# FLUID MOVERS

## Pumps, Compressors, Fans and Blowers

# FLUID MOVERS

## Pumps, Compressors, Fans and Blowers

Edited by
Jay Matley
and
The Staff of Chemical Engineering

McGraw-Hill Publications Co., New York, N.Y.

Printed in the United States of America.

**Library of Congress Cataloging in Publication Data**

Main entry under title:

Fluid movers—pumps, compressors, fans, and
blowers.

1. Pumping machinery. 2. Compressors.
3 Fans (Machinery) I. Chemical Engineering
TJ900.F65 621.6 79-20472
ISBN 0-07-010769-6 (case)
07-606631-2 (paper)

# Contents

# FOREWORD

This book contains a wealth of practical information on the most common fluid movers—pumps, compressors, fans and blowers—for engineers who work in process operating plants.

*For the project engineer*—whose combination of managerial and technical skills ensures the viability of "grass roots" plants and the effective major modification of existing ones—there is comprehensive guidance on the specification and selection of fluid-moving equipment, as well as coverage of its installation. Not only is this guidance anchored in thermodynamic and hydraulic fundamentals, it is also channeled by step-by-step procedures that—together with information on the capabilities and limitations of particular fluid movers—will direct the project engineer to the most efficient and economical fluid-moving equipment and circulation system.

*For the plant process engineer*—the invaluable caretaker whose skill and ingenuity bring about ever-increasing efficiency in plant operation—there are practical insights into how pumps and compressors can be made to function more effectively, and ideas on how this equipment can be modified for changed process demands, such as (to cite only one example) how a compressor can be operated or altered so as to achieve higher, or lower, capacity than that for which it was originally designed. With the understanding that can be gleaned from this book of the intricacies of the wide range of fluid-moving equipment and of the systems in which it operates, the process engineer can more ably solve plant problems that involve the movement of gases, liquids and slurries.

*For the operations engineer or supervisor*—upon whom depends the continuous, profitable operation of the equipment at his disposal—there are detailed discussions and descriptions of fluid-moving equipment that will enable him to fully comprehend its functioning. There are also insights into operating characteristics, limitations and most-suitable control schemes—all of which will help him achieve the troublefree operation that is the measure of his success. Still other information will enable the operations engineer to quickly pinpoint problems involving fluid movers, to determine the causes of malfunctionings and to arrive at the most practical solutions. For example, a checklist guide to centrifugal pump problems includes eighty-nine probable causes, and is accompanied by discussions on how the malfunctionings might be remedied. Presented also is information on how to cope with troublesome fluids, such as those that are highly viscous, corrosive or abrasive, and how to achieve the closely controlled flows vital to high-yielding operations.

*For the maintenance engineer*—among whose expertise and efficiency depends the online availability of fluid-moving equipment—there is a vast amount of diagrams and discussion that will provide the familiarity with the external and internal construction of the large variety of such equipment that will promote the expeditious repair of it. There is also practical guidance on how to align shafts and care for bearings, seals, couplings, speed gears, and lubrication and control systems. Such knowledge can be critical. For example, in the modern high-speed compressor, good lubrication is vital to reducing friction, transferring heat, sealing against air leakage and flushing away contaminants. Among the many trouble spots addressed in the book are those that arise from mechanical seals and packing, which particularly plague the maintenance engineer, and a variety of means are presented by which he can avoid or cure such problems, and so contribute immeasurably to the important goal of holding down maintenance costs.

*For the plant design engineer*—from whose technical expertise must come long-range solutions to persistent process problems—there is detailed information on the systems in which fluid movers perform that will both widen and deepen his understanding of them and so improve his approach to redesigning them. And there is information that is directly applicable, such as how to achieve better equipment layouts and piping arrangements, as well as how to alter the fluid mover itself so as to suit changed process conditions. On this last point, for example, ideas are given on the best mechanical means of revising the output from a pump or a compressor, considering the restrictions imposed by practical reality. Also, because the saving of energy is of paramount importance, particularly valuable to the plant design engineer is an examination of what an inefficient pumping system can cost in terms of energy, and the actions that can be taken to reduce these costs. For many such problems, the book provides guidance to the optimum solutions.

*For the safety engineer*—upon whom rests the responsibility for guiding equipment design, process operations and maintenance procedures so as to ensure the security of life and property—information vital to the safe operation and maintenance of fluid movers is interspersed through the book. For example, the matter of access for safety is dealt with in the discussion of the optimum layout of pumps and compressors. When, however, the operation and maintenance of a fluid mover bears critically and immediately upon safety, as in the case of oxygen compressors, then safety is focused upon directly.

All the foregoing, of course, only provides the briefest summary of the information on fluid movers contained in the book that can be of value to the process-plant engineer. Understandably, the fluid movers most emphasized are centrifugal, because this type prevails in industry. In the chemical process industries, it is clearly the mainstay of the fluid movers. Its popularity largely stems from its relatively simple design and construction, which causes it to be less demanding in terms of maintenance, and enables it to operate continuously for long periods.

# Section I
# Introduction

# Compressors and Pumps: the principal fluid movers

Here's a guide to the structure and operating characteristics of centrifugal and positive-displacement compressors and pumps. It provides information necessary to ensure proper selection and long, trouble-free operation.

ROBERT W. ABRAHAM, The Badger Co.

In the chemical process industries, the trend is to build larger and larger plants with larger, more reliable, single-component equipment.

Reliability of rotating equipment for such plants must always be defined in terms of the expected life of a plant and the associated payout time required to earn profits for the company. Many chemical plants may have a life expectancy of five years or less, with the process being outdated at that time, whereas refinery or petrochemical plants may have a payout time of 10 to 15 years or more.

Several, seemingly unrelated, questions are of prime importance when evaluating, selecting and installing rotating equipment. Is the plant to have a continuous or batch process? What premium is placed on operating cost versus capital cost? Are adequate maintenance personnel available or is it the intent to minimize labor and provide more-automatic control of the process?

With such matters in mind, we can now try to evaluate and make use of equipment existing in the field.

The heart of most processes, and potentially the most troublesome, is the compressor. When selecting a type of compressor, it is most important to have all the process conditions available for review. If there are specialists in the company, they should be informed of all these conditions. Failure to do so has been a source of many problems in the past.

Fig. 1 shows the operating range of the most widely used types of compressors in the CPI. Care should be taken when using Fig. 1 because there are no definite boundaries for considering operating conditions where two or more types of compressors can be used. In this case, alternatives must be looked at. The first step is to define the types, and principles of operation, of compressors.

## Centrifugal Compressors

A centrifugal compressor develops pressure by increasing the velocity of gas going through the impeller, and then recovering the velocity in a controlled manner to achieve the desired flow and pressure. Fig. 2 shows a typical impeller and diffuser. The shape of the characteristic curve depends on the angle of the impeller vanes at the outer diameter of the impeller and also on the type of diffuser. For technical details on the theory of operation of different impeller types, see Ref. 1. These compressors are normally installed as a single unit, unless flow is too high or process requirements dictate otherwise.

Originally published October 15, 1973

Most impellers furnished in the CPI are the backward-leaning type. This makes the compressor more suitable for control because of a steeper performance curve. The tip speed of a conventional impeller is usually 800 to 900 ft./sec. This means that an impeller will be able to develop approximately 9,500 ft. of head (the head depends on the gas being compressed). Multistage compressors are needed if duties exceed this value. Heavy gases such as propane, propylene or Freon require a reduction in tip speeds due to lower sonic velocities of these gases when compared with air. For these gases generally, the relative Mach number at the side of the impeller is limited to 0.8.

An excellent summary is given in Ref. 2 that discusses in everyday language the reason for the shape of characteristic curves. When evaluating a centrifugal compressor, close attention should be paid to the percent increase in pressure from the normal operating point to the surge point. The surge point is defined as the location where a further decrease in flow will cause instability in the form of pulsating flow, with possible damage resulting from overheating, failure of bearings due to thrust reversals, or excessive vibration.

Because of the high speeds used in centrifugal compressors, greater care must be taken with rotor balance. The following formula is now accepted by the industry generally for allowable vibration limits on compressor shafts:

$$Z = \sqrt{12{,}000/n}$$

where $Z$ is the maximum allowable vibration limit, peak to peak, mil (1 mil = 0.001 in.) and $n$ is speed, rpm. $Z$ has a maximum limit of 2.0 mil at any speed. Because of high speeds, most users specify that vibration monitors of the noncontacting type be provided to sense excessive shaft vibration.

Depending on the type of process system, various anti-surge controls are required to prevent the compressor from reaching the surge limit. Usually, a safety margin of from 5 to 10% should be allowed for automatic controls. Simple resistance circuits may not need any antisurge controls because the surge line could never be reached (see Fig. 3).

When a fixed backpressure is imposed on the compressor, special care must be given to selecting a steep performance curve (i.e., a rise in head of approximately 10 to 15% from rated point to surge point) so that a small variation in pressure will not cause a large change in flow and possible surging. Fig. 4 illustrates this process. When recycling gas in the antisurge loop, the gas must be cooled before it is returned to the inlet of the compressor. Furthermore, if variable speed is desired, a pressure control will regulate the speed of the driver.

When both a fixed backpressure and friction drop are required, an antisurge system will be needed—especially if wide variations in flows and pressures can exist (see Fig. 5). Head rise from rated point to surge should be at least 10% for good stability. The control scheme is the same as that shown in Fig. 4, and will usually be based on measurement of flow through the compressor. Again, the bypass flow must be cooled before it is returned to the compressor.

From a process viewpoint, the centrifugal compressor offers the advantage of oilfree gas and no wearing parts in the compression stream. There are several choices of end-seals available. The selection of seals depends on the suction pressure of the compressors since most compressors have the discharge end balanced back to the suction pressure (i.e., both the inlet and discharge ends of the compressor are under suction pressure). The following table gives types of seals and the common range of pressures. For configuration, see Fig. 6.

| Type of Seal | Approximate Pressure Psig. |
|---|---|
| Labyrinth | 15 |
| Carbon ring | 100 |
| Mechanical contact | 500 |
| Oil film | 3,000 and higher |

There are several variations of these seals. For example, if the process gas contains a "sour" component such as $H_2S$, a sweet gas such as nitrogen could be used to buffer the area between the mechanical-contact or oil-film seal and the process gas. (See Fig. 6.) An eductor could be used in conjunction with injection of a sweet gas in order to ensure that the outward leakage is in the direction of eduction.

The advantage of the labyrinth seal is that it is a clearance-type seal with no rubbing parts and is by far the simplest of all seals. The type is also used between stages for multistage compressors. Its disadvantage is that high leakage losses may be encountered which, for valuable gases such as nitrogen or oxygen, cannot be tolerated.

Carbon-ring seals are not commonly used except where the compressed gas is clean or there is a clean buffering medium containing a lubricant. Since these are close-clearance seals, they are subject to wear. They are

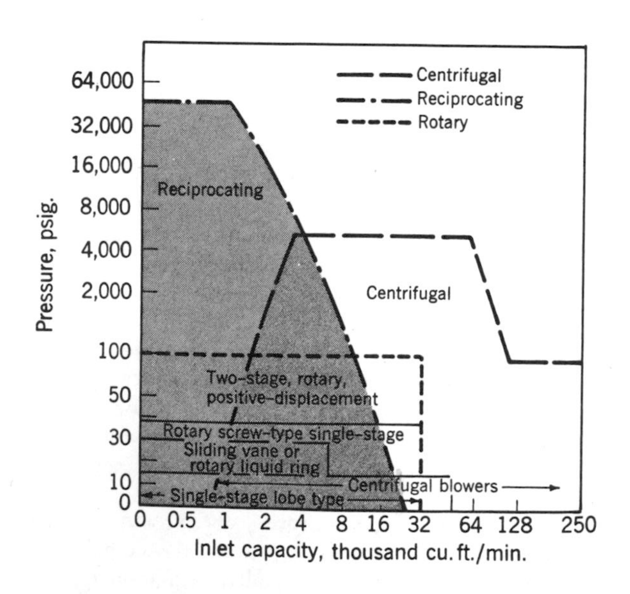


**COMPRESSORS** have wide ranges for process use—Fig. 1

**GAS FLOW** through a centrifugal compressor—Fig. 2

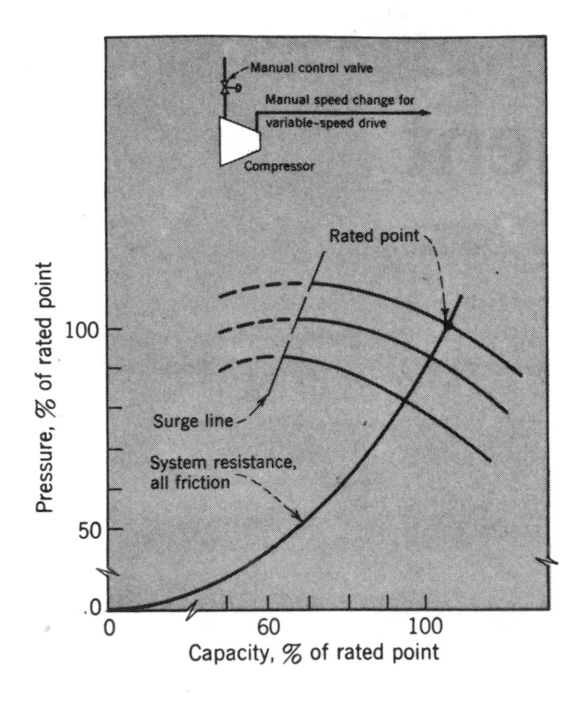


**RESISTANCE** to flow is due only to friction—Fig. 3

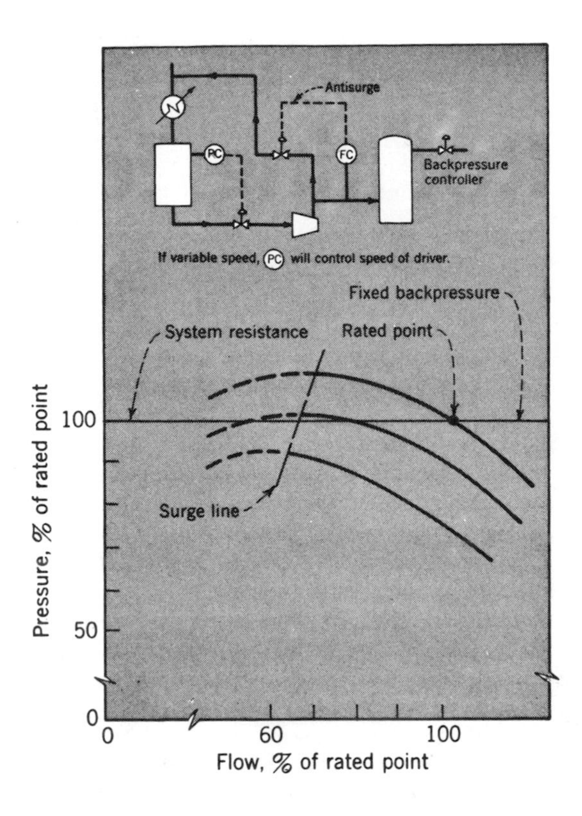


**FIXED** backpressure imposes careful control—Fig. 4

relatively inexpensive compared with oil-film or mechanical-contact seals but still have the advantage of limiting the outward leakage of the compressed gas.

The mechanical-contact seal has an oil film that is maintained between the running and stationary faces. This seal has the advantage of minimizing oil leakage toward the gas side. It is also relatively insensitive to the differential pressure between the compressed-gas suction pressure and the sealing-oil pressure. The disadvantage of the seal is the possible loss of the oil film, which may cause serious damage to the mating faces.

The oil-film seal, like the mechanical-contact seal, relies on an oil film to seal the compressed gas from the atmosphere. Unlike a mechanical-contact seal, however, it is a clearance type and requires a very close differential between suction pressure and sealing pressure in order to minimize inward leakage of oil. When the seal oil is common to the lube system, this could amount to excessive oil losses and become a maintenance problem both in disposing of the contaminated oil and in replenishing the oil for the lubricating system. This type of seal is used for the highest suction pressures commonly found in the CPI.

The disadvantage of both the oil-film and mechanical-contact oil-seal systems is the requirement for sophisticated controls, plus extra pumps and a seal-oil cooler and filter if a separate seal-oil system is required. For further details on seal-oil systems and lube-oil systems, see Ref. 2 and 3.

Casings for centrifugal compressors can be either horizontally or vertically split with relation to the shaft.

From a maintenance viewpoint, it is easier to get at the rotor assembly for a horizontally split casing than for a vertically split one. However, the horizontally split casing is limited in pressure because of the large sealing area of the joint. The American Petroleum Institute's Subcommittee on Mechanical Equipment has recently adopted a guideline that calls for a vertical sealing joint. The proposed guideline for changing to a vertically-split or barrel casing is:

| Mol Fraction $H_2$ % | Maximum Case Working-Pressure, Psig. |
|---|---|
| 100 | 200 |
| 90 | 222 |
| 80 | 250 |
| 70 | 295 |

Where vertical-type split casings are used, space must be allowed to withdraw the inner casing and rotor.

Material selection for casings and rotors depends on the gas being compressed. Recent studies indicate that gases containing $H_2S$ cause stress corrosion in highly stressed areas. In order to overcome this, softer impeller materials are required. This necessitates reduced impeller tip speeds. In some cases, because of this reduction in speed, the next-larger compressor may have to be selected. What this means is that the compressor manufacturer should be made aware of all gas components as well as all operating conditions.

Advantages of using a centrifugal compressor:

1. In the range of 2,000 to 200,000 cu.ft./min. (and depending on pressure ratio), the centrifugal compressor is economical since a single unit can be installed.

2. The centrifugal offers a relatively wide variation in flow with relatively small change in head.

3. Lack of rubbing parts in the compression stream

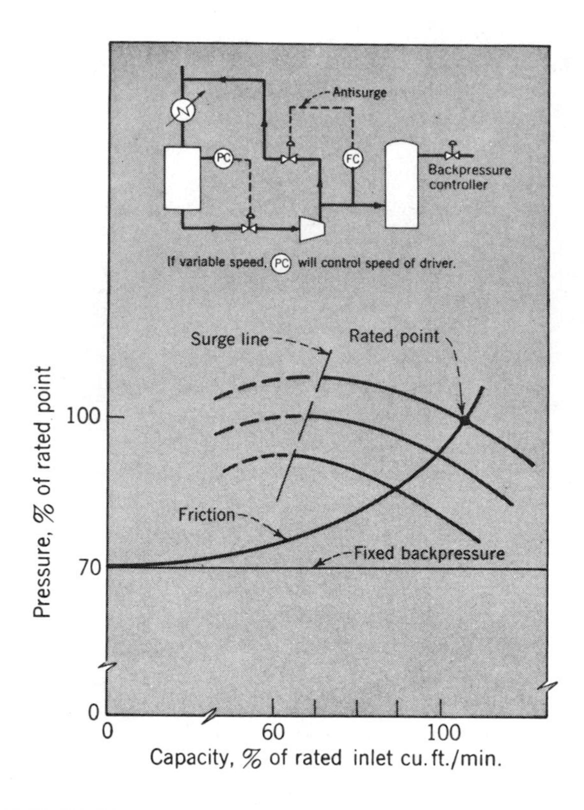


**ANTISURGE** control handles fixed backpressure—Fig. 5

enables long runs between maintenance intervals, provided that auxiliary systems such as lube-oil and seal-oil are designed properly.

4. Large throughputs can be obtained with relatively small plot size. This can be an advantage where land is valuable.

5. When enough steam is generated in the process, a centrifugal compressor will be well matched with a direct-connected steam-turbine driver.

6. Smooth, pulsation-free flow is characteristic.

Disadvantages:

1. Centrifugals are sensitive to the molecular weight of the gas being compressed. Unforeseen changes in molecular weight can cause discharge pressures to be very low or very high.

2. Very high tip speeds are required to develop the pressures. With the trend to reduce size and increase flow, much greater care must be taken in the balancing of rotors and materials used for highly stressed components.

3. Relatively small increases in process-system pressure drops can cause very large reductions in compressor throughput.

4. A complicated lube-oil system and sealing system is required.

## Positive-Displacement Compressors

Positive-displacement compressors can be categorized as rotary and reciprocating for the more common process applications. Unlike the centrifugal, these are essentially constant-capacity machines having varying discharge pressures. Fig. 7 shows a typical performance curve. For this diagram, suction pressure and temperature, and discharge pressure, are assumed constant. Capacity is changed by speed or suction-valve unloading. Furthermore, there is only a slight variation in flow over a wide pressure change.

Reciprocating compressors operate on an adiabatic principle whereby the gas is drawn into the cylinder through inlet valves, is trapped in the cylinder and compressed, and is then passed through the discharge valves against discharge pressure. These compressors are seldom used as single units unless the process is such that intermittent operation is required. For instance, if catalyst must be regenerated every two or three months, or a backup supply is available from another source, this would allow time to make valve or piston-ring repairs or replacement if required. Reciprocating compressors have contacting parts such as piston wear-rings mating to cylinders, valve springs and plates or disks mating to valve seats, and piston-rod packing mating to piston rods. These parts are all subject to wear due to friction.

Reciprocating compressors can be furnished in a lubricated or nonlubricated design. If the process permits, it is usually better because of increased life of wearing parts to have a lubricated compressor. However, care must be taken to avoid overlubricating since carbonizing of oil on valves can cause sticking and overheating. Also, oil-saturated discharge lines are potentially a fire hazard, so an adequate separator should be placed downstream to remove the oil. The biggest problems for lubricated-cylinder compressors are dirt and moisture, both serving to destroy the oil film created in the cylinder.

The best way to prevent dirt is to start up with a clean system by using temporary suction strainers. Moisture or condensables carrying over into the compressor suction can be avoided by using an efficient separator located as close to the compressor as possible. Steam tracing or preheating of the inlet gas should be considered downstream of the separator if a wet gas is being compressed.

For nonlubricated machines, dirt is usually the most severe problem. Other problems can arise from the gas itself. For example, a bone-dry gas can cause severe ring wear. The manufacturer should be consulted in cases such as this, since new test data are constantly being developed. For nonlubricated machines, the piston and wear rings are usually made of Teflon-filled, bronze, glass or carbon materials, depending on the gas being compressed. Honing of the cylinder to 12 $\mu$ (rms.) finish usually prolongs ring life. (See Ref. 4.) Packing is subject to the same wear as the piston rings.

If the gas being compressed is a sour one, or the lubrication used for the cylinder is not compatible with the oil used in the crankcase, or vice versa, then an extra-long distance piece should be specified. See Fig. 8 for various distance-piece configurations. In cases where the gas is a safety hazard, a double-distance piece should be specified, and the distance-piece next to the cylinder should be purged with an inert gas.

Packing leakage should be vented to a flare system or returned to suction. Lubricated compressors may require separate feedlines for lubricating packing, though smaller-size cylinder diameters may not require them. Teflon, nonlubricated, packing usually requires water-

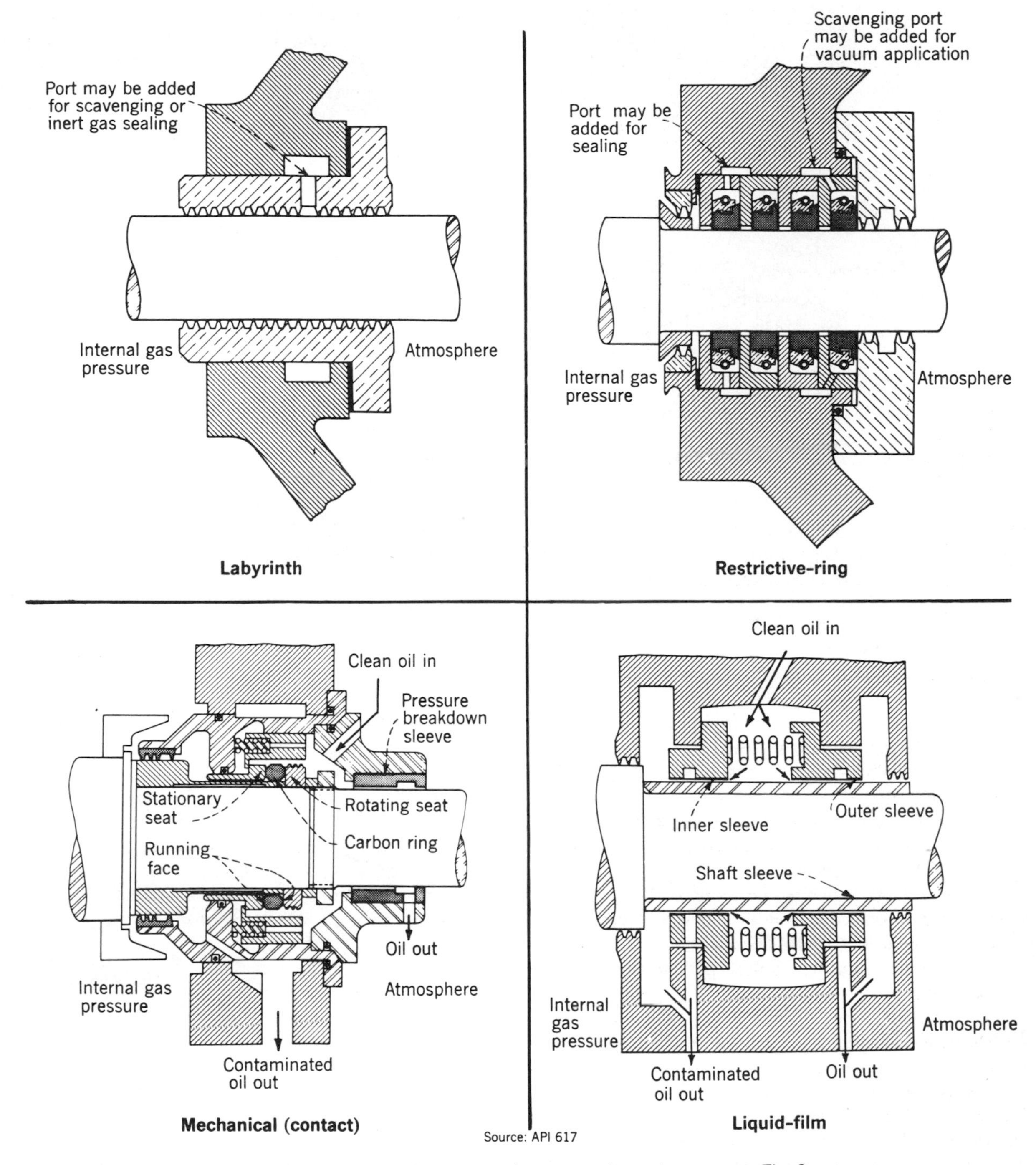


**SHAFT** end-seals for centrifugal compressors handle a range of pressures for various gases—Fig. 6

cooling, since its thermal conductivity is very low. If low-temperature gases are being handled below 10 F., the manufacturer should calculate the amount of preheat in the gas from internal bypassing. This will mean that a slightly larger cylinder is required to pass the same weight of flow.

Reciprocating compressors are best suited to low-speed, direct-drive motors, especially above 300 hp. These machines are normally constant-speed. Capacity control is accomplished by unloading valves. These valves should be either the valve-plate-depressing type or the plug-unloader type. Unloaders that lift the entire valve off its seat can introduce sealing problems. Unloading can be done automatically or manually. Normal unloading steps are: 0-100%, 0-50-100%, 0-25-50-75-100%. Intermediate steps can be obtained by clearance pockets or bottles. However, these pockets should not be used if polymerization can take place, unless adequate precautions are taken.

## Cylinder Cooling

If pressure ratios are low, resulting in discharge temperatures of 190 F., or lower, a static closed system or

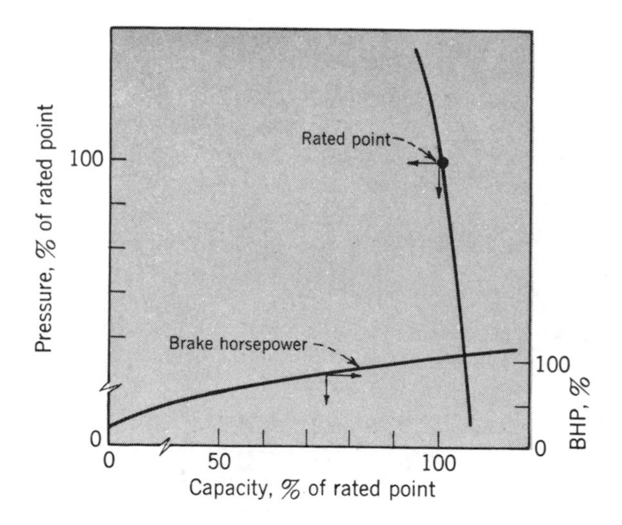


**POSITIVE** displacement compressor curve—Fig. 7

thermosyphon cooling system can be used. Here again, care should be exercised not to run at prolonged periods of no load. Otherwise, a forced, closed-loop system should be used. The inlet temperature of the cooling water should always be maintained at least 10°(F.) above the suction temperature of the incoming gas in order to prevent condensation from forming in the compressor cylinder.

Discharge temperature for nonlubricated machines should be held to a maximum of 350 F. for process compressors. Lubricated-compressor discharge temperatures should be held to 300 F. With synthetic lubricants, these temperatures can be raised to 350 F.; however, care must be taken to see that these lubricants do not cause problems such as paint removal.

The above limitations may be reduced. For example, oxygen in nonlubricated service should be limited to a discharge temperature of 300 F., while chlorine compressors should be limited to 225 F. to prevent fouling (see Ref. 9).

## Compressor Loadings, Speeds and Pulsations

Compressor ratings are based on rod loadings. Longer strokes usually mean higher rod-load ratings and a greater differential-pressure and horsepower capability. Most manufacturers have established proper frame sizes. It is important that the frame and rod loadings are not exceeded even during relief-valve operation.

Average piston speeds for nonlubricated compressors should be approximately 700 ft./min. maximum, while lubricated compressors could run at approximately 850 ft./min. maximum. Rotative speeds for heavy-duty compressors should be held to below 600 rpm., and even lower for high horsepower (over 400 hp.).

Inherent with the reciprocating compressor is pressure-pulsation. This is caused by the reciprocating motion of the piston. In order to avoid it, dampers are fitted to the compressors as closely as possible. The following formula is presented as a guide for maximum limitation of

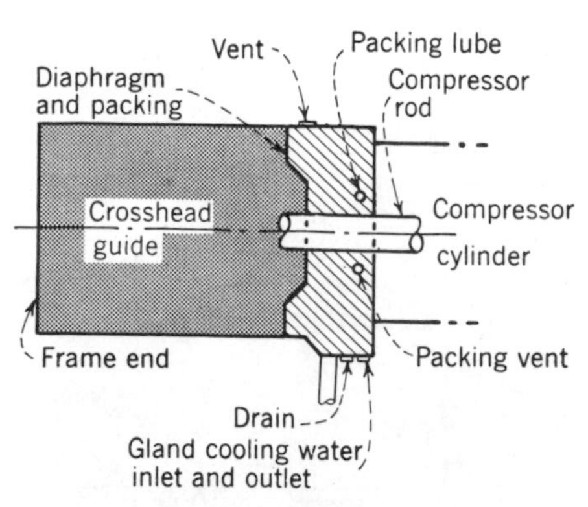


**Standard cylinder mounting**

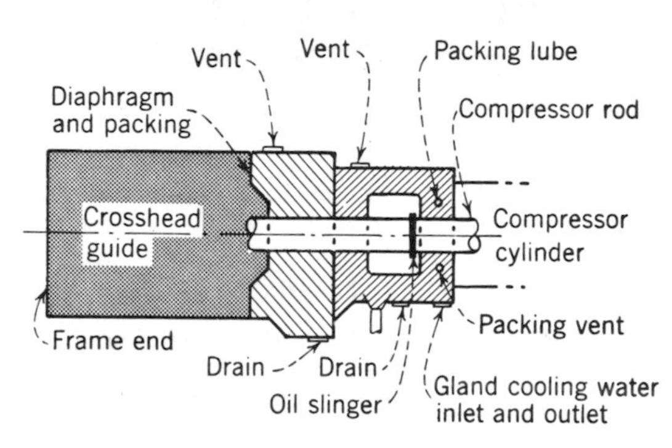


**Single compartment**

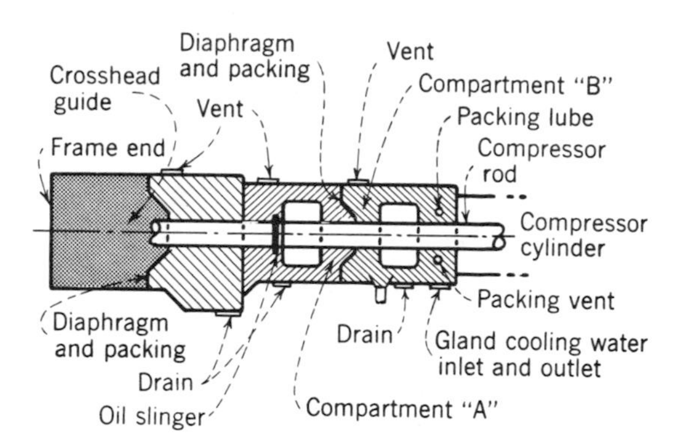


**Long two compartments**

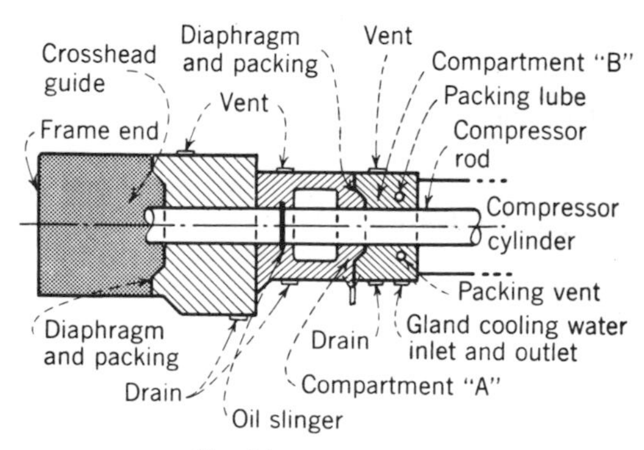


**Short two compartments**

**DISTANCE-PIECES** protect area from leakage—Fig. 8

peak-to-peak pulsations in compressor suction and discharge lines:

$$P_1 = \frac{1.5}{\sqrt[3]{0.001p}}$$

where $P_1$ is maximum allowable pulsation, %; and $p$ is mean effective line pressure, psia. The value for $P_1$ is that obtained from the formula, or 1%, whichever is greater. Compliance with these limits not only ensures reduced pulsations but also better valve performance from the compressor.

In order to ensure that the whole compressor system is adequate, including piping and associated vessels, an analog study should be done by the manufacturer. In case of complicated systems, Southwest Research Institute[5] has the capability of doing both a mechanical and acoustical vibration check.

An excellent source of information on reciprocal compressors is API 618.[6] Following the API recommendations does bring the cost of the equipment up; however, these specifications represent many years of experience and can mean reduction of costly startup or onstream repairs.

## Rotary Positive-Displacement Compressors

Various types of rotary positive-displacement compressors are available. Among these: the lobe-type blower (such as the Rootes design), SRM rotary-screw type, water-ring design, and sliding vane. All of these have the same type of performance curve as a reciprocating compressor; i.e., they are essentially fixed-capacity machines with varying backpressure. Rotary compressors are most readily adaptable to variable-speed drives, such as steam turbines, than are reciprocating compressors. Normally, these compressors are limited to a maximum of about 25,000 cu. ft./min. with present technology. This applies to the rotary-screw and lobe types. The water-ring design offers the advantage of not having the compressed gas in contact with the rotating metal parts. The critical areas are the vapor pressure of the incoming gas vs. the vapor pressure of the liquid forming the water ring, and the temperature rise of the liquid ring. The vapor pressure of the fluid used for sealing should be well below the boiling point, otherwise the sealing ring will evaporate and cause loss of capacity and possibly serious damage from overheating.

Since sliding-vane compressors require lubrication, they are limited to applications where the process can tolerate the lubricant. The oil in the compression chamber does tend to lower discharge temperatures. Oil consumption is fairly high compared to a reciprocating compressor. The sliding-vane compressor is very compact—although it has the same disadvantage as a reciprocating compressor in that rubbing parts are required in the gas stream. Loss of lubricant could cause overheating of the cylinder. High-water-temperature and high-air-temperature switches are necessary for these compressors. Speed reduction is limited to about 60% of normal, since decrease in centrifugal force results in loss of sealing efficiency.

The more common types of rotary, positive-displacement compressors in the CPI are the rotary screw and rotary lobe. These units offer the advantage of oilfree gas, since contact is not made with any parts in the compression area. Their rotary design gives them a throughput capability much greater than the reciprocating compressor, and without pulsation problems.

Timing gears are used to maintain rotor separation and to prevent contact with rotors. For the Rootes-type blower, these gears transmit about 30% of the torque, whereas the rotary-screw type transmits about 10% of the torque. Because these compressors are positive-displacement types, a relief valve should be placed between the compressor and the block valve.

The Rootes lobe-type has a low differential-pressure capability, usually limited to about 15 psig. The rotary-screw compressor can take differentials of much higher values. Both have a fixed slippage rate, causing internal bypassing and a preheat of suction gas. The slower the speed for a given compressor size, the greater the internal bypass that will occur. If the speed is too low, overheating will occur, with possible damage to the rotors. The vendor should stipulate a minimum speed of operation in anticipation of this.

If the discharge temperature of rotary-screw compressors exceeds 350 F., oil-cooled rotors should be specified. It is also well to determine whether the vendor has a minimum backpressure requirement, in order to prevent backlash of timing gears. Another wise precaution is to require the vendor to make a torsional analysis of the compressor and driver.

The first lateral critical speed for these types of machines is usually above the operating speed. This critical speed must be established for both compressor and driver, and should be at least 20% above the highest operating speed and always above the trip speed if a turbine drive is used.

Since noise levels are of increasing interest, it should be noted that these compressors are quite noisy, and are without acoustical provisions such as inlet and discharge silencers, or complete acoustical enclosures. More details can be found in Ref. 8.

## Pumps for Processes

In most processes, liquid is transferred from one vessel to another by means of a pump. The type most common to the CPI is the centrifugal pump that operates on the same principle as centrifugal compressors except that the fluid handled is essentially incompressible. The large clearance and lack of rubbing parts, except for bearings and seals, have been responsible for the wide use of the centrifugal pump.

Some guidelines for specifying and evaluating centrifugal pumps are:

1. In the range of 20 to 500 gpm., standard AVS horizontal pumps[10] are used. Vendors making this type of pump have extended the range to about 1,500 gpm., though no official standards have been approved as of this writing. These pumps are normally limited to about 500 F. maximum. Pumps below 20 gpm. will probably require a maximum-flow bypass. Vertical, inline pumps

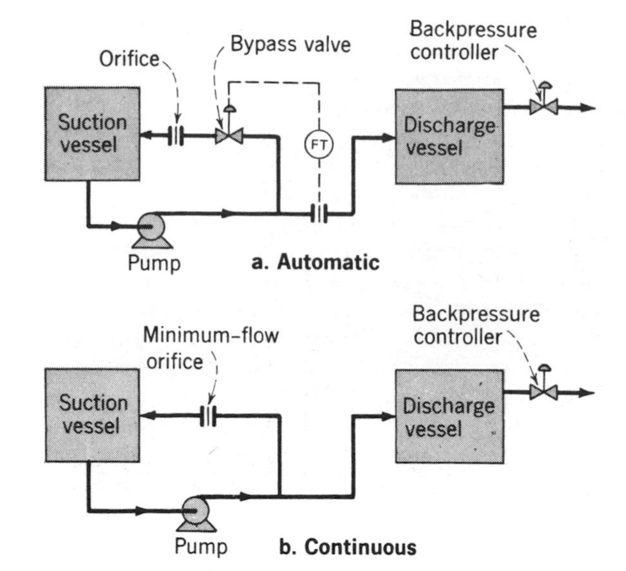


**BYPASS** keeps pump operating at low flows—Fig. 9

offer the advantage of requiring no foundation in the smaller sizes (less than 50 hp.) and no special need of piping flexibility. Above this size, foundation pads should be installed for the pumps to rest on. There is also a proposed standard[10] for these pumps, which will standardize the flange-to-flange dimension if accepted.

2. Depending on pump size, viscosity is limited to a maximum of about 3,000 to 5,000 SSU. Above this range, positive-displacement pumps should be considered. Where possible, heating the liquid is recommended in order to lower its viscosity.

3. High-temperature and high-pressure pumps should be specified per API 610.[12]

4. It is important to ensure that the *NPSH* (net positive suction head) available is above the *NPSH* required by the pump. *NPSH* available is defined as the net positive suction head available over the vapor pressure of the liquid, plus the suction head, or minus the suction lift, minus any friction losses. Some companies use a safety margin of 2 ft. or more between the required and available *NPSH* to ensure no problems after field installation.

5. Undissolved gases in liquids affect the capacity of centrifugal pumps. These should be limited to a maximum of 5%.

6. Some processes require the pump to operate under low-flow conditions. All centrifugal pumps have a minimum flow at which they are capable of operating satisfactorily. If run below this flow, overheating of the pump, and reduction of bearing life, may result. The way to resolve this problem is to install a minimum-flow bypass valve (see Fig. 9). This minimum flow can be continuous or automatic. Where the bypass valve is under automatic control, it will open when minimum flow is reached. If the differential head exceeds 200 ft. in the minimum-flow line, an orifice is required or a control valve can be installed. Use the continuous-flow system when low horsepower and relatively low discharge pressures are required. Flow through the minimum-flow orifice must be added to process requirements when sizing the pump. Orifices should be multiple-step type if the differential head equals 300 ft. or more, or when flashing occurs. For both systems, the minimum flow is normally 20 to 25% of flow at best efficiency point. Head rise should be at least 10% from rated point to shutoff point. To avoid heating the inlet flow, minimum flow should always be taken back to the suction vessel, not directly into the pump's suction line. These systems are needed only if the process flows can vary. If process flow is fixed near rated point, bypass valves are not required.

7. Where possible, it is wise to use two-pole electric motors. The higher the speed, the more efficient (and least expensive) the pump; however, a greater *NPSH* is required by the pump. The cost of the pump must always be weighed against the cost of raising a vessel to obtain adequate *NPSH*.

8. Pumps should always be specified with continuously rising curves. Pumps operating in parallel should have a 15% rise from rated point to shutoff.

9. Care must be taken with pumps that have mixed-flow impellers—usually in the higher capacities above 2,000 gpm.—since some of these pumps may have shutoff pressures as high as 200% of the rated point. This can cause a lot of grief with downstream equipment, so the equipment must either be designed for the shutoff pressure or a relief valve installed.

10. Centrifugal pumps are in themselves not self-priming and must either have a flooded suction or, in case of a suction lift, have a vacuum device to lower the pump-casing pressure in order to have the liquid pushed into the pump. However, there are special centrifugal pumps that can prime themselves.

11. For high-temperature service (above 350 F.) on large single-stage pumps, or multistage pumps with temperature above 150 F., a warmup line should always be provided to ensure uniform thermal expansion.

12. First critical speeds for single-stage pumps are usually above the operating speeds. However, for multistage pumps, this may be below the first critical speed.

13. Most process centrifugal pumps have a specific speed range below 2,000 (see Ref. 13). Larger pumps such as cooling-tower or loading pumps may range as high as 5,000 rpm.

$$N_S = n\sqrt{Q/H^{3/4}}$$

where $N_s$ is the specific speed, $n$ is speed, rpm.; $Q$ is flow, gpm.; and $H$ is head at best efficiency point, ft.

Centrifugal pumps are generally flow-controlled or level-controlled. Unlike the characteristic curve for the centrifugal compressor, the curve for the centrifugal pump normally rises from the rated point to the shutoff.

If the pump is completely oversized and the system pressure is low, the pump may operate at the end of the curve. This can cause excessive vibration, with increased loads on bearings. If added capacity is desired for the future, then a flow-controller should be included with a minimum-flow bypass for current operating conditions.

Pumps handling liquids with viscosity over 100 SSU. should be checked for possible viscosity correction, as noted in Ref. 13. Fig. 10 gives an example of how the performance curve of a pump may be affected by viscosity. Note the dropoff in capacity and head when oper-

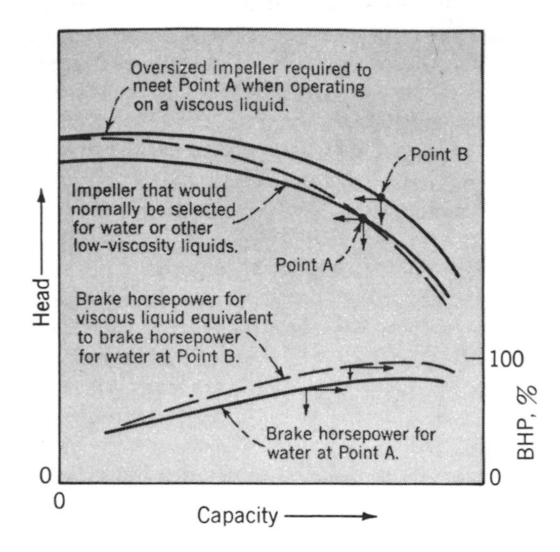


**VISCOSITY** of liquid affects centrifugal pump—Fig. 10

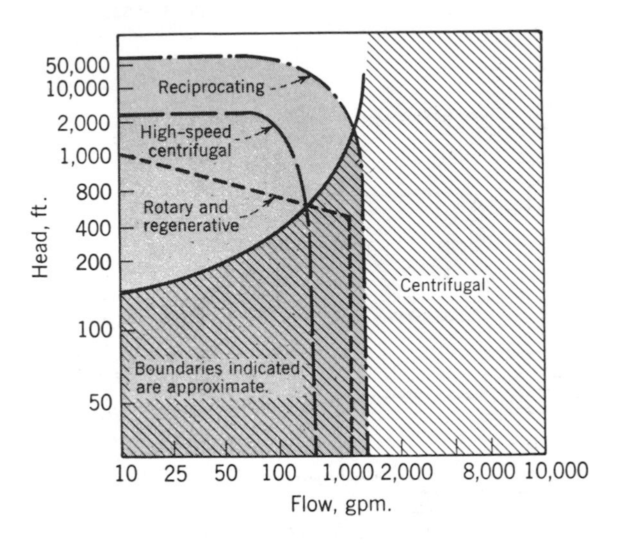


**PUMPS** have many ranges for process use—Fig. 11

ating with a viscous fluid. The amount of head reduction and flow depends on the viscosity of the fluid and the size and type of pump. In the curve (Fig. 10), the impeller selected would have to be increased to Point B if operation at Point A were desired. Sizing the motor without taking into account the decreased efficiency may result in overloading the motor at the operating Point A.

Most pumps used in the CPI have mechanical seals. These are unbalanced or balanced, depending on the pressure in the stuffing box and the peripheral speed of the mating faces of the seals. Single seals are normally used unless the liquid pumped is hazardous, abrasive, etc., in which case a double seal is used.

## Positive-Displacement Pumps

Positive-displacement pumps are generally suited to low-flow, high-head applications, as noted in Fig. 11. Where possible, centrifugal pumps should be used. For viscosities above 3,000 SSU., rotary pumps should be considered first. Lower flows (about 100 gpm.) with viscosities of 100 SSU. and higher require positive-displacement pumps. The general types encountered in the CPI are reciprocating or rotary of either gear- or screw-type. These pumps operate on the same principle as the positive-displacement compressors except that the liquid is usually noncompressible in the lower pressure range, say below 3,000 psi. Above this range, compressibility must be dealt with. These pumps have very close clearances and require a clear fluid in order to minimize maintenance. Fig. 12 shows a typical drawing of each type. As with the positive-displacement compressor, capacity stays essentially constant while backpressure varies.

Reciprocating pumps are normally used to handle low flows. As with the reciprocating compressor, rotary motion is changed to reciprocating by means of a crankshaft and crosshead piece. The liquid is drawn into the cylinder and then pressurized against the system discharge valve. These pumps also produce pulsation flow. To limit this, velocities on the suction side should be kept low, 2 to 3 ft./sec. or less, and discharge velocities should be held to 3 to 4 ft./sec. Pulsations may also be reduced by the addition of an accumulator.

Larger reciprocating pumps are normally specified in triplicate to reduce pulsations. Fig. 13 shows the difference between curves for a single and double plunger versus three plunger (triplex) 120° apart. Note the much smoother curve for this pump. When calculating *NPSH* available, acceleration load and peak velocity should be considered when using a triplex pump. Ref. 14 gives a detailed procedure based on the theory behind calculations of *NPSH* available for reciprocating pumps.

Reciprocating pumps use a packing or seal such as a V-ring chevron. This means that leakage from the pumps must be disposed of, and this should be considered when installing this equipment. Smaller reciprocating pumps (below 25 hp.) have variable-speed drives (such as Reeves). Larger pumps (over 25 hp.) can be variable-speed electric motors with eddy-current coupling or fluid coupling. Normal drives for reciprocating pumps in the CPI are motor driven through a gear box. The gear box should have a 2.0 minimum service factor (Ref. 17). The motor should be sized to operate at the relief-valve setting to avoid overloading.

Rotary pumps normally used in the CPI are either gear type or rotary-screw type with or without timing gears. They are used for normally viscous materials. Abrasive or low-viscosity fluids are not handled readily by the pumps. Special rotors must be used for abrasive fluids. Bearings and timing gears should be external.

Rotary pumps can handle up to about 500 gpm. If viscosity is low (below 100 SSU.), timing gears should be provided. With external timing gears, however, at least four stuffing boxes will be required, which means more potential leakage. For clean lubricating fluids (such as lube and fuel oils below 200 F., below 100 psi., with viscosities above 100 SSU.), gear-type pumps can probably be used.

When liquids handled by these pumps solidify at ambient temperatures, the pumps should be jacketed or

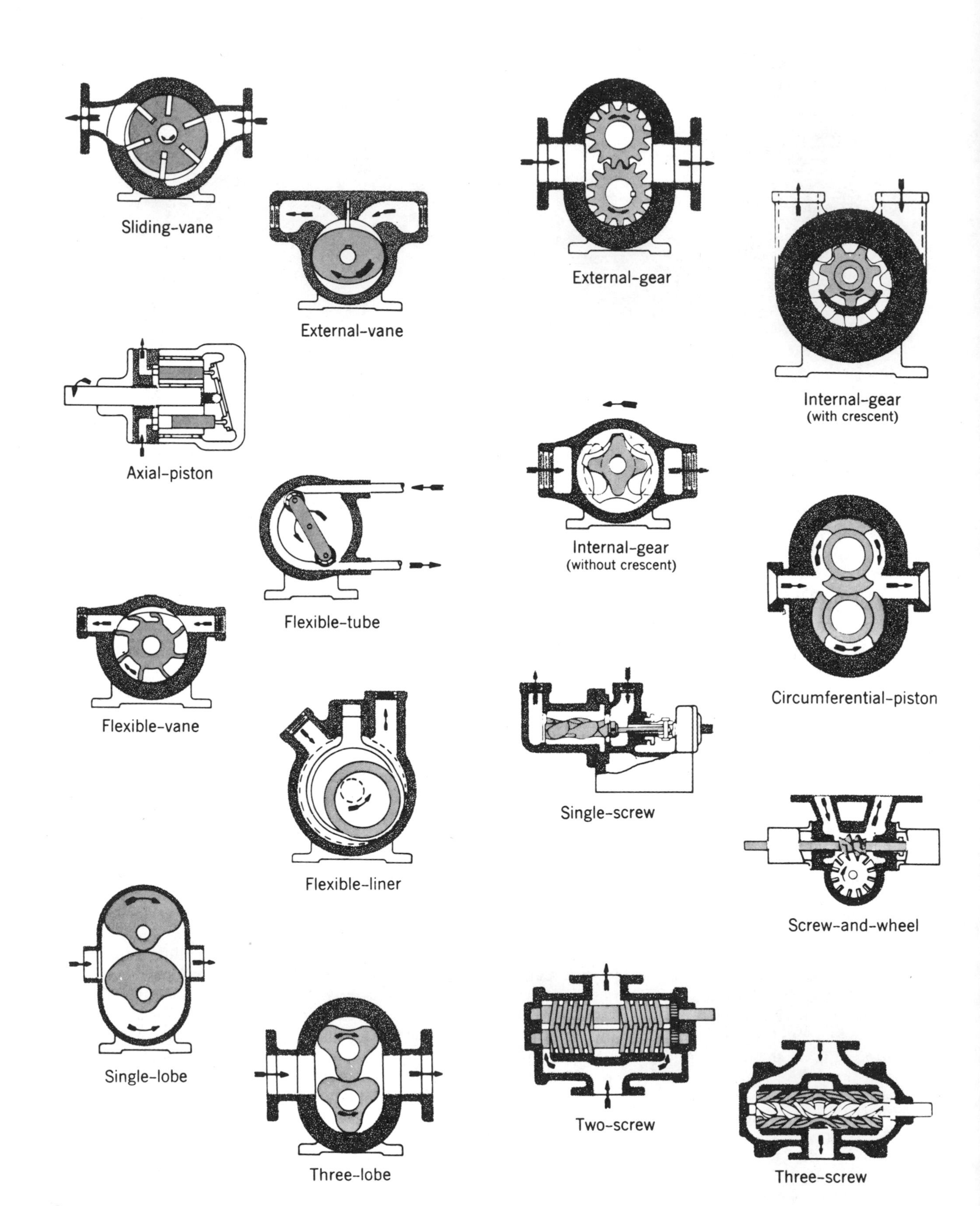


Source: Hydraulic Institute Standards

**ROTARY** postitive-displacement pumps handle many liquids, usually viscous, at flows up to about 500 gpm.—Fig. 12

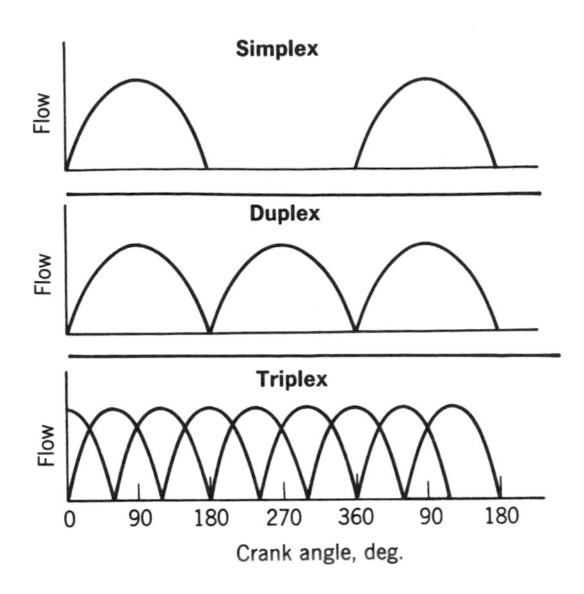


**DISCHARGE** flow of reciprocating pumps—Fig. 13

steam-traced. When specifying this modification, however, care should be taken that the steam temperature does not exceed the rating of either the pump or seal materials. It is important to specify a relief valve between the pump and the first downstream block valve.

## References

1. Stepanoff, A. J., "Centrifugal Pumps and Blowers," Wiley, New York, 1965.
2. Hallock, D. C., Centrifugal Compressors, The Cause of the Curve, *Air Gas Eng.*, Jan. 1968.
3. "Centrifugal Compressors in General Refinery Service," API 617, American Petroleum Institute, Washington, D. C.
4. Theberge, J. and Lomax, J., Design Properties of Filled TFE Compounds, *Compressed Air*, Dec. 1971.
5. Southwest Research Institute, San Antonio, Tex.
6. "Reciprocal Compressors for General Refinery Service," API 618, American Petroleum Institute, Washington, D. C.
7. "Engineering Data Book," Natural Gas Processors Suppliers Assn., Tulsa, Okla., 1973.
8. Abraham, R. W., Selection of Rotary Screw Compressors, *Oil Gas J.*, June 12, 1972, pp. 91-93.
9. Jones, C. A., Chlorine: Glamor Product of the CPI, *Power and Fluids From Worthington*, 10, No. 1 (1967).
10. "American National Standard Specification for Centrifugal Pumps for Process Use," ANSI B123.1, American National Standards Institute, New York, 1971.
11. "Vertical Inline Centrifugal Pumps for Process Use," Manufacturing Chemists Assn., Washington, D. C.
12. "Centrifugal Pumps for General Refinery Service," API 610, American Petroleum Institute, Washington, D. C.
13. "Hydraulic Institute Standards," Hydraulic Institute, New York, 1972.
14. Hattiangadi, U. S., Specifying Centrifugal and Reciprocating Pumps, *Chem. Eng.*, Feb. 23, 1970, pp. 101-108.

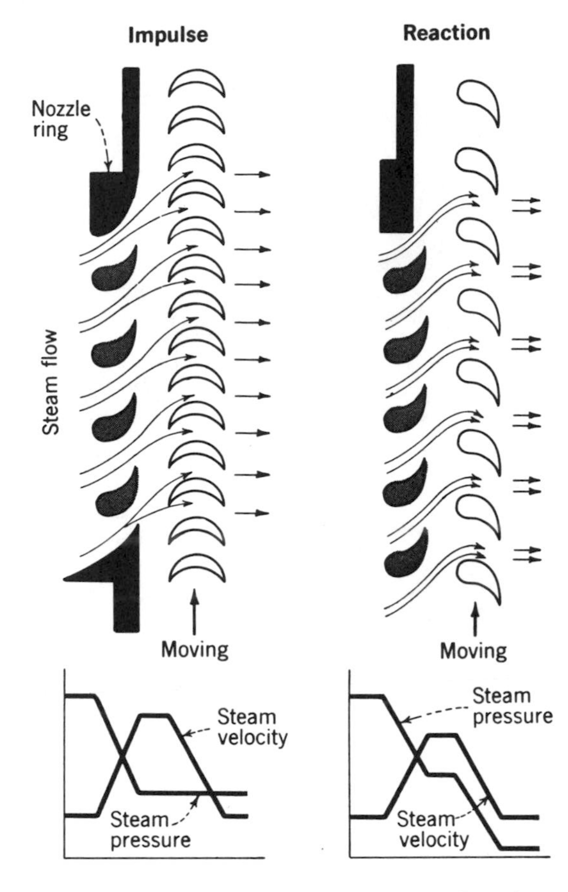


**STEAM** flow in impulse and reaction turbines—Fig. 14

## Meet the Author

**Robert W. Abraham** is supervisor of rotating equipment for The Badger Co., One Broadway, Cambridge, MA 02142. He works directly with project groups on all types of rotating equipment for the chemical, petrochemical and refinery industries. He has a B.S. in mechanical engineering from Lowell Technological Institute, is a member of ASME and the American Petroleum Institute Subcommittee on Mechanical Equipment, and a registered professional engineer in Massachusetts.

# Section II

# Gas Movers: Compressors, Fans and Blowers

# Keys to Compressor Selection

Handling gases in process plants ranges from very high pressures to vacuum for many flow conditions. Here is an analysis of equipment characteristics in order to make a preliminary selection of the most suitable type and size of compressor.

RICHARD F. NEERKEN, The Ralph M. Parsons Co.

The chemical process industries use compressors of all types and sizes for increasing the pressure of air or gases. In this article, we will provide a general description of all types of compressors, with specific examples to show how preliminary selections may be made by the project engineer. Such selections follow the basic fundamentals of thermodynamics and should not be considered so difficult or so complicated that only compressor manufacturers can make the initial choice of compressor for given process conditions.

Some typical applications are:

- Air compressors to provide service or instrument air for almost any plant.
- Simple air blowers in sulfur-recovery plants.
- Large air blowers in catalytic-cracking units.
- Low-temperature refrigeration compressors in ethylene, polyethylene or *p*-xylene units.
- High-pressure feed-gas, booster and recycle-gas compressors in hydrocarbon, ammonia and methanol-synthesis plants.

Compressors are either dynamic or positive-displacement types (Fig. 1). The dynamic types include radial-flow and axial-flow centrifugal machines; and to a limited extent, partial-emission types for low flows. Positive-displacement types exist in two basic categories: reciprocating and rotary. The reciprocating compressor consists of one or more cylinders, each with a piston or plunger that moves back and forth, displacing a positive volume with each stroke. Rotary compressors include lobe-type, screw-type, vane-type, and liquid-ring type,

Originally published January 20, 1975

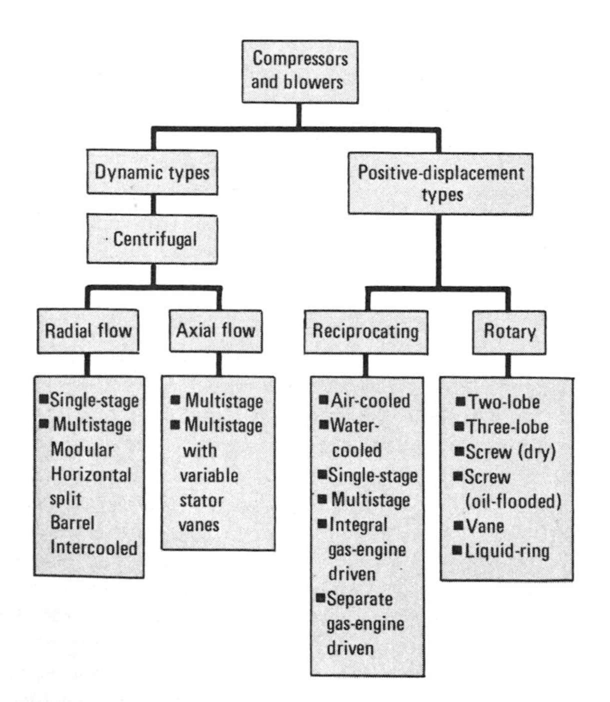


**TYPES** of compressors for chemical processes—Fig. 1

each having a casing with one or more rotating elements that either mesh with each other (such as lobes or screws) or that displace a fixed volume with each rotation.

## Operating Conditions

We must have certain information concerning (a) the operating conditions of any compressor and (b) the properties of the air, gas or gas mixture that is to be compressed.

Gas analysis is usually given in percent by volume. A molal analysis can be readily converted into a mol-percent analysis for use in determining the properties of the gas mixture. For air compressors, the inlet relative humidity or wet-bulb temperature is required, from which the amount of moisture present in the air can be determined.

The ratio of specific heats, $k$, where ($k = c_p/c_v$), may be given at suction temperature. For a more accurate calculation, $k$ should be at the average temperature during the compression cycle.

Compressibility factors (indicating deviation from an ideal gas) are given or calculated at suction and discharge conditions. For air or a pure gas, charts of compressibility

**Gas Analysis, Other Data and Computations for Example 1—Table I**

| Gas Mixture | | Molecular Weight, $M_w$ | Contribution,* $\%(M_w)/100$ | Specific Heat at 150°F, $c_p$, Btu/(Lb-Mol)(°F) | Contribution,* $\%(c_p)/100$ | Critical Pressure, $P_c$, Psia | Contribution,* $\%(P_c)/100$ | Critical Temperature, $T_c$, °R | Contribution,* $\%(T_c)/100$ |
|---|---|---|---|---|---|---|---|---|---|
| Component | Mol% | | | | | | | | |
| Hydrogen | 85 | 2.016 | 1.714 | 6.94 | 5.899 | 327 | 278 | 83 | 71 |
| Methane | 9 | 16.04 | 1.444 | 8.95 | 0.805 | 673 | 61 | 344 | 31 |
| Ethane | 3 | 30.07 | 0.902 | 13.77 | 0.413 | 708 | 21 | 550 | 17 |
| Propane | 2 | 44.09 | 0.882 | 19.53 | 0.390 | 617 | 12 | 666 | 13 |
| Isobutane | 0.5 | 58.12 | 0.291 | 25.75 | 0.128 | 529 | 3 | 735 | 4 |
| n-Butane | 0.5 | 58.12 | 0.291 | 25.81 | 0.129 | 551 | 3 | 766 | 4 |
| Total | 100 | 208.5 | 5.524 | | 7.764 | | 378 | | 140 |

* By multiplying the composition of each component in mixture by the property for that component, we obtain the contribution of the property for the amount of that component in the mixture.

factors as functions of actual pressure and temperature are available. Unless such charts are provided for mixed gases, it is normal to use the generalized compressibility charts [*1,2,3,4*], which require calculation of the reduced pressure, $P_r$, and reduced temperature, $T_r$. These terms are defined by: $P_r = P/P_c$ and $T_r = T/T_c$—where $P_r$ and $T_r$ are the reduced pressure and temperature, respectively; $P$ and $T$ are pressure, psia, and temperature, °R, at actual operating conditions, respectively; and $P_c$ and $T_c$ are critical pressure, psia, and critical temperature, °R, of the mixture. To demonstrate the several relations, let us review the procedure for a gas mixture.

*Example 1*—A typical hydrogen-hydrocarbon gas mixture has the composition shown in Table I. For this mixture, let us find the molecular weight, ratio of specific heats, critical pressure and critical temperature. The computations for the several components of the mixture are summarized in Table I, along with the pertinent data for each pure component. The specific-heat ratio, $k$, is computed as follows:

$$k = \frac{c_p}{c_v} = \frac{c_p}{c_p - 1.986} = \frac{7.764}{7.764 - 1.986} = 1.343$$

For this example, the molal specific heat, $c_p$, was taken at 150°F (assumed as a typical average temperature during the compression cycle for 100°F suction temperature). If the average temperature varies greatly from this value, the molal specific heat for the average temperature during compression must be used.

These computations may be done manually or by computer. If computerized, the standard values for all common gases for molecular weight, molal specific heat, critical pressure and temperature are stored in the computer memory.

Pressures and temperatures must be given at suction or inlet conditions, and pressure at discharge or outlet conditions, including the pressure of any side-load or intermediate requirement in the total compression cycle. The discharge temperature is not given, but is calculated to include the effects of heat rise during compression. Pressures are normally expressed in lb/in² gauge, psig, or lb/in² absolute, psia.

Capacity may be expressed in any of several ways:

- Weight flow, $W$, lb/h or lb/min.
- Volume flow referred to standard conditions (usually 14.7 psia and 60°F in the chemical process industries) as:

  (*SCFM*), standard cubic feet per minute.
  (*SCFH*), standard cubic feet per hour.
  (*MMSCFD*), million standard cubic feet per 24-hour day.

- Volume flow referred to inlet conditions usually expressed as:

  (*ICFM*), ft³/min, or ft³/s.
  $Q$ or $Q_s$, ft³/min, or ft³/s.

Regardless of how the capacity is given, it is necessary to convert to capacity at the inlet conditions to properly select or size a compressor. This conversion can be made by using any or all of the following relations:

$$\frac{P_1 V_1}{T_1 z_1} = \frac{P_2 V_2}{T_2 z_2} \qquad (1)$$

where $V$ is volume, $P$ is absolute pressure, $T$ is absolute temperature, and $z$ is the compressibility factor. In Eq. (1), the compressibility factor, $z_1$, may be assumed as 1.0 if $P_1$ and $T_1$ are at standard conditions of 14.7 psia and 520°R.

$$(ICFM) = Q_s = W\bar{v} = W/\rho \qquad (2)$$

where $W$ is flow, lb/min, $\bar{v}$ is specific volume, ft³/lb, and $\rho$ is density, lb/ft³. Specific volume, $\bar{v}$, may be calculated from:

$$\bar{v} = z\left(\frac{1{,}545}{M_w}\right)\left(\frac{T}{144P}\right)$$

where $M_w$ is the molecular weight.

$$(SCFM) = 379.46M/60 \qquad (3)$$

where $M$ is flow, mol/h.

$$W = M(M_w) \qquad (4)$$

where $W$ is weight flow, lb/h, $M$ is flow, mol/h and $M_w$ is molecular weight.

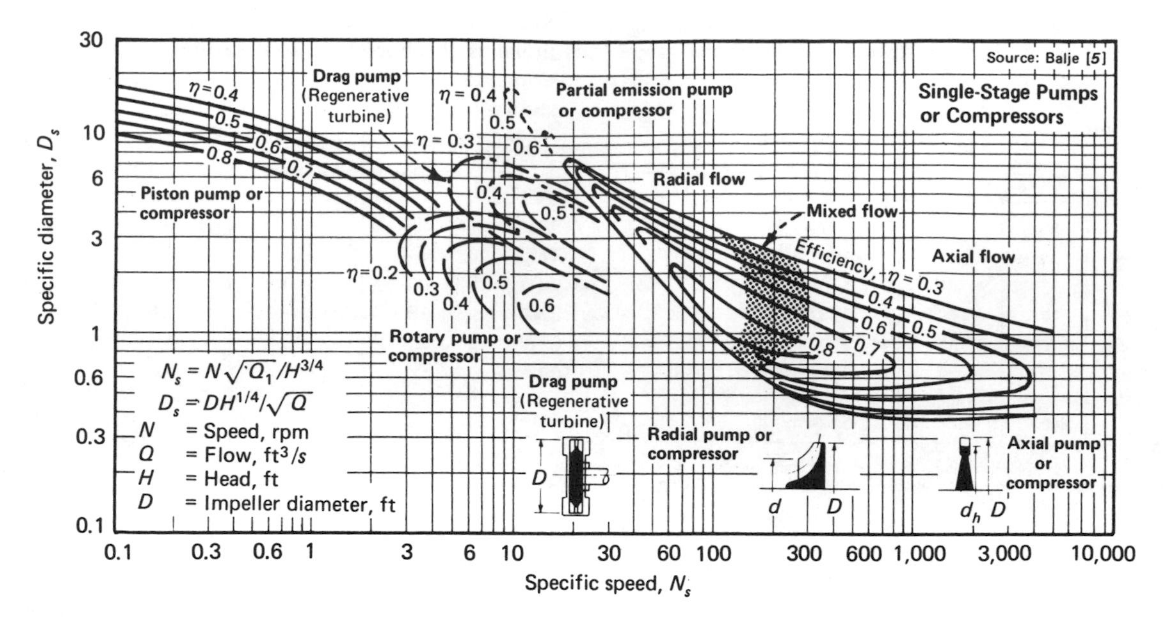


**SPECIFIC** speed and specific diameter enable the initial selection of a definite type of single-stage compressor—Fig. 2

$$Q_s = (ICFM) = \left[\frac{(MMSCFD) \times 10^6}{60 \times 24}\right]\left(\frac{14.7}{P_s}\right)\left(\frac{T_s}{520}\right)\left(\frac{z_s}{1.0}\right) \quad (5)$$

where the subscript, $s$, denotes properties at suction or inlet conditions.

## Compressor Head and Horsepower

For any compressor, the horsepower required is:

$$(HP)_{g(ad)} = WH_{ad}/33{,}000\eta_{ad} \quad (6)$$

$$H_{ad} = \left(\frac{z_s + z_d}{2}\right)\left(\frac{1{,}545}{M_w}\right)T_s\left[\frac{r_c^{(k-1)/k} - 1}{(k-1)/k}\right] \quad (7)$$

where $(HP)_{g(ad)}$ is adiabatic gas horsepower, hp; $W$ is weight flow, lb/min; $H_{ad}$ is adiabatic head, (ft-lb)/lb; $\eta_{ad}$ is adiabatic efficiency; $z_s$ is compressibility factor at suction conditions, $z_d$ is compressibility factor at discharge conditions; $M_w$ is molecular weight; $T_s$ is suction temperature, °R; and $r_c$ is the ratio of compression, i.e., $P_d/P_s$.

Adiabatic discharge temperature, $T_{d(ad)}$, °R is:

$$T_{d(ad)} = T_s r_c^{(k-1)/k} \quad (8)$$

Certain types of compressors perform closely to the adiabatic conditions; many others deviate significantly from adiabatic, and the compression cycle must be considered polytropic. In this case, the necessary relations are:

$$(HP)_{g(poly)} = WH_{poly}/33{,}000\eta_{poly} \quad (9)$$

$$H_{poly} = \left(\frac{z_s + z_d}{2}\right)\left(\frac{1{,}545}{M_w}\right)T_s\left[\frac{r_c^{(n-1)/n} - 1}{(n-1)/n}\right] \quad (10)$$

where $(HP)_{g(poly)}$ is polytropic gas horsepower, hp; $W$ is weight flow, lb/min; $H_{poly}$ is polytropic head, (ft-lb)/lb; $\eta_{poly}$ is polytropic efficiency; $z_s$ and $z_d$ are compressibility factors for suction and discharge conditions, respectively; $M_w$ is molecular weight; $T_s$ is suction temperature, °R; and $r_c$ is the ratio of compression.

Polytropic discharge temperature, $T_{d(poly)}$, is calculated from:

$$T_{d(poly)} = T_s r_c^{(n-1)/n} \quad (11)$$

The value of the quantity $n$ in the several polytropic relations is obtained from:

$$\left(\frac{n-1}{n}\right) = \left(\frac{k-1}{k}\right)\left(\frac{1}{\eta_{poly}}\right)$$

When using tables of gas properties or Mollier diagrams for performing compressor calculations, the adiabatic head, $H_{ad}$, is obtained from:

$$H_{ad} = 778\Delta h \quad (12)$$

where $h$ is enthalpy, Btu/lb.

Relation of adiabatic efficiency to polytropic efficiency:

$$\eta_{ad} = \left[\frac{(r_c^{(k-1)/k} - 1)}{(r_c^{(n-1)/n} - 1)}\right] \quad (13)$$

## Specific Speed

Specific speed, $N_s$, is an index number for the impellers or rotors of various types of pumps and compressors. The definition is the same for both pumps and compressors:

$$N_s = N\sqrt{Q}/H^{3/4} \quad (14)$$

When using Eq. (14) for compressors, it is common to express speed, $N$, in rpm; capacity, $Q$, in ft³/s at inlet conditions; and head, $H$, in (ft-lb)/lb.

Another dimensionless quantity for impellers or rotors is termed the specific diameter, $D_s$, defined by:

$$D_s = \frac{DH^{1/4}}{\sqrt{Q}} \quad (15)$$

where $D$ is diameter of impeller or rotor, ft.

Balje [5] presented a specific-speed chart combining the

relations of Eq. (14) and (15), which is shown in Fig. 2. If Fig. 2 is used, it must be on a head per stage basis—i.e., each impeller or stage must be chosen with regard to the inlet capacity and head for that stage. Although past experience with existing types and designs of compressors will often make reference to a diagram such as Fig. 2 unnecessary, it does provide a logical correlation for selecting the type of compressor that can be used for any given application. The following example will illustrate a typical use of Fig. 2.

*Example 2*—Let us make a preliminary selection of a compressor to handle 90,000 ICFM of air when inlet conditions are 14.3 psia, 90°F, and 70% relative humidity. Discharge pressure will be 22.3 psia, molecular weight = 28.59, $k = c_p/c_v = 1.395$. We will assume an impeller diameter, $D$, of 55 in, and a rotating speed, $N$, of 3,550 rpm.

In order to use Fig. 2, we must find the specific speed and specific diameter from Eq. (14) and (15). To do so, we first calculate inlet air flow, $Q_s = 90{,}000/60 = 1{,}500$ ft³/s, and adiabatic head from Eq. (7), keeping in mind that compressibility factors are unity for these conditions. Hence:

$$H_{ad} = \left(\frac{1{,}545}{28.59}\right)(550)\left[\frac{(22.3/14.3)^{0.283} - 1}{0.283}\right] = 14{,}072$$

$$N_s = \frac{3{,}550\sqrt{1{,}500}}{(14{,}072)^{3/4}} = 106.4$$

$$D_s = \frac{(55/12)(14{,}072)^{1/4}}{\sqrt{1{,}500}} = 1.29$$

By using these values in Fig. 2, we find that a centrifugal compressor with a single radial-flow impeller will give a selection in the optimum-efficiency range.

## Centrifugal Compressor Selection

Centrifugal compressors have become the most popular type for use in the chemical process industries largely because their relatively simple, maintenance-free designs provide long periods of continuous operation.

The simplest style of centrifugal compressor is the single-stage overhung design. Current commercial designs are available from flows of about 3,000 ICFM to as high as 150,000 ICFM. The conventional closed or shrouded impeller (Fig. 3) will be used for adiabatic heads to about 12,000 (ft-lb)/lb. The open, radial-bladed impeller (Fig. 3) will develop more head for the same diameter and speed. Variations of this type (Fig. 3), having inducer or three-dimensional-type blading, will develop up to 20,000 (ft-lb)/lb of head.

Similar designs are used with higher-strength materials, at higher speeds, for specialized applications such as the integral-gear plant-air compressor, or for aerospace applications, engine turbochargers, expander load-compressors, etc.

## Multistage Centrifugal Compressors

When the head requirement is too great for one impeller, the obvious solution is two or more impellers in series. These form the multistage compressors that are found in most process applications. The most conventional is the horizontal-split casing with three to eight impellers in series, with or without intercooling such as shown on p. 79.

Designs are available in flow ranges from 1,000 to 100,000 ICFM, at total polytropic heads from 20,000 to 100,000 (ft-lb)/lb, based on the number of impellers or stages per casing. Such a casing is sometimes arranged with opposed impellers to partially equalize thrust, and to simplify design problems relating to thrust bearings, balance drums and shaft seals.

Similar impeller arrangements are used in vertical-split, barrel-type casings made of welded, cast or forged steel. The vertical-split casings are more suitable for higher pressures than the horizontally split designs.

The current "Standard for Centrifugal Compressors," [*6*] specifies that barrel-type casings must be used for pressures above 200 to 250 psig if the hydrogen content of the gas mixture is 70% or greater, to help ensure against leakage. Flow ranges are from about 1,000 to 100,000 ICFM. Such compressor barrels have been built for pressures as high as 10,000 psig.

The standard machine for compressed-air service today is the three-stage or four-stage intercooled design such as shown on p. 78, that is built in sizes ranging from 500 to 70,000 ICFM, based on atmospheric air compressed to about 125 psig. At present, this type has not been widely applied to gas service—especially if the gases are dirty, corrosive or toxic. Closer examination of this design reveals that the impellers are attached to pinion shafts that rotate at different speeds for succeeding stages. This arrangement enables the machine designer to achieve optimum dimensional sizing and efficiency on a volume of air or gas that is continually being reduced due to compression. This results in a more efficient machine than the conventional design for single-shaft gas or air compressors.

A popular derivative of the multistage compressor is the external-bolted casing or modular type developed for low-pressure air or gas service. This machine is widely used for flows ranging from 400 to 20,000 ICFM, at heads as high as 18,000 to 20,000 (ft-lb)/lb. Here, the casing is assembled in modules (doughnut-shaped rings each containing one diffuser section and one impeller). The unit runs at speeds of 3,000 to 4,000 rpm, which allows the use of inexpensive ring-oiled or grease-lubricated ball bearings. Also, the low tip-speeds permit the use of cast-aluminum or fabricated-aluminum impellers rather than the more expensive forged-steel types in higher-speed machines.

Higher-speed designs of the modular type are also available. These cover flows from 500 to 15,000 ft³/min, and heads to about 60,000 ft in a single casing. This modular type has most of the high-speed features regarding bearings, seals, shaft and impeller designs but costs much less than the horizontal split-case multistage unit.

All of these types have mechanical limitations, due to shaft and bearing stiffness, shaft deflection, critical speed, and rotor dynamic problems. When the process requires more head than can be developed with the maximum number of impellers in one casing, two or three casings may be used in series to obtain as many as 25 or 30

## Example 3: Overall Polytropic-Head Method—Table II

**Centrifugal-Compressor Calculation**

| Identification | Recycle Compressor | Alternate Selection | Source or Explanation |
|---|---|---|---|
| Capacity, MMSCFD | 80 | Same | Given |
| Capacity, $W$, lb/h | — | — | Given (sometimes) |
| Inlet capacity, $Q$, ICFM | 2,961 | ← | Eq. (5) |
| Suction pressure, $P_s$, psia | 300 | ← | Given |
| Suction temperature, °F | 100 | ← | Given |
| Suction temperature, °R | 560 | ← | Given |
| Relative humidity, % | — | — | Given (if air) |
| Discharge pressure, $P_d$, psia | 450 | ← | Given |
| Molecular weight, $M_w$ | 5.524 | ← | Given |
| Gas constant, $R = 1{,}545/M_w$ | 279.69 | ← | Calculated |
| Specific heat ratio, $k$ | 1.343 | ← | Given, or calculated. See Table I |
| Compressibility, suction, $z_s$ | 1.01 | ← | Given, or calculated. See Table I |
| Compressibility, discharge, $z_d$ | 1.022 | ← | Given, or calculated. See Table I |
| Compressibility, average, $(z_s = z_d)/2$ | 1.016 | ← | Calculated |
| Specific volume, $\bar{v}$, ft³/lb | 3.66 | ← | See Eq. (2) |
| Weight flow, $W_s$, lb/min | 809 | ← | See Eq. (2) |
| Specific heat exponent, $(k - 1)/k$ | 0.255 | ← | Calculated |
| Acoustic velocity at inlet, $U_a$, ft/s | 2,616 | ← | $U_a = \sqrt{kgRT_s z_s}$ |
| Ratio of compression, $r_c = P_d/P_s$ | 1.5 | ← | Calculated |
| Head coefficient, $\mu$ | 0.49 | ← | From Table IV |
| Polytropic efficiency, $\eta_{poly}$, % | 73 | ← | From Fig. 4 |
| Nominal impeller diameter, $D$, in | 18 | ← | From Table IV |
| Polytropic exponent ratio, $(n - 1)/n = Y$ | 0.349 | ← | $Y = \dfrac{(k-1)/k}{\eta_{poly}}$ |
| $(r_c)^Y$ | 1.152 | ← | Calculated |
| Discharge temperature, polytropic, $T_d$, °R | 645 | ← | $T_d = T_s(r_c)^Y$ |
| Discharge temperature, polytropic, $t_d$, °F | 185 | ← | |
| Polytropic head, $H_{poly}$, (ft-lb)/lb | 69,307 | ← | Eq. (10) |
| Gas horsepower, $(HP)_{g(poly)}$, hp | 2,328 | ← | Eq. (9) |
| Friction horsepower, bearings, hp | 28 | 34 | Select from Fig. 6 |
| Friction horsepower, seals, hp | 27 | 35 | Select from Fig. 6 |
| Friction horsepower, gear | 0 | 0 | None (use steam turbine) Estimate at 2% of gas horsepower |
| Total brake-horsepower, BHP | 2,383 | 2,397 | |
| Impeller tip-speed, maximum, $U$, ft/s | 900* | | $U \leqq 0.9$ to $1.0(U_a)$ |
| Impeller tip-speed, actual, $U$, ft/s | 807 | 871 | $U = \sqrt{\dfrac{H_{poly}g}{N_{st}\mu}}$ |
| Number of stages, $N_{st}$ | 7 | 6 | From above relation |
| Frame size or designation | #2 | #2 | From Table IV or manufacturer |
| Rotating speed, $N$, rpm | 10,267 | 11,081 | $N = 229U/D$ |
| Flow coefficient at inlet, $\phi_s$ | 0.0346 | 0.0321 | Eq. (17) |
| Flow coefficient at outlet, $\phi_d$ | 0.0269 | 0.0249 | Eq. (17) |

*In this example, the maximum impeller top-speed is set by stress limitations in a conventional backward-curved-blade impeller, not by comparison with acoustic velocity. Conclusion: The preliminary selection is a 7-stage or 6-stage centrifugal compressor, without intercooling, requiring approximately 2,400 BHP, and operating at 10,267 or 11,081 rpm.

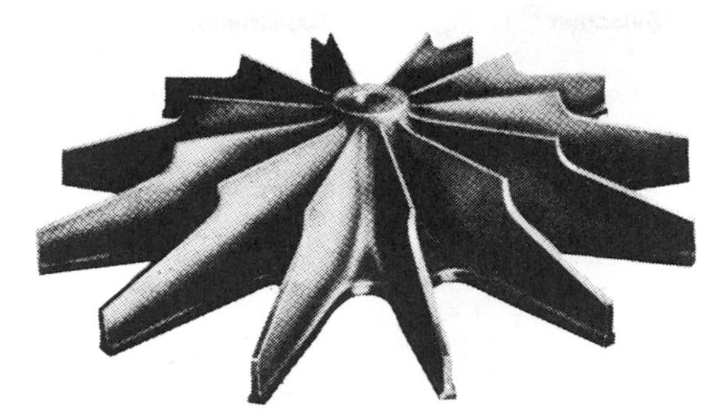

Open radial impeller

Open inducer impeller

Closed impeller

**SINGLE-STAGE** impellers for compressors—Fig. 3

impellers in series. Selection of such combinations still comes back to fundamentals, i.e., calculating (a) the ICFM (inlet ft$^3$/min) at whichever casing is being considered, (b) total adiabatic or polytropic head across that stage or section, and then (c) matching available casing and impeller sizes and speeds, with or without speed-change gears, to obtain a complete series of casings.

Before proceeding to select or specify a compressor with multiple stages, the temperature rise during compression must be carefully considered. If the resulting discharge temperatures are above approximately 350°F, some means of cooling the gas must be included in order to avoid dangers from hot discharge gases or problems

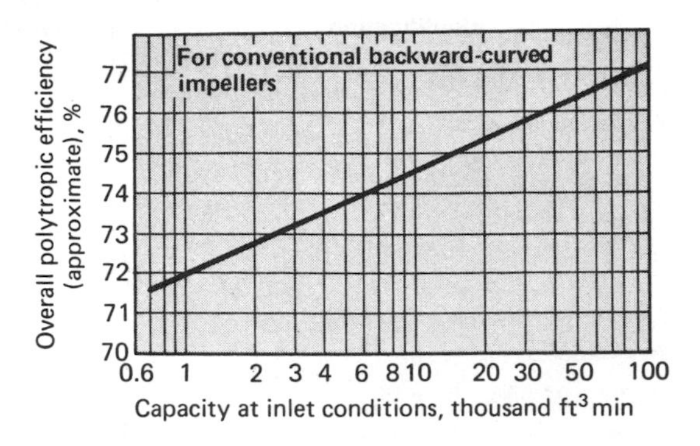


**EFFICIENCY** of multistage centrifugal compressors—Fig. 4

associated with materials of construction at higher temperatures. Intercoolers are usually required to cool the gases before further compression after each stage (in some designs) or after several stages.

## Calculation Methods

Weight flow, inlet capacity, adiabatic or polytropic head, and the approximate horsepower may be quickly estimated by the basic relations thus far described. Two examples summarize the necessary procedures.

*Example 3*—Let us make a preliminary selection of a centrifugal compressor to handle 80 MMSCFD of a recycle gas having a molecular weight of 5.524. We will use the methods for an overall polytropic head. Other pertinent data and the necessary computations are shown in Table II.

*Example 4*—Let us now make a preliminary selection of a centrifugal compressor to handle a mainstream of 64,200 lb/h and an additional sidestream of 42,300 lb/h of propane. We will use the method requiring a Mollier chart in this problem. Other pertinent data and the necessary computations are shown in Table III.

In performing the computations for Examples 3 and 4, we will need some additional information and explanations of thc material shown in Tables II and III.

Some representative values for polytropic efficiency, flow range, impeller diameter, and head coefficient for currently manufactured industrial machines are shown in Fig. 4 and Table IV. More-specific data must be obtained from manufacturers of compressors.

The head coefficient, $\mu$, and the flow coefficient $\phi$, are dimensionless values that are used to describe the performance of any single compressor impeller or group of impellers. The relation can be expressed as a performance curve (Fig. 5). The value of $\mu$ at or near peak efficiency is chosen for the preliminary selection. The head and flow coefficients are defined as:

$$\mu = H_{st}g/U^2 \tag{16}$$

$$\phi = 700Q_s/ND^3 \tag{17}$$

where $H_{st}$ is head per stage, ft; $g$ is the gravitational constant, 32.2 ft/$s^2$; $U$ is impeller tip-speed, ft/$s$; $Q_s$ is inlet capacity, ft$^3$/min; $N$ is impeller speed, rpm; and $D$ is impeller diameter, ft.

The actual values of $\mu$ and the shape of the curve are

## Example 4: Mollier-Chart Method—Table III

### Centrifugal—Compressor Calculation

| Section or Stage | First | Side Load | Second | Source or Explanation |
|---|---|---|---|---|
| Gas | Propane | ← | ← | Given |
| Molecular weight, $M_w$ | 44 | ← | ← | Given |
| Ratio specific heats, $k$ | 1.13 | ← | ← | Given |
| Compressibility, $z_s$ | 0.95 | --- | 0.915 | Given or from Tables or Mollier chart |
| Load, $W_1$, lb/h | 64,200 | --- | --- | Given |
| Load, $W_1$, lb/min | 1,070 | --- | --- | |
| Added load, $W_2$, lb/h | --- | 42,300 | --- | Given |
| Added load, $W_2$, lb/min | --- | 705 | --- | |
| Total load, $W_1 + W_2$, lb/min | 1,070 | --- | 1,775 | |
| Suction pressure, $P_s$, psia | 24 | 56 | 56 | Given |
| Suction temperature, $t_s$, °F | −20 | +20 | +35* | Given (* or calculated) |
| Suction temperature, $T_s$, °R | 440 | 480 | 495 | |
| Suction enthalpy, $h_1$, Btu/lb | 104.5 | 115 | 122.2 | From Tables or Mollier chart |
| Suction specific volume, $\bar{v}_s$, ft³/lb | 4.25 | | 1.96 | From Tables or Mollier chart |
| Suction capacity, $Q_s$, ft³/min | 4,548 | | 3,479 | See Eq. (2) |
| Discharge pressure, $P_d$, psia | 56 | | 215 | Given |
| Discharge enthalpy, $h_{2(ad)}$, Btu/lb | 123 | | 151.2 | From Tables or Mollier chart |
| $\Delta h$ (adiabatic), Btu/lb | 18.5 | | 29 | $\Delta h = h_{2(ad)} - h_1$ |
| Head (adiabatic), (ft-lb)/lb | 14,393 | | 22,562 | Eq. (12) |
| Efficiency (polytropic), $\eta_{poly}$, % | 73.5 | | 73.3 | From Fig. 4 |
| Efficiency (adiabatic), $\eta_{ad}$, % | 72.5 | | 71.5 | Eq. (13) |
| $\Delta h$ (polytropic), Btu/lb | 22.5 | | 40.6 | $\Delta h_{poly} = h_{ad}/\eta_{ad}$ |
| Enthalpy (polytropic), Btu/lb | 127 | | 162.8 | $h_1 + \Delta h_{poly}$ |
| Discharge temperature, °F | 48 | | 162 | From Tables or Mollier chart |
| Discharge specific volume, $\bar{v}_d$, ft³/lb | 2.0 | | 0.58 | From Tables or Mollier chart |
| Discharge capacity, $Q_d$, ft³/min | 2,140 | | 1,030 | $Q_d = w_1\bar{v}_d$ |
| Gas horsepower, $(HP)_{g(ad)}$, hp | 644 | | 1,697 | Eq. (6) |
| Bearing and seal loss, hp | | 63 | | |
| Gear loss (if used), hp | | 47 | | Estimate 2% of gas horsepower |
| Total brake-horsepower, BHP | | 2,451 | | |
| Frame size | #2 | | #2 | |
| Number of stages, $N_{st}$ | 2 | | 3 | Estimate |
| Preliminary impeller diameter, $D$, in | 17.5 | | 18 | From Table IV or manufacturers' data |
| Preliminary head coefficient, $\mu$ | 0.50 | | 0.49 | From Table IV or manufacturers' data |
| Preliminary tip-speed, $U$, ft/s | 681 | | 703 | $U = \sqrt{H_{ad}g/N_{st}\mu}$ |
| Acoustic velocity, $U_a$, ft/s | 731 | | 761 | $U_a = \sqrt{kgRTz}$ |
| Ratio, $U/U_a$ | 0.932 | | 0.924 | $\cong$ 0.9 to 1.0 |
| Shaft speed, $N$, rpm | 8,943 | | 8,943 | $N = 229U/D$ |

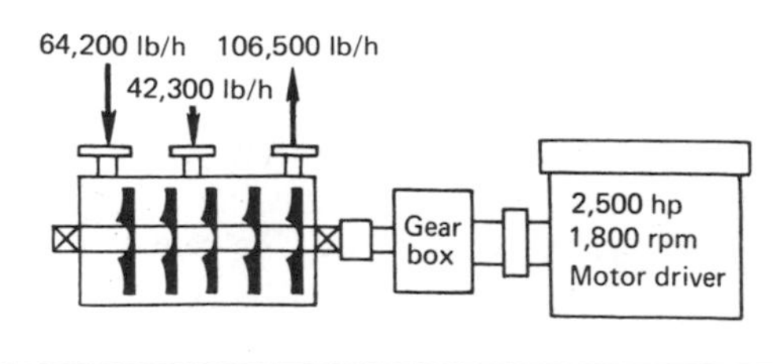

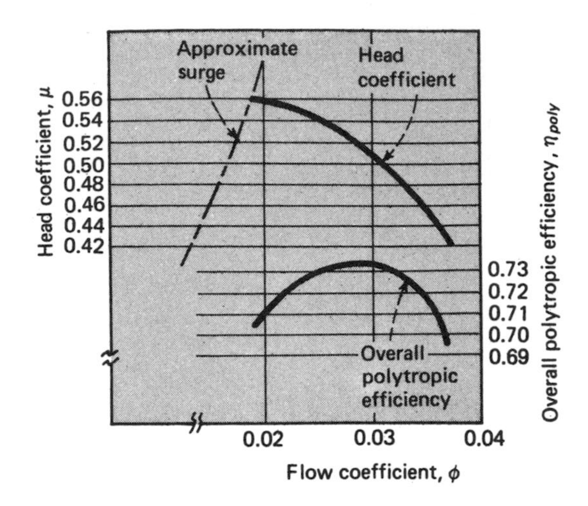


**PERFORMANCE** of a centrifugal compressor—Fig. 5

dependent on the design of the impeller [7]. More-specific information, regarding surge point and rise in head, is necessary before attempting to estimate the shape of an actual curve.

After the gas horsepower has been determined by either method, horsepower losses due to friction in bearings, seals and speed-increasing gears must be added. Fig. 6 shows suggested losses for conventional multistage units, based on the use of oil-film seals [8].

All of these calculations give only preliminary results. More-accurate determinations are made by the compressor designer through the use of individual impeller data, making possible the "wheel by wheel" selection, in which the performance of each impeller is determined on the basis of its specific inlet conditions, then added together to give the total performance of the compressor.

## Control of Centrifugal Compressors

When any or all of the following parameters—molecular weight, ratio of specific heats, suction or discharge pressure, or temperature—change with respect to flow, a different point is reached on the head-capacity curve for any given compressor because the compressor develops head and *not* pressure.

Centrifugal compressors and blowers follow the "fan laws" or "affinity laws" regarding variation in capacity and head as a function of speed:

$$\frac{N_1}{N_2} = \frac{Q_1}{Q_2} = \frac{\sqrt{H_1}}{\sqrt{H_2}} \tag{18}$$

where $N$ is speed, $Q$ is volumetric capacity, and $H$ is head.

Thus, the most effective way to match the compressor characteristic to the required output is to change the speed in accordance with Eq. (18). This is one of the principal advantages of using steam or gas turbines as drivers for compressors; they are inherently suited to variable-speed operation. With such drivers, the speed can be controlled manually by an operator adjusting the speed governor on the turbine. Or, the speed adjustment can be made automatically by a pneumatic or electronic

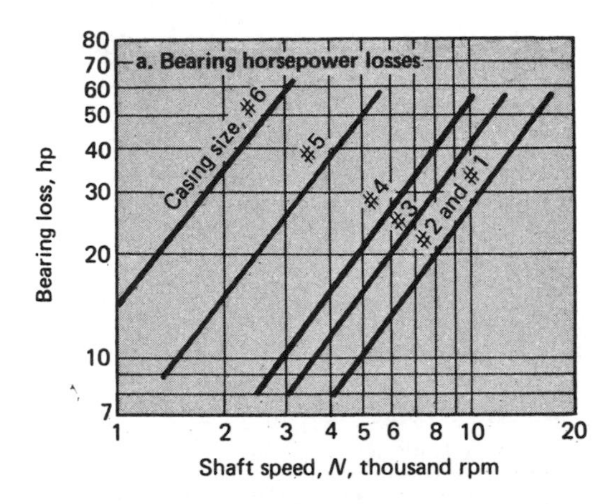


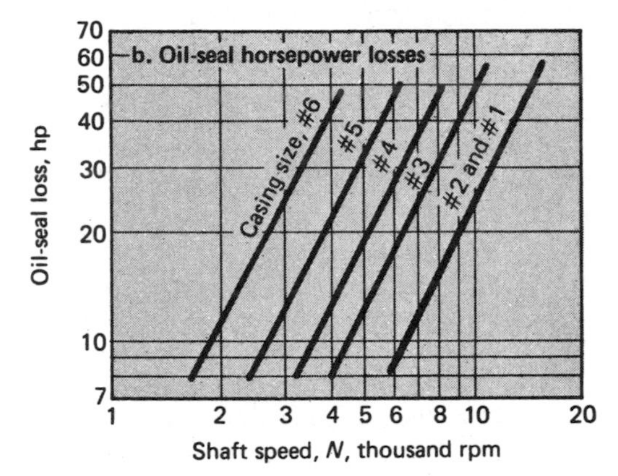


**LOSSES** due to friction in bearings and seals—Fig. 6

controller that changes the speed in response to a flow or pressure signal.

For constant-speed drives such as electric motors,* the compressor must be controlled in one of three ways:

1. Inlet-guide vanes (most efficient).
2. Throttling of suction pressure.
3. Throttling of discharge pressure (least efficient).

Inlet-guide vanes are manually or automatically adjustable stationary vanes in the inlet to the first stage (sometimes to succeeding stages) that cause the angle of approach of the incoming gas to change relative to the rotating impeller. This changes the flow characteristic in response to varying load requirements. Fig. 7 illustrates the effect of such control on head and capacity. Although most efficient, the economics of inlet-guide vanes must be carefully studied, as they are expensive, complex on some types of machines, and constitute additional mechanisms requiring maintenance and adjustment.

A compromise to achieve simplicity and efficiency is usually made by throttling the suction. This results in a slightly lower suction pressure than the machine was designed for, and thus yields a higher total head if the discharge pressure remains constant. This can be matched to the compressor's head-capacity curve, i.e. higher head at reduced flow. In throttling the inlet, the density of the

*__Note:__ Variable-speed electric-motor drivers are also available but seldom used in the chemical process industries. Two-speed or multispeed wound-rotor motors present problems in hazardous areas. The use of electrical or hydraulic variable-speed clutches or couplings often presents mechanical problems and potential inefficiencies at off-design conditions.

**Preliminary Selection Values for Multistage Centrifugal Compressors—Table IV**

| Nominal Size | Flow Range, $Ft^3$/Min | Head Coefficient,* Average, $\mu$ | Impeller Diameter, Nominal, $D$, In |
|---|---|---|---|
| 1 | 800 to 2,000 | 0.48 | 14 to 16 |
| 2 | 1,500 to 7,000 | 0.49 to 0.50 | 17 to 19 |
| 3 | 4,000 to 12,000 | 0.50 to 0.51 | 21 to 22 |
| 4 | 6,000 to 17,000 | 0.51 to 0.52 | 24 |
| 5 | 8,000 to 35,000 | 0.51 to 0.52 | 32 |
| 6 | 35,000 to 65,000 | 0.53 | 42 to 45 |
| 7 | 65,000 to 100,000 | 0.54 | 54 to 60 |

*Based on impellers with backward-curved blades. Impellers with radial blades have higher values.

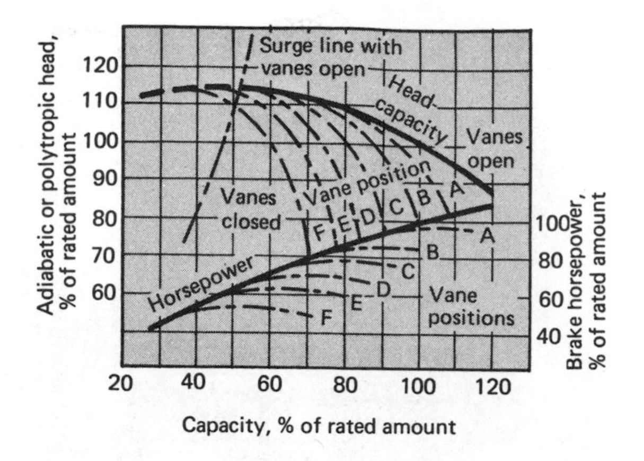


**INLET** guide vanes affect performance—Fig. 7

gas is reduced, resulting in a matching of the required weight flow to the compressor's inlet-volume capabilities at other points on the head/capacity curve.

The least-efficient method of control is to throttle the discharge. At reduced flow, the compressor develops more head (and pressure) than the process requires. This extra head or pressure is throttled before passing to the process equipment, but the horsepower consumed in compressing it is wasted—hence, the relative inefficiency. This method has the advantage, however, of being very simple, and is often applied on small-horsepower machines where the inefficiency can be neglected.

## Surge Control of Centrifugal Compressors

All dynamic compressors have a limited range of capacity for a given selection of impellers at a fixed speed. Below the minimum value (usually 50 to 70% of the rated flow), the compressor will surge, i.e. become unstable in operation. Excessive vibration and possibly sudden failure or shutdown may then occur.

It is essential that all compressor systems be designed to avoid possible operation in surge. This is usually done by providing some type of antisurge control. The simplest form is used on air compressors, and consists of a blowoff valve, automatically controlled, to open and blow off excess capacity to the atmosphere if the process-flow requirement is too low. More-efficient methods with suction-control valves are sometimes used.

For a gas that cannot be wasted to atmosphere, the most common antisurge control is bypass control (i.e., bypassing unwanted flow back to the suction source). Since this gas has already been compressed, its temperature has increased. Therefore, it must be cooled before entering the compressor a second time. A bypass cooler may be required. In systems where the suction source is large enough, or far enough away, so that the heat is dissipated by mixing or radiation, the cooler may not be required.

Several well-engineered surge-control systems are also available from companies specializing in process controls. It may be preferable to purchase such a control system than to design your own [*9*].

## Shaft Seals on Rotating Machines

Shaft seals are required for every rotating compressor shaft to contain the gas being compressed or to allow a controlled leakage. Seals are generally of four basic types: (1) labyrinth, (2) restrictive ring (carbon ring), (3) oil-film and (4) mechanical contact. Fig. 8 illustrates these types Ref. [*6*].

Labyrinth or restrictive-ring types are used only when some leakage of air or gas can be tolerated. The oil-film types are normally used on process gases—especially on gases containing impurities that are dangerous or toxic, such as hydrogen sulfide. The mechanical-contact type can also be applied to most gases, but finds its biggest use on clean, heavier hydrocarbon gases, refrigerant gases, etc.

Occasionally, a buffer gas will be required to form a buffer between the compressed gas and the atmosphere, and is often found when compressing dangerous or toxic gases that must not be allowed to leak out. Such a system has the disadvantage of requiring an external gas supply at a pressure higher than the compressor's suction pressure. The system also requires a buffer gas that is clean and compatible with the gas being compressed (since some of the buffer gas may leak inward), and is available in an uninterrupted supply. If this last is not possible, a backup gas such as bottled nitrogen may be required as well—making the overall system very complicated and expensive.

## Lubrication and Seal-Oil Systems

When oil-film or mechanical-contact seals are used, seal oil is required. This is usually a self-contained, recirculated system, supplying oil under pressure to the seals; and draining oil away from the seals in one or two separate streams, depending on the contamination of the oil by contact with the gas.

For example, if the gas contains hydrogen sulfide, the oil that leaks inward toward the gas will be contaminated. This oil will be drained separately into contaminated-oil drain traps, possibly to be discarded or at least reconditioned before reuse. If the gas is not toxic, the inner drain

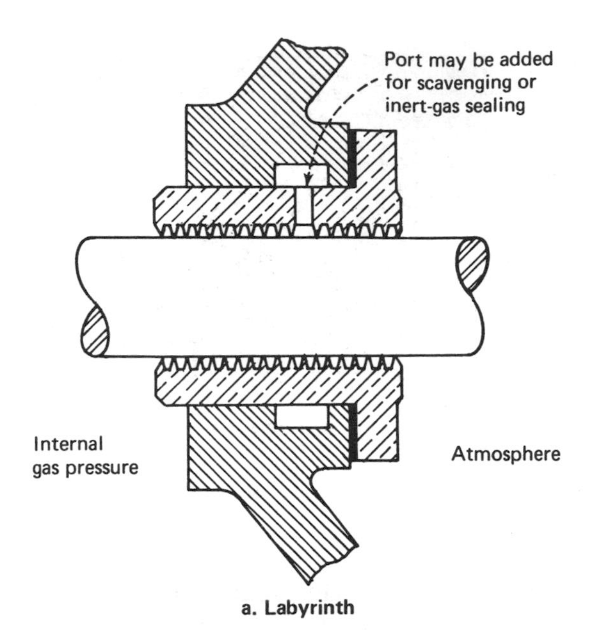


a. Labyrinth

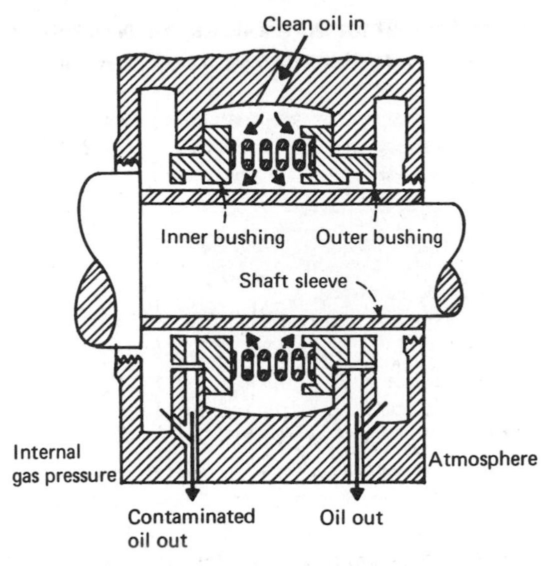


d. Liquid film with cylindrical bushing.

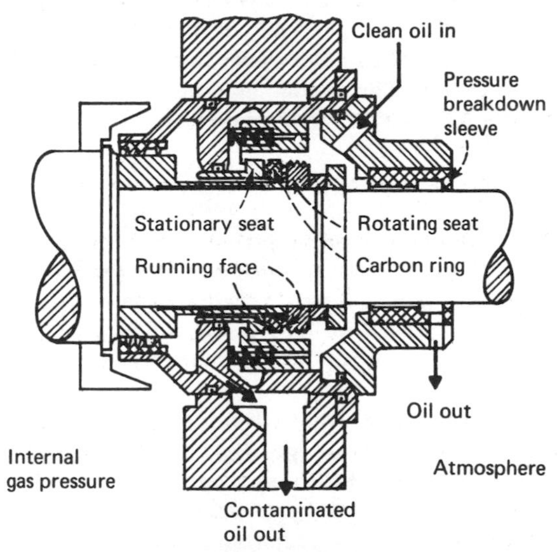


b. Mechanical (contact)

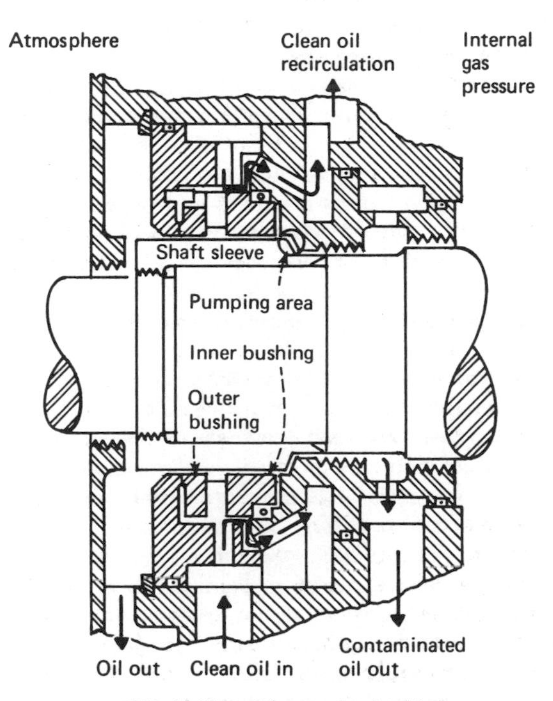


e. Liquid film with pumping bushing

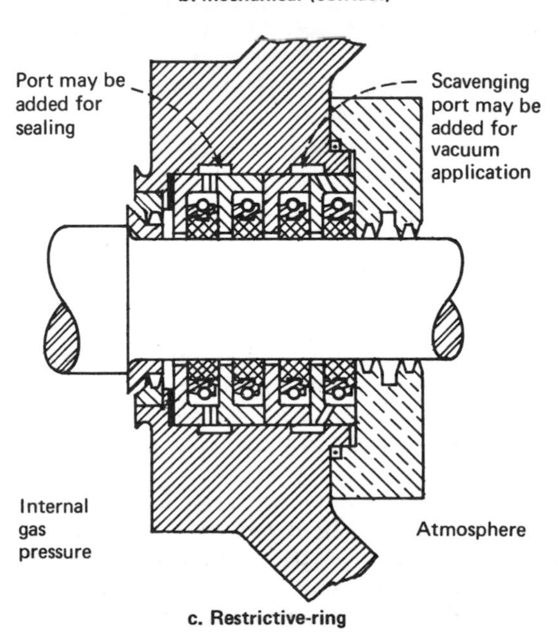


c. Restrictive-ring

**SHAFT** seals for the rotating compressor shaft either contain the gas being compressed or allow for its controlled leakage—Fig. 8

traps may be connected directly to the main-seal oil-return system, and the oil reused.

On all centrifugal compressors that have pressure-lubricated sleeve bearings, a lubrication-oil supply system is required. On any compressor requiring seal oil, a similar seal-oil supply system is required. These may be combined into one system—or into one lubrication system having booster pumps to increase the pressure of only the seal oil to the required sealing pressure. On more-complex installations, separate seal-oil and lube-oil systems will be required.

Each such system is normally furnished by the supplier of the compressor or the driver. The system may supply oil for lubricating both compressor and driver bearings. In some special cases, a driver will have its own lube-oil supply. The oil systems are usually specified to be mounted by the supplier on a console or baseplate, located adjacent to the compressor. Occasionally, on small simple, oil systems, the oiling equipment may be mounted on the same base as the compressor or turbine-driver. For more information on lube-oil and seal-oil systems, see Ref. [*10*].

## Axial-Flow Compressors

In axial-flow types of dynamic compressors, the flow of gas is parallel to the compressor's shaft and does not change direction as in radial-flow centrifugal types. The capacity range for axial machines as shown in Fig. 2 is to the right of radial-flow centrifugals, indicating the use of axials for higher flows than centrifugals. The head per stage of the axial machine is much lower (less than one-half) than a centrifugal type. Therefore, most axials are built with many stages in series. Each stage consists of rotating blades and stationary blades. In a 50% reaction design, one-half of the pressure rise occurs across the rotor blades and one-half across the stator blades.

Axial-flow machines are available from about 20,000 ICFM to over 400,000 ICFM, developing pressures up to 65 psig in typical 12-stage industrial designs; or slightly over 100 psig in current gas-turbine 15-stage air compressors. These types are used in combustion gas-turbines, and aircraft jet engines of all but the smallest sizes. They are also widely used in process applications requiring air flows or gas flows above 75,000 or 100,000 ICFM—especially as they exhibit greater efficiency than comparable multistage centrifugals. The axial is usually higher priced than a centrifugal. In smaller sizes, it can only be justified on the basis of higher efficiency.

The characteristic curve of an axial machine is much steeper than a centrifugal. Due to the flow characteristics of the rotor, and the large number of stages, the axial machine has a very narrow stability range (Fig. 9). It is most readily controlled by a variable stator-blade control, usually on the first several stages of any axial machine (partial stator-blade control); and sometimes on every stage for greatest stability, range and efficiency.

Mechanical details regarding bearings, shaft seals, lubrication and seal-oil systems, and also regarding control and performance at varying speeds, are all similar to those for centrifugal compressors previously reviewed.

Less information is available regarding preliminary

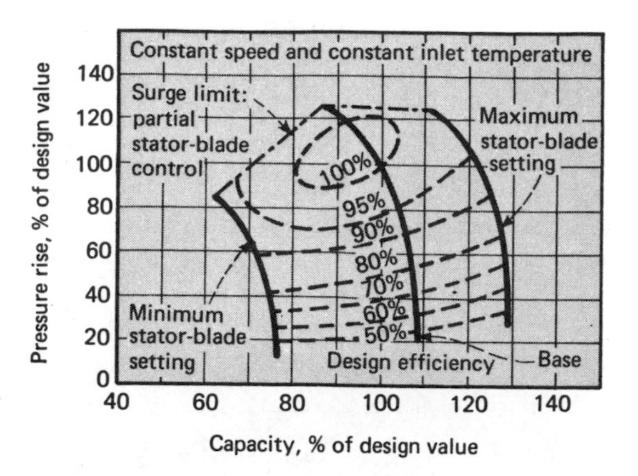


**STATOR** blade control in axial compressor—Fig. 9

selection methods for axials than for centrifugals. Although the axial compressor follows the same basic laws regarding adiabactic and polytropic head, weight flow, peripheral speed (at pitch line of rotor blades), etc., the project engineer should consider this type in a preliminary way only, before calling in a manufacturer's representative to obtain accurate estimating data.

## Positive Displacement: Reciprocating

Reciprocating compressors cover the range from the smallest capacity requirements through approximately 3,000 ICFM. Larger sizes are available but generally are not desired today for process service—centrifugal types being preferred. High pressures and relatively low-volume flows usually require reciprocating machines. The number of stages or cylinders must be chosen with relation to discharge temperature, available cylinder size, and compressor frame load or rod load.

The smallest sizes (to about 100 hp) may have single-acting cylinders, are air cooled, and may allow oil vapors from the compressor crankcase to mix with the compressed air or gas. Such types are only desired for process service in certain specialized, modified designs.

Small single-cylinder process types (25 to 200 hp) have a water-cooled cylinder, double-acting piston, separate stuffing box (allowing controlled leakage), and may be furnished with nonlubricated construction (no lubricant contacts the compressed air or gas). These serve for instrument air, or for small process-gas applications.

Larger gas- or air compressors will require two or more cylinders. On most process installations, these cylinders will be arranged horizontally and in series, forming two or more stages of compression. Typical size-ratings currently available for process applications are shown in Table V. The number of stages of compression is largely a function of the temperature rise across a stage (usually limited to about 250°F), the frame or rod loading that the compressor can handle, and occasionally the total pressure rise across a stage as it relates to compressor-valve design (usually limited to less than 1,000-psi).

The total ratio of compression is determined in order to obtain a first approximation of the number of stages.

**Size Ratings for Reciprocating Compressors—Table V**

| Type | Typical Stroke, *L*, In | Typical Speed, *N*, Rpm | Approximate Horsepower Range, Hp |
|---|---|---|---|
| Single-crank frames | 5, 7 | 600 to 514 | to 35 |
| | 7, 9 | 450 | 30 to 60 |
| | 9, 11 | 400 | 50 to 125 |
| | 11, 13 | 300 to 327 | 100 to 175 |
| Slow-speed frames | 9, 9½ | 600 to 514 | 200 to 800 |
| Horizontal | 10, 10½ | 450 | 400 to 1,200 |
| Opposed cylinders | 11, 12 | 450 to 400 | 800 to 2,000 |
| (two or more) | 14 | 327 | 1,000 to 2,500 |
| | 15, 15½, 16 | 327 to 300 | 1,500 to 4,000 |
| | 17, 18<br>19, 20 | 277 to 257 | 3,000 to 10,000 |
| Medium-speed and | | | |
| high-speed frames | 5 | 1,000 | 150 to 400 |
| Horizontal | 6, 8 | 720 to 900 | 1,000 to 4,500 |
| Opposed cylinders | | | |
| (two or more) | 9 | 600 | 4,000 to 8,000 |

If the ratio is too high for one stage (about 3.0 to 3.5), then the square root of the total ratio will equal the ratio per stage for two stages; the cube root for three stages, etc. Actual interstage pressures and ratio per stage will be modified after allowances are made for pressure drops through intercoolers, interstage piping, separators, and pulse dampers when used.

## Selecting Reciprocating Compressors

A quick and reasonably accurate method to determine the approximate horsepower requirement for each stage of a reciprocating compressor is to use the "horsepower per million" curves, as shown in Fig. 10.* For accurate results on gases much lighter or heavier than air, a correction factor must be applied to reflect the changes in valve losses resulting from the molecular weight of the actual gas (Fig. 11). The basic relationship is:

$$(HP)_{st} = \frac{(BHP)}{(MMCFD)}(MMCFD)F_{sg}\left(\frac{z_s + z_d}{2}\right) \quad (19)$$

where $(HP)_{st}$ is power per stage, hp; $(BHP)/(MMCFD)$ is the power required for a given compression ratio, brake hp/$10^6$ ft$^3$/d at 14.4 psia and suction temperature; $(MMCFD)$ is required capacity, $10^6$ ft$^3$/d at 14.4 psia and suction temperature; $F_{sg}$ is factor for specific gravity of gas; and $z_s$ and $z_d$ are the compressibility factors for the gas at suction and discharge conditions, respectively. If nonlubricated construction is specified, increase the horsepower obtained from Eq. (19) by approximately 5%.

Curves such as shown in Fig. 10 are industry-accepted for preliminary selection. More-accurate curves can often be obtained from compressor manufacturers, which will aid in obtaining somewhat more accurate results.

*This chart covers only a portion of the range of compression ratios. The complete set of curves contains ratios up to 6.0.

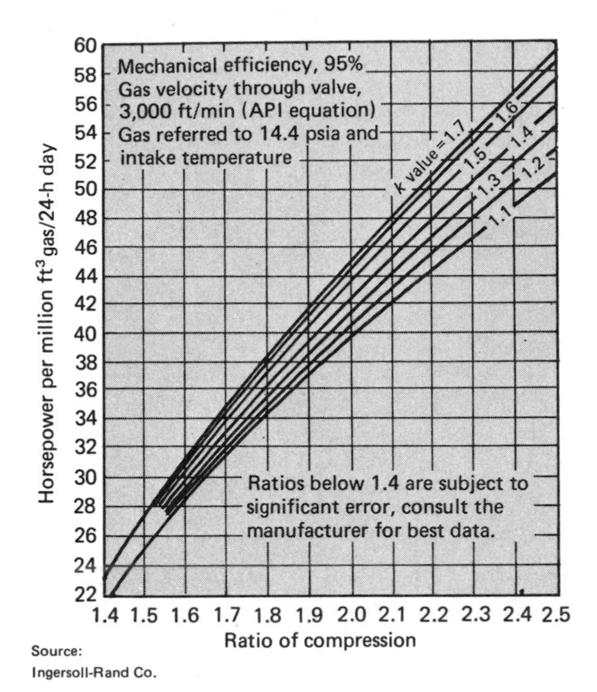


**POWER** needed for reciprocating compressors—Fig. 10

## Cylinder Sizing

When the interstage pressure and temperatures on a multistage application have been established, the capacity at inlet conditions for each stage can be found. Due to the clearance necessary to permit operation and allow valve passages to be designed, the piston does not sweep the entire volume of the cylinder. Hence, the actual cylinder capacity is somewhat lower than the cylinder displacement. Expressed as a volumetric efficiency of the cylinder, this relation is:

$$E_v = Q/C_{dis} \quad (20)$$

where $E_v$ is the volumetric efficiency; $Q$ is the capacity at inlet conditions, ICFM; $C_{dis}$ is the cylinder displacement, ft$^3$/min.

$$C_{dis} = \left(\frac{A_{he} + A_{ce}}{144}\right)\left(\frac{L}{12}\right)N \quad (21)$$

where $L$ is piston stroke, in; $A_{he}$ is area of head end of

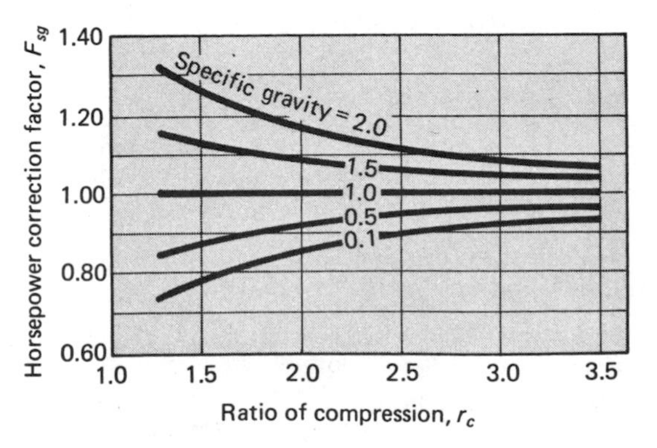


**CORRECTION** factor for specific gravity—Fig. 11

## Example 5: Horsepower per Million Method—Table VI

**Reciprocating-Compressor Calculation**

| | | | | |
|---|---|---|---|---|
| Capacity, MMSCFD | 41.3 | | | Given |
| Capacity, lb/h | — | | | Given (sometimes) |
| Gas | Hydrogen + hydrocarbon | | | Given |
| Molecular weight, $M_w$ | 2.925 | | | Given or calculated |
| Ratio specific heats, $k$ | 1.40 | | | Given or calculated |
| Suction pressure, $P_s$, psia | 208 − 2 = 206 | | | Given (− pulse damper loss) |
| Suction temperature, °F | 100 | | | Given |
| Suction temperature, °R | 560 | | | |
| Discharge pressure, $P_d$, psia | 1,885 + 19 = 1,904 | | | Given (+ pulse damper loss) |
| Overall compression ratio, $r_c$ | 1,904/206 = 9.24 | | | |
| Number of stages, $n$ | | | | Assume ($r_c \approx 2.0$ to 3.0/stage) |
| Stage | 1 | 2 | 3 | For identification |
| Approximate ratio/stage | 2.10 | 2.10 | 2.10 | $(r_c)^3$ |
| Approximate discharge pressure, $P_d$, psia | 433 | 909 | 1,909 | $r_cP_s$ |
| Interstage pressure drop, psi | 7 | 12 | — | $\approx 0.1(P_d)^{0.7}$ |
| Pulsation damper loss, psi | 2 Suction | — | 19 Discharge | 1% of absolute pressure |
| Suction pressure, $P_s$, psia | 206 | 433 | 909 | Given or calculated |
| Suction temperature, $T_s$, °F | 100 | 100 | 100 | Given or assume perfect intercooling |
| Suction compressibility, $z_s$ | 1.01 | 1.018 | 1.035 | Given or calculated |
| Discharge pressure, $P_d$, psia | 440 | 921 | 1,904 | $P_d$ above + interstage loss |
| Discharge temperature, $T_d$, °F | 236 | 235 | 231 | $T_s(r_c)^{(k-1)/k}$ |
| Discharge compressibility, $z_d$ | 1.016 | 1.03 | 1.062 | Given or calculated |
| Actual ratio/stage, $r_c$ | 2.14 | 2.13 | 2.09 | Calculated: $P_d/P_s$ (above) |
| Inlet capacity, $Q_s$, ICFM | 2,226 | 1,067 | 517 | Eq. (5) |
| Inlet capacity at $T_s$ and 14.4 psia, MMCFD | 45.40 | 45.40 | 45.40 | $(MMCFD) = (ICFM)P_s/(10^4)z_s$ |
| BHP/MMCFD | 46.5 | 46.25 | 45.2 | From Fig. 10 |
| Specific gravity correction, $F_{sg}$ | 0.875 | 0.87 | 0.865 | From Fig. 11 |
| Average compressibility, $z_{avg}$ | 1.013 | 1.024 | 1.049 | $(z_s + z_d)/2$ |
| BHP/stage | 1,871 | 1,871 | 1,862 | $(BHP/MMCFD)F_{sg}(MMCFD)z_{avg}$ |
| Total brake-horsepower, BHP | | 5,604 | | Sum of all stages |
| Number of cylinders | 1 | 1 | 1 | Assume, based on knowledge of available frames |
| Cylinder clearance, $C_c$, % | 15 | 15 | 15 | Use 10 to 15% for estimate |
| Volumetric efficiency, $E_v$, % | 86.3 | 86.6 | 87.3 | Eq. (22) |
| Displacement required, ft³/min | 2,579 | 1,233 | 593 | $(ICFM)/E_v$ |
| Stroke, $L$, in | 18 | ← | ← | From Table V or manufacturer |
| Speed, $N$, rpm | 277 | ← | ← | From Table V or manufacturer |
| Piston-rod size, $d$, in | 5 | ← | ← | From Table V or manufacturer |
| Cylinder diameter, $D$, in | 24.12 use 24¼ | 16.87 use 17 | 11.97 use 12 | Eq. (21) |
| Area, head end, $A_{he}$, in² | 461.86 | 226.94 | 113.10 | |
| Area, crank end, $A_{ce}$, in² | 442.23 | 207.35 | 93.47 | $A_{he} - A_{rod}$ |
| Frame load, compression, $F_c$, lb | 112,119 | 119,266 | 130,375 | Eq. (23) |
| Frame load, tension, $F_t$, lb | −99,438 | −92,687 | −75,156 | Eq. (24) |
| Frame load limit, lb | 150,000 | ← | ← | Based on available frame |

**Summary:** Preliminary selection is a three-cylinder, 18-in stroke machine, running at 277 rpm, with one 24¼-in cylinder (first stage), one 17-in cylinder (second stage) and one 12-in cylinder (third stage). All cylinders are double-acting. Driver size is 6,000 hp.

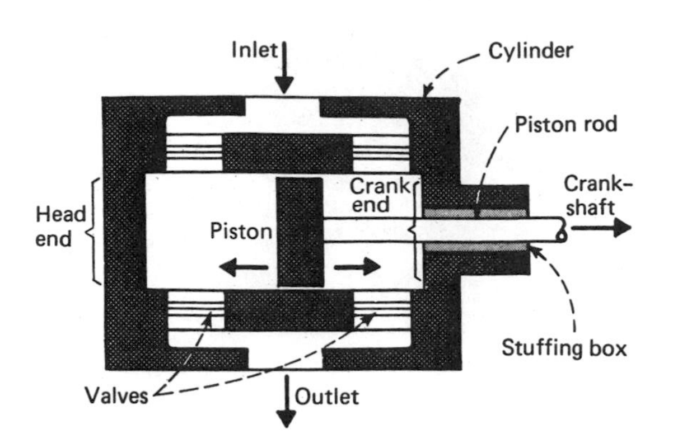


**MACHINE** components for double-acting cylinder—Fig. 12

piston, in²; $A_{ce}$ is area of crank end of piston, in² and $N$ is rpm. Note that area of crank end is the area of head end minus the area of the piston rod.

Many formulas for volumetric efficiency are in use. For preliminary estimates, the following is usually sufficient:

$$E_v = 0.97 - C_c\left[\frac{(r_c)^{1/k} - 1}{z_d/z_s}\right] \quad (22)$$

where $C_c$ is cylinder clearance, $r_c$ is ratio of compression, $k$ is ratio of specific heats, and $z_d$ and $z_s$ are the compressibility factors of the gas at discharge and suction conditions, respectively.

## Frame Load or Rod Load

Every compressor frame has a limit to the forces that can be applied during compression. In the simplest manner, this load can be computed from the known cylinder diameter and the pressures acting upon the piston. This is sometimes called the "gas-rod loading," and neglects the loading due to the reciprocating weights and motion of the machine.

For a double-acting cylinder when the piston is moving inward toward the crankshaft, the frame load in compression, $F_c$, is:

$$F_c = P_d A_{he} - P_s A_{ce} \quad (23)$$

and, for the piston moving outward away from the crankshaft, the frame load in tension, $F_t$, is:

$$F_t = P_s A_{he} - P_d A_{ce} \quad (24)$$

where $F_c$ and $F_t$ are in lb; $P_s$ and $P_d$ are suction and discharge pressure, psi; $A_{he}$ is piston area of head end, in²; and $A_{ce}$ is piston area of crank end, in.² References to Fig. 12 will make these relations clear.

Every compressor frame has maximum values for frame load that must not be exceeded during normal operation. The actual calculated values are usually preferred to be 60% to 75% of the maximum capabilities of the frame.

Limits on rotating speed, $N$, and average piston speed, $U_p$, should be specified to avoid selecting a design that operates too fast and has excessive wear and maintenance.

$$U_p = 2N(L/12) \quad (25)$$

where $U_p$ is piston speed, ft/min; $N$ is rotating speed, rpm; and $L$ is stroke, in. The general limit of piston speed for compressors in process applications is 800 to 850 ft/-min, and is somewhat lower for nonlubricated types, e.g. 700 ft/min.

*Example 5*—Let us make a preliminary selection of a typical multistage reciprocating compressor to handle 41.3 MMSCFD of a hydrogen/hydrocarbon gas mixture having a molecular weight of 2.925. The pertinent data and the necessary computational steps are shown in Table VI.

## Controlling Reciprocating Compressors

If sufficient power is supplied to a positive-displacement compressor, the machine will continue to increase the pressure over its rated value until some limit is reached. This may occur by the opening of a relief valve, the actuating of a high-discharge-temperature switch, or by mechanical failure of the machine. None of these alternatives is desired for process control. Hence, the compressor must be equipped with cylinder control devices or with bypass valves, or must respond by having its speed changed for variations in capacity.

Most reciprocating compressors are driven by constant-speed electric motors. Therefore, the controls are based on a machine running at the constant speed. Capacity control of a fixed-speed machine can be accomplished by:

1. External bypass of gas or air around the compressor back to the suction source or to atmosphere.
2. Cylinder unloaders.
3. Cylinder clearance pockets.
4. A combination of the above.

Cylinder unloaders are manually or automatically operated devices on one or both ends of a cylinder. They are designed to unload (i.e., hold open) the inlet valves. Thus, the compressor does no work on that portion of that stroke. For example, inlet-valve unloaders could be placed on the head end of a cylinder, which when actuated would cause the cylinder's net output to be reduced by approximately one-half. Problems of frame loading, excessive pulsation or decreased valve life may arise if unloaders are used as a permanent control means. Unloaders are usually always supplied as a means of unloading the compressor for startup.

Clearance pockets are additional capacity that is built in or bolted on to the head end or crank end of one or more cylinders to increase the cylinder clearance. This reduces the volumetric efficiency, and results in a lower net output for a given size cylinder. By referring to Fig. 13, we can see how increased cylinder clearance reduces volumetric efficiency.

Regardless of the capacity control used, relief valves must always be provided in the piping adjacent to any positive-displacement machine. The relief valves should be sized for the full output capacity of the cylinder.

Variable-speed gas-engine drivers are sometimes applied to reciprocating compressors, and provide the obvious advantage of capacity variation directly as a function of speed. For additional flexibility, clearance pockets and cylinder unloaders are often used. Higher first cost

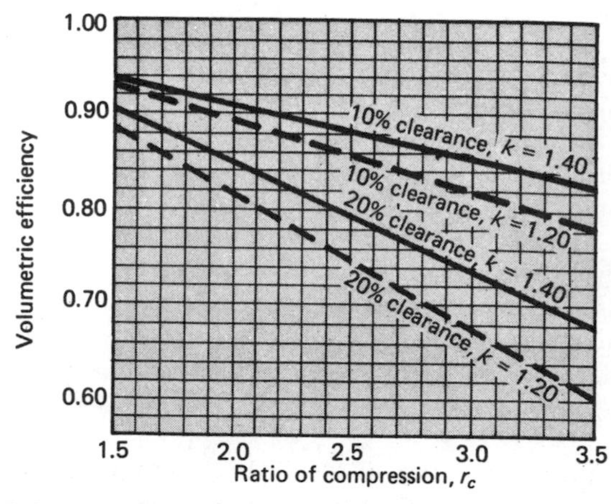


**CLEARANCE** affects volumetric efficiency—Fig. 13

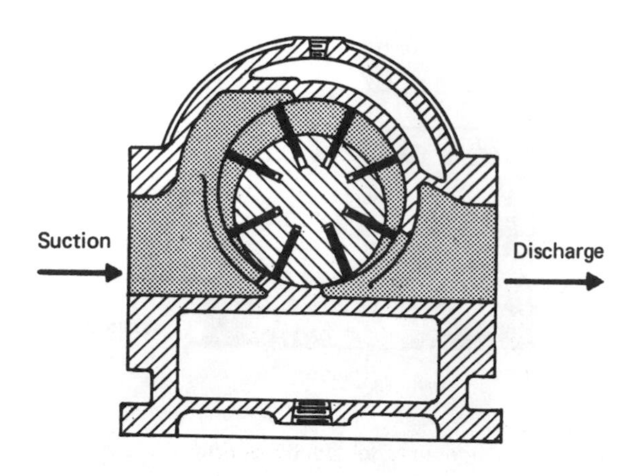


**ROTARY** compressor has sliding vanes—Fig. 14

and higher maintenance make this type of driver less popular, except in certain special applications.

Steam or gas turbines can also be used to drive reciprocating compressors, but must be carefully applied because speed-reducing gears, flywheels, torsional couplings and special analyses are usually required.

## Rotary Compressors

Rotary blowers, vacuum pumps and compressors are all positive-displacement machines in which a rotating element displaces a fixed volume during each revolution.

The many different styles that exist can be generally grouped into four basic types. Oldest and most widely known is the lobe-type, in which two or three figure-8-shaped rotors, meshing together, are driven through timing gears attached to each shaft. Lobe types range from very small, mass-produced machines (from approximately 2 ft$^3$/min) through the largest sizes (up to 20,000 ICFM). They are used primarily as low-pressure blowers, compressing air or gases from atmospheric to 5 to 7 psig, or higher (up to 25 psig) for special types. They also find wide application as vacuum pumps, which are really compressors operating with suction pressures below atmospheric, and discharge pressures at or slightly above atmospheric.

A second style is the sliding-vane type. This has one offset rotor containing slots in which vanes slide in and out during each revolution. The vanes trap air or gas and gradually reduce its volume and raise its pressure, until it is exhausted through ports in the casing (Fig. 14). This type of compressor can produce up to 50 psig per stage and is also available in two-stage arrangements for pressures up to 125 psig. Capacities range up to 1,500 to 2,000 ft$^3$/min. This type is also used as a vacuum pump.

For the chemical process industries, both the lobe-type and vane-type are limited in their application because of the low pressure-rise obtainable, and because they are generally available only with cast-iron casings. This makes them unsuitable for certain corrosive or dangerous gases.

A third type is the rotary-screw compressor, which has become moderately popular in recent years for higher pressures and is available in large sizes. A twin-screw concept was developed in Europe in the 1930s, together with the specialized equipment to machine the complex rotors, by A. J. R. Lysholm.

These designs are now available in both oil-cooled and dry constructions. Capacities range from about 50 ICFM to 3,500 ICFM in oil-flooded designs, and about 1,000 to 20,000 ICFM for the dry types. The latter will operate at speeds up to 10,000 or 12,000 rpm, and may be obtained for discharge pressures as high as 250 to 400 psig, or approximately 50-psi pressure rise per casing.

The dry type is generally preferred in the chemical process industries because it does not cause oil carryover into process streams. However, oil-flooded models have found wide acceptance as plant-air, service-air, and portable-air compressors, and as refrigeration compressors for fluorocarbon refrigerants.

The fourth version is known as the liquid-ring compressor (or liquid-ring pump), and is correctly classified as a rotary machine. However, it has a unique operating principle unlike that of any other rotary. A circular vaned-type rotor turns in a circular or oval-shaped casing in which water or other sealing liquid is also present (Fig. 15). Centrifugal force causes the liquid to form a ring around the periphery of the casing while in operation. Air or gas travels inward toward the center of the vaned rotor, and gradually decreases in volume and increases in pressure until passing discharge ports, where it leaves the casing. The liquid, present in the discharged air or gas, is then separated and either cooled and recirculated, or discarded in a once-through system.

The liquid-ring type finds greatest application as a vacuum pump to as low as 3 or 4 in Hg absolute. It may also be applied as a low-pressure blower (to about 25 psig), or an intermediate-pressure instrument-air compressor (to about 100 psig). Sizes range from very small models of about 10 ft$^3$/min, and up, through the largest single casings for about 10,000 ft$^3$/min. Liquid-ring compressors have found wide use on certain difficult gases such as chlorine, acid gas, hydrogen-sulfide-laden gases, carbon dioxide, etc. Stainless-steel construction is available in most designs.

In general, the rotary compressor serves only special

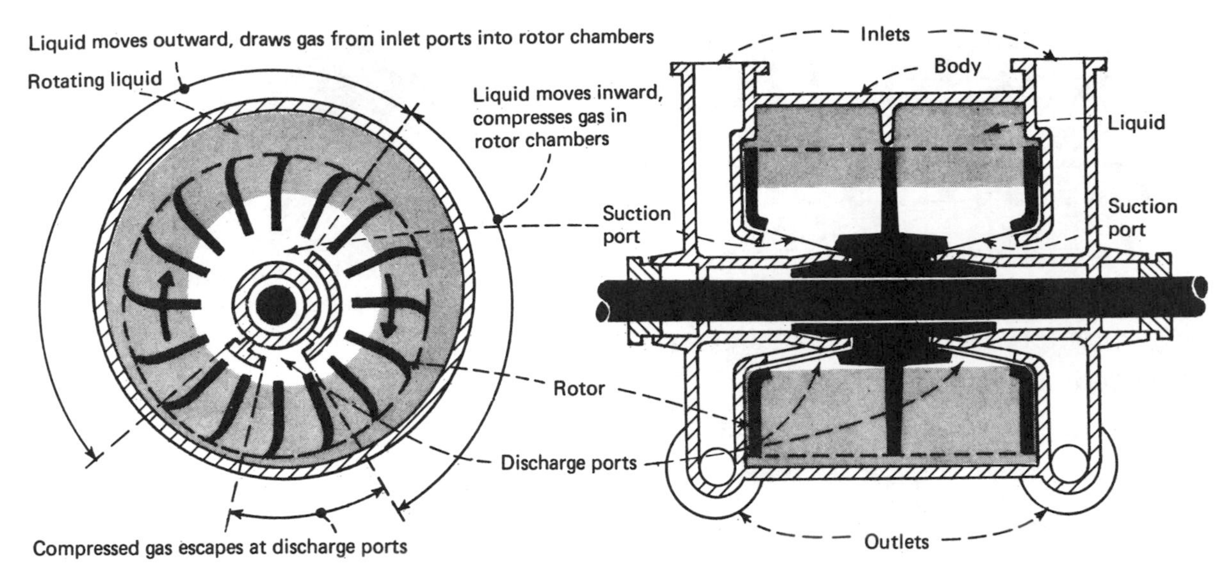


**LIQUID-RING** rotary compressor finds wide use for handling acid-laden gas mixtures or corrosive gases—Fig. 15

areas, such as low-pressure-rise, low-capacity requirements. But such compressors should not be ignored when attempting to choose the correct machine for an application. Reference to the specific-speed chart (Fig. 2) may sometimes prove helpful in attempting to apply a rotary compressor. Adiabatic head, capacity and horsepower determinations can be made in much the same way as for a centrifugal machine. However, a widely recognized source of data for efficiency is not available, hence, preliminary estimates for selecting a machine must usually be made by using equipment manufacturers' data.

## Summary

The project engineer need not feel apprehensive about making preliminary selections for a compressor or blower of any type for any process-plant application if he carefully follows the basic relations and laws discussed in this report. Obviously, the final detailed performance and design are the responsibility of the compressor manufacturer, from whom much valuable assistance is available. It should prove helpful and time-saving in every case if the project engineer will first study his needs and make his own preliminary determination of compressor size and type.

## Acknowledgements

The following firms have supplied background information and/or illustrative material: Allis-Chalmers; Atlas Copco Inc.; Cooper Bessemer Co.,; DeLaval Turbine Inc.; Dresser Industries; Elliot Div., Carrier Corp; Gardner-Denver Co.; Hoffman Air & Filtration Div., Clarkson Industries Inc.; Ingersoll-Rand Co.; Joy Manufacturing Co.; Kellogg-American Inc.; Nash Engineering; Sulzer Bros.; Sundstrand Fluid Handling; Vilter Manufacturing Corp.; White Superior Div., White Motor Corp; and Worthington-CEI Inc. #

## References

1. "Engineering Data Book," 9th ed., Natural Gas Processors Suppliers Assn., Tulsa, 1972.
2. "Gas Properties and Compressor Data," Form 3519-C, Ingersoll-Rand Co., Woodcliff Lake, N.J., 1967.
3. Scheel, L. F., "Gas Machinery," Gulf Publishing, Houston, 1972.
4. Compressibility Charts and Their Application to Problems Involving Pressure-Volume-Energy Relations for Real Gases, Bulletin P-7637, Worthington-CEI Inc., Mountainside, N.J., 1949.
5. Balje, O. E., A Study on Design Criteria and Matching of Turbomachines—Part B, *Trans. ASME, J. Eng. Power,* Jan. 1962.
6. Centrifugal Compressors for General Refinery Service, 3rd ed., API Standard 617, American Petroleum Institute, Washington, 1973.
7. Hallock, D. C., Centrifugal Compressors—The Cause of the Curve, *Air and Gas Eng.,* Jan. 1968.
8. Centrifugal Compressors, Bulletin 8282-C, Ingersoll-Rand Co., Woodcliff Lake, N.J., 1972.
9. Magliozzi, T. L., Control System Prevents Surging in Centrifugal Flow Compressors, *Chem. Eng.,* May 8, 1967, pp. 139–142.
10. Lubrication, Shaft-Sealing, and Control-Oil Systems for Special-Purpose Applications, API Standard 614, American Petroleum Institute, Washington, 1973.
11. Reciprocating Compressors for General Refinery Service, 2nd ed., API Standard 618, American Petroleum Institute, Washington, 1974.

## Meet the Author

**Richard F. Neerken** is Chief Engineer, Rotating Equipment Group, for The Ralph M. Parsons Co., Pasadena, CA 91124. He joined Parsons in 1957 and has worked continuously with rotating machinery such as pumps, turbines, compressors and engines on all projects for the company. Previously, he spent over 11 years as an applications engineer for a major manufacturer of pumps, compressors and turbines. He has a B.S. in mechanical engineering from California Institute of Technology and is a member of the Contractors Subcommittee on Mechanical Equipment for the American Petroleum Institute.

# Guide to Trouble-Free Compressors

Having a compressor with minimal operating problems does not only depend on the selection of the right type and size for the job. Detailed specifications of all auxiliary equipment and operating conditions, as well as keeping constant vigilance over the engineering and installation phases, are also essential.

SIDNEY A. BRESLER, Consulting Engineer, and J. H. SMITH, American Cyanamid Co.

A compressor is frequently an expensive device, a critical item in a process, and a rather complicated piece of equipment to specify and purchase because of the many alternatives open to the engineer. Faced with this wide choice, on what basis can a selection be made so that the chosen compressor will do the intended job?

Since each type of compressor has specific characteristics, these must first be understood. Then, the many factors that enter into the final decision to purchase a compressor and auxiliaries are considered.

## Compressor Types

There are two basic mechanical methods of increasing the pressure of a gas: reducing its volume, and increasing its velocity so that the velocity energy may be converted into pressure.

Positive-displacement machines that increase pressure by reducing volume are:

• Reciprocating compressors, which have a piston moving within a cylinder (Fig. 1).

• Rotary-screw compressors, in which gas is squeezed between two rotating intermeshed helices and the casing in which they are housed.

• Rotary-lobe compressors, through which gas is pushed by intermeshing lobes.

• Sliding-vane compressors, where an eccentric

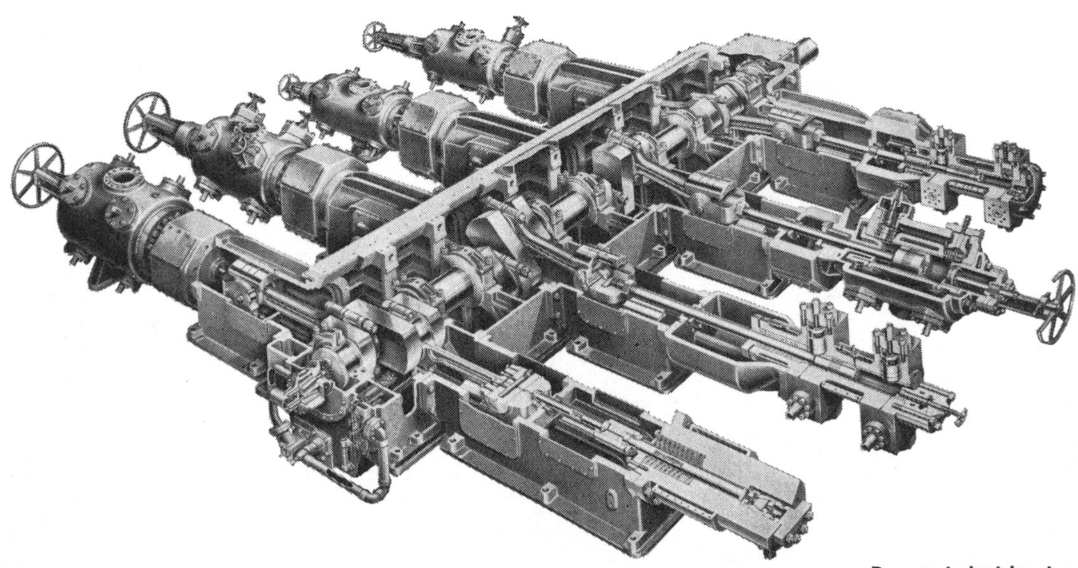

Dresser Industries, Inc.

**PROCESS RECIPROCATING COMPRESSOR** operates from vacuum to over 36,000 psi.—Fig. 1

Originally published June 1, 1970

cam (into which sealing vanes slide) rotates inside a housing.

• Liquid-piston type, in which a partially liquid-flooded case creates the equivalent of sliding vanes.

• Diaphragm compressors, in which a flexible diaphragm is pulsed inside a concave housing.

The two types of compressors that convert velocity into pressure are:

• Radial-flow compressors, generally called "centrifugal compressors."

• Axial-flow compressors, known as "axial compressors."

In centrifugal compressors, the gas enters the eye of the impeller, and rotative force moves the fluid to the rim of each wheel or stage. Diffusers convert the velocity head into pressure, and return passages are then used to lead the gas to the compressor discharge or to the next impeller stage.

In axial compressors, flow occurs through a series of alternating rotating and stationary blades, and in a direction basically parallel to the compressor shaft. Each passage through the rotating blades increases the velocity of the fluid, and each passage through the stationary diffuser blades converts the velocity head into a pressure head.

## Selection Considerations

Not all types of machines are made in all pressure-volume ranges. Fig. 1 indicates, in a very general manner, the capacities of reciprocating, centrifugal, rotary-screw, and axial compressors available. The more common usage is indicated by the deeper shad-

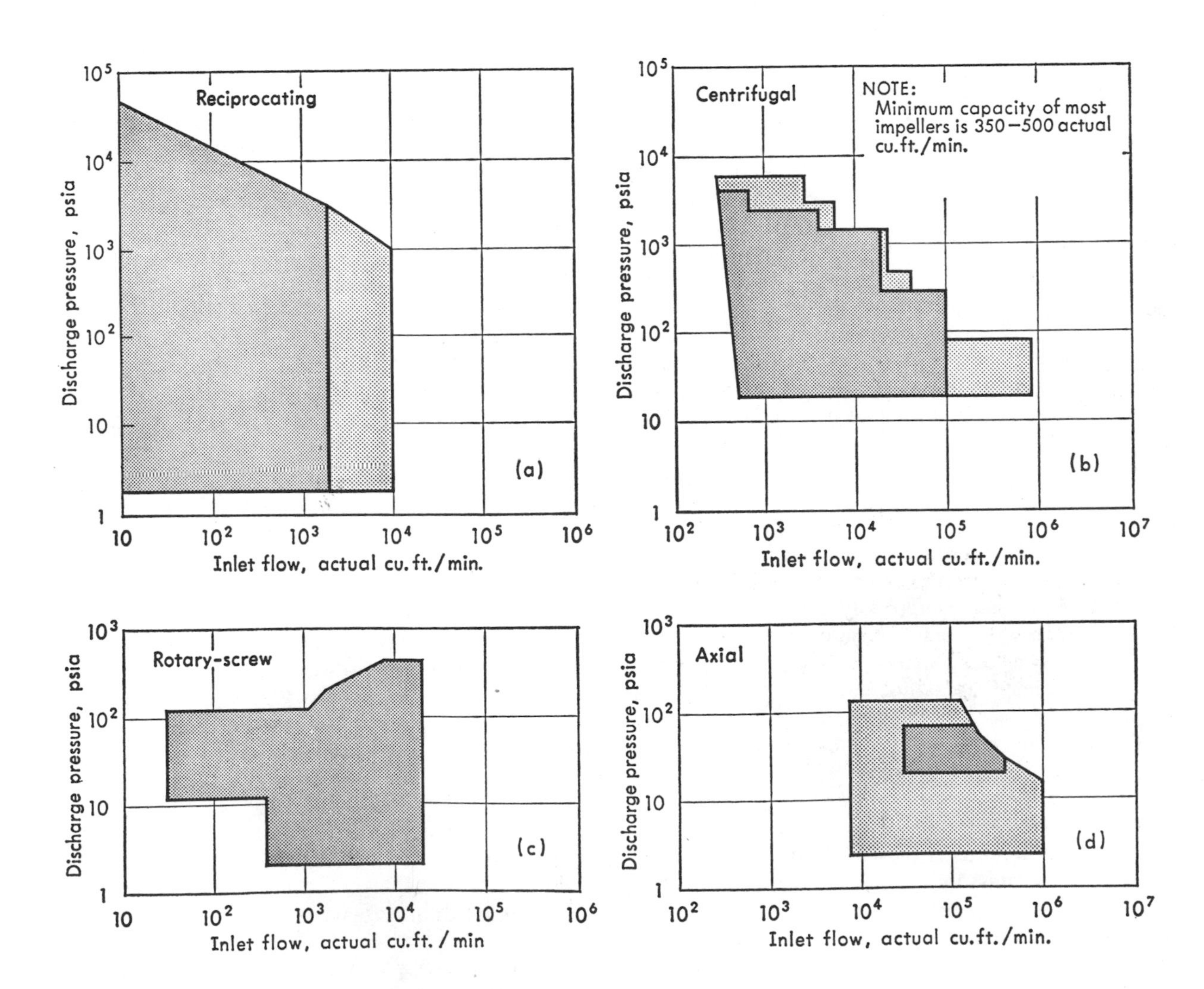


**PRESSURE-VOLUME** operating range (discharge pressure vs. feed volume) of various compressors—Fig. 2

ing. Although these figures do not indicate the theoretical or engineering limits of any design (the limits are continually being expanded), they may be used as guides to current technology.

Since sealing systems for axial compressors are not as versatile as for other types, normally only those gases whose leakage to the atmosphere can be tolerated should be handled by this type of machine.

Rotary-lobe, sliding-vane, liquid-piston and diaphragm compressors have relatively small capacities and, typically, atmospheric-pressure suction. Of these four types, the rotary-lobe can deliver the most gas, as its maximum suction volume is about 30,000 actual cu.ft./min. A maximum discharge pressure of about 40 psia. can be attained. However, rotary-lobe compressors are most competitive at capacities of 17,500 actual cu.ft./min. or less, and discharge pressures of about 22 psia.

Maximum inlet capacities of sliding-vane units are about 3,000 actual cu.ft./min., or double this amount if a duplex compressor is used. The latter consists of two compressors attached to a single drive. Maximum discharge pressures of standard machines are about 65 psia. in a single stage, and 140 psia. in two stages.

The liquid-piston compressor has a maximum capacity of about 10,000 actual cu.ft./min., and can deliver this amount of air (or gas) at about 30 psia. Volumes of 300 actual cu.ft./min. or less can be compressed to about 115 psia.

The three foregoing compressor types can produce moderate to high vacuum, particularly if multiple-staged.

Diaphragm compressors have much smaller volumetric capacities, with maximum flows ranging from 40 to perhaps 200 actual cu.ft./min. These machines, however, can develop pressures up to 40,000 psi.

Before selecting a compressor type, one must decide how many machines will be needed to handle the process load. In former years, reciprocating machines were used for almost all process applications. Since compressor capacity was low, large plants would require trains of machines. As machine reliability and capacity increased, the tendency became to install two machines, each with 55 or 60% of capacity, perhaps with a third unit as a spare.

The spare unit ensured operation at full capacity, but at an increased compressor cost of about 50%. If the spare compressor were omitted, but two half-size machines installed, one could still be reasonably sure of continued operation at all times. This was particularly important when the process included equipment that could not be shut down frequently, such as furnaces. Later, to take advantage of larger machine capability, several services were placed on the same frame.

Today, the situation is somewhat different, as more centrifugal compressors are being used (Fig. 3). For one thing, downtime of rotating equipment generally is appreciably less than that of reciprocating equipment. Therefore, in many instances a single centrifugal compressor may be satisfactory. However, it must be recognized that when a compressor is down, it will usually take longer to repair or overhaul a centrifugal unit than a reciprocating one—unless a complete spare rotor is available.

Cooper-Bessemer Co.
**TRAIN** of three barrel-type centrifugal compressors in 1,400-ton/day ammonia plant—Fig. 3

Also, the pricing structure of centrifugal compressors is quite different from that of reciprocating ones. As a first very rough approximation, one may assume that halving the size of a reciprocating compressor will halve its cost. Yet, halving the size of a small centrifugal compressor may only decrease its cost 20%, and halving the size of a large machine may only reduce its cost 30%.

Furthermore, because of their flat operating characteristics, the running of centrifugal compressors in parallel may result in surging unless very careful attention is given to avoiding unstable operation. Therefore, in many process applications for which one centrifugal will have adequate capacity, an installed spare is not provided. In these instances, a complete spare rotor may be bought.

The choice between reciprocating and centrifugal compressors is not always simple, particularly for high-head, medium-capacity service such as gasfield repressuring. If several reciprocating compressors are used, each can be multiple-staged to develop the desired head. The shutting down of one machine would merely cause a decrease of plant output. But if several centrifugal compressors are used in series, failure of one would stop the entire operation.

## Operating Characteristics

A positive-displacement compressor is characterized by a pressure-rise - volume curve that is almost vertical. (It is not completely vertical because there is mechanical clearance, and slip and leakage from the discharge to the suction, the slip increases as the compression ratio rises.) The compressor delivers its gas against any pressure head up to the limit of its mechanical strength and drive capability. Capacity is almost directly proportional to speed.

The characteristics of a centrifugal compressor are appreciably different. Generally, the pressure-rise-volume curve is quite flat (Fig. 4a). (It may be somewhat steeper if a heavier gas is being compressed.) A small change in the compression ratio produces a marked effect on compressor output. As the discharge pressure increases, the flow decreases, and if the flow decreases too much, the machine will start to surge.

Surging occurs when the velocity of gas leaving an impeller wheel is too low to move the fluid through the machine. With no gas leaving the impeller, the discharge pressure may drop. Should this occur, the machine will again start to compress gas, and the cycle will be repeated. Such intermittent operation may severely damage a compressor. The characteristic curve can be modified by the installation of adjustable inlet guide vanes (Fig. 4b). These are most effective on machines having few stages. Adjustable diffuser vanes have been used on some machines.

In some installations, process requirements may dictate that the compressor be run at the far right of the characteristic curve, where it is very steep. Operating in this area requires careful control and is accomplished at some penalty of compressor efficiency.

The volumetric capacity of a centrifugal compressor is almost directly related to its speed; its developed head, to the square of the speed. (The horsepower requirements are thus related to the cube of the speed.) The efficiency of centrifugal compressors is lower than that of reciprocating machines by perhaps 5 to 20%.

These characteristics establish the sensitivity of the compressor to variations of flow conditions. For example, a change in the density of the fluid being compressed will have little effect on either the volume of gas pumped or the discharge pressure developed by a reciprocating machine, although one would have to be sure that no component parts of the compressor were being mechanically overstressed. Any variation in the density of a gas being compressed will result in a proportionate change in the weight of gas pumped.

On the other hand, because the head developed by a centrifugal compressor depends only on the velocity developed, a change of gas density will be directly reflected by a proportionate change in the developed discharge pressure. However, at a given density, if the discharge pressure can be permitted to change slightly, one can obtain large variations in volumetric gas flow through the compressor.

The axial compressor has a very steep characteristic curve (Fig. 4c). The unit's surge capacity is thus close to its operating capacity. However, by providing a method of adjusting the angle of stator blades and inlet guide vanes, a greater operating range can be obtained (Fig. 4d).

Generally, the efficiency of an axial compressor exceeds that of a multistage centrifugal machine by perhaps 5 to 10%. The axial compressor does not contain diaphragms that expand radially as the compressed gas gets hot. This mechanical factor, combined with higher efficiency, leads to greater freedom from temperature limits, and permits a higher compression ratio per case than do centrifugal units.

## Speed Considerations

The type of mechanical drive (including gears) that is used may influence the choice of compressor.

### Speed Range for Compressors and Drives—Table I

| Compressor Types | Usual Speed Range, Rpm. | Remarks |
|---|---|---|
| Large reciprocating compressors | 300-600 | Some even 1,000 to 1,500 rpm. |
| Small reciprocating air and refrigeration machines | 1,000-1,500 | |
| Rotary-screw | 3,000-10,000 | |
| Process centrifugal units | 3,000-12,000 | Some large-horse-power machines up to 17,000 rpm. |
| Special, small-volume high-head air centrifugals | 30,000-50,000 | |
| Axial compressors | 3,000-6,000 | Some up to 16,000 rpm. |
| Large internal-combustion engines and reciprocating gas expanders | 300-600 | |
| Small rotary and radial engines | 3,000-8,000 | |
| Mechanical-drive gas turbines and centrifugal expanders (over 1,000 hp.) | 10,000 or less | Small gas-turbine compressor drives have operated at up to 50,000 rpm. |
| Mechanical-drive, back-pressure steam turbines (3,000 to 40,000 hp.) | 16,000 or less | Condensing turbines have lower maximum speeds. |
| Electric motors | 3,600 or less | |

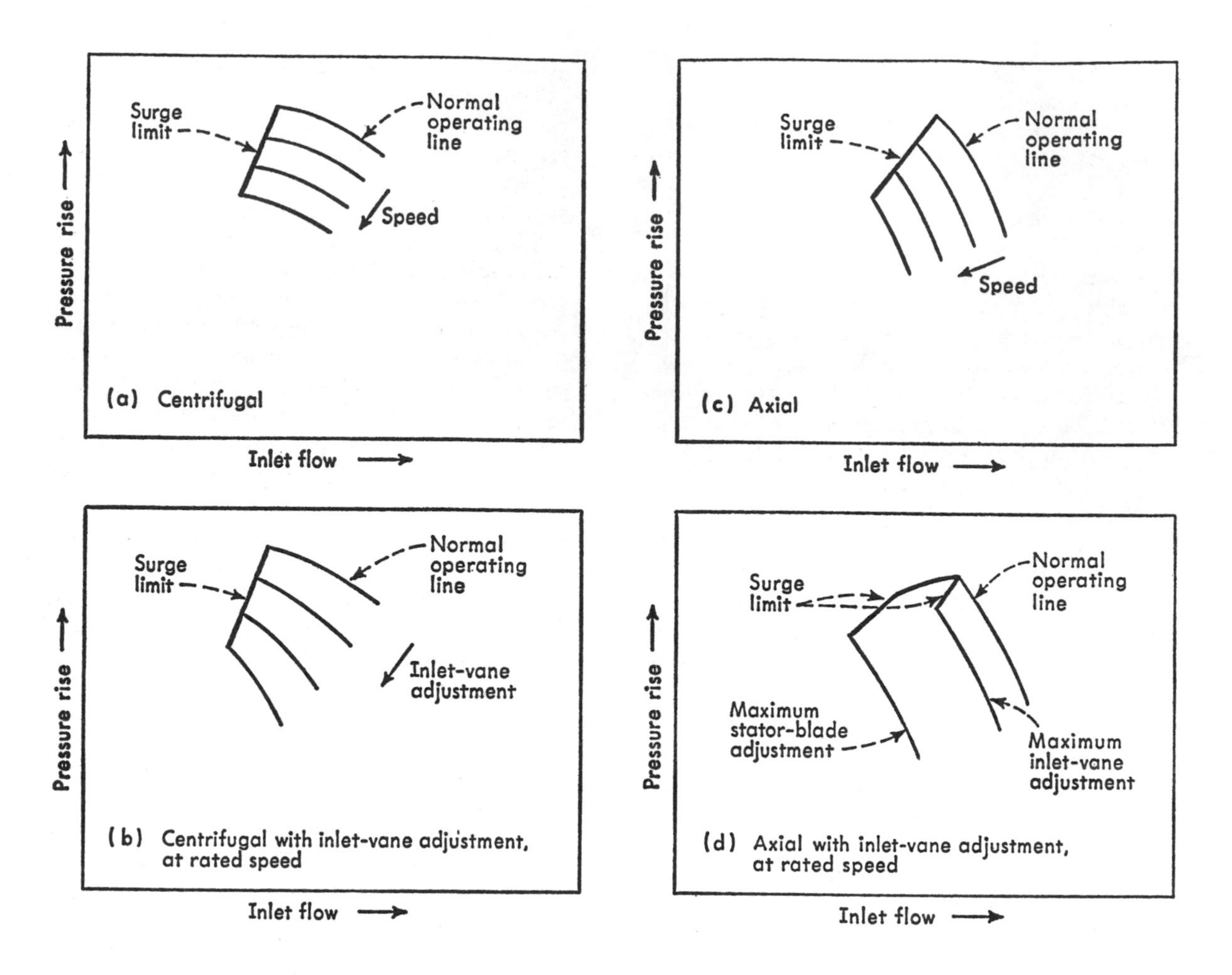


**OPERATING CHARACTERISTICS CURVES** for centrifugal and axial compressors—Fig. 4

Compressor and drive speeds are very pertinent if one wishes to avoid gearing. The accompanying table provides speed ranges of the most common types of compressors and drives. There are specially designed units, however, that do not fall within the ranges listed. One of these, for example, is a carbon dioxide compressor with a suction volume of approximately 50 actual cu.ft./min. at the last wheel, which rotates at 25,000 rpm. and delivers gas at 5,000 psi. The tip speed of this compressor's impeller is approximately 650 ft./sec. The compressor itself is directly driven by a specially designed 1,000-hp. steam turbine.

## Cost Comparisons

In very general terms, at low pressures and large flows, the purchase cost of a reciprocating compressor may be estimated to be perhaps twice that of a centrifugal machine of the same capacity (Fig. 5, 6, 7). The cost differences narrow as pressure increases or actual flow decreases. At high pressures and at low flows, costs may be quite close to each other. A reciprocating compressor will need a more massive foundation, more protection from the environment, and a more careful piping design to avoid vibration and pulsation.

On the same rough basis, one may estimate the costs of rotary-screw and axial compressors to be about the same or less than that of centrifugal units. In their most suitable applications, the costs of the screw and of the axial compressors may be considerably lower.

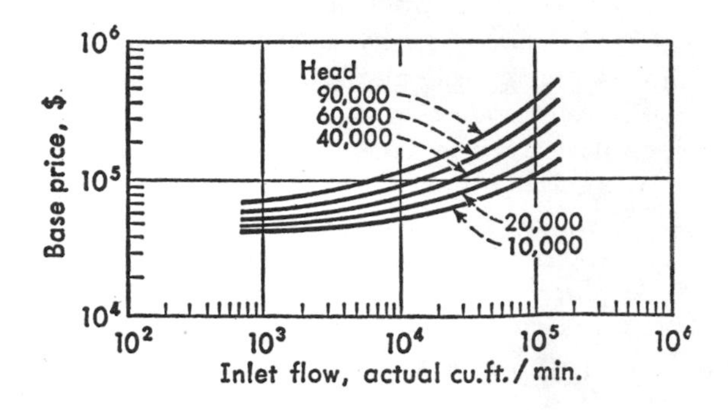


**COST OF** centrifugal compressors according to inlet volume and head output (head is ft. of gas)—Fig. 5

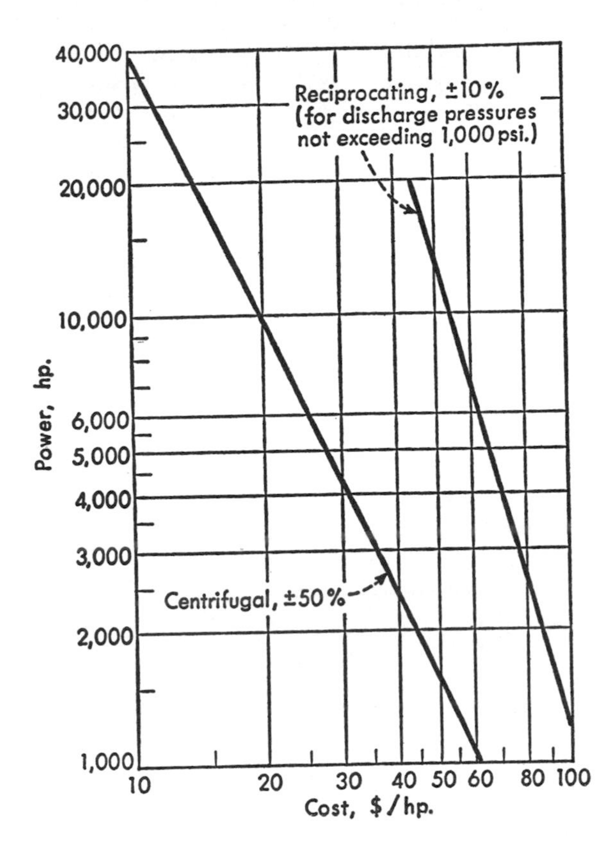


**BASE PRICE** of reciprocating and centrifugal compressors according to horsepower of unit—Fig. 6

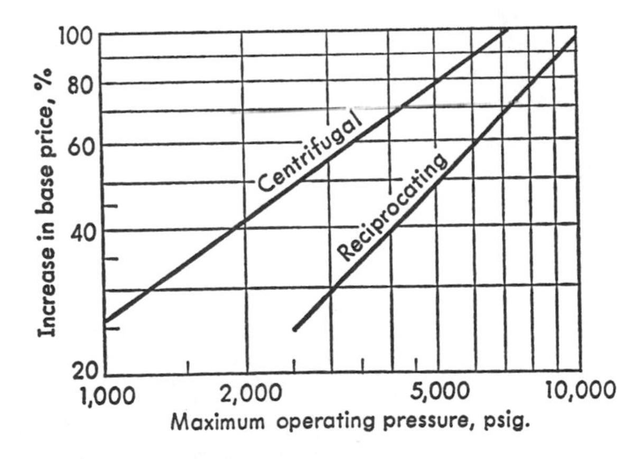


**EFFECT OF OPERATING PRESSURE** on cost of centrifugal and reciprocating compressors—Fig. 7

## Combinations of Compressors

At times, combining compressors may be worth considering. For example, in compressing to a very high pressure, it may be possible to use a centrifugal machine or a rotary-screw machine for the lower pressure, and then pipe the gas to a reciprocating unit. In some instances, axial and centrifugal impellers may be placed on the same shaft. In addition, one might also resort to placing axial- and centrifugal-compressor cases in a common drive-train.

As an alternative to an axial compressor, three or four single-stage centrifugal compressors may be connected by a gear train to a single drive. With the gas cooled after each stage of compression—and gears designed to permit each stage to be run at its optimum speed—the efficiency of these centrifugals is comparable to that of an axial compressor, while their operating characteristics are those of a centrifugal machine. Extensive gearing, however, is a distinct disadvantage.

Additional significant characteristics of each type of compressor will be discussed in a forthcoming article.

## Driving Mechanisms

Smaller units are usually electric-motor driven (direct or belt); for medium to large units there is a wide choice of drives. These include motors (synchronous, induction, low- or high-speed); steam turbines (back-pressure, condensing or controlled-extraction); internal-combustion engines (integral or direct-connected); gas turbines (single or double-shaft); and expanders.

The selection of a drive depends to some extent on the compressor service, but more important are the overall energy balance, energy utilization and availability, and heat-rejection methods. Within the limits imposed by these criteria, the selection should stress a drive system that is simple, dependable and straightforward. The compressor is the reason for the drive, not the other way around.

The drives of internal-combustion engines and steam turbines can ordinarily be operated over a fairly large range of speeds. This may not be the case, however, if gas-turbine or electric-motor drives are used.

Let us first consider gas turbines, almost all of which have axial-type air compressors for pressure-air supply to the compressors. For single-shaft units (air compressor, gas turbine and driven unit on one shaft), the speed range is most often determined by the steep performance curve of the axial rather than by the much flatter curve of the centrifugal process compressor. Double-shaft machines permit constant speed for the axial air-compressor and variable speed for the process compressor. The selection of sizes, speeds and horsepower outputs of commercially available gas turbines is limited. Very often, there is not a wide freedom of choice as to single or double-shaft units.

Motor drives are usually of constant speed. In a limited number of cases, variable-speed couplings, wound-rotor or multipole (PAM) motors may be used. Large motors may be of the synchronous or induction kind. For a unit driven at above-synchronous speeds (3,600 rpm. for 60 Hz.), the choice should be based on the total cost of the motor and the speed increaser. Thus, an 1,800-rpm. induction motor and its speed increaser may cost less (including operating costs) than a 1,200-rpm. synchronous motor with its speed increaser.

Constant-speed centrifugals in process plants tend to operate at a high-enough average load so that the economic rewards of power-factor correction—obtained by the use of a synchronous motor—are minor.

Fossil-fuel drives are used when: initial and operating costs are more attractive than steam or motor drives; sufficient electric power is not available; and electric or steam sources are not reliable. In this last case, the entire system must be carefully specified to ensure that minor items such as cooling-water pumps, pressure switches, control air, etc., are independent of any source of power not as dependable as the compressor drive.

Internal-combustion engines are usually turbo-supercharged, and may be two- or four-cycle, integral with or separate from the compressor. The type of engine can normally be selected on the basis of drive features, including accessories and costs (purchase order, installation, fuel consumption, spare parts, maintenance) independent of the compressor. It is best, however, to include the drive as part of the compressor system.

The gear mechanical rating, including the American Gear Mfrs. Assn. (AGMA) service factor, should be selected so that the gear rating does not become the limiting factor in the compressor and drive train. Steam-turbine drives combined with a gear (with the turbine at a lower speed than the compressor) are sometimes lower in cost than higher-speed turbines. The policy of the user and his insurance carrier on warehouse spares for gears affects the choice, since the gears and additional couplings increase the probability of outage.

## Marketing of Compressors

Compressor systems and their drives range from small through medium to large. Marketing methods range from those suitable for catalog items handled by distributors to engineered systems oriented to specific market areas such as chemical processing, gas distribution, petroleum refining and electric-power generation. Engineers holding discussions with equipment suppliers, manufacturer representatives, or suppliers of package or skid-mounted units should recognize these elements of supplier organizations; such knowledge will aid in establishing the scope and detail most useful to include in purchase requisitions and specifications.

Catalog items require little more than hardware description, as performance is specified in published information. Engineered systems, on the other hand, require definition of performance requirements for the overall compressor system. Hardware definitions also are needed to establish the quality level of the system and its components, and to define the number and type of auxiliary elements such as oil coolers, governor, pumps, etc. Other hardware items include controls, heat-rejection systems, drives, gears, piping, ducts and electrical wiring (cable, conduit, trays, etc.).

The environment (indoor or outdoor) should also be made clear so that due allowances can be made for access for construction and maintenance, sound control and isolation, and area electrical classification.

## System Specifications

The form of a purchase specification should be the one most familiar to, and most commonly used within, the issuing organization; typical forms are given in API standards 617 and 618. Here we shall consider the content of the specification.

Performance criteria must be carefully defined for the end-use that the compressor must have within the overall system. Included in the specification should be:

- Range of mass and volumetric flow as influenced by variations in inlet temperature, pressure, molecular weight, gas composition (vapor loading, compressibility factor, etc.), discharge pressure, temperature and flow of cooling fluids (water, air, etc.).
- Startup, standby and shutdown conditions of the compressor and of the entire system.
- Mention of even traces of vapors, liquid droplets, dusts or gases that may be minor items for the chemistry of the process but may cause fouling, gunking, seal problems, etc., either by themselves or when mixed with lubricants or sealing fluids. Items such as these may appreciably influence the choice of compressor type.
- Range of ambient temperature.
- Altitude.
- Area electrical classification.
- Applicable codes and standards from such organizations as Tubular Exchanger Mfrs. Assn. (TEMA) and American Soc. of Mechanical Engineers (ASME).

Purchase specifications must define quality requirements for auxiliary equipment such as seals, piping systems (material and arrangement), type and quality of control elements and systems, level of redundancy, and shop testing (if any). Check lists for such items may be prepared based on available knowledge within an organization, as well as on accepted references on inspection of completed installations, such as Chapter X of the API "Guide for Inspection of Refinery Equipment." Purchase specifications—or those prepared by the customer for use by an engineer-constructor—should not limit the bidders from using their knowledge and experience.

Required controls cover a very wide range of supply. The compressor specification should include all elements that directly measure and control any part of the compressor system. This includes local panels, receivers from external inputs, and any items to provide outputs to external devices. Devices for volume control as such—or those used to control mass flow and provide anti-surge control—must be carefully defined as to which elements are supplied as part of the compressor system and which elements are external. Thus, such items as inlet guide vanes are best included in the compressor system, while anti-surge and recycle devices are usually best considered as external to the system.

An increasing area of interest for controls consists of types of diagnostic devices used to measure, indicate, alarm, and to record vibration (velocity and displacement), axial movement, bearing temperatures, and drive-motor copper temperatures. Axial-movement and motor-copper temperature indicators are best used for both alarm and then shutdown. Other instruments are most suitable for alarm only and as trend indicators. The compressor supplier is in the best position to select the points of pickup and recommend types of pickup and readout devices. Controls for units to be attended only by remote or occasional local surveillance require very careful attention.

Job cost and completion time is improved with proper use of shop-assembled units. Typical packages, including skid-mounted units, comprise refrigeration (chilled water and low-temperature brines), and complete instrument and plant-air units. Purchase specifications should therefore call for or permit packaged units to be offered where feasible. The units should be such that they need merely to be set on simple foundations, and have power, cooling-water and supply and discharge piping connected.

## Factors That May Be Overlooked

Perhaps the most important single factor to determine is whether a proposed piece of machinery has been used in similar service, and what its history has been. This by no means implies that one should never install a newly designed compressor, or use an older design for a new application. The first to use a new design may enjoy an advantageous position before competitors follow suit. Furthermore, if the decision to begin the construction of a plant cannot be delayed, it may be a matter of either installing something new or installing a proven design that one realizes may soon be outmoded. Features of a design that have not been proven should be reviewed in detail, and perhaps consideration should be given to courses of action to be taken if unexpected difficulties occur. Attempting to foresee possible failures and developing corrective courses of action is very time-consuming, very difficult and usually not very rewarding. But when trouble does occur, such planning may more than compensate for the effort spent.

If a compressor and its drive are investigated separately, one should not overlook the direction of rotation (which may not always be changeable) and its effect on gear requirements. One supplier should be given the responsibility of completing a combined torsional and vibration analysis of the entire system. An agreement as to who will do this work should be reached as soon as possible, so that any required design changes can be made with a minimum of difficulty.

Similarly, it is advisable to have one supplier assume responsibility for collecting and correlating design data pertaining to noise emission, and for making the final recommendation for noise suppression. The best procedure is to select one supplier and have him assume responsibility for the overall unit.

When reviewing vendors' proposals, there are several items, in addition to price and energy requirements, that should not be overlooked.

One may question what spares the supplier will generally have in his shop. If the purchaser does not buy a spare rotor, how long will it take to obtain a new one in case of emergency? Is there another company that might share the cost of a standby spare? Such sharing has not been generally accepted because even though costs are reduced, the risk of extended outage is increased. Nevertheless, this course of action may be worth considering.

When a centrifugal compressor is bought, one should check the closeness of the operating point to the surge point. On low-capacity, high-head wheels, these points may be quite near each other. It may not be possible to reduce speed very much without resorting to bypass control or to the installation of suction or discharge valves, etc. Critical speeds should also be reviewed to be sure they are far enough away from any desired operating speed, particularly if operation at reduced capacities may be considered.

Allowable noise levels are a function of frequency as well as intensity, and levels must be lower at higher pitches. The amount of noise generated depends on the type of compressor, its horsepower rating, compression ratio, speed, etc. Silencers or acoustic coverings may be used to reduce emitted noise to levels acceptable by the user and by regulatory authorities. Methods of estimating compressor noise levels, and the effect of various kinds of silencers are in the literature.[1, 2]

Drive and compressor characteristics must be evaluated. Operating at the maximum continuous speed of the centrifugal compressors—which is generally 5% greater than the speed at the compressor rated point—may call for a power expenditure 15% greater than the rated horsepower. If the compressor is driven by a turbine, increasing the latter's capacity by 15% may impose a significant increase in the cost of the drive and its auxiliary facilities, as well as impose a penalty of poorer efficiency when running the drive at the rated compressor speed. Normally, the permissible maximum mechanical speeds of the drive and driven unit should be the same.

It may also be well to review guarantees. In general, the following applies (API Std. 617, 618) unless other representations are made: For centrifugal compressors operating at constant speed, the capacity is guaranteed; the head may vary within +5 and −0% of that specified; the horsepower (when corrected to the specified head-capacity conditions) may vary by not more than 4% of the stated horsepower. For centrifugal compressors operating at a variable speed (i.e., gas- or steam-turbine drive in most instances), the capacity and the head are guaranteed but the speed is not. Horsepower may vary ±4%.

For reciprocating compressors, one may specify a guaranteed capacity with no negative tolerance, as well as a guaranteed maximum horsepower and a specified speed. However, process industries frequently accept a manufacturer's capacity guarantee of ±2 to 3% rather than pay more for the no negative tolerance. When motor drives are to be used, obviously one must review compressor-speed guarantees much more carefully than when turbine drives are involved.

## Installation Engineering

After the compressor unit is selected and a purchase order issued and accepted, the next steps require continued vigilance. This is not simple because many more people in the engineering and supplier organizations now become involved. Follow these guidelines to prevent certain items from being neglected:

1. For all but simple catalog units, prepare a process and instrumentation diagram or engineering flow-diagram for the complete compressor system.

2. Establish the layout requirements including those determined by operator assignment—i.e., the number of operators assigned to the compressor during normal operations, or during startup only, etc. Provide terminals for remote control if an operator will not be in attendance at all times.

3. If a large, complex compressor is involved, hold meetings with the supplier's engineering group to establish schedules for submission, review and approval of the supplier's engineering data and drawings, and tentative plans for use of his servicemen during installation and startup.

4. Review the compressor manufacturer's drawings and those of his suppliers to ensure that quality and performance criteria are being met.

5. Review the torsional-vibration analysis and lateral critical studies completed by the compressor and drive supplier to make certain that no contemplated operating condition will cause the machine to operate at a hazardous speed.

6. Review unpriced supplier orders. (Priced orders would not be made available and are unnecessary.)

7. Review control plans, including startup, normal operation, scheduled and forced shutdowns, protective and safety devices for alarm and shutdown, and the duties to be assigned to the operators.

8. Submit to the supplier, for his comments and review, design bases and installation drawings for foundations, piping and pipe supports. Such information should include the calculated forces and moments (hot and cold) exerted by the piping on the equipment flanges. Guidelines for allowable values are established by the compressor and turbine suppliers on their outline drawings.

9. Review requirements for shop and field pressure- and performance-testing. In most applications, test procedures established by the supplier are sufficient. Establish procedures for shop erection and match marking of prefabricated pipe to be furnished by the compressor supplier. Establish what shop tests are to be witnessed.

10. Establish requirements for such items as operating and maintenance access (cranes, monorails, etc.); noise control within buildings and other enclosures; protection from fumes and dust; winterizing; and piping systems, including drains, vents, and access for field flushing and cleaning.

11. Provide methods so that the installed dimensional accuracy of piping right at the compressor is high, and thus compatible with the level of dimensional exactness required by the machinery. Neglect of this may require field changes in the piping arrangement to secure and maintain acceptable compressor alignment. Arrange major piping so that supports can be taken from concrete substructures rather than from elevated steel structures. This is most important on reciprocating compressors because it helps in attenuating vibration.

12. Make provisions for shop inspection during fabrication and assembly, as well as during shop testing.

13. Obtain copies of expediting and inspection reports. Monitor delivery schedule.

## Vigilance During Installation

Job pressures during installation, run-in and startup create many hazards to the achievement of quality results from the compressor system. To minimize the hazards, follow these guidelines:

*Manufacturer's Representatives*—Err on the side of too much participation in the field of representatives of the supplier's service organization and of his major subsuppliers (turbines, gears, motors). Most suppliers include a specific number of days for such representatives in the original proposal price (and thus in the purchase order), with a per diem rate beyond this limit. Do not save these "free" days for when troubles may be encountered. Use the days to avoid trouble. Do not hestitate to obtain such service beyond the limit of days included in the purchase order for such items as machine setting and grouting, alignment, initial run-in and actual startup and demonstration.

*Field Checking*—During field checking of all kinds,

not only must construction drawings be consulted but also flowsheets, operating manuals, etc.

*Foundations and Superstructures*—Check these vital elements of the system, using supplier drawings and installation recommendations, as well as the engineer's construction or working drawings. Dimensional accuracy and quality of construction are both vital.

*Materials of Construction*—Check to see whether certified mill certificates have been received and whether they are acceptable.

*Piping, Ductwork and Supports*—Ascertain accuracy and structural adequacy, so that excessive loads are not imposed on equipment flanges. Also make sure that: provisions are made for controlled movement due to thermal expansion; proper line slopes are maintained; noise and vibration are attenuated; and resonant conditions are avoided.

Conduct hydrostatic and leak tests, remove all temporary blanks and install rugged line strainers.

Flush, degrease, mechanically clean and, when pertinent, clean with chemicals. Chemical cleaning (pickling) of carbon steel piping for such services as lubricating oil should also be done. Chemical cleaning of field-assembled systems should be considered very carefully because the results can be very hazardous if any of the chemical solution enters the compressor during startup operations. Proper drainage and venting provisions are essential for effective and safe removal of cleaning fluids. Chemical cleaning, moreover, does not prevent trouble from mud, stones, welding rods and slag.

Small piping systems such as sealing, venting, drain and control connections must be checked for continuity and completeness. Frequently, these systems are not given enough attention. This kind of piping (usually field run) must be arranged so as to permit access to the compressor, as well as for lubrication and maintenance. The supports must be sturdy to avoid leakage or rupture from vibration during normal and upset operating conditions.

*Instrumentation and Controls*—For the proper functioning of instruments and controls: (1) conduct completeness and continuity checks—hydraulic, pneumatic, electrical; (2) commission, field-calibrate, and establish setpoints; (3) check for accessibility for operator use and vision, and for adjustment and maintenance; (4) check mounting locations, and methods to minimize vibration pickup—avoid mounting directly on the compressor, light platforms or hand rails; (5) check the supply of instrument air and control power for adequacy and reliability.

*Electrical Power and Lighting*—For power, test for continuity, electrical-insulation soundness, proper grounding and settings of relays (including correctness of thermal overload trip devices), tightness and quality of all connections, sealing of fittings, and use of flexible connections for equipment and instruments.

As for lighting, check to see that it is adequate for those areas frequently missed by general lighting such as control stations, instruments, and lubrication points.

*Thermal Insulation and Painting*—Check to see that proper insulation will: (1) avoid hazard to personnel, (2) prevent thermal shock to piping, compressor and turbine from rain; (3) prevent fire from occurring as a result of oil spillage on a hot metal surface.

Avoid "overcompleteness" such as (1) paint on valve stems, instruments, etc., (2) unnecessary insulation on flanges and flange bolting, (3) external insulation on internally insulated brick or refractory piping and ducts, (4) insulation restraint on expansion joints and (5) insulation that will limit freedom of required pipe movement.

## References

1. Heitner, I., How to Estimate Plant Noise, *Hydrocarbon Process.*, Dec. 1968, pp. 67-74.
2. Golden, B. G., Ways to Reduce Plant Noises, *Hydrocarbon Process.*, Dec. 1968, pp. 75-78.

## Acknowledgments

The authors are grateful to Struthers Energy Systems for authorizing the presentation of Fig. 2 and 5 to 7. They are also grateful to the individuals and companies who reviewed the manuscript of this article and were kind enough to offer their comments and suggestions, particularly C. C. Kirby of American Cyanamid Co., J. Dzuback of Dresser Industries, Inc., J. Gooch of Cooper-Bessemer Co., and engineers at Allis-Chalmers Mfg. Co.

Fig. 5 is based, in part, on data from *Hydrocarbon Process.*, Nov. 1965, p. 120.

## Meet the Authors

◀ **Sidney A. Bresler** is a chemical engineering consultant, P.O. Box 86, Cathedral Station, New York, N.Y. 10025. He has worked on the design and economic evaluation of petrochemical and fertilizer plants for many years, is an experienced project manager, and has presented a number of papers that deal with technical and financial aspects of the process industries. Holder of an M.Ch.E. degree from Brooklyn Polytechnic Institute, he also has an M.B.A. from Columbia University.

**John H. Smith** is senior mechanical engineer with American Cyanamid Co., Engineering and Construction Div., Wayne, N. J. 07470, where he is responsible for the application, specification and selection of mechanical equipment for process and utility services. Prior to joining Cyanamid, he worked as a designer for Manning, Maxwell and Moore, Inc., and as a thermodynamicist for Northrop Corp. He holds B.S. and M.S. degrees in mechanical engineering from Purdue University and is a member of the American Soc. of Mechanical Engineers and the Connecticut Soc. of Professional Engineers. ▶

# Compressor Efficiency: Definition Makes a Difference

Compressor efficiency ratings can be as misleading as interest rate figures. To be certain of getting the most for your money when buying compressors, you must be sure that efficiency comparisons are made on the same basis.

EDWARD R. LADY, Los Alamos Scientific Laboratory, University of California

The compression of air and other gases consumes a significant amount of power in the chemical process industry. In the production of industrial gases, such as oxygen, nitrogen and helium, or in the rapidly growing field of natural-gas liquefaction, compression power accounts for more than 80% of total energy requirements. Clearly, compressor efficiency has a direct impact on product costs.

This article compares isothermal and adiabatic efficiencies, shows the effect of intercooling between stages of compression, and provides convenient ways for obtaining preliminary figures on compression power.

## Isothermal Compression

The basic equation used to calculate the work of gas compression in a steady flow process (and flow through a reciprocating compressor may be considered steady, despite the pulsations of the individual compression strokes) is:

$$w = \int_1^2 v dp \qquad (1)$$

The work on which this article is based was performed under the auspices of the U.S. Atomic Energy Commission.

Originally published August 10, 1970

with $w$ the work required per lb.-mole, $v$ the molal volume, and $p$ the pressure.

For preliminary calculations, the ideal-gas equation of state may be used to relate pressure, volume and temperature:

$$pv = RT \qquad (2)$$

In fact, this equation will yield quite accurate results at pressures below 10 atm. and at temperatures well above the critical.

If the gas could be compressed without friction and isothermally, the work required to raise the pressure from $p_1$ to $p_2$ is:

$$w_{iso} = RT_1 \ln (p_2/p_1) \qquad (3)$$

In this equation, $R$ is the gas constant, 1.986 Btu./(lb.-mole)(°R): $T_1$ the initial temperature, °*R;* and work has the units Btu./lb.-mole.

The isothermal efficiency of a compressor is defined as the ratio of work calculated by Eq. (3) to the actual work required, or:

$$\eta_{iso} = w_{iso}/w_a \qquad (4)$$

In many cases, the numerical value of isothermal efficiency may appear to be low, e.g. 65%, and yet the actual work required be less than that of a compressor with an adiabatic efficiency of 80%. We shall see how this apparent discrepancy comes about.

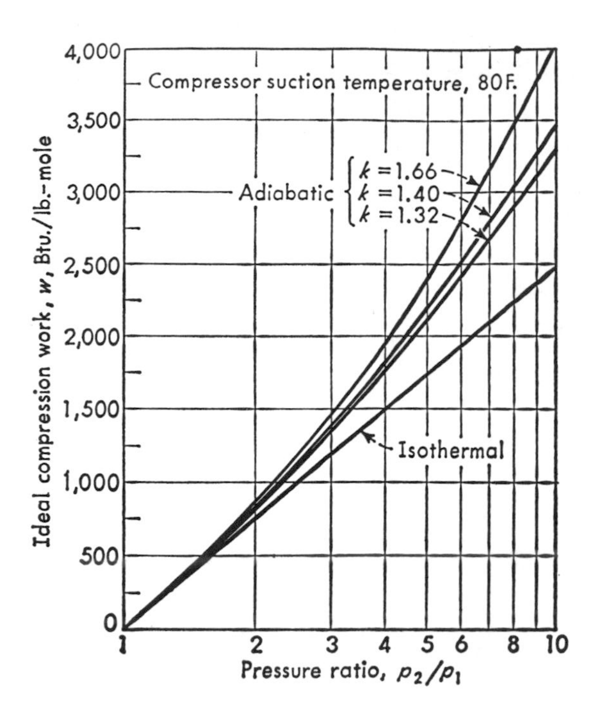


**ADIABATIC** is greater than isothermal work—Fig. 1

## Adiabatic Compression

An ideal gas compressed adiabatically and without friction requires work according to the expression:

$$w_{ad} = RT_1\left(\frac{k}{k-1}\right)\left[\left(\frac{p_2}{p_1}\right)^{\frac{k-1}{k}} - 1\right] \tag{5}$$

with $k$ the ratio of specific heats, $C_p/C_v$, and all the other terms defined as before.

Adiabatic efficiency is defined similarly to isothermal efficiency:

$$\eta_{ad} = w_{ad}/w_a \tag{6}$$

It is clear that the two definitions of compression efficiency vary only in the standard of comparison, i.e., ideal isothermal work vs. ideal adiabatic work. Fig. 1 shows the variation of ideal adiabatic work as a function of pressure ratio, $p_2/p_1$, and specific heat ratio, $k$. In all cases, this work is greater than the isothermal work shown by the lowest curve. The influence of specific heat ratio is pronounced. Monatomic gases, such as helium and argon, have $k = 1.66$, and therefore the adiabatic work for compression of such gases is significantly greater than the isothermal work. Diatomic gases, such as nitrogen, oxygen, hydrogen and air, have $k = 1.4$. More-complex gaseous molecules have a lower value of $k$, such as 1.32 for methane.

As an example, consider the compression of air at 14.0 psia. and 80 F. to 56 psia. From the appropriate equation, or from Fig. 1 with $k = 1.40$ and $p_2/p_1 = 4.0$, we find: $w_{iso} = 1{,}487$ Btu./lb.-mole and $w_{ad} = 1{,}824$ Btu./lb.-mole.

If the actual work required by a compressor is 2,280 Btu./lb.-mole, the compressor efficiency may be expressed as: $\eta_{iso} = (1{,}487/2{,}280)(100) = 65\%$ and $\eta_{ad} = (1{,}824/2{,}280)(100) = 80\%$. Both of these efficiency definitions are equally valid, although the adiabatic efficiency is usually used when no effort is made to cool the gas during or between stages.

## Equal-Work Efficiency Ratio

We have seen in the preceding example that for the same actual work of compression, the adiabatic and isothermal efficiencies vary by a factor of 1.23. From Eq. (3-6), the equal actual work of compression, $w_a$, results when:

$$\frac{\eta_{ad}}{\eta_{iso}} = \frac{\left(\frac{k}{k-1}\right)\left[\left(\frac{p_2}{p_1}\right)^{\frac{k-1}{k}} - 1\right]}{\ln(p_2/p_1)} \tag{7}$$

The equal-work efficiency ratio given by Eq. (7) is plotted as a function of pressure ratio in Fig. 2. With the same conditions as in the earlier example, $k = 1.4$ and $p_2/p_1 = 4.0$, the equal-work efficiency ratio is 1.23.

To illustrate how the equal-work efficiency ratio may be used, consider the problem of evaluating compressor bids. Again, we shall compress air from 14.0 psia. and 80°F. to 56 psia. If one vendor guarantees 80% adiabatic efficiency and another

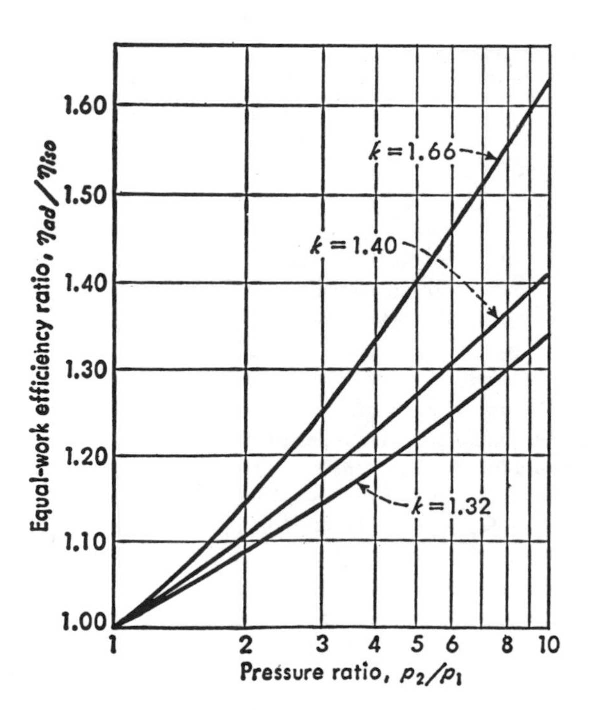


**RATIOS** for actual work of compression—Fig. 2

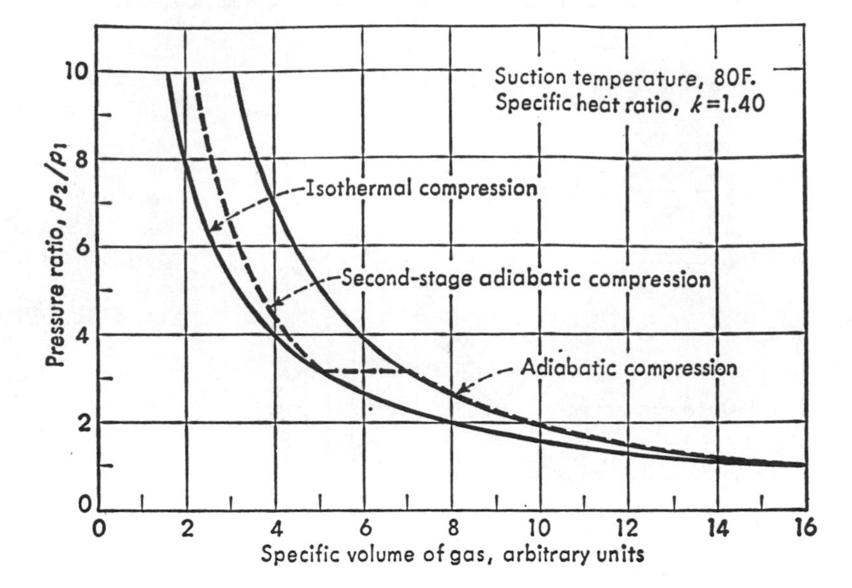


**DASHED-CURVE** represents two-stage adiabatic compression with intercooling to the suction temperature; solid curves depict adiabatic and isothermal compression of gases—Fig. 3

70% isothermal efficiency, the quotation of the latter must indicate a lower power requirement, despite the lower efficiency. We found that the equal-work efficiency ratio for these conditions is 1.23. The ratio of quoted efficiencies is 1.14. Whenever the actual ratio is less than the equal-work ratio, the compressor whose efficiency is based on isothermal efficiency will require less power. Above the equal-work curve, the adiabatic efficiency machine requires less power.

## Multistage Compression With Intercooling

Whenever the overall pressure ratio exceeds four, consideration must be given to multistage compression with intercooling. This is necessary to keep the temperature of the compressed gas at a safe value, e.g., 365 F. for air compressors with hydrocarbon lubricants, as well as to reduce the overall power requirement.

Ideal intercooling would reduce the temperature between stages to the suction temperature. The effect of staging with intercooling is shown in Fig. 3. The ideal work of compression in steady flow, given by Eq. (1), is represented by the areas to the left of the curves. The isothermal curve represents the minimum work requirement, whereas the adiabatic curve encloses a much larger area. The intermediate dashed-curve represents two-stage compression with intercooling to the suction temperature, 80 F.

Now let us consider a two-stage compressor, compressing air from 14.0 psia. and 80 F. to 140 psia., with intercooling to 80 F. Should the efficiency of a compressor be based on the ideal isothermal or the ideal adiabatic compression, as shown by the solid curves on Fig. 3? Actually, we should refer to stage efficiency but keep overall efficiency clearly in mind.

Because, in this example, the overall pressure ratio is 10.0, we may compute the ideal work of compression as: $w_{iso}$ = 2,470 Btu./lb.-mole and $w_{ad}$ = 3,480 Btu./lb.-mole.

These numbers may also be found from Fig. 1. For each stage of compression, with a pressure ratio of 3.16, the ideal adiabatic work of compression is 1,460 Btu./lb.-mole, or a total, for two stages, of 2,920 Btu./lb.-mole. Using a realistic adiabatic stage efficiency of 80%, the actual work of compression is 2,920/0.80 = 3,650 Btu./lb.-mole. We now have three equally accurate, but quite confusing, compressor efficiencies:

Adiabatic compressor efficiency: (3,480/3,650)(100) = 95%
Adiabatic stage efficiency: (2,920/3,650)(100) = 80%
Isothermal compressor efficiency: (2,470/3,650)(100) = 68%

Efficiency is a word of many meanings. We have seen that a higher numerical value of efficiency based on one definition may actually represent poorer performance than a lower numerical value based on another definition. The equations and graphs presented in this discussion are designed to refresh practicing engineers on these points. There is no doubt, however, that for a given compression job, the compressor that requires the least horsepower is the most efficient.

## Meet the Author

**Edward R. Lady** is a visiting staff member at the University of California's Los Alamos Scientific Laboratory (P.O Box 1663, Los Alamos, N. M. 87544) on leave from the University of Michigan (where he is an associate professor). His industrial experience of 12 years was with Union Carbide Corp. and Air Products and Chemicals Co. A member of ASME and American Soc. for Engineering Education, he is a registered engineer in the states of Pennsylvania and Michigan.

# How to achieve online availability of centrifugal compressors

*Here is practical information that will ensure the best selection and proper maintenance of centrifugal compressors and their drivers, bearings, seals, couplings, speed gears, and lubrication and control systems.*

**M. P. Boyce,** *Boyce Engineering International, Inc.*

☐ Centrifugal compressors are an integral part of the chemical process industries (CPI). They are used extensively because of their smooth operation, large tolerance to process fluctuations, and higher reliability than other types of compressors.

Centrifugal compressors range in size from pressure ratios of 1.3:1 per stage to as high as 12:1 on experimental models. We will limit our discussion to compressors having pressure ratios below 3.5:1 because these are the ones used extensively in the CPI.

Proper selection of these compressors involves making complex decisions, since the successful operation of many CPI plants depends upon the smooth and efficient functioning of such units. To ensure the best selection and proper maintenance of centrifugal compressors, the engineer must have a wide knowledge of many engineering disciplines.

## Specification criteria

Detailed compressor specifications can vary from customer to customer. Some provide only basic information, such as pressure, flowrate, type of gas, driver, and site conditions. Others present a lengthy document detailing types of bearings, rotor response, lubrication system, acceptable tolerance on performance, etc. The latter requires that the engineer be very conversant with compressors and their total support systems. A starting point for the specifications is to be found in the publications of the American Petroleum Institute (API) for turbomachinery. Some of the applicable ones are listed in Table I.

Specifications for the API Standards are written by user engineers, with the input of manufacturers and engineering contractors. Thus, the standards represent a wealth of experience and are a very good base from which to start turbomachinery specifications.

Many decisions, regardless of details contained in specifications, have to be made in advance by the engineer. Some of these decisions may involve company philosophy about various units; others could be strictly job-oriented. In writing a job specification, we must have an understanding of the major problem areas. These will now be covered.

## Compressor layout

The general topography of the plant must be known so that the proper site can be established. Whether the unit will be grade-mounted or mezzanine-mounted is important in determining the foundation characteristics. Enough space should exist for the ducting, so that the inlet conditions to each stage allow the flow to enter without large distortions of velocity and pressure.

Accessibility requirements should be kept in mind so that repair and maintenance work on the unit can be performed with relative ease. Location of the oil system for the unit is also an important aspect. It is advisable to locate the oil reservoir away from the base plate, with the bottom sloped toward the low drain-point. Enough space should be provided so that the return-oil lines can enter the reservoir away from the oil-pump suction. This would greatly reduce disturbance of the pump suction, and also help in keeping the reservoir-retention time to around 10 min.

## Compressor environment

The environment in which a machine operates is as important as any other factor. In many cases, this detail is often overlooked, or described in a phrase such as "extreme cold climate." The vendor needs to know much more. He must know what extreme cold means (usually below −25°F), and what the transition weather is.

Weather creates many problems—though few precautions are taken because operating problems are either not recognized or are glossed over. All cold-weather applications need to have some icing protection, espe-

Originally published June 5, 1978

**Sources for compressor specifications** **Table I**

| Standard | Title |
|---|---|
| API 611 | General Purpose Steam Turbine for Refinery Services |
| API 612 | Special Purpose Steam Turbine for Refinery Services |
| API 613 | High-Speed Special Purpose Gear Units for Refinery Services |
| API 614 | Lube and Seal Oil Services |
| API 616 | Combustion Gas Turbines for General Refinery Services |
| API 617 | Centrifugal Compressors for General Refinery Services |
| API 670 | Noncontacting Vibrating and Axial Position Monitoring Systems |

Available from: American Petroleum Institute, 2101 L St., N.W., Washington, DC 20037.

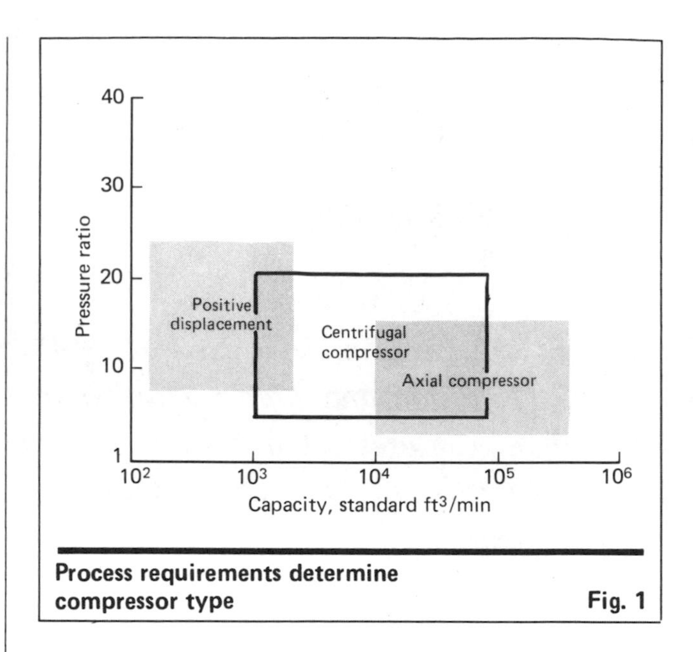


**Process requirements determine compressor type** **Fig. 1**

cially in the air-inlet supplies. Fuel-supply ventilation, pneumatic controls and actuators need to have some degree of deicing. Many types of deicing systems exist. The two most common use exhaust gases or compressor bleed-air.

Tropical climates present their own problems, such as excessive corrosion, high moisture content, and high ambient temperatures that increase both the horsepower required and the required cooling capacity of the lube-oil systems. Desert locations require special filtration systems to prevent erosion of the blades, and special sealing on joints to prevent micron-sized sand particles from entering the lube system, etc.

Machine usage is an environmental consideration because intermittent operation can be the most severe kind of service. The aggressiveness of the gas to be handled and its temperature determine the materials of construction to be used.

## Compressor types and arrangements

It is not always obvious what type of compressor is needed for an application. Of the many kinds, some of the more significant are the centrifugal, axial, rotary and reciprocating. Fig. 1 will aid in the selection of a compressor. For very high flows and low pressure ratios, an axial-flow compressor would be best. Axial-flow compressors usually have a higher efficiency but a smaller operating region than a centrifugal machine. Centrifugal compressors operate most efficiently at medium flowrates and high pressure ratios. Rotary and reciprocating compressors (positive-displacement machines) are best used for low flowrates and high pressure ratios.

The general configuration for compressors and their drive trains must fit the location, environment and type of compressor. A decision must be made as to whether the units are to operate in series or parallel, and it requires a knowledge of the necessary discharge pressure and flow. In most cases, a number of casings are connected together to form a "compressor train." This is nothing more than connecting various compressors in series. The limit to the number of casings so connected is due to the rotor dynamics of the coupled train.

A decision must also be made as to what type of mounting is desirable for a given arrangement. Most turbomachinery is mounted on structural-steel platforms, referred to as base plates or skids. These platforms are then positioned on a mass of concrete at the job site by installing them on sole plates or through direct grouting. Platforms should be considered part of the foundations, with great care exercised in their design. Insufficient rigidity can allow the rotating machinery to excite these platforms.

## Driver selection

The three main types of drives for centrifugal compressors are (1) steam turbines, (2) gas turbines and (3) electric motors. The decision of which drive is best is not always easy. Selection depends on many factors, such as location, process, and unit size.

For remote locations, gas turbines are mostly used, due to their low maintenance and the ability to prepackage the units. Their light weight makes them a must for offshore platforms. For chemical plants, steam turbines are widely used, due to the needs for process steam. In this manner, CPI plants can use energy more efficiently. For smaller flows, electric motors drive the compressor, usually through speed-increasing gears.

Typical ranges for the various drives are shown in Fig. 2, from which we note that the higher the flow, the lower the speed. At high flows, the compressor diameter must be large; therefore, the speed must be reduced to maintain the same stress levels in the machinery.

## Compressor design and configuration

To properly design a centrifugal compressor, we must know the operating conditions—the type of gas, and its pressure, temperature and molecular weight. We must also know its corrosive properties so that proper metallurgical selection can be made. Gas fluctuations due to process instabilities must be pinpointed so that the compressor can operate without surging.

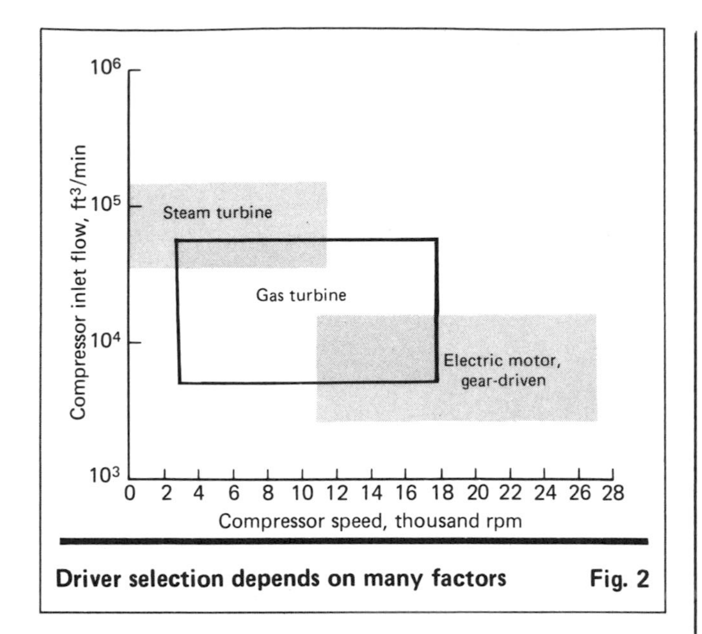


**Driver selection depends on many factors** **Fig. 2**

Centrifugal compressors for industrial applications have relatively low pressure-ratios per stage. This is necessary so that the compressors can have a wide operating range, and stress levels can be kept at a minimum. Due to the low pressure-ratios for each stage, a single machine may have a number of stages in one "barrel" to achieve the desired overall pressure ratio. Fig. 3 shows some of the many configurations. Some of the factors to be considered when selecting a configuration to meet plant needs are:

1. Intercooling between stages can considerably reduce the power consumed.
2. Back-to-back impellers allow for a balanced rotor thrust, and minimize overloading the thrust bearings.
3. Cold inlet or hot discharge at the middle of the case reduces oil-seal and lubrication problems.
4. Single inlet or single discharge reduces external piping problems.
5. Balance planes that are easily accessible in the field can appreciably reduce field-balancing time.
6. Balance piston with no external leakage will greatly reduce wear on the thrust bearings.
7. Hot and cold sections of the case that are adjacent to each other will reduce thermal gradients, and thus reduce case distortion.
8. Horizontally split casings are easier to open for inspection than vertically split ones, reducing maintenance time.
9. Overhung rotors present an easier alignment problem because shaft-end alignment is necessary only at the coupling between the compressor and driver.
10. Smaller, higher-pressure compressors that do the same job will reduce foundation problems but will have greatly reduced operational range.

## Impellers and their fabrication

Centrifugal-compressor impellers are either shrouded or unshrouded. The blading for the impellers can have one of three configurations—most commonly radial, followed by backward-curved and, in some rare cases, forward-curved. Backward-curved blades have the

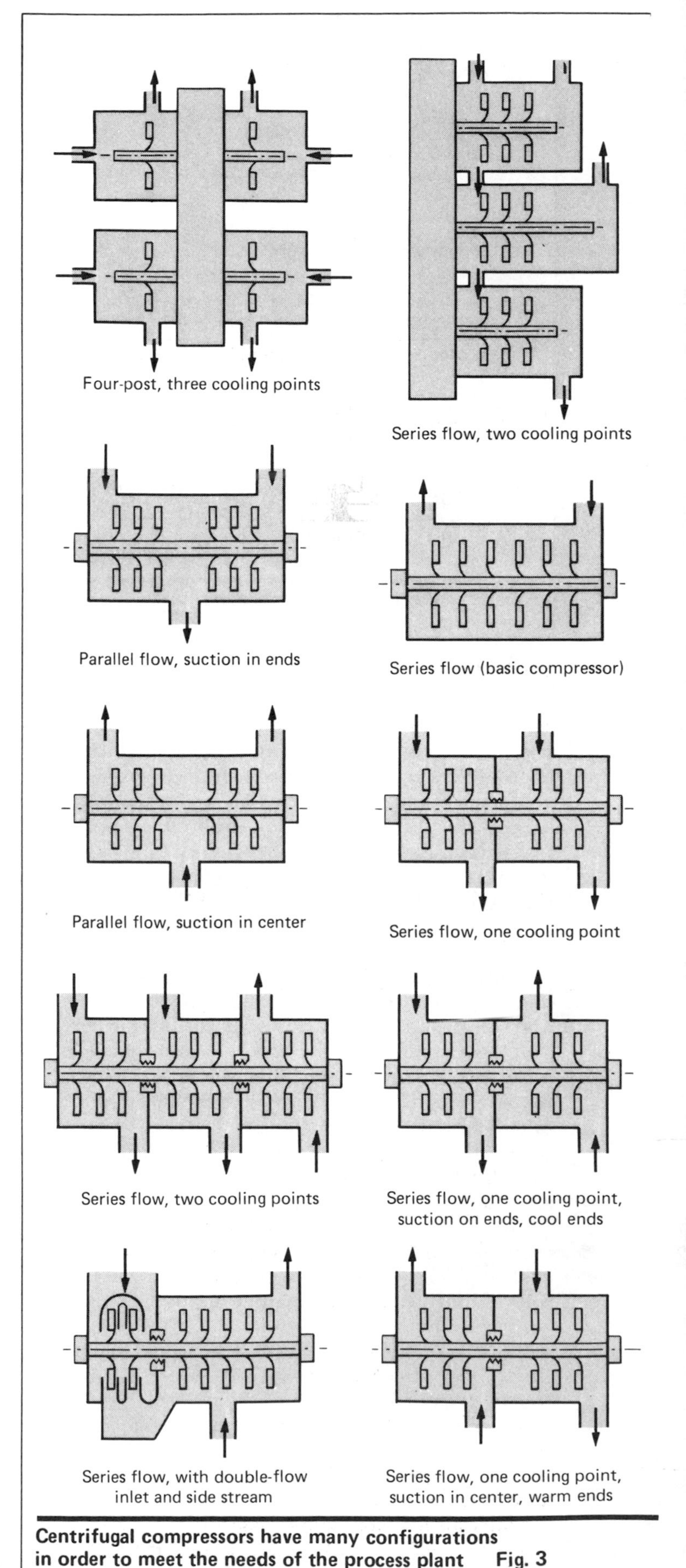


**Centrifugal compressors have many configurations in order to meet the needs of the process plant** **Fig. 3**

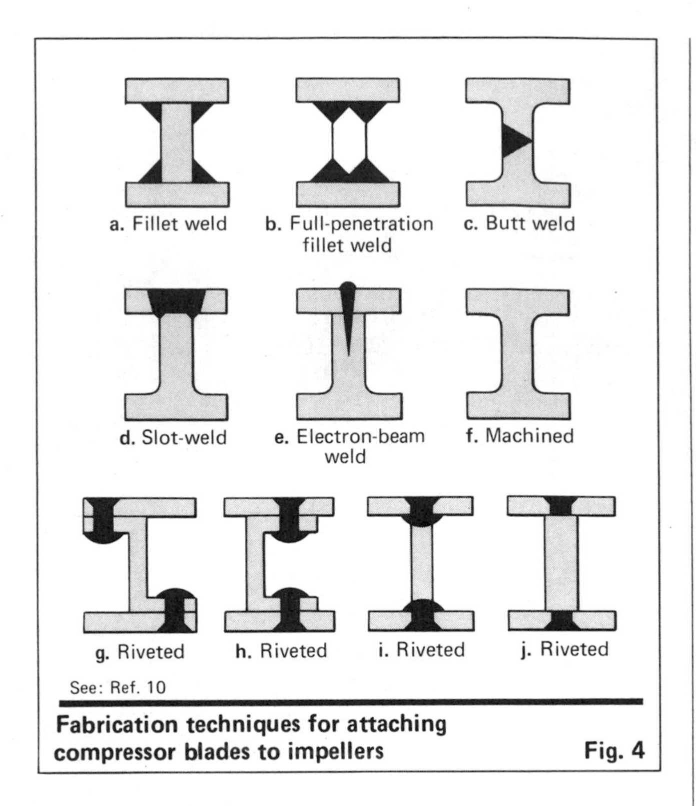


**Fabrication techniques for attaching compressor blades to impellers** **Fig. 4**

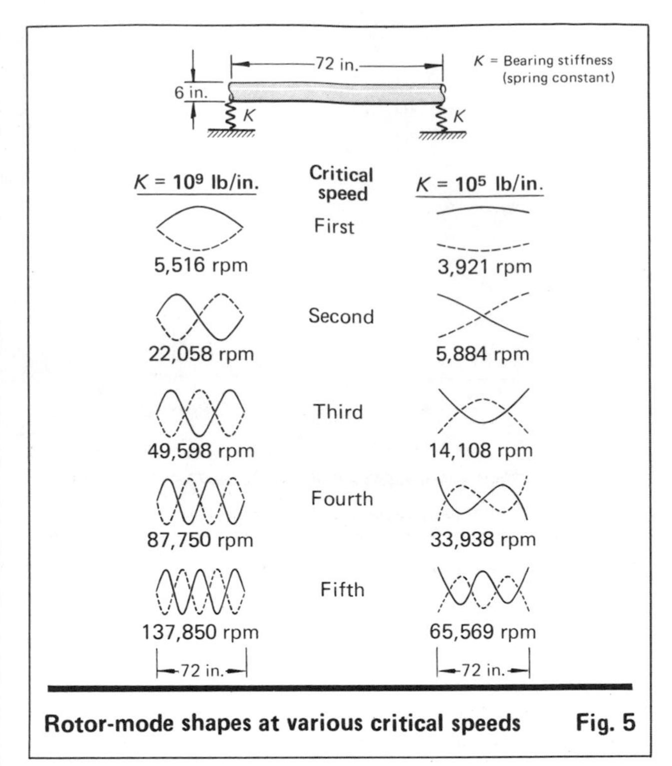


**Rotor-mode shapes at various critical speeds** **Fig. 5**

largest operating range and are usually higher in efficiency. The radial-bladed impeller produces a higher head but has a smaller operating range and a slightly lower efficiency than one with backward-curved blades.

Open, shrouded impellers that are mainly used in single-stage applications are made by investment-casting techniques or by three-dimensional milling. Such impellers are used, in most cases, for the high-pressure-ratio stages. In the CPI, the shrouded impeller is the most common.

Fig. 4 shows several fabrication techniques. The most-common type of construction is seen in Fig. 4a and 4b, where the blades are fillet-welded to the hub and shroud. In Fig. 4b, the welds are full penetration. The disadvantage of this type of construction is the obstruction to the aerodynamic passage. In Fig. 4c, the blades are partially machined with the covers, and then butt-welded down the middle. For backward-lean-angle blades, this technique has not been very successful, and there has been difficulty in achieving a smooth contour around the leading edge.

Fig. 4d illustrates a slot-welding technique and is used where blade-passage height is too small, or the backward-lean angle too high, to permit conventional fillet welding. In Fig. 4e, an electron-beam technique is used to weld on the shroud or the hub. This technique is still in its infancy and work needs to be done to perfect it. Its major disadvantage is that electron-beam welds should preferably be stressed in tension, but for the configuration of Fig. 4e, they are in shear. The configurations of Fig. 4g through 4j use rivets. Where the rivet heads protrude into the passage, aerodynamic performance is reduced.

Materials for fabricating these impellers are usually low-alloy steels, such as AISI 4140 or AISI 4340. AISI 4140 is satisfactory for most applications; AISI 4340 is used for larger impellers requiring higher strengths. For corrosive gases, AISI 410 stainless steel (about 12% chromium) is used. Monel K-500 is employed in halogen-gas atmospheres and in oxygen compressors because of its resistance to sparking. Titanium impellers have been applied to chlorine service. Aluminum-alloy impellers have been used in great numbers—especially at lower temperatures (below 300°F). With new developments in aluminum alloys, this range is increasing. Aluminum and titanium are sometimes selected because of their low density. This can cause a shift in the critical speed of the rotor, which may be advantageous.

## Rotor dynamics

Rotor movement and its effect on the performance of the entire unit is the most important aspect of centrifugal-compressor design. Most compressors for the CPI are built in accordance with API 617 specifications. The natural frequency of the rotor should not occur in the variable-speed range of the compressor, so as not to excite any of these frequencies. In newer high-speed compressors, many operate above their first critical speed. Shafts that operate above the first critical are said to be "flexible shafts."

It is desirable that the first critical not be around half the design speed. Otherwise, a problem known as "oil whirl" may be induced. Oil whirl is a major cause of instability in turbomachines. It may occur in the journal bearings or in the seals in which the shaft and the stationary seal are separated by a film of fluid.

Many types of whirling motions occur in turbomachinery and are generated by various factors, such as:

■ Hydrodynamic or "oil whip" or half-frequency whirl is caused by instability of the fluid film in the bearing.

■ Aerodynamic whirl is usually induced by the tip

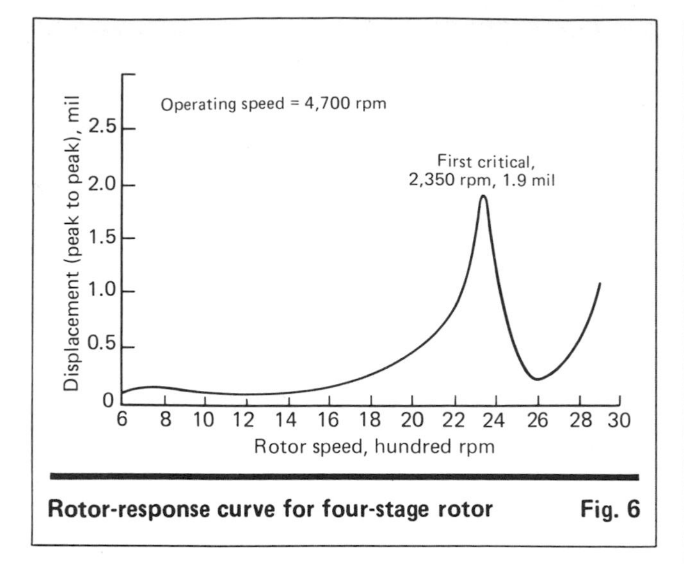


**Rotor-response curve for four-stage rotor** **Fig. 6**

clearances on compressors and turbines. It cannot be eliminated on balancing.

- Dry-friction whirl is usually initiated by rubbing caused by a large unbalance force.
- Torque-induced whirl occurs in large flexible rotors under large power levels.
- Gyroscopic-induced whirl may be caused by bowed shafts and skewed disks.
- Asymmetric whirl is caused by the difference in shaft stiffness, and is most violent near the first critical.

Balancing is a major problem for the newer flexible rotors that operate, in many cases, above the first critical and, often, above the second or third critical. Fig. 5 shows the various modes that the rotor shaft undergoes as it passes through these criticals. High-speed balancing of these rotors is sometimes a must for smooth operation. This inevitably means field balancing, because there are only a few test-stands that can balance these rotors at design speed.

Fig. 6 is a typical rotor-response curve for a four-stage rotor. Here, the rotor is operating above the first critical, but the steepness of the curve near the design point is a cause for concern. Modification of the rotor and a change in bearing stiffness will move the slope from the design point.

## Bearings for high-speed machines

Journal and thrust bearings are among the most important components to assure maintenance-free running of high-speed turbomachines. Bearings in these machines range from simple journal bearings and flat thrust bearings to multiwedge designs for both thrust and journal bearings. Some of the many factors that enter into the selection of such bearings are:

- Speed range of shaft.
- Maximum misalignment that can be tolerated by the shaft.
- Critical-speed analysis, and influence of bearing stiffness on this analysis.
- Loading of the compressor impellers.
- Oil temperature and viscosity.
- Foundation stiffness.
- Axial movement that can be tolerated.

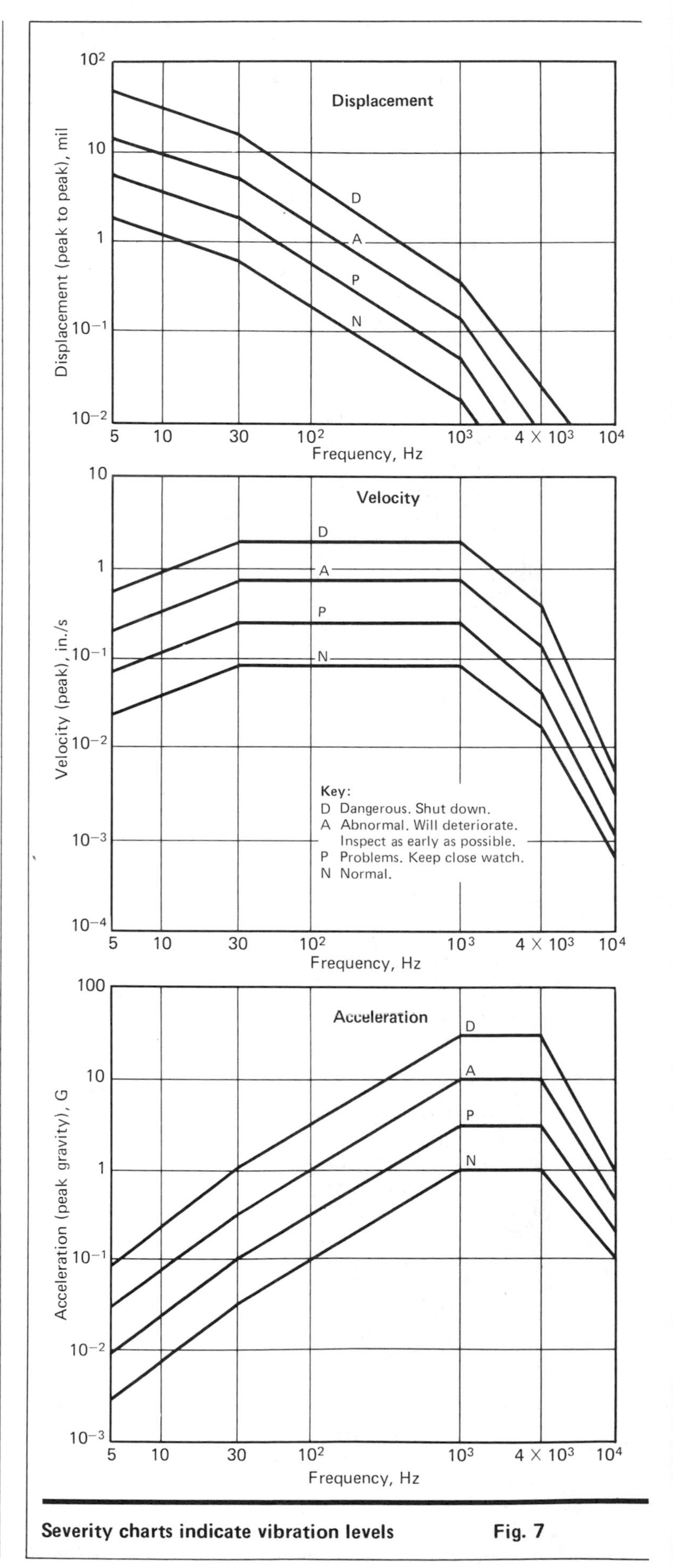


**Severity charts indicate vibration levels** **Fig. 7**

- Type of lubrication system and its contamination.
- Maximum vibration levels that can be tolerated.

All rotating machines vibrate when operating, but failure of the bearings is mainly due to their inability to resist cyclic stresses. The level of vibration that a unit can tolerate is shown in the severity charts (Fig. 7). These charts are modified by many users to reflect the critical values for their machines.

## Journal bearings

The journal bearing for turbomachinery has a fluid film that carries the load. Film thickness in most applications ranges from 0.0003 in. for gases to 0.008 in. for hydrostatic oil-lubricated bearings.

In this article, we will only discuss journal bearings in which a positive supply of lubricant is fed to the bearing at all times. Fig. 8 shows a number of such bearings. The circumferential-grooved bearing normally has the oil groove at half the bearing length. This provides better cooling, but reduces load capacity by dividing the bearing into two parts. The cylindrical bearing, used in turbines, has a split construction with two axial oil-feed grooves at the split. The pressure-dam bearing is used where bearing stability is required.

The most common bearing is the tilting-pad type, whose most important feature is self-alignment when the bearing is used with spherical pivots. This bearing offers the greatest increase in fatigue life because of these advantages:

1. Self-aligning for optimum alignment and minimum limit.
2. Thermal-conductivity backing material to dissipate heat developed in the oil film.
3. Thin babbitt layer, centrifugally cast to a uniform thickness of about 0.005 in. Thick babbitts greatly reduce bearing life. Babbitt thickness of about 0.01 in. reduces bearing life by more than half.
4. Oil-film thickness can be varied by changing the number of pads, directing the load onto or in between the pads, or changing the axial length of the pad. Oil-film thickness is critical when making bearing-stiffness calculations.

| Bearing type | Load capacity | Suitable direction of rotation | Resistance to half-speed whirl | Stiffness and damping |
|---|---|---|---|---|
| Cylindrical bore | Good | | Worst | Moderate |
| Cylindrical bore with dammed groove | Good | | | Moderate |
| Lemon bore | Good | | | Moderate |
| Three-lobe | Moderate | | to | Good |
| Offset halves | Good | | | Excellent |
| Tilting-pad | Moderate | | Best | Good |

Journal bearings for turbomachinery **Fig. 8**

## Thrust bearings

The most important function of a thrust bearing is to resist the unbalanced force developed in the working fluid of the machine, and to maintain the rotor in its position within the prescribed limits. A complete analysis of the thrust load must be conducted. Compressors with back-to-back rotors greatly reduce the load on thrust bearings.

Fig. 9 shows a number of different types of such bearings. When properly designed, the tapered-land thrust bearing (Fig. 9a) can take and support a load equal to that of a tilting-pad thrust bearing. With perfect alignment, it can match the load of even a self-equalizing tilting-pad thrust bearing. Fig. 9b is a nonequalizing tilting-pad thrust bearing that pivots on the back of the pad along a radial line. Fig. 9c is a nonequalizing tilting-pad bearing whose pads are supported on spherical-pivot points. Since this allows the pads to pivot in any direction, alignment is not as serious a problem as in the other two types. Fig. 9d is the Kingsbury-type self-equalizing thrust bearing. This bearing virtually eliminates the problem of misalignment. The major drawback is that standard designs require more axial space than do a nonequalizing type.

## Misalignment

The amount of misalignment that can be tolerated in high-speed turbomachines depends on the types of journals and thrust bearings used. Fig. 10 shows misalignment for such bearings. In a journal bearing, misalignment will cause the shaft to contact the end of the bearing. Thus, journal length is a criterion in the amount of misalignment that a bearing can tolerate. Obviously, a shorter length bearing can tolerate more misalignment. In a thrust bearing, misalignment will cause the loading up of one segment of the thrust-bearing arc and the unloading of the opposite segment. This effect is more pronounced at higher loads and with less-flexible bearings.

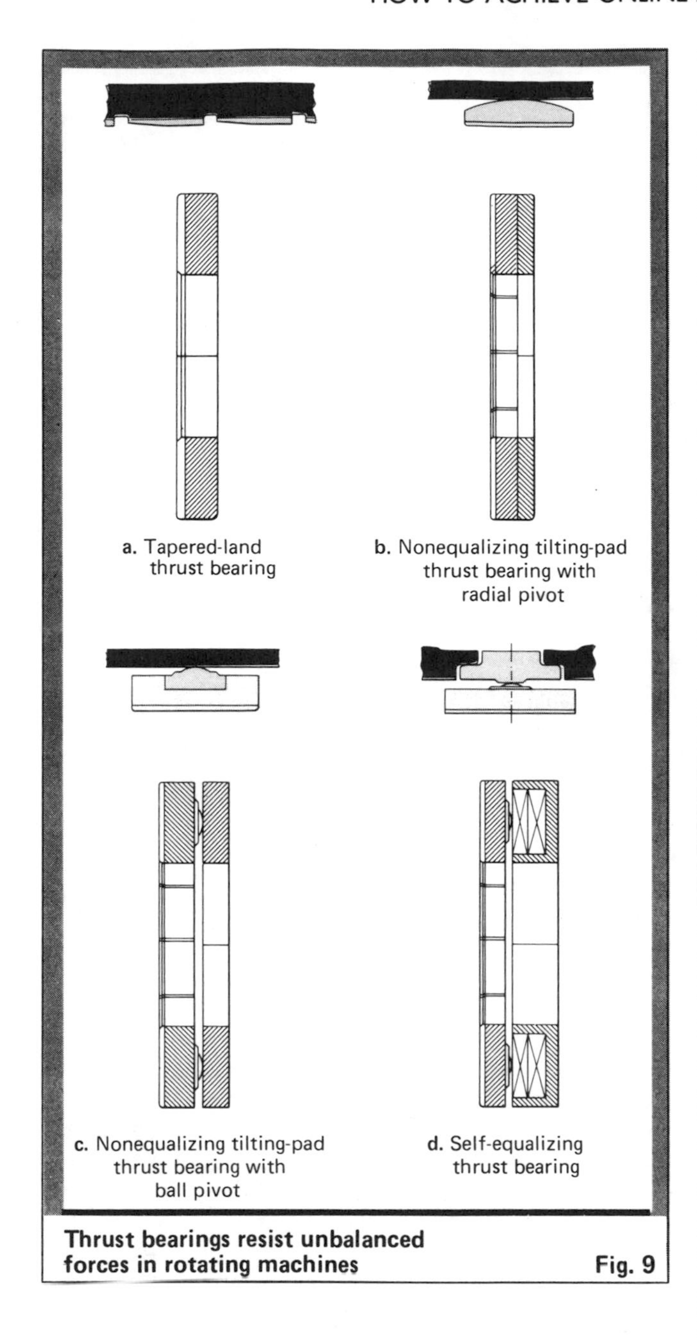


**Thrust bearings resist unbalanced forces in rotating machines** Fig. 9

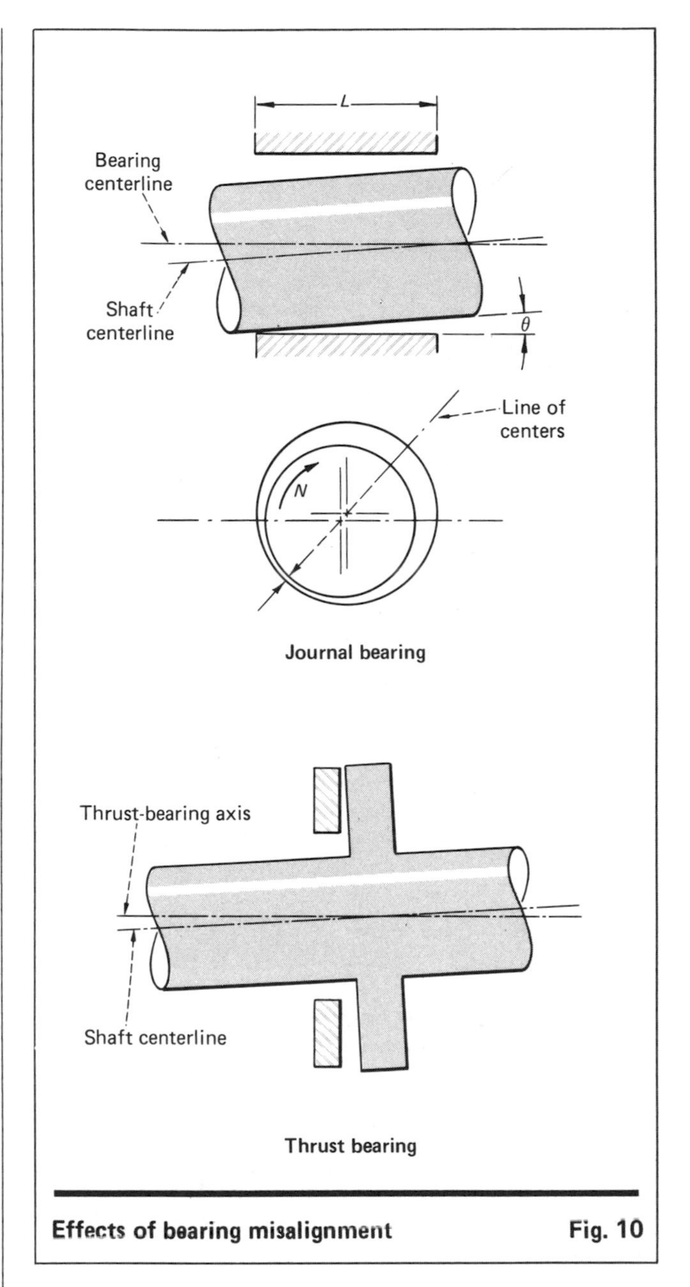


**Effects of bearing misalignment** Fig. 10

To adjust for misalignment other than by using tilting-pad bearings, various techniques must be employed. The simplest and most common is the so-called "cold alignment" (also known as base alignment) method. Once this is done, hot alignment is carried out. Hot-alignment techniques measure the changes when the unit is operational and temperature growth is stabilized, so that accurate alignment data are obtained. There are many ways to perform these tests. The recommended procedure is to do cold alignment by using the "reverse-indicator graphical plotting," and the hot alignment by using an optical alignment technique.

Reverse-indicator graphical plotting is normally obtained when the unit is cold. This is done by first laying out the desired hot operating line on graph paper (Fig. 11). This line shows the desired equilibrium operating conditions. Then, the desired cold position of the shaft is plotted, based on thermal-growth information that must be supplied by the manufacturer. Actual positions of the shaft in the field are taken. This information is plotted, the difference computed, and shims added to the supports. (The major assumptions here are that the computation measures actual vertical growth, and that growth in the horizontal direction is zero.) This procedure is then repeated, after hot-checking the alignment.

For hot-checking, an optical alignment procedure is recommended. The heart of this technique is the optical instrument with built-in optical micrometers to measure the displacement from a precise and referenced line of sight. After the initial cold alignment, the train is ready for startup. Optical reference points are established at each end and each side of each unit of the train, and as close to the couplings as possible. On most units, this is at the bearing housing.

A jig transit is readied, a line of sight established, and readings taken at each reference point in the compressor train in the vertical plane. For the horizontal plane, a

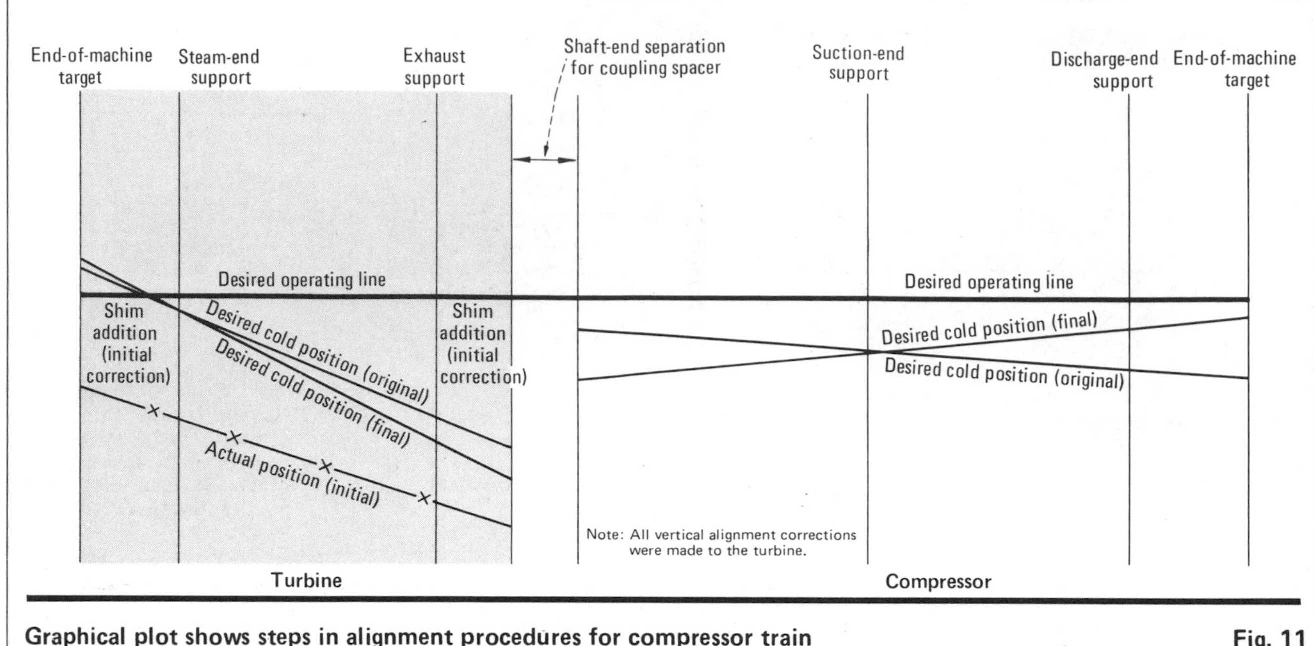


**Graphical plot shows steps in alignment procedures for compressor train** **Fig. 11**

somewhat similar procedure is followed, and data are recorded for reference points. The train is then started up and operated at design or near-design conditions, and temperatures are allowed to stabilize. Then, another set of optical readings is taken. In this manner, when a comparison is made with the cold readings, actual thermal growth can be calculated. Thus, the actual thermal growth can be plotted on the graph, and new corrections can be computed.

The technique outlined here is used on new machines. On old machines, a reverse technique can be applied. First, a hot alignment check is made, then a cold optical check, followed by the mechanical reverse-indicator readings. The information is plotted, and realignment measurements are taken.

The outline of the technique given here is simple, but in practice one must develop this skill. Some of the major problems encountered in alignment are due to pipe strain. This is caused by the piping being offset from the intake or exhaust nozzles from a few thousandths of an inch to several inches. Many engineers cannot understand how a "small" pipe is able to move a large, heavy piece of machinery. Tension on pipe hangers can change the vibration level considerably.

Another contributor to the alignment problem is the gear casing. Thermal growth in the new fabricated cases has been unpredictable. In some instances, gear cases rise with a twist to create another problem that is very difficult to correct. Short couplings also present a problem and magnify misalignment. Some users are now specifying that the coupling spacer will be at least 18 in. long.

The preceding discussion has presented a general outline of alignment techniques. Alignment of high-speed machinery must be accurate; otherwise, major problems will arise. Sometimes, to prevent major shutdowns to correct misalignment, heaters are added to one or the other legs of the unit to align them while running. This technique is not a cure but rather a temporary relief when shutdown is impossible.

## Compressor seals

The internal seals that prevent leakage around the impellers are usually labyrinth type, as shown in Fig. 12. They have a series of circumferential knife edges that are positioned closely to the rotating impeller. If damaged by rubbing, erosion or corrosion, these knife edges will lose their effectiveness. In some cases, the knife edges are machined onto the rotating part, while a sleeve of soft material is positioned on the stationary part. The rotating part then cuts a groove into the stationary sleeve, reducing leakage considerably and thus increasing the compressor efficiency. The stationary sleeve can be manufactured in babbitt-lined steel, ceramic, compressed steel-fiber material, etc., while the knives are made of a high-quality steel.

Shaft seals are usually mechanical-contact or liquid-film types, as seen in Fig. 13. In some cases where a small amount of leakage can be tolerated, labyrinth seals are used. The oil or liquid-film seal consists of two stationary bushings that surround the shaft with a clearance of a few thousandths of an inch. Oil at a nominal rate of 10 gpm is introduced between the bushings at a positive pressure higher than that of the process gas, and leaks in both directions along the shaft. The oil is retained in the seal housing by "O" rings. To limit the inward oil leakage, the differential pressure across the inner bushing is only a few pounds per square inch. The inner leakage rate varies from 1 to 4 gph, depending on the size of the seal, but is independent of the gas pressure being contained.

This leakage is collected in a chamber that is usually separated from the gas stream by a labyrinth seal. Overflow of oil from the leakage chamber and its subsequent entering the compressor is the biggest problem.

Mechanical-contact seals have two major elements

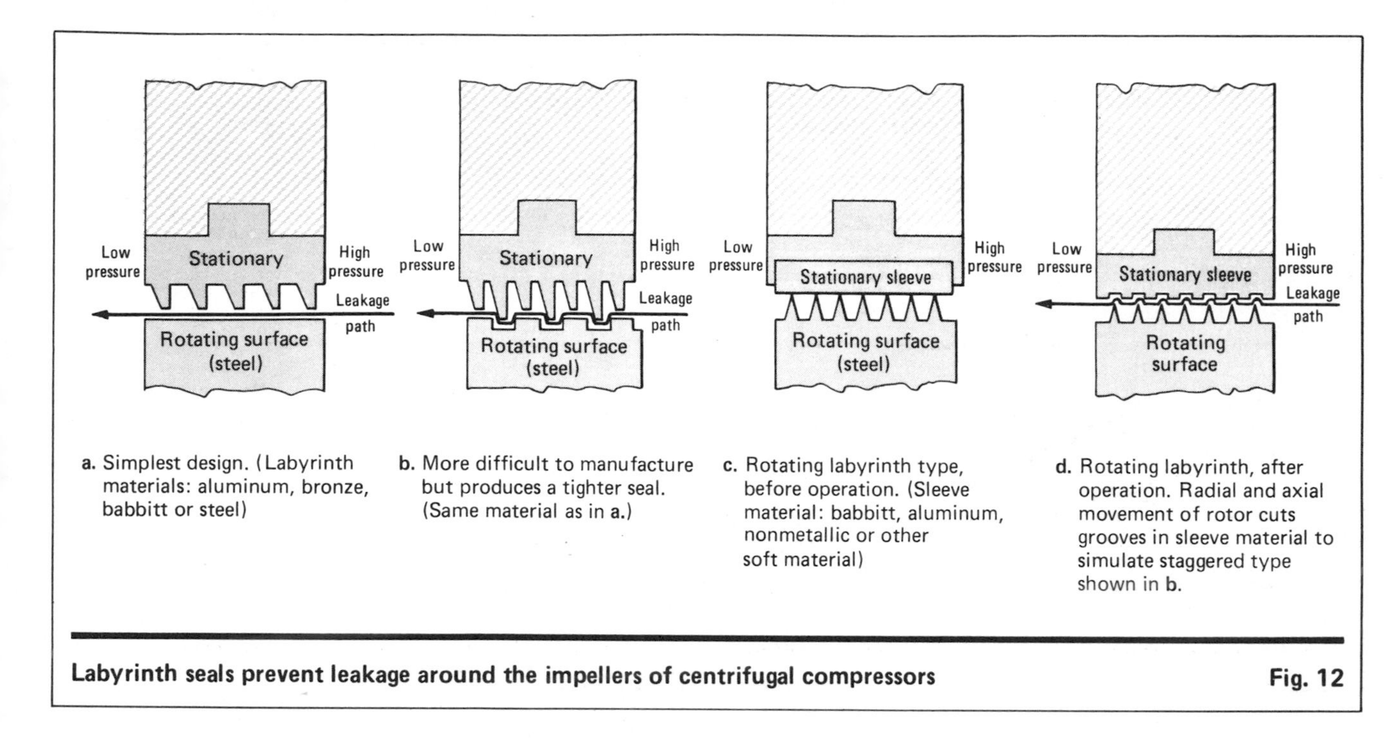


**a.** Simplest design. (Labyrinth materials: aluminum, bronze, babbitt or steel)

**b.** More difficult to manufacture but produces a tighter seal. (Same material as in **a.**)

**c.** Rotating labyrinth type, before operation. (Sleeve material: babbitt, aluminum, nonmetallic or other soft material)

**d.** Rotating labyrinth, after operation. Radial and axial movement of rotor cuts grooves in sleeve material to simulate staggered type shown in **b.**

**Labyrinth seals prevent leakage around the impellers of centrifugal compressors** **Fig. 12**

(Fig. 14). These are the oil-to-process gas seal, or carbon ring, and the oil-to-uncontaminated-seal-oil drain seal, or breakdown bushing. This seal can maintain a lower inner-leakage ratio with higher oil-to-gas differential pressure.

In operation, the seal-oil pressure is maintained at about 25 to 50 psia over the process-gas pressure against which the seal is sealing. High-pressure oil enters the seal cavity, completely filling it. Some of the oil (ranging from 2 to 8 gph) is forced across the carbon-seal face, and flows out the contaminated-oil drain. The mechanical seal's great advantage over the oil-film seal is that it has a minimum effect on rotor dynamics. On the other hand, when the oil-film bushings lose their free-floating feature, they can upset the stability of the rotor when operating at high speeds.

## Balancing

Balancing of high-speed rotating machinery is very crucial in the smooth running of the machine. Machines designed to operate higher than their first-critical speed need even more care. Presently in most vendors' shops, each wheel is balanced separately and then shrink-fitted to the shaft. During this process, shaft runout must be closely watched, since large changes could indicate that the impeller is not installed square to the shaft. When all rotors are fitted on the shaft and the final balancing is done, correction is usually made to the last impeller installed if an unbalance exists.

The best technique for high-speed rotors is to balance them at their rated speed, and not in low-speed machines. This is not always possible in the shop, and so is often done in the field. To achieve best results, balancing should be performed at all rotating planes. This is done with the influence-coefficient technique. Here, the effect of unbalance on each plane, and its influence on the other planes, is noted. Once the influence coefficients of each plane are known, balancing can be performed at each plane and a smooth rotor obtained. Fig. 15 shows rotor-response curves before and after balancing.

## Couplings

Flexible gear couplings have been the most widely used. In many new machines, they are being replaced by disk-type couplings. Gear-type couplings consist of a hub gear and a sleeve gear. In most cases, male teeth are integral with the hub, but some couplings have male teeth integral with the sleeve. Most of these couplings are mounted on a tapered-shaft end. In many instances, a key and keyway are provided to transmit torque; in others, the coupling is shrink-fitted on the tapered shaft.

Lubrication of the gear coupling is accomplished by a continuous or batch-packed technique. With continuous lubrication, nozzles spray oil into the coupling's teeth. The oil is recirculated through the machine's lubricant system, which dissipates the heat generated in the coupling, and thus maintains a relatively constant temperature. With batch-packed or seal lubrication, a recommended grease or oil is sealed in the coupling, which is changed as necessary or during scheduled shutdown. The major advantages of the batch-packed technique are that the lubricant is the best available for the application and that the lubricant does not get contaminated. For high-temperature applications, continuous lubrication is recommended. Such lubrication requires a good filtration system; otherwise, a centrifuging of contaminants in the gear teeth may occur.

The disk-type or metal-flexible coupling, shown in Fig. 16, consists of two hubs rigidly mounted by interference fit or by flange bolting to the driven and driving shafts. Flexible elements, one attached to each hub, compensate for misalignment. They are connected by a spacer (usually tubular) to span the gap between shafts. The major problem of this coupling is its potential to vibrate when excited at its resonance frequency. The

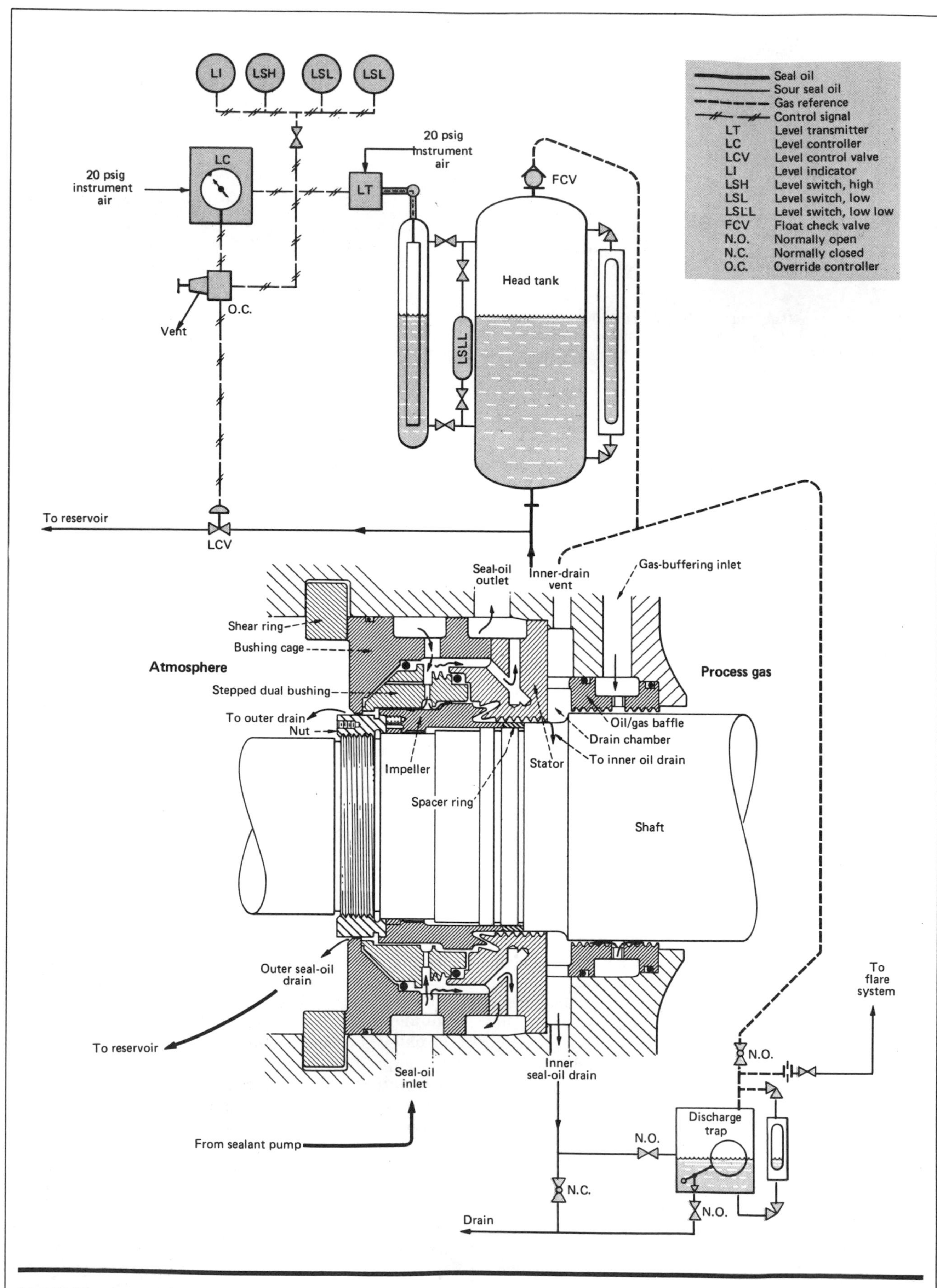


Liquid-film seal for compressor shaft provides continuous flow of seal oil **Fig. 13**

major advantage is that it needs no lubrication system and can tolerate a higher degree of misalignment.

## Lube-oil systems

API Standard 614 covers in detail the minimum requirements for lubrication systems, oil-type shaft-sealing systems, and control-oil supply systems for special-purpose applications.

The base system consists of a reservoir that should be separately placed from the base plate. Working capacity of the oil tank should be at least 5 min, based on normal flow. Reservoir retention time should be 10 min, based on normal flow and total volume below the minimum operating level.

Arrangements for heating the oil should also be provided. If an immersion heater is used, maximum watt density should be 15 W/in.$^2$ If steam heating is used, the heating element should be external to the reservoir.

The oil system should be equipped with a main oil pump, a standby pump and, for critical machinery, an emergency pump. Power sources for the main and standby pumps should be different. For example, if steam is used to power the main pump, the standby pump should be electrical. The emergency pump is also usually electrical, but will be driven by either a d.c. supply or a completely separate a.c. supply.

Twin oil coolers should be provided and piped in parallel, using a single multiport transfer valve to direct the flow to the coolers. The coolers should carry water on the tubeside and oil on the shellside. Oilside pressure should be greater than the waterside. Twin full-flow filters should be located downstream of the oil coolers. Do not pipe the filters with separate inlet and outlet block-valves. This could cause loss of oil flow from the possible human error of blocking the flow during filter switching.

Oil for turbomachinery should be of correct viscosity, and formulated with the required chemical additives to prevent rusting, resist oxidation and sludging, be non-corrosive to machine parts, resist foaming, and separate rapidly from water. The oil should be checked periodically to detect changes in viscosity, pH and neutralization number, and precipitation.

## High-speed gears

In many compressor applications, high-speed gears are used between the drive (especially, electric motors) and the compressor. The application of gear drives to large trains involving high speeds and power levels has never been an easy task. Currently, high-hardness gearing is used, with tooth loads in the range of 1,500 to 2,000 lb/in. of face, at pitchline velocities of 20,000 to 30,000 ft/min. Normally, the gear drives used in such an application are of the single-helical or double-helical type, with rotors carried in sleeve-type bearings. The double-helical gear is the first choice for high-reliability applications.

Proper alignment of gears to the shafts is very important, and hot checks on alignment are usually a must for high-speed gearing. When possible, gears should be run-in on initial startups. Speed and load should be increased percentagewise. Lube-oil pressure and temperature should be closely observed, and adjustments

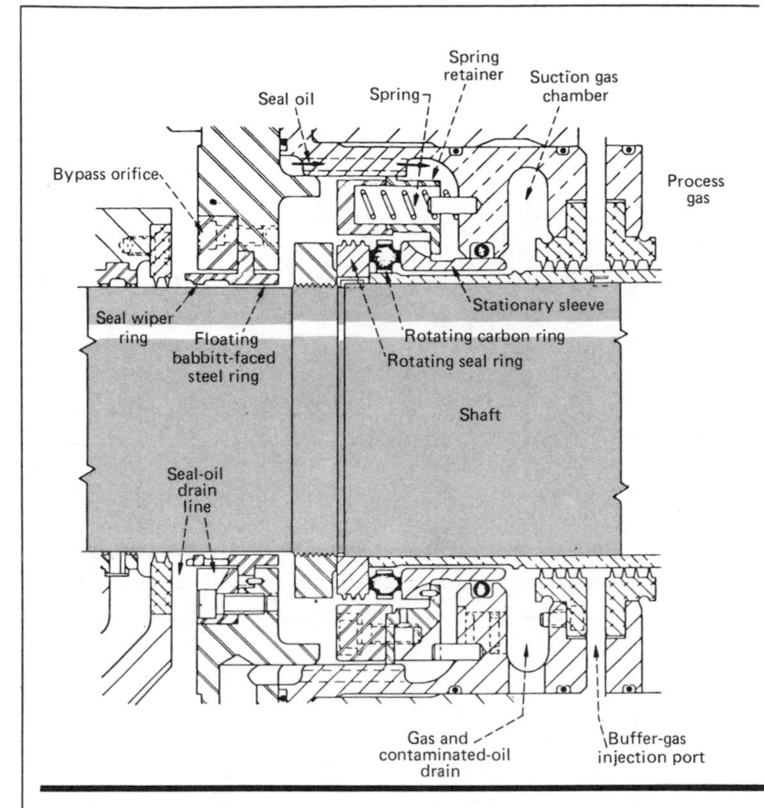


**Mechanical-contact seal for compressor shaft Fig. 14**

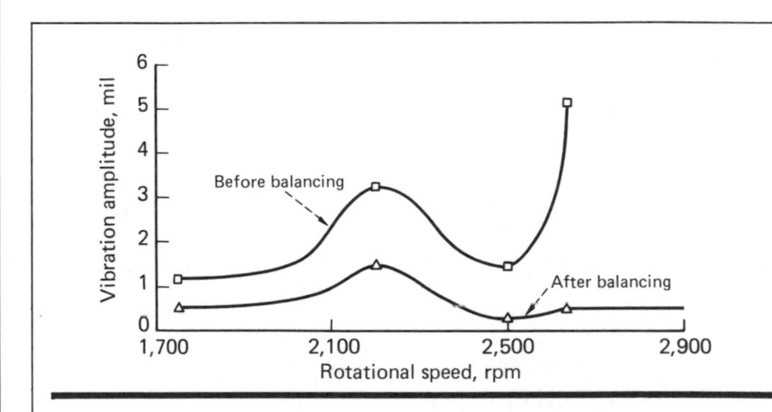


**Rotor-response curves show effects of balancing Fig. 15**

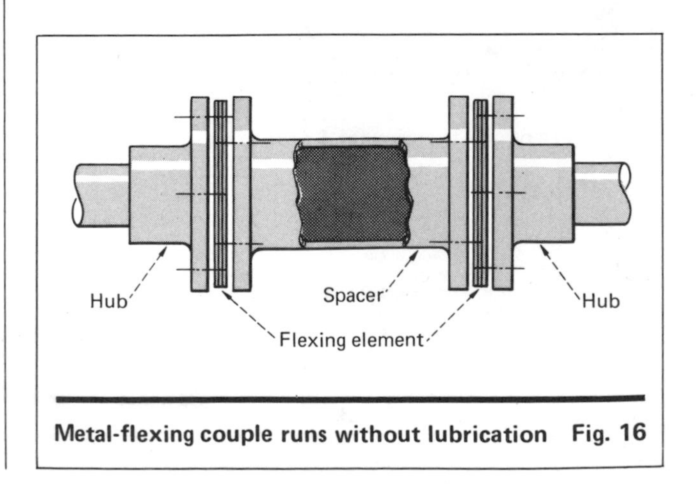


**Metal-flexing couple runs without lubrication Fig. 16**

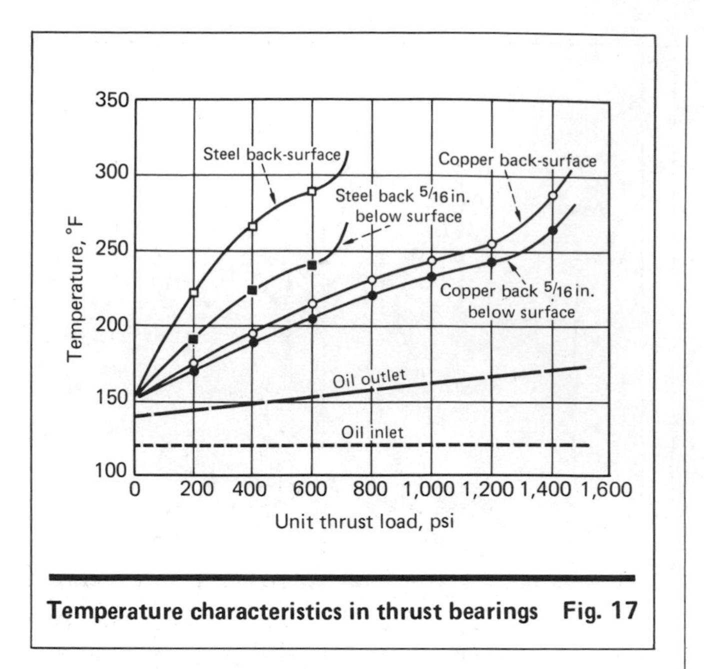


**Temperature characteristics in thrust bearings Fig. 17**

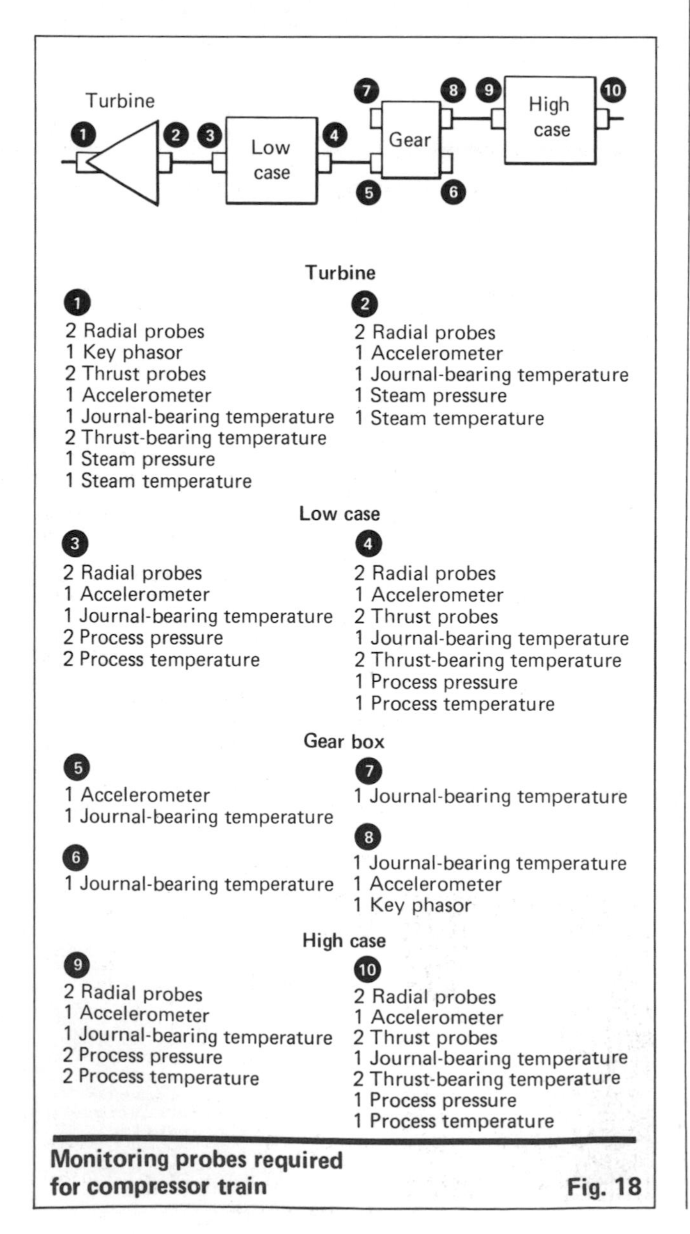


**Monitoring probes required for compressor train Fig. 18**

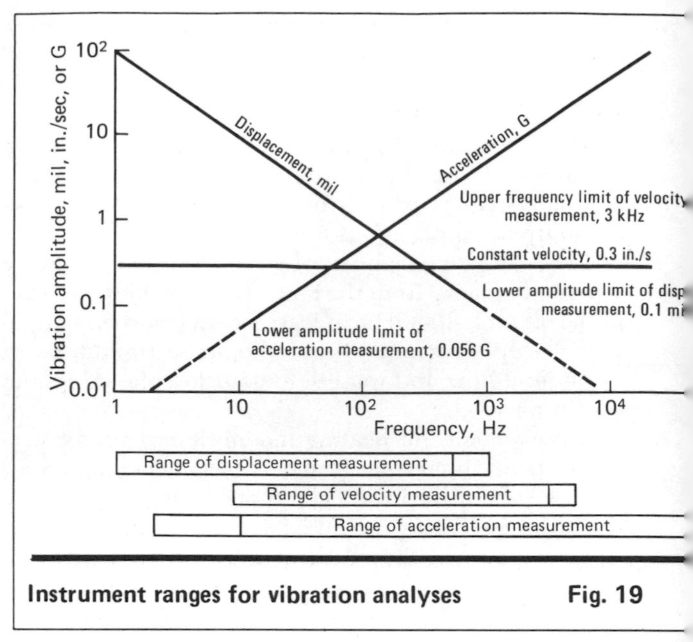


**Instrument ranges for vibration analyses Fig. 19**

made to the lube system, as required. Oil pressure must be maintained at all costs. High-speed gearing must be treated with care.

## Control systems

The controls for most compressor trains consist of two major systems: one for the lubrication system, and one for the compressor. For the lubrication system, minimum alarms are: low oil pressure, low oil-pressure trip (at some point lower than the alarm point), low oil level in the reservoir, high oil-filter differential pressure, high thrust-bearing metal temperature, and high oil temperature. Each pressure and temperature sensing switch should be in a separate housing. The switch type should be single-pole, double-throw, and furnished as open (deenergized) to alarm and close (energize) to trip. Pressure switches for alarms should be installed, with a "T" connection for a pressure gage and bleeder valve to test the alarms. Temperatures should be monitored in the oil piping to and from the coolers, and at the outlet of each radial and thrust bearing. Bearing-metal temperature should also be measured, since problems will show up much faster in the metal temperature than in the oil temperature. Fig. 17 shows that the change in metal temperature is much higher than the change in oil temperature as load is increased.

Pressure gages should be provided at the discharge of the pumps, bearing header, control-oil line, and seal-oil line. Each atmospheric oil-drain line should be equipped with steel nondestructive bull's-eye flow indicators, positioned for viewing through the sides.

For the compressor, a complete instrumentation package should be provided, as shown in Fig. 18. In this manner, a full analysis can be made for both vibration and performance characteristics. For vibration analysis, both proximity probes and accelerometers have been suggested. Fig. 19 shows the ranges for which the various probes are applicable. The reason for two different transducers is that proximity probes show only the movement of the shaft at the location of the probes. The

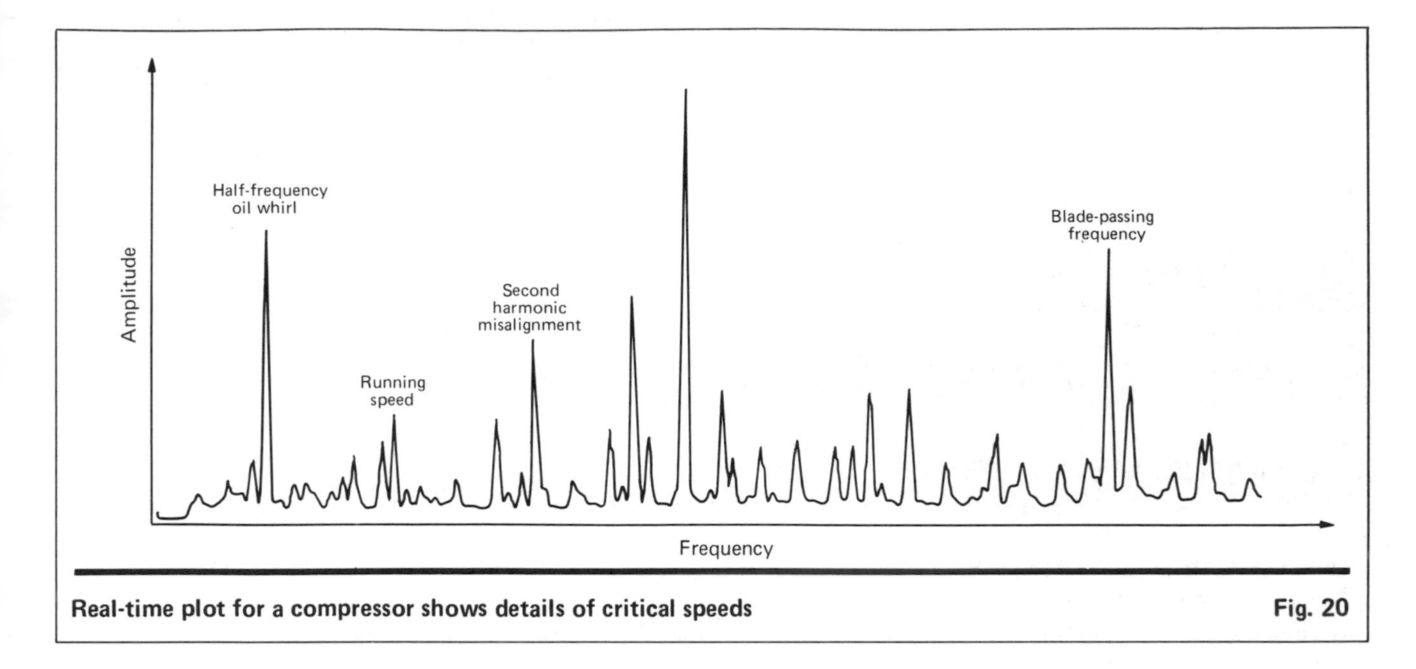


**Real-time plot for a compressor shows details of critical speeds** **Fig. 20**

probes do not pick up high-frequency problems such as blade-passing or gear-mesh frequencies.

Fig. 20 is a real-time plot that shows the blade-passing frequencies, dominating as the compressor approaches surge. High subharmonic frequencies indicate oil whirl, and second harmonics indicate some misalignment. In some cases, catastrophic failure has occurred even though no indication was given by the proximity probes. When properly used, proximity probes are excellent in locating subharmonic and misalignment problems. Thus, by using a combination of proximity probes and accelerometers, a complete picture of rotor dynamics is obtained.

Surge-control instruments on the market are basically static devices. These measure the pressure rise across the compressor, and sense the flowrate from pressure drop across an orifice. The pressure rise and flowrate are then compared with values programmed into the device; and when they exceed these values, the signal is given for a bypass valve to open.

The biggest drawback of this system is that it does not provide for any changes in the molecular weight of the gas or the degradation of the compressor itself. Presently, programs are underway that measure the dynamic change in the pressure head in the boundary layer. When flow-reversal is sensed, the bypass valve is opened. This program is still being evaluated.

Until now, monitoring systems have consisted of no more than vibration monitors. With the advent of real-time analyzers and minicomputers, a whole new era is opening in monitoring and diagnostic systems. These new systems indicate problem areas, give expected life of various components, and schedule maintenance. They correlate both process and mechanical parameters to fully diagnose the compressor train and run it at its most efficient point. These systems use a performance matrix in combination with trending data and real-time analysis to properly project and diagnose problems. A good monitoring and diagnostic system can save thousands of dollars in downtime and energy.

## References


1. Housman, J. G., Turbomachinery Specifications, *Proc. First Turbomachinery Symp.*, pp. 77–78, Texas A&M University, College Station, Tex., 1972.
2. Davis, H. M., Centrifugal Compressor Operation and Maintenance, *Proc. First Turbomachinery Symp.*, pp. 10–25, Texas A&M University, College Station, Tex., 1972.
3. Wilcock, D. E. S. and Booser, E. R., "Bearing Design and Application," McGraw-Hill, New York, 1961.
4. Gunter, E. J., "Dynamic Stability of Rotor Bearing Systems," NASA SP-113, 1966.
5. Herbage, B. S., High Speed Journal and Thrust Bearing Design, *Proc. First Turbomachinery Symp.*, pp. 55–61, Texas A&M University, College Station, Tex., 1972.
6. Boyce, M. P. and Hanawa, D. A., Development of Techniques for Monitoring Turbomachinery, *Proc. Gas Turbine Operations and Maintenance Symp.*, Edmonton, Canada, 1974.
7. Boyce, M. P. and Hanawa, D. A., Parametric Study of a Gas Turbine, *J. Eng. Power,* Dec. 1975.
8. Clapp, A. M., Fundamentals of Lubrication Relating to Operation and Maintenance of Turbomachinery, *Proc. First Turbomachinery Symp.*, pp. 67–74, Texas A&M University, College Station, Tex., 1972.
9. Cameron, J. A. and Danowski, F. M., Some Metallurgical Considerations in Centrifugal Compressors, *Proc. Second Turbomachinery Symp.*, pp. 116–128, Texas A&M University, College Station, Tex., 1973.
10. Jackson, C. J., Cold and Hot Alignment Techniques of Turbomachinery, *Proc. Second Turbomachinery Symp.*, pp. 1–7, Texas A&M University, College Station, Tex., 1973.
11. Lesiecki, G., Evaluation of Liquid Film Seals: Associated Systems and Process Considerations, *Proc. Sixth Turbomachinery Symp.*, pp. 145–148, Texas A&M University, College Station, Tex., 1977.
12. Lewis, R. A., Mechanical Contact Shaft Seal, *Proc. Sixth Turbomachinery Symp.*, pp. 149–151, Texas A&M University, College Station, Tex., 1977.
13. Boyce, M. P., "Turbomachinery Notes," Texas A&M University, College Station, Tex., 1977.
14. Boyce, M. P., Morgan, E. and White, G., Simulation of Rotor Dynamics of High Speed Rotating Machinery, *Proc. First Int. Conf. Centrifugal Compressor Technol.*, pp. E6–32, Madras, India, 1978.


## The author

**Meherwan P. Boyce** is president of Boyce Engineering International, Inc., Suite 520, 2990 Richmond Ave., Houston, TX 77098, which does engineering consulting in turbomachinery. He is also director of the Gas Turbine Laboratories, and a full professor in the Mechanical Engineering Dept., Texas A&M University, College Station, TX 77840. He has a B.S. in mechanical engineering from S.D. School of Mines and Technology, an M.S. in mechanical engineering from the State University of N.Y., and a Ph.D. in mechanical engineering from the University of Okla. He is a member of several professional and honor societies and a registered engineer in Texas.

# Basics of Surge Control for Centrifugal Compressors

Surging is an unstable operating condition that causes erratic compressor performance. Here is a basic antisurge control scheme for centrifugal compressors, along with suggestions on how to cope with variations in operating conditions and how to handle multi-compressor arrangements.

M. H. WHITE, The Foxboro Co.

Compressors, like other major processing equipment, should be controlled as effectively as possible to ensure efficient plant operation. In devising a control scheme, the designer must consider how to avoid an unstable operating condition known as surging.

Simply stated, surging occurs when the compressor throughput is reduced to a point sufficiently below design conditions that erratic performance results. The job of the antisurge system then, is to detect the potential upset and automatically compensate for it by maintaining a flow through the compressor in excess of the surge condition.

Before getting into the details of how this system works, we must first understand the phenomenon of surging from the standpoint of compressor-system variables.

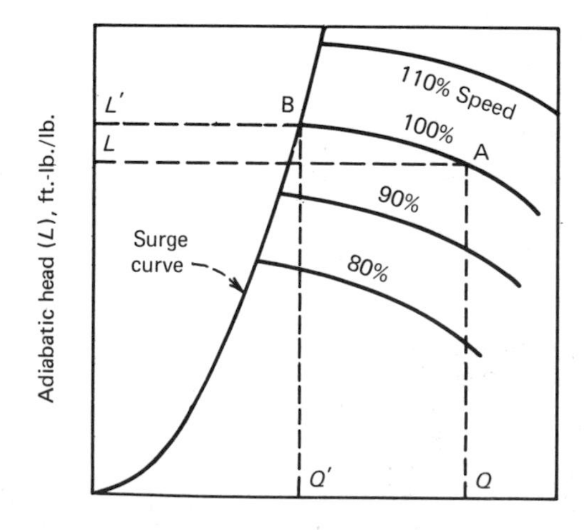


**CHARACTERISTIC CURVES** for typical centrifugal compressor are based on Eq. (3)—Fig. 1

Originally published December 25, 1972

## Surging Characteristics

A typical family of characteristic curves for a variable-speed centrifugal compressor is shown in Fig. 1. These curves depict adiabatic head as a function of actual inlet-volume flow, with a characteristic curve for each rotational speed.

If we assume that the compressor is operating at Point A on the 100%-speed curve, the inlet flow is $Q$ and the head is $L$. Now if the external load resistance gradually increases while the speed remains constant, the flow decreases and the operating point will move to the left along the 100%-speed characteristic curve. When it reaches Point B, the flow decreases to $Q'$, and the head increases to $L'$, the maximum head the compressor can produce at this speed.

At this point, the characteristic curve is practically flat and the operation of the compressor becomes unstable. This condition is called surging and it appears as rapid pulsation in the flow and discharge pressure, producing high-frequency reversals in the axial thrust on the compressor shaft. In some machines, surging can become severe enough to cause mechanical damage. To avoid this danger, a control system must be devised to prevent the compressor from operating in this unstable area.

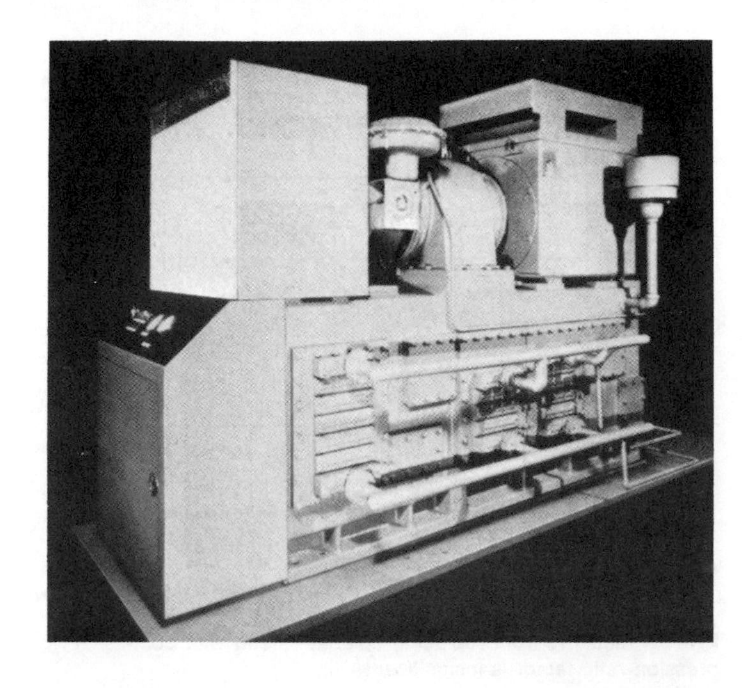

## Basic Control Theory

In all centrifugal machines (both pumps and compressors), the head produced is proportional to the square of the speed, while the flowrate is linearly proportional to the speed. These relationships are described equations:

$$Q = CN \qquad (1)$$

$$L = CN^2 \qquad (2)$$

Substituting for $N$ yields:

$$L = CQ^2 \qquad (3)$$

This equation is used to calculate the parabolic surge curve usually provided by the compressor manufacturer.

Since there is no way to measure adiabatic head directly, we must find another quantity to use in its place. The one most easily related to head is compression ratio, $R_c$.

$$R_c = P_2/P_1 \qquad (4)$$

Eq. (5) is used to relate compression ratio to adiabatic head:

$$R_c = (1 + Lm\phi/1{,}545T_1Z_1)^{1/\phi} \qquad (5)$$

where

$$\phi = (K - 1)/K$$

When the compressor handles gas of only one composition, and when the suction temperature is nearly constant, the quantities $m$, $\phi$, $T_1$ and $Z_1$ are constant, and Eq. (5) reduces to:

$$R_c = (1 + CL)^{1/\phi} \qquad (6)$$

Although Eq. (6) shows that the relation between $R_c$ and $L$ is not linear, when this equation is plotted for most of the gases commonly encountered (e.g., air, natural gas, etc.), the curves approach straight lines over the normal operating ranges. The quantity $(R_c - 1)$ can then be substituted for $L$ without producing significant distortion of the surge curves.

Eq. (3) shows that the relation between adiabatic head and $Q^2$ is linear. Fig. 2 is a plot of $L$ and the corresponding values of $(R_c - 1)$ vs. $Q^2$ for the same range of heads for natural gas to show the magnitude of the error introduced by substituting $(R_c - 1)$ for $L$ in Eq. (3). At greater heads, this error increases and may require an adjustment in the system to correct it, but in most cases, the error is small enough to be neglected.

This permits us to write the equation:

$$R_c - 1 = CQ^2 \qquad (7)$$

The differential pressure across the compressor is:

$$\Delta P = P_2 - P_1 \qquad (8)$$

From Eq. (4), $\quad P_2 = P_1R_c$

Then: $\quad \Delta P = P_1R_c - P_1$

$$\Delta P = P_1(R_c - 1) \qquad (9)$$

$$R_c - 1 = \Delta P/P_1 \qquad (10)$$

Substituting Eq. (10) into Eq. (7) provides:

$$\Delta P/P_1 = CQ^2 \qquad (11)$$

### Nomenclature

| | |
|---|---|
| $C$ | Constant (not necessarily the same in all equations) |
| $h$ | Differential pressure across primary flow device, in. of water |
| $K$ | Specific heat ratio |
| $L$ | Adiabatic head, ft.-lb./lb. |
| $m$ | Molecular weight |
| $N$ | Compressor speed, rpm. |
| $P$ | Pressure, psia. |
| $Q$ | Actual inlet-volume flow, cfm. |
| $R_c$ | Compression ratio |
| $T$ | Absolute temperature, °R. |
| $V$ | Specific volume, cu. ft./lb. |
| $W$ | Mass flowrate, lb./min. |
| $Z$ | Supercompressibility factor |

**Greek letters**

| | |
|---|---|
| $\Delta$ | Differential |
| $\phi$ | Ratio of $(K - 1)$ to $K$ |

**Subscripts**

| | |
|---|---|
| $D$ | Design |
| 1 | Suction |
| 2 | Discharge |

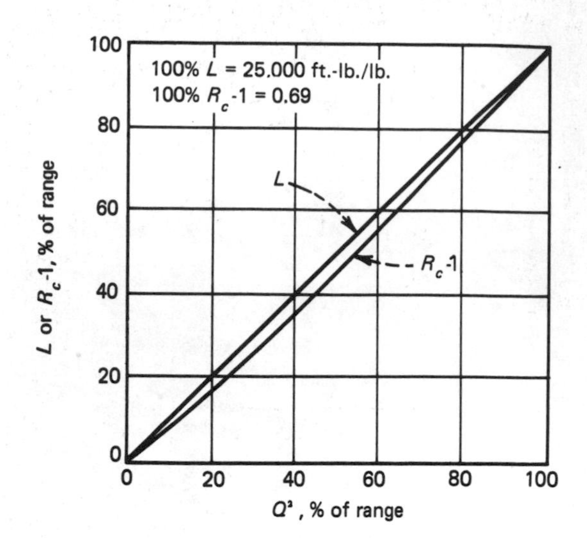


**ERROR** resulting from replacing adiabatic head with compression-ratio factor is insignificant—Fig. 2

To measure $Q$, we must have a primary device, such as an orifice or venturi tube, in the compressor suction line. If we let $h$ represent the differential pressure across this primary device in inches of water, we can calculate the mass flow, $W$, to the compressor from the equation:

$$W = C\sqrt{hP_1/T_1} \tag{12}$$

$$V = C(T_1/P_1) \tag{13}$$

$$Q = WV \tag{14}$$

Substituting Eq. (12) and (13) in Eq. (14) produces:

$$Q = C\sqrt{hP_1/T_1} \times T_1/P_1$$

$$Q = C\sqrt{hT_1/P_1} \tag{15}$$

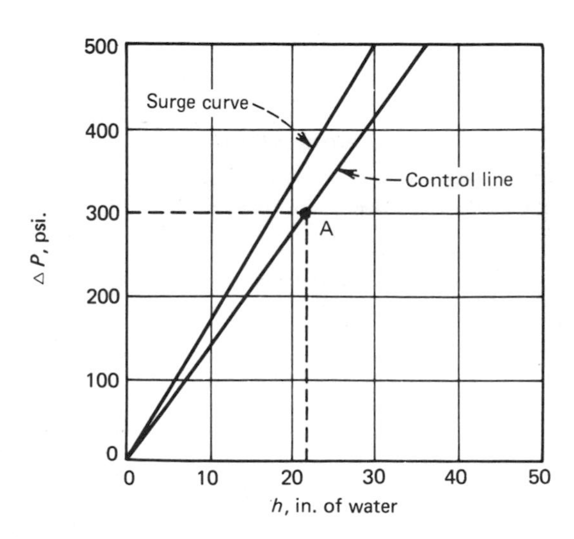


**CONTROL LINE** is generally displaced about 10% of $Q$, or 20% of $\Delta P$, from the surge curve—Fig. 3

If, as before, we assume the temperature to be constant, then:

$$Q = C\sqrt{h/P_1}$$

$$Q^2 = C\,h/P_1 \tag{16}$$

Substituting Eq. (16) in Eq. (11) yields:

$$\Delta P/P_1 = C\,(h/P_1)$$

$$\Delta P = Ch \tag{17}$$

Eq. (17) is used to calculate the surge curve, which will be our basic design equation. It verifies two important points:

1. There is a linear relationship between $\Delta P$ and $h$, as seen in Fig. 3; and
2. The surge curve is not affected by variations in suction pressure, $P_1$.

Fig. 3 also shows the control line, displaced to the right of the surge curve. Moving the control line to the right reduces the possibility of reaching surge conditions in case a rapid decrease in flow causes an overshoot to the left of the control line. However, if the control line is placed too far to the right, gas will be bypassed unnecessarily and power wasted.

Consequently, location of the control line represents a compromise based on the actual operating conditions of each individual system. As a general rule, the control line is displaced about 10% of flow, or 20% of differential pressure. However, there are cases where the compressor must operate close to the surge curve and it is necessary to reduce this margin.

## Effect of Temperature Changes

In deriving Eq. (17), we assumed that the compressor inlet temperature was constant. Since this is not always true in actual practice, it is necessary to investigate the effect of changes in this variable on the performance of the control system.

In order to determine the change in $\Delta P$ resulting from a change in $T_1$, we start with Eq. (5), which expresses the relationship between compression ratio ($R_c$) and temperature. If it is assumed that the head and gas composition are constant, the factors $L$, $m$ and $\phi$ in the equation will be constant. The factor $Z_1$ may be somewhat affected by the temperature; but for this part of the analysis, it will be considered constant.

Eq. (5) reduces to:

$$R_c = (1 + C/T_1)^{1/\phi} \tag{18}$$

Fig. 4 shows a plot of this equation for two gases, air and natural gas, that have widely different molecular weights and specific heat ratios. The curves show the magnitude of the change in compression ratio for a 120 deg. F. change in inlet temperature in a compressor producing a head of 40,000 ft.-lb./lb.

From Eq. (9) above, $\Delta P = P_1\,(R_c - 1)$, it is evident that, for any particular value of suction pressure ($P_1$), $\Delta P$ will vary directly with $R_c - 1$; and the shape of the curves for $\Delta P$ vs. $T_1$ is the same as those for $R_c$ vs. $T_1$.

Now, we will consider the effect of changes in the differential pressure across the orifice, $h$, resulting from variations in suction temperature. Starting with Eq. (15),

$$Q = C\sqrt{hT_1/P_1}$$

$$Q^2 = C(hT_1/P_1)$$

$$h = C(Q^2P_1/T_1) \qquad (19)$$

Then, for specific values of $Q$ and $P_1$,

$$h = C/T_1 \qquad (20)$$

The curves in Fig. 4 also show that both $\Delta P$ and $h$ vary inversely with the absolute inlet temperature, $T_1$. If these temperature effects were identical, both sides of Eq. (17) would be changed equally and the system would be exactly self-compensating. However, this is not the case—although both variables change in the same direction, the magnitudes are different, with the result that there is a shift in the slope of the surge curve.

In most cases, this problem can be solved simply by placing the control line to the right of the surge curve for the highest temperature, and accepting the fact that, when minimum temperature conditions exist, some gas may be bypassed unnecessarily.

If temperature variations are too great, or if the compressor must operate close to the surge curve, it will be necessary to provide temperature compensation in the control system. The instrumentation required to perform this function will be described later.

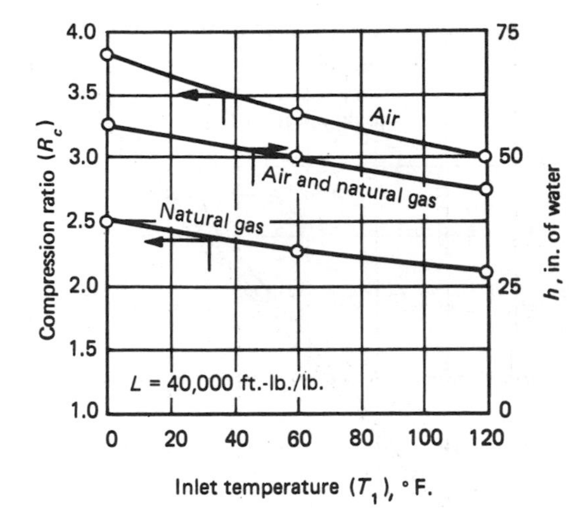


**CURVES** for both compression ratio, $R_c$, and $h$ vary inversely with gas inlet temperature—Fig. 4

## Molecular-Weight Changes

In most processes, each compressor handles the same gas all the time. Typical examples are air, refrigerant and natural-gas pipeline compressors. In some applications, however, the compressor may be called on to handle gases of varying composition.

Since a variation in composition almost always results in a change in molecular weight, it is important to determine the effect of this change on the surge curve. Since the surge curve is plotted with $\Delta P$ vs. $h$, we must determine how each of these variables is affected.

Referring back to Eq. (5):

$$R_c = (1 + Lm\phi/1{,}545\ T_1Z_1)^{1/\phi}$$

for fixed values of $L$, $T_1$ and $Z_1$, this equation reduces to:

$$R_c = (1 + Cm\phi)^{1/\phi} \qquad (21)$$

This equation indicates that, as with temperature, there is a nonlinear relation between molecular weight, $m$, and differential pressure across the compressor, $\Delta P$. This is further complicated by the fact that a change in molecular weight is accompanied by a change in the specific heat ratio, $K$, and consequently the factor $\phi$.

In some cases, such as the lighter saturated hydrocarbons (i.e., methane, ethane and propane, and mixtures of these gases), there is a definite relationship between $m$ and $K$, as shown in Fig. 5. This permits accurate calculation of the change in $R_c$ and $\Delta P$ for a given change in $m$.

However, in some processes, the compressor may be handling two or more gases where there is no correlation between $m$ and $K$. An example of this would be a methane compressor that is also periodically used to purge the system with nitrogen. Under such a condition, each value of $R_c$ and $\Delta P$ must be calculated individually to determine the magnitude and direction of the change.

To compute the effect of changes in $m$ on the value of $h$, we must include $m$ in Eq. (15), which then becomes:

$$Q = C\sqrt{hT_1/P_1m} \qquad (22)$$

$$h = C\,Q^2P_1m/T_1 \qquad (23)$$

Then, for specific values of $Q$, $P_1$, and $T_1$,

$$h = Cm \qquad (24)$$

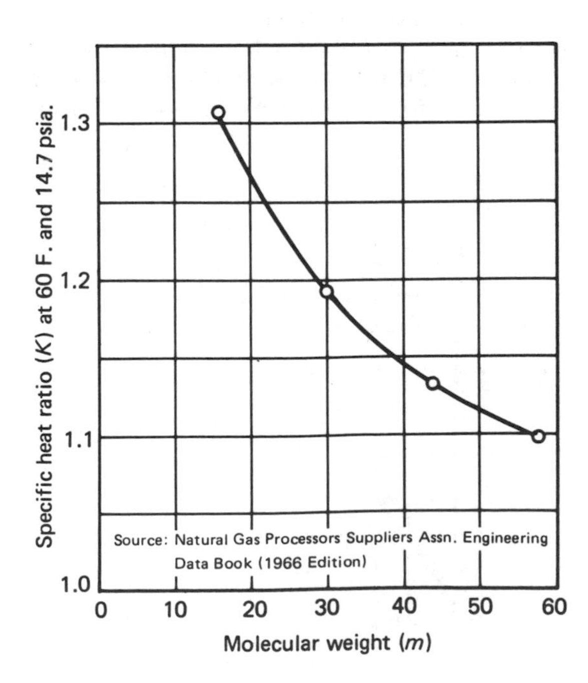


**DEFINITE RELATIONSHIP** exists between molecular weight and specific heat ratio for lighter hydrocarbons—Fig. 5

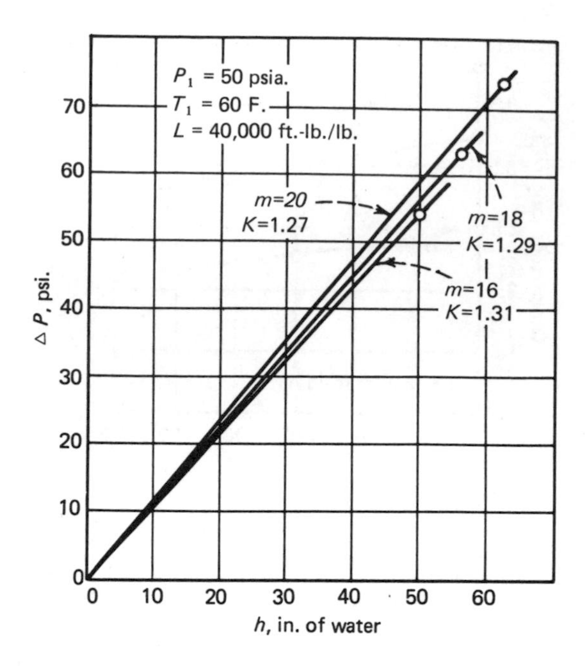


**INCREASING** $m$ shifts surge curve to left—Fig. 6

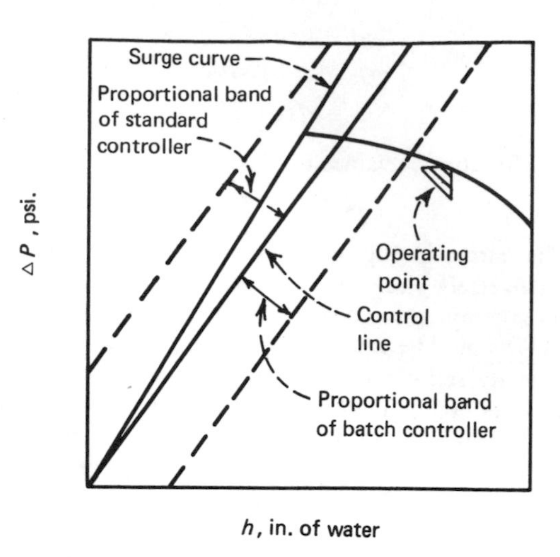


**"BATCH"** controller provides anti-windup protection—Fig. 8

By comparing Eq. (21) and (24), it is apparent that the system will not be exactly self-compensating for changes in molecular weight. For lighter hydrocarbons, changes in $m$ produce changes in the same direction in both $\Delta P$ and $h$, but in different magnitudes. Fig. 6 shows how the surge line shifts in slope for changes in $m$ from 16 to 20.

It is not possible to generalize, however, concerning the overall effect of molecular-weight changes because of the influence of the specific-heat-ratio factor. As mentioned earlier, when two or more unrelated gases are handled by the same compressor, the slope of the surge line must be calculated for each condition. And if there is too great a divergence to be covered by a single control line, provision must be made to adjust the slope as required. This is usually done manually.

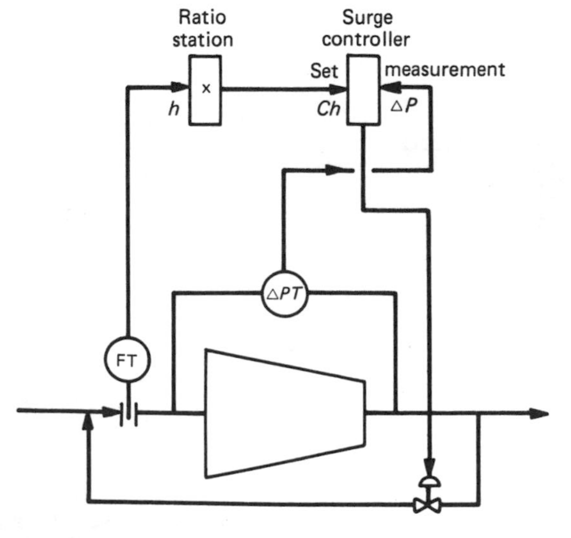


**BASIC SCHEME** to provide antisurge control—Fig. 7

## Instrumentation

Instrumentation required to use Eq. (17) is shown in Fig. 7. This equipment, of course, may be either pneumatic or electronic. A high-range differential-pressure transmitter measures $\Delta P$ and sends an output signal that becomes the measurement signal to the surge controller. The flow transmitter is connected across a primary device such as an orifice or venturi in the compressor suction line, and its output signal, $h$, is fed to a ratio station. There it is multiplied by the constant, $C$, and becomes the setpoint of the surge controller.

The surge controller should have, in addition to proportional and reset functions, an anti-windup function, sometimes referred to as the "batch" feature.

To understand the need for a "batch" feature, consider Fig. 8. Under normal conditions, the compressor operates in an area some distance from the control line. This results in an offset between the measurement and the setpoint of the controller. In a standard proportional-plus-reset controller, this causes the output signal to wind up to either its high or low limit.

In this condition, the proportional band and the operating point will be on opposite sides of the setpoint or control line, and no control action will be obtained until the measured operating point reaches the control line. If the measurement approaches the control line rapidly, it will overshoot before the controller can unwind, and the compressor may surge.

The anti-windup or "batch" function is arranged so that when the controller output reaches its limit, the reset loading is adjusted to shift the proportional band to the same side of the control line as the measurement, as indicated in Fig. 8. Then, if it approaches the control line rapidly, the measurement enters the proportional band, and control action starts before it reaches the control line; overshoot is thus reduced or eliminated. If the measurement change is slow—that is, no faster than the reset rate—the controller will perform as a standard instrument.

### Instrument Ranges

The basic antisurge-control system shown in Fig. 7 includes two differential-pressure transmitters—one measuring the differential pressure across the primary flow device, $h$, and the other the differential pressure across the compressor, $\Delta P$.

If flow measurement is used only for antisurge control, the primary device can be designed for optimum conditions. The preferred design for maximum flow is about 25–30% above the maximum volume flow indicated by the surge curve. To keep pressure loss as low as possible, the maximum differential pressure, $h$, should be as small as practical.

If flow measurement is used in the normal operation of the compressor, the primary device will probably be designed for a considerably higher maximum flow. As a result, the value of $h$ at the surge line will be a much smaller percent of full scale and may require an undesirably high setting of the ratio station. Under these circumstances, it is better to install a second transmitter with a lower range for use only with the antisurge system.

Calculations required to establish the range of the $\Delta P$ transmitter depend on the variables used by the compressor manufacturer to plot the surge curve. When the curve is plotted with discharge pressure vs. flow for a given suction pressure, a simple subtraction produces the maximum $\Delta P$ at surge.

If the surge curve is plotted using compression ratio vs flow, the maximum value of $\Delta P$ can be obtained from Eq. (9). However, if the manufacturer has used adiabatic head vs. flow to draw the surge curve, calculations become a little more involved, and it is necessary to use Eq. (5) to calculate the compression ratio, and then compute $\Delta P$ from Eq. (9).

Having determined the maximum $\Delta P$, the transmitter range should be set at some reasonable amount above this value (e.g., 20–25%).

In order to plot the surge curve in the form shown in Fig. 3, which is expressed by Eq. (17), about four points should be selected from the manufacturer's surge curve, and the corresponding values of $\Delta P$ and $h$ computed.

The $\Delta P$ values can be calculated in the same manner as the range of the $\Delta P$ transmitter is determined. To compute the corresponding values of $h$, we use Eq. (15):

$$Q = C\sqrt{hT_1/P_1}$$

$$h = C\,Q^2P_1/T_1 \tag{25}$$

Using the subscript $D$ to indicate the design conditions for the primary flow device, the equation for full-scale flow is:

$$h_D = C\,Q_D^2P_D/T_D \tag{26}$$

and dividing Eq. (25) by Eq. (26) yields:

$$h = (h_D)\,Q^2P_1T_D/Q_D^2P_DT_1 \tag{27}$$

If the flowing pressure and temperature are the same as the design conditions, Eq. (27) reduces to:

$$h = h_D\,(Q/Q_D)^2 \tag{28}$$

Points plotted using $\Delta P$ and $h$ will usually fall very close to a straight line. The control line in Fig. 3 is located by adding 5 to 10% to the flow values and calculating the corresponding values of $h$. This provides the necessary safety margin to keep the compressor out of the surge area.

Some compressors with very high compression ratios have parabolic surge curves in the lower range that straighten out, or even bend to the right, at the upper end. These curves, when replotted as $\Delta P$ vs. $h$, obviously do not produce a straight line. However, in almost every case, a control line can be drawn that will provide adequate protection over the normal operating range of the machine. If this control line does not pass through the origin, it is necessary to add bias between the ratio station and the controller.

### Ratio Setting

With the control line established, as in Fig. 3, the slope of the line (the factor $C$ in Eq. 17) represents the setting of the ratio station. Rearranging Eq. (17) produces:

$$C = \Delta P/h \tag{29}$$

In calculating $C$, values of $\Delta P$ and $h$ must be expressed in terms of percent of full scale of their respective transmitters. This calculation can best be shown by an example. If we assume, as in Fig. 3, that the transmitters have the following ranges:

$$\Delta P = 0\text{–}500 \text{ psi.}$$

$$h = 0\text{–}50 \text{ in. of water}$$

and if we select an arbitrary point, A, on the control line where:

$$\Delta P = 300 \text{ psi.}$$

$$h = 22 \text{ in. of water}$$

then: $\Delta P = (300/500)\,100 = 60\%$

and: $h = (22/50)\,100 = 44\%$

Using these values in Eq. (29) gives:

$$C = 60/44 = 1.36$$

and the equation becomes:

$$\Delta P = 1.36h$$

### Automatic Temperature Compensation

As we mentioned earlier, the slope of the surge curve changes with variations in the compressor inlet temperature. When these changes in inlet temperature are large and occur frequently, and when the compressor must operate near the surge curve, it may be wise to provide automatic temperature compensation in the antisurge control system.

This can be done by replacing the manually adjusted ratio station in Fig. 7 with an analog dividing computer,

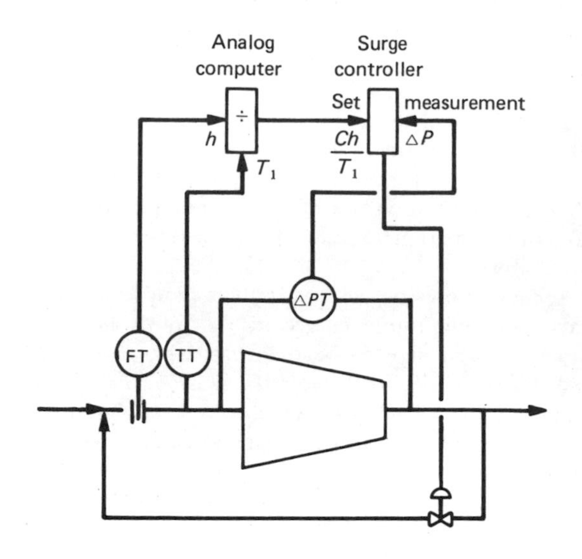


**CONTROL** arrangement to provide automatic temperature compensation uses analog dividing computer—Fig. 9

as shown in Fig. 9. By properly scaling the computer, the equation:

$$\Delta P = Ch/T_1 \tag{30}$$

can be solved. This equation indicates that the temperature correction is inversely proportional to the first power of the absolute temperature. Although, in theory, this is not precisely correct, the error introduced by this assumption is negligible.

When the largest temperature changes occur over long periods of time (e.g., seasonally), changes in the slope of the control line can be made manually to keep the system as simple as possible (as in Fig. 7).

## Inlet-Guide Vanes

Constant-speed centrifugal and axial compressors are frequently equipped with adjustable inlet-guide vanes to control output flow. In addition, the axial compressor may have adjustable stator blades. Besides influencing the compressor output, moving these vanes also changes the slope of the surge curve. In other words, there is a surge line for each vane position, just as there is for each inlet temperature.

Since the magnitude of this change in the slope of the surge line is different for each compressor design, there is no convenient way to express it mathematically. However, it is generally true that moving the vanes in the counter-rotation direction will reduce the slope of the surge line (i.e., it moves the surge line to the right). It is also generally true that the change in slope is not a linear function of vane position.

There are several solutions to this problem. If the change in slope is reasonably small, it may be possible to use the surge line for the maximum counter-rotation vane position and set the control line accordingly, as in Fig. 3. This places the control line on the safe side of the surge line for all vane positions. But it also provides an unnecessarily wide safety margin when the vanes are moved in the pre-rotation direction. This could result in bypassing gas at times when it is not required. This drawback must be weighed against the advantage of using the simple control system shown in Fig. 7.

When vanes are positioned from a manual station, and the position is not frequently changed, it may be satisfactory to use the basic control system and adjust the setting of the ratio station manually. This can be done by providing the operator with a table showing the correct ratio setting for each vane position.

When frequent changes in vane position produce large variations in the slope of the surge line (as when they are operated by a controller), it is usually desirable to make the changes in the ratio setting automatically. A system like the one for automatic temperature compensation (Fig. 9) can be used. The signal from the controller to the vane operator is a measure of vane position and, instead of the temperature measurement, can be fed into the analog computer to adjust the ratio setting. This system produces a linear relation between ratio setting and vane position that is not precisely correct. However, the errors introduced are usually small enough to be neglected.

## Discharge-Line Flow Measurement

The antisurge control system based on Eq. (17) uses a primary flow-measuring device in the compressor suction line. This system offers the very important advantage of being self-compensating for changes in suction pressure and, at least partially, for variations in suction temperature. These benefits are not realized when the flow-measurement device is located in the discharge line.

In some installations, because of the piping configuration or the size of the line, it is not feasible to put the primary flow device in the suction line. This may also be true when the compressor operates with a very low suction pressure and any additional pressure drop in the line cannot be tolerated. It may be possible to resolve the problem by using the inlet eye of the compressor as an orifice. However, it must first be established that this measurement is a usable one (i.e., noise-free, repeatable and representative). It also requires individual calibration.

If this cannot be done, it is necessary to install the primary flow device in the discharge line and modify the antisurge control system accordingly. This modification consists of adding analog computers to calculate the value of $h$ for use in Eq. (17). The instrumentation to perform this computation is shown schematically in Fig. 10.

Under steady-state conditions, the weight flow (in lb./min. or in scfm.) is the same at both suction and discharge; that is:

$$W_1 = W_2 \tag{31}$$

Using Eq. (12) to substitute for $W$ provides:

$$C_1 \sqrt{h_1 P_1/T_1} = C_2 \sqrt{h_2 P_2/T_2}$$

If the primary flow devices are designed for the same maximum values of weight flow, and if we make

$$\sqrt{h_1 P_1/T_1} = \sqrt{h_2 P_2/T_2}$$

then: $$h_1 = h_2 (P_2T_1/P_1T_2) \qquad (32)$$

With these relative maximum values of $h_1$ and $h_2$, the same relation will hold true at any point on the scale and we can use Eq. (32) to calculate $h_1$ under all conditions. The surge curve and control line can be plotted in the conventional manner, and the setting of the ratio station can be computed by the method described earlier.

Although the inlet temperature may vary considerably, in some compressor applications the ratio $T_2/T_1$ will be nearly constant, (remember these are absolute temperatures).

If the temperature ratio is assumed constant, Eq. (32) reduces to:

$$h_1 = Ch_2\, P_2/P_1 \qquad (33)$$

This greatly simplifies the control system by eliminating the temperature transmitters as well as two analog computers (i.e., the one to divide $T_2$ by $T_1$ and the one to multiply the temperature ratio by the pressure ratio).

When the compressor operates at constant speed, the instrumentation can be further simplified. Under this condition, the compression ratio is constant at the surge point, so Eq. (33) becomes:

$$h_1 = Ch_2 \qquad (34)$$

This makes it possible to eliminate two more analog computers from Fig. 10 and use the standard basic control system, by replacing $h_1$ with the corrected value of $h_2$.

## Compressors in Series

When two or more compressors are connected in series, using the control system shown in Fig. 7 on each individual machine will provide the best protection against surge. This, of course, requires a recirculation control valve for each compressor.

In some designs, two compressors in series are driven by the same prime mover, and consequently, they both always run at the same speed. The machines are sometimes considered as a single unit: the manufacturer may provide one surge curve for the combination and the installation usually includes only one recirculation control valve. For surge control, the system shown in Fig. 7 is used.

When compressors are independently driven at variable speed, but have a single recirculation control-valve, surge is controlled by the pneumatic system shown in Fig. 11. Each compressor is equipped with its own ratio station and controller, but the two are combined into an autoselector system to enable either controller to operate the common recirculation valve as required.

If, at times, either compressor operates singly, the three-way valves in the controller output lines should be installed to ensure a positive air supply to the selector relay of the controller in operation. If both compressors will always operate together, the three-way valves may be eliminated altogether.

## Compressors in Parallel

When centrifugal compressors are operated in parallel, the problem of properly dividing the load is always pres-

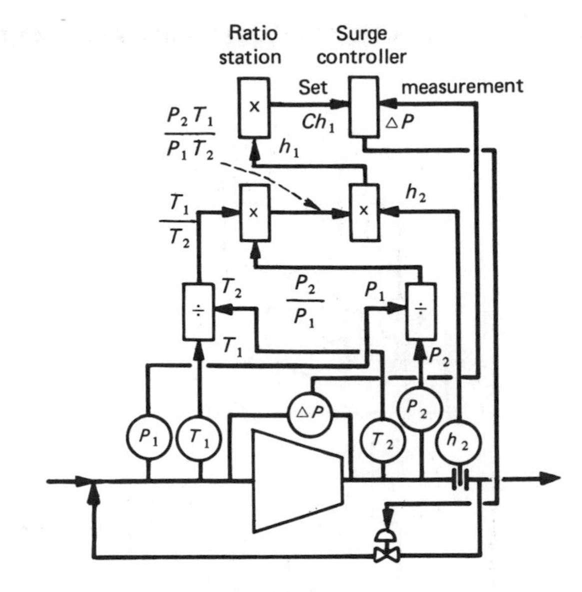


**SCHEMATIC DIAGRAM** shows instrumentation required when primary flow-measuring device is located in centrifugal-compressor discharge line—Fig. 10

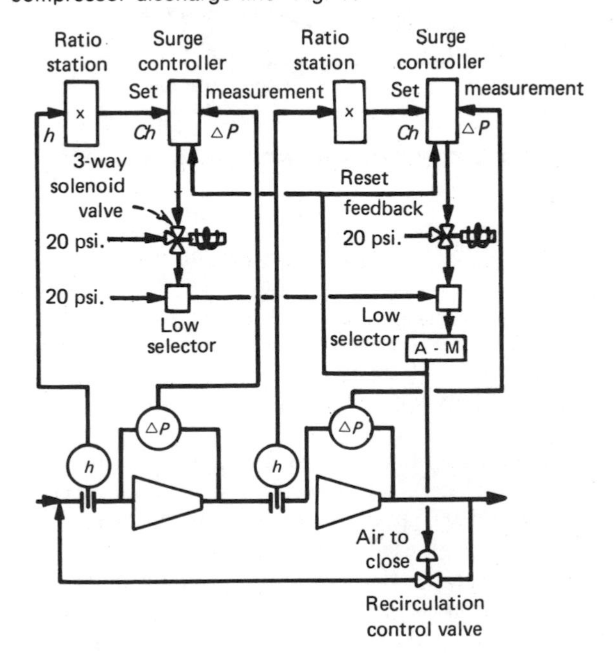


**SYSTEM** for controlling compressors in series—Fig. 11

**LOAD DISTRIBUTION** for parallel compressors can be seriously affected if they run at different speeds—Fig. 12

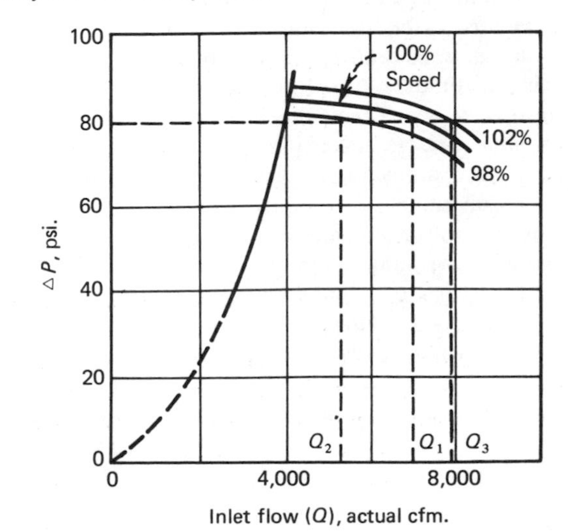

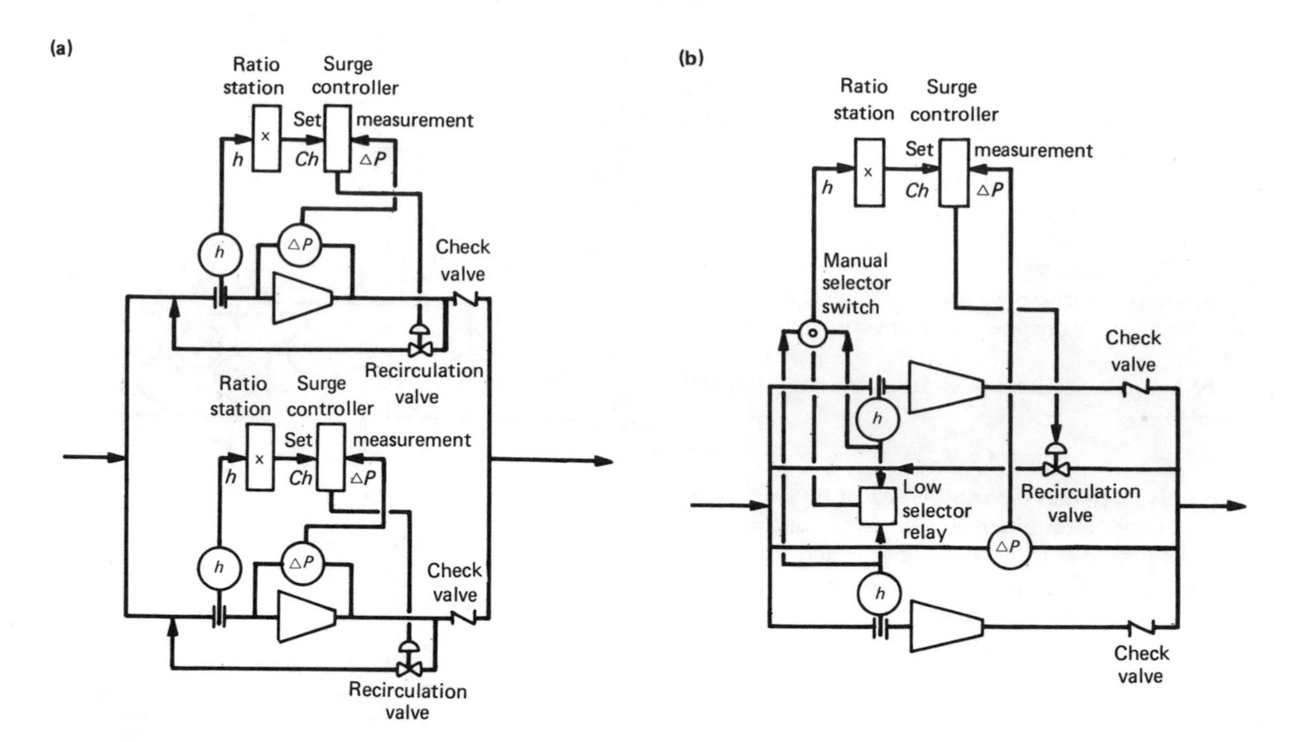


**ALTERNATE CONTROL ARRANGEMENTS** for centrifugal compressors operating in parallel—Fig. 13

ent. The solution is particularly difficult when the compressors have dissimilar characteristics. But even when the characteristics are identical, some means must be provided to ensure that the desired load distribution is obtained. Controls to accomplish this will not be discussed here, except as they affect the antisurge control system.

Fig. 12 shows the effect of inadequate control. With a differential pressure of 80 psi. and both compressors running at 100% speed, the load is equally divided and the flow through each machine is $Q_1$.

If, however, for some reason such as dissimilar governor characteristics, one compressor is operating at only 98% speed and the other is at 102%, the flows will be $Q_2$ and $Q_3$, respectively. If the difference in speeds is large enough, the flow through one machine will decrease sufficiently to cause surging.

From Fig. 12, it can be seen that the flatter the characteristic curves, the more difficult the problem becomes. For this reason, centrifugal compressors require more-precise methods of load distribution than axial-flow compressors, which have much steeper curves. This susceptibility to surge makes it even more essential to provide positive protection for centrifugal machines during parallel operation.

The control arrangement shown in Fig. 13(a) allows maximum operating flexibility with minimum supervision. It consists of a standard surge-control system for each compressor, which provides complete protection at all times—whether the machines are operating singly or together. It is even possible to start a compressor with the surge controller in automatic while the other compressor is running. As the newly-started machine approaches its operating speed, the surge controller holds the recirculation valve open just far enough to bring the compressor up along the control line and keep it from surging. And this system can be used for any number of machines.

Fig. 13(b) shows an alternate surge-control system that requires less equipment. It uses only one ratio station and controller, one $\Delta P$ transmitter and a single recirculation control valve. When both compressors are operating, the low selector relay chooses the lower flow signal and sends it to the ratio station. If only one machine is running, its flow signal is connected to the ratio station by a manual selector switch. This selection can also be performed automatically with relay logic, if desired.

The cost savings resulting from the use of this system, however, are not obtained without some sacrifice. For example, each compressor must be equipped with a manual recirculation valve for starting. With one machine operating, the recirculation control valve cannot be used for starting the second compressor—so the operation must be performed manually and without automatic surge protection.

## Meet the Author

**M. H. White** recently retired as an Oil and Gas Industry consultant after more than 20 years service with The Foxboro Co., Foxboro, MA 02035. A specialist in instrument-applications engineering for pipelines and refineries, he holds a B.S. in electrical engineering from the University of Maryland. Before joining Foxboro, he was affiliated with the Atlantic Refining Co.

Mr. White was active in the formation of the Instrument Society of America, having served as chairman of the constitution committee and on the board of directors. He is also a member of the American Petroleum Institute and the Petroleum Electrical Supply Assn.

# Improved surge control for centrifugal compressors

Surge may occur without being detected by conventional control systems, cutting compressor and process efficiency and causing compressor damage. Excess flow recirculation or blowoff can help avoid some surges, but energy costs are high. Here is a control system that copes with all surges without incurring such penalties.

*Naum Staroselsky and Lawrence Ladin, Compressor Controls Corp.*

☐ The operation of a centrifugal compressor can become unstable due to changes in many conditions such as flowrate, pressure, and the molecular weight of gas. This causes rapid pulsations in flow, called surge. No system is immune to sudden upsets at one time or another.

We have surge-tested air compressors, gas compressors, centrifugal compressors and axial compressors, at high and low pressures. The amplitude of the flow drop and the frequency of surge cycles vary. However, there are certain common results: the pressure change is less than the flow change in every case that we have recorded; flow usually drops extremely fast just before surge, and then always drops precipitously during surge, at least in the authors' experience—a time of 0.05 s from setpoint flow to reverse flow being common.

The speed of some surges is such that conventional control systems cannot detect them, let alone react to them. So, often even when records indicate that no surge has taken place, stripping down a compressor will reveal surge damage, ranging from changes in clearances, which exact a penalty in compressor efficiency, to destruction of parts.

Unstable operation—whether it is detected or not—affects the operation of the process that relies on the unit.

Typically, surge is prevented by recirculating some of the flow or by blowing off the excess flow. Constant recirculation of flow—often at 30–40% of the amount needed by the process—is common. For instance, this is found frequently in units used to compress chlorine and wet gases, among others. Obviously, this excess flow costs energy. And it does not necessarily avert all surges.

## Improved surge control

Here, we will present an improved surge-control strategy and instrumentation system. This system copes with even very rapid upsets, minimizing disturbances to process operations. The amount of recirculation is reduced sharply, saving significant amounts of energy. Furthermore, the system lends itself to actual field calculation of the surge limit.

With conventional closed-loop control schemes, once surge begins, the compressor control-system oscillates—and the only way to stop this is by manual override. Here, if surge occurs, the compressor will automatically be brought out of it during the first surge cycle, and the control system will reset itself to prevent such oscillations from recurring.

Conventional systems can be set into oscillation by operator error—such errors are minimized by this system. Also, fluctuations in the pressure and flowrate of the compressed gas are kept to a minimum, resulting in relatively minor disturbances to the process. (The reasons for these advantages will be explained later on.)

In order to develop this control system, we will first relate surge to compressor operation, then determine where surge begins and where to set the control system to prevent it. Next, we will discuss instrumentation, control strategies and calibrating the equipment. Compressors in series and parallel will not be discussed here, due to limitations of space.

## Surge characteristics

Fig. 1a shows characteristic curves for a typical single-stage centrifugal compressor. Each rotational speed ($N_1$, $N_2$, $N_3$, $N_4$) has its own characteristic curve.

If the compressor is operating at speed $N_1$ at Point $A$ (mass flowrate $W_A$) and the flow is decreased, surge will occur at $W_B$. The pressure has increased from $P_{D,A}$ to $P_{D,B}$. Point $B$ lies on the surge limit line, a naturally occurring line that is peculiar to each compressor geometry. The area to the left of this line is the surge zone, where there are oscillations of flow and pressure.

Originally published May 21, 1979

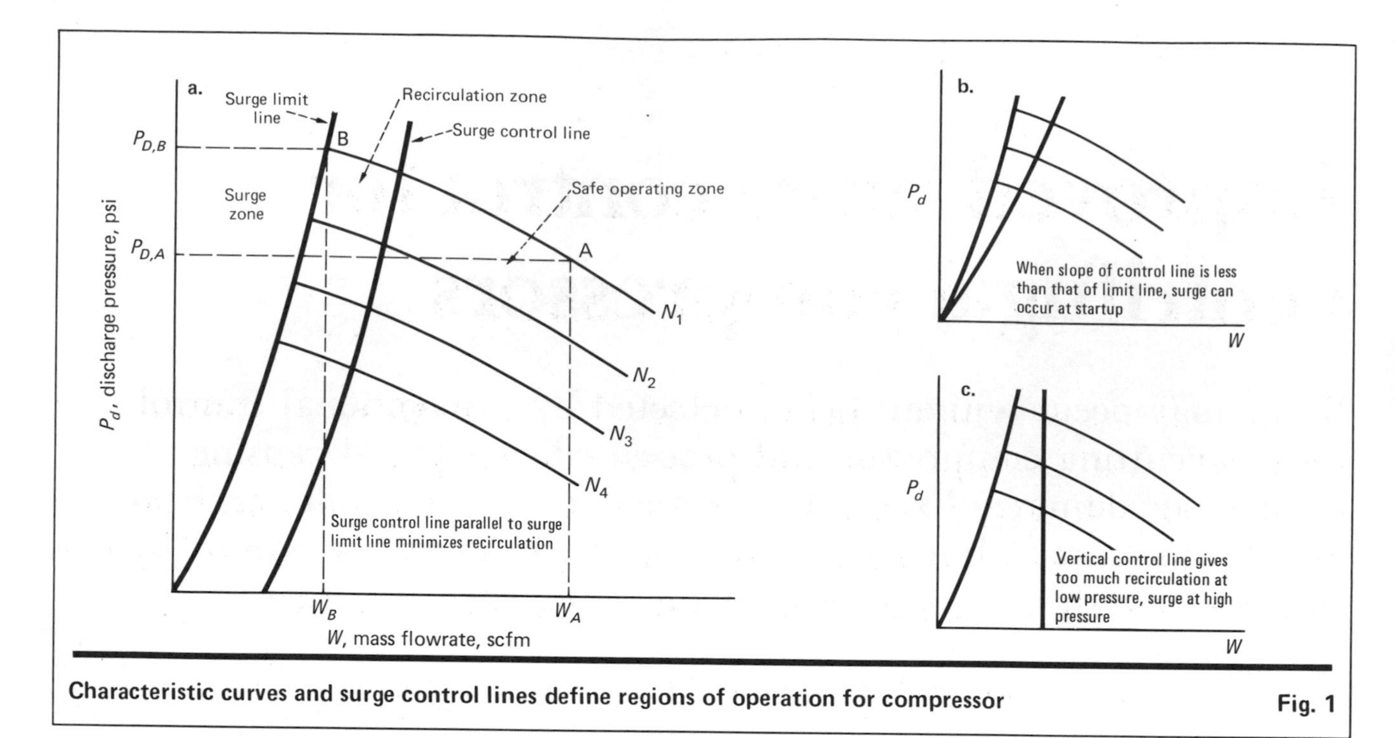


**Characteristic curves and surge control lines define regions of operation for compressor** **Fig. 1**

The line to the right of the surge limit line is the surge control line, which is an artificial line set by an engineer. These lines are the boundaries of the recirculation zone, where recirculation or blowoff is used to prevent surge. To the right of the control line is the safe operating zone, where recirculation is considered unnecessary. In order to develop the control strategy, we must first develop improved equations for the control and limit lines.

## Surge limit line

Here, we will formulate an equation for the surge limit line that does *not* assume constant inlet temperature and molecular weight. Typical equations assume that these parameters are constant. Corrections must be applied when conditions change. Such is the case with White [*1*], whose method is widely used.

The head of a centrifugal compressor is a function of angular velocity, speed of rotation, inlet volumetric flowrate and impeller diameter. Using a form of the kinetic energy of the compressor, given by Davis and Corripio [*2*], and performing a dimensional analysis:

$$\frac{g_c H_p}{N^2 D^2} = f\left(\frac{Q_s}{ND^3}, Ma, Re\right) \tag{1}$$

At the high velocities at which centrifugal compressors operate, the variation of Reynolds number, *Re*, with the velocity of the gas is negligible. The variation of Mach number, *Ma*, with velocity also is negligible. Mach number can be assumed constant as long as the gas velocity does not approach the speed of sound. We will assume that operation is well below this speed in the region close to surge.

For a given compressor geometry, under these conditions, if $H_p/N^2$ is plotted against $Q_s/N$, then the performance curves for different speeds will reduce to one curve, called the universal performance curve (Fig. 2).

The surge limit line is reduced to a single point, the surge limit. Since the surge limit is fixed for a particular geometry, the values of this point are constant:

$$\frac{H_p}{N^2} = C_1 \tag{2}$$

$$\frac{Q_s}{N} = C_2 \tag{3}$$

Thus, the curves of the surge limit line follow the fan law, which states that inlet volume is proportional to speed, and head is proportional to the square of speed. The surge limit may be defined by either Eq. (2) or (3). These equations apply not only to single-stage centrifugals but also to multistage units without intercoolers.

While Eq. (2) and (3) can be used to set the surge limit (and the surge limit line), this is impractical since $H_p$ and $Q_s$ depend upon a measurement of molecular weight.

Variations in molecular weight cannot be continuously measured at the high speeds present in compressors, and therefore must be eliminated from the equation of the surge line. Also, $H_p$ and $Q_s$ depend upon variations in inlet temperature and pressure. We will eliminate molecular weight and temperature from the equation of the surge limit line. Eq. (2) and (3) are combined to eliminate $N$:

$$C_3 H_p = Q_s^2 \tag{4}$$

Polytropic head is given by:

$$H_p = \frac{C_4 Z_{av} T_s}{M} \frac{(R_c^{\sigma} - 1)}{\sigma} \tag{5}$$

Substituting $H_p$ in Eq. (4):

$$\frac{C_5 Z_{av} T_s}{M} \frac{(R_c^{\sigma} - 1)}{\sigma} = Q_s^2 \tag{6}$$

## Nomenclature

| | |
|---|---|
| $b_a$, $b_b$, $b_n$ | Distance between surge control line and surge limit line ($n=1,2,3$), in. $H_2O$ |
| $C_n$ | Constant ($n=1,2,3,\ldots$). Units vary from equation to equation |
| $D$ | Impeller diameter, ft |
| $d_1$ | Bias added by summing device in No. 5, Fig. 7, psi |
| $d_2$ | Bias added by summing device in No. 7, Fig. 7, psi |
| $g_c$ | Acceleration of gravity, ft/s$^2$ |
| $H_p$ | Polytropic head, (ft-lb)/lb |
| $k$ | Ratio of specific heat at constant pressure to specific heat at constant volume |
| $k_1$ | Slope of surge control line, dimensionless |
| $M$ | Molecular weight |
| $Ma$ | Mach number |
| $N$ | Speed of rotation, rpm |
| $\Delta P_c$ | Differential pressure across compressor, psi |
| $\Delta P_{or,s}$, $\Delta P_{or,d}$ | Orifice pressure at suction; discharge psi |
| $P_s$, $P_d$ | Suction pressure; discharge pressure, psi |
| $Q_s$ | Volumetric flowrate in suction, actual cfm |
| $R_c$ | Pressure ratio across compressor |
| $Re$ | Reynolds number |
| $T_s$, $T_d$ | Temperature at suction; discharge, R |
| $W$ | Mass flowrate, scfm |
| $Z_{av}$ | Average compressibility = $(Z_s + Z_d)/2$ |
| *Greek letters* | |
| $\sigma$ | $\dfrac{k-1}{k\eta_p}$ |
| $\eta_p$ | Polytropic efficiency |

Volumetric flowrate equals mass flowrate divided by density:

$$Q_s = \frac{W}{\text{density}} = \frac{C_6 W}{(P_s M/Z_s T_s)} \tag{7}$$

Substituting $Q_s$ in Eq. (6):

$$C_7 Z_{av}\frac{(R_c{}^{\sigma} - 1)}{\sigma} = \frac{W^2 Z_s{}^2 T_s}{P_s{}^2 M} \tag{8}$$

Using an equation for gas flow across an orifice in the suction of the compressor:

$$W = C_8\sqrt{\frac{\Delta P_{or,s} P_s M}{T_s Z_s}} \tag{9}$$

Eq. (8) becomes:

$$C_9 Z_{av}\frac{(R_c{}^{\sigma} - 1)}{\sigma} = \frac{\Delta P_{or,s} P_s M Z_s{}^2 T_s}{Z_s T_s P_s{}^2 M} \tag{10}$$

Or:

$$C_9 \frac{Z_{av}}{Z_s} P_s \frac{(R_c{}^{\sigma} - 1)}{\sigma} = \Delta P_{or,s} \tag{11}$$

Eq. 11 may be written for the discharge side of the compressor. Since the mass flowrate at any instant in

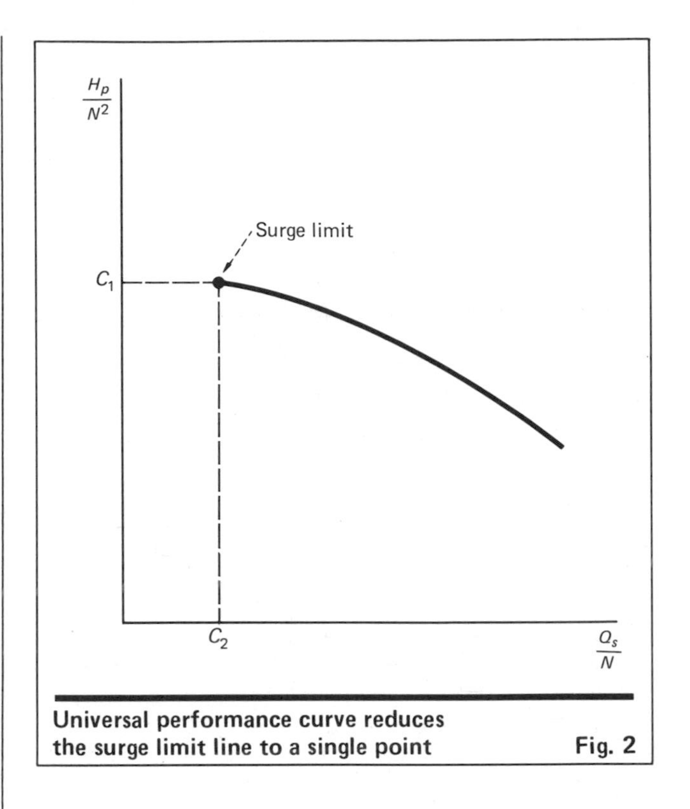


**Universal performance curve reduces the surge limit line to a single point** **Fig. 2**

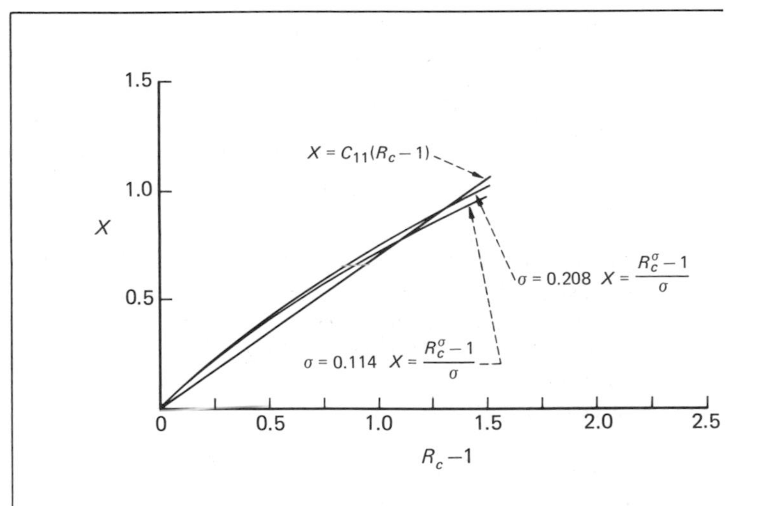


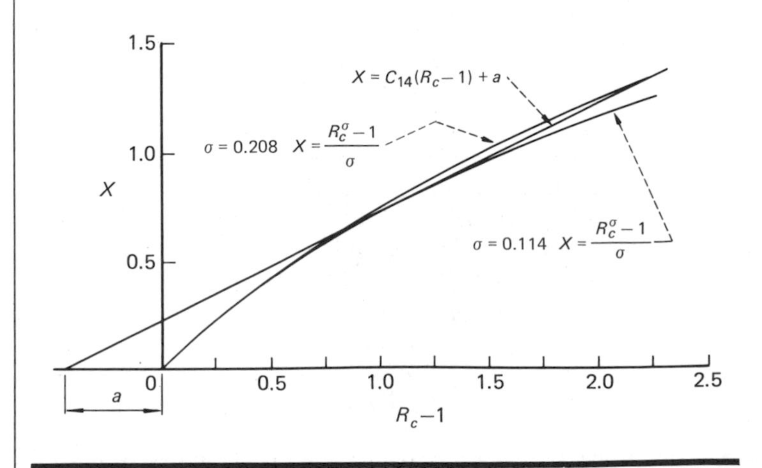


**Approximating $\frac{(R_c^{\sigma} - 1)}{\sigma}$ with a linear function** **Fig. 3**

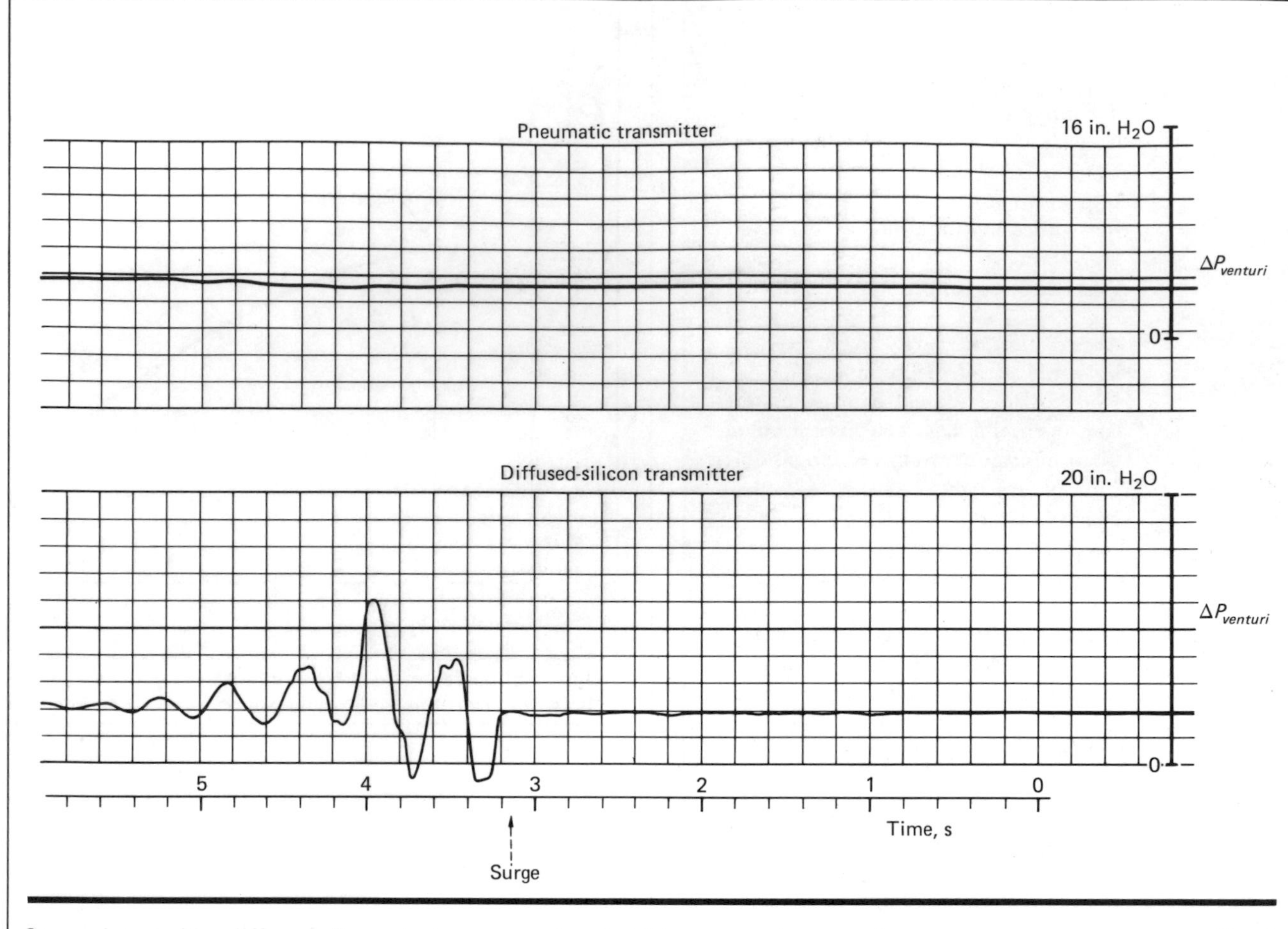


**Surge, detected by diffused-silicon device, occurs too fast for pneumatic transmitter** **Fig. 4**

the compressor is the same at suction and discharge, Eq. (9) can be written as:

$$W = C_8\sqrt{\frac{\Delta P_{or,s}P_sM}{T_sZ_s}} = C_8\sqrt{\frac{\Delta P_{or,d}P_dM}{T_dZ_d}} \quad (12)$$

Or:

$$\Delta P_{or,s} = \Delta P_{or,d}\frac{P_d}{P_s}\frac{T_sZ_s}{T_dZ_d} \quad (13)$$

Eq. (11) becomes:

$$C_{10}\frac{Z_{av}Z_dT_dP_s^{\ 2}}{Z_s^{\ 2}T_sP_d}\frac{(R_c^{\ \sigma}-1)}{\sigma} = \Delta P_{or,d} \quad (14)$$

If we assume that changes in compressibility are small, then $Z_{av} = Z_d = Z_s$. The term $(R_c^{\ \sigma} - 1)/\sigma$ can be approximated by a linear function (Fig. 3). If $R_c$ is less than 2.5, the approximation $C_{11}(R_c - 1)$ can be used. Eq. (11) and (14) become:

$$C_{12}(P_d - P_s) = C_{12}\Delta P_c = \Delta P_{or,s} \quad (15)$$

$$C_{13}\frac{P_sT_d}{P_dT_s}(P_d - P_s) = C_{13}\frac{P_sT_d}{P_dT_s}\Delta P_c = \Delta P_{or,d} \quad (16)$$

Eq. (15) is the same as one developed by White [1]. However, White arrived at the same result by assuming that inlet temperature and molecular weight were constant; he later offered corrections for these. Here, we have shown that such corrections are unnecessary. As in White's case, here the surge limit line is unaffected by variations in suction pressure.

If $R_c$ is greater than 2.5, $(R_c^{\ \sigma} - 1)/\sigma$ may be approximated by $C_{14}(\Delta P_c - 1) + a$ and the surge limit line becomes:

$$C_{15}\Delta P_c + a_2P_s = \Delta P_{or,s} \quad (17)$$

When changes in inlet pressure are negligible:

$$C_{15}\Delta P_c + b_1 = \Delta P_{or,s} \quad (18)$$

where $b_1$ is a constant.

Note that the above equations apply to centrifugal compressors without intercoolers. With intercoolers, it is necessary to consider each section as a separate compressor, or use an approximation that considers the unit as a whole.

Eq. (15), (17) and (18) are recommended for calculating the surge limit line, as they contain the fewest number of variables, all of which can be easily measured. Sometimes, the surge limit line is obtained from Eq. (3). When this is done, it is necessary to neglect variations in molecular weight, and this can lead to inaccuracies, even on air compressors, since the humidity of ambient air varies.

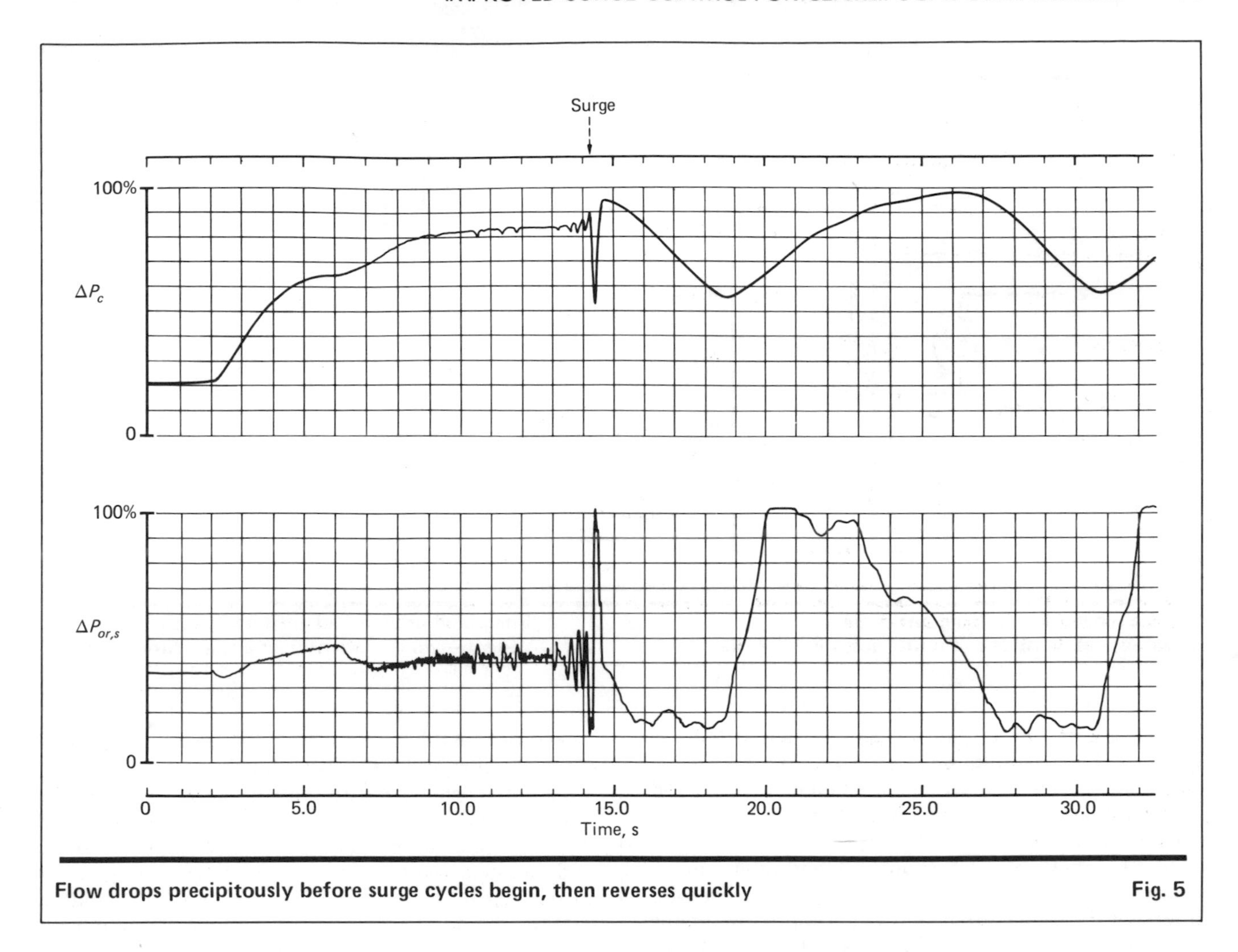


**Flow drops precipitously before surge cycles begin, then reverses quickly** **Fig. 5**

## Surge control line

The three common forms of the surge control line are shown in Fig. 1. The optimal position of this line is parallel to the surge limit line (Fig. 1a). To minimize recirculation, the surge control line should be set as close to the surge limit line as possible. Setting the control line with a slope less than that of the limit line (Fig. 1b) can lead to excess recirculation at high pressures, and surge at low pressures during stopping and startup. The third method is to select a minimum safe volumetric flow, and set a vertical control line (Fig. 1c). This can lead to excess recirculation at low pressures, and surge at high pressures. Many systems measure flow in the discharge without correcting for suction conditions. This gives maximum recirculation with minimum surge protection and is not recommended.

If it is desired to keep the control line parallel to the limit line, how close together can the two lines be?

This depends upon how accurately the surge control line is set and how well it accounts for changes in inlet temperature and pressure, and molecular weight.

Also critical to the location of the control line is the effectiveness of the antisurge system in handling upsets. This effectiveness depends on the control strategy chosen and the nature of the antisurge system, its transmitters, controller and antisurge valve.

To set the surge control line equidistant from the limit line, Eq. (15) can be used. The control line is displaced to the right by some fixed amount, $b_a$:

$$C_{12}\Delta P_c + b_a = \Delta P_{or,s} \quad (19)$$

Eq. (19) applies when $R_c$ is less than 2.5. For values of 2.5 and greater, Eq. (17) may be used. The control line is displaced by some fixed amount, $b_b$:

$$C_{15}\Delta P_c + a_2 P_s + b_b = \Delta P_{or,s} \quad (20)$$

## Instrumentation and valves

How fast must the antisurge system be to detect the onset of surge and effectively stop it?

It is often thought that a very rapid response is not necessary; also, many believe that protection against large upsets is unnecessary. However, upsets that are both fast *and* large are not rare, and these can be caused by factors such as jammed check valves, operator errors and shutdowns of process equipment triggered rapidly by protective systems.

We have found that surge cycles can occur faster than is thought possible, and that often surge goes undetected. This is because conventional pneumatic controllers are too slow to detect this phenomenon. Only fast transmitters can cope with the high speeds of compressor transients. We recommend a diffused-silicon electronic transmitter. We also recommend pressure-differen-

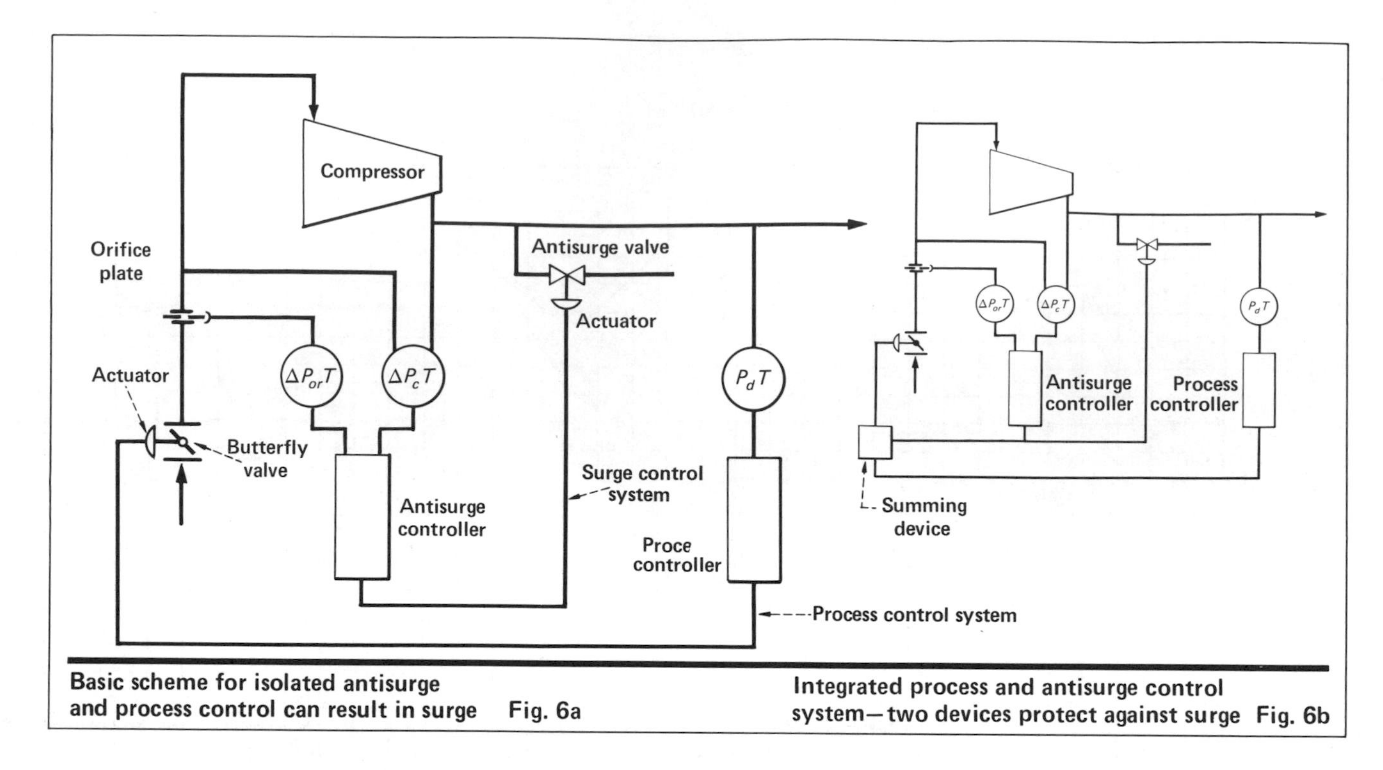


**Basic scheme for isolated antisurge and process control can result in surge** Fig. 6a

**Integrated process and antisurge control system—two devices protect against surge** Fig. 6b

tial transmitters with a lag not greater than 0.035 s as necessary to maintain surge control and to calibrate the surge limit line. Also, recorders (used for calibrating and testing the antisurge system) must have a chart speed not less than 25 mm/s.

Fig. 4 compares the response of a diffused-silicon transmitter with that of a pneumatic device. Pressure differential was measured across a venturi. At about 3.2 s, surge began, and the pneumatic transmitter was too slow to detect flow fluctuations.

The slowness of most transmitters and recorders explains why records will show no surge, yet, upon inspection, compressors reveal changed clearances or damaged impellers, seals or bearings. Such undetected surges will gradually alter the compressor, resulting in decreased efficiency. Some electronic transmitters are dampened and are too slow for use here; others are not stable at high speeds.

Fig. 5 shows how quickly surge can begin. Tests were made on a centrifugal air compressor that supplied a blast furnace. Here, surge occurred at about 14 s. Occasionally, there are pre-surge oscillations (incipient surge). Flow drops quickly before surge occurs. As surge begins, flow drops precipitously—typically from a setpoint level to a reverse condition in about 0.05 s, regardless of pressure or compressor speed (see Fig. 4).

## Antisurge valve and controller

For proper surge control, the antisurge valve must be large and fast-acting. It should be capable of recirculating 100% of design flowrate.

When flow is decreasing and operation is moving close to surge, it is necessary to effect a quick increase in flow. The rate at which flow can be increased depends upon the response of the valve and its size. The larger the valve, the greater the effect of its opening. The valve should be able to recirculate the entire flowrate because sometimes complete blockage of the system occurs, and all of the flow must be recirculated.

Field tests show that the full stroke of the antisurge valve should be from 0.5 to 1.5 s. New valves can meet this requirement; existing valves can be speeded up with boosters. Boosters increase the time of response by increasing the pneumatic signal to the valve actuator.

Why should the transmitter have a maximum response time of 0.035 s, while the antisurge valve has a maximum time of 1.5 s? The surge control line is set close to the surge limit line—typically, the distance between them is set at 15% of design flow. Thus, a partial valve stroke will be enough to stop movement toward surge; further opening of the valve will restore the operating point to the surge control line.

The controller must be fast as well. Pneumatic controllers are too slow to reliably prevent surge. Digital controllers should be used with caution. Since these devices look in sequence at each control loop, scanning time may not be short enough to detect the transients typical of compressor surge cycles. Microprocessors must have a scanning time less than 0.1 s.

## Control: isolated antisurge loop

First, we will look at the antisurge control system without considering its interaction with the process controller. Then we will integrate the two systems. In developing our control system, two typical systems will be studied first: a proportional-plus-reset controller and a relay (on-off) controller.

Consider an electronic analog antisurge controller that has a proportional-plus-reset response, with an antiwindup device (Fig. 6a).

The antiwindup device is necessary due to the nature of the proportional and reset functions. Normally, the compressor operates in an area some distance from the control line, resulting in an offset between the measurement and the setpoint of the controller. As a result, the output signal winds up to its high or low limit.

The proportional band and the operating point will be on opposite sides of the setpoint or control line—control will not be affected until the measured operating point reaches the control line. Surge can result if the measurement approaches the control line quickly, since overshoot will occur before the controller can unwind.

Antiwindup adjusts the reset loading to shift the proportional band to the same side of the control line that the measurement is on when the controller reaches its output limit. Then, if the control line is approached rapidly, the measurement enters the proportional band, and control starts before the value reaches the control line. Overshoot is reduced.

Derivative control is not recommended, for it can open the antisurge valve far from the surge control line and can cause system oscillations. Rapid oscillations in flow, even in the safe operating zone, can cause the valve to open because of the nature of the derivative response.

The response speed of this controller depends upon the proportional-band width and reset time. These parameters influence the stability of the system. Decreasing the proportional band or increasing the reset time increases the speed of the controller's response, but past a certain point system stability will be disturbed—all closed-loop control systems have a stability limit.

The speed of an antisurge controller's response is limited mainly by the inertia of the compressor and its networks, the transmitters, and the antisurge valve and its actuator.

The slower these elements are, the slower the controller must be set. The inertia of the compressor and its piping network cannot be changed, but the inertia of the transmitter and the valve and its actuator can be decreased by a proper selection of faster elements.

Once fast transmitters and an antisurge valve are selected, the proportional band level and reset time are set, based on the parameters of the compressor and its network. These parameters include the volume of the system, length of the pipes, and inertia of the compressor. This limitation is common to all closed control loops with feedback. Thus, modulated control responds well to slow upsets and give good control. However, the valve cannot be opened quickly for fast upsets, due to limitations of system stability.

Now consider using a relay for control. The relay is part of an open-loop system that opens the antisurge valve to a pre-established level after the compressor's operating point reaches the surge control line. The output of a relay device can change from minimum to maximum at any speed, without upsetting the system's stability. This is because (1) the relay is an on-off device and there is no limit as to how fast the device can be turned on and off and (2) the system is open-loop and oscillations cannot be set up in the control system.

However, the relay system has certain disadvantages for antisurge controllers. The preadjusted output level may be either lower or higher than that required for protection. If it is lower, the compressor's operating point will cross the surge limit; if it is higher, the compressor will be operated with large amounts of recirculation or blowing off when the antisurge valve is opened. Also, the operator can be unsure as to when to reset the relay without endangering the compressor.

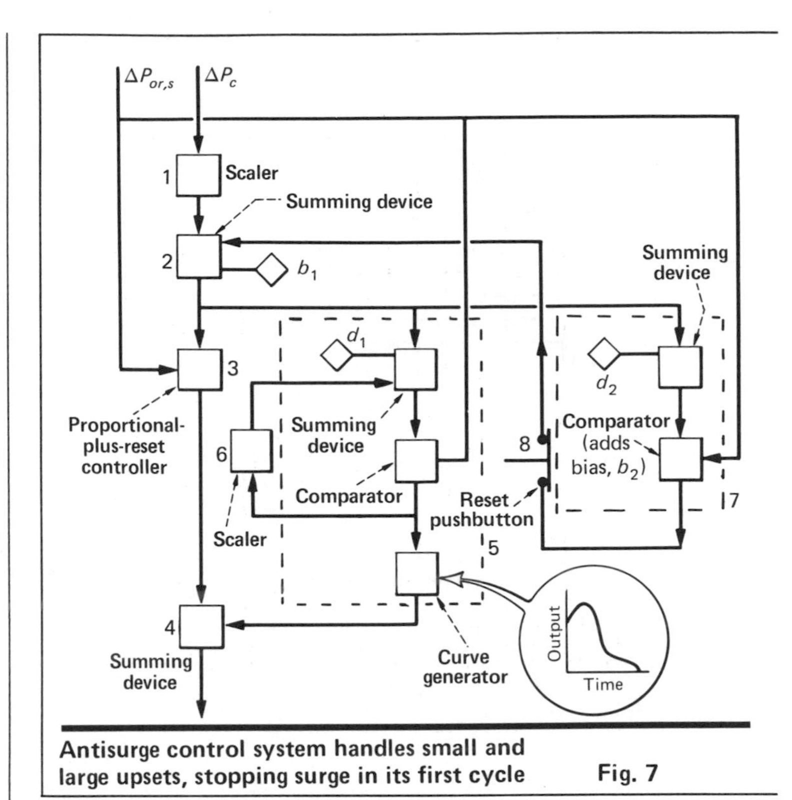


**Antisurge control system handles small and large upsets, stopping surge in its first cycle** **Fig. 7**

Comparing the closed-loop system to the open-loop system has led to a patented method that combines the advantages of both, while limiting their shortcomings.

Fig. 7 shows the controller with inputs for Eq. (18), one form of the surge control line. The input corresponding to $\Delta P_c$ passes through a scaler (No. 1) and summing unit (No. 2) that transform it to the required form:

$$k_1 \Delta P_c + b_1 \tag{21}$$

where $k_1$ is the slope of the surge limit line; and $b_1$ is the normal distance between the surge limit line and the surge control line.

The value of $k_1\Delta P_c + b_1$ is compared with $\Delta P_{or,s}$ by the proportional-plus-reset controller (No. 3). If $\Delta P_{or,s}$ is greater, then the controller's output is set at zero.

When $\Delta P_{or,s}$ is less than the signal given by Eq. (21), the controller begins producing a signal that increases until $k_1\Delta P_c + b_1$ becomes equal to $\Delta P_{or,s}$.

As a result, the compressor's operating point will be restored to the surge control line. This system will work well if disturbances are small and slow, but cannot open the antisurge valve quickly enough to prevent surge when disturbances are large or fast.

To handle large disturbances, the output from summing device (No. 2) is decreased by a fixed amount, $d_1$, which is typically set at one half the value of $b_1$. This addition is done by the summing device in element 5. The comparator in element 5 compares the value of $\Delta P_{or,s}$ with $k_1\Delta P_c + b_1 - d_1$. When $\Delta P_{or,s}$ is smaller, a signal is sent to the curve generator. The output of the curve generator follows the shape shown in Fig. 7—it

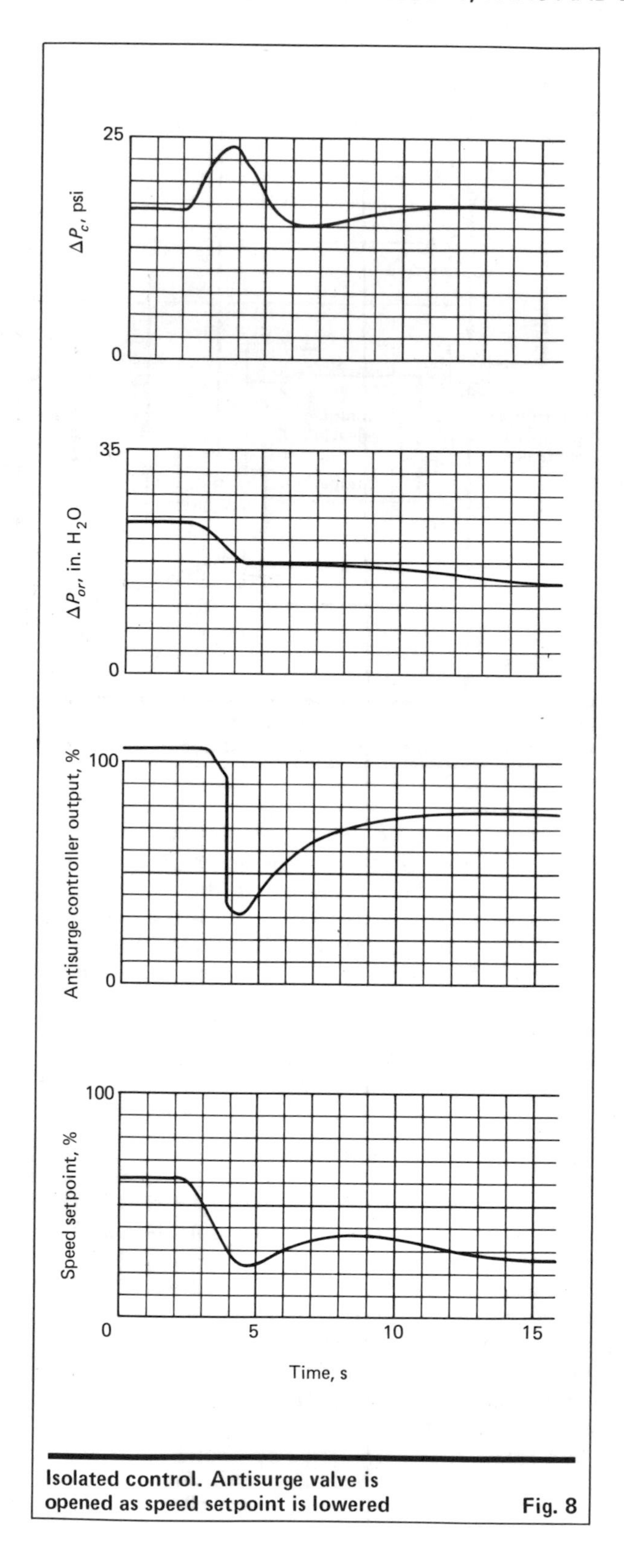


**Isolated control. Antisurge valve is opened as speed setpoint is lowered** **Fig. 8**

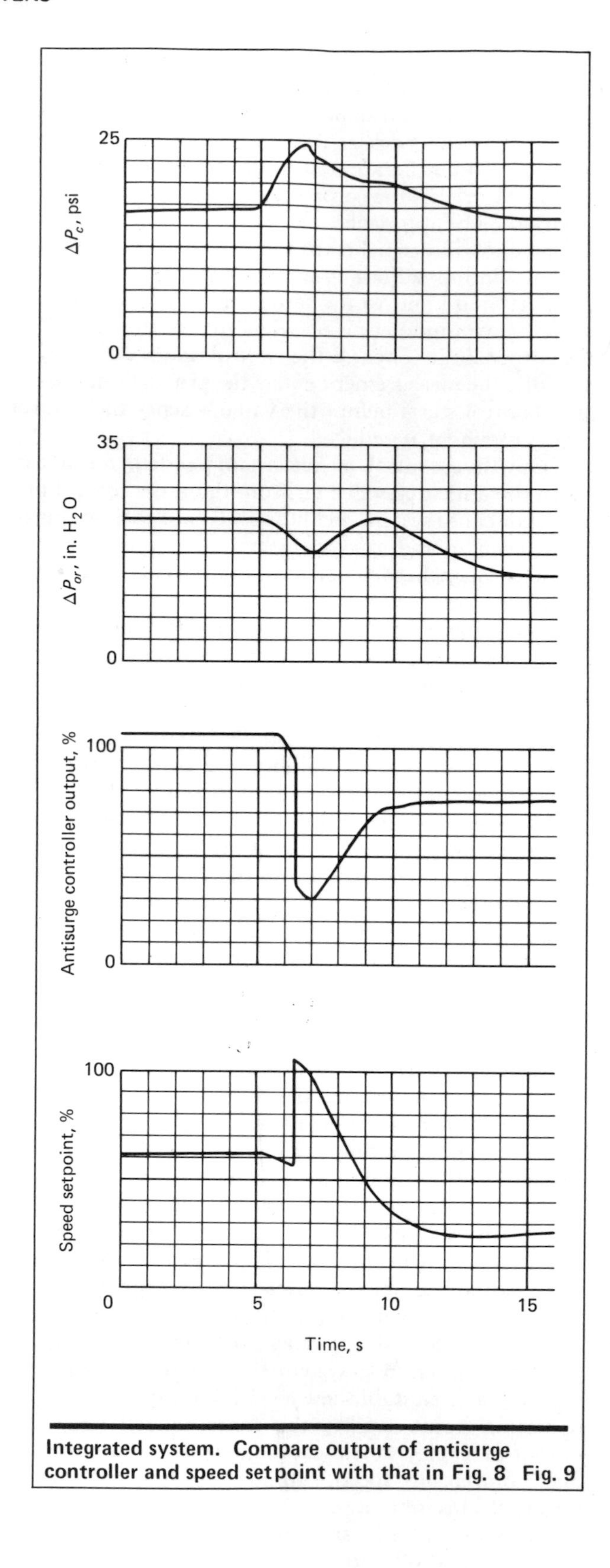


**Integrated system. Compare output of antisurge controller and speed setpoint with that in Fig. 8** **Fig. 9**

increases nearly instantaneously, then exponentially decreases to zero. This signal goes to a summer (No. 4), overriding the signal from the proportional-plus-reset controller.

Thus, for slow upsets, the valve will be opened slowly by the action of the proportional-plus-reset controller; for fast upsets, No. 5 will quickly open the valve, overriding the slow response of the controller.

The output from No. 5 decreases exponentially, allowing the proportional-plus-reset element to take over.

Thus, the operating point is restored to the surge control line as in any other antisurge closed loop. After the output of No. 5 decreases to zero, No. 6 automatically resets the override system.

Field tests have shown that, if the surge control line is set far enough from the surge limit line, then the combination of proportional-plus-reset response with the relay override signal is sufficient to prevent surge.

However, if the surge control line is set too close to the limit line, the operating point of the compressor can cross the limit line before the override signal appears. As a result, surge may begin, causing an almost instantaneous flow decrease. Under such circumstances, the deviation of the compressor's operating point from the surge control line increases.

Here, it is necessary to move the surge control line to the right—the task of No. 7. This element, like No. 5, subtracts a fixed amount ($d_2$) from the signal from No. 2 and compares the sum with $\Delta P_{or,s}$. When the sum is greater than $\Delta P_{or,s}$, a preselected bias, $b_2$, is added:

$$k_1 \Delta P_c + b_1 + b_2 = \Delta P_{or,s} \tag{22}$$

Bias $b_2$ moves the surge control line to the right.

These three elements (3, 5 and 7) can stop surge during the first cycle and keep the operating point at a safe distance from the surge limit line, even if gas consumption decreases to zero. This system protects the compressor against changing operating conditions, not just at design conditions. The reset pushbutton (No. 8) restores the surge control line to its original position.

Test results on this system are shown in Fig. 8. Here, rather than using a butterfly valve, the compressor was controlled by changing its speed. In Fig. 8, the speed setpoint is a record of how the process controller changes the setpoint of the speed governor. The action of the antisurge system is seen in the graph of the antisurge controller output. The horizontal part corresponds to the antisurge valve's being closed; the negatively sloped line, to the opening of the valve by the proportional-plus-reset controller (No. 3 in Fig. 7); the vertical drop, to the action of the override controller (No. 5 in Fig. 7).

## Interaction with process control

A process control system is shown in Fig. 6a. This consists of a pressure transmitter, pressure controller, and butterfly valve with actuator.

By closing the butterfly valve, the process controller forces the compressor's operating point to move toward surge. If the point crosses the surge control line, the antisurge controller will open the antisurge valve.

With both control systems, the deviation of the compressor's operating point from the surge control line depends on the speed of response of the antisurge controller and of the process controller, and on the dead times of the antisurge valve and its actuator.

If the process controller is faster than the antisurge controller, then this difference may interfere with surge control. The dead time of the antisurge valve can further add to this problem. Until the output of the antisurge controller overcomes the dead time, the process controller may push the compressor's operating point over the surge limit line.

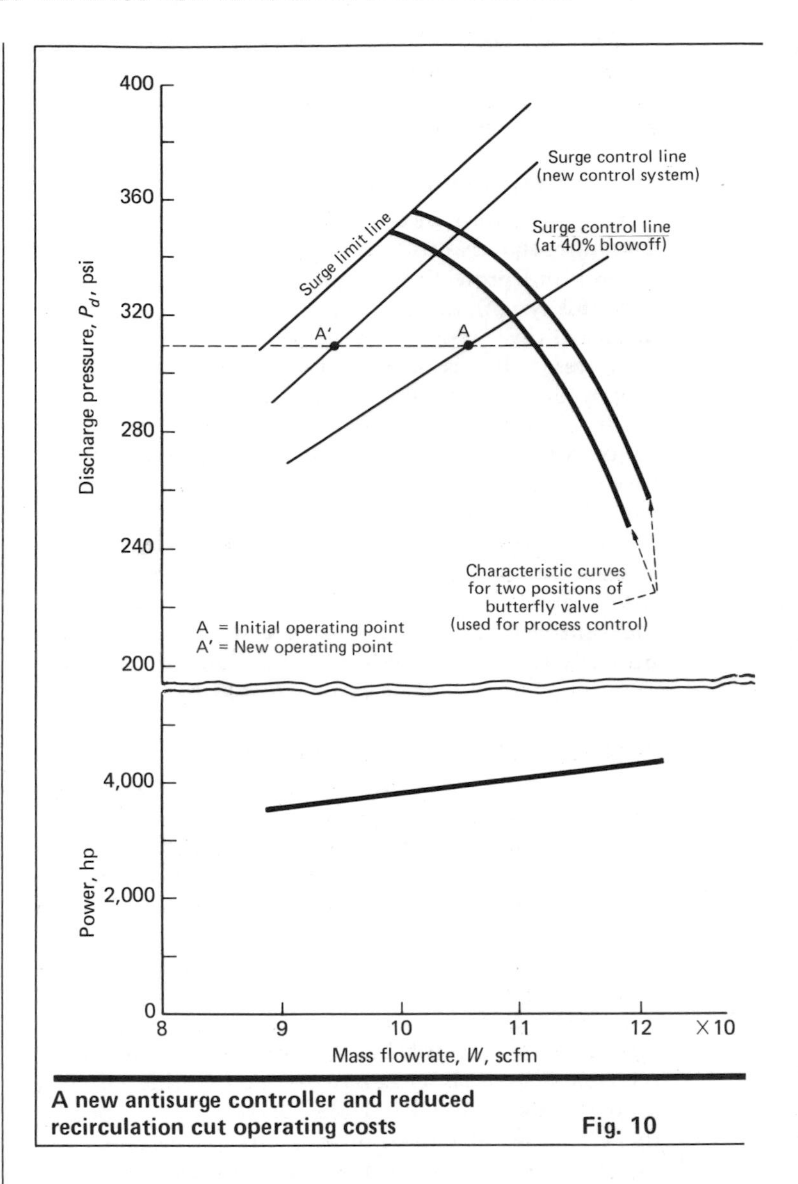


**A new antisurge controller and reduced recirculation cut operating costs** **Fig. 10**

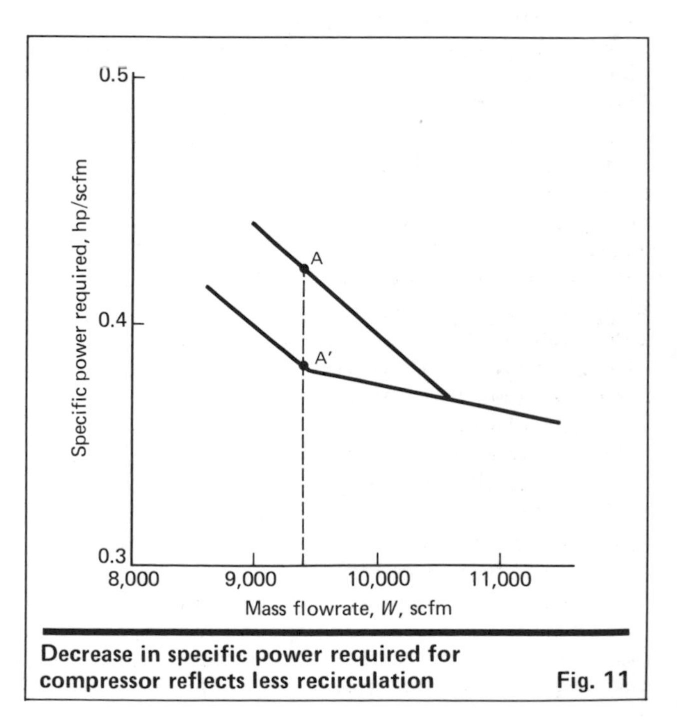


**Decrease in specific power required for compressor reflects less recirculation** **Fig. 11**

This interaction between an isolated surge-control system and the process controller occurs not only with control of discharge pressure but also with control of suction pressure and flow.

## Integrated system

System performance can be improved by changing from an isolated antisurge loop to an integrated process control and protective system (Fig. 6b).

In this system, the butterfly valve is controlled by the output of the summing device. This applies equally to a compressor that is under variable speed control. The output of this device depends on the output of the antisurge controller, as well as on that of the process controller. The operating point will, after reaching the surge control line, follow this line rather than the line of constant pressure.

With integrated process control and antisurge protection, the effect of the dead time of the antisurge valve will be minimized, since two devices are used to protect the compressor from surge—the antisurge valve and the butterfly valve. Test results of the integrated system are shown in Fig. 9, with a variable-speed compressor.

## Setting and testing equipment

The surge limit line can be located precisely, since the control system stops surge during its first cycle.

To calibrate the surge limit line, the slope of the surge control line, $k_1$, is set at zero to avoid response by the proportional-plus-reset controller. The distance between the surge limit and surge control lines, $b_1$, is set at 15% of flowrate, and $b_2$ is set at 75%. This is because the control line is vertical and a value of $b_1 = 15\%$ will induce surge. The distance between the surge control line and the control line of the override antisurge relay-element, $d_1$, is set at zero, so that override will start immediately. The level of the override signal opening the antisurge valve must be set at 100%; the time of its exponential decrease must be not less than 3 min. This ensures that the valve is kept open.

If the discharge line of the compressor is closed completely, surge will be induced. At the beginning of the first surge cycle, the override antisurge relay-element opens the antisurge valve at maximum speed. At the same time, the comparator (No. 7; Fig. 7) moves the surge control line to the right, making the distance between the surge control and surge limit lines $b_1 + b_2$ equal to 90%, which is safely out of the surge zone. This will keep the antisurge valve open, and surge will be stopped. The outputs of the transmitters, $\Delta P_c$ and $\Delta P_{or,s}$, will be obtained at the moment surge occurs.

The ratio $\Delta P_{or,s}/\Delta P_c$ yields $k_1$. If the control line is not a straight line, this test can be repeated to generate different values of $k_1$.

Setting $d_1$ depends on the speed of response of the proportional-plus-integral part of the antisurge controller. If the discharge line is closed slowly, the override antisurge relay should not operate. Experience indicates that $d_1$ may be between 2% and 5% of $\Delta P_{or,s}$.

The distance, $b_1$, between the surge limit and surge control line is selected so that the severest disturbances (for example, the fastest possible complete closing of the discharge line) will not cause surge. This distance can be narrow, because the improved antisurge controller operates at such high speed. In most cases, $b_1$ is set between 5% and 10% of $\Delta P_{or,s}$. Careful selection of $k_1$ and $b_1$ will reduce the energy expense for any compressor that operates with recirculation or blowoff.

## Example of energy savings

A 4,000-hp air compressor was being operated at constant 40% blowoff (see Fig. 10). The specific power required by this machine is shown in Fig. 11. The flowrate was 9,400 scfm at 310 psi.

The unit had a pneumatic surge-control system that was supplied by the manufacturer. In addition to constant blowoff, damage was frequent—the machine had to be rebuilt twice during one year. The control system described here was installed at a cost of about $15,000. About half the cost was for instruments, the other half for installation. After installation, the antisurge valve was closed and recirculation was kept to a minimum. A valve booster was not needed.

Assuming operation on a yearly basis, the savings in energy costs are calculated, with electricity costs of $0.024/kWh:

$$\text{Savings} = 0.746\,\frac{\text{kW}}{\text{hp}} \times 8{,}760\,\frac{\text{h}}{\text{yr}} \times (0.4225 - 0.3825)\,\frac{\text{hp}}{\text{scfm}} \times \frac{\$0.024}{\text{kWh}} \times 9{,}400\ \text{scfm} = \$59{,}000/\text{yr.}$$

## References


1. White, M. H., Surge Control for Centrifugal Compressors, *Chem. Eng.*, Vol. 79, No. 29, Dec. 25, 1972, p. 54.
2. Davis, Frank G., and Corripio, Armando, Dynamic Simulation of Variable Speed Centrifugal Compressors, "Instrumentation in the Chemical and Petroleum Industries," Vol. 10, 1974, Instrument Soc. of America, p. 15.


## The authors

**Naum Staroselsky** is director of engineering for Compressor Controls Corp., P.O. Box 1936, Des Moines, IA 50306. Telephone: (515) 244-1180. He designs automatic controllers and control systems for turbomachinery. The holder of a Ph.D. degree in mechanical engineering from Leningrad Polytechnic Institute, Staroselsky has received seven U.S. patents and has taught courses in control theory sponsored by Instrument Soc. of America. Control installations based on his inventions are in chemical, petrochemical and steel plants. He has designed numerous installations in the U.S. and holds a dozen patents in the U.S.S.R. He is a member of ISA and ASME.

**Lawrence Ladin** is business manager of Compressor Controls Corp. and president of its parent company, Ladin Industries, Inc. For the past twenty years, he has been involved with sales, design and marketing for Ladin. He holds a B.A. degree in mathematics and chemistry from the University of Minnesota and is a member of ISA.

# Can You Rerate Your Centrifugal Compressor?

Do you have a centrifugal compressor that you would like to use above its capacity or pressure rating? Here is how to determine if it is possible.

RONALD P. LAPINA, Elliott Co.

Compressor users often want to obtain increased production—which can take the form of increased capacity, increased pressure rise, or both.

One way to obtain an increase in production lies in the purchase of new equipment; however, one can often rerate (i.e., rebuild) an existing compressor to provide the required increase. Even if the compressor requires a new rotor and all internal stationary hardware, one can usually still salvage the compressor casing.

Most often, compressors are rerated to obtain greater throughput, with only a minor change in pressure levels. Under these conditions, one can probably reuse a significant amount of existing hardware.

One problem always appears when considering a rerate. No way exists to change the casing size or the bearing centerline distance. One cannot change the nozzle locations, and only a limited number of impellers will fit in any section of the compressor.

When he designs new equipment, the manufacturer starts with a "clean sheet of paper" and can do almost anything to satisfy the requirements of the process. However, the manufacturer must forego this luxury when rebuilding an existing unit.

When considering a rerate, the project engineer must first study the process and determine the flow required to handle the scheduled production increase. What pressure levels will he require of the compressor? How will the compressor's operation affect the overall process? He must determine the answers to these and other questions before considering the compressor for possible rerate.

A frustrating thing can happen to a project engineer. He designs the process, determines the new required capacity and associated pressure levels at which he expects his compressor to operate—only to learn from the manufacturer that the compressor cannot meet intended requirements. He then heads back to the drawing board and attempts to answer the pertinent questions until he arrives at a set of numbers compatible with both the compressor and the process.

The project engineer could often save a great deal of time if he understood the rerate capabilities of his compressor. If he could figure out for himself what his compressor could deliver, he could determine the probability of reaching his scheduled production increase—without communicating with the manufacturer—until he derived a feasible set of numbers. He might save days, or even months, depending on the extent of the process change and its relation relative to the compressor casing. He would know better than to try to put "10 pounds of coal into a 5-pound bag."

Actually the engineer need possess only a minimal amount of compressor knowledge, and spend a minimal amount of calculation time in order to determine the rerate feasibility of a compressor. He must consider:

1. Capacity—Will the size of the nozzle accept the projected increase in flow?
2. Horsepower—Will the motor, motor-gear, turbine or turbine-gear handle the increased horsepower?
3. Pressure—Can the casing accommodate the intended pressure levels? Can the compressor aerodynamically meet these levels?
4. Speed—Will the compressor handle the required speed within the API (American Petroleum Institute) critical-speed limitations?

We will now review these four considerations, presenting them, as above, in the order of their importance.

## Rerate Feasibility: Capacity

The most important consideration for capacity lies in nozzle sizes. One must determine whether the nozzle size will pass the required volume-flow with a reasonable pressure-drop. An existing compressor possesses fixed nozzle sizes—the geometric size of the nozzle sets the maximum feasible volume flow possible. Hence we can reduce the analysis to the consideration of inlet velocities.

Although inlet-velocity limits vary with conditions, a good rule-of-thumb sets the limit at a maximum of 140 ft/s for air and lighter gases.

Owing to the inverse proportionality between the inlet

Originally published January 20, 1975

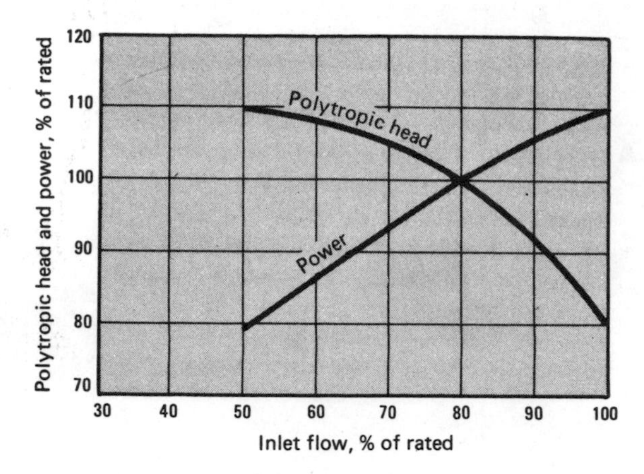


**CENTRIFUGAL COMPRESSOR:** typical performance—Fig. 1

velocity limit and the square root of the molecular weight of the gas, one should generally limit heavier hydrocarbons such as propane (mol. wt. = 44.06) to approximately 100 ft/s.

Eq. (1) determines the actual inlet velocity:

$$V_a = 3.06Q/D^2 \quad (1)$$

Since velocity limitations set the maximum capacity, the maximum velocity of 140 ft/s can limit one when rerating a compressor. With geometrically sound inlet piping, the engineer may find higher inlet velocities possible. Use Ref. 1 as a practical guide to the maximum inlet velocity tolerable. (Although written as a design guideline for compressor inlet piping, the design engineer will find it useful for determining the maximum inlet velocity a nozzle will tolerate.) Essentially, one must provide as even a flow distribution through the nozzle as possible. Consult Ref. 1 if you expect to exceed the above inlet-velocity criteria.

Manufacturers often furnish compressors with inlet nozzles other than the main inlet. They design some compressors with side-load nozzles, some with interstage cooling nozzles, some with a combination of both. Use the above inlet-velocity criteria for all inlet nozzles, and check all inlet nozzles for gas velocities.

Once the design engineer decides that the compressor can pass the desired capacity, he can investigate the effect of the increase on the horsepower required.

## Rerate Feasibility: Horsepower

The project engineer must include power requirements as the second major consideration in a rerate-feasibility analysis. This holds especially true in the case of motor-driven compressors, since customers usually do not purchase oversized motors along with the original equipment. However, one can usually rebuild gears and turbines to provide greater power capacity. Motor drives may require buying a new motor, which may lead, in some cases, to foundation problems.

The compressor will require power approximately proportional to the increase in weight flow desired. This means that an increase in weight flow of 20% will dictate an increase in horsepower of at least 20% (or 1.2 times the original power). Moreover, the manufacturer will usually suggest at least an additional 10% horsepower availability to take care of overload should this occur. Therefore, a 20% weight-flow increase will result in an increase of approximately 32% in power requirements (1.2 + 10%(1.2) = 1.32), should the project engineer wish to design a bit conservatively (see Fig. 1). Note also that in the case of a motor-gear drive, general practice requires the allowance of an additional 2% horsepower for gear losses.

If the driver cannot deliver the required horsepower to the compressor, the engineer must either buy a new driver or relax the scheduled production increase until he can reduce the required weight flow to within the capabilities of the driver. In cases where critical power requirements prevail, driver capabilities formulate the starting point of a rerate feasibility analysis.

## Rerate Feasibility: Pressure

The project engineer must next consider the hydrostatic test pressure. The compressor maker, at the time of manufacture, hydrostatically tests the casing to 150% of the maximum expected operating pressure. Compressor aerodynamics may set the maximum operating pressure, or the process may set it by the existence of a relief valve in the system.

If the expected pressure levels of the new process exceed the nameplate maximum pressure, one should make a new hydrostatic test. CAUTION: Sometimes, in a compressor made up of two or more sections (as with interstage cooling, side loads or extractions), the test engineer will test the compressor sectionally, with the inlet section at a lower pressure than the outlet section. In such a case, the project engineer must review both sections for the possibility of exceeding the maximum, sectional, safe-operating-pressure.

He must then investigate the compressor's aerodynamic ability to deliver the required head. He can more easily do so in terms of polytropic head—the amount of work done by the compressor (which differs from the input work by the polytropic efficiency). Eq. (2) relates the polytropic head to the desired pressure ratio:

$$Hp = ZRT[n/(n-1)][r_p^{(n-1)/n} - 1] \quad (2)$$

Although the polytropic efficiency can vary from 68% on small impellers to 83% on ideal impellers, for a feasibility analysis one can assume a polytropic efficiency of 70%. If the engineer analyzes the role of the polytropic efficiency in Eq. (2), he will find that the error in the "$n/(n-1)$" multiplier tends to counterbalance the error in the "$(n-1)/n$" exponent, thus considerably reducing error in the calculated polytropic head.

Eq. (2) can give both the original polytropic head and the rerate polytropic-head requirement. With the original rated speed known, the "Fan Law," Eq. (3), determines the approximate rerate speed:

$$N_{rerate} = N_{original}(Hp_{rerate}/Hp_{original})^{1/2} \quad (3)$$

The speed calculated from Eq. (3) may turn out too high for safe operation (see the considerations of rotational

## Nomenclature

| | |
|---|---|
| $a$ | Number of impellers in original rotor |
| $b$ | Number of "blank" stages in original configuration |
| $d$ | Tip dia. of largest impeller, in. |
| $D$ | Inside dia. of nozzle flange, in. |
| $GHP$ | Gas horsepower |
| $Hp$ | Polytropic head, ft-$lb_f/lb_m$ |
| $K$ | Ratio of specific heats, dimensionless |
| $N$ | Rotational speed, rpm |
| $n/(n-1)$ | $[K/(K-1)]\eta_P$ |
| $Q$ | Flange volume flow, $ft^3$/min, based on rated inlet conditions (i.e., inlet pressure, temperature, compressibility factor and molecular wt. |
| $r_p$ | Pressure ratio, $P_{discharge}/P_{inlet}$ |
| $R$ | Gas constant, 1,545/mol. wt. |
| $T$ | Inlet temperature, °R |
| $u$ | Mechanical tip speed, ft/s |
| $V_a$ | Actual inlet velocity, ft/s |
| $W$ | Weight flow, $lb_m$/min |
| $Z$ | Average compressibility factor |
| $\eta_P$ | Polytropic efficiency |

speed discussed in the next section). However, the manufacturer can sometimes obtain increased polytropic head by adding an impeller, or replacing the existing impellers with ones of lower backward lean. The manufacturer will ultimately have to determine the feasibility of obtaining the required head.

The project engineer can make a quick estimate of the maximum head capability of his compressor. If the compressor contains a "blank" stage (stage space with no impeller), Eq. (4) will show the approximate maximum head capability with no change in speed:

$$Hp_{max} = Hp_{original}[(a + b)/a] \qquad (4)$$

One can then apply the "Fan Law" along with Eq. (4), to determine the approximate polytropic-head capability of the compressor. This procedure should yield a polytropic head within 10% of the maximum compressor capability.

The above procedure for estimating the required polytropic head, although presented for straight through units, will also work for units with side loads, extractions or interstage cooling, provided one uses the sum of the head requirements for each section as the total head.

Determine the head for Section 1, based on the operating parameters of Section 1 (i.e., the inlet temperature, inlet pressure, discharge pressure, etc.); the head for Section 2, based on the operating parameters for Section 2, and so on. The sum of all sectional head requirements yields the total head requirement for the rerate. One can determine the total head requirement for the original rated point in a similar manner, and derive the approximate rerate speed from Eq. (3) as before.

One important effect of an increase in head requirement remains to be analyzed. That is, the head effect on the power requirement. Horsepower is directly proportional to polytropic head:

$$GHP = \frac{WHp}{33{,}000\eta_P} \qquad (5)$$

As can be seen, an increase of 20% in required polytropic head will raise the power requirement 20% if the efficiency remains constant. For example, if an increase of 20% in weight flow is coupled with one of 20% in polytropic head, the power requirement will rise by 44% ($1.20 \times 1.20 = 1.44$). To this figure, an additional 10% horsepower should be added for overload. A 2% gear loss should also be included, where appropriate.

### Rerate Feasibility: Speed

The final consideration is that of rotational speed. Two major criteria must be satisfied:

1. The rotational speed must be slow enough to not overstress the impellers.
2. The rotational speed should fall within certain limits of the compressor's first and second critical speeds, as specified in Ref. [2].

The first criterion above is best viewed from the standpoint of mechanical tip-speed limitations. The mechanical tip-speed can be determined by:

$$u = \pi dN/720 \qquad (6)$$

Mechanical tip-speed limits vary with manufacture, size, impeller material and type of construction. A typical limit on mechanical tip-speed might be 900 ft/s. When using the "Fan Law" of Eq. (3), the rotational speed that results in a tip-speed of 900 ft/s for the largest impeller should be considered the upper limit (providing that this speed falls within API critical-speed limitations as specified in Ref. [2].

The project engineer will not know what effect the rebuild will have on the compressor critical speeds when conducting the feasibility analysis, except possibly through experience with such effects. Since the bearing centerline distance is fixed and most rebuilds can be accomplished without a change of bearings, it can generally be assumed that the critical speeds will not be affected (at least, as a first approximation).

### Sample Problem

For the purpose of discussion, let us consider a "straight through" centrifugal compressor on a dry-air process, having the following nameplate data: inlet capacity = 11,000 icfm; inlet temperature = 90°F; rated inlet pressure = 14.5 psia; rated discharge pressure = 55 psia; rated power input = 1,700 hp; rated speed = 8,100 rpm; max. continuous speed = 8,500 rpm; first critical speed = 4,800 rpm; rated molecular weight = 28.97; $K = C_p/C_v = 1.4$; max. discharge pressure = 65 psia.

An investigation of the data files on the compressor reveals that the second critical speed = 10,800 rpm, and that the smallest wheel diameter is 22 in. Also the cross-sectional drawing of the compressor indicates that the inlet nozzle diameter is 20 in.

The desired rerate is an increase of the inlet capacity to 12,300 icfm and an increase of the discharge pressure to 60 psia, with all other inlet conditions unchanged:

1. Compute the inlet velocity, based on the new inlet

volume flow using Eq. (1):

$$V_a = 3.06\frac{Q}{D^2} = 3.06\left[\frac{12{,}300}{(20)^2}\right] = 94 \text{ ft/s}$$

Since this is an acceptable inlet velocity, the proposed capacity is feasible.

2. Since the rated inlet conditions have not changed, the increase in weight flow will be proportional to the increase in volume flow, and therefore the power requirement due to the change in volume flow will increase by the same proportion:

$$\frac{GHP_{\text{rerate}}}{GHP_{\text{original}}} = \frac{W_{\text{rerate}}}{W_{\text{original}}} = \frac{Q_{\text{rerate}}}{Q_{\text{original}}} = \frac{12{,}300}{11{,}000} = 1.12$$

$$GHP_{\text{rerate}} = 1.12GHP_{\text{original}} = 1.12(1{,}700) = 1{,}910 \text{ hp}$$

Note that, up to this point, the driver will have to be capable of:

$$(1.1)(1{,}910) = 2{,}100 \text{ hp}$$

plus 2% excess horsepower if a gear is involved.

3. Since the nameplate maximum-discharge-pressure is 65 psia, the compressor will not have to be hydrostatically retested, providing that the process will not allow the value of 65 psia to be exceeded.

4. The approximate polytropic head can now be calculated for both the original and the rerate conditions from Eq. (2):

Original:

$$n/(n-1) = [K/(K-1)]\eta_P = [(1.4/1.4-1)](0.76) = 2.66$$

$$H_p = ZRT(n/n-1)[(P_2/P_1)^{(n-1)/n} - 1]$$

$$= (1.0)(1{,}545/28.97)(550)(2.66)[(55/14.5)^{1/2.66} - 1]$$

$$= 50{,}700$$

Rerate:

$$H_p = (1.0)(1{,}545/28.97)(550)(2.66)[(60/14.5)^{1/2.66} - 1]$$

$$= 55{,}000$$

The new required speed can be determined from the "Fan Law," Eq. (3):

$$N_{\text{rerate}} = N_{\text{original}}\sqrt{H_{p_{\text{rerate}}}/H_{p_{\text{original}}}}$$

$$= 8{,}100\sqrt{55{,}000/50{,}700} = 8{,}440 \text{ rpm}$$

The smallest wheel diameter is 22 in, therefore, from Eq. (6):

$$u = \frac{\pi dN}{720} = \frac{(\pi)(22)(8{,}440)}{720} = 810 \text{ ft/s}$$

The new rotational speed results in a satisfactory mechanical tip-speed. API states that the second critical speed must be 20% above the highest operating speed. Assuming that the new rerate speed is the highest for the new process, the second critical speed must be at least:

$$(1.2)(8{,}440) = 10{,}130 \text{ rpm}$$

The second critical speed (10,800 rpm) is higher than that required and therefore the rotational speed is feasible.

5. The total increase in gas horsepower can now be determined. The new horsepower will be proportional to the increase in polytropic head and weight flow (in this case, volume flow):

$$\frac{GHP_{\text{rerate}}}{GHP_{\text{original}}} = \left(\frac{H_{p_{\text{rerate}}}}{H_{p_{\text{original}}}}\right)\left(\frac{Q_{\text{rerate}}}{Q_{\text{original}}}\right)$$

$$= \left(\frac{55{,}000}{50{,}700}\right)\left(\frac{12{,}300}{11{,}030}\right) = 1.21$$

$$GHP_{\text{rerate}} = (1.21)(1{,}700) = 2{,}060 \text{ hp}$$

The driver must therefore be capable of:

$$(1.1)(2{,}060) = 2{,}270 \text{ hp}$$

plus 2% if a gear is involved.

Since the inlet velocity, maximum operating pressure and required rotational speed are within satisfactory limits, the rerate is feasible.

## Summary

We have analyzed the four major considerations of a rerate-feasibility analysis—(1) capacity, (2) horsepower, (3) pressure and (4) speed. The nozzles must be large enough to pass the required flow—or, to state this another way, the maximum capacity will be limited by the size of the inlet nozzle. The driver must be capable of delivering the required power. The casing must be capable of handling the pressure levels, both mechanically and aerodynamically. The required speed must be within certain mechanical limitations and should meet the API critical-speed requirements. Once the Project Engineer is satisfied that the above criteria can be met, a request for a quotation can be forwarded to the manufacturer with greater confidence.

## References

1. Hackel, R. A., and King, R. F., Jr., "Centrifugal Compressor Inlet Piping—A Practical Guide" *CAGI* **4,** No. 2.
2. "API Standard 617 for Centrifugal Compressors and General Refinery Services," American Petroleum Institute, Washington, D.C. 3rd ed., 1973.

## Meet the Author

**Ronald P. Lapina** is an application engineer with the Elliott Co. Div. of Carrier Corp., Jeannette, Pa 15644, who has primary responsibility for the rebuilding of centrifugal compressors. Previously, he was employed by Pratt and Whitney Aircraft in West Palm Beach, Fla., where he was involved in turbine-blade-cooling research. He has a B.S. in aerospace engineering and an M.S. in mechanical engineering, both from the University of Pittsburgh.

# EASY WAY TO GET COMPRESSION TEMPERATURES

BILL SISSON, Nipak, Inc.

An engineer or operator can quickly find the temperature of gas discharging from a compressor with this nomograph. All he needs to know is the suction pressure and temperature, the discharge pressure, and the ratio of specific heats for the gas.

The basic equation for compressor discharge temperature is:

$$T_2 = T_1(p_2/p_1)^{(n-1)/n}$$

where: $T_2$ = absolute discharge temperature, °F + 460
$T_1$ = absolute suction temperature, °F + 460
$p_2$ = discharge pressure, psia.
$p_1$ = suction pressure, psia.
$n$ = ratio of specific heat at constant pressure to specific heat at constant volume
$p_2/p_1$ = compression ratio

The ratio of specific heats, $n$, can be calculated, read from tables, or estimated from the chart.

If the compressor suction or discharge pressure is beyond the range in the nomograph, the nomograph can still be used by calculating the compression ratio and taking that as the starting point. It should be noted that the absolute values of temperature and pressure in the equation have been converted to °F and psig.

*Example:* A hydrocarbon gas whose molecular weight has been given as 21 is fed to a compressor at 60 psig and 70°F, then compressed to 350 psig. What is its temperature leaving the compressor? Read $n = 1.26$ on the plot of molecular wt. versus $n$. Connect 60 psig on the $P_1$ scale with 350 on the $P_2$ scale (note that these pressures are psig on the scale and not psia) and read the compression ratio, $p_2/p_1$, as 4.88. Connect this with 1.26 on the $n$ scale, note the intersection on the pivot line, and align this intersection with 70°F on the $T_1$ scale to read 275°F on the $T_2$ scale. #

**Nomograph Determines Gas Compression Temperature**

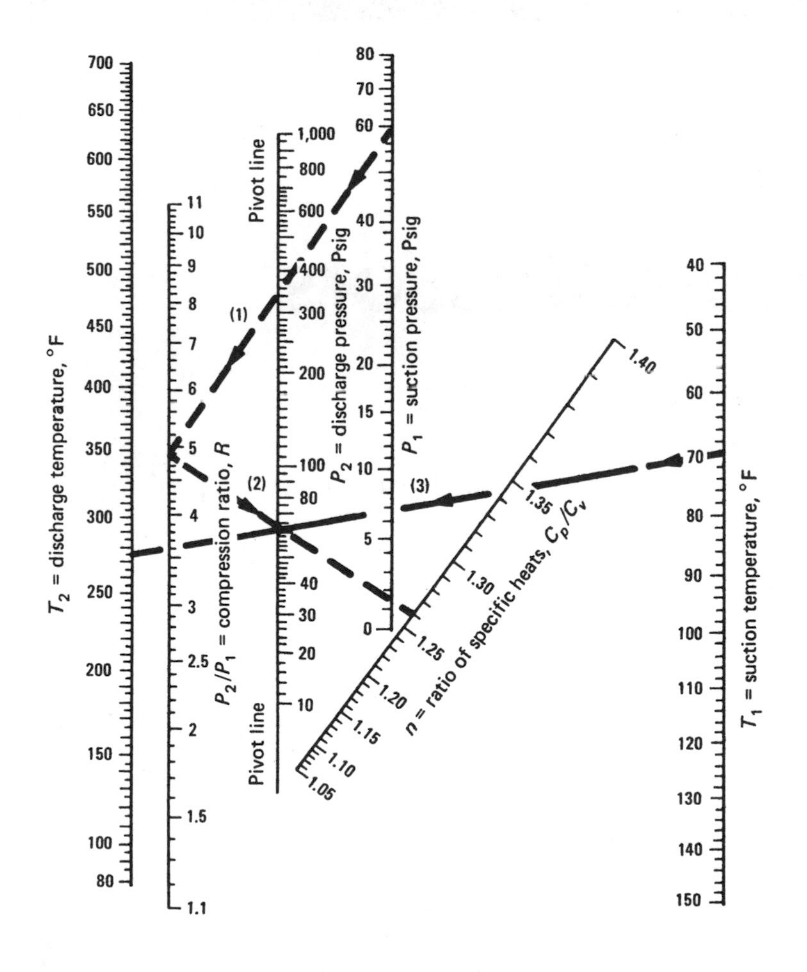


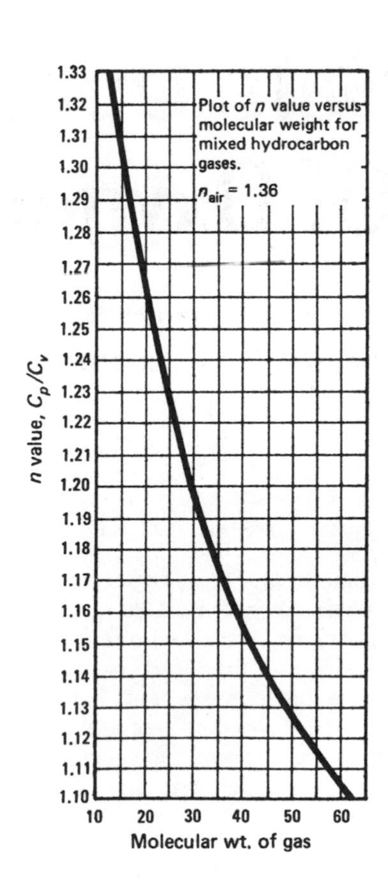

# Safe Operation Of Oxygen Compressors

When only fingerprints on a cylinder can cause a violent fire, what special design features, fabrication precautions and maintenance procedure are required to keep oxygen compressors operating safely?

WILLIAM M. KAUFFMANN, Consulting Engineer

Supervisors and engineers who operate oxygen compressors are usually aware of the safety precautions required. Too often, however, because compressors are the heart of oxygen operations, they are overhauled on a crash basis. When this happens, supervisors become tempted to expedite repairs by taking shortcuts on the seemingly tedious procedures and precautions that are always recommended by manufacturers for the servicing of oxygen-handling compressors.

Continual shifting of operations personnel and turnover in millwrights and mechanics also create situations that result in dangerous relaxations of established safety practices.

Oxygen equipment should not be treated casually. Although oxygen serves man in the air he breathes and the water he drinks, in its pure form it is a potentially violent oxidant.*

Once, a vendor representative, called into a plant to supervise the reassembling of an oxygen com-

* Bulletin No. 259, which has been issued by the U. S. Dept. of Labor in its Environmental and Chemical Hazard Series and which deals with the use and handling of compressed gasses, offers valuable information on the hazards of oxygen; copies of it can be obtained from the U. S. Government Printing Office, Washington, D. C.

pressor, arrived to find millwrights installing the piston with bare hands. When his protests went unheeded, he left abruptly. Shortly afterwards, a telegram arrived at his office informing him that the compressor had burned and was a total loss. Fingerprint grease on the piston had contaminated the cylinder sufficiently to cause ignition.

## Safe Maintenance and Operation

Engineers and maintenance personnel should know not only what to guard against in oxygen compressor maintenance and operations, but also what manufacturers provide in the way of a safe compressor, from both the standpoint of design and fabrication.

Specifically, the following questions will be answered:

- What design features and fabrication procedures that contribute to safe, reliable compressor operation should operating and maintenance personnel know about especially?
- What should supervisors know to keep oxygen compressors operating efficiently and safely?
- What kind of repair program will contribute to safe and efficient compressor operation?

A repair program and procedures for safe startup and operation will be reviewed. This will cover preparation for overhaul, inspection procedure and work schedule, contaminant detection and evaluation, assembly procedure, and starting up and running in the compressor.

## Design for Safety

Compressors for oxygen service differ from other types because they incorporate the following major features: (1) nonlubricated components, including TFE* piston rings, piston-rider rings and rod packing; (2) extended frame housing with a sealed and vented packing section and double oil-scraper rings; (3) special nonlubricated valve assemblies; (4) TFE-fitted diaphragm unloaders; and (5) TFE-honed cylinders.

The cylinder of a modern oxygen compressor is designed to operate with components made of self-lubricating materials such as glass-filled TFE.

For additional protection, the cylinder is also separated from the compressor frame by an extra-long distance piece. Also, an oil slinger, clamped on the piston rod, blocks off contaminants that may pass the scraper rings of the crankcase (Fig. 1). The distance piece is long enough that the oil slinger travels full stroke in the housing. A double distance piece with an oil slinger in each provides additional safety.

Finishes are finer and tolerances closer than in lubricated compressors because mating surfaces of self-lubricating materials cannot otherwise seal off small extrusion leakage. Cylinder bores are honed

* Tetrafluoroethylene, also Teflon.

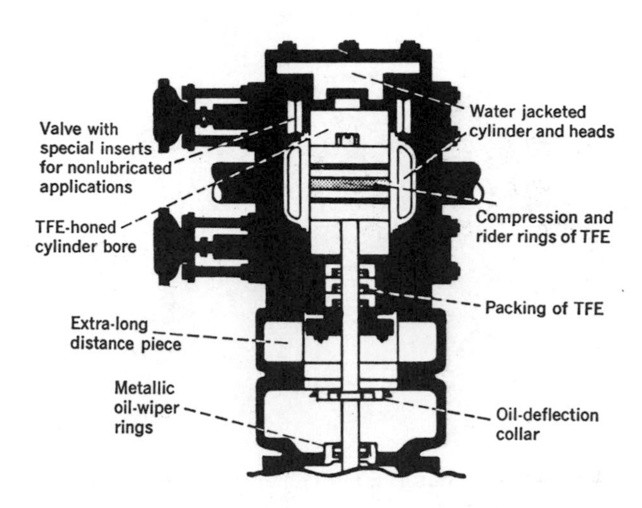


**INTERNALS** of nonlubricated oxygen compressor—Fig. 1

to a 10-to-20 microin. finish, then with TFE to provide a compatible wearing surface.

Cylinders are rust-proofed with a coating of manganese phosphate, which permeates the iron surface and renders it impervious to oxidation. Piston rods are normally surface hardened in the packing area (to a nominal hardness of 55 Rockwell C), and ground to an 8-to-15 microin. finish.

## Ring and Rider-Ring Assembly

Wearing parts of cylinders for oxygen service are mostly of glass-filled TFE. The trend is to solid (one piece) piston-rider, or wear, rings. Two types of jointless rider rings are now mostly used, both appropriately nicknamed "rubberband." One type is expanded and forced over the piston and into the ring grove by means of a taper mandrel. If correctly installed, this ring contracts fully into the groove in an hour or less. The other type, already expanded on a mandrel whose inside diameter is slightly larger than that of the piston, is pressed off the mandrel onto the piston and into the groove. If the ring contracts too slowly, heat is applied to it.

The most widely accepted nonlubricated piston-ring configuration is the step-cut (or butt-cut) one-piece piston ring. This type is usually selected for oxygen service because of ease of assembly, low cost and simplicity of parts replacement. Because spring expander rings are not required, the possibility of metal-to-metal contact is eliminated if ring wear is excessive. For pressures above 1,500 psi., special designs that provide pressure balancing extend ring life.

## Valves and Unloaders

Valves for nonlubricated operation are equipped with replaceable TFE nubs at the ends of the strip profiles. Valve seats and guards, at both suction

**VALVE** assembly has TFE buttons at strip ends—Fig. 2

and discharge, are also chemically treated to resist oxidation (Fig. 2).

Load control in oxygen compressors is accomplished by external diaphragm-operated unloading elements, which consist of a springloaded finger assembly that prevents the strips from seating during unloading. The sliding members of the actuating elements are guided in TFE sleeves, which provide lubrication.

The external diaphragm casing is mounted above the unloading assembly, and the actuating stem is sealed with TFE packing. Because the seal gland is open to the atmosphere, it can be easily inspected while the compressor is operating.

## Building in Safety

The key to operational safety is the "clean room" of the compressor manufacturer. It is here that all compressor parts that will come in contact with oxygen are cleaned (with a special detergent solution), inspected, assembled and protectively wrapped. Cylinders, ports and passages are rustproofed, and openings sealed and bores protected with a special vapor-inhibiting paper.

Protection often extends to double wrapping of parts in the special vapor-inhibiting paper, and covering everything with an outer-wrap of polyethylene, which is held in place with a special waterproof tape suitable for oxygen service.

Before shipment, cylinders and openings are charged with a special vapor-inhibiting powder that prevents the accumulation of moisture and condensation in them.

## Installation for Safety

Setting up a compressor in a plant requires continuing checks for cleanliness. Parts that will come in contact with oxygen should be thoroughly chipped or shot-blasted to remove all weld slag, scale and other deleterious substances. Acid cleaning to remove rust should follow dipping in a stripping tank to take off paint or other coatings.

After acid cleaning and phosphatizing, the parts should be dried and protective coating applied. Flanged openings should be blanked off with heavy gaskets. Each pipe and pressure vessel prefabricated by the manufacturer should be tagged to show that it has been cleaned and protected for oxygen service. Such protection should not be taken off until after the last connection has been made.

Intercoolers, aftercoolers and interstage piping should be similarly handled.

## No Shortcuts in Preparation

More time will be needed to properly service and overhaul an oxygen compressor than, for instance, an air compressor, except for the crankcase, which can be handled routinely. This is because the gear in the crankcase does not come in contact with oxygen, which is why it is oil lubricated.

However, special procedures must be followed when stripping down cylinders, removing pistons and piston rods, valves and unloader assemblies.

Preparations for servicing critical parts should be made before disassembly, including the following:

- Work tables should be clean and completely covered with the special vapor-inhibiting paper, which should be secured with a tape suitable for oxygen service.
- Cleaning tanks should be provided for two separate rinses of an inhibited 1,1,1-trichloroethane solution. A tank containing fluid for blacklight inspections for contamination should also be available.
- Tools should be thoroughly cleaned in the aforementioned tanks and placed in a clean location. Before the tools are used, they should be checked for contamination under the blacklight, and periodically afterwards.
- All compressor parts that will contact oxygen should only be handled by millwrights who wear clean, white cotton-canvas gloves, especially after the parts have been cleaned and when they are being assembled.
- Clothing worn by the maintenance crew should be clean and free of grease or oil.
- A 3,200-3,800 angstrom blacklight should be on hand for examining parts after they have been cleaned.
- Dry nitrogen should also be available for blowing off or drying parts. Compressed air from the plant system should never be used for cleaning or drying parts.

## Inspections at Overhauls

Systematic inspection is mandatory before and during the assembling of the cylinder and its associated parts. For finding contamination, a 3,200-3,800 angstrom blacklight is indispensable. The light will reveal hydrocarbon or dirt contamination as a fluorescent smear, smudge or blot. Contaminated parts must be cleaned until the fluorescence disappears.

The cleaning fluid should also be checked periodically. This can be done by placing a sample of it on a blotter and examining it under the blacklight. Contaminants in excess of 100 ppm. will cause the fluid to glow, in which case it should be replaced.

Parts and tools should be inspected under a blacklight during the assembling of the compressor, and contamination removed by cleaning and scrubbing. Brushes should be of bronze or stainless steel, never of a synthetic material. (A typical overhauling of an oxygen compressor consists of the steps shown in the box on this page.)

### Oxygen Compressor Overhauling Checklist

- ☐ Removing valves, covers and gaskets, which must be protected from dirt and moisture.
- ☐ Taping unloader-pipe openings with special tape.
- ☐ Removing and tagging water lines.
- ☐ Removing the cylinder head.
- ☐ Opening the frame to expose the piston rod and crosshead.
- ☐ Tramming the rod to the crosshead and checking the crosshead guide clearance; tramming is not necessary if the piston rod is flanged-bolted to the crosshead.
- ☐ Loosening the lock and crosshead nut and removing the rod and piston.
- ☐ Removing packing, cases and oil scraper rings.
- ☐ Inspecting and measuring the cylinder bore.
- ☐ Checking piston grooves for wear, and the rings for end gaps.
- ☐ Checking the piston rod for runout and wear at packings.
- ☐ Servicing valves and replacing worn parts.

## Starting Up Is Critical

Oxygen compressors are normally started up on dry nitrogen. Atmospheric air should never be used. The nitrogen should not only be dry, but also should have been compressed by a machine that also does not require lubrication so as to eliminate the possibility of hydrocarbon contamination.

During the startup, the upper and lower valve should be left out of each cylinder. Water jackets should also be filled, then the water flow stopped. Next, the compressor should be turned over by hand several times. The crankcase oil should be checked for proper type and level.

Before starting the compressor, make sure the suction and discharge valves have been properly placed in the cylinder. Serious damage and loss of capacity can result if they are improperly installed. Also, check that the valve locking screws are tight and the gaskets correctly in place, both for the valve seats and valve covers.

Bump the motor by power and check for proper rotation. If it is correct, bump the motor again; count three seconds; then press the stop button. All this time, be checking the oil pressure gage. By the time the compressor stops, oil pressure should register on the gage. If it does, start the compressor again. If the driver is a synchronous motor, be sure the exciter is on when it synchronizes. Check that proper field d.c. amperage is applied either through fixed taps or manual rheostat control. Check the motor nameplate for the a.c. voltage and amperage and the d.c. voltage and amperage required to properly operate the motor.

Run the compressor for five minutes, observing all the gages. Then shut it down by tripping the oil pressure switch. This tests whether it is properly wired into the stop circuit. Also check bearings, connecting rods and piston rods for abnormal heat.

Next, start the compressor again and run it for 10 minutes. Then shut it down again for the aforementioned checks. If everything appears to be satisfactory, start it again and run it for 20 minutes before shutting it down and repeating the heat check. If the temperatures are acceptable, start it up and run it for one hour before shutting down and repeating the heat check.

The piston rod temperature should not rise over 140 F. They usually run fairly hot in nonlubricated packing. By now, the water jackets will show signs of warming. Flow should be adjusted as required until pressure begins to build up.

If the compressor is two stage, the first stage valves should now be installed and the machine started up again. After the second stage has been blown through for 10 minutes, the compressor should be shut down and the second stage valves installed.

The final discharge bypass to the first stage should now be opened and dry nitrogen admitted. The compressor should then be started again and nitrogen circulated around the loop for about two hours or until the pressure rises to approximately 10 psi. Then the compressor should be shut down and the heat check repeated.

Finally, start the compressor again and begin closing the bypass of the suction valve to build up pressure. Admit additional nitrogen to hold 2-to-6 psi. suction pressure. After checking operating conditions, build up the load 10%/hr. of the final discharge, while at the same time maintaining the proper suction pressure.

After the proper loading at the final discharge has been reached and the compressor is running satisfactorily, unload and load it to check out the unloader elements. Be sure interstage pressure is correct for multistage machines.

Adjust the water flow through the cylinders so that the outlet temperature from the final stage will be from 110-to-115 F. Crankcase oil temperatures should be about 120-to-140 F. After operating con-

**CENTRALIZED** stations improve safety control—Fig. 3

ditions are satisfactory on dry nitrogen, oxygen can be introduced into the compressor.

It is important to continue a close watch of all the operational data, including especially the pressures and temperatures of water and oil and at the interstage coolers and the final discharge.

## Continuing Maintenance Assures Safety

Once an oxygen compressor has been put into operation, it should be the object of a rigorous preventive maintenance program. Vendor representatives who specialize in overhauling oxygen compressors are usually available to train plant personnel in preventive maintenance practices. Often overlooked is the valuable guidance in the proper maintenance of oxygen compressors offered in manufacturers' instruction booklets.

Safe compressor operations demand alert and conscientious maintenance. Compressor fires have resulted because valves and packing rings were not correctly assembled or installed. Inoperable safety devices have also led to considerable damage. Such devices should be checked frequently to ensure their proper operation.

During normal operation, the following items should be vigilantly checked: cooling water flow; crankcase oil level, pressure and temperature; operation of controls and control pressure; suction and discharge pressures and temperatures; unusual noises; and motor loading and temperature.

If a portable blacklight is available, checking the piston rod between the frame and the cylinder will reveal the presence of contamination in the distance piece, such as that which would appear if the scraper rings became worn. Such inspections should be scheduled according to length of service and to accommodate work schedules.

Fires have been caused by cylinder hot spots when cylinder-jacket water passages have become clogged with sludge or scale. Jackets, coolers and the packing-case cooling system should be inspected each time the compressor is overhauled. Clean surfaces will result in cooler operation and greater safety. Water treatment is essential to control scale and sediment deposits.

A daily log of compressor operations is essential for maintenance efficiency, especially with multistage units. At least the following should be recorded: (1) suction, discharge and interstage temperatures and pressures; (2) inlet, outlet and interstage water jacket temperatures; (3) temperature and pressure of lubricating oil to frame bearings; (4) motor load, amperage and voltage; (5) ambient temperature; and (6) date and time.

From such a log, a supervisor can note changes in pressure or temperature that signal a malfunction in the system. Prompt corrective action often can prevent serious trouble later.

Frequent checks of the open housing between the cylinder and crankcase with a blacklight for possible contamination and oil carryover from the crankcase into the housing should also be continued.

## Safety Devices Offer Insurance

Compressor safety controls are rapidly being incorporated into centralized control stations. Such controls offer the means for the rapid assessment of a malfunction and its location. They may also automatically initiate alarms and shutdowns. One set of points may be set up to set off an alarm and trigger an amber warning light; then, if the malfunction continues, the alarm sounds a second time, actuates a red light and shuts down the compressor. Another set of more-critical points may actuate an alarm and red light and shut down the compressor immediately (Fig. 3).

The following conditions are generally arranged to lead to only a delayed shutdown: excessive interstage gas temperature; low oil level in crankcase; high final-discharge gas temperature or pressure; and excessive frame oil temperature.

An immediate alarm and shutdown is usually arranged for the following conditions: failure of frame oil pressure; excessive vibration; and lack of cooling water flow.

## Meet the Author

**William M. Kauffmann,** now a consultant (2577 Nottingham Dr., Parma, Ohio 44134), was formerly a regional representative in customer service for Worthington Corp. During his 31 yrs. with Worthington, he had also been manager of research and chief engineer of its engine division.

A holder of a mechanical engineering degree from Armour Tech (now Illinois Institute of Technology), he is a member of ASME, National Society of Professional Engineers and Society of Automotive Engineers.

# Lubricating air compressors

**Choosing the proper lubricant for plant air compressors pays off in terms of long life and trouble-free operation.**

*R. G. Winters, Ingersoll-Rand Co.*

☐ The lubrication system in a compressor is a critical necessity in keeping the machine running. It does this by reducing friction, transferring heat to the cooling system, sealing against air leakage, and flushing away dirt and debris.

The lubrication systems of present-day air compressors must be able to provide these vital functions in atmospheres that are usually polluted with dirt particles as well as with corrosive fumes that pass right through the inlet air filter and enter the compression chamber.

In addition, modern air compressors are much smaller, and run at speeds undreamed of a few decades ago, making lubrication requirements all the more critical.

The packaged, integral-gear, centrifugal, plant air compressor has become important over the past 15 years. Here, the lubrication system has to deal with individual pinion gears that run at speeds of 30,000 to 50,000 rpm.

A more recent compressor is the oil-flooded rotary-screw machine. It has brought forth still a different set of requirements, because the oil and air pass through the compression cycle together, with the oil picking up all the dirt and contaminants that come through the inlet air filter.

Suppliers of lubricants have had to introduce more-sophisticated products in recent years to meet the requirements of this new generation of air compressors, operating under increasingly adverse environmental conditions. The oil industry has not only developed new lubricants to meet these requirements but has actually been able to extend the service life between oil changes in most cases.

## Lubrication problems

Many factors should be considered when your oil suppliers visit your plant to advise you on the proper lubricant for a particular compressor. Among the significant factors are inlet air temperature, air contamination, condensation, operating temperature, discharge temperature and pressure.

Inlet air contamination is a major factor in compressor wear, oil oxidation, and discharge-valve deposits. Recent years have seen an amazing increase in the rate of buildup of industrial air contaminants, which are often odorless and invisible. Even on the clearest day, considerable amounts of vapors and fumes come through an inlet air filter in a typical industrial plant. It surprises most people to learn that the human body can tolerate far more air contamination than can a modern air compressor.

Condensation seldom occurs in a compressor during normal operation. However, when the relative humidity of inlet air is high, and the cylinder temperature during the suction stroke is below that of the inlet air, condensation will occur. For instance, to prevent condensation, the cylinder walls in the first stage of a two-stage unit have to be kept about 100°F when inlet air is 80°F. The inexperienced operator who runs too much cold water through the cooling jackets may cause water to condense within the cylinders and wash away the oil film. This promotes the formation of rust, which in turn causes abrasion during compressor operation.

High operating temperatures reduce the viscosity of an oil, thereby reducing the oil-film thickness. This film reduction creates a marginal condition where metal-to-metal contact can occur. This is why the viscosity index of the oil purchased is so important. Prolonged operation at temperatures in the range of 140° to 160°F also increases the oxidation rate of the lube oil. Oxidation can change a thin oil to a thick, carbonaceous, sticky mass in a relatively short time. Hence, the oil-change period must be strictly maintained.

Air discharge temperature directly affects the quantity of deposits on the discharge valves and downstream piping. High discharge temperatures cause rapid oil oxidation; the oxidized residue plates out on the valves. This buildup eventually interferes with normal valve motion, causing valve leakage. Then, preheated or recompressed air increases the discharge temperature, further speeding oxidation and promoting more deposits. This cycle can lead to component failure and, at the extreme, to incandescent hot spots.

Another warning: If someone claims he has the ideal oil for all reciprocating compressors, don't believe him. A single-stage reciprocating machine has a theoretical discharge temperature of 400 to 450°F, while a two-stage unit would discharge at about 250 to 300°F. Some reciprocating units are splash-lubricated; others are force-feed lubricated. Some are air-cooled; others, water-cooled. Some use one grade of oil for all moving parts, while others use different oils for cylinders and running gear. There are many variables.

## Rotary-screw compressors

Oil-flooded rotary-screw compressors have different requirements from reciprocating ones. Because air and

Originally published August 14, 1978

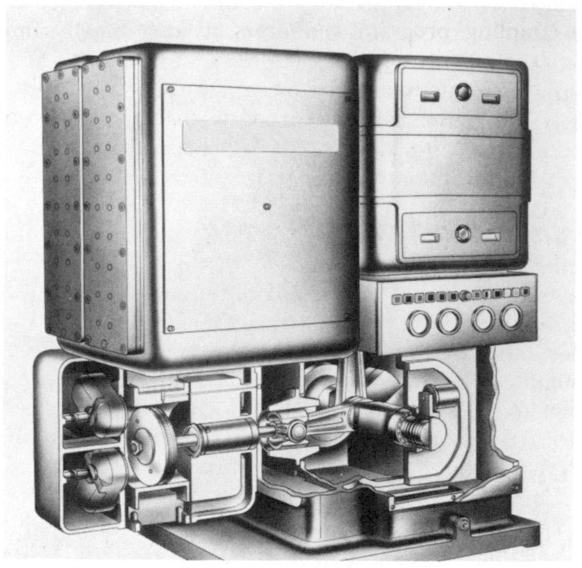

**Packaged compressor using pressurized lube system**

**Bull gear and pinions of centrifugal compressor**

moisture sucked in at the inlet carry through with the oil, and are separated after the compression cycle, the oil must have an excellent water-separation characteristic. Since the oil is continually cooled in a screw compressor, the internal temperatures can be kept quite low. However, this is dangerous; as has been said, condensation presents a severe hazard. Thus, most manufacturers recommend that discharge air temperature be maintained between 170 and 200°F. This concept of deliberately increasing the discharge temperature of a screw machine by reducing oil cooling is difficult for some operators to accept, but it must be insisted upon to prevent condensation and consequent bearing damage.

## Types of lubricating oils

Most of the premium-grade lubricants advertised to the public are the so-called Pennsylvania-crude-based oils, which are excellent for automotive service. They come from a large natural pool under parts of Pennsylvania, Ohio and West Virginia, and are unique in the U.S. These are paraffin-based oils, as opposed to the naphthenic-based oils produced elsewhere in the country. However, paraffin-based lubricants form a hard, varnish-like residue when used in compression cylinders; naphthenic oils form a lighter, fluffy, carbon residue that is more easily cleaned off.

Discharge-valve deposits can be the principal problem in some reciprocating compressor installations. When this is the case, it is essential to use a naphthenic oil and minimize discharge temperature. If the air compressor must also be operated under high humidity, high ambient temperatures, and intermittent-duty conditions, then condensation and resulting sticky valves or rings may occur. A compounded or oxidation-inhibited naphthenic-based oil is best suited for condensation conditions where valve deposits are a chief lubrication problem.

Paraffinic oils can be used in air compressors that do not have to contend with valve deposits. They are more resistant to thinning out as temperature increases than are straight naphthenic oils. That is, they have a better viscosity index.

Bearing protection is one of the primary functions of a frame or crankcase lubricant. The critical period for any bearing lubricant is when bearing failure begins. The bearing surfaces during critical periods become irregular, and high localized pressures occur. Naphthenic oils under high pressure become viscous. This provides an effective medium for carrying the bearing metal until its surface is smooth and the load is evenly distributed.

Synthetic lubricants were developed to overcome the two most frequent complaints by users of petroleum oils—carbonaceous deposits that necessitate frequent valve cleaning, and danger of fire.

## Danger of fire

The so-called "fire triangle"—oxygen, fuel and an ignition source—exists in any air compressor. Air provides the oxygen, and petroleum oils provide the fuel. Normal discharge temperatures are never high enough to trigger a fire or cause an explosion. The ignition temperature of most air-cylinder lubricating oils is above normal compressor operating temperatures.

However, carbonaceous deposits from lube oils do collect on valves, heads, and discharge piping. As has been stated, this may cause leaking discharge valves, so that hot air bypasses and continues to build up heat until auto-ignition can occur. This is practically always the result of inadequate maintenance, such as dirty fins on air-cooled compressors, dirty intercoolers, water jackets clogged with scale, broken or leaking valves, and the like. Therefore, the best oils should be used in a limited amount. Keep the compressor clean. Inspect valves frequently. Remove scale formations in the cooling-water jackets.

Phosphate-ester synthetic lubricants were developed to minimize compressor fires. Synthetics have distinct fire-resistant qualities in that they have an auto-ignition point that is about 50% higher than most petroleum oils, and have excellent film strength.

There are some disadvantages to synthetics: Their viscosity indexes are generally not as high as those of petroleum lubes, and they are more vulnerable to water washing in the presence of condensate. They are much

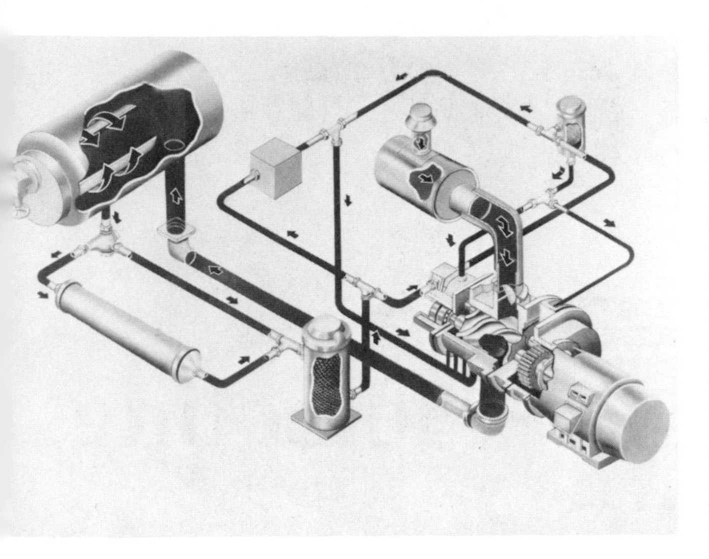

Oil-flooded rotary-screw compressor

more expensive than petroleum products, and the cylinder feedrate must usually be substantially higher. These synthetics are powerful solvents. They can remove all paint on the inside of air passages, and this carries into the valves, clogs oil lines, fouls intercoolers, and raises general havoc inside the compressor.

These synthetics attack normal gasketing material, and fumes must be directed away from an electric motor, for they can destroy electrical insulation. Never use a phosphate-ester synthetic lubricant without first checking with the compressor manufacturer.

Diester-based lubricants, while not offering quite the same explosion protection as phosphate esters, have many other properties that are advantageous, particularly a long service life between oil changes. Then, there are the so-called "super oils," which are petroleum oils with additives. They provide somewhat longer service life than standard petroleum oils, at somewhat lower cost than the synthetics.

## Which type of oil is best?

The next task is selecting the right oil for your needs.

Consider first the centrifugal compressor. This probably presents the least demanding lubrication problem. Remember, since the centrifugal air compressor is a nonlubricated machine, there is no oil in the air compression chambers; lubrication is confined to the driving gear.

Here the requirement is for good maintenance of film strength under load, with added inhibitors to prevent rust, sludge formation and foaming.

Oil-flooded screw compressors, on the other hand, are very sensitive to both the quality and condition of the lube oil. There are no high-temperature conditions to contend with, but the oil is repeatedly exposed to the air being compressed, plus all the contaminants that pass through the inlet filter. Since a part of the oil that floods the compression chamber is bypassed to go through the bearings, moisture collecting in the unit can be critical to bearing life.

Since operators are not always alert to the dangers of condensation-contamination or oil degradation in an oil-flooded screw compressor, one manufacturer offers an oil-sampling program and urges its use. Small sample bottles of oil from the compressor are mailed to the compressor manufacturer on a regular basis, and an analysis is immediately returned to the user. This gives him adequate warning and enables an operator to schedule a maintenance check before a bearing failure shuts him down.

Because of the very high temperatures incurred in a piston compressor, sometimes 300 to 400°F, this may present the toughest lubrication requirement. Synthetic lubricants and the super oils have been mentioned. Their selection, compared with a conventional petroleum oil, becomes largely a matter of economic choice; that is, are the long-life advantages worth the extra cost?

## Setting the oil feedrate

Cylinder lubricators on piston compressors have adjustable feedrates. The oil droplets can easily be counted as they pass through a sight glass. The question most often raised is, how many drops per minute should be used? Attempts have been made to relate drops per minute to cylinder diameter, piston speed, etc. However, it must be noted that such an approach may be dangerous; the correct oil quantity for one machine may be too much or too little for another.

A new compressor should be broken in with a very heavy feedrate (three or four times normal) for the first 500 hours of operation, using a cylinder oil two SAE grades heavier than normal, until the piston and cylinder have acquired a glaze. This break-in routine should be followed using only petroleum oil, even if there is to be a switch to synthetic cylinder oils for regular operation.

Then the lubricator feedrate should be reduced in small steps, shutting down and inspecting the cylinder bore after a few hours of operation at each step, until the top inside surface has just a slight oil film. Puddles of excess oil lying in the bottom of the cylinder indicate too heavy a feedrate; dry spots on the top are a signal of too little. An experienced operator knows that there is much more art than science in arriving at just the right lubricator setting.

Lubricating oil is the lifeblood of a compressor. Its selection and use should not be left to chance. Manufacturers' publications, such as "Maintenance of Reciprocating Compressors" and the "Compressed Air & Gas Data" book, as well as operating manuals for the particular compressor at hand, should be studied. Then, call in the expert—your oil supplier.

## The author

**Robert G. Winters** is manager of compressor sales for the Air Compressor Group of Ingersoll-Rand Co., Woodcliff Lake, NJ 07675. Mr. Winters joined Ingersoll-Rand in 1950 and has held various positions of increasing responsibility with the company. In 1975, he became regional manager of air power sales, and he was appointed to his present position in 1978. Mr. Winters has a B.S. in mechanical engineering from the University of Pennsylvania and has done graduate work at New York University. He is a member of ASME.

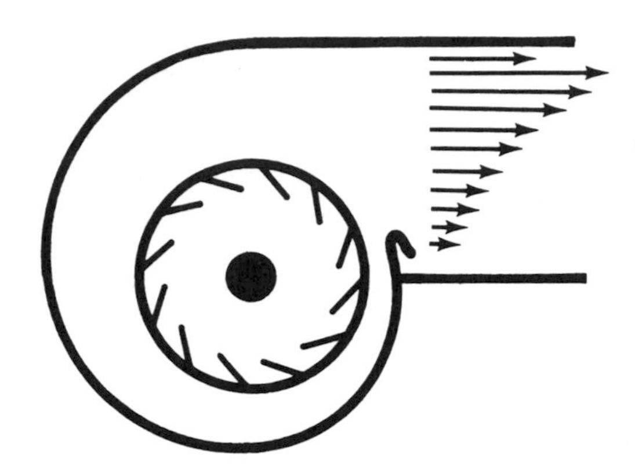

# Selecting

This discussion of available types of fans and blowers, and of the factors that should be considered in their selection, maintenance and installation, should help you choose the most adequate unit for your application.

ROBERT POLLAK, Bechtel, Inc.

Few pieces of equipment have as wide a range of application in the chemical process industries as do fans and blowers. Considering that they have such diverse uses as exhausting or introducing air or other gases into process reactors, dryers, cooling towers and kilns; assisting combustion in furnaces; conveying pneumatically; or simply ventilating for safety and comfort, these machines can well be regarded as basic pieces of equipment.

In the last few years, fan-assisted, air-cooled heat exchangers also have made considerable inroads into the CPI, as engineers have sought to solve thermal water-pollution problems.

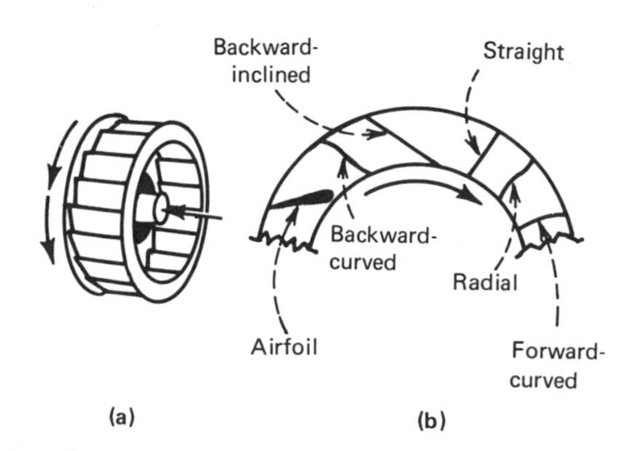


**CENTRIFUGAL FAN:** (a) entering air is turned 90 deg. as it is discharged; (b) blade types—the airfoil kind is the most efficient—Fig. 1

Originally published January 22, 1973

Because of an increasing demand for smaller, more-reliable fans and blowers, and due to the new impetus on occupational health and safety, these machines are now receiving increasing attention. At the same time that user requirements have forced manufacturers to build fans for higher pressures—with resulting higher speeds—environmental considerations have pressed for lower noise levels and shorter noise-exposure times.

Because fan manufacturers are supplying machines at higher compression ratios and at lower and higher flowrates than ever before, an in-depth engineering evaluation of fans or blowers may be justified before selecting one or the other. For this, a basic knowledge of what the various types of fans and blowers can and cannot do is essential.

## Classification of Fans and Blowers

The word *fan* is ordinarily used to describe machines with pressure rises up to about 2 psig. Between this pressure and approximately 10 psig., the name applied to the machine is *blower*. For higher discharge pressures, the term used is *compressor*.

Fans are normally classified as *axial* (where air or gas moves parallel to the axis of rotation), or *centrifugal* (air or gas moves perpendicular to the axis). The National

# Fans and Blowers

Assn. of Fan Manufacturers has established two general categories of axial-flow (AF) fans: *tube-axial* and *vane-axial*.

AF units are usually considered for low-resistance applications because of their ability to move large quantities of air at low pressure.

Centrifugal-flow (CF) fans are used for jobs requiring a greater head, where moving air encounters high frictional resistance. CF fans are classified by blade configuration: *radial, forward-curved, backward-curved* or *inclined,* and *airfoil* (Fig. 1).

Blowers are generally single-stage, high-speed machines, or multi-stage units that operate at pressures close to, or in the range of, compressors (Fig. 2). The term *blower* is also applied to rotary (positive-displacement) compressors that can handle relatively low flows at high compression ratios.

## Characteristics of Axial Fans

Classified into tube-axial and vane-axial types, the characteristics of these machines are as follows:

*Tube-Axial Fans*—Designed for a wide range of volumes at medium pressures, these consist primarily of a propeller enclosed in a cylinder that collects and directs air flow. A helical or screwlike motion is the typical air-discharge pattern (Fig. 3).

*Vane-Axial Fans*—These are characterized by air-guide

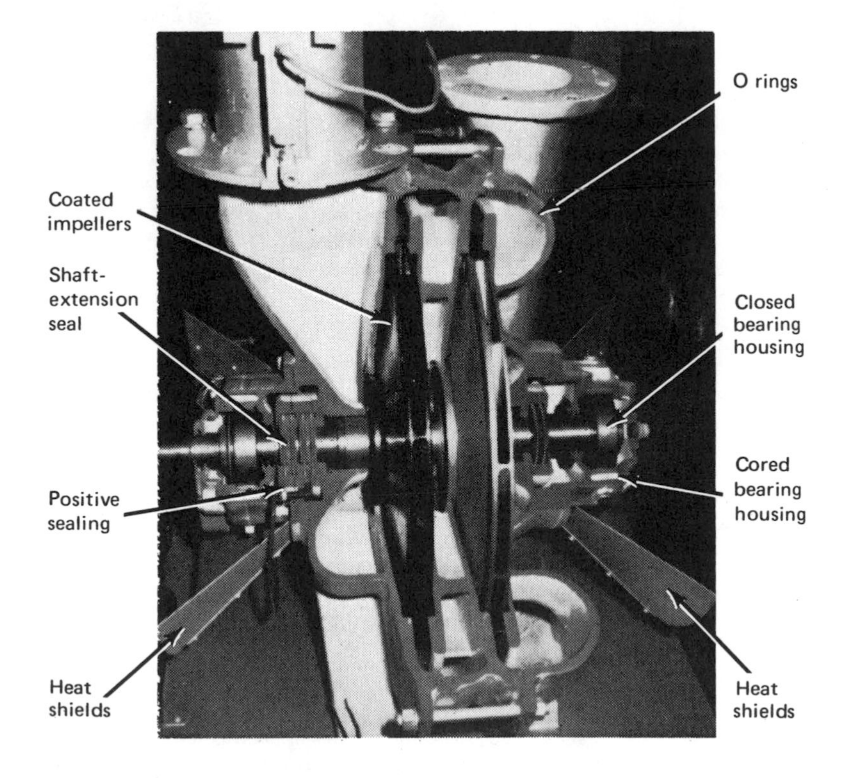


**AIR-TIGHT PRESSURE BLOWER** can handle air, natural gas, organic vapors, helium, nitrogen, etc.—Fig. 2

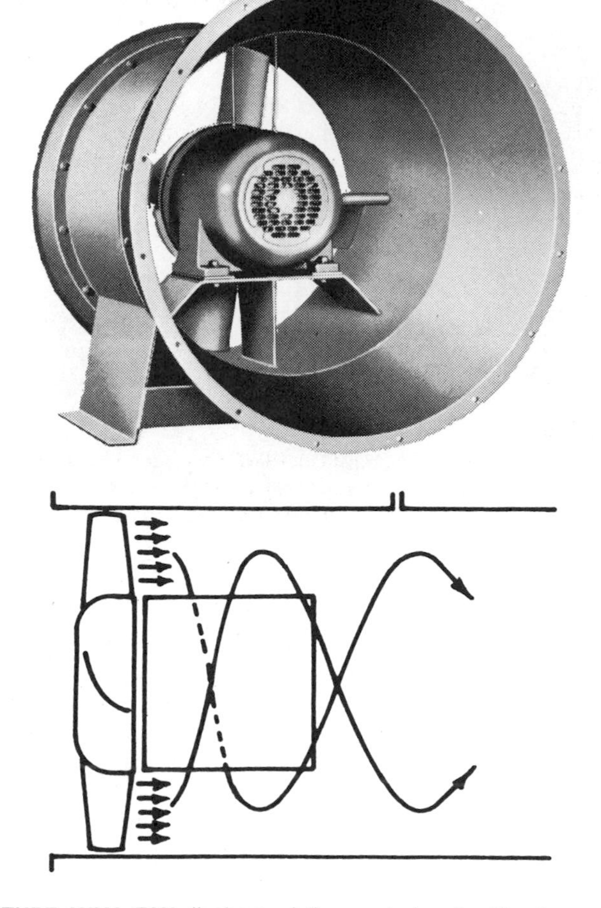

**TUBE-AXIAL FAN** discharge follows spiral path—Fig. 3

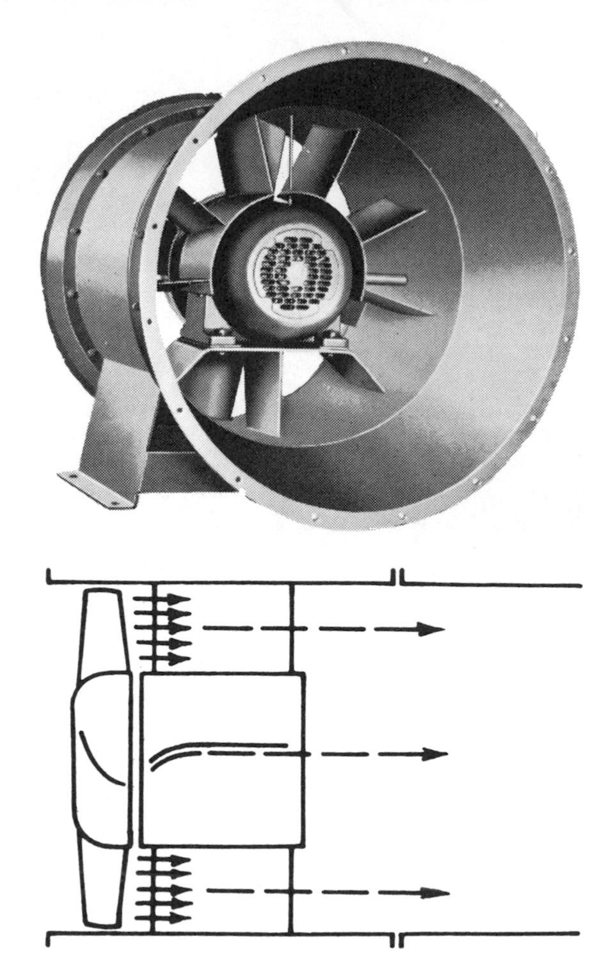

**VANE-AXIAL FAN** discharges straight-line flow—Fig. 4

vanes on the discharge side, which is what differentiates this type of fan from the tube-axial kind. By combining the wheel construction of the tube-axial unit with the directional guides, the air-flow pattern produced is that of a straight line (Fig. 4). Turbulence is thus reduced, which improves efficiency and pressure characteristics.

Vane-axial fans can develop pressures up to 20 in. of water; modified, they can go even higher. Usually, these fans are nonoverloading; i.e., they can be powered by a driver to take care of any horsepower required. These units are available with adjustable fan-blade pitch, which provides performance variation. In some instances, this design feature enables direct connection of the fan wheel to the motor shaft, thus eliminating certain disadvantages of V-belt drives.

## Centrifugal Fans

As already stated, centrifugal fans are classified as radial-blade, forward-curved, backward-curved or inclined, and airfoil types.

*Radial-Blade Type*—This wheel design performs well in many applications, ranging from pneumatic conveying to exhausting process air or gas in high-resistance systems. The chief feature is flexibility in proportional width construction, which allows the fan to achieve high static pressure with relatively low capacity.

When large-horsepower motors are required, the radial blade is frequently connected at synchronous-motor speed. Usually, the machine also provides stable service, regardless of the percentage of wide-open capacity.

This fan can develop high pressures at high speeds.

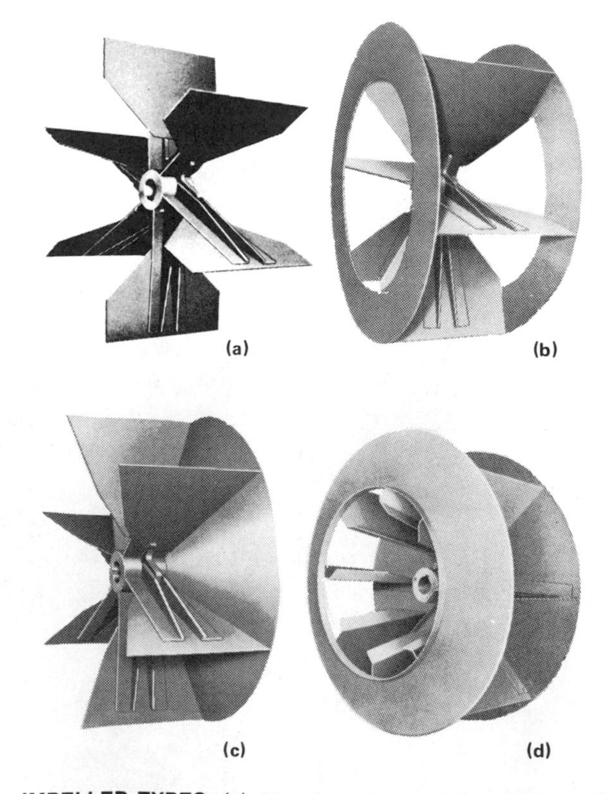


**IMPELLER TYPES:** (a) open type is general-purpose, self-cleaning; (b) type closed on one side is for use with stringy materials; (c) rim type is good for severe duties; (d) back-plate type creates good draft but is not suitable for chunky or fibrous materials—Fig. 5

**FORWARD-CURVED FAN WHEEL** has large-volume capacity at low speed and operates fairly quietly_Fig. 6

**BACKWARD-INCLINED WHEEL** develops much of its energy directly as pressure—Fig. 7

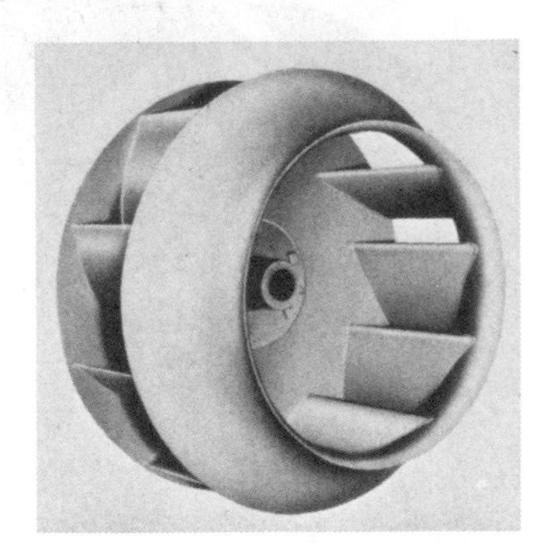

**AIRFOIL FANS** have backward-inclined blades with airfoil cross-section for less air turbulence—Fig. 8.

Blades tend to be self-cleaning, and can be of high structural strength. Typical impeller types are shown in Fig. 5. Normally, the machine is not used for ventilating purposes.

*Forward-Curved Type*—This fan imparts a greater velocity to the air leaving the blade than a backward-inclined blade running at the same tip speed. Although the machine discharges high-velocity air, it runs at slower speeds than the other types, which makes it suitable for process equipment requiring long shafts. The machine operates fairly quietly and requires little space (Fig. 6).

*Backward-Curved and Backward-Inclined Types*—These feature blades that are curved or tilted backward to the optimum angle to develop much of the energy directly as pressure (Fig. 7). This makes the units efficient ventilators.

These fans operate at medium speed, have broad pressure-volume capabilities, and develop less velocity head than forward-curved units of the same size. Another advantage of these backward-inclined fans is that small variations in system volume generally result in small variations of air pressure, which makes the units easy to control.

*Airfoil Centrifugal Fans*—These are backward-curved-blade units that have been given an airfoil cross-section to increase their stability, efficiency and performance. While operating, airfoil fans are also generally quieter, and do not pulsate within their operating range, because the air is able to flow through the wheels with less turbulence (Fig. 8).

*Tubular Centrifugal Fans*—These are enclosed in a duct so that air enters and leaves axially, and all changes in direction of flow are within the fan (Fig. 9). Their design produces a steeply rising pressure over a wide range of capacity (Fig. 10). Being nonoverloading, these fans are good for general building ventilation and air conditioning, as well as for fume removal, humidifying, drying, motor cooling, and supplying combustion air.

## Axial Versus Centrifugal Fans

In general, centrifugal fans are easier to control, more robust in construction, and less noisy than axial units. Their efficiency does not fall off as rapidly at off-design conditions.

Inlet boxes* can sometimes be used without impairing the pressure or efficiency of centrifugal fans, but they are generally not recommended with axial-flow machines. If possible, axial-flow fans should have about two diameters of axial distance upstream and downstream without obstructions or changes in direction.

Centrifugal fans are less affected by miter elbows at

*Devices used to turn the air 90 deg. at the fan inlet in a space close to one diameter in the axial direction.

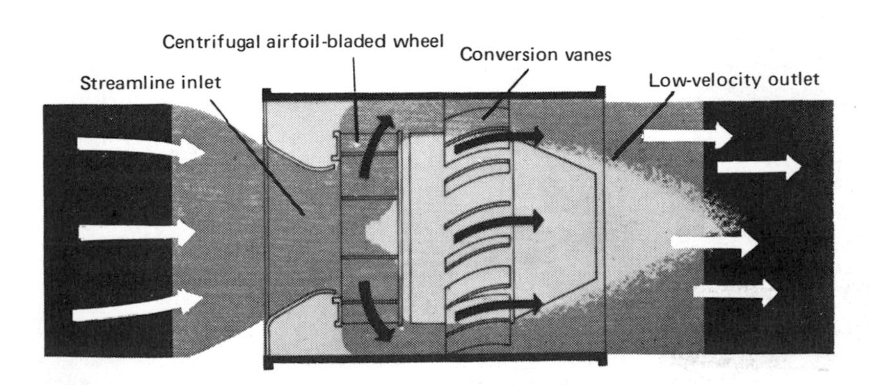


**TUBULAR CENTRIFUGAL FAN** is enclosed in a duct for air to enter and leave axially—Fig. 9

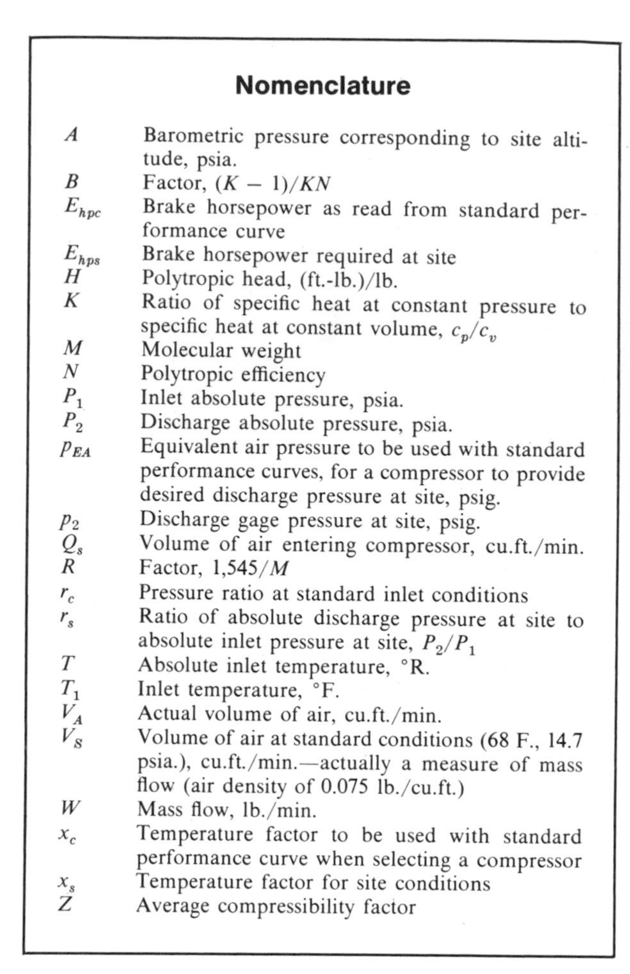

**Nomenclature**

| | |
|---|---|
| $A$ | Barometric pressure corresponding to site altitude, psia. |
| $B$ | Factor, $(K - 1)/KN$ |
| $E_{hpc}$ | Brake horsepower as read from standard performance curve |
| $E_{hps}$ | Brake horsepower required at site |
| $H$ | Polytropic head, (ft.-lb.)/lb. |
| $K$ | Ratio of specific heat at constant pressure to specific heat at constant volume, $c_p/c_v$ |
| $M$ | Molecular weight |
| $N$ | Polytropic efficiency |
| $P_1$ | Inlet absolute pressure, psia. |
| $P_2$ | Discharge absolute pressure, psia. |
| $p_{EA}$ | Equivalent air pressure to be used with standard performance curves, for a compressor to provide desired discharge pressure at site, psig. |
| $p_2$ | Discharge gage pressure at site, psig. |
| $Q_s$ | Volume of air entering compressor, cu.ft./min. |
| $R$ | Factor, $1{,}545/M$ |
| $r_c$ | Pressure ratio at standard inlet conditions |
| $r_s$ | Ratio of absolute discharge pressure at site to absolute inlet pressure at site, $P_2/P_1$ |
| $T$ | Absolute inlet temperature, °R. |
| $T_1$ | Inlet temperature, °F. |
| $V_A$ | Actual volume of air, cu.ft./min. |
| $V_S$ | Volume of air at standard conditions (68 F., 14.7 psia.), cu.ft./min.—actually a measure of mass flow (air density of 0.075 lb./cu.ft.) |
| $W$ | Mass flow, lb./min. |
| $x_c$ | Temperature factor to be used with standard performance curve when selecting a compressor |
| $x_s$ | Temperature factor for site conditions |
| $Z$ | Average compressibility factor |

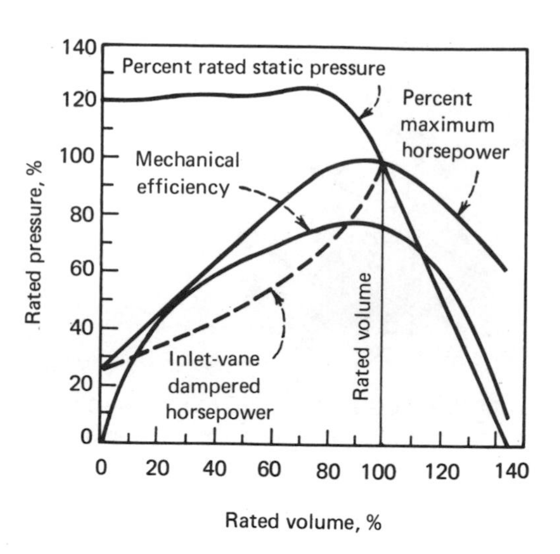


**STEEPLY RISING PRESSURE** is produced by tubular centrifugal fan over wide range of capacity—Fig. 10

the inlet than vane-axial fans, but losses in efficiency up to 15% can be expected when abrupt changes in air-flow direction occur at the fan inlets.

Inlet guide vanes usually provide smooth control down to less than 30% of normal flow, but there have been instances of vibration problems on large, induced-draft and forced-draft fans when their inlet guide vanes have been closed between 30 and 60%.

When high duct velocities are present with a fan equipped with inlet guide vanes, extra consideration should be given to having smooth air-flow patterns in the inlet and outlet ducts, as well as making ducts as strong as necessary to avoid vibration damage. Vibration is aggravated by turbulence and improper inlet-guide-vane setting.*

Axial fans have a narrower operating range at their highest efficiencies (Fig. 11), which makes them less attractive when flow variations are expected. The hump on the axial-fan-performance curve (Fig. 12)—at about 75% of flow—corresponds to the stall point. Operation

*Ref. 5 provides a good general treatment of how fans work.

## Typical Industrial Applications for the Various Types of Fans—Table I

| Application | Type of Fan | | | | | |
|---|---|---|---|---|---|---|
| | Tube-Axial | Vane-Axial | Radial | Forward-Curved | Backward-Inclined | Airfoil |
| Conveying systems | | | X | | X | |
| Supplying air for oil and gas burners or combustion furnaces | X | X | X | X | X | X |
| Boosting gas pressures | | | X | | X | X |
| Ventilating process plants | X | X | | | X | X |
| Boilers, forced-draft | | X | | | X | X |
| Boilers, induced-draft | | | X | X | | |
| Kiln exhaust | | | X | X | | |
| Kiln supply | | X | | | X | X |
| Cooling towers | X | | | | | |
| Dust collectors and electrostatic precipitators | | | X | X | | |
| Process drying | X | X | X | | X | X |
| Reactor off-gases or stack emissions | | | X | X | | |

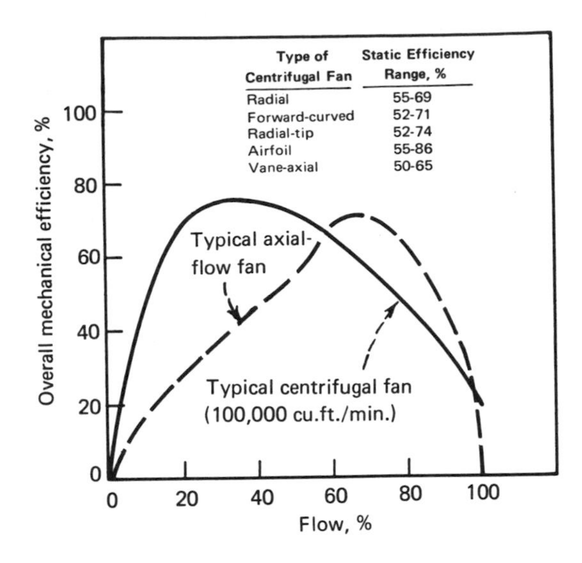


**EFFICIENCY CURVES** for centrifugal and axial fans—Fig. 11

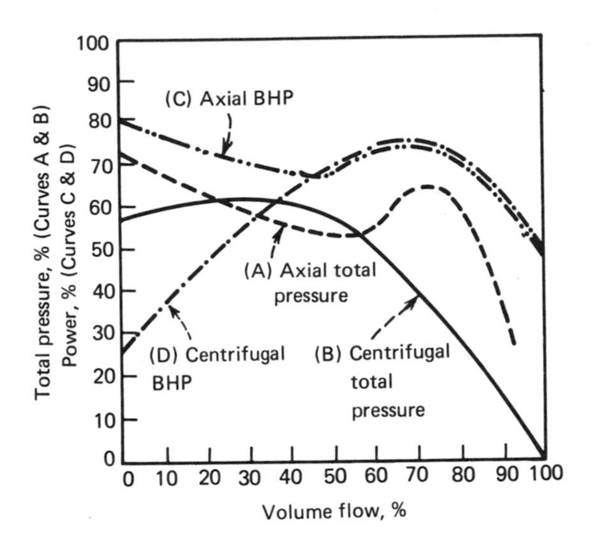


**PERFORMANCE COMPARISON:** total pressure and brake horsepower of axial versus centrifugal fans—Fig. 12

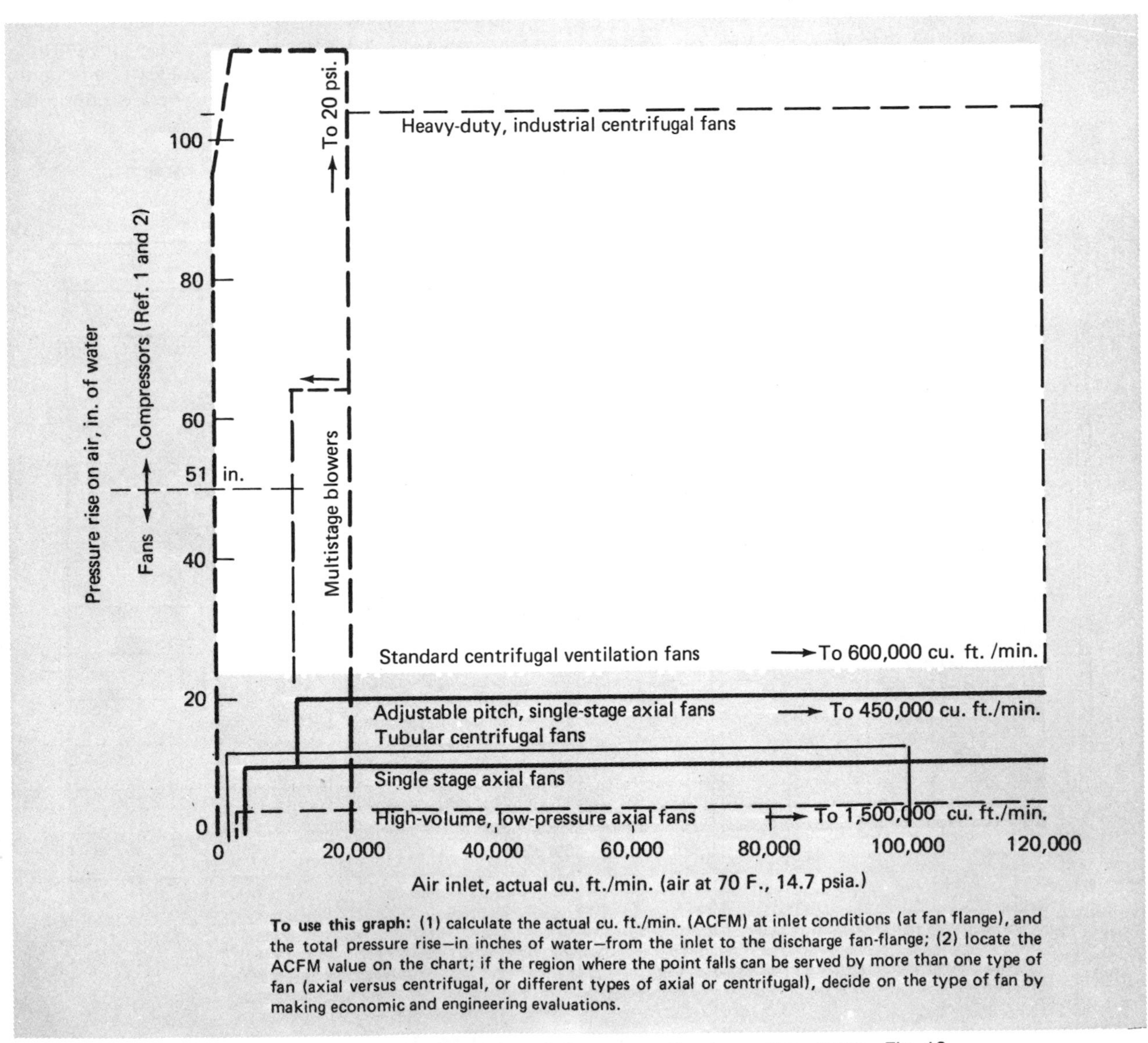


**FAN SELECTION GUIDE,** based on pressure rise versus air flow, according to catalog ratings—Fig. 13

of an axial-flow fan between this point and no-flow is not desirable; performance is difficult to predict.

Fig. 11 also shows the efficiency curve for centrifugal fans (CF). Bear in mind that both these curves are general and are not intended to imply that axial fans are less efficient than centrifugal ones.

Process applications, in general, are more apt to use centrifugal fans, although there is a considerable amount of overlap between centrifugal and axial units at the lower end of the flow-pressure range. A performance comparison of centrifugal versus axial machines is shown in Fig. 12. Table I lists typical applications.

Fig. 13 shows the range of centrifugal and axial machines. This chart is based on catalog ratings. The standard, centrifugal ventilation fans operate up to approximately 22 in. of water. Beyond this, heavy-duty centrifugal fans—with higher compression ratios at some flows—may be made to specifications. The only area where no fan is available is above 100 in. of water at extremely low air flows.

When an application is outside the standard range for fans, it is advisable to consult manufacturers to see if a special heavy-duty unit can be built. At higher pressure, it may be difficult to decide initially whether the process requires a compressor or a fan. When this is the case, it may become necessary to obtain estimated prices from manufacturers of both types of equipment before making a selection.

## Sizing Procedure

To estimate the air-horsepower requirements of fans—when density changes between inlet and outlet can be neglected—the following formula can be used with air:

$$\text{Air hp.} = (144 \times 0.0361)Qh/33{,}000 \qquad (1)$$

where $Q$ = inlet volume, cu.ft./min.; and $h$ = static-pressure rise, in. of water.

For estimating brake horsepower (BHP), an efficiency value—obtained from Fig. 11—can be used with the above formula (efficiency = output-air horsepower/input horsepower). The actual efficiency will depend on the type of fan. The driver horsepower is usually selected so that a power margin of safety of at least 10% exists at the expected operating point, and the required horsepower at any flow is less than the driver horsepower. This permits operation at other-than-design conditions.

Manufacturers' catalogs are usually arranged to show a tabulation of standard cu.ft./min. versus pressure rise across the machine. When air is not at standard conditions, volume, pressure and horsepower corrections must

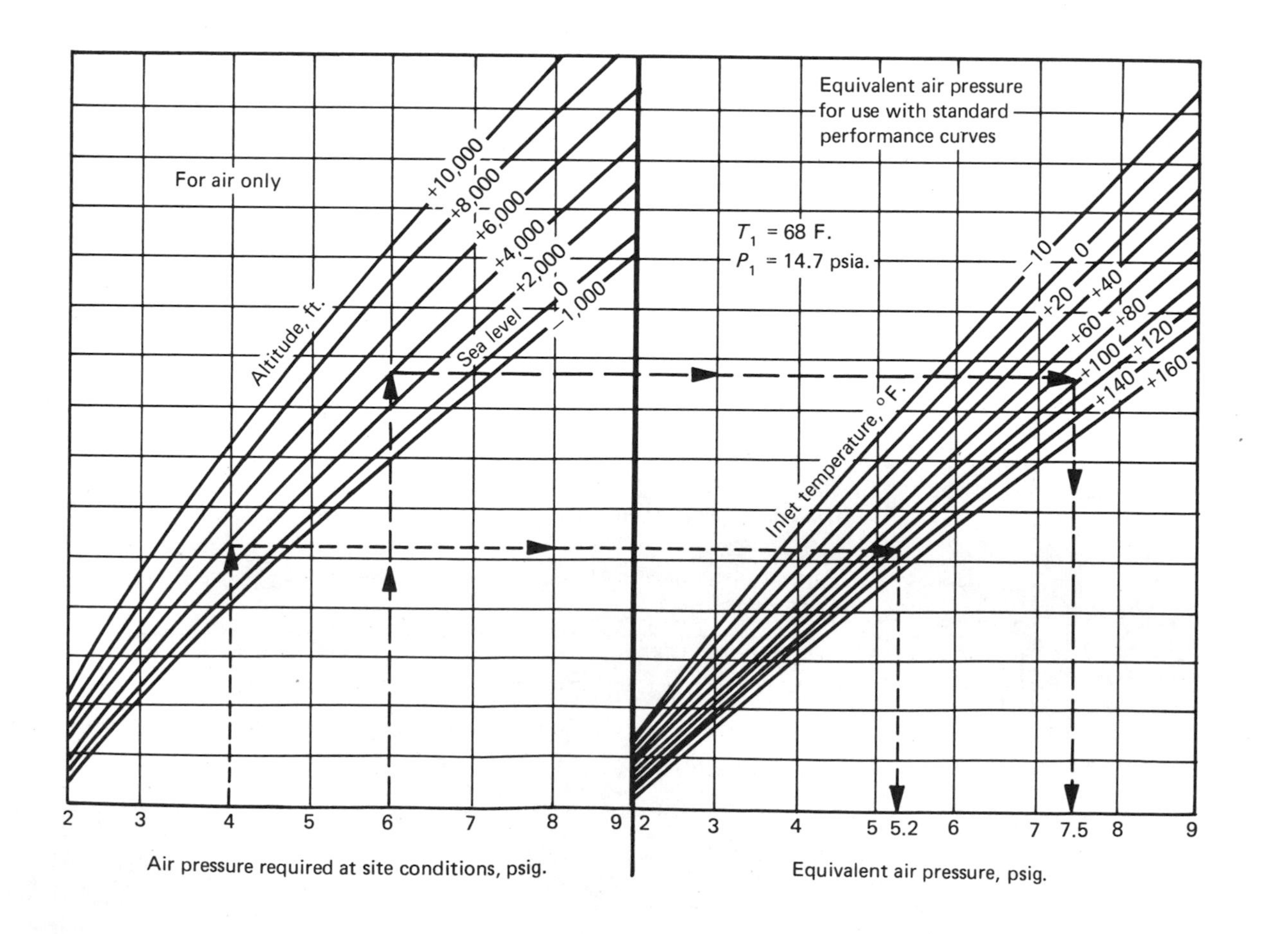


**PRESSURE CORRECTION CURVES** to be used for altitude and inlet temperature of air—Fig. 14

**Value of $x$ for Air Only*—Table II**

| $r$ | 0 | 1 | 2 | 3 | 4 | 5 | 6 | 7 | 8 | 9 |
|---|---|---|---|---|---|---|---|---|---|---|
| 1.0 | 0.0000 | 0.0028 | 0.0056 | 0.0084 | 0.0112 | 0.0139 | 0.0166 | 0.0193 | 0.0220 | 0.0247 |
| 1.1 | 0.0273 | 0.0300 | 0.0326 | 0.0352 | 0.0378 | 0.0404 | 0.0429 | 0.0454 | 0.0480 | 0.0505 |
| 1.2 | 0.0530 | 0.0554 | 0.0579 | 0.0603 | 0.0628 | 0.0652 | 0.0676 | 0.0700 | 0.0724 | 0.0747 |
| 1.3 | 0.0771 | 0.0794 | 0.0817 | 0.0841 | 0.0864 | 0.0886 | 0.0909 | 0.0932 | 0.0954 | 0.0977 |
| 1.4 | 0.0999 | 0.1021 | 0.1043 | 0.1065 | 0.1087 | 0.1109 | 0.1130 | 0.1152 | 0.1173 | 0.1195 |
| 1.5 | 0.1216 | 0.1237 | 0.1258 | 0.1279 | 0.1300 | 0.1321 | 0.1341 | 0.1362 | 0.1382 | 0.1402 |
| 1.6 | 0.1423 | 0.1443 | 0.1463 | 0.1483 | 0.1503 | 0.1523 | 0.1542 | 0.1562 | 0.1581 | 0.1601 |
| 1.7 | 0.1620 | 0.1640 | 0.1659 | 0.1678 | 0.1697 | 0.1716 | 0.1735 | 0.1754 | 0.1773 | 0.1791 |
| 1.8 | 0.1810 | 0.1828 | 0.1847 | 0.1865 | 0.1884 | 0.1902 | 0.1920 | 0.1938 | 0.1956 | 0.1974 |
| 1.9 | 0.1992 | 0.2010 | 0.2028 | 0.2045 | 0.2063 | 0.2080 | 0.2098 | 0.2115 | 0.2133 | 0.2150 |
| 2.0 | 0.2167 | 0.2184 | 0.2202 | 0.2219 | 0.2236 | 0.2253 | 0.2269 | 0.2286 | 0.2203 | 0.2320 |

*The table is used as in these examples: if $r$ = 1.00, $x$ = 0.0000; if $r$ = 1.01, $x$ = 0.0028; if $r$ = 1.86, $x$ = 0.1920.

be applied to be able to select a machine at an "equivalent" volume and pressure.

Make the following corrections when inlet conditions are not the standard 68 F. and 14.7 psia.:

### Volume Correction

$$Q_s = (14.7/A)\left(\frac{460 + T_1}{528}\right) V_{SCFM} \qquad (2)$$

### Pressure Correction

Method A: Use Fig. 14

Method B: $r_s = (A + P_2)/A$

$$x_s = r_s^{0.283} - 1 \text{ (see Table II for value of } x)$$

$$x_c = x_s\left(\frac{T_1 + 460}{528}\right)$$

$$r_c = (x_c + 1)^{3.53} \text{ (see Table II for value of } r_c)$$

$$p_{EA} = 14.7\ (r_c - 1) \qquad (3)$$

### Horsepower Correction

$$HP_s = \left(\frac{A}{14.7}\right)\left(\frac{528}{460 + T_1}\right) HP_c \qquad (4)$$

When making calculations, bear in mind the following:

1. Make appropriate substitutions in Eq. (2) through (4) if manufacturers' catalogs have been prepared for conditions other than the standard 68 F. and 14.7 psia.

2. When an approximate value of the equivalent air pressure ($p_{EA}$) is needed, enter Fig. 14 on the left-hand graph at the proper pressure, and read up to the corresponding site elevation. From this point, draw a line to the maximum inlet temperature expected (right-hand graph) and proceed down from this intersection to the equivalent air pressure on the X axis. For instance, to obtain 6.0 psig. with inlet conditions of 4,000-ft. altitude and 100 F., a blower must be selected that will develop 7.5 psig. under standard conditions.

3. For an accurate $p_{EA}$ value, use pressure-correction Method B above—for Eq. (3)—with $x$ factors shown in Table II.

4. The brake horsepower needed for job-site conditions is determined from standard performance curves.

5. For gases other than air, use Eq. (5) to calculate head, and select a compressor that will develop this same head *on air*. Brake horsepower may then be calculated by means of Eq. (6). For these applications, it is advisable to consult a manufacturer's representative.

## Sample Calculations

*Example 1*—Calculate the brake horsepower required for these conditions: suction flow = 10,000 standard cu.ft./min.; $P_1$ = 12.7 psia. at 4,000-ft. elevation; $p_2$ = 4 psig.; $T_1$ = 120 F.

$$V_A = \left(\frac{14.7}{12.7}\right)\left(\frac{460 + 120}{528}\right) = 12{,}700 \text{ cu.ft./min.}$$

$$r_s = 16.7/12.7 = 1.318$$

$$x_s = 1.318^{0.283} - 1 = 0.805$$

$$x_c = 0.0805\left(\frac{580}{528}\right) = 0.0884$$

$$r_c = (1 + 0.0884)^{3.53} = 1.35$$

$$p_{EA} = 14.7 \times 0.353 = 5.2 \text{ psi. (check value with Fig. 14)}$$

Entering catalog rating tables with a 5.2-psi. pressure rise, and 12,700 ACFM (actual cu.ft./min.):

$$E_{hps} = \left(\frac{12.7}{14.7}\right)\left(\frac{528}{580}\right) E_{hpc} = (1/1.27)\ E_{hpc}$$

Although, in general, fan-manufacturers' representatives should be contacted when sizing for gases other than air, the procedure that follows may be used to estimate equivalent fan horsepower and flow. Fig. 15, which is used in this procedure, was prepared by means of the equation for polytropic head (similar to column height in liquids), which applies to given speeds and inlet flows, regardless of the type of gas:

$$H = \frac{ZRT(P_2/P_1)^B - 1}{B} \qquad (5)$$

Fig. 15 can be used with little error for efficiencies between 0.60 and 0.80. Note that at lower compression ratios, air (or gas) compressibility can be neglected.

To determine the horsepower required by a fan, Eq. (6) can be used:

$$E_{hps} = \frac{HW}{33{,}000\ N} \qquad (6)$$

Although $N$ in this equation is polytropic efficiency,

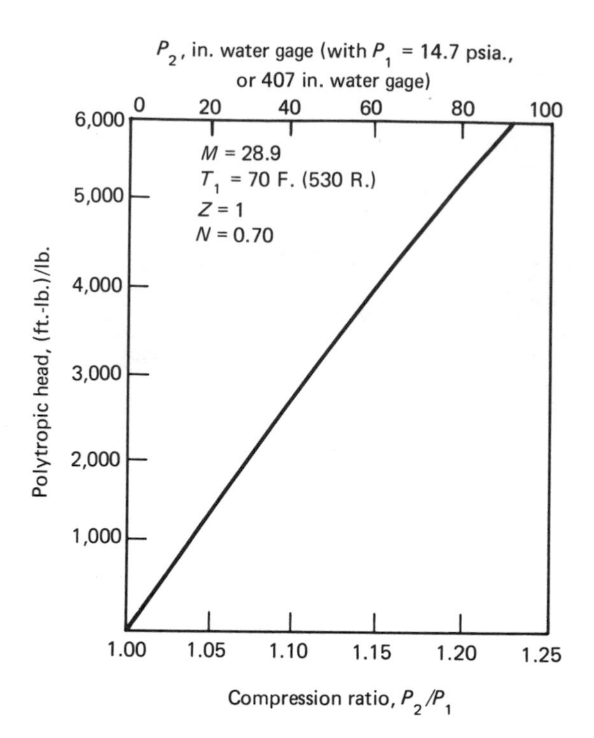


**POLYTROPIC HEAD** versus compression ratio—Fig. 15

static efficiencies may be used as first approximations.

*Example 2*—Calculate the brake horsepower ($E_{hps}$) required and the mass flow ($W$) attainable for a dry carbon dioxide system, for a fan whose air-handling characteristics are: $P_1 = 14.7$ psia.; $T_1 = 70$ F.; actual cu.ft./min. ($V_A$) = 26,000; discharge pressure = 18 in. water gage; brake horsepower for air ($E_{hps}$) = 103 (assume 98 hp. for air + 5 hp. for bearing losses); speed = 960 rpm. Pertinent data for the $CO_2$ system are: $K = 1.3$; molecular weight ($M$) = 44; $T_1 = 100$ F.

$$18 \text{ in. water gage} = 18 \times 0.03613 = 0.65 \text{ psi.}$$

$$P_2 = 14.7 + 0.65 = 15.35 \text{ psia.}$$

$$P_2/P_1 = 15.35/14.7 = 1.045$$

$$\text{At } R = 1.045, \quad H = 1{,}260 \text{ (ft.-lb.)/lb.} \quad \text{(from Fig. 15)}$$

$$W_{air} = \frac{26{,}000 \times 144 \times 14.7 \times 28.9}{1{,}545 \times 530} = 1{,}940 \text{ lb./min.}$$

$$N = \frac{1{,}940 \times 1{,}260}{33{,}000 \times 98} = 0.756$$

With carbon dioxide at a head of 1,260 (ft.-lb.)/lb., the $B$ factor is:

$$B = \left(\frac{K-1}{K}\right)\left(\frac{1}{N}\right) = \left(\frac{1.3-1}{1.3}\right)\left(\frac{1}{0.756}\right) = 0.305$$

And the pressure ratio:

$$(P_2/P_1)^B = 1 + \left(\frac{BH}{RT}\right) = 1 + \frac{0.305 \times 1{,}260}{(1{,}545/44)(560)} = 1.196$$

$$\frac{P_2}{P_1} = \sqrt[B]{1.196} = 1.80$$

Therefore, at 26,000 actual cu.ft./min., and 100 F., the mass flow ($W$) and the brake horsepower ($E_{hps}$) for the $CO_2$ system are:

$$W = \frac{26{,}000 \times 144 \times 14.7 \times 44}{1{,}545 \times 560} = 2{,}800 \text{ lb. } CO_2\text{/min.}$$

$$E_{hps} = \frac{2{,}800 \times 1{,}260}{33{,}000 \times 0.756} = 141.2 \text{ hp. (plus 5 hp. for bearing losses)}$$

Checking the discharge temperature:

$$T_2 = T_1\left(\frac{P_2}{P_1}\right)^B = 560 \times 1.196 = 670 \text{ deg. R. (210 F.)}$$

Before the fan in this example is used on carbon dioxide, the fan manufacturer must be consulted to determine whether the equipment is suitable for the new service. He could suggest changing the speed or restricting the flow to bring down the required power.

A lower speed would reduce the pressure ratio produced by the machine (fan laws may be used to estimate the new performance). The flow would have to be restricted within the stable flow for the fan, and a new performance curve obtained from the manufacturer.

Ordinarily, fans are not switched from one service to another, but the methods outlined above can be used for estimating required power and—using general vendor literature—selecting a fan size.

## Specifications, Data Sheets

An essential part of correct sizing is an accurate definition of operating conditions and requirements. When a fan is to be purchased, the usual procedure is to issue a data sheet and specifications to fan manufacturers. The data sheet should contain not only information to enable the manufacturer to size the fan, but also a list of necessary accessories; and sufficient space should be provided to enter data supplied by the manufacturer. This aids in evaluating the mechanical and aerodynamic characteristics of the fan. A typical data sheet includes the items listed in Table III.

Gas characteristics and operating conditions must be defined as accurately as possible. Included should be the widest expected range of gas components, pressures and temperatures. For instance, a forced-draft fan for a boiler in northern Canada may have to draw air at temperatures from −50 to +90 F. It may therefore be necessary to drive this fan with a motor that is nonoverloading at any air-inlet temperature.

If fans are to be used in outside, unprotected areas, the motor driver and other electric and control equipment should be specified with enclosures suitable for the environment (such as a totally enclosed, fan-cooled motor). The fan itself can be protected with paint.

One should also keep in mind that the Air Moving and Conditioning Assn. (AMCA)[4] has standardized fan and blower designations for spark-resistant construction, wheel diameters, outlet areas, sizes, drive arrangements, inlet-box positions, rotation and discharge, motor positions and operating limits.

All the foregoing points are covered in AMCA Standards 2401 through 2410. Reference to these standards permits specifying fan characteristics in a precise manner. Process plants generally use Class IV construction, which covers fans for greater than 12.25 in. water-gage total-pressure-rise.

## Investment Costs

Costs of centrifugal fans are difficult to estimate accurately because of the many different fan types, classes and arrangements available. Recent purchases indicate about $30 to $40/brake horsepower (BHP) for fans of about 50,000 cu.ft./min., 45-in. water-gage pressure (500 BHP), and about $60/BHP for fans of 25,000 cu.ft./min., 40-in. water-gage pressure (250 BHP).

The costs just mentioned include: fan; totally-enclosed, fan-cooled (TEFC) motor drive; coupling; coupling guard; baseplate mounting for the small units; and standard materials of construction. The costs do not include starters, accessories or controls. A generalized guide for fan costs—up to about 20 in. water-gage pressure and 1,000 BHP—can be found using a nomograph prepared by J. R. F. Alonso.[6]

The price of small, high-pressure, single-stage fans (up to 100 cu.ft./min.) is comparatively high. For example, a 65-cu.ft./min. fan at 14-in. water gage recently cost over $1,000/BHP. Blowers, in general, range from $50 to $60/BHP. When stainless steel construction is required by the process, prices may be two to three times as high as those for standard materials.

## Drives and Couplings

Whenever a fan is to be driven by a turbine or other variable-speed device, it is important to ascertain that the integrity of the rotating parts is assured up to the trip speed of the driver. In the case of steam turbines, trip speed is about 10 to 15% above normal running speed. It is advisable to include in the specifications a rotating assembly test at the trip speed. Advantages and disadvantages of several variable-speed drives for fans are listed in Table IV.

### Pros and Cons of Variable-Speed Drives for Fans—Table IV

| Advantages | Disadvantages |
|---|---|
| **Direct-Current Motor** | |
| Wide range of adjustable, stepless, speed variation. | High initial cost; requires a.c.-to-d.c. conversion equipment; presents maintenance and installation problems. |
| **A.C. Variable-Speed Motor** | |
| All the advantages of d.c. variable-speed drives; many do not have commutators or brushes to maintain. | High initial cost. |
| **Two-Speed A.C. Motor** | |
| Simple speed change. | Limited choice of only two speeds—in which category are included single-winding, consequent-pole machines (with a 2:1 speed ratio), and pole amplitude-modulated motors, which have a speed ratio of 3:2 to 3:1. |
| **Hydraulic Drives** | |
| Low-cost; simple; they allow motor to start against low torque; generally trouble-free. | Inefficient at other than full speed; some hydraulic clutches are difficult to control near the full-speed position; an auxiliary lube-oil system is required. |

### Information That Should Be Provided on a Data Sheet—Table III

**Gas Characteristics**
- Composition
- Molecular weight
- Flow required

**Operating Conditions and Characteristics**
- Suction pressure and temperature
- Discharge pressure and temperature
- Required power
- Speed of fan
- Rotation of fan
- Diameter of impeller
- Number of stages
- Type of fan
- Starting torque
- Moment of inertia

**Bearings and Lubrication**
- Type of bearings (radial and thrust)
- Lubrication system and recommended lubricant

**Connections**
- Size and rating
- Location
- Drain connections

**Accessories Required**
- Driver (motor, steam turbine, hydraulic turbine, other)
- Coupling information (supplier, type, size, etc.)
- Gear required
- Control (dampers, inlet guide-vanes, variable-speed drive, variable-pitch blades—axial, actuators)
- Safety devices (pressure, temperature, vibration)
- Filters (inlet), or screens
- Cleanout holes
- Noise-attenuation equipment and lagging

**Construction and Materials Specifications**
- For case, impeller, shaft and other parts
- Type of seals
- Shaft diameter

**Testing and Miscellaneous**
- Required testing
- Inspection
- Witnessing tests
- Test driver
- Weights

When a gear is used between the fan and the driver, a torsional analysis of the entire train (including drive, couplings, gear and fan) should be made. This analysis can be done by the seller of the fan or driver, and should be purchased with the unit.

The American Gear Mfrs. Assn. (AGMA)[8] recommends that a service factor be used with gears. The following factors are usually applied to the power capability of the driver to obtain the rated horsepower of the gear unit:

| Type of Fan | Motor | Turbine | Internal Combustion Engine (Multi-Cylinder) |
|---|---|---|---|
| Centrifugal units, including blowers and forced-draft fans | 1.4 | 1.6 | 1.7 |
| Induced-draft fans | 1.7 | 2.0 | 2.2 |
| Industrial and mine fans | 1.7 | 2.0 | 2.2 |

Gear losses of about 2 or 5%—depending on the type and quality of the gear unit—are added to the power required from the driver.

Accessories for the gear, depending on its size, may be bearing-temperature gages, vibration detectors, type of thrust bearing (tapered-land,* tilting-pad, antifriction, shoulder, etc.), and a type of lube-oil system. For fans

*Defined in Ref. 1, pp. 8-170, 8-171.

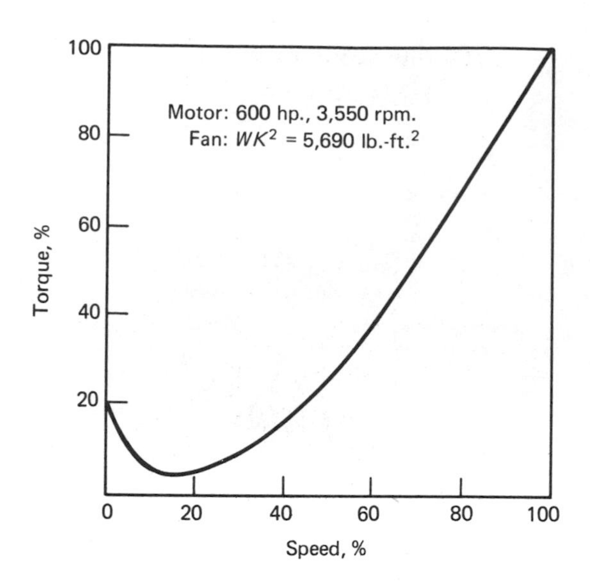


**SPEED-VERSUS-TORQUE CURVE** to estimate fan horsepower of air, or gas, flow—Fig. 16

in refinery applications, American Petroleum Institute Standard 613 can be applied.[9]

In addition to the above considerations, a decision as to whether to purchase the driver separately or with the fan must be made. If the fan to be purchased is large, it is advisable to buy the motor driver with the fan, to avoid the coordination problems encountered in the selection of the motor and coupling.

The fan manufacturer must determine an expected speed-torque curve (Fig. 16), as well as the moment of inertia of the fan. This will enable him to select a motor to suit the electrical and area classification of the application. The coupling, which must suit both the fan and driver shaft, should be supplied by the fan vendor.

Other items subject to coordination are sole plates, or a baseplate under fan and motor, and coupling guard.

Some of the complexities of coordination are illustrated by this actual example:

An industrial, forced-draft fan, for approximately 150,000 cu.ft./min. at 38-in. water gage, and requiring 1,300 BHP on air, was to have a dual drive (motor and turbine), with one-way clutches so that either motor or turbine could be serviced with the fan running.

The fan and motor were to be purchased overseas from different manufacturers; the turbine, gear, clutches and coupling in the U.S., through the turbine manufacturer. With so many vendors involved, none could be held responsible for the unit, with the result that the fan could not be run with either motor or turbine in the factory.

It is much simpler, and probably less expensive overall, to purchase all the equipment through one vendor. This is especially desirable when problems are encountered in the field, and responsibility for repairs is difficult to pinpoint.

A good treatment of fan motors, and how moment of inertia ($WR^2$), weight of the fan, and other factors affect motor selection can be found in Ref. 7.

## Fan Controls

The throughput of centrifugal or axial fans may be changed by varying the speed of the fan, or by changing pressure conditions at the inlet and/or outlet with dampers or with inlet guide-vanes. Axial-flow fans may be controlled also by varying the pitch of the blades.

Of these methods, the most efficient is changing the speed. Since, however, this feature is not generally available—because fans are ordinarily driven by constant-speed motors—other means of varying flow must be resorted to. The next best way of accomplishing this is by means of variable inlet guide-vanes, which must be purchased with the fan.

The most common of the controls used with constant-speed centrifugal fans is the inlet damper. As the damper closes and reduces inlet pressure, the pressure ratio across the machine increases, so that the operating point on the fan curve moves in the direction of lower flow.

Sometimes, the extra pressure drop is taken by a discharge damper, but the power wasted is greater than with inlet-damper control. Partially closed dampers on axial fans may increase power as they decrease flow, in accordance with the general performance characteristics of the axial fan.

Surge, which is a condition of unstable flow in dynamic-type compressors, can also occur in fans. This happens at less-than-normal flowrates, when the fan (or compressor) can no longer develop the required pressure ratio. On fans or blowers of more than about 2 psi. (55 in. of water) and 150 BHP, surge can be damaging. Some type of antisurge control should therefore be considered.

Occasionally, on high-head fans in services other than air, it may be necessary to bypass some of the gas from the discharge back to the suction side, to keep the flow above the minimum needed to avoid surge. This gas must be cooled, and it should be taken from a point in the discharge line upstream of the discharge backflow preventer (if used). On air service, flow can be maintained above the surge point by dumping air to the atmosphere or by bleeding some air into the suction side (on induced-draft fans).

To avoid possible reverse rotation after shutdown, some kind of backflow preventer should be considered on fans exhausting gas from a closed system.

## Vibration

Vibration limits depend on speed. A maximum peak-to-peak amplitude (measured on the bearing caps), as follows, would be classified as "good." Vibrations 2.5 times larger than the following values would be "slightly rough," but still acceptable after some use.

| Rpm. | "Good" Vibration Amplitude, In. |
|---|---|
| 400 | 0.003 |
| 800 | 0.002 |
| 1,200 | 0.0013 |
| 1,800 | 0.0008 |
| 3,600 | 0.0005 |

At the lower speeds—say, less than 800 rpm.—acceptable vibration-amplitude values taken from charts may not be a good criterion. It is then better to limit shaft vibratory velocity to 0.1 in./sec.

Vibration monitoring should be considered for fans in critical service, to provide an automatic warning when the machine's vibration reaches a trouble-indicating level. Fan vibration due to imbalance can be minimized by asking the factory to balance the entire rotating assembly (fan and shaft). With the larger fans, the impeller may be shipped disassembled to the user. Fan manufacturers should be responsible for field balancing to a level agreed upon with buyers.

## Coping With Noise

Noise-attenuation equipment must be considered for fans that exceed established noise limits. It is, however, most difficult to specify a fan's maximum noise level.

The sound power generated by a fan depends on the flow, fan-pressure level, and impeller type and configuration. It is not possible to design a quiet fan at high-pressure levels; for fans of 2 to 3 psi., a sound-power level in the range of 110 to 130 db. is not uncommon. Obviously, this type of fan must either be installed in an unmanned area of a plant, or be suitably modified with sound attenuators to bring the noise level within acceptable limits. The Walsh-Healy Act[10] and the Occupational Safety and Health Act (OSHA)[14] specify permissible sound levels in working areas.

To bring the noise level down, silencers, insulation around ducts, and lagging—or a housing around the unit—can be considered. Fan-pressure losses in cylindrical attenuators are generally 2 in. water gage or less.

Silencing equipment can be placed in the inlet or discharge ducts near the fan, or around the fan casing. Fan suppliers can ordinarily furnish data on the sound level generated by a particular fan. These data are usually taken from tests at the factory on typical field installations of similar fans.

When rated under operating conditions, inlet and discharge silencers usually provide the required noise attenuation. They are made for insertion in round or rectangular ducts, in standard or special materials, and with special acoustic fills for corrosive atmospheres.*

Sometimes, when an application is in the range of a compressor manufacturer's equipment, the cost of a fan plus associated noise-reducing accessories can be less than that of a compressor that does not exceed the specified maximum sound level.

## Flange Loading, Shaft Seals

Fan manufacturers generally require no load to be transmitted to the fan casing due to attached ducts. This is desirable, but when it is unavoidable to impose loads—due to thermal expansion or weight—it may be possible to strengthen the fan casing to avoid distortion and misalignment.

Concerning leakage, a certain amount is usually tolerable on fans and shaft seals because a prime consideration in selecting fans is ease of seal replacement. Seals can be made of felt, rubber, asbestos or other packing.

* To estimate noise levels, consult Ref. 11 and 12.

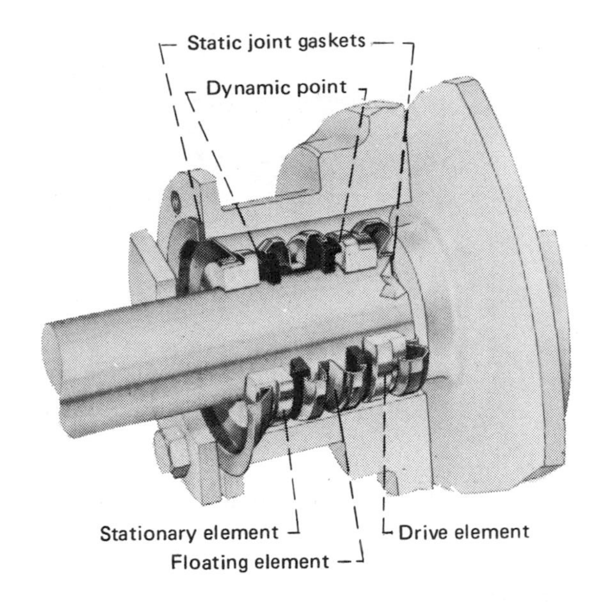


**CONTACT-TYPE SEAL** holds all seal faces in constant contact with shaft to prevent leakage—Fig. 17

When leakage cannot be tolerated, a contact-type seal may be considered. One of these (Fig. 17) has a centrally located, annularly compensating feature, preloaded to hold all seal faces in constant contact. This seal is claimed to be suitable for liquids, gases, vapors and fine solids in the chemical, petroleum, pharmaceutical and food industries.

## Systems Analysis

For a fan to function properly, one must understand the effects of the fan system on the fan itself; otherwise, neither the system nor the fan will work well. A fan system consists of the whole air path—usually a combination of pipes or ducts, coils, filter, flanges and other equipment.

In a fixed system, the volume flowrate in cu.ft./min. will have an associated pressure loss, which is caused by the system's resistance. Head loss for a fan system is calculated similarly to the head loss for flow of fluids in a process piping system. First, the complex system is broken down into its component parts, with known pressure-drop values. The summation of all these resistances yields the total resistance of the system.

A system's total resistance would include the resistance in the main duct to the fan inlet; the one in the main duct from the fan discharge to the end of the duct; and the ones in branch pipes (or ducts), filters, dust collectors, grilles, or other pieces of equipment. A novice tackling an extensive project would do well to consult a specialist in the field.

In a typical fan-system curve (Fig. 18), the static pressure ($P_s$) of the system is a parabolic function. The point of operation (PO) is located at the intersection of the fan static pressure and the system's $P_s$.

Occasionally, fans operating at other than the design PO are unstable and cause pulsation. This can damage

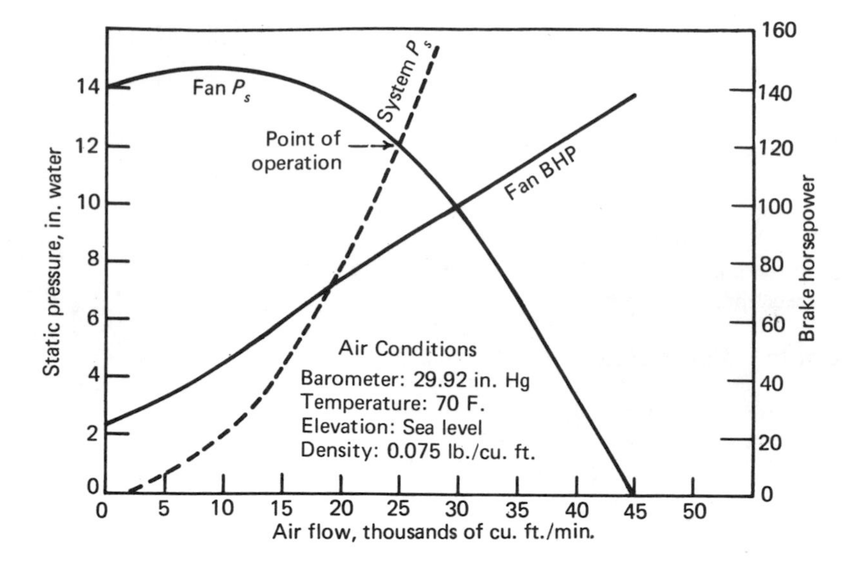


**FAN-SYSTEM CURVE** locates operating point at intersection of fan's static pressure and system's $P_s$-Fig. 18

the fan, the system or both. To overcome the problem, a fan should be selected so that its PO always falls in the stable range—i.e., in the down-sloping portion of the flow-versus-pressure-rise curve, and preferably at some flow that corresponds to only one pressure-rise point. On Fig. 18, for instance, this corresponds to flows exceeding 17,500 cu.ft./min.

Another important factor in a system's design is the choice of fan blade. For example, according to Fig. 19, there is less likelihood of particulate buildup on the blade if a forward-curved blade is used, but there is a tradeoff in fan stability. The backward-inclined-blade fan is inherently more stable; forward-curved blades must be carefully matched to the duct system.

## Materials of Construction

The materials of construction and the types of seals used in a fan depend on the composition of the gas handled. Standard materials include cast iron and carbon steel for casings; aluminum and carbon steel for impellers; and carbon steel for shafts. In some cases, other materials may be required. For instance, if the fan is required to move a wet mixture of ammonia, carbon dioxide and air, stainless-steel (304 or 316) construction for all parts in contact with the gas may be necessary.

Plastics reinforced with fiber glass (FRP) are now also accepted materials of construction for fans and blowers, even though FRP units have pressure limitations. For example, a $7\frac{1}{2}$-in.-dia. fan at 120,000 standard cu.ft./min. was built for only a 2-in. maximum static pressure.

With backward-inclined blades, FRP fans can handle flows of 65,000 cu.ft./min. at 3-in. static pressure (8,200 ft./min. tip speed). With special supports and reinforcement, FRP fans with radial blades can handle pressures up to 20 in. of water, at flowrates up to 45,000 cu.ft./min. (16,500 ft./min. tip speed).

Corrosion resistance can be enhanced with special coating materials, which are often readily available from fan manufacturers at lower costs than special materials. However, the successful application of coatings depends largely on experience. A coating's suitability is usually proven by its previous use in a similar service.

Coatings are generally classified as *air dry*—such as special paints, asphalt, epoxy, air-dry phenolic, vinyl, silicone, or inorganic zinc—and *baked*—such as polyester (with or without fiber-glass reinforcement), baked polyvinyl chloride, baked epoxy, and baked phenolic.

Whenever a coating is specified, its extent and thickness must be clearly indicated. The surface preparation and method of application must be in accordance with recommendations of the coating manufacturer. Generally, sand-blasted or shot-blasted surfaces may be included in the price quoted by the fan manufacturer, but special preparations are not.

Coating the entire inside and outside surfaces may be impossible with baked coatings. In some instances, just coating the airstream surfaces may be satisfactory, and much less costly. For example, in an application for a blower to compress 100 actual cu.ft./min. of ammonia and hydrogen sulfide from 18 to 21 psia., the blower manufacturer recommended lining the internal parts only with Heresite (Heresite and Chemical Co., Manitowoc, Wis.) at about $1,000 per blower.

Allowable temperatures of coatings should be higher than the expected operating temperatures by an appreciable margin. Rubber, also sometimes used, is limited to about 180 F. The tip speed of wheels lined with rubber is about 13,000 ft./min. (or lower for thick coats).

As a rule, the upper limit of tip speed for modern, large, industrial fans is approximately 40,000 ft./min. At such a high speed, pressure increases of 25% are attainable on air. Whenever the wheel is coated with any material, it is necessary to use slower speeds. Thus, the pressure ratio of the machines becomes limited.

Linings of epoxy resins, such as Coroline (Ceilcote Co., Berea, Ohio), or polyester resins, such as Flakeline (Ceilcote Co.), are also used successfully to protect fan surfaces from corrosive gases. These and similar coatings can be used up to tip speeds of 20,000 to 28,000 ft./min.

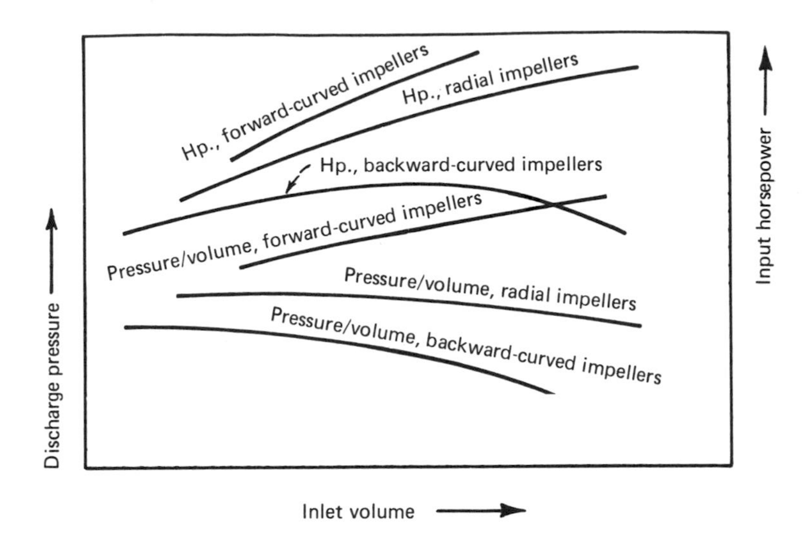


**BLOWER-PERFORMANCE** for different impeller types—Fig. 19

However, if abrasive particles, dust or liquid droplets are present in the gas stream, such coatings may fail. In such instances, fans must be made of satisfactory metals.

To reduce costs, a manufacturer may recommend coating the shaft and some slower-moving portions of the fan with a lining, and using bolted, riveted or sprayed-on metal surfaces on the high-speed portions. Colmonoy 5 (Wall Colmonoy Corp., Detroit, Mich.), Stellite 3 and 5 (Stellite Div., Cabot Corp., Kokomo, Ind.), and Inconel X (International Nickel Co., Huntington, W. Va.) may be used satisfactorily in environments subject to hydrogen sulfide stress-corrosion. When the construction of the fan allows, plates of Inconel X, Hastelloy (Stellite Div., Cabot Corp.) or other suitable material may be bolted or riveted on to minimize erosion.

Fan construction methods are as numerous as fan manufacturers. Casings and impellers may be riveted, welded, cast or bolted. Bearings may be sleeve-type or antifriction and—depending on speed, load and temperature—may be self-lubricated or require a lube system. Minimum life and maximum allowable temperature and antifriction bearings—according to American National Standards Institute standards—should be specified. Generally, 30,000 hr. minimum is a satisfactory life, but for heavy-duty fans 50,000 hr. is a conservative figure. Bearing temperature, as measured internally, should not exceed 180 F.

## Performance Tests

In the factory, fans are tested with open inlets and smooth, long, straight discharge ducts. Since these conditions are seldom duplicated in the field, the result often is reduced efficiency, impaired performance and—in some extreme cases—failure of the fan or overloading of the driver. Although fans are affected more by inlet than by discharge conditions, care should be exercised in both inlet and outlet ducts to provide proper flow patterns.

Axial-flow fans are more prone to be affected by inlet conditions than centrifugal ones. Consider the case of an actual 33-in.-dia., 1,000-rpm., vane-axial fan, which had a 70% efficiency and a total pressure of 1.0-in. water gage, when the inlet was connected to a smooth 90-deg. elbow (outside to inside radius ratio of 2).

At constant flow, when a miter elbow with turning vanes was used instead of the smooth elbow, the efficiency dropped to 54%, and the total pressure to 0.8-in. water gage. With a miter without turning vanes, the efficiency was 45%, and the total pressure 0.6-in. water gage.[13]

Although shop testing a fan can disclose its mechanical integrity or its aerodynamic performance, the ability to perform tests may be governed by the size of the factory test stand. For units in critical service, this may be an important factor in fan selection.

A shop mechanical test—which should last for at least 2 hr. at maximum continuous speed—should certainly be obtained if at all possible. This should be a witnessed test to obtain readings of bearing temperatures, oil flows and vibration amplitudes. For turbine-driven fans, an over-speed test should also be included.

Performance tests are recommended when:

- The fans in question are large centrifugal or axial units, or fans whose design has not been previously manufactured, or is a scaled-up version of an existing fan.
- The quoted efficiency is at the upper end of the scale for the type of fan, and utility costs are high.
- The fan is to be in critical service, and missing the guaranteed operating point by any margin would be costly.

Because testing codes (Ref. 2 and 4) outline only test methods (no penalties are assessed for not meeting performance promises), buyers should specify limits of acceptable performance in their purchase orders.

Field-performance tests are very difficult to make with any degree of accuracy, but if the performance in the field is not satisfactory, the purchaser should have the option of requiring the seller to supervise such a test. Even though factory conditions for measuring pressure, temperature, system air humidity, and power consumed cannot be duplicated in the field, major fan-performance deficiencies can be demonstrated and corrective action initiated. Sufficient instrumentation, or space for its location, should be provided in the layout of fans that may need field testing.

## Installation Guidelines

Installing the fan on a heavy base is essential for a good, long, trouble-free life. Fans mounted on concrete slabs at ground level are ideally placed. If a fan must be mounted on an elevated structure, such as on top of a furnace, extra care in balancing must be taken to avoid shaking the structure. For critical installations, vibration analysis of the entire structure is necessary.

A rule of thumb for fans installed on concrete slabs at grade level is to use a weight of concrete about six times the mass of the rotating elements of the unit.

As the fan is installed on its foundation, shims and sole plates should be used to aid in the alignment of driver and fan (and gear, if used). Alignment is especially critical on induced-draft fans running at elevated temperatures. Here, allowances should be made for movement as the casing, shaft and impeller reach operating temperature. If possible, vibration should be continuously monitored while the unit is heated to its normal running temperature. A gradual increase in vibration is a good indication of poor alignment due to temperature rise.

When ball or roller bearings are used on V-belt-driven shafts, caution should be exercised to prevent excessive preloading of the bearings (which could bend the shaft) while the V belts are tightened. Units that have V-belt drives should have the sheaves mounted with the impeller at the factory, at the time that balancing is done.

If the bearing temperature exceeds 180 F., special lubricants may be used. If, however, the temperature is lower than about −30 F., not only will special lubricants have to be used but antifriction bearing metals may have to be specially processed by the bearing manufacturer.

Safety devices available for fans are the same as those used on centrifugal compressors. When a fan requires a separate lubricating oil system, adequate pressure and temperature protection must be provided to avoid running the fan unit without lubrication (even during coast-down times due to power failure). Vibration switches are recommended on high-speed fans that are in hot or dirty service, as well as on most axial-type units.

A survey of plant startup problems for the last seven years indicates that the incidence of failures attributed to fans and their drives has been very small. The failures on centrifugal fans were minor and were easily resolved. The ones on axial fans were more severe but caused only minor damage to other equipment.

## Acknowledgments

The following companies provided information and/or illustrative material for this report: American Standard, Inc., Industrial Products Div., Detroit, Mich. (Fig. 3, 4, 9, 10); Buffalo Forge Co., Buffalo, N.Y. (Fig. 8); Castle Hills Corp., Piqua, Ohio; Clarkson Industries, Inc., Hoffman Air Systems Div., New York, N.Y. (Fig. 2, 14); Ernest F. Donley's Sons, Inc., Cleveland, Ohio (Fig. 17); Dresser Industries, Inc., Franklin Park, Ill.; Fuller Co., Lehigh Fan & Blower Div., Catasauqua, Pa.; Garden City Fan & Blower Co., Niles, Mich.; Joy Mfg. Co., Pittsburgh, Pa.; Lau, Inc., Lebanon, Ind.; The New York Blower Co., Chicago, Ill. (Fig. 6, 7, 18); Niagara Blower Co., Buffalo, N.Y.; Westinghouse Electric Corp., Westinghouse Sturtevant Div., Boston, Mass.; Zurn Industries, Kalamazoo, Mich. (Fig. 5).

## References

1. "Mark's Standard Handbook for Mechanical Engineers," 7th ed., T. Baumeister, ed., McGraw-Hill, New York (1967).
2. "ASME Standard PTC-11," Test Code for Fans, American Soc. of Mechanical Engineers, New York.
3. "API Standard 617," American Petroleum Institute, New York.
4. "AMCA Standard 210-67," Test Code for Air Moving Devices, Air Moving and Conditioning Assn., Park Ridge, Ill.
5. Fans, A Special Report, *Power,* Mar. 1968.
6. Alonso, J. R. F., Estimating the Costs of Gas-Cleaning Plants, *Chem. Eng.,* Dec. 13, 1971, p. 86.
7. Rajan, S., and Ho, T. T., Large Fan Drives in Cement Plants, *IEEE Transactions IGA,* **Vol IGA-7,** No. 5, Sept.-Oct. 1971.
8. American Gear Mfrs. Assn., Washington, D.C.
9. "API Standard 613," High-Speed, Special-Purpose Gear Units for Refinery Service, 1st ed., American Petroleum Institute, New York (1968).
10. Walsh-Healy Act, *Federal Register,* **Vol. 34,** No. 96, May 20, 1969; revised, Jan. 24, 1970.
11. "ASHRAE Guide and Data Book," Chapter 31, American Soc. of Heating, Refrigerating and Air Conditioning Engineers, New York (1967).
12. Graham, J. B., How To Estimate Fan Noise, *Sound and Vibration,* May 1972.
13. Christie, D., Fan Performance as Affected by Inlet Conditions, *ASHRAE Transactions,* **Vol. 77,** Part 1, 1971, pp. 84-90, American Soc. of Heating, Refrigerating and Air Conditioning Engineers, New York.
14. Occupational Safety and Health Act. *Federal Register,* **Vol. 36,** No. 105, May 29, 1971.
15. "Fan Engineering," Buffalo Forge Co., 1970 ed., Buffalo, N.Y.

## Meet the Author

**Robert Pollak** is engineering specialist with Bechtel, Inc., Refinery and Chemical Div., P. O. Box 3965, San Francisco, CA 94119, where he is engaged in the specification and selection of centrifugal and reciprocating compressors and fans, as well as of their motors, steam turbines and other drivers. A graduate of the University of Illinois, he holds an M.S. degree in mechanical engineering. He is a member of the American Soc. of Mechanical Engineers.

# Selecting and maintaining reciprocating-compressor piston rods

**Here is information on material selection, finish, thread design, piston fit, coatings and rod runout—all important points in rod selection. Also included are tips that will help the user to avoid rod breakage.**

***Jim Messer,*** *Dresser Industries Inc.*

☐ Piston rods, a critical part of any reciprocating compressor, must be designed for maximum safety and long life. (Some engineers believe that such rods are intentionally designed to be a "weak link" in the compressor, but this is untrue.)

Here are some considerations for selecting and maintaining reciprocating-compressor piston rods.

## Material selection for gas atmospheres

A piston rod must be compatible with the gas atmosphere in which it will perform. In addition, the rod has to be designed to resist wear, so a compromise is often made that is satisfactory to both the customer and manufacturer. The most common piston-rod materials and their properties are discussed below and summarized in Table I.

*Heat-treated SAE-4140*—this is the most widely used material for compressor piston rods. It can be heat-treated for maximum strength and is easily induction-hardened on the surface to provide superior wearing qualities. However, SAE-4140 is not suitable for corrosive atmospheres. Also, since it rusts easily, it should not be used in nonlubricated applications. It performs best in lubricated, noncorrosive atmospheres such as sweet natural gas, air or nitrogen.

*Annealed SAE-4140*—Although this material has superior resistance to hydrogen sulfide attack in sour-gas atmospheres, the soft annealed piston rods do not wear well. Though they may not break, they are likely to require shorter service intervals.

To improve this, the manufacturer may be asked to induction-harden the surface. This tends to defeat the original purpose for the softer material. Hence, expensive oversized piston rods are needed to keep stress levels low, with little gain in corrosion resistance. The use of annealed SAE-4140 piston rods should be avoided.

*17-4PH stainless steel*—17-4 PH is a martensitic precipitation-hardening stainless steel that, when heat-treated to the H-900 condition, becomes the strongest rod material

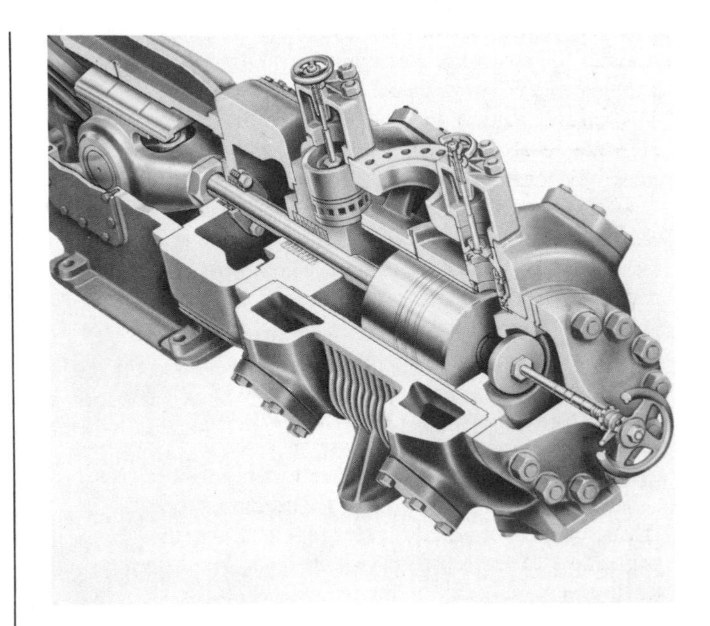

in use today. It is easily through-hardened to 40-45 Rockwell-C hardness, and can be run with all types of packing rings except for cast iron. For some unknown reason, cast iron wiper and packing rings do not perform as well on 17-4 rods as they do on SAE-4140 rods.

The 17-4 stainless rods possess superior corrosion resistance in all types of gas atmospheres—including hydrogen sulfide gas—and published data state that the material is as good as types 302 and 304 stainless steels. Yet, while 17-4 possesses all these good qualities, it has yet to be tried in a sour-gas application. Metallurgists usually agree that 17-4 looks good, but are reluctant to recommend it because it has not been proven as yet. It is hoped that some day this material will be thoroughly tested in sour gas (heat-treated to H-1025 condition). The results should be gratifying.

17-4 stainless steel rods are considered standard for most other corrosive atmospheres, and also for nonlubricated compressors. They run very well with Teflon rings, either lube or nonlube and, in the last 15 years, this material has performed excellently.

Heat-treated SAE-8620—This is not often used, but it does make an excellent base for Colmonoy overlay. SAE-8620 should be used when Colmonoy-coated rods are required.

*K-500 Monel*—This has only one application: in recip-

Originally published May 21, 1979

**Materials in common use for piston rods** **Table I**

| | Heat-treated SAE-4140 | Annealed 4140 | Heat-treated 17-4 PH stainless steel | Heat-treated SAE-8620 | K-500 Monel |
|---|---|---|---|---|---|
| Ultimate tensile strength, psi | 120,000 | 95,000 | 200,000 | 115,000 | 140,000 |
| Minimum yield strength, psi | 100,000 | 60,000 | 185,000 | 70,000 | 100,000 |
| Endurance limit used for design, psi | 55,000 | 40,000 | 75,000 | 50,000 | 43,000 |
| Hardness, Rockwell-C | | | | | |
| Core | 40 max. | 22 max. | 40-45 | 22 max. | 30 max. |
| Surface | 50 min. | 50 min. | 40-45 | 22 max. | 30 max. |
| Requires overlay | No | No | No | Yes | Yes |
| Recommended overlay | — | — | — | Colmonoy | Linde LW-1 Carbide |
| Maximum allowable design stress, psi | 8,500 | 6,000 | 10,000 | 7,500 | 6,500 |
| Maximum allowable compressor pinload at: | 100% rated | 30% reduction | 100% Rated | 15% reduction | 25% reduction |

rocating oxygen compressors. Many oxygen compressors use this base material with some form of overlay such as Metco #439 or Linde LW-1. K-Monel material will not support combustion and is one of the safest materials available for this application.

Other materials—Over the years, several other rod materials have been tried, but none has had the impact on the industry that SAE-4140 and 17-4 PH stainless steel have had.

## Allowable stress levels

Stress levels in various compressor rods should always be kept as low as possible, but not so low as to add undue cost. Large, slow-speed compressor rods are generally not stressed higher than 8,500 psi for heat-treated 4140 stainless steel, and are proportionately stressed up to 10,000 psi or higher. In either case, when these stress levels are exceeded, the risk associated with rod breakage increases drastically. Size the rod as large as possible, within reason, if there is a choice.

## Finish, and packing-ring compatibility

Rod finish should not exceed 16 $\mu$in. rms (root mean square), which is an acceptable level for lubricated compressors handling most gases above a molecular weight of 10. For low-molecular-weight gases (running lubricated) and for all nonlubricated applications running against Teflon, this maximum level is dropped to 8 rms. There are also special applications such as in ethylene gas, where a 4-rms maximum limit is set, because of the tendency for the gas to dissolve the lubricant, thus making proper lubrication more difficult.

The hardness of rod surfaces is also important. In general, the harder the surface, the better the wear. Because the packing rings are pressure-actuated against the rod, the harder surface resists these forces and counteracts wear.

Finally, consider packing ring compatibility. Heat-treated 4140 rods that run lubricated perform well with cast iron, bronze, Micarta or Teflon rings. For nonlubricated applications, heat-treated 4140 rods are not used.

For lubricated 17-4 PH stainless steel rods, bronze or Teflon rings are most acceptable. Cast iron rings are not normally used. For nonlubricated applications, rings made of Teflon are recommended.

When nonmetallic piston-ring and packing-ring materials are used, care must be exercised so as not to thermally insulate the piston-and-rod assembly. For example, Micarta piston rings and rod packings should never be used together because, since they insulate so well, there is no way to remove frictional heat from the rod. When using Micarta piston rings, longer life can be realized if metallic rod-packing rings are used with them. If Teflon rings and packings are employed, use a metallic backup ring as part of the rod packing-ring set in order to remove the heat of friction. Too frequently, compressors have suffered from excessive ring and packing wear brought on by frictional heat buildup that cannot be detected by normal instrumentation.

## Thread design

Historically, most piston-rod breakage problems have occurred in the threads at the crosshead end. Sometimes this failure could be attributed to a lost lock nut or an improperly tightened nut—but most of the time it has been traced to fatigue of the metal, starting at the thread root. Most piston rods formerly used a conventional 60-deg thread form, made by cutting the thread onto the rod by means of a cutting tool. When viewed under magnification, the threads were seen to be ragged and torn, with sharp corners at the thread root. With this type of thread, it was almost impossible for designers to provide sufficiently low stress in the part to prevent breakage.

With the introduction of thread-grinding equipment, a giant step was achieved for users of the standard 60-deg thread form. This new thread was so clean in comparison to the cut thread that it reduced fatigue-type failures by over 50%. However, too many piston-rods still failed.

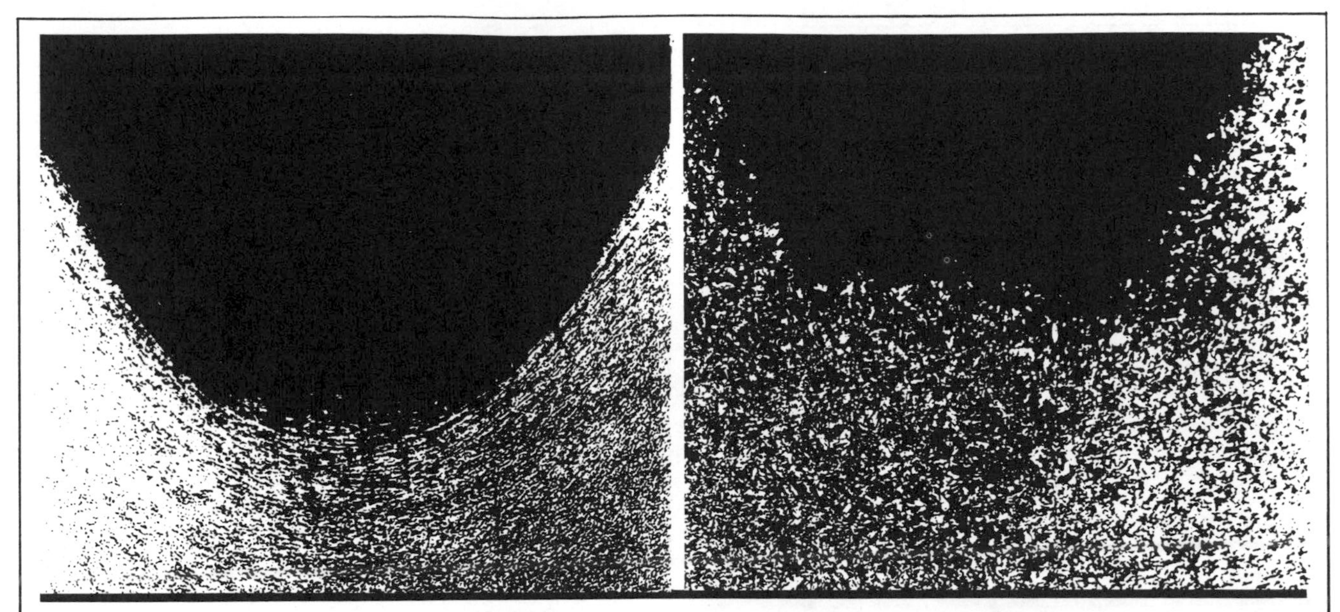

**Rolled thread (left) eliminates minute tears and galls such as appear in root of typical milled thread (right) Fig. 1**

Shortly thereafter, thread-rolling machines were being designed and marketed, and most compressor builders immediately tooled-up to roll conventional 60-deg threads on their piston rods. Laboratory tests proved that rolled threads reduced fatigue stresses by one half when compared with cut threads, and field results confirmed this. The rolled-thread piston rod was a tremendous improvement in technology and is now standard.

The next step was thread redesign; the place to start was at the thread root, since it provided the source for stress concentration. In 1960, a new thread form was designed that had a full-radius thread root, called the "high fatigue strength" or HFS thread. It was adopted, in 1963, as standard on all our company's piston rods 2¼ in. dia. and over. It has worked so well that today, 15 years later, the company has yet to see the first failure caused by fatigue in the piston rod threads.

The HFS thread form in Fig. 1 clearly shows the revised grain structure occurring in the rod itself as a result of the rolling process, as well as the full root radius, which is the largest that can be designed into a thread of this type.

## Piston fits

In order to efficiently minimize rod failures near the piston end, care must be taken in fitting and securing the piston to the rod. Certain compressor pistons have a slight interference fit between the piston hub and the piston rod, which helps to keep the piston tight on the rod and to keep the two parts from rotating when the piston nut is tightened. The piston is generally assembled onto the rod, heated with steam, and the rod inserted until it shoulders. It is very important that the assembly then be allowed to cool to room temperature before tightening the piston nut. The piston nut should not be tightened while the piston is still hot, for as the piston cools, it will decrease in length and the prestress in the rod introduced by the piston nut tightening will be lost. Once the prestress is lost, the piston can be considered "loose" on the rod, and eventually the micro-movement will generate a pounding action that will cause a fatigue failure of the rod in the area of the piston fit.

How much should the piston nut be tightened? While the answer is simple, obtaining desired results is more difficult, especially if the proper tools are not used. Pistons with solid hubs should be torqued to prestress the piston rod to a 15,000-psi level. Pistons that do not have solid hubs should not be torqued greater than a 10,000-psi prestress level, in order to prevent piston collapse. However, 10,000 psi is the lowest prestress level that should be considered.

How then is the prestress level achieved? For piston rods up through 2½ in. dia., a torque wrench will guarantee that the job is done right. On larger-diameter rods, the only acceptable method of tightening the piston nut is by means of hydraulics. Some compressor units have a special hydraulic-ram piston-nut-tightening arrangement that correlates hydraulic pressure versus rod prestress as a function of rod diameter. These special tools must be used whenever a piston is reassembled to a rod (Fig. 2). Failure to properly torque the piston nut will most likely result in a broken rod. Rods assembled by using impact wrenches or sledge hammers usually break.

## Coatings

There are as many opinions on this subject as there are people who provide coating products and services. We will only attempt here to pass on some of the things learned over 20 years of experience.

*Carbides*—Though expensive, Union Carbide's Linde Div. LW-1 carbide has an excellent track record. This coating can be applied over most rod materials and the finished surface is compatible with most pressure packing materials. Furthermore, it stands up either lubricated or nonlubricated, and has the lowest wear rate of any material we have used. Once applied to a new rod, or when reclaiming an old rod, the finished product will outwear any original noncoated rod by at least a factor of

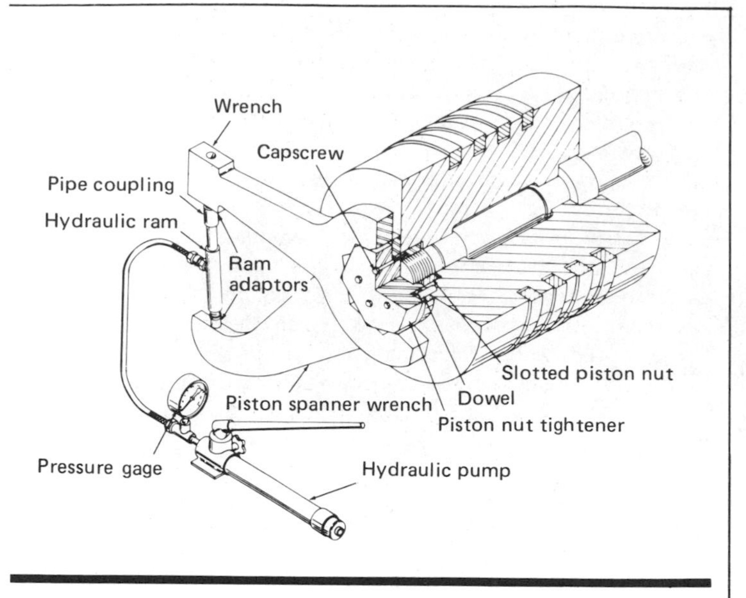


**Tightening piston nut with hydraulic ram tightener** **Fig. 2**

two. It is used on K-Monel piston rods used in oxygen compressors and on some high-pressure-polyethylene hyper-compressor plungers. It is also available on any new rod for any noncorrosive service. There are other carbide coatings marketed that cost less to apply and that also have had excellent results.

*Chrome plating*—Some chrome-plated rods run quite well, whereas others have failed miserably. We have never been able to predict which will be successful. It may be that some chrome surfaces are such that the lube oil cannot wet them, and when this happens the packing rings cause excessive wear that peels the chrome off in just a few strokes of the compressor. Also, Teflon packing rings should not be run against chrome-plated piston rods, unless there is no other choice of material for the application. This applies to both lubricated and nonlubricated applications.

In general, if you anticipate using chrome-plated piston rods, remember that there is a difference in "chrome." The platings are not all alike, and there is definitely some element of risk.

*Colmonoy*—This material has been around a long time and is especially good in corrosive atmospheres. However, to apply it the base metal must be heated to approximately 1,600°F and allowed to cool slowly. This process completely anneals the base metal, and any previous heat treatment applied to the rod for the purpose of obtaining more-desirable physical properties is lost. Annealed or soft rods must then be derated to keep within safe stress limitations.

Colmonoy overlay cannot be regularly applied over standard SAE-4140 rods without surface cracking and should not be used. However, Colmonoy can be applied over SAE-8620 rod material, if the rod is oversized by 15% in order to maintain safe stress levels.

## Allowable rod runout

On compressors that have adjustable crossheads, piston rods should be adjusted to run true in both horizontal and vertical directions within 0.004 in. This is important so as not to impose undue bending stresses into the rod where it fastens to the crosshead. Only persons skilled in operating reciprocating compressors should attempt to make this adjustment, and they should be thoroughly familiar with the unit being worked on. Although most crossheads appear to be quite heavy, they sometimes do ride the top of the guide—depending on rotation and load—and the runout adjustment must take this into consideration. Failure to adjust for proper runout will usually lead to a broken piston rod somewhere close to the crosshead.

Many smaller compressors are designed with nonadjustable barrel-type crossheads, where the maximum rod runout sometimes does exceed 0.004 in. Here, the user must exercise care to assure that the compressor is assembled as close as possible to a theoretical centerline. Many such compressors are in operation and, although in some cases the runout may get as high as 0.007 in., rod breakage has not been a major problem.

## Reasons for piston-rod breakage

*High tensile stress*—We have already pointed out the need to keep tensile stresses within certain design limitations. The resulting safety factor usually compensates for some degree of bending imposed by runout, but more importantly it sets limitations on pressure differentials that can be safely tolerated across the piston. New compressors, as shipped from the factory, have these limitations clearly spelled out. Frequently, however, compressors are revamped without regard to pinload limitations.

Recently, I witnessed a compressor operating at 50% over the allowable rodload rating. The compressor had been revamped by others, and the user registered a complaint about excessive cylinder movement. The problem was corrected by a readjustment of pressure ratios. So, under normal operating conditions, tensile stress usually will not cause rod breakage of itself.

*Bending stresses*—Bending fatigue is a common cause for rod breakage and can usually be traced back to excessive rod runout. This might be caused by cylinder-liner wear, or by wornout piston wear-bands that allow the piston to drop in the cylinder. It is an important part of any preventive-maintenance program to periodically check rod runouts.

*Scored rods*—Proper lubrication of rod packings is important in order to keep piston rods from wearing or scoring. While it does not happen often, rods that run scored also run hot. The combination of tensile stresses, heat, and stress-concentration in the scored grooves can cause the rod to break in the pressure-packing area.

*Loose piston nut*—A loose piston on a rod, if not corrected immediately, will almost always break the rod, usually in the area of the piston. Loose pistons normally produce a knocking sound and are easily detected. Piston nuts should always be properly torqued and locked.

*Loose piston-rod nut*—The piston-rod nut locks the rod to the crosshead and should not be confused with the piston nut. The purpose of this nut is to prestress the rod in the crosshead. Thus, the threads must be in good condition and the faces of the nut and crosshead must be square with one another and free of any surface roughness (in order to prevent bending stresses).

Small piston-rod nuts can be tightened by sledging;

however, larger nuts must be torqued by hydraulics. Failure to tighten the piston-rod nut properly can result in rod breakage—usually close to the point where the nut mates to the crosshead. A loose rod-nut can also allow the piston to turn and use up end-clearances. Hence, the piston could end up striking one of the heads. This can also cause rod breakage through pounding action.

*Insufficient clearance between piston and heads*—Many compressor cylinders are designed to provide minimum clearance. This means that the running clearance between the piston and heads at the end of the stroke is minimal—0.062 in. each end, or 0.125 in. total. When installing a compressor piston rod, this clearance should be divided 65-60% in the head-end, and 35-40% in the crank-end. In this example, 0.075 in. clearance in the head-end with 0.050-in. clearance in the crank-end would be considered good distribution. This is to compensate for the expansion of the piston rod when it warms up and gets longer, so that in normal operation the actual clearances will be somewhere near 0.060 in. on each end.

Failure to make this adjustment could result in thermal growth, allowing the piston to strike the head-end head. If it does not strike the head hard enough to produce an audible knock, and the problem is therefore not discovered and corrected, the repeated pounding can cause excessive compressive stresses in the rod that could propagate a fatigue failure of some sort.

## Liquid slugs

Compressors are built to pump gas, and ingested liquids are always a cause for concern. If a noncompressible liquid gets into a compressor cylinder in enough volume to fill up the clearance space between the piston and heads, the old rule that "no two things can occupy the same space at the same time" prevails, and at that instant something is going to break. It will not always be the piston rod.

If the jam-up occurs in the head end of the cylinder, the piston rod is placed in compression, so the bolting that holds the cylinders to the distance pieces, and the distance pieces to the frame, are in jeopardy. If the jam-up occurs on the crank-end of the cylinder, the piston rod is then in tension, as are the connecting rod and connecting rod bolts. Here, a broken piston rod would probably cause the least amount of overall damage to the compressor.

Liquids cannot be tolerated in a reciprocating gas compressor. A good operator must not allow his compressor to run without taking some action to prevent the liquids from entering.

## The author

**Jim Messer** is head project engineer for all reciprocating compressors manufactured by the Dresser Clark Div. of Dresser Industries, Inc., P. O. Box 560, Olean, NY 14760. At Dresser, he is responsible for the design of air, gas and polyethylene-type compressors. He holds an engineering degree from the University of Cincinnati and is a registered professional engineer.

# Section III

# Liquid and Slurry Movers: Centrifugal and Positive-Displacement Pumps

# Keys to Pump Selection

Pump selection requires a careful analysis of the hydraulic system and the pump's location and function within the system. With a knowledge of pump capabilities and limitations, a step-by-step procedure will direct the engineer to the most efficient and economical equipment.

WILLIAM H. STINDT, Brown & Root, Inc.

Proper pump selection can reduce start-up and operating problems, as well as future maintenance costs. A reasonably detailed preselection also provides an opportunity to issue a more complete specification. This serves to facilitate proposals for a final selection without continual revisions.

Some selections, however, are best made by evaluating proposals based on pump hydraulic requirements rather than complete specifications. This is true in situations that could use a number of pump designs equally well; a rigid specification would tend to restrict proposals to one design. Usually, though, the importance of a complete specification and proposal evaluation can not be over emphasized.

## Fluid Characteristics

The normal variables in pump selection are location, function and type of hydraulic service. The most common objectives are low operating and maintenance costs, although it is not unusual to have these goals superseded by low initial cost. National purchase agreements, spare parts interchangeability and working relationships with local service personnel are other factors that influence selections.

A detailed listing of fluid characteristics is usually the first step in selection. The most important characteristics are:

1. The chemical identity of the fluid pumped, including such items as pH, dissolved oxygen concentration and historic data acquired during previous handling.
2. Absolute viscosity at pumping temperature and 60 F., usually given in centipoises. A substantial viscosity increase at ambient temperature could indicate that the liquid is unsuited for centrifugal pumps without steam jacketing. Kinematic viscosity, which is usually given in centistokes, is equal to absolute viscosity divided by density. The term is also often expressed as Saybolt Seconds Universal (SSU).
3. Specific gravity of the fluid at the pumping temperature and 60 F. (or ambient).
4. Pumping temperature during normal operation and vapor pressure of the fluid measured at the pumping temperature.
5. A number of items that make up the "personality" of the fluid. These include toxicity, explosive nature, crystallizing or polymerizing tendencies, and the presence of entrained gases. If there are over 3% solids by volume the fluid may qualify as a slurry and require special pump design such as rubber or hard-alloy linings.

## Hydraulics of the System

After the nature of the fluid has been recorded, the hydraulics of the system must be studied. Pertinent factors are discussed below.

*Pressure.* Barometric pressure is the atmospheric pressure at the pump location. It varies with altitude and weather conditions. Gage pressure is the pressure above atmospheric (or below in the case of vacuum). Absolute pressure is defined as the sum of atmospheric and gage pressures.

*Head.* The head measured in feet is the height of a fluid column required to produce the gage pressure at the bottom of the column. Equal pressures created by two different fluids require different heads. The relationship is:

$$\text{Head, ft.} = 2.31\ (\text{pressure, psi.})/(\text{Sp. Gr.})$$

Originally published October 11, 1971

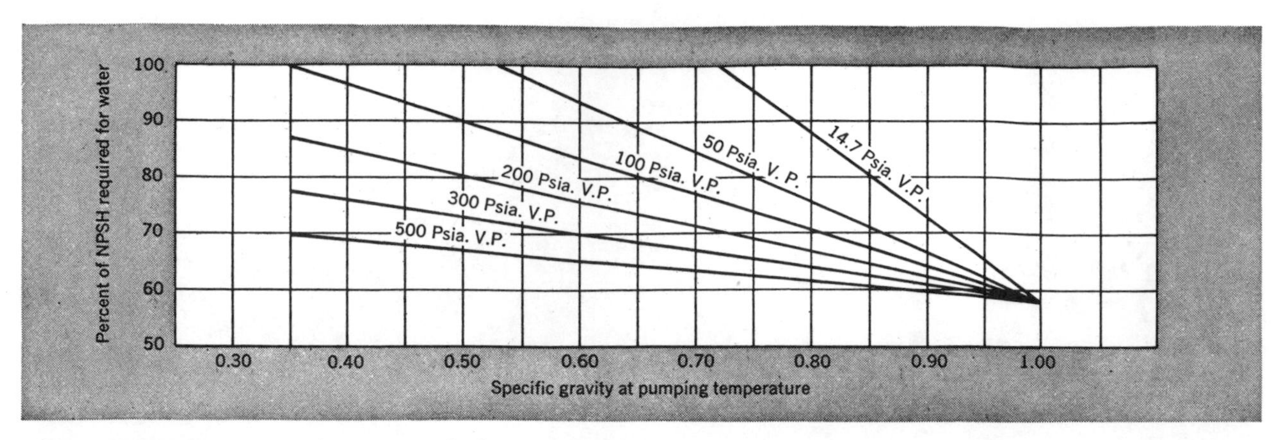


**NPSH CORRECTION** factors for pumping hydrocarbons—Fig. 1

*Static Head.* Fluid elevation above or below the datum line is the static head. For horizontal centrifugals, static head is usually measured from the pump centerline; with vertical pumps it is taken from the eye of the first stage impeller.

*Velocity Head.* The velocity head is the kinetic energy of a flowing fluid. It can be calculated in feet of liquid by dividing fluid velocity in ft./sec.$^2$ by 2g.

*Suction Lift.* This term expresses the suction head when it is below atmospheric pressure. It is found by taking a manometer reading in feet of liquid at the pump suction (corrected to datum), minus the velocity head.

*Total Discharge Head.* This is the pressure reading at the pump discharge in feet of liquid, corrected to datum, plus the velocity head.

*Total Head.* The algebraic difference between the discharge and suction heads is called the total head. A suction lift is additive to the total discharge head; a positive suction head should be subtracted from the discharge head.

*Head Losses.* These are the frictional head losses caused by fluid flow through pipes, fittings, valves and nozzles.

*NPSH.* The net positive suction head is probably the most critical factor in a pumping system. Adequate NPSH is essential, whether working with centrifugal, rotary or reciprocating pumps.

Marginal or inadequate NPSH will cause cavitation —the formation and rapid collapse of vapor bubbles in a fluid system. This occurs at points such as the pump impeller entrance, trailing edges of valve gates, other low pressure areas, and voids that demand immediate fluid replacement. Collapsing bubbles place an extra load on pump parts and can remove a considerable amount of metal from impeller vanes.

Cavitation is often taking place before the symptoms become evident. Factors that indicate cavitation are increased noise, loss of discharge head and reduced fluid flow.

Fig. 1 is reprinted from Hydraulic Institute standards, 12th Edition, copyright 1969 by the Hydraulic Institute, 122 East 42nd St., New York, NY 10017.

## Calculating NPSH

NPSH requirements are usually established with water. Cold water has a high vapor-to-liquid volume ratio, which makes it difficult to pump under cavitating conditions. NPSH requirements will be lower with fluids that have a smaller vapor-to-liquid ratio. Correction factors supplied by the Hydraulic Institute are shown in Fig. 1.

Absolute pressure values are used to calculate NPSH. The formula is:

$$\text{NPSH, ft.} = (H_A - H_{VP}) \pm (H_S - H_F)$$

where: $H_A$ pressure on the fluid in the suction tank, ft.
$H_{VP}$ vapor pressure of the fluid at the pumping temperature, ft.
$H_S$ static head or lift from the fluid level to the pump datum, ft.
$H_F$ friction head for the design flow rate at the pump suction, ft.

The NPSH for an existing installation can be measured as follows:

$$\text{NPSH, ft.} = P_A \pm P_S + (V_S^2/2g) - H_{VP}$$

where: $P_A$ atmospheric pressure, ft.
$P_S$ gage pressure measured at the suction flange corrected to datum, ft.
$V_S^2/2g$ velocity head
$H_{VP}$ vapor pressure of the fluid, ft.

## Other Hydraulic Factors

The acceleration head is the force required to accelerate the fluid in the suction line. This term is used with reciprocating pumps and is the head normally required in addition to the NPSH at the suction flange. The acceleration head is equal to:

$$H_a = LV_nC/gK$$

where: $L$ length of pipe, ft.
$V_n$ fluid velocity in the suction pipe, ft./sec., at $n$ rev./min. of the pump crankshaft
$g$ gravity acceleration, ft./sec.$^2$
$C$ 0.20 —simplex double-acting pump
0.115—duplex double-acting pump
0.066—triplex double-acting pump
$K$ varies from 2.5 for hot oil to 1.4 for deaerated hot water. See reference 4, p. 14–6.

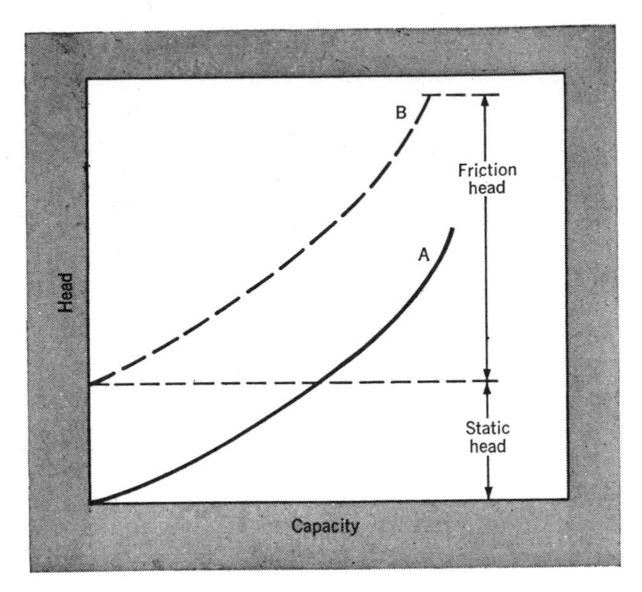


**SYSTEM HEAD** curves for a typical application—Fig. 2

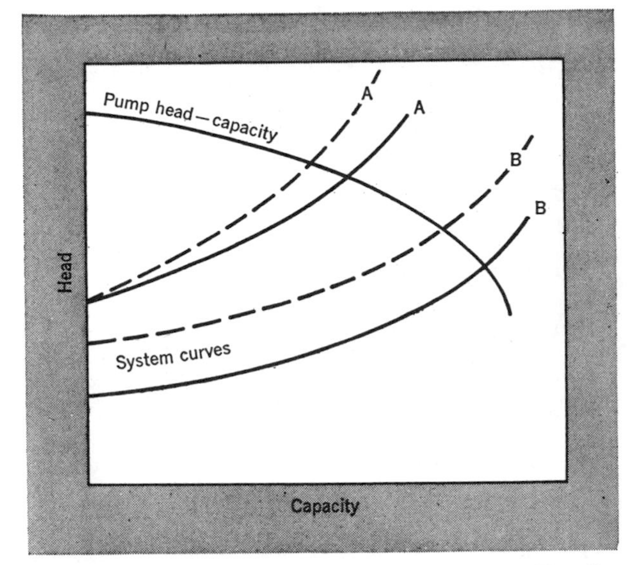


**EFFECT** of changes in the hydraulic system—Fig. 3

## System Head Curves

To make the best application of a particular pump, it is necessary to identify the system's head-capacity relationships. This is done most frequently for centrifugal pumps, although an analysis of the system should also be considered for positive displacement units.

In Fig. 2, curve A represents friction losses from fluid motion, while curve B includes total static head. It should be noted that even where friction losses are a minor part of the system curve, pump sizing could be missed considerably if they are not considered. With high volume, low head pumps (mixed flow), the friction head becomes a major percentage of the total head requirement.

Curves for another common system are shown in Fig. 3. The effect of throttling a valve on the pump discharge is shown by curves A and A′, while curves B and B′ represent the effect of changes in the static head. When studying a system at various conditions, two or more points should be obtained on the curve near the operating point.

Centrifugal pumps are often arranged in series or parallel. With series operation of two pumps, the heads are additive while the capacity remains constant. Thus, a combined curve shows twice as much head output at each capacity than the curve for a single pump, assuming the two pumps are identical.

The reverse is true with parallel operation. That is, pump capacities are additive while the head remains constant. Most systems involve flow control, and a method for determining the head throttling value that will give half the desired capacity is shown in Fig. 4.

## Special Problems

Entrained gases may present a problem in some pumping applications. More than 5% by volume can not be handled by most centrifugals. Regenerative turbine pumps can operate with up to about 50% entrained gas, and there is virtually no limit with reciprocating or rotary types.

Handling viscous fluids with centrifugal pumps reduces efficiency and increases horsepower requirements. Correction factors are available to predict performance at various viscosities. When there is a borderline application, pump manufacturers can be consulted about special impeller designs. *(For more in-*

Slurries present another situation that calls for careful pump selection. There are actually two separate problems in handling solid/liquid mixtures: mechanical clogging of the impeller and shaft sealing; and reduced performance of the pump.

Solids in a slurry only possess or transmit kinetic energy. Energy imparted to the solids by the pump is virtually lost since it is not converted to pressure energy. Therefore, pump efficiency drops as the solids concentration increases because there are additional hydraulic losses due to the relative difference in motion between solids and the liquid.

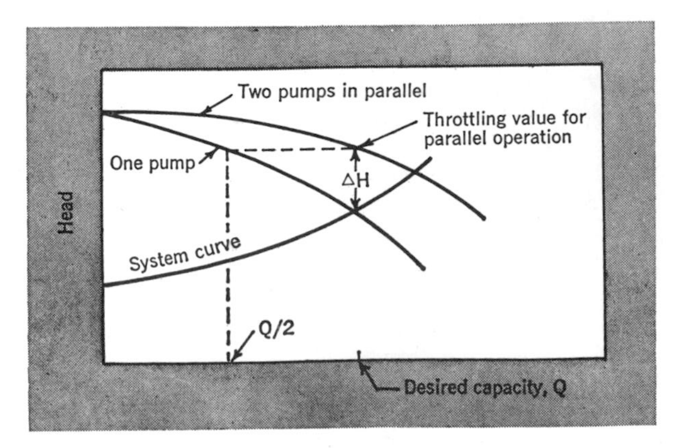


**THROTTLING** with parallel pumps—Fig. 4

Data on handling specific slurries have not been well organized. The actual "personality" of the solid particle, such as shape, size and coagulative tendencies, probably has much to say about its successful handling. Centrifugal pumps, other than regenerative turbine types, are best suited for slurries, and mixtures with over 50% by weight solids have been pumped successfully.

## Function and Installation

Before proceeding to the actual pump selection, it is necessary to have knowledge of the location and basic job to be performed. Such information will reduce the number of specific types to be considered. The following are questions that should be answered at this point in the selection process:

1. Will the pump function as a continuous service unit, or will it operate intermittently in a remote area?
2. Does company policy require work permit authorizations before entry can be made by maintenance men and equipment? Restricted accessibility can influence future maintenance spending.
3. Does the installation require a vertical or horizontal pump. If a vertical pump, will a dry or wet pit be used?
4. Will repair parts have to be transported unusual distances?
5. Should space and weight be considered?
6. Will the installation be indoors or outdoors, above or below grade, or subject to any unusual climate conditions?
7. Does construction or performance have to meet certain company specifications or standard preferences?
8. Will special pump design or construction be required to handle the fluid?

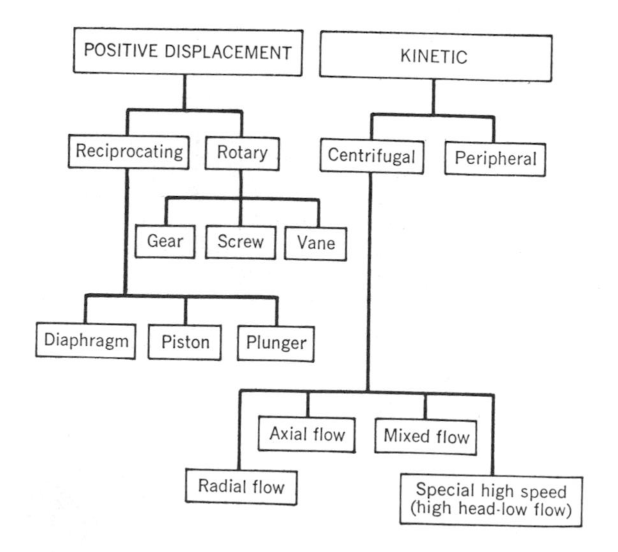


**GUIDE** to pump classifications—Fig. 5

## Steps in Pump Selection

Most pumps can be classified into two major categories for selection purposes: positive displacement and kinetic. Various types are represented schematically in Fig. 5. Firsthand experience with identical or related services is always the best guide to selection. The following discussion presents a step-by-step method for arriving at a reasonable choice.

The first steps are to analyze the pump's function and installation, and develop system characteristics, as presented earlier in this article. If the application requires a controlled volume, this immediately dictates a positive displacement metering pump.

The curves in Fig. 6 show approximate capabilities of the major pump classes. Note that the graph is scaled to emphasize the lower ranges of heads and flows. If performance requirements fall in the area of normal centrifugal capabilities, the next consideration is the fluid's viscosity. Above 500 SSU, a positive displacement pump (discussed below) is probably the best choice. This type is also indicated if there are more than 5% entrained gases. When handling slurries with centrifugal pumps, speeds should be 1,800 rpm. or lower when the solids volume is over 3 to 5%. A special slurry pump may be required.

Specific speed—a correlation of pump capacity, head and speed at optimum efficiency—should be analyzed. Higher speed pumps are more economical and should be selected if possible.

The next factor in centrifugal selection is the choice between horizontal or vertical designs. Most vertical pumps are specified for obvious applications such as lift stations, sumps, and inherent mixed or axial flow. They are also used to obtain an adequate NPSH by submerging the impeller at the required depth in an outer case.

A vertical pump may be the answer where unusually high heads are needed. About 12 to 14 stages are the practical limit with a horizontal pump. Special high-speed, single-stage pumps can be used as an alternative to multistage vertical units. These are mounted above ground. Their approximate performance limitations can be seen in Fig. 6.

At this point a decision is required between a lighter-duty chemical pump or those built to meet the American Petroleum Institute standards for refinery service. The major advantage of chemical or "AVS" pumps is that all manufacturers follow the same external dimensions, model for model.

A refinery service pump usually costs two to three times as much as a corresponding chemical pump. Pressure and temperature capabilities are generally about double, and each type is also available as a vertical in-line unit. In-line models are becoming more popular as experience is gained in electric motor designs, specifications and quality control. The motor is a major design component since the pump relies on the motor's bearings.

## Materials of Construction

A detailed discussion of pump materials and their resistance to various environments may be found in the Standards of the Hydraulic Institute. It should be kept in mind that the manufacturer's standard materials should be used if possible. In some applications it can be cheaper to replace the entire pump rather than specify costly special alloys.

Ductile iron is standard on many chemical pumps, although cast iron is usually cheaper. Cast steel, often required in refinery or high pressure services, is 2 to 2.5 times the cost of cast iron.

A wide range of materials is usually available in chemical-service centrifugals. Pump cases may be of cast or ductile iron, steel, bronze, chrome and chrome-nickel stainless steels. Special materials used in centrifugals include Hastelloy and Monel alloys, rubber and Teflon plastic.

Cast iron has good corrosion resistance to mild caustic and cold non-corrosive solutions. If the product is flammable or toxic, cast steel may be preferred because it is more resistant to fracture when sprayed with water during a fire. For high temperatures and where corrosion or erosion is a problem, the chrome and chrome-nickel alloy steels are often specified. These are generally used for acid solutions, although the bronzes can serve equally well at mild temperatures. However, a tougher material is needed for centrifugal pump impellers.

Concentrated, hot, caustic solutions or fluids with high chloride content require nickel-base alloys. In all cases, it is advisable to select compatible case and impeller materials to avoid galvanic corrosion in solutions that are good electrolytes.

Wear rings in centrifugal pumps may use cast iron against cast iron for fluid viscosities of 50 SSU and above. A cast iron or bronze ring against chrome stainless steel is often used with less viscous liquids. Chrome steels should be hardened to about 450 BHN, and chrome-nickel alloys should be hard faced to prevent galling. There should be a hardness differential of at least 50 BHN between the two parts.

With positive displacement pumps, material selections are usually dictated by pressure requirements, unless fluid characteristics call for a special material. Manufacturers' standard compliments are good selections in most cases. Extensive metallurgical data are required to design these pumps and this work is passed on to the user as part of the product.

In fact, the supplier can offer worthwhile knowledge on positive displacement pump metallurgy, and this information should be obtained before final specifications are set. This allows a mechanical review that produces a better specification before cost bidding for the equipment.

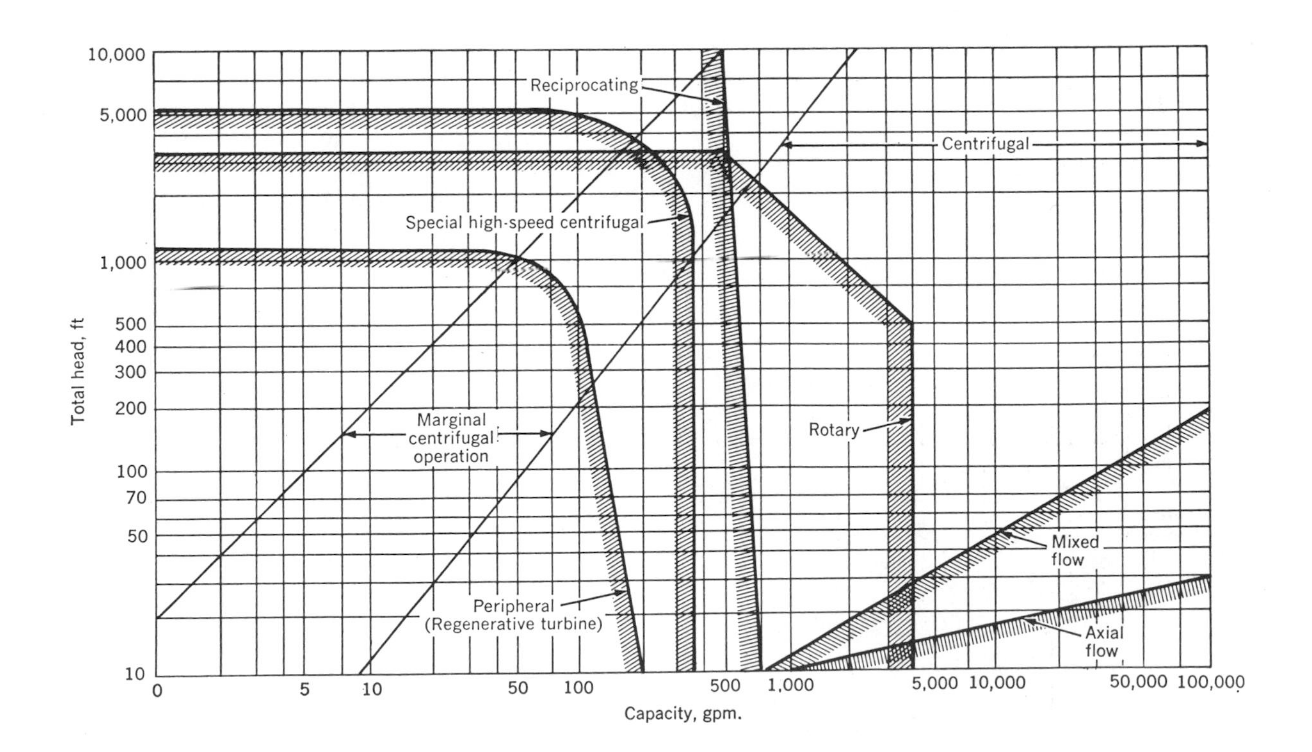


**OPERATING** ranges of common process pumps—Fig. 6

## Positive Displacement Pump Selection

If the factors discussed thus far point to a positive displacement pump as the best choice, then the relative merits of rotary and reciprocating types must be evaluated. Gear, screw, vane and other rotary designs have a number of advantages such as:

- They can operate under low NPSH requirements and produce high suction lifts.
- They have relatively high efficiencies when pumping viscous liquids.
- Hydraulic characteristics are good, with high heads at a wide range of capacities.
- They have an extensive speed range, generally limited only by the fluid's viscosity. The Hydraulic Institute recommends speed-viscosity relationships.
- Rotary pumps are inherently self-priming since they can virtually operate as compressors. They can produce a vacuum of 29 in. Hg.

The final selection of a particular type of rotary will usually be based on a cost evaluation. However, each type has particular capabilities that should be understood.

▶ **Vane** types operate by differential pressures, usually at 50 psi., and have capacities up to about 375 gpm. The practical limit on liquid viscosity is around 100,000 SSU. This type is subject to wear and should not be specified when the liquid has poor lubricating qualities.

▶ **Gear** pumps are used to about 650 gpm. and 350 psi. They can handle viscosities approaching 5 million SSU. Internal bearings and timing gears are preferred because this requires only one shaft sealing area. They may have to be mounted externally, however, if the fluid's lubricating characteristics are not good. Vane and gear pumps usually have a top speed of 1,200 rpm. when pumping liquids in the range of 100 to 500 SSU viscosity.

▶ **Screw** pumps are applied to large flows since they have capacities up to 4,000 gpm. and 3,000 psi. Viscosity capability is about 100 million SSU. Screw pumps have bearing and timing gear requirements similar to gear pumps.

Screw pumps can usually operate at higher speeds because of relatively lower fluid velocities. The reason is that in this type the flow is axial, as opposed to flow around the periphery of a gear or vane pump.

## Reciprocating Pump Capabilities

The diaphragm type of reciprocating pump is generally found in controlled volume applications. Standard models have capacities up to 600 gal./hr. at discharge pressures to 3,500 psi. These figures are for simplex units; higher capacities can be obtained by adding an extra plunger. (Fig. 9)

Plunger pumps, a second type of reciprocating machine, are also used for metering. They have larger capabilities than diaphragm pumps—up to 1,000 gal./hr. and 7,500 psi. Large plunger types, often called power pumps, turn out 600 gpm. at pressures up to 10,000 psi. See Fig. 8.

Piston pumps are double acting in that fluid discharges on both the forward and reverse strokes. Maximum performance is in the range of 600 gpm. and 750 psi. Piston pumps are generally less expensive than plunger types, but also have shorter service life. One example is shown in Fig. 7.

This type is usually steam driven and thus is also called a steam pump. Air or gases can also be used. The liquid side of steam pumps have a "valve plate" for pressures up to 350 psi. and a "valve pot" for services to 750 psi. Cast iron piston rings should be used above 350 F.

Pulsation is inherent with reciprocating pumps. Dampeners on the pump discharge will help correct the uneven pressure pattern, or multiple ends can be arranged so that one discharge stroke begins before another has ended. This overlap eliminates sudden drops in capacity and pressure.

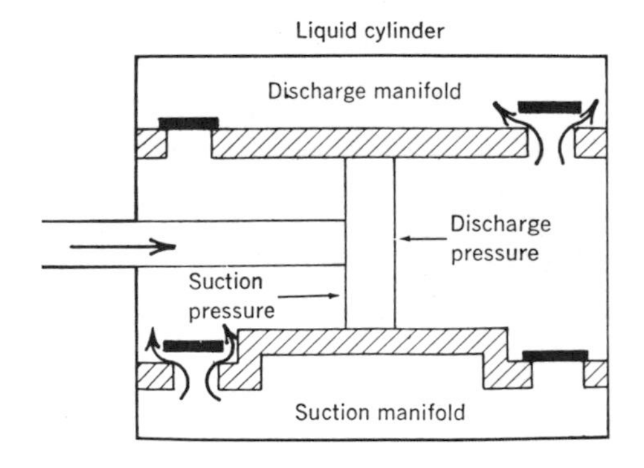


**PISTON PUMPS** are double acting—Fig. 7

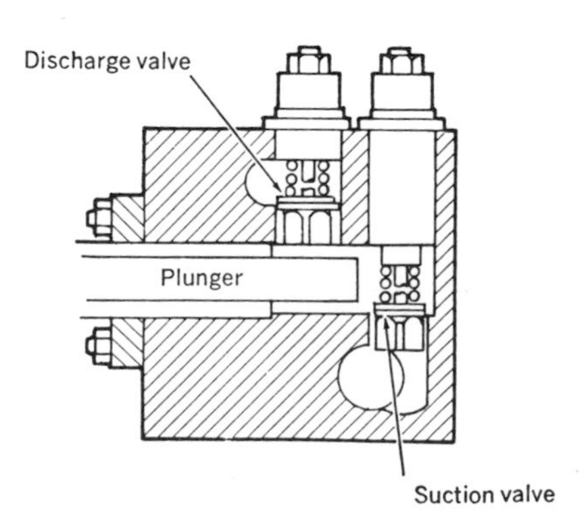


**PLUNGER** type pumps—Fig. 8

The Hydraulic Institute recommends piston speeds and gives correction factors for viscosities up to 10,000 SSU. Viscosities under 200 SSU do not require an appreciable speed correction. The Institute's recommendations are very liberal by some manufacturers' standards, but can be used as a guide during preliminary selection.

## Final Pump Evaluation

Specifications that will be used in obtaining bids should be drawn up based on the pump merits outlined here. Individual company preference and requirements should, of course, be included. Catalogs can provide useful information, but this should not be applied in a manner that would restrict other manufacturers from complying with the intent of the specification.

After proposals are received, the final selection will be based on economics, a mechanical review and a hydraulic review. Hydraulic requirements of the application must be satisfied before making any further evaluation of a particular quotation.

Mechanical review of the proposals should study such items as basic design, pressure limits, speed and balancing requirements, materials of construction, control mechanisms and other accessory equipment. At this stage, one often discovers why there is a large discrepancy between some quotes.

The economic evaluation should consider initial, operating and installation costs on a comparable basis. The review may include a maintenance cost history of comparable equipment already in service. An economic analysis can reach complex proportions and this article only intends to present some of the factors to be considered.

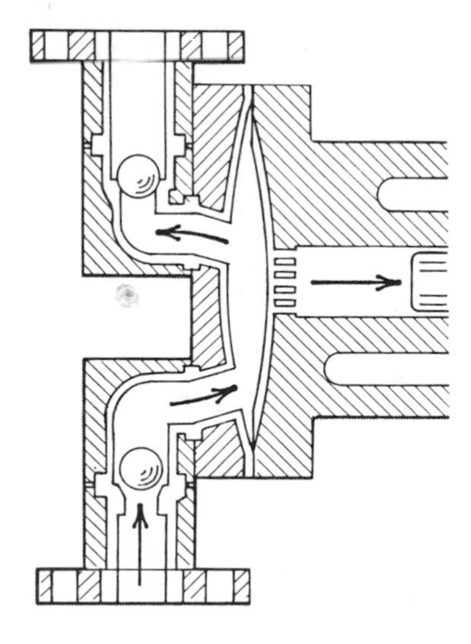

**DIAPHRAGM** pump for metering—Fig. 9

In the absence of actual cost data, an assumption should be made that maintenance costs will be equal to those for pumps of similar design. It should only be necessary to compare the capital costs of pumps and drivers since accessory equipment should be essentially equal for all pumps being considered for the same services. Naturally, if a steam-rod pump is being compared with a centrifugal, accessory equipment becomes a major consideration.

A complete analysis would also include company philosophy on operating vs. capital costs, the cost of capital, depreciation allowables and taxes. Most companies have a minimum rate-of-return on capital that must be satisfied. In the following simplified example involving four different pumps, we will assume this figure is 20%.

| Pump | Capital Cost | Operating Cost |
|---|---|---|
| A | $39,500 | $20,600 |
| B | $38,500 | $21,000 |
| C | $38.900 | $21,400 |
| D | $41,500 | $21,000 |

First look at the lowest priced pump. If its operating costs are also the lowest, then this is the economic choice. In this case, pump A has lower operating costs than pump B, so the rate-of-return on capital must be considered. (Choices C and D are, of course, eliminated since their capital and operating costs are equal to or higher than B).

Pump A requires additional capital of $1,000, which will reduce annual operating costs by $400.

$$\text{Rate-of-return} = 400/1{,}000 \times 100 = 40\%$$

Therefore, pump A is the economic choice because the additional capital required for its purchase will bring an annual return of 40%.

## Bibliography


1. "Flow of Fluids," Crane Co., Chicago, 1965.
2. Hicks, Tyler G., "Pump Selection and Application," McGraw-Hill Book Co., New York City, 1957.
3. Hicks, Tyler G. and Edwards, Theodore, "Pump Application Engineering," McGraw-Hill Book Co., New York City, 1971.
4. Marks, Lionel S., "Mechanical Engineers Handbook," McGraw-Hill Book Co., New York City, 1958.
5. "Power Manual of Practical Engineering Data," McGraw-Hill Publishing Co., New York City.
6. "Standard of the Hydraulic Institute," The Hydraulic Institute, New York City, 1969.
7. Stepanoff, A. J., "Centrifugal and Axial Flow Pumps," John Wiley and Sons, 1964.


## Meet the Author


**William H. Stindt** is a mechanical engineer with Brown & Root, Inc., Houston, Tex. 77001. His primary duties are mechanical pump design for petrochemical and power plants. Previously, he worked with Monsanto as a plant engineer involved with pump, compressor and associated plant equipment problems. Mr. Stindt has a B.S. degree in mechanical engineering from Texas A&M University, and has 10 years experience with process plant equipment.

# Selecting the right pump

For any application, picking the proper pump from the multitude of available styles, types, and sizes can be a difficult job for the user or the contracting engineer. The best approach is to make some preliminary searches, come to some basic decisions and preliminary selections, and then discuss the application with the pump supplier.

***Richard F. Neerken,*** *Ralph M. Parsons Co.*

☐ The key to making the correct pump selection lies in understanding the system in which the pump must operate. The engineer specifying a pump may make a poor selection because he has not investigated total system requirements and completely understood how the pump should perform. Moreover, when the responsibility of pump selection falls entirely on the supplier's representative, it may be difficult or impossible for him to determine overall operating requirements.

So, with the first rule in pump selection being full understanding of the system, how is this achieved? In the chemical process industries, the starting place is

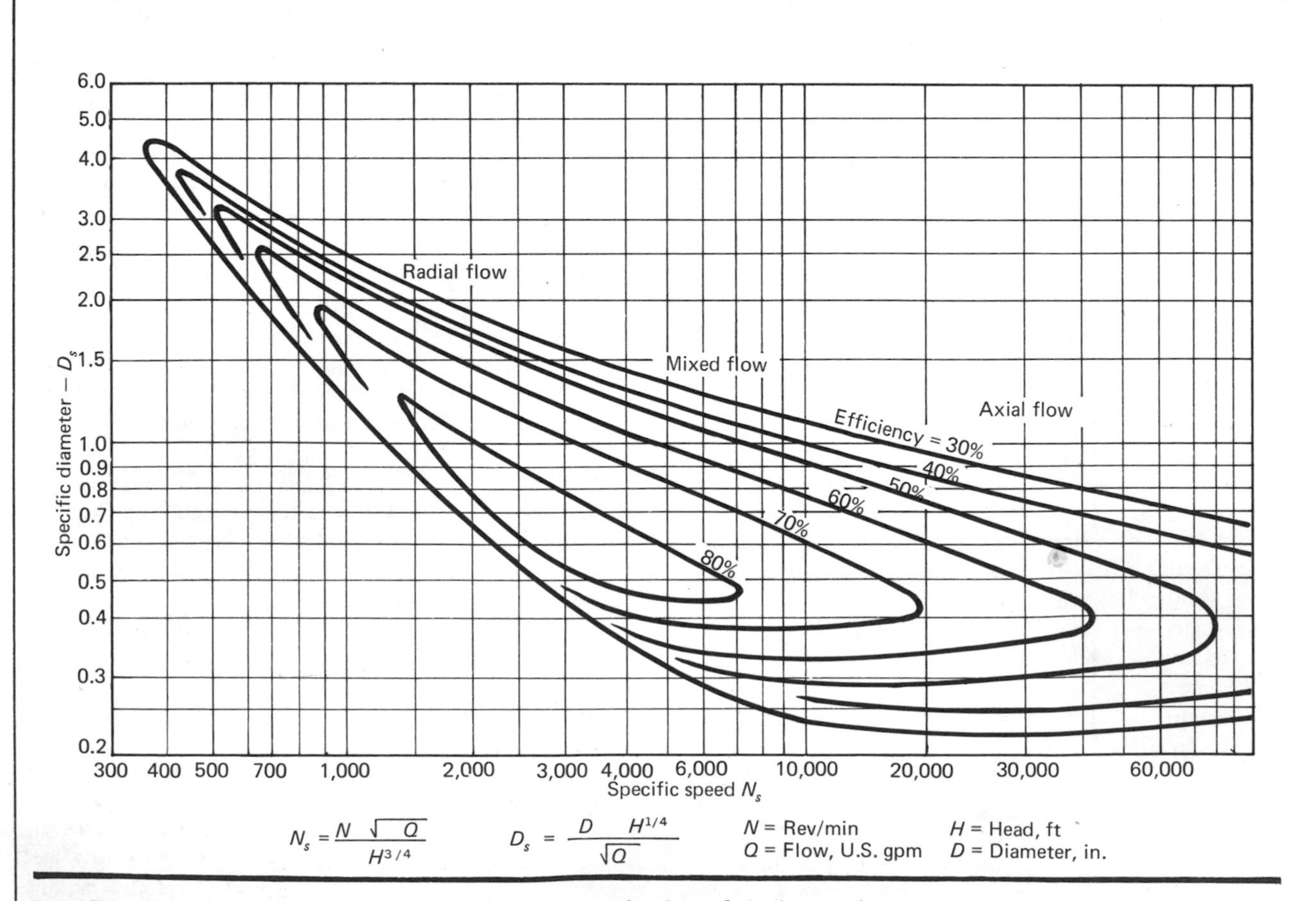


**Specific-speed chart is of great use in the preliminary selection of single-speed centrifugal pumps** **Fig. 1**

Originally published April 3, 1978

**Pump selection for problem shown in Fig. 1** **Table I**

Capacity, gpm 500 Sp. gr. @ temp. 0.88
Total head, ft 350 Viscosity @ temp. 0.8 cP
Temperature, °F 110 NPSH available, ft 20

| Manufacturer | A | B | C |
|---|---|---|---|
| Pump model or size | 3×4×10½ | 4×6×10½ | 3×4×11 |
| No. of stages | 1 | 1 | 1 |
| Speed, rpm | 3570 | 3570 | 3550 |
| Efficiency, % | 71 | 61 | 69 |
| Brake horsepower @ rated point | 54.8 | 72.5 | 56.4 |
| @ end of curve | 63 | 95 | 70 |
| NPSH required, ft. | 18 | 9 | 13 |
| Impeller dia. rated/max.,in | 9⅞/10½ | 9¼/10½ | 9½/11 |
| Cost – pump with motor driver, $ | 6,000 | 6500 | 5,500 |
| Power evaluation | 0 | + 6,338 | + 573 |
| Power cost basis 3 ¢ per kWh 8,000 h/yr 2 years, $ | 19,623 | 25,961 | 20,196 |
| Recommendation | Based on highest efficiency ↑ | | |

**Pump selection for higher NPSH problem described in text** **Table II**

Capacity, gpm 500 Sp. gr. @ temp. 0.88
Total head, ft. 350 Viscosity @ temp. 0.6 cP
Temperature, °F 250 NPSH available, ft Various

| Manufacturer | X | Y | Z |
|---|---|---|---|
| Pump model or size | 3×4×10½ | 3×4×11 | 4×6×10½ |
| No. of stages | 1 | 1 | 1 |
| Speed, rpm | 3,570 | 3,570 | 3,570 |
| Efficiency, % | 71 | 69 | 61 |
| Brake horsepower @ rated point | 54.8 | 56.4 | 72.5 |
| @ end of curve | 63 | 70 | 95 |
| NPSH required, ft. | 18 | 13 | 9 |
| Impeller dia. rated/max.,in | 9⅞/10½ | 9½/11 | 9¼/10½ |
| Cost – pump with motor driver, $ | 6,000 | 5,500 | 6,500 |
| Power evaluation | 0 | +573 | + 6338 |
| Power cost basis 3 ¢ per kWh 8,000 h/yr 2 years, $ | 19,623 | 20,196 | 25,961 |
| Recommendation | Cost of raising suction drum by 5 ft. exceeds power saving. Will not pay out. | Based on adequite NPSH & near to best efficiency. ↑ | Lower NPSH results in too-low efficiency — will not pay out. |

process flow sheets and schematics such as piping and instrument diagrams.

Where pumps take suction from vessels or drums, with the height above the pump changeable, the pump engineer must find the optimum height and coordinate pump requirements in collaboration with other engineers who are designing vessels or foundations. If the pump is installed in a sump or a pit, essential factors include correct sizing of the pit, flow requirements as the liquid approaches the pump, and positioning of the pump in the pit—with adequate spacing and baffles, if required.

Where friction loss through apparatus or piping forms a significant part of the total head, the pump engineer should have some influence in the selection of allowable pressure drop. Often—for example, in an attempt to save on initial cost—the piping designer may select a pipe size which results in high pressure drop. This will require a pump of far more horsepower than a larger-size line would call for. The horsepower consumed by this higher head should be carefully evaluated, as it results in a constant higher cost during the operating life of the pump.

Volatile liquids, hot liquids, viscous liquids, slurries, and crystalline solutions all require special thought and selection methods. Horizontal-shaft or vertical-shaft pumps must be considered along with pump type—centrifugal, reciprocating, rotary, turbine, high-speed, or low-speed. Specification of materials compatible with the pumped liquids is an obvious requirement; not so obvious to some is that a particular type or style of pump may not be available, or may not be economical, in certain special materials. Types of drivers, drive mechanisms, couplings, gears, and seals also enter into the final decision. Much of this is a joint development between purchaser and seller, as the two get closer together on requirements and availability.

## Specific speed as a guide

Referenced in the Hydraulic Institute Handbook [1] and many well-known texts [2,3] also in the writer's earlier article on pump selection [4] is the dimensionless number, "specific speed":

$$N_s = \frac{N\sqrt{Q}}{H^{3/4}}$$

where $N_s$ = specific speed, $N$ = rotating speed, $Q$ = capacity, and $H$ = head.

This helps in rating all centrifugal pumps.

A new chart (Fig. 1), derived from the earlier ones [5] but made more useable for pump work, expresses capacity in gal/min, the customary rate-of-flow unit used in the U.S. today. The same type of chart can be converted to SI units whenever American industry truly goes metric. Some of the examples that follow will refer to Fig. 1, and show how to use it.

**Pump selection for high-pressure problem** **Table III**

Capacity, gpm 250 — Sp. gr. @ temp. 0.88
Total head, ft. 2,625 — Viscosity @ temp. 0.8 cP
Temperature, °F 110 — NPSH available, ft 20

| Manufacturer | A | B | C | D |
|---|---|---|---|---|
| Pump model or size | Horizontal 10-stage | Vertical multistage | Vertical high speed | Reciprocating plunger type |
| No. of stages | 10 | 12 | 1 | 5 cylinders |
| Speed, rpm | 3,550 | 3,550 | 16,200 | 320 |
| Efficiency, % | 67.5 | 68 | 62 | 90 |
| Brake horsepower @ rated point | 216 | 215 | 235 | 162 |
| @ end of curve | 238 | 254 | 250 | Max. 200 |
| NPSH required, ft. | 10 | 8 | 10 | 29* |
| Impeller dia. rated/max., in | $8\frac{1}{4}/8\frac{3}{8}$ | $7\frac{11}{16}/7\frac{13}{16}$ | 5 | — |
| Cost – pump with motor driver, $ | 50,000 | 70,000 | 35,000 | 65,000 |
| Power evaluation | +25,920 | +25,440 | +35,040 | 0 |
| Power cost basis 3¢ per kWh 8,000 h/yr 2 years, $ | 103,680 | 103,200 | 112,800 | 77,760 |
| Recommendation | Most likely choice of conventional style pump. ↑ | Very small power saving does not warrant added cost. | Lowest first cost. Might be seriously considered. ↑? | Although highest in efficiency this would probably not be chosen for a modern plant |

*Too high.

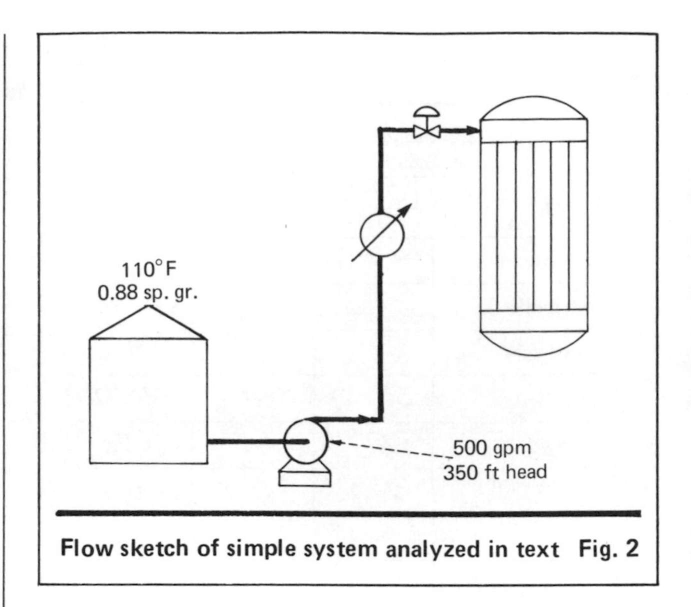


**Flow sketch of simple system analyzed in text Fig. 2**

## Selection for best efficiency

Most process pumps today are of the centrifugal type. High on the list of important things to consider is pump efficiency. In an effort to save on first cost, engineers have frequently chosen pumps that do not represent the most efficient designs available for a given service. Should selection of efficiency be left entirely to the pump manufacturer? He should certainly be given some guidance by the user regarding energy costs and payout methods.

Fig. 2 shows a typical feed pump, taking suction from a storage tank, pumping through a heat exchanger and a control valve into a reactor or process vessel. Assume atmospheric temperature, clean liquid, non-volatile or non-toxic, ample Net Positive Suction Head (NPSH), no solids, viscosity about like water—in other words, about as simple a system as possible. In theory, we could begin by assuming a 60-Hz motor speed of 3,550 rpm and finding the specific speed from Fig. 1 of such a pump (981). Similarly, the specific diameter, $D_s$, and estimated efficiency (72%) could be read from Fig. 1, which would result in a single-stage centrifugal pump at 3,550 rpm, having an impeller of 8.53 in. diameter and an overall efficiency of 72%.

Actually, most of this would be obvious to the experienced user or contractor engineer and such calculations would be unnecessary. A single-stage pump could be specified with confidence, and a preliminary estimate of horsepower, using a rule-of-thumb estimate of 70% efficiency, would not be far wrong. Table I shows how three manufacturers might choose a pump for such conditions. Variations in size and efficiency result from each manufacturer's effort to choose, from his standard line of pumps, the one which most closely meets the purchaser's required conditions. The lower part of the table shows how these selections could be evaluated, based on an assumed cost of 3¢ per kWh and a payout of two years.

## Pump selection for volatile liquids

Using a similar example, but assuming the liquid to be at or near its vapor pressure and stored in a sphere or drum rather than an atmospheric tank, let us look at pump selection based on NPSH.

Usually in such a process flow scheme, it is assumed that there will be liquid/vapor equilibrium in the suction drum. This most conservative approach results in a completely safe calculation. The formula for NPSH available to the pump (NPSHA) is:

$$\text{NPSHA (ft of liquid)} = \frac{(p_s - p_{vp})2.31}{\text{sp. gr.}} + h_s - h_{fr} \quad (1)$$

where $p_s$ = suction pressure, psi; $p_{vp}$ = vapor pressure, psi; sp. gr. = specific gravity of the liquid at pump temperature; $h_s$ = static height, ft; and $h_{fr}$ = friction loss in suction line, ft.

So, NPSHA is entirely a function of static height of the vessel above the pump, less pipe friction in the suction line, since we have assumed that $p_s$ equals $p_{vp}$. See previous references for more on NPSH.

The user must specify the NPSH available to the pump. The pump manufacturer cannot know all details of the user's system, nor reply to a purchase inquiry with alternatives on different pumps requiring different NPSH values. The engineer must look at the economics of setting the suction drum higher, or perhaps increasing the size of the suction line to reduce the friction loss, to reach a realistic value of NPSHA for the given system.

Suppose the engineer suggests 10 feet from grade

level to the lowest liquid level in the suction drum. NPSHA is about 6 feet, based on a pump with impeller centerline two feet above grade level and 1.7-foot piping loss in the 6-inch suction line. He sees that this value of NPSHA appears low, so he also considers heights of 12 and 14 feet, and also considers use of an 8-inch suction line, resulting in several higher values of NPSHA. Pump manufacturers might respond as shown in Table II, where it appears obvious that the higher value of NPSHA makes possible a more efficient pump selection, which can surely pay out the higher initial construction cost in a short time.

Fig. 3 is offered as a guideline for the engineer in determining how much NPSH should be made available to get good pump selections. This guide is based on suction specific speed ($N_{ss}$), an index of pump-suction capabilities, or NPSH required (NPSHR):

$$N_{ss} = \frac{N\sqrt{Q}}{H_s^{3/4}} \qquad (2)$$

where $N$ = rotative speed; $Q$ = capacity (gpm); and $H_s$ = NPSHR (ft).

In these units, modern centrifugal pumps are available with $N_{ss}$ values from 7,000 to 13,000 or higher. Values above about 15,000 will require use of an inducer-type impeller. Double-suction impellers, really equivalent to two single-suction impellers back-to-back and cast in one piece, will give lower NPSHA for the same flow and speed than single-suction types. When the curves in Fig. 3 are used for double-suction impellers, the value for $Q$ must be divided in half.

## Selection of pumps for large capacities

Suppose our example in Fig. 2 requires a flow 10 times as large, with head remaining the same by increasing the size of the piping and apparatus in the system. Suppose also that this pumping system is handling a volatile liquid and that available NPSH consists only of static height minus suction-line friction loss. Checking Fig. 3 for a flow of 5,000 gpm we can see that selecting a single-suction pump at 3,550 rpm is not realistic.

An NPSHR of 50 feet would be unreasonable and unacceptable in a process unit, and probably no such designs exit with manufacturers' standard products. Single suction at 1,760 rpm may be satisfactory, but best of all appears to be the double-suction type, where for a suction specific speed of 11,000 and rotating speed of 1,760, NPSH-required is 16 ft. Cross-checking with manufacturers' standard curves shows that such a pump is definitely available. The user or contractor-engineer can proceed with confidence on such a selection, allowing a reasonable margin of safety in setting vessel or suction-drum heights so that NPSHA exceeds the pump NPSH-required by at least two or three feet.

Certain special services may require a greater margin between NPSHA and NPSHR. A hot vacuum-column bottoms-pump in a typical crude-oil distillation unit becomes a potential troublemaker. Vortex breakers in the bottom of the column, and adequate layout of the suction line to the pump, will help assure a successful operation.

Boiler-feed pumps handling hot water from deaerators will usually require a greater margin, because of alternative operating conditions, or upset conditions that affect equilibrium conditions of water [3]. It is good insurance to make this extra NPSH available in the original system design, because an adequate suction design will eliminate many costly pump troubles.

Similarly, on even larger flows a 1,760 rpm pump may not be adequate, and lower speeds will be required. A pump operating at 1,180 rpm, for example, while perfectly feasible from an NPSH standpoint, may not be available to meet the total head requirements in a single stage. While a multi-stage pump might be used on very large flow services, splitting the total flow into two or more units, each delivering a portion of the total, may solve the problem. Otherwise, a low-speed booster

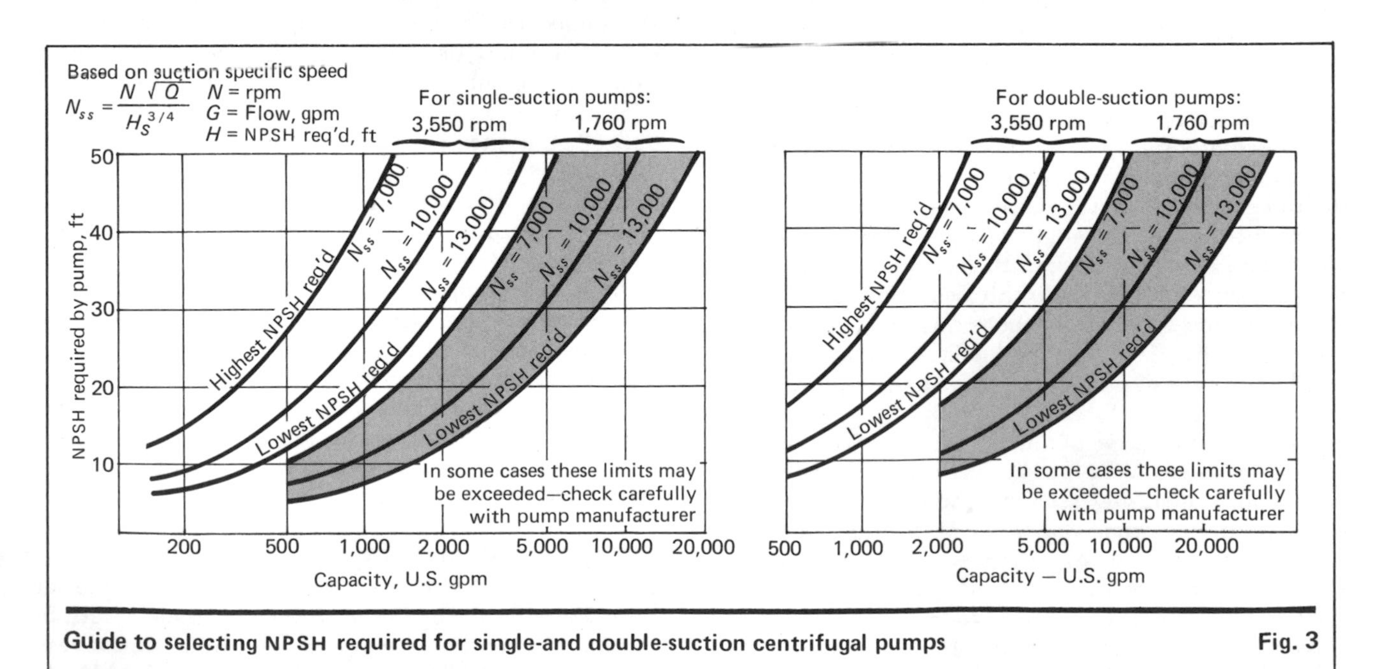


Guide to selecting NPSH required for single-and double-suction centrifugal pumps — Fig. 3

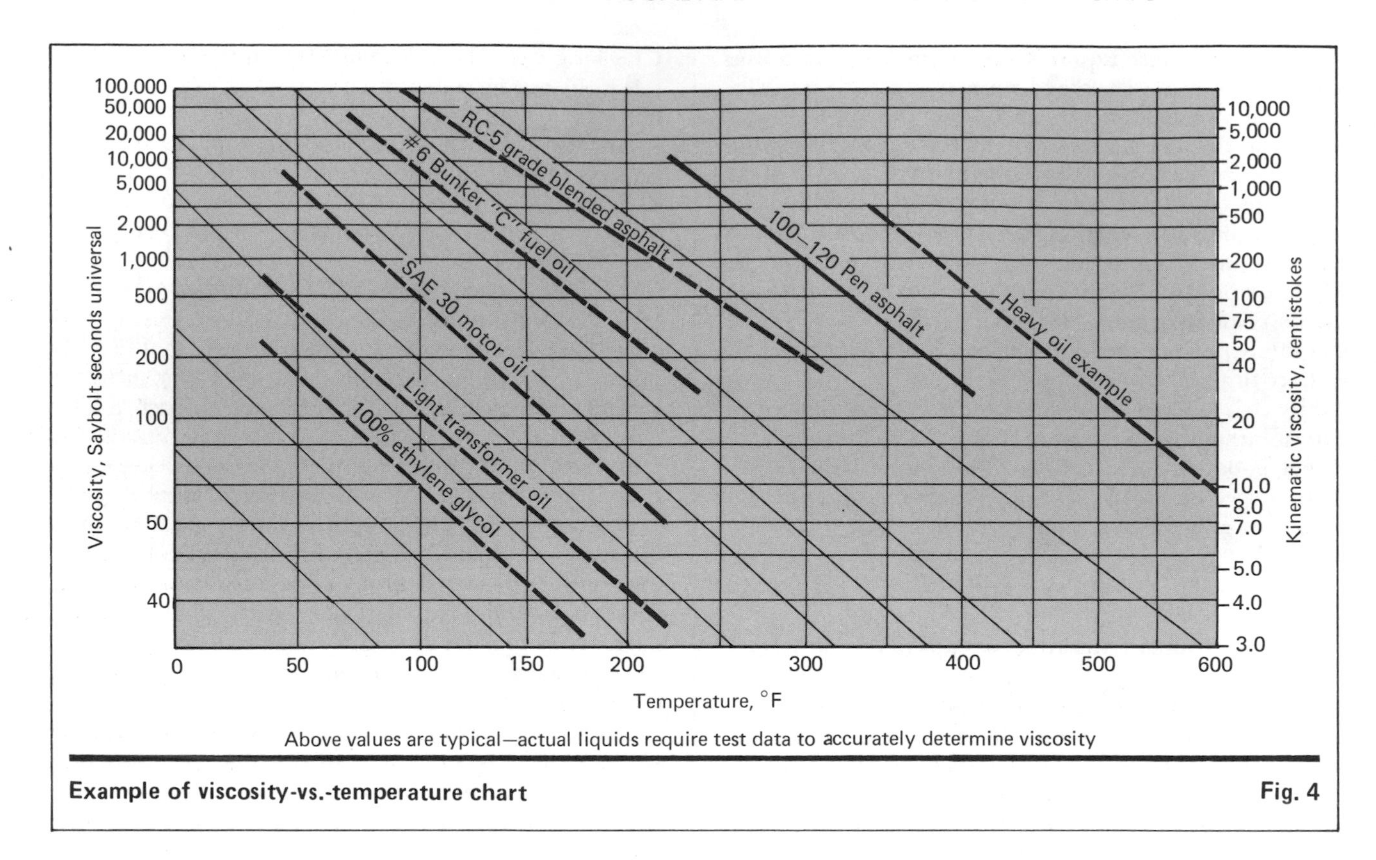


**Example of viscosity-vs.-temperature chart** **Fig. 4**

pump selected for low NPSHR will be used with a conventional multistage pump at higher speed.

## Pumps for higher pressures

If we take, as another example, a reactor-charge pump, as shown in Fig. 2, but assume the flow is 250 gpm and the discharge pressure is 1,000 psi (= 2,625 feet head), how should we approach this pump selection? The multi-stage horizontal centrifugal pump would be the first to look at. From the specific-speed chart (Fig. 1), assume a 10-stage pump and find the required specific speed, impeller diameter, and efficiency. ($N_s = 860$, $D = 7.07$ in., eff. = 70%). Note that all specific-speed charts (including this one) are based on head per stage, not total head of a multistage pump.

A horizontal pump with 10 or more stages may present problems to the designer regarding to shaft design, shaft deflection, interstage clearances, or critical speeds. A vertical-shaft pump might be used, where it is possible to have more stages, since the vertical shaft does not pose the same problems of shaft deflection and critical speed. Assume 12 stages for the vertical type, and again find $N_s = 987$, $D = 6.78$ and efficiency = 71% from Fig. 1. This type of pump may prove to be slightly more efficient.

Another possibility, becoming more popular in the process industries today, is the vertical- or horizontal-shaft pump operating at higher speeds. With the restriction on speed removed, a one-stage pump, with an inducer-type impeller to keep NPSH requirements low, may work under these conditions. (Assume $N_s = 700$, find $N = 16{,}236$ and efficiency = 62% from Fig. 1.)

Last but not least, consider the use of a reciprocating pump. With greater attention focused on energy usage, it takes careful examination of each pumping problem to find the most efficient unit available. The reciprocating pump for these conditions will undoubtedly work most efficiently. Other factors will tend to offset the higher efficiency, however; e.g., higher maintenance required for valves, packing rings, plungers or pistons; and power-frame drive-assemblies. A multi-cylinder reciprocating pump will cause flow pulsations, which call for the use of accumulators or dampers. NPSH requirements of a reciprocating pump may be satisfactory for a pump selected to operate at a reasonable speed. Table III summarizes comparative information on the four suggested types of pumps. Arrangements A and C would be the only commercially attractive alternatives.

Should the head be 2,625 feet, as above, but the flow only 50 gpm, then either the reciprocating pump or the high-speed partial-emission-type centrifugal would be the only solutions, for—as Fig. 1 calculations show—a reasonable multi-stage centrifugal pump at 3,550 rpm could not be designed.

Again, if the flow were 5,000 gpm, such as a large boiler-feed pump might require, a horizontal multistage centrifugal pump becoms the only viable alternative. Thus each situation for high-pressure pumping will be somewhat different, and will require individual attention. The user or contractor-engineer should investigate several types of pumps before choosing one for any given high-pressure service.

## Pumps for viscous liquids

Selecting pumps for viscous liquids requires special considerations. First, the user must accurately specify the actual viscosity of the pumped liquid. Handbooks give viscosities for standard liquids, but special blends, mixtures, etc., may require some special calculations or tests to determine viscosity accurately.

Viscosity is usually expressed in one of three standard units—centipoise, centistokes, or Saybolt Second Universal (SSU). The latter two kinematic viscosities differ from the centipoise, an absolute viscosity. Relation between absolute viscosity and kinematic viscosity is given by:

$$\text{Kinematic viscosity (cSt)} = \frac{\text{absolute viscosity (cP)}}{\text{specific gravity of liquid}}$$

Also useful (above 250 SSU) is the approximate conversion, SSU = 4.62 × cSt. Below 250 SSU, see tables in standard handbooks.

Viscosity of a given liquid will vary with temperature. ASTM provides a chart [8], similar to logarithmic scales, on which viscosities may be plotted against temperature (Fig. 4). Customarily, when making laboratory tests for viscosity, taking two or more points at different temperatures will describe the liquid completely.

The most likely choice for pumping viscous liquids will be a positive displacement pump, either rotary or reciprocating. Rotary-gear, screw, or lobe types actually perform best when applied to viscous liquids, and for the highest viscosities, they are the only usable types.

On the other hand, the *minimum* viscosity must also be known when choosing a rotary pump for a viscous liquid. With low viscosity, slip will be considerably greater in a rotary pump. This reduces the pump capacity to less than the rated capacity at the higher viscosity. The Hydraulic Institute Handbook, pp. 133–134, shows many available positive-displacement rotary pumps. Some of these have definite limits in maximum pumping temperature, maximum working pressures, or choice of available materials of construction.

A reciprocating pump operating at a reduced speed can give excellent performance with viscous liquids. Again, because it is a positive-displacement type, it requires different control methods from those used with centrifugal pumps. If the required discharge pressure is

**Viscous performance of rotary pumps** **Table IV**

| IV-1 A | IV-1 B | Operating case | IV-2 A | IV-2 B |
|---|---|---|---|---|
| 313 | 313 | Required capacity, gpm | 313 | 313 |
| 50 | 50 | Differential pressure, psi | 50 | 50 |
| 8 | 150 | Viscosity, cP | 8 | 150 |
| 600 | 400 | Temperature, °F | 600 | 400 |
| Screw-type rotary (external bearing) 6 × 4 in. | | Type and size of pump | 1-stage centrifugal 3 × 4 × 8 1/2 in. | |
| 1,760 rpm | | Operating speed | 3,550 rpm | |
| 313 | 334 | Delivered capacity, gpm | 325 | 313 |
| 61% | 38% | Approx. efficiency | 66% | 44% |
| 15 | 26 | Approx. horsepower required | 15 | 21 |

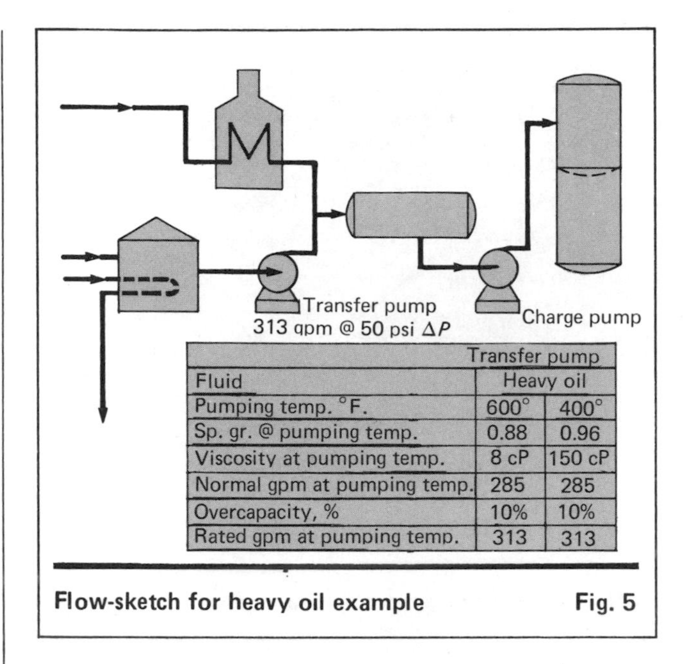


| Transfer pump | | |
|---|---|---|
| Fluid | Heavy oil | |
| Pumping temp. °F. | 600° | 400° |
| Sp. gr. @ pumping temp. | 0.88 | 0.96 |
| Viscosity at pumping temp. | 8 cP | 150 cP |
| Normal gpm at pumping temp. | 285 | 285 |
| Overcapacity, % | 10% | 10% |
| Rated gpm at pumping temp. | 313 | 313 |

**Flow-sketch for heavy oil example** **Fig. 5**

high (500 psi or above), reciprocating pumps probably represent the best choice. Manufacturers' data or Hydraulic Institute methods will help determine how much to derate the capacity of a reciprocating pump for viscous liquids.

Centrifugal pumps are regularly used on liquids of moderate viscosity—up to about 1,000 SSU, sometimes higher. The Hydraulic Institute Handbook carries a chart (p. 104) that is widely accepted in the pump world for derating centrifugal-pump performance in relation to viscosity. This chart shows that, above certain viscosities, centrifugal pumps are not desirable.

## Pump selection for viscous liquids

Consider the system shown in Fig. 5. At first glance, the conditions appear to describe a rather straightforward high-temperature, centerline-supported type of single-stage centrifugal pump. Shown as an alternate operating condition, however, is a much higher viscosity corresponding to a lower temperature. Assuming the engineer has received accurate information about the viscosity at both temperatures, he can select a centrifugal pump for both the highest and lowest temperatures, using the Hydraulic Institute method for derating. Table IV-1 tabulates expected performance; the efficiency has been considerably reduced at the increased viscosity. The pump driver will have to be large enough to accommodate this lower pump efficiency.

All of the above assumes that a centrifugal pump would be chosen. Why not choose the more obvious type, a positive-displacement rotary pump? Many rotary types would be unsuitable for this example because of the high pumping temperature, since most rotaries have an upper operating limit of about 400°F. Most are available only with cast iron or ductile iron casings. For this application a steel or alloy steel casing would undoubtedly be required. The proper high-temperature rotary pump for this service is a screw pump with external timing gears. (Fig. 6). The approximate performance for the rated and alternative conditions are

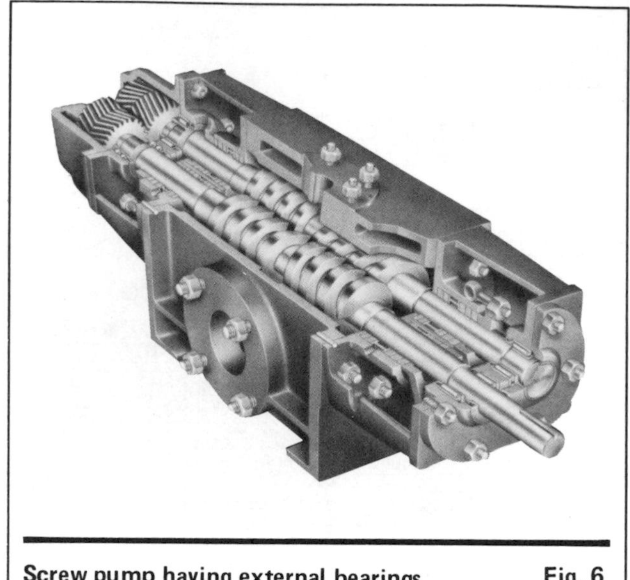

**Screw pump having external bearings** **Fig. 6**

shown in Table IV-2. Horsepower from manufacturer's data has been back-calculated to obtain efficiencies at both points. The pump must be oversized for the viscous condition in order to give sufficient capacity at the lowest viscosity, due to the increased slip at the lower viscosity.

Objections usually raised to the use of this type pump include its having four stuffing boxes instead of one. Also, the flow or pressure controller used downstream of the pump must open a valve in a bypass line since the output of a positive displacement pump running at a fixed speed cannot be throttled, as with a centrifugal.

There is one circumstance that makes the use of a rotary pump almost mandatory. Suppose the viscosity data for the pumped liquid is not well known, and the value of 150 cP at 400°F is the process engineer's best estimate. In further consultation with him, the pump engineer learns that viscosity might run as high as 300 cP, or as low as 100 cP. That higher value would almost certainly preclude the use of a centrifugal pump of any sort (see Hydraulic Institute curve), leaving the rotary pump as the only answer.

## Pumps for slurry service

Either centrifugal or positive-displacement pumps can handle a mixture of solids and liquids, sometimes called two-phase flow, or slurry pumping. Centrifugal pumps, by far the most common for comparatively low head requirements, are normally available in single-stage designs only. Two or more arranged in series can provide higher heads. Pump casings and impellers can be lined with natural or synthetic rubber, or made of hard metal such as alloy iron, 28%-chrome alloys, Ni-hard, etc. Certain processes may require stainless steel.

Chemical-type pumps made of suitable materials are commonly used for light, non-abrasive crystalline slurries. The heavy-duty slurry pump, available in both horizontal and vertical shaft orientation, will perform for more difficult applications, as found in mining and metallurgical processing. The horizontal pumps have end-suction design, and should be lined with rubber for fine abrasive slurries, and with hard metal for coarse slurries. Both types must be designed for easy disassembly (to replace worn parts) with features such as two-piece casings having slotted casing-assembly bolts, plus adjustable wear plates in the hard-metal type. Pump-out vanes on impellers will keep solids build-up away from stuffing box or packing areas.

Vertical slurry pumps operate when submerged in a sump, tank, flotation cell, etc. V-belt drives, often applied to both types, permit the pump speed to match the required service conditions. This works better than attempting to run at fixed motor speeds and to meet pump heads by cutting impellers.

Reciprocating slurry pumps have been used as mud pumps in oil-field drilling, high-pressure slurry pipelines, and high-pressure process applications such as carbamate service in urea processes.

Rotary pumps of the single-screw type or twin-screw type (Fig. 6) have served in relatively nonabrasive-slurry applications, and especially with semi-solids—sludge, thixotropic materials, paste, resin, etc.

Centrifugal pumps in slurry service follow the basic laws and principles as for pumping clear liquids. However, the effects of the solids in the mixture must obviously be taken into account in making a proper pump selection. Some of the considerations are:

■ The correct **specific gravity** of the solids/liquid mixture must be ascertained—determining the concentration by volume ($C_v$) or the concentration by weight

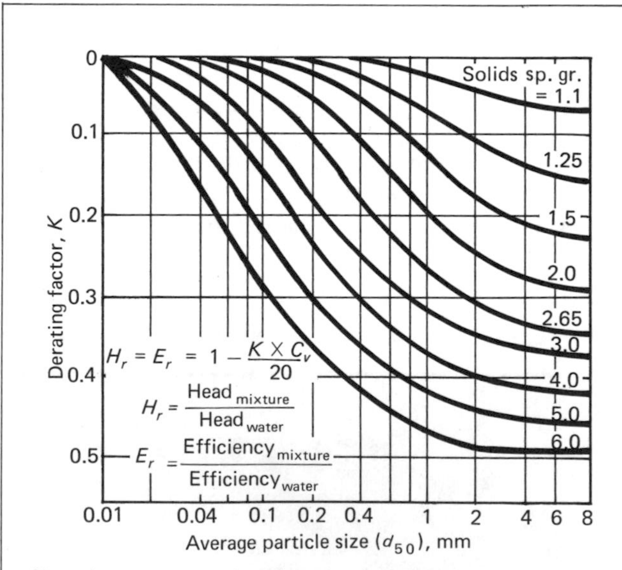


Note these basic equations for mixtures of water and solids:

$$S_m = 1 + \frac{C_v}{100}(S_s - 1) \qquad C_w = \frac{100\, S_s}{\frac{100}{C_v} + (S_s - 1)}$$

$$S_s \times \frac{C_v}{C_w} = S_m$$

$$H_r = E_r$$

where:
$S_s$ = sp. gr. of solids
$S_m$ = sp. gr. of mixture
$C_w$ = % solids in mixture by wt.
$C_v$ = % solids in mixture by vol.

**Slurry factor chart for head and efficiency** **Fig. 7**

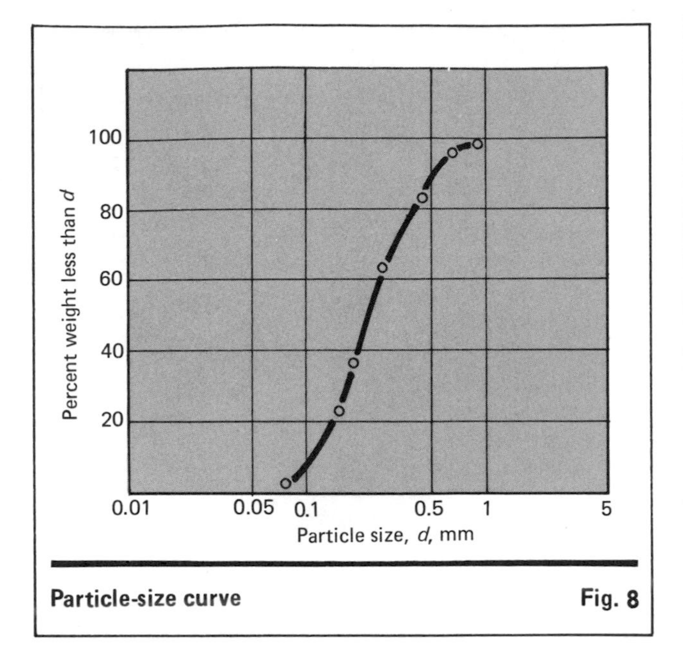


**Particle-size curve** **Fig. 8**

($C_w$) of the solids, the specific gravity of the solids, and the specific gravity of the fluid (usually water), and by utilizing basic relations or nomographs to find the mixture's specific gravity.

■ Correct **materials** for pumps must be chosen, to resist abrasion and wear, with regard to the nature of the solids (sharp or round, hard or soft, crystalline, etc). Sharp particles in rubber-lined pumps will tear the rubber. Fracture of crystals may be avoided or reduced by choosing proper speeds. Pump design must make provision for replacement of worn parts. Use of multiple or spare pumps is often a worthwhile precaution.

■ Pump **performance** on clear water must be derated to compensate for specific gravity, concentration of solids, viscosity of mixture, and added slip due to the mixture.

■ Pumps must run slower than when pumping clear liquids, mainly to reduce wear and abrasion and to give longer life.

Slurry pumps also follow the "affinity laws" regarding change in capacity, head, and horsepower with changing speed:

$$\frac{rpm_1}{rpm_2} = \frac{Q_1}{Q_2} = \sqrt[3]{\frac{H_1}{H_2}} = \sqrt[3]{\frac{hp_1}{hp_2}}$$

where *rpm* is speed, *Q* is flow, *H* is head and *hp* is power.

Variable speed, and provision for selecting any exact speed, are extremely important for slurry pumps.

Knowing details of the solids and carrier fluid makes it possible to determine whether the slurry is non-settling or settling. Non-settling slurries require corrections for viscosity of the pumped mixture, which can be made according to the Hydraulic Institute method [*1*].

Settling-type slurries call for even more arbitrary corrections, since no industry-wide correction method is universally accepted. The reduction in pump performance, due to solids present, is caused not only by the viscosity of the mixture but primarily by losses in the form of slip between the fluid and the solid particles as the mixture accelerates through the pump impeller. Obviously, this slip and the reduction in performance would seem greatest in mixtures with the highest settling velocities.

One method is illustrated (Fig. 7) for obtaining the estimated factors for head reduction ($H_r$) and efficiency reduction ($E_r$) as functions of the solids concentration by volume ($C_v$) and average particle size ($d_{50}$; i.e., the size 50% of the particles pass and 50% do not).

## Slurry-pump selection: example

Problem: to pump 1,000 gpm of sodium carbonate crystalline slurry in water.

Specific gravity of solids ($S_s$) = 2.46 and concentration by weight ($C_w$) = 25%. The total head required is 47 feet. Particle sieve analysis and average particle size determinations are as follows:

| U.S. mesh | % by weight solid | Particle size | Accumulated % passing |
|---|---|---|---|
| +200 | 3 | 0.074 mm | 3 |
| +140 | 9 | 0.105 mm | 12 |
| +100 | 10 | 0.149 mm | 22 |
| + 80 | 15 | 0.177 mm | 37 |
| + 60 | 27 | 0.250 mm | 64 |
| + 40 | 20 | 0.42 mm | 84 |
| + 30 | 13 | 0.59 mm | 97 |
| + 20 | 3 | 0.84 mm | 100 |

(1) Plot on semi-logarithmic paper and read $d_{50} = 0.2$ mm (Fig. 8)

(2) Determine concentration of solids by volume ($C_v$) (Fig. 7)

$$C_w = 25\% = \frac{(100)(2.46)}{\frac{100}{C_v} + (2.46 - 1)}\ ;\ C_v = 11.9\%$$

(3) Determine specific gravity of mixture by reference to Fig. 7

$$S_m = 1 + \frac{11.9}{100}(2.46 - 1) = 1.173$$

(4) Determine head and efficiency reduction, $H_r$ and $E_r$ (Fig. 7)

For $d_{50} = 0.2$ and $S_s = 2.46$, read $K = 0.08$ calculate $H_r = 0.952 = E_r$

(5) Determine head required on clear water

$$H_w = \frac{H_{mix}}{H_r} \text{ or } \frac{47}{0.952} = 49 \text{ ft}$$

(6) Select a suitable slurry pump from manufacturer's printed curves, running at slowest speed with maximum-diameter impeller (6- × 8-in. pump, 880 rpm, 14.75-in. impeller, 78% efficiency, per Fig. 9)

## Speed variation for slurry pumps

If the required speed for a constant-speed application cannot be matched with an induction motor at full-load speed (1,750, 1,160, 875, 705, etc. on 60 Hz, or 1,450, 975, 730, 585 etc. on 50 Hz), a direct-connected driver cannot be used. V-belt drives, commonly used to obtain any intermediate speed, have the advantage that the ratio of the drive can be changed if pumping requirements change. V-belt drives can be overhead-mounted or side-mounted (either side) on horizontal pumps, or arranged vertically to drive vertical pumps.

Where the speed of the pump may be changed during operation, variable-speed drives are also regularly used with slurry pumps. These include:

- Variable-pitch V-belt or chain drives.
- Variable-speed electric motors, such as d.c. motors, or two-speed or multi-speed wound-rotor a.c. motors.
- Eddy-current couplings, where speed of the output shaft to the pump will be varied electrically.
- Hydraulic couplings (fluid couplings), where the hydraulic-fluid mechanism varies the output-shaft speed.

Variable speed drives are desirable for many applications. Although they represent a higher initial investment than a fixed-speed drive, they may be required, or may prove economically attractive, for many reasons, including:

- Variations in flow rate.
- Variations in head, as caused by changing the length of a discharge pipe leading to tailings disposal, as the plant operation is continued.
- Deterioration of pump performance due to wear, making it possible to continue only for a while to pump at required rate of flow and head.
- Correction of errors, made in initial system calculations, due to insufficient or inaccurate data regarding a particular slurry.

## Limitations on slurry pumps

- Head developed per impeller is limited to about 180 to 220 feet, except for certain special designs.
- Impeller-tip speed (the speed of the periphery of the impeller) is limited by some users to 3,500 to 4,500 ft/min. especially on severe-abrasive service. (Centrifugal pumps handling clear liquids may operate at two to three times these speeds.)
- Highest head, or pressure, requirements will probably require reciprocating-type pumps (e.g., long-distance slurry pipelines, requiring 500 to 1,500 psi pressure drop between pumping stations).
- Thickest unpredictable mixtures that cannot be pumped by centrifugal pumps may require air-operated or motor-operated diaphragm pumps.
- Even with proper centrifugal pump selection, many rubber-lined or hard-metal pumps will last only a few weeks in constant service before replacement or adjustment of wear parts is required.

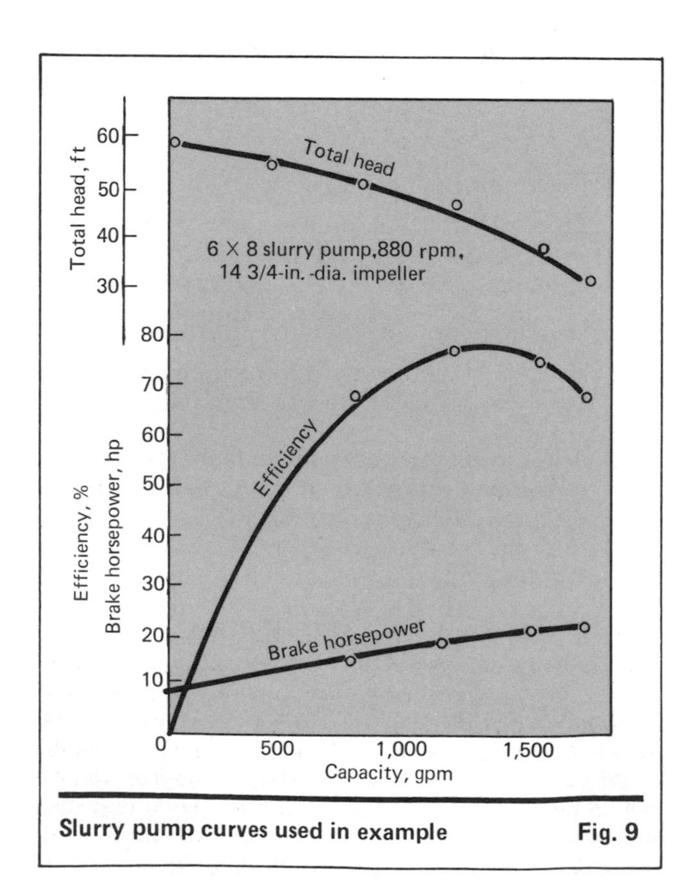

**Slurry pump curves used in example** **Fig. 9**

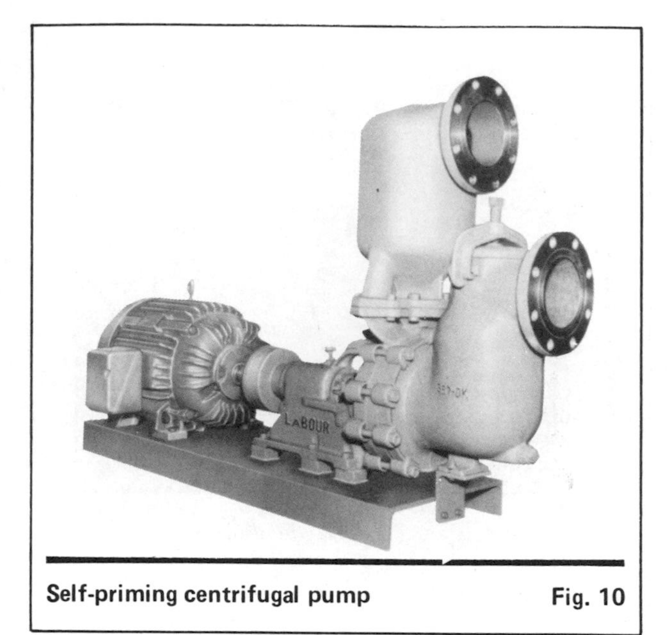

**Self-priming centrifugal pump** **Fig. 10**

## Sealing of slurry pumps

Centrifugal slurry pumps, horizontal type, will have a shaft passing through the pump casing, which must be sealed to prevent leakage. Mechanical seals, such as are used on clean liquids, have not been used for solid/liquid mixtures until recently. Packed stuffing boxes have customarily been used to seal the shafts. Such packing requires a clean liquid (usually water) to flush the pumped material away, providing a film of clean liquid between the packing rings and the shaft or shaft sleeve.

Flushing liquid must come from an external source. In most slurry pump designs, the pump impeller will have some type of back-vanes, on the back side of the impeller, which act as pump-out (or expeller) vanes, to pump liquid away from the stuffing-box region. This makes the pressure at the stuffing box essentially the same as the suction pressure to the pump, so sealing fluid need be supplied at a pressure only 5 to 10 psi above this suction pressure.

On certain designs, however, such vanes are not present and discharge pressure affects the stuffing box. With an excessively worn pump, the pressure may rise in the stuffing box region even if it originally had pump-out

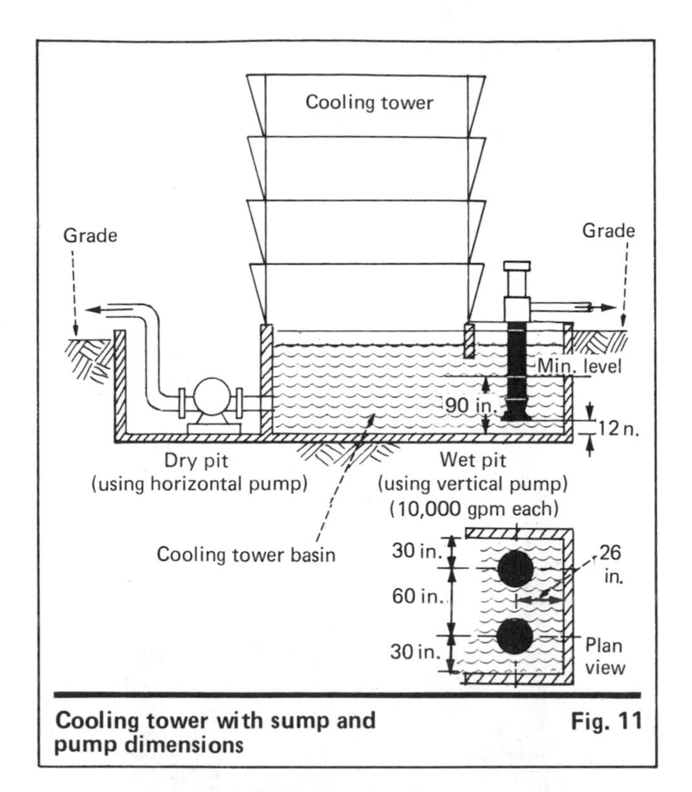


**Cooling tower with sump and pump dimensions** **Fig. 11**

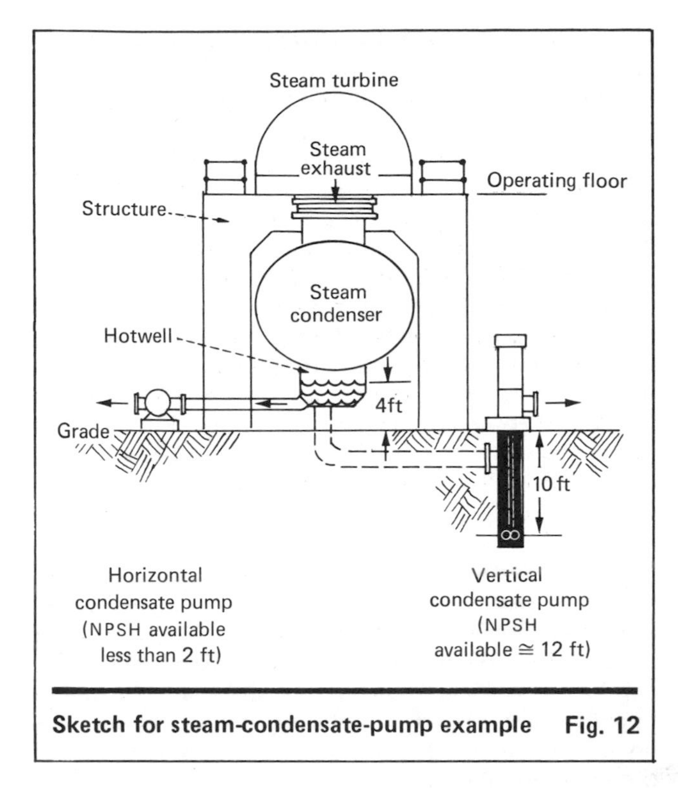


**Sketch for steam-condensate-pump example** **Fig. 12**

vanes. Hence, it is not uncommon to provide seal fluid at 5 to 10 psi above discharge pressure, especially after examination of the particular pump design to determine what the estimated pressure will be in the stuffing box.

## Selection of vertical pumps

Centrifugal pumps, both volute and diffuser types (and some rotary positive-displacement types), are also available with vertical shafts. The specifying pump engineer often must decide which type he wants before contacting a manufacturer's representative.

Many things can enter into his decision. Sometimes the decision appears obvious, such as the use of a vertical pump to pump water from a well or to pump from a pit or sump where the lift is above 34 feet. (Any horizontal pump will have a certain capability for suction lift, and sometimes a horizontal can pump from a basin, a pit, or a river. But such suction lift is limited to less than 34 feet—equal to atmospheric pressure at sea level.) Although horizontal pumps are sometimes used for low-suction lifts, the problem of priming the pump presents itself each time the pump is started. Eductors, vacuum pumps, or other air-removal apparatus can be used; foot valves can retain the liquid in the suction pipe; or, within limited size ranges, self-priming pumps may be used (Fig. 10). The self-priming pump has a built-in chamber in front of the impeller that traps a sufficient volume of liquid to allow the pump to make repeated starts indefinitely after the initial filling of the chamber.

Unless there is a compelling reason not to submerge the pump in the pit, such as the presence of a very dirty or corrosive liquid, the vertical pump is often considered best in spite of several disadvantages: first, it is more difficult to maintain a vertical, as the pump must be completely removed from its location to get to the pumping elements; second, the lineshaft that supports the pump impeller will require bearings, usually at about five-foot intervals, which in turn require lubrication and regular maintenance; third, vertical drivers must be used, generally more expensive than horizontal drivers; also, coupling alignment may be more critical, to ensure absolute concentricity for the overall shaft assembly.

In spite of these objections (all of which have been successfully overcome by vertical-pump technology), the vertical pump is the preferred choice for many applications. Two examples illustrate this preference:

Consider pumps required to circulate cooling water from a cooling-tower basin throughout a process plant. Fig. 11 shows two possible arrangements, based on the assumption that the cooling-tower basin is located at grade level, with the depth of water variable as a function of pit design, cooling-tower size, and pump requirement.

Obviously, the vertical pump, which can be submerged directly in the basin, permits lower construction costs for the basin. The horizontal, located in a dry pit adjacent to the cooling-tower basin to avoid the problem of priming, requires added plot area, higher construction costs, installation of access ladders or stairs, and usually a drainage sump pump to keep the pit dry. The main advantage of the horizontal—easier maintenance—is usually outweighed by the lower installed cost of the vertical type. Maintenance requirements for either type should be rather low for clean, clear liquids.

When considering vertical pumps in pits or sumps, the engineer must not overlook the very important matter of where to place the pumps in the pit. Proper velocities must be maintained in the approach channels to the pumps. A sufficient distance must be allowed between two or more pumps in a sump. To avoid vortexing and consequent pump trouble, optimum distances for pumps from the walls or floors of pits or sumps are essential to satisfactory operation. The Hydraulic Institute has published recommendations on this subject in every recent edition of its Handbook [*1*]. Based on current recommended figures, the typical installation shown in Fig. 11 for flow of 10,000 gpm would have pumps located approximately as shown.

The vertical pump may often be preferred for volatile liquids where low NPSHA is a problem. Consider a condensate pump, pumping hot water from a hotwell of a steam condenser in a steam power plant (Fig. 12). Obviously, NPSHA is a potential problem. To get adequate values for a conventional horizontal pump, the condenser must be raised, and also the entire structure above it.

Special horizontal low-NPSH condensate pumps are available, with slow speeds and a large impeller eye-area, resulting in NPSH-required values of from $1\frac{1}{2}$ to 4 ft. But the modern way to handle this pumping job employs a vertical "can" pump in which the pump impellers are placed a sufficient distance below the hotwell to ensure adequate NPSHA with no need to raise the structure above.

Another form of vertical-shaft pump, the vertical in-line type (Fig. 13), has become very popular in recent years. Hydraulically this type may be similar to, or even the same as, its horizontal counterpart. But it can offer some big advantages in overall plant constructed-cost, because of the saving in space and the simplified piping required.

Smaller-size in-line pumps can be installed without any baseplate or foundation, held only by pipe supports on the adjacent piping. High-speed models, incorporating built-in gear-speed increasers, make it possible to achieve much higher heads with a single-stage pump than can be generated at 3,600 rpm.

The user or contractor should decide whether a given pumping system should be horizontal or vertical. It is not realistic, or even fair to suppliers, to compare bids on one type against the other. First, make the choice, then compare and recommend-for-purchase from available similar types or styles.

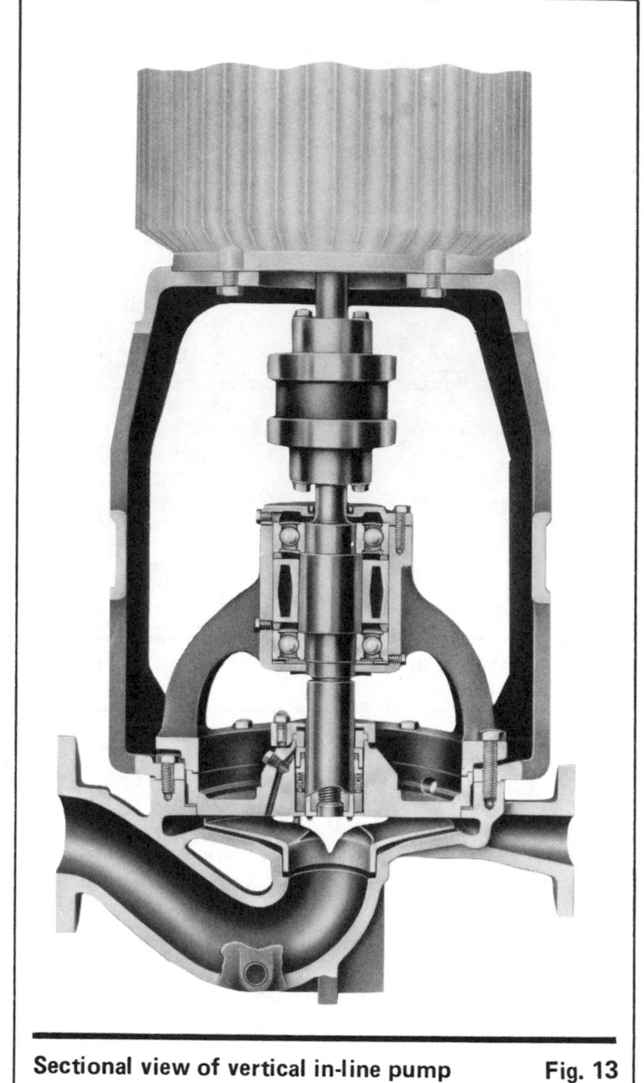

**Sectional view of vertical in-line pump** **Fig. 13**

## Selection of controlled-volume pumps

A special category of pumps, known variously as controlled-volume, metering, or proportioning, is really a type of positive-displacement pump, where motion is transmitted from a driver through cranks, wobble-plates, or various mechanisms to one or more reciprocating plungers. These, in turn, either pump the actual fluid or pump hydraulic oil that actuates a diaphragm to do the pumping. The special feature of these pumps is their capability for stroke adjustment, either manually or automatically, to permit metering an exact amount of fluid into the system.

The pumps have the characteristic of repetitive accuracy. This means that—under fixed conditions of speed, pressure, and stroke-setting—they will continue to deliver the same quantity of fluid on every stroke. Also, such a pump has reset accuracy, meaning that it can be set to an original capacity, then to a different capacity, then back to the original setting where it will again deliver the original amount within a stated minimal tolerance.

When the engineer finds an application requiring this kind of accuracy, it usually involves a comparatively small total flow (often expressed in gallons per hour or cubic centimeters per hour rather than gpm). The controlled-volume pump is the only type that will give the needed accuracy. Pressure requirements may range from very low pressures—such as injecting chemicals into a cooling-water system to control pH of the total system—to extremely high pressures (10,000 psi or higher), where plunger types can readily meet the requirements.

The specifying engineer should make the choice between a plunger pump, where the liquid is pumped directly by the plungers, and a diaphragm pump, where the pumped liquid is entirely separate. For extremely corrosive, toxic, or lethal liquids, or those otherwise dangerous in event of leakage, the diaphragm type will probably be mandatory. Even for non-toxic or non-dangerous liquids, the diaphragm type may be pre-

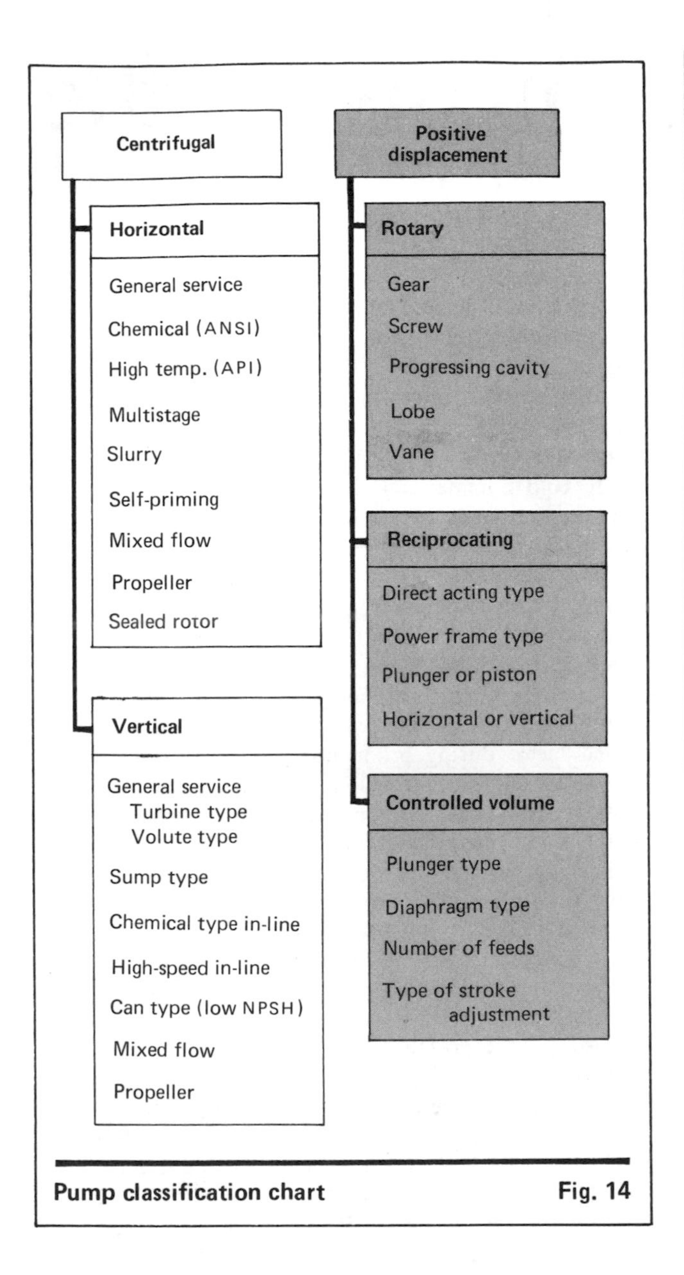


**Pump classification chart** **Fig. 14**

ferred, depending on size and pressure requirements. Many special liquid-end materials, now available, suit the needs of the pumped fluid. The user should examine pump speeds to avoid choosing a unit that runs too fast, and may need excessive maintenance.

When several different streams must be metered in the same location, pumps with two or more plungers may be called for. In most designs, the plungers can be controlled and adjusted independently. On some complex systems, as many as ten pump feeds are connected to one driver.

The subject of NPSHA, and also acceleration head in suction-piping systems, is significant, even though these pumps are usually small in size. Several references provide guides for controlled-volume pump application and selection [6,7].

## Sealed pumps

Another specialty category, important to the chemical process industries, is the sealed pump, which has no external seal or potential leakage. The two major types are the canned-rotor and the magnetic. Sealed pumps are used where no leakage can be tolerated, or where pump-seal failure might cause major trouble or disaster.

Such pumps are available in a limited range of sizes, most are low-flows, and all are of single- or two-stage construction. They have been used for both high-temperature and very low-temperature liquids. High-suction-pressure applications avoid the need for a troublesome high-pressure stuffing box. The centrifugal-type pumps follow the same hydraulic performance rules as conventional centrifugal pumps. Because of their small size, these pumps show a rather low efficiency but, in dangerous applications, efficiency must often be sacrificed for safety.

## Summary

Fig. 14 will help the selecting engineer to choose the proper pump for any system. Use the chart as a starting point, or a reminder of the many types available, and proceed logically to one or more suitable types for a specific duty. Call on the pump manufacturer for help, but don't leave the whole job to him. The user must make certain important decisions before the manufacturer or supplier can do his job properly.

## Acknowledgment

I would like to thank the following companies that provided information for this article. (Numbers in parentheses indicate the figure numbers of illustrations provided by that company.) Bingham Pump Co.; Borg-Warner Corp., Byron Jackson Div.; Crane Co., Chempump Div.; The Duriron Co. (14); The Galigher Co.; Hills-McCanna Co.; Ingersoll-Rand Co.; La Bour Pump Co. (10); Pacific Pumps Div., Dresser Industries; Sundstrand Fluid Handling Div.; Warman International, Inc.; Warren Pumps, Inc. (6); Worthington Pump Inc.

## References

1. Hydraulic Institute Standards for Centrifugal, Rotary and Reciprocating Pumps, 13th ed. Hydraulic Institute, Cleveland, Ohio.
2. Stepanoff, A. J., "Pumps and Blowers; Two Phase Flow," Wiley, N.Y., 1965.
3. Karassik, I. J., and Krutzsch, W. C., "Pump Handbook," McGraw-Hill, New York, 1976.
4. Neerken, R. F., Pump Selection for the Chemical Process Industries, *Chem. Eng.,* Feb. 18, 1974, pp. 104–115.
5. Baljo, O. E., A Study on Design Criteria and Matching of Turbomachines, *Trans. ASME (Am. Soc. Mech. Engs.) Ser. A: J. Eng. Power,* Jan. 1962.
6. Hernandez, L. A., Jr., Controlled-Volume Pumps, *Chem. Eng.,* Oct. 21, 1968.
7. Hefler, John, Figure NSPH for Proportioning Pumps, *Pet. Ref.,* June 1956.
8. ASTM D341-77, Viscosity-Temperature Charts for Liquid Petroleum Products, ASTM Standards, Part 23, 1977.

## The author

**Richard F. Neerken** is chief engineer of the Rotating Equipment Group, The Ralph M. Parsons Company, Pasadena, California, 91124. He joined Parsons in 1957 and has worked continuously with rotating machinery such as pumps, turbines, compressors, and engines on many projects for the Company. He directs a group of over thirty engineers doing similar work on Parsons' projects throughout the world. He has a B.S. degree in mechanical engineering from California Institute of Technology, is a Registered Professional Engineer in California, and is a member of the Contractors' Subcommittee on Mechanical Equipment for API.

# Pump Requirements for the

JOHN R. BIRK and JAMES H. PEACOCK, The Duriron Co.

Pumps for the chemical process industries differ from those used in other industries primarily because of the materials of which they are made.

While cast iron, ductile iron, carbon steel, and aluminum or copper-base alloys will handle a few chemical solutions, most chemical pumps are made of stainless steels, nickel-base alloys, or more-exotic metals such as titanium and zirconium. Pumps are also available in carbon, glass, porcelain, rubber, lead and whole families of plastics, including phenolics, epoxies and fluorocarbons.

Each of these materials has been incorporated into pump designs for just one reason—to eliminate or reduce the destructive effect of the chemical on the pump parts.

Since the type of corrosive liquid will determine which of these materials will be most suitable, a careful analysis of the chemical to be handled must first be made.

## Major and Minor Constituents

Most important in studying any solution will be a knowledge of its constituents. This includes not only the major constituents but the minor ones as well, for in many instances the minor constituents will be the more important. They can drastically alter corrosion rates and, therefore, a detailed analysis is most critical.

Closely allied, and directly related to the constituents, is the concentration of each. Merely stating "concentrated", "dilute" or "trace quantities" is basically meaningless because of the broad scope of interpretation of these factors.

For instance, some interpret "concentrated" as meaning any constituent having a concentration greater than 50% by weight; whereas others interpret any concentration above 5% in a like manner. Hence, it is always desirable to cite the percentage by weight of each constituent in a given solution. This eliminates multiple interpretation and permits a more accurate evaluation.

It is also recommended that the percentage by weight of any trace quantities be cited, even if this involves only parts per million. For example, high-silicon iron might be completely suitable in a given environment in the absence of fluorides. If, however, the same environment contained even a few parts per million of fluorides, the high-silicon iron would suffer catastrophic failure.

## Properties of the Solution

Generalized terms such as "hot," "cold," or even "ambient" are ambiguous in that they can be interpreted in different ways. The preferred terminology would be the maximum, minimum and normal operating temperature in degrees, either F or C. Chemical reactions, in general, increase in rate of activity approximately two to three times with each increase of 18°F in temperature. Corrosion can be considered a chemical reaction, and with this in mind, the importance of temperature or temperature range is obvious.

A weather-exposed pump installation is a good illustration of the ambiguity of the term ambient for there could be as much as a 150°F difference between an extremely cold climate and an extremely warm one. If temperature cannot be cited in actual degrees, the ambient temperature should be qualified by stating the geographic location of the pump. This is particularly important for pump materials that are subject to thermal shock in addition to increased corrosion rates in high-temperature environments.

More often than not, little consideration is given to the pH of process solutions. This can be a critical and well-controlled factor during production processing, and it can be equally revealing in evaluating solution characteristics for material selection. One reason the pH may be overlooked is that generally it is obvious whether the corrosive is acidic or alkaline. However, this is not always true, particularly on process solutions that may have the pH adjusted so that it will always be either alkaline or acidic. When this situation exists, the precise details should be known so that a more thorough evaluation can be made.

It is also quite important to know when a solution is alternating between acid and alkaline conditions because this can have a pronounced effect on materials selection. Some materials, while entirely suitable for handling a given alkaline or acid solution, may not be suitable for alternately handling the same solutions.

Erosion-corrosion, velocity, and solids in suspension are closely allied in pump services for the chemical process industries.

Pump design is a very critical factor when solids are in the solution. It is not uncommon for a given alloy to range from satisfactory to completely unsatisfactory in a given chemical application with hydraulic design being the only variable factor. Failure to cite the presence of solids on the data sheet for a solution is a common occurrence. This undoubtedly is the cause of catastrophic erosion-corrosion failures experienced in many pumps when the presence of solids is omitted.

The presence of air in a solution can be quite significant. In some instances, it is the difference between success and failure, in that air can conceivably render a reducing solution into an oxidizing one. Under these circumstances, an altogether different material may be required. A good example of this would be a self-priming, nickel-molybdenum-alloy pump for handling commercially pure hydrochloric acid. This alloy is excellent for the commercially pure form of this acid, but any condition that can induce even slightly oxidizing tendencies renders this same alloy completely unsuitable.

Originally published February 18, 1974

# Chemical Process Industries

## Other Factors in Pump Operation

When a pump is used for transferring or recirculating a solution, there is a possible buildup of corrosion products or contaminants that can influence the service life of the pump. Such a buildup can have a beneficial or deleterious effect; and for this reason, the possibility of such buildup should be an integral part of evaluating solution characteristics.

Inhibitors and accelerators can be intentionally or unintentionally added to the solution. Inhibitors reduce corrosivity, whereas accelerators increase it. Quite obviously, no one would add an accelerator to increase the corrosion rate on a piece of equipment; but the addition of a minor constituent, as a necessary part of a given process, may serve the same purpose. Thus, the importance of knowing whether such constituents are present.

Where purity of product is of absolute importance, particular note should be made of any element that may cause contamination problems, whether it be discoloration of product or solution breakdown. In some environments, pickup of only a few parts per billion of certain elements can create severe problems. This is particularly important in pumps, where velocity effects and the presence of solids can alter the end-result, as opposed to other types of process equipment where the velocity and/or solids may have little or no effect. When a material is basically suitable for a given environment, it follows that purity of product should not be a problem. However, this cannot be an ironclad rule, particularly with chemical pumps.

Depending on the process liquid, continuous or intermittent contact can affect service life. Intermittent duty in some environments can be more destructive than continuous duty if the pump remains half full of corrosive during periods of downtime, and causes accelerated corrosion at the liquid interface. Perhaps of equal importance is whether or not the pump is flushed and/or drained when not in service.

## Corrosives and Materials

Materials for pump applications can, in general, be divided into two very broad categories; metallic and nonmetallic. The metallic category can be further subdivided into ferrous and nonferrous alloys, both of which have extensive application in the chemical industry. The nonmetallics can be further subdivided into natural and synthetic rubbers, plastics, ceramics and glass, carbon and graphite, and wood.

Of the nonmetallic categories, wood of course has little or no application for pump services. The other materials cited have definite application handling heavy corrosives. Plastics, in particular, are noted for their corrosion resistance, and see wide service in chemical environments.

For a given application, a thorough evaluation of not only the solution characteristics but also the materials available should be made to ensure the most economical selection.

## Sources of Data

To evaluate materials for chemical pump services, various sources of data are available. The best is practical experience within one's own organization. It is not unusual, particularly in large organizations, to have a materials group or corrosion group whose basic responsibilities are to collect and compile corrosion data pertaining to process equipment in service at the company's various plants. These sources should be consulted whenever a materials-evaluation program is made.

A second source of data is laboratory and pilot-plant experience. Though it quite logically cannot provide as valuable and detailed information as actual plant experience, it can certainly be indicative and serve as an important guide.

The knowhow of suppliers can be a third source of information. Though suppliers cannot hope to provide comments on the specific details of a given process and the constituents involved, they normally can provide assistance and corrosion coupons to facilitate a decision.

Technical journals, handbooks and periodicals are a fourth source of information. A wealth of information is contained in them, but if an excellent information-retrieval system is not available, it can be very difficult to obtain the desired information.

## Types of Pump Corrosion

The corrosion encountered in chemical pumps may appear to be unique compared to that found in other process equipment. Nevertheless, like other types of chemical process equipment, pumps will experience eight forms of corrosion, with some being more predominant in them than in other types of equipment. It is not the intent here to describe in detail these forms of corrosion, but only a brief description, so that the various forms can be recognized when they occur.

1. General, or uniform, corrosion is the most common type, characterized by essentially the same rate of deterioration over the entire wetted or exposed surface. General corrosion may be very slow or very rapid, but is of less concern than the other forms because of its predictability. However, predictability of general corrosion in a pump can be difficult because of the varying velocities of the solution within.

2. Concentration-cell, or crevice, corrosion is a localized form resulting from small quantities of stagnant solution in areas such as threads, gasket surfaces, holes, crevices, surface deposits, and under bolt and rivet heads. When concentration-cell or crevice corrosion occurs, a difference in concentration of metal ions or oxygen exists in the stagnant area compared to the main body of the liquid. This causes an electrical current to flow between

the two areas, resulting in severe localized attack in the stagnant area. Usually, this form of corrosion does not occur in chemical pumps except perhaps in misapplications, or in designs where the factors known to contribute to concentration-cell corrosion have been ignored.

3. Pitting corrosion is the most insidious, destructive form of corrosion, and very difficult to predict. It is extremely localized; is manifested by small or large holes (usually small); and the weight loss due to the pits will be only a small percentage of the total weight of the equipment. Chlorides in particular are notorious for inducing pitting that can occur in practically all types of equipment. This form of corrosion, in some instance, can be closely allied to concentration-cell corrosion, as pits may initiate in the same areas where such corrosion is manifested. Pitting is commonplace at areas other than stagnant ones; whereas, concentration-cell corrosion is basically confined to areas of stagnation.

4. Stress-corrosion cracking is localized failure caused by the combination of tensile stresses and a specific medium. Undoubtedly, more research and development has been conducted on this form of corrosion than any other. Nevertheless, the exact mechanism of stress-corrosion cracking is still not well understood. Fortunately, castings, due to their basic overdesign, seldom experience stress-corrosion cracking. Corrosion fatigue, which can be classified as stress-corrosion cracking, is of concern in chemical-pump shafts because of repeated cyclic stressing. Failures of this type occur at stress levels below the yield point, due to the cyclic application of the stress.

5. Intergranular corrosion is a selective form of corrosion at, and adjacent to, grain boundaries. It is associated primarily with stainless steels but can also occur with other alloy systems. In stainless steels, it results from subjecting the material to heat in the 800 to 1,600°F range. Unless other alloy adjustments are made, this form of corrosion can be prevented only by heat treating. It is easily detectable in castings, because the actual grains are quite large compared to wrought material of the equivalent composition. In some instances, uniform corrosion is misinterpreted for intergranular corrosion because of the etched appearance of the surfaces exposed to the environment. Even in ideally heat-treated stainless steels, very slight accelerated attack can be noticed at the grain boundaries because these areas are more reactive than the grains themselves. Caution is required to avoid misinterpreting general and intergranular corrosion. Stainless-steel castings will never encounter intergranular corrosion if they are properly heat treated, after being exposed to temperatures in the 800 to 1,600°F range.

6. Galvanic corrosion occurs when dissimilar metals are in contact or otherwise electrically connected in a corrosive medium. Corrosion of the less-noble metal is accelerated, and the more corrosion-resistant metal is decreased compared with their behavior when not in contact. The further apart that the metals or alloys are in the electromotive series, the greater the possibility of galvanic corrosion.

When it is found necessary to have two dissimilar metals in contact, caution should be exercised to make certain that the total surface area of the least-resistant metal far exceed the more corrosion-resistant material. This will tend to prevent premature failure by simply providing a substantially greater area of the more corrosion-prone material. This form of corrosion is not common in chemical pumps, but may be of some concern with accessory items that may be in contact with the pump parts, and are subjected to the environment.

7. Erosion-corrosion is characterized by accelerated attack resulting from the combination of corrosion and mechanical wear. It may involve solids in suspension and/or high velocity. It is quite common with pumps where the erosive effects prevent the formation of a passive surface on alloys that require passivity to be corrosion resistant. The ideal material to avoid erosion-corrosion in pumps would possess the characteristics of corrosion resistance, strength, ductility and high hardness. Few materials possess such a combination.

Cavitation is considered a special form of erosion-corrosion that results from the collapse or implosion of gas bubbles against the metal surface in high-pressure regions. The stresses created are high enough to actually remove metal from the surface and destroy passivity. The same material qualities mentioned above are desirable for cavitation resistance, but a change in piping or an increase in suction pressure will eliminate the problem in most cases.

8. Selective-leaching corrosion involves removal of one element from a solid alloy in a corrosive medium. Specifically, it is typified by dezincification, dealuminumification, and graphitization. This form of attack is not common to chemical pumps, because the alloys in which it occurs are not commonly used in heavy chemical applications.

## Typical Materials of Construction

The most widely used metallic materials of construction for chemical pumps are the stainless steels. Of the many available, the most popular are the austenitic grades such as Type 304 and Type 316 because of their superior corrosion properties compared to the martensitic or ferritic grades.

The stainless steels are used for a wide range of corrosive solutions. They are suitable for most mineral acids at moderate temperatures and concentrations. The notable exceptions are hydrochloric and hydrofluoric acid. In general, the stainless steels are more suitable for oxidizing as opposed to reducing environments. Organic acids and neutral-to-alkaline salt solutions are also handled by stainless-steel pumps.

Carbon steel, cast iron, and ductile cast iron are also frequently used for the many noncorrosive applications found in most plants.

For the more-severe or critical services, the high-alloy stainless steels such as Alloy 20 are frequently specified.

Nickel-base alloys, because of their relatively high cost, are generally used only where no iron-base alloy is suitable. This family of corrosion-resistant materials includes: pure nickel, nickel-copper, nickel-chromium, nickel-molybdenum, and nickel-chromium-molybdenum alloys.

Copper-base alloys such as bronze or brass, aluminum and titanium are the most frequently used nonferrous metals for chemical pumps after the nickel-base alloys. Zirconium has also found application in a few very-special areas.

Both natural-rubber and synthetic-rubber linings are used extensively for abrasive and/or corrosive applications. Soft natural rubber generally has the best abrasion resistance but cannot be used at as high a temperature as semihard natural rubber or the synthetic rubbers such as Neoprene and butyl. In most cases, the hard rubbers and synthetic rubber also possess better chemical resistance.

Plastics are one of the fastest growing families of pump materials. A multitude of new plastics have been introduced in the last 10 to 15 years. For the ultimate in chemical resistance, the fluorocarbon resins such as polytetrafluoroethylene (PTFE) and fluorinated ethylene-propylene (FEP) are finding wide application. Where strength and chemical resistance is needed, a variety of fiber-reinforced plastics (FRP) are available. Epoxy, polyester, and phenolic are three of the more popular FRP materials. Polyvinyl chloride, polyethylene and polypropylene are also finding application. Plastics are gaining in popularity because they offer the corrosion resistance of the more expensive metals at a fraction of the cost. However, it is doubtful that plastics will ever completely replace metals.

Ceramic or glass construction is avoided whenever possible, because of the poor mechanical properties of these materials. However, for many extremely corrosive services at elevated temperatures, glass or ceramic are the most suitable materials because of their extreme chemical inertness.

Carbon or graphite construction is generally used for the same kind of service. The primary reason for using carbon or graphite instead of glass or ceramic is that the latter are unsuitable for services where HF or alkalis are handled.

## Types of Chemical Pumps

The second step in selecting a chemical pump is based on the characteristics of the liquid and on the desired head and capacity. It should be noted that not all types of pump are available in every material of construction, and the final selection may depend on availability of designs in the proper material.

Centrifugal pumps are used extensively in the chemical process industries because of their suitability for practically any service. They are available in an almost unending array of corrosion-resisting materials. While not built in extremely large sizes, pumps with capacity ranges of 5,000 to 6,000 gpm are commonplace. Heads range as high as 500 to 600 ft at standard electric-motor speeds. Centrifugals are normally mounted in the horizontal position; but they may also be installed vertically, suspended into a tank, or hung in a pipeline. Disadvantages include reduced performance when handling liquids of more than 500 SSU viscosity, and the tendency to lose prime when comparatively small amounts of air or vapor are present in the liquid.

Rotary pumps such as the gear, screw, deforming-vane, sliding-vane, axial-piston, and cam types are generally used for services requiring 500 to 1,000-psi discharge pressure. They are particularly adept at pumping liquids of high viscosity or low vapor pressure. Their constant displacement at a set speed makes them ideal for metering small quantities of liquid. Since they operate on the positive-displacement principle, they are inherently self-priming. When built of materials that tend to gall or seize on rubbing contact, the clearances between mating parts must be increased, resulting in decreased efficiency. The gear, sliding-vane and cam units are generally limited to use on clear, nonabrasive liquids.

Reciprocating pumps have, to a great extent, been replaced by centrifugal or rotary units, except for special applications. They are still used extensively where their variable-speed and variable-stroke capabilities are important process considerations. This characteristic, in combination with their inherent ability to handle volatile and very viscous liquids, makes them particularly suitable for metering and injection systems where low capacity and high head are normal duty parameters. These pumps are available for discharge pressures as high as 50,000 psi. Disadvantages include comparatively high *NPSH* requirements, the vulnerability of available check-valve materials to chemical liquids, and relatively poor performance where solids, abrasives or dirt are present in the process liquid. Most commercially available reciprocating pumps use multiple cylinders, i.e., duplex, triplex, or quintuplex, to smooth the pulsating flow generated by the reciprocating motion.

Diaphragm pumps are also classed as positive-displacement pumps because the diaphragm acts as a limited-displacement piston. Pumping action is obtained when the diaphragm is forced into reciprocating motion by mechanical linkage, compressed air, or oil from a pulsating external source. This type of construction eliminates any connection between the liquid being pumped and the source of energy, and thereby eliminates the possibility of leakage. This characteristic is of great importance when handling toxic or very expensive liquids. Disadvantages include a limited selection of corrosion-resistant materials, limited head and capacity range, and the necessity of using check valves in the suction and discharge nozzles. Construction details are shown in Fig. 1.

Regenerative-turbine pumps easily handle flowrates up to 100 gpm and heads up to 700 ft. When used for chemical service, the internal clearances must be increased to prevent rubbing contact, resulting in decreased efficiency. These pumps are generally unsuitable for solid-liquid mixtures of any concentration.

## Design Considerations for Chemical Pumps

Practically all the major components of chemical pumps are castings. Needless to say, it is a fruitless venture to thoroughly evaluate the detailed characteristics of the pumped liquid and the material to be used if the component castings do not satisfy the quality needed to provide good service life. This is probably of more concern in chemical-pump applications than in any other type of service because leakage, loss of product, and downtime can be extremely costly, and leakage can be very dangerous.

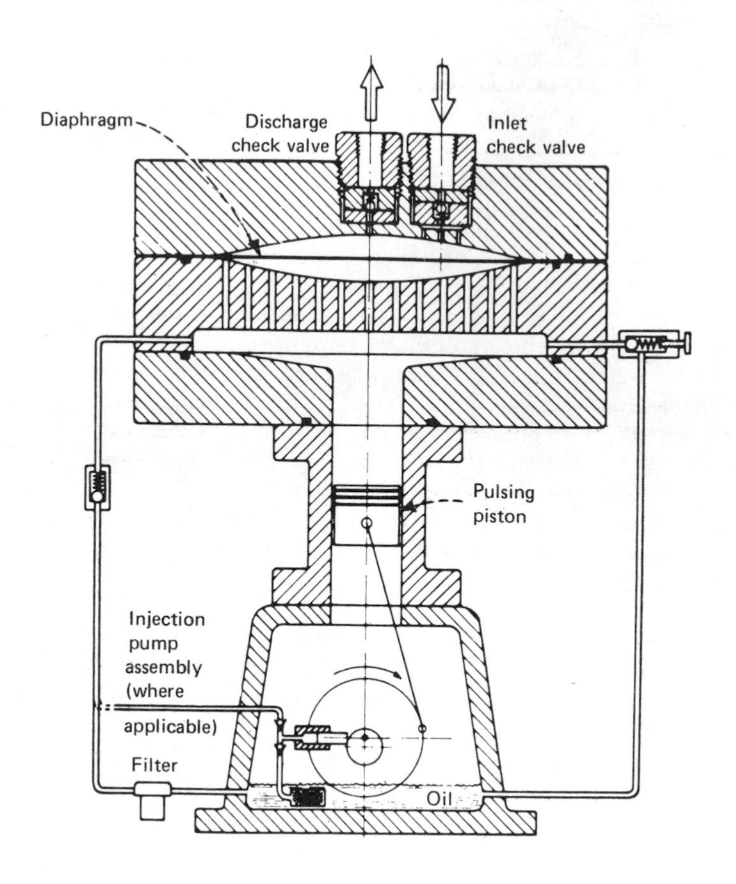


**DIAPHRAGM** pump provides positive displacement—Fig. 1

Of the several factors that determine whether or not a certain material can be used for a particular pump design, mechanical properties are the most important. Materials may possess outstanding corrosion resistance but may be completely impossible to produce in the form of a chemical pump because of limited mechanical properties. Hence, awareness of these properties is essential for any material being considered in a corrosion-evaluation program. Such a program gives a relatively good indication of whether or not a particular design may be available. Since most materials are covered by ASTM or other specifications, such sources can be used for reference purposes. A table of mechanical properties and other characteristics of proprietary materials not included in any standard specification should be readily available from the manufacturer of the material.

Weldments or welded construction should impose no limitation, providing the weldment is as good as, or better than, the base material. Materials requiring heat treatment in order to achieve maximum corrosion resistance must be heat treated after a welding operation, or other adjustments must be made, to make certain that corrosion resistance has not been sacrificed.

Wall sections in pumps are generally increased over the mechanical-design requirements, so that full pumping capability will be maintained even after the loss of some material to the corrosive environment. Parts that are subject to corrosion from two or three sides, such as impellers, must be made considerably heavier than their counterparts in water or oil pumps. Pressure-containing parts are also made thicker so they will remain serviceable after a specified amount of corrosive deterioration. Areas subject to high velocities, such as the cutwater of a centrifugal casing, are further reinforced to allow for the accelerated corrosion caused by high velocities in the liquid.

Threaded construction of any type within the wetted parts must be avoided whenever possible. The thin thread is subject to attack from two sides, and a small amount of corrosive deterioration will eliminate the holding power of the threaded joint. Pipe threads are also to be avoided because of their susceptibility to attack.

Gasket materials must be selected to resist the chemical being handled. Compressed asbestos, lead, and certain synthetic rubbers have been used extensively for corrosion services. In recent years, the fluorocarbon resins have come into widespread use due to their almost universal corrosion resistance.

The power end of pumps consists of the bearing housing, bearings, oil or grease seals, and the bearing lubrication system. This assembly is normally made of iron or steel components and thus must be designed to withstand the severe environment of the chemical plant. For example, when venting of the bearing housing is required, special means of preventing the entrance of water, chemical fumes or dirt must be incorporated into the vent design.

The bearing that controls axial movement of the shaft is usually selected to limit shaft movement to 0.002 in, or less. End-play values above this limit have been found detrimental to mechanical-seal operation.

Water jacketing of the bearing housing may be necessary under certain conditions to maintain bearing tem-

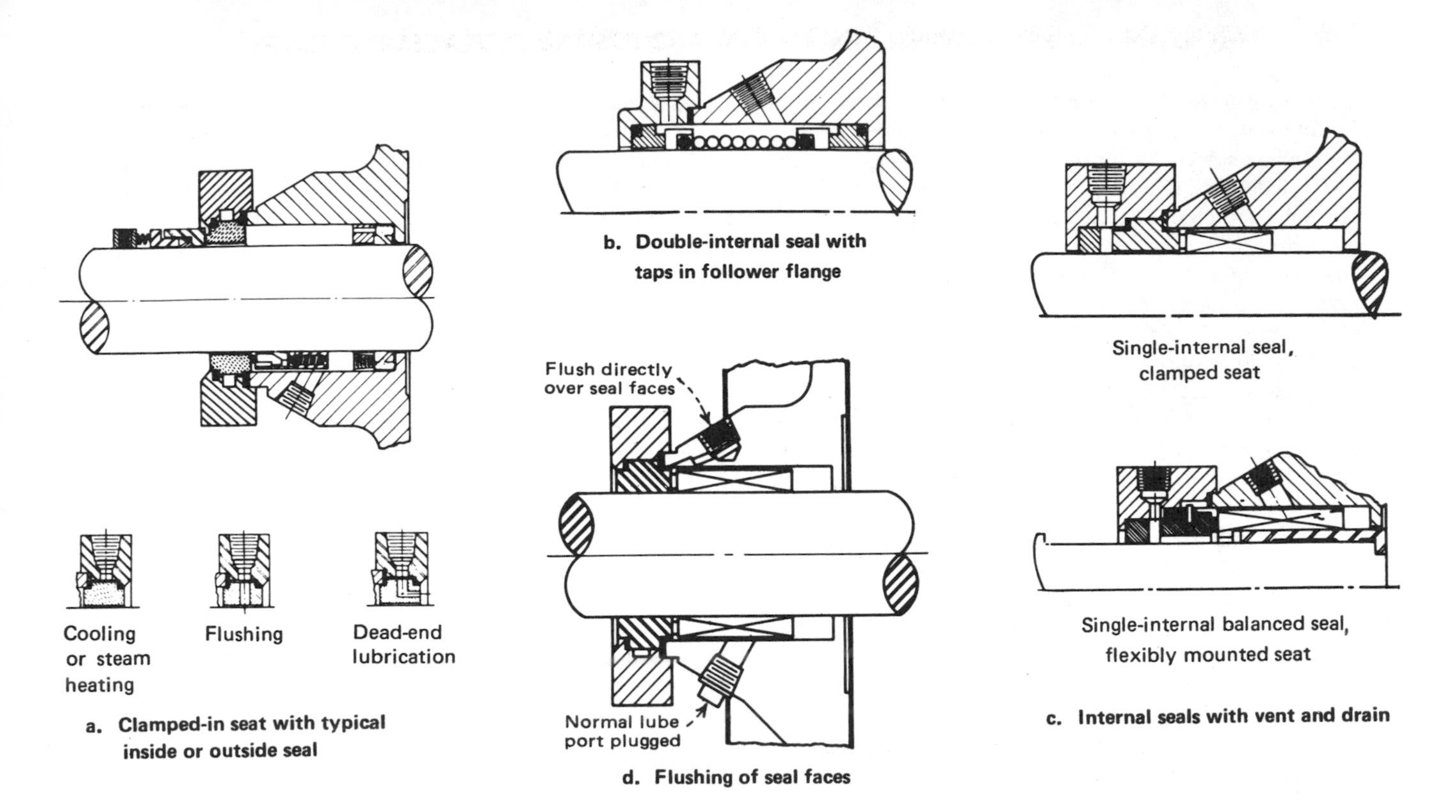


**MECHANICAL** seals have many designs and accessories for handling process liquids at operating conditions—Fig. 2

peratures below 225°F, the upper limit for standard bearings.

Maintenance of a chemical pump in a corrosive environment can be very costly and time consuming. It can be divided into two categories—preventative and emergency. When evaluating materials and design factors, maintenance aspects should be high on the priority list. The ease and frequency of maintenance are very critical factors in any preventative-maintenance program.

## Stuffing-Box Design

The area around the stuffing-box probably causes more failures of chemical pumps than all other parts combined. The problem of establishing a seal between a rotating shaft and the stationary pump parts is one of the most intricate and vexing problems facing the pump designer.

Packings of braided asbestos, lead, fluorocarbon resins, aluminum, graphite and many other materials, or combinations of these materials, have been used to establish the seal. Inconsistent as it seems, a small amount of liquid must be allowed to seep through the packing to lubricate the surface between packing and shaft. This leakage rate is hard to control, and the usual result is overtightening of the packing to stop the leak. The unfortunate consequence is the rapid scoring of the shaft surface, making it much harder to adjust the packing to the proper compression. Recommendations as to the type of packing to be used for various chemical services should come from the packing manufacturer.

Mechanical shaft seals are used extensively on chemical pumps. Once again, the primary consideration is selection of the proper materials for the type of corrosive being pumped. Stainless steels, ceramics, graphite and fluorocarbon resins are used to make the bulk of the seal parts. Several large manufacturers of this equipment have very complete files on seal designs for various chemical services. Typical seal installations are shown in Fig. 2.

The operating temperature is one of the most important factors affecting the stuffing-box sealing medium. Most packings are impregnated with a grease for lubrication, but these lubricants break down at temperatures above 250°F, resulting in further temperature increases because of friction. One of the less obvious results of this temperature increase is corrosive attack on the pump parts in the heat zone—many materials selected for the pumping temperature will be completely unsuitable in the presence of the corrosive at elevated stuffing-box temperatures. Another source of heat is the chemical solution itself. These liquids often are in the 300°F range, and some go as high as 700°F.

The best answer to the heat problem is removal of the heat by means of a water jacket around the stuffing box. While heat conductivity is rather low for most chemical-pump materials, the stuffing-box area generally can be maintained in the 200°F to 250°F range. This cooling is of further benefit in that it prevents the transfer of heat along the shaft to the bearing housing, thereby eliminating other problems around the bearings.

Stuffing-box pressure varies with suction pressure, impeller design, and the degree of maintenance of close-fitting seal rings. Variations in impeller design would include those using vertical or horizontal seal rings in combination with balance ports, as opposed to those using back vanes or pumpout vanes. All impeller designs depend upon a close-running clearance between the impeller and the stationary pump parts. This clearance must be kept as small as possible to prevent excessive recirculation of the liquid, and resulting loss of efficiency. Unfortunately, most chemical pump materials tend to seize when subjected to rubbing contact. Therefore, running clearances must be increased considerably above those used in pumps for other industries.

At pressures above 100 psi, packing is generally unsatisfactory unless the stuffing box is very deep, and the operator is especially adept at maintaining the proper gland pressure on the packing. Mechanical seals incorporating a balancing feature to relieve the high pressure are the best means of sealing at pressures above 100 psi.

The pump shaft can create additional stuffing-box problems. Obviously, a shaft that is out-of-round or bent will form a larger hole in the packing than the shaft can fill, thereby allowing liquid to escape. Lack of static or hydraulic balance in the impeller produces a dynamic bend in the shaft, resulting in the same condition. Undersize shafts, or those made of materials that bend readily, will deflect from their true center in response to radial thrust on the impeller. This action produces a secondary hole in the packing, and again allows the liquid to escape.

Mechanical-seal performance is also impaired when the shaft is bent or deflects during operation. Since the flexible member of the seal must adjust with each revolution of the shaft, excessive deflection results in shortened seal life. If the deflection is of more than nominal value, the flexible seal member will be unable to react with sufficient speed to keep the seal faces together, allowing leakage at the mating faces.

An arbitrary limit of 0.002 in has been established as the maximum deflection or runout of the shaft at the face of the stuffing box.

In the stuffing-box region, the shaft surface must have corrosion resistance at least equal to, and preferably better than, that of the wetted parts of the pump. In addition, this surface must be hard enough to resist the tendency to wear under the packing or mechanical-seal parts. Further, it must be capable of withstanding sudden the temperature changes often encountered.

Since it is not economically feasible to make the entire pump shaft of stainless alloys, and physically impossible to make functional carbon, glass or plastic shafts, chemical pumps often have carbon-steel shafts with a protective coating or sleeve over the steel in the stuffing-box area. Cylindrical sleeves are sometimes made so that they may be removed and replaced when they become worn. Other designs use sleeves that are permanently bonded to the shaft to obtain lower runout and deflection values.

Another method of obtaining a hard surface in this region is the welded overlay or spray coating of hard metals onto the base shaft. Materials applied by these techniques are generally lacking in corrosion resistance and have not been widely accepted for severe chemical service. Ceramic materials applied by the plasma-spray technique possess excellent corrosion resistance but cannot achieve the complete density required to protect the underlying shaft.

Composite shafts using carbon steel for the power end and a higher alloy for the wet end have been used extensively where the high-alloy end has acceptable resistance. Since the two ends are joined by various welding techniques, the combination of metals is limited to those that can be easily welded. On such assemblies, the weld joint and the heat-affected zone must be outside the wetted area of the shaft.

Elimination of the stuffing box and its problems has been the object of several designs for chemical pumps other than the diaphragm pump previously described.

Vertical submerged pumps use a sleeve-type bearing in the area immediately above the impeller, to limit the flow of liquid up the shaft. For chemical service, the problem of materials associated with this bearing and its lubrication have been major disadvantages.

Canned pumps, wherein the motor windings are hermetically sealed in stainless-steel "cans," also avoid the use of a stuffing box. The pumped liquid is circulated through the motor section, lubricating the sleeve-type bearings that support the rotating assembly. Disadvantages again center on the selection of bearing materials compatible with the corrosive liquid; the lubrication of these bearings when handling nonlubricating liquids; and the probability of the liquid paths through the motor section becoming clogged when solid-liquid mixtures are handled.

## Designing With Special Materials

As described earlier, a number of low-mechanical-strength materials have been used extensively in chemical-pump construction. While breakage problems are inherently associated with these materials, their excellent corrosion resistance has allowed them to remain competitive with higher-strength alloys. Of course, their low tensile strength and brittleness makes them sensitive to tensile or bending stresses, requiring special pump designs. The parts are held together by outside clamping means, and braced to prevent bending. The unit must also be protected from sudden temperature changes and from mechanical impact from outside sources.

Although produced by very few manufacturers, high-silicon iron is the most universally corrosion-resistant metallic material available at an economic price. This resistance, coupled with a hardness of approximately Brinell 520, provides an excellent material for handling abrasive chemical slurries. The material's hardness, however, precludes normal machining operations, and the parts must be designed for machine grinding. The hardness also eliminates the possibility of using drilled or tapped holes for connecting piping to the pump parts. Therefore, special designs are required for process piping, stuffing-box lubrication, and drain connections.

Ceramics and glass are similar to high-silicon iron in regard to hardness, brittleness and susceptibility to thermal or mechanical shock. Pump designs must, therefore, incorporate the same special considerations.

Glass linings or coatings on iron or steel parts are sometimes used to eliminate some of the undesirable characteristics of solid glass. While this usage provides for connecting process piping, the dissimilar expansion characteristics of the two materials generate small cracks in the glass, allowing corrosive attack.

Thermosetting and thermoplastic materials are used

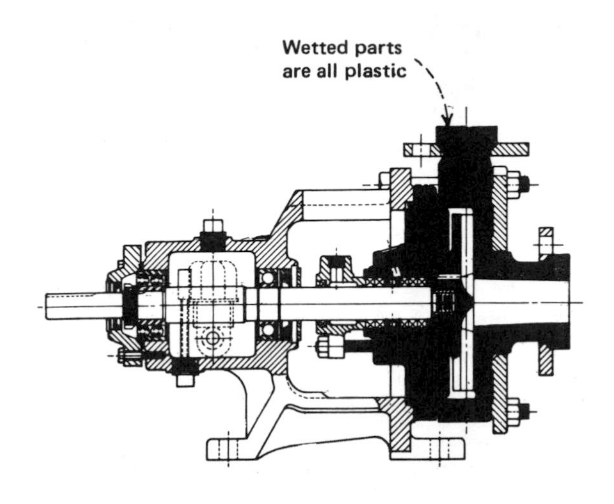


**PLASTIC** pump has all-plastic liquid end—Fig. 3

extensively in services where chlorides are present. Their primary disadvantage is loss in strength at higher pumping temperatures. Phenolic and epoxy parts are subject to gradual loss of dimensional integrity because of the material's creep characteristics. The low tensile strength of the unfilled resins again dictates a design that will place these parts in compression, and eliminate bending stresses. Typical construction details are shown in Fig. 3.

Polytetrafluoroethylene and hexafluoropropylene possess excellent corrosion resistance. These resins have been used for gaskets, packing, mechanical-seal parts, and flexible-piping connectors. Several pumps made of these materials have reached the market in recent years. Problems associated have centered around these materials' tendency to cold flow under pressure, and their high coefficient of expansion compared to the metallic components of the unit. Pumps may be made of heavy solid sections, as illustrated in Fig. 3, or may use more-

## Dimensions for Horizontal Chemical Process Pumps — Table I

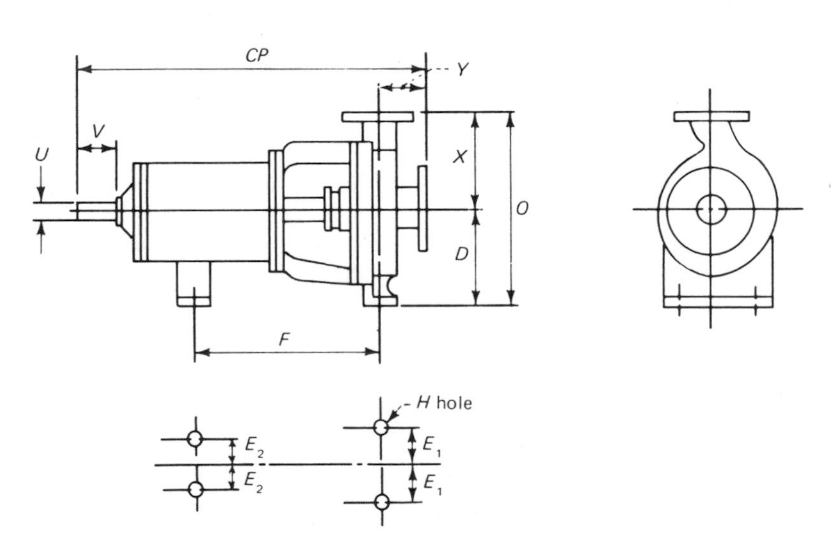


Dimensions in Inches

| Dimension Designation | Size, Suct. x Disch. x Nom. Impeller | CP | D | $2E_1$ | $2E_2$ | F | H | O | U Dia. | U Keyway | V Min. | X | Y |
|---|---|---|---|---|---|---|---|---|---|---|---|---|---|
| AA | 1-1/2 x 1 x 6 | 17-1/2 | 5-1/4 | 6 | 0 | 7-1/4 | 5/8 | 11-3/4 | 7/8 | 3/16 x 3/32 | 2 | 6-1/2 | 4 |
| AB | 3 x 1-1/2 x 6 | 17-1/2 | 5-1/4 | 6 | 0 | 7-1/4 | 5/8 | 11-3/4 | 7/8 | 3/16 x 3/32 | 2 | 6-1/2 | 4 |
| A10 | 3 x 2 x 6 | 23-1/2 | 8-1/4 | 9-3/4 | 7-1/4 | 12-1/2 | 5/8 | 16-1/2 | 1-1/8 | 1/4 x 1/8 | 2-5/8 | 8-1/4 | 4 |
| AA | 1-1/2 x 1 x 8 | 17-1/2 | 5-1/4 | 6 | 0 | 7-1/4 | 5/8 | 11-3/4 | 7/8 | 3/16 x 3/32 | 2 | 6-1/2 | 4 |
| A50 | 3 x 1-1/2 x 8 | 23-1/2 | 8-1/4 | 9-3/4 | 7-1/4 | 12-1/2 | 5/8 | 16-3/4 | 1-1/8 | 1/4 x 1/8 | 2-5/8 | 8-1/2 | 4 |
| A60 | 3 x 2 x 8 | 23-1/2 | 8-1/4 | 9-3/4 | 7-1/4 | 12-1/2 | 5/8 | 17-3/4 | 1-1/8 | 1/4 x 1/8 | 2-5/8 | 9-1/2 | 4 |
| A70 | 4 x 3 x 8 | 23-1/2 | 8-1/4 | 9-3/4 | 7-1/4 | 12-1/2 | 5/8 | 19-1/4 | 1-1/8 | 1/4 x 1/8 | 2-5/8 | 11 | 4 |
| A05 | 2 x 1 x 10 | 23-1/2 | *8-1/4 | 9-3/4 | 7-1/4 | 12-1/2 | 5/8 | 16-3/4 | 1-1/8 | 1/4 x 1/8 | 2-5/8 | 8-1/2 | 4 |
| A50 | 3 x 1-1/2 x 10 | 23-1/2 | 8-1/4 | 9-3/4 | 7-1/4 | 12-1/2 | 5/8 | 16-3/4 | 1-1/8 | 1/4 x 1/8 | 2-5/8 | 8-1/2 | 4 |
| A60 | 3 x 2 x 10 | 23-1/2 | 8-1/4 | 9-3/4 | 7-1/4 | 12-1/2 | 5/8 | 17-3/4 | 1-1/8 | 1/4 x 1/8 | 2-5/8 | 9-1/2 | 4 |
| A70 | 4 x 3 x 10 | 23-1/2 | 8-1/4 | 9-3/4 | 7-1/4 | 12-1/2 | 5/8 | 19-1/4 | 1-1/8 | 1/4 x 1/8 | 2-5/8 | 11 | 4 |
| A20 | 3 x 1-1/2 x 13 | 23-1/2 | 10 | 9-3/4 | 7-1/4 | 12-1/2 | 5/8 | 20-1/2 | 1-1/8 | 1/4 x 1/8 | 2-5/8 | 10-1/2 | 4 |
| A30 | 3 x 2 x 13 | 23-1/2 | 10 | 9-3/4 | 7-1/4 | 12-1/2 | 5/8 | 21-1/2 | 1-1/8 | 1/4 x 1/8 | 2-5/8 | 11-1/2 | 4 |
| A40 | 4 x 3 x 13 | 23-1/2 | 10 | 9-3/4 | 7-1/4 | 12-1/2 | 5/8 | 22-1/2 | 1-1/8 | 1/4 x 1/8 | 2-5/8 | 12-1/2 | 4 |
| A80 * | 6 x 4 x 13 | 23-1/2 | 10 | 9-3/4 | 7-1/4 | 12-1/2 | 5/8 | 23-1/2 | 1-1/8 | 1/4 x 1/8 | 2-5/8 | 13-1/2 | 4 |

*Suction connection may have tapped bolt holes.

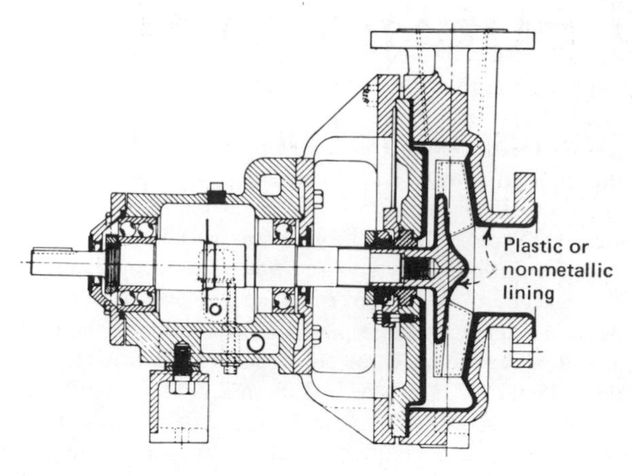


**LINED** pump has metallic components that contain a liner usually made of fluorocarbon materials—Fig. 4

conventional metallic components lined with the fluorocarbon material, as shown in Fig. 4.

## Chemical-Pump Standards

Early in 1962, a committee of the Manufacturing Chemists Assn. (MCA) reached agreement with a special committee of the Hydraulic Institute on a proposed standard for chemical-process pumps. This document was referred to as the American Voluntary Standard (AVS), or MCA Standard. In 1971, it was accepted by the American National Standards Institute (ANSI), and issued as ANSI Standard B123.1. Many pump manufacturers in the U.S. and a few in foreign countries are building pumps that meet these dimensional and design criteria.

It is the intent of this Standard that pumps of similar size from all sources of supply shall be interchangeable with respect to mounting dimensions, size and location of suction and discharge nozzles, input shafts, base plates and foundation bolts. Table I lists the pump dimensions that have been standardized.

It is also the intent of this Standard to outline certain design features that will minimize maintenance problems. The Standard states, for instance, that the pump shaft should be sized so that the maximum shaft deflection, measured at the face of the stuffing box when the pump is operating under its most adverse conditions, will not exceed 0.002 in. The Standard does not specify shaft diameter, since impeller diameter, shaft length, and provision for operation with liquids of high specific gravities would determine the proper diameter.

The Standard also states that the minimum bearing life, again under the most adverse operating conditions, should not be less than two years. Bearing size is not specified but is to be determined by the individual manufacturer and will depend on the load to be carried.

Additional specifications in the Standard include: hydrostatic test pressure, shaft finish at rubbing points, and packing space.

Other dimensional standards are in use in foreign countries on both horizontal and vertical pumps. In 1971, the International Organization for Standardization (ISO) reached agreement on a set of dimensional standards for horizontal, end-suction, centrifugal pumps. This document, ISO 2858, is in metric units, and describes a series of pumps of slightly higher capacity range than described in B123.1. It does not include design criteria as to minimum shaft deflection, minimum bearing life, or other characteristics required to reduce maintenance.

The British Standards Institution issued BS4082 in 1966 to describe a series of vertical inline centrifugal pumps. While dimensional interchangeability was the primary reason for this standard, it also includes requirements for hydrostatic testing of the pump parts. It is made up of two sections: Part 1 covers pumps wherein the suction and discharge nozzles are in a horizontal line (the "I" type); and Part 2 covers pumps wherein the nozzles are on the same side of the pump and parallel to each other (the "U" configuration).

## References

1. "Corrosion Data Survey," National Assn. of Corrosion Engineers, Houston, 1967.
2. Fontana, M. G. and Greene, N. D., "Corrosion Engineering," McGraw-Hill, New York, 1967.
3. Lee, J. A., "Materials of Construction for Chemical Process Industries," McGraw-Hill, New York, 1950.
4. "Proceedings, Short Course on Process Industry Corrosion," National Assn. of Corrosion Engineers, Houston, 1960.

### Meet the Authors

◀ **John R. Birk** is Vice-President of Engineering for The Duriron Co., Dayton, OH 45401, where he was responsible for the development of the standard AVS pump and TFE-lined equipment. He has a degree in mechanical engineering from the University of Cincinnati, and is a member of ASME, American Soc. of Professional Engineers, Dayton Engineers' Club and the Hydraulic Institute. He is also on the ANSI Committee B73 on centrifugal pumps, and is a professional engineer in Ohio.

**James H. Peacock** is Manager of the Machine Div. of The Duriron Co. He is a frequent lecturer and has published articles on the applications of materials and equipment in corrosive environments. He has a degree in metallurgical engineering from Purdue University, and is a member of the National Assn. of Corrosion Engineers, American Soc. for Metals, and American Soc. for Testing and Materials. ▶

# Select pumps to cut energy cost

Specifications that are out-of-date and too restrictive can prevent engineers from selecting energy-efficient pumps. A guide based on pump specific speed indicates the type of pump to choose.

*John H. Doolin,* *Worthington Pump Inc.*

☐ Considerable energy savings can be made in pump systems. Of course, the first place to look for these savings is in system design [*1*]. However, even after system requirements have been reduced to the minimum and hydraulic conditions have been fixed, attention should be given to selecting the most efficient pumps for the system.

Most engineers fully regard the efficiencies quoted by vendors. This, however, may not be enough, because company specifications—such as on operating speed, number of stages and impeller configuration—may preclude the supplier from offering the most efficient pump.

Although pump technology has advanced considerably during the past 25 years, many specifications are still based on experience prior to then. Such specifications can lead to the selection of inefficient pumps in terms of power consumption at a time when a savings of 1 hp can justify an investment of $1,000 [*2*].

## A guide for selecting efficient pump type

The most efficient type of pump for a particular application could be single-stage, multistage, high-speed, or even reciprocating. Too many specifications, especially those based on out-of-date experience, limit the possibility of energy-efficient pumps being selected

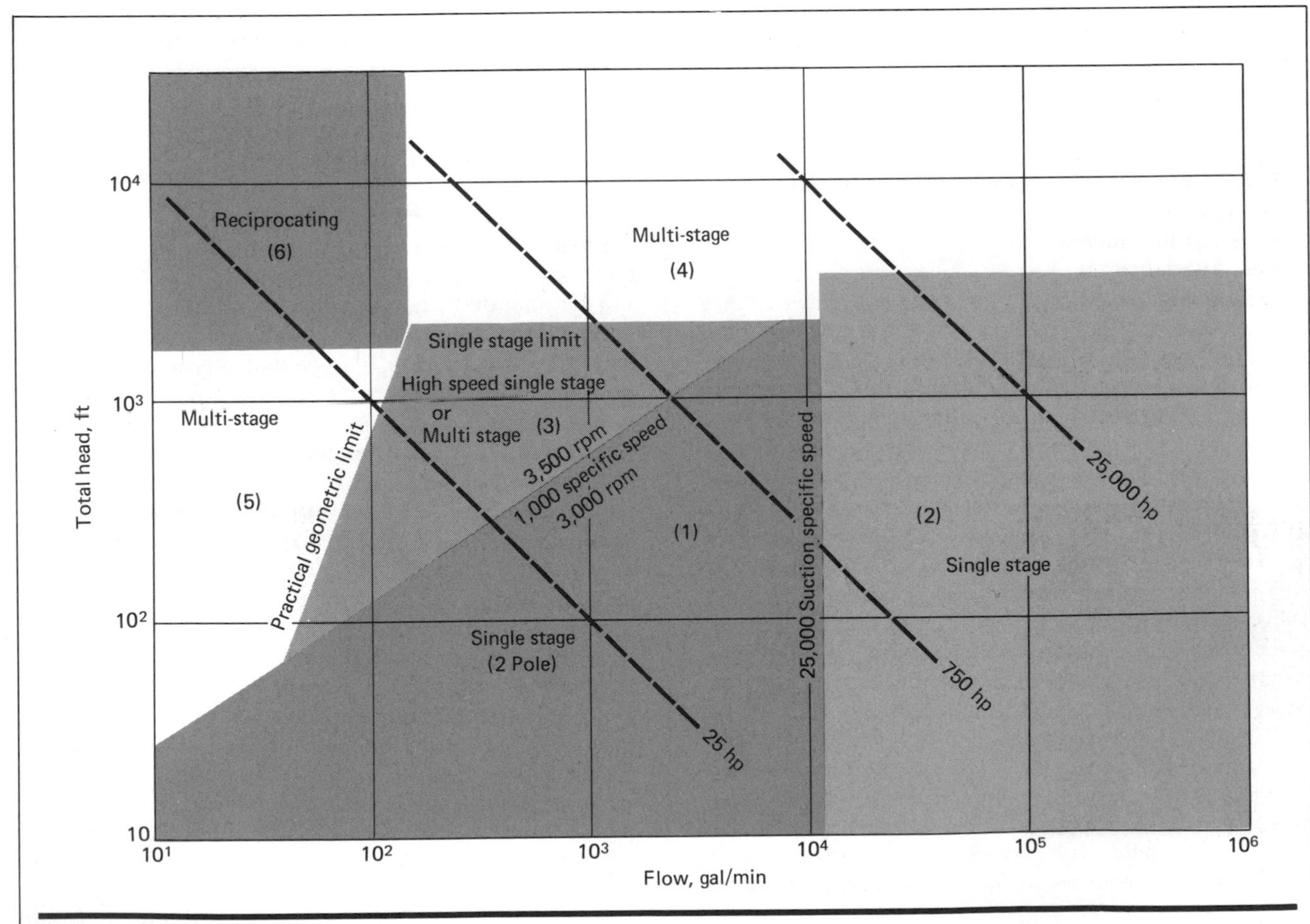


**Selection guide is based mainly on specific speed, which number gives an indication of impeller geometry** **Fig. 1**

Originally published January 17, 1977

Inducer fits into suction opening of impeller Fig. 2

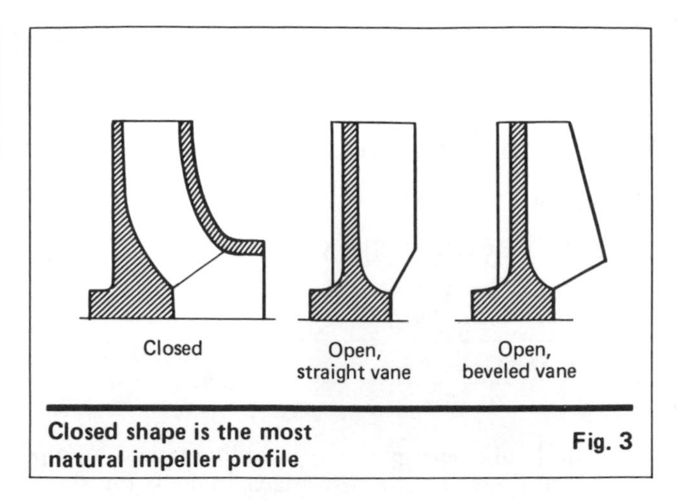


Closed shape is the most natural impeller profile Fig. 3

by restrictive specifying, such as "single-stage centrifugal," "two-stage centrifugal," or "multistage."

Fig. 1 offers a guide to efficient pumps ranging in capacity to 100,000 gal/min and in total head to 10,000 ft. The selection guide is based primarily on the characteristic of pump specific speed: $N_s = NQ^{1/2}/H^{3/4}$. (In this equation, $N$ = rotating speed, rpm; $Q$ = capacity, gal/min; and $H$ = total head, ft.)

Fig. 1 is mapped into six areas, each of which indicates the type of pump that should be selected for the highest energy efficiency, as follows:

**Area 1:** single-stage, 3,500 rpm.

**Area 2:** single-stage, 1,750 rpm or lower.

**Area 3:** single-stage, above 3,500 rpm, or multistage, 3,500 rpm.

**Area 4:** multistage.

**Area 5:** multistage.

**Area 6:** reciprocating.

When the value of $N_s$ for any condition is less than 1,000, the operating efficiency of single-stage centrifugal pumps falls off drastically, in which case either multistage or higher-speed pumps offer the best efficiency.

Area 1 is the densest, crowded both with pumps operating at 1,750 rpm and 3,500 rpm, because years ago 3,500-rpm pumps were not thought to be as durable as 1,750-rpm ones. Since the adoption of the AVS standard in 1960 (superseded by ANSI B73.1), pumps with stiffer shafts have been proven reliable.

Also responsible for many 1,750 rpm pumps in Area 1 has been the impression that the higher (3,500 rpm) speed caused pumps to wear out faster. However, because impeller tip speed is the same at both 3,500 and 1,750 rpm (as, for example, a 6-in. impeller at 3,500 rpm and a 12-in. one at 1,750 rpm), so is the fluid velocity, and so should be the erosion of metal surface. Another reason for not limiting operating speed is that improved impeller inlet design allows operation at 3,500 rpm to capacities of 5,000 gal/min, and higher.

## Evaluate limits on suction performance

Choice of operating speed may also be indirectly limited by specifications pertaining to suction performance, such as that fixing the top suction specific speed directly, or indirectly by choice of Sigma constant or by reliance on Hydraulic Institute charts.

Suction specific speed is defined as: $S = NQ^{1/2}/H_s^{3/4}$. (In this equation, $N$ = rotating speed, rpm; $Q$ = capacity, gal/min; and $H_s$ = net positive suction head, ft.)

Values of $S$ below 8,000 to 10,000 have long been accepted for avoiding cavitation. However, since the development of the inducer (Fig. 2), $S$ values in the range of 20,000 to 25,000 have become commonplace, and values as high as 50,000 have become practical.

The Sigma constant, which relates NPSH to total head, is little used today, and Hydraulic Institute charts (which are being revised) are conservative.

In light of today's designs and materials, past restrictions due to suction performance limitations should be re-evaluated or eliminated entirely.

## Consider off-peak operation

Even if the most efficient pump has been selected, there are a number of circumstances in which it may not be operated at peak efficiency. Today's cost of energy has made these considerations more important.

A centrifugal pump, being a hydrodynamic machine, is designed for a single peak operating-point of capacity and total head. Operation at other than this best efficiency point (bep) reduces efficiency. Specifications now should account such factors as:

1. A need for a larger number of smaller pumps. When a process operates over a wide range of capacities, as many do, pumps will often work at less than full capacity, hence at lower efficiency. This can be avoided by installing two or three pumps in parallel, in place of a single large one, so that when operations are at a low rate one of the smaller pumps can handle the flow.

2. Allowance for present capacity. Pump systems are frequently designed for full flow at some time in the future. Before this time arrives, the pumps will operate far from their best efficiency points. Even if this interim period lasts only two or three years, it may be more economical to install a smaller pump initially and to replace it later with a full-capacity one.

3. Inefficient impeller size. Some specifications call for pump impeller diameter to be no larger than 90 or 95% of the size that a pump could take, so as to provide reserve head. If this reserve is used only 5% of the time,

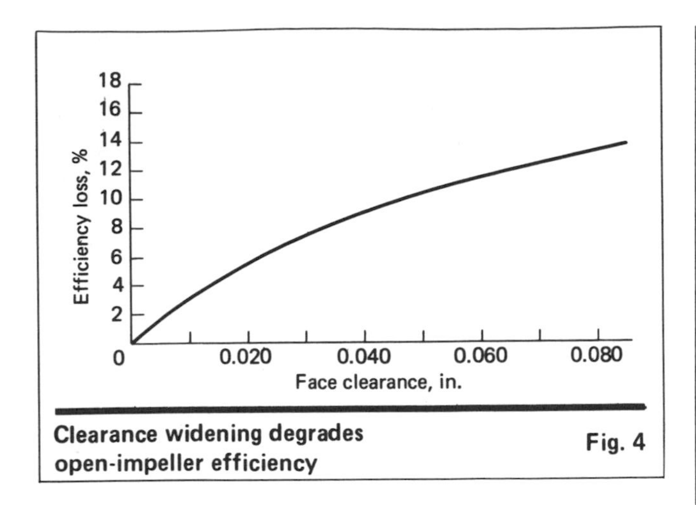


**Clearance widening degrades open-impeller efficiency** **Fig. 4**

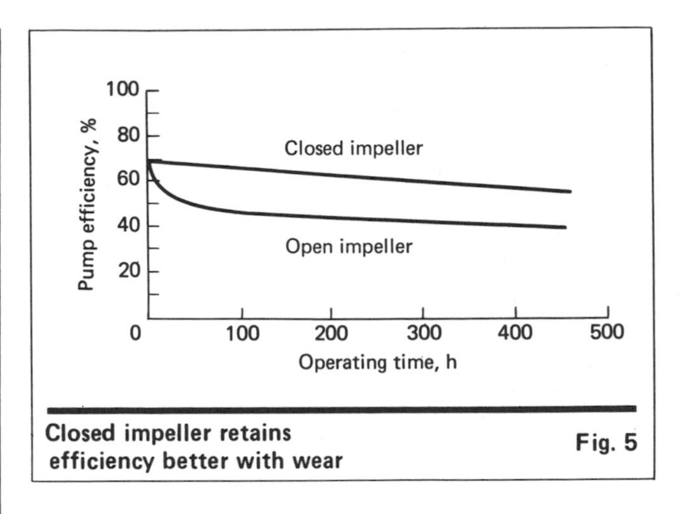


**Closed impeller retains efficiency better with wear** **Fig. 5**

all such pumps will be operating at less than full efficiency most of the time.

4. Advantages of allowing operation to the right of the best efficiency point. Some specifications, the result of such thinking as that which provides reserve head, prohibit the selection of pumps that would operate to the right of the best efficiency point. This eliminates half of the pumps that might be selected, and results in oversized pumps operating at lower efficiency.

If an open impeller is carefully machined, it can be as efficient as a closed one. Because of the manufacturing problems of getting smooth hydraulic profiles (even though numerically controlled machine tools now make contoured open impellers), closed-impeller pumps tend to be more efficient.

The closed shape is the most natural profile for impellers (Fig. 3), being the easiest to achieve with cast surfaces. Open impellers are typically straight-sided or beveled, because these shapes are simple to machine.

Closed impellers are also more efficient because the efficiency of open impellers is very dependent on the amount of face clearance between the impeller and casing wall. Although an open-impeller pump can be built with only a 0.015-in. clearance, it is not uncommon for this clearance to enlarge to 0.050 in. after a short time in service. Many studies have revealed this, including an NASA report, which shows efficiency drops 10% at a 0.050-in. clearance (Fig. 4) [3].

Of course, closed-impeller clearances also widen with service. However, the loss in efficiency is less. An accelerated wear test of open and closed impellers of otherwise identical geometry showed that when the clearances of both impellers opened to 0.050 in., the efficiency of the open impeller dropped 28%, whereas that of the closed impeller fell only 14% (Fig. 5) [4].

## If possible, avoid special pumps

Sometimes, the unusual requirements of a system dictate the selection of special pumps, which are often inherently lower in efficiency. Such requirements should be carefully evaluated to determine if a special pump is really necessary and worth the loss of efficiency. The following are examples of such special pumps:

■ Self-priming pumps are designed with suction and discharge chambers that generate significant friction loss. Recirculation also reduces efficiency. A vertical wet-pit pump could be considered as a substitute.

■ Canned motor pumps, which are often installed when zero leakage is mandatory, are less efficient because the magnetic gap must be wider to allow for the chamber that seals the motor.

■ Hydrodynamically-sealed pumps, another form of zero-leakage pump, prevents leakage by the back-pumping of a second impeller. The pumping of the second impeller penalizes power efficiency.

■ Pumps for handling solids are usually specified to be oversized or of low-efficiency design so as to handle fairly large solids without plugging. Rather than select such a pump, it may be more economical to unplug a more efficient pump that clogs occasionally.

■ Pumps of proprietary mechanical design are sometimes chosen because the special design may best fit the unusual requirements of a system. Nevertheless, the likely loss of efficiency should be carefully evaluated before such a pump is specified.

In general, avoid restrictive specifications that exclude more-efficient pumps. Allow manufacturers latitude in offering pumps by following such a guide as Fig. 1. In fact, take an affirmative stance and preface specifications with a statement that efficiency is given primary consideration and that inefficiency is penalized at $1,000 per horsepower.

## References

1. Karassik, I. J., Design and Operate Your Fluid System for Improved Efficiency, *Pump World*, Summer, 1975, Worthington Pump Inc.
2. Reynolds, J., Saving energy and costs in pumping systems, *Chem. Eng.*, Jan. 5, 1976.
3. NASA Report No. CR-120815.
4. Pumping Abrasive Fluids, *Plant Eng.*, Nov., 1972.

## The auth

**John H. Dooli**
Development
Pump Div. (
NJ 07017). H
on pump de
articles on c
Worthingto
Hydraulic
on ANSI
Chairman
end sucti
and Mas
Newark
is a lice
New Je

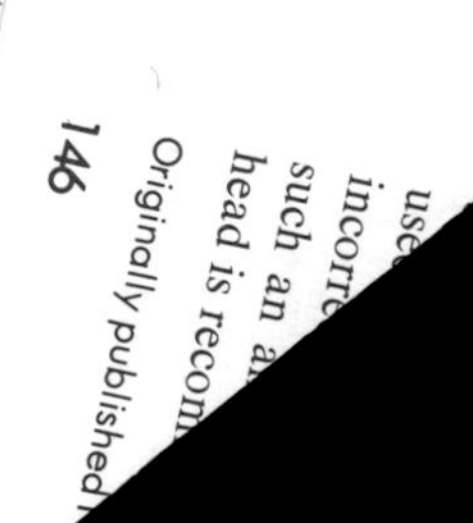

# Pump Selection for the

An evaluation of the hydraulic requirements of a pumping system coupled with a knowledge of the basic performance and operating characteristics of different types of pumps enable the selection of the proper pump for a given application.

RICHARD F. NEERKEN, The Ralph M. Parsons Co.

Pumps may be classified into two general types, dynamic and positive-displacement. Dynamic pumps, such as centrifugal pumps, are those in which energy is imparted to the pumped liquid by means of a moving impeller or propeller rotating on a shaft. Velocity energy, imparted to the fluid by the moving impeller, is converted to pressure energy as the liquid leaves the impeller and moves through a stationary volute or diffuser casing. Obviously, for a greater velocity (which is obtained either by higher rotating speed or larger impeller diameter, or both), a higher head can be achieved.

Positive-displacement pumps are those in which energy is imparted to liquid in a fixed displacement-volume such as a casing or cylinder, by the rotary motion of gears, screws or vanes, or by reciprocating pistons or plungers.

Our aim will be to examine the basic types that exist today, and explore some of the ways for making a rational approach to pump selection.

## Hydraulic Analysis

It is first assumed that a proper hydraulic analysis has been made of the system in which the pump is to be d. Unfortunately, this is not always done, and many ct pump selections have resulted from the lack of alysis. The use of a form to calculate pump mended, especially in the chemical process industries, because it provides a valuable worksheet, checklist and reference during the course of the project. A typical form, filled out, is shown in Fig. 3.

What does the engineer do when such hydraulic calculations have been completed? The next task is to select the best or most suitable pump for the service. Although pumps fall into the two general categories previously described, there are hundreds of varieties from which a choice must be made. Let us examine ways of doing so.

## Specific Speed of Pumps

Specific speed is a useful index in getting a general idea of the type of pump to be chosen. All pumps can be classified with a dimensionless number, called specific speed, $N_s$, and defined as:

$$N_s = \frac{N\sqrt{Q}}{H^{3/4}}$$

where $N$ is speed, rpm; $Q$ is capacity or flow; and $H$ is head. When capacity is expressed in gpm (U.S.) and head in ft, centrifugal pumps have specific speeds that range from about 400 to over 10,000, depending on the type of impeller design (Fig. 1).

Smaller dynamic types such as the drag pump (regenerative-turbine pump) and the partial-emission type pump, are in the specific speed ranging from about 100

# Chemical Process Industries

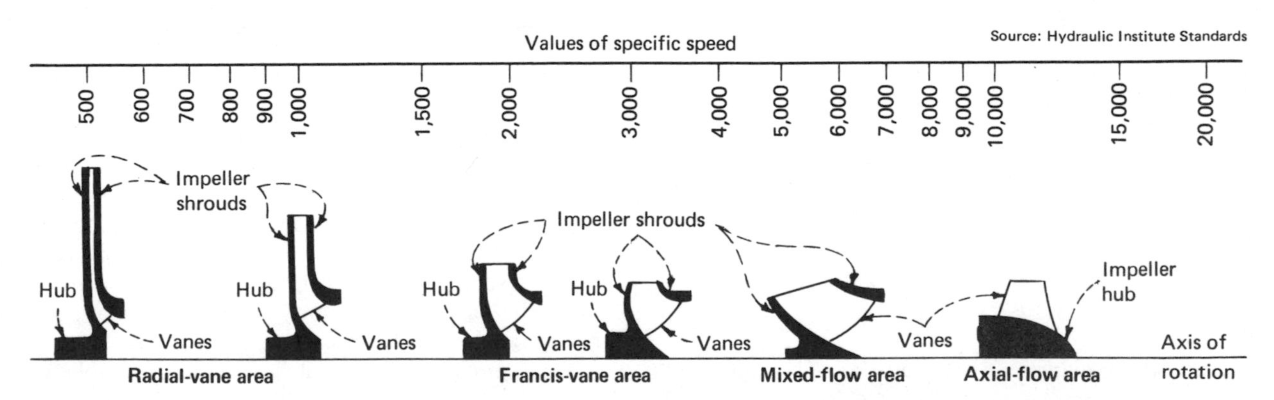


**SPECIFIC** speed, a dimensionless number for different impeller designs, is a useful index for choosing a pump type—Fig. 1

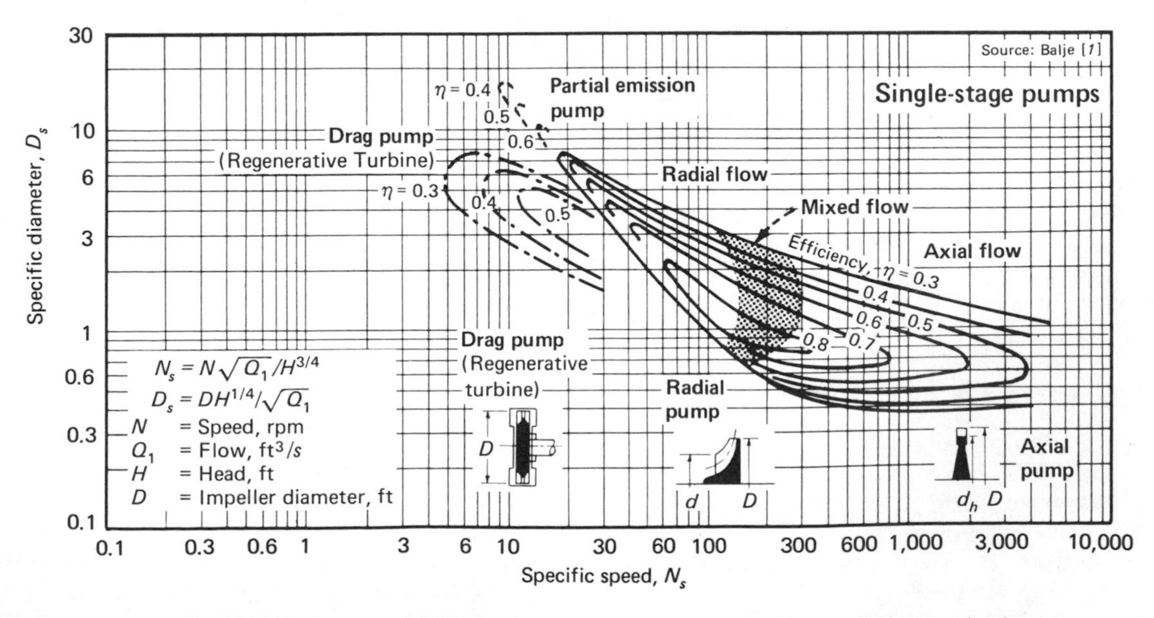


**CHART** of specific speed vs specific diameter for single-stage centrifugal pumps enables preliminary selection of pump—Fig. 2

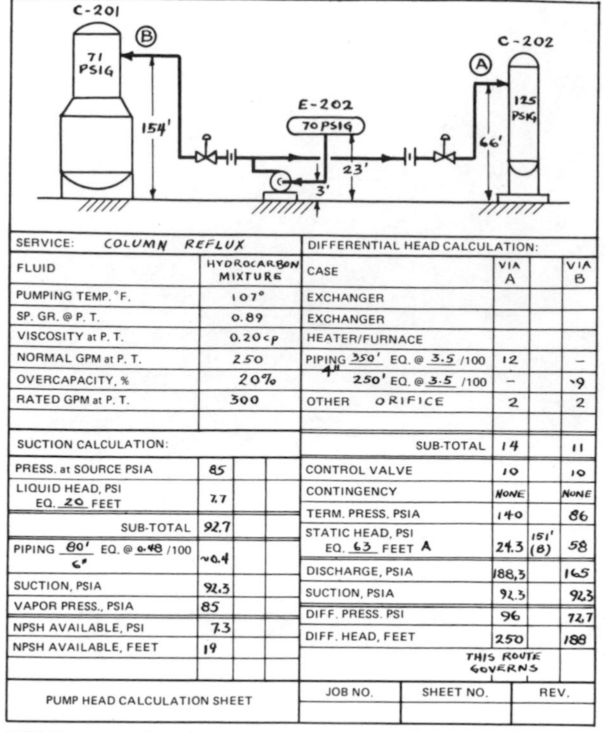


| SERVICE: COLUMN REFLUX | |
|---|---|
| FLUID | HYDROCARBON MIXTURE |
| PUMPING TEMP. °F. | 107° |
| SP. GR. @ P. T. | 0.89 |
| VISCOSITY at P. T. | 0.20cp |
| NORMAL GPM at P. T. | 250 |
| OVERCAPACITY, % | 20% |
| RATED GPM at P. T. | 300 |

| SUCTION CALCULATION: | | | |
|---|---|---|---|
| PRESS. at SOURCE PSIA | 85 | | |
| LIQUID HEAD, PSI EQ. 20 FEET | 7.7 | | |
| SUB-TOTAL | 92.7 | | |
| PIPING 80' 6" EQ. @ 0.48 /100 | ~0.4 | | |
| SUCTION, PSIA | 92.3 | | |
| VAPOR PRESS., PSIA | 85 | | |
| NPSH AVAILABLE, PSI | 7.3 | | |
| NPSH AVAILABLE, FEET | 19 | | |

| DIFFERENTIAL HEAD CALCULATION: | | | |
|---|---|---|---|
| CASE | VIA A | | VIA B |
| EXCHANGER | | | |
| EXCHANGER | | | |
| HEATER/FURNACE | | | |
| PIPING 350' EQ. @ 3.5 /100 | 12 | | – |
| 4" 250' EQ. @ 3.5 /100 | – | | ~9 |
| OTHER ORIFICE | 2 | | 2 |
| SUB-TOTAL | 14 | | 11 |
| CONTROL VALVE | 10 | | 10 |
| CONTINGENCY | NONE | | NONE |
| TERM. PRESS. PSIA | 140 | | 86 |
| STATIC HEAD, PSI EQ. 63 FEET A | 24.3 | 151' (B) | 58 |
| DISCHARGE, PSIA | 188,3 | | 165 |
| SUCTION, PSIA | 92.3 | | 92.3 |
| DIFF. PRESS. PSI | 96 | | 72.7 |
| DIFF. HEAD, FEET | 250 | | 188 |
| | THIS ROUTE GOVERNS | | |

| PUMP HEAD CALCULATION SHEET | JOB NO. | SHEET NO. | REV. |
|---|---|---|---|

**HEAD** calculation for process-pump requirement—Fig. 3

to 1,200. Rotary and reciprocating pumps have even lower values.

A useful method of visualizing this relationship was offered some years ago by Balje [1] in the form of a chart plotting specific speed vs. specific diameter. The latter is defined as:

$$D_s = \frac{DH^{1/4}}{\sqrt{Q}}$$

where $D$ is the diameter of the impeller. The units used in this chart (Fig. 2) are flow, $Q$, ft$^3$/s equal to [gpm/(60)(7.48)]; head, $H$, ft; diameter of impeller, $D$, ft; and speed, $N$, rpm.

As an example of how this approach is used, let us consider a typical process-pump requirement for a chemical or petrochemical plant, as shown on the head-calculation form of Fig. 3. Assuming that the pump driver is a 60-Hz motor at a speed of 3,550 rpm, the specific speed is:

$$N_s = 3{,}550\sqrt{0.669}/(250)^{3/4} = 46.5$$

Note that a flow of 300 gpm = 0.669 ft$^3$/s. From the data in the chart of Fig. 2, it is seen that a radial-flow single-stage centrifugal pump appears as the correct choice. The approximate best efficiency to be expected is 72%. We will compare this with the selection shown later in Fig. 8.

## Net Positive Suction Head

All pumps require a certain net positive suction head (*NPSH*) to permit the liquid to flow into the pump casing. This value is determined by the pump designer, and is influenced by the speed of rotation, the inlet area or eye-area of the impeller in a centrifugal pump, the type and number of vanes in the impeller, etc. On reciprocating pumps, it is a function largely of speed and valve design. Most pump curves show the values of *NPSH* required at a given speed, which vary with flow.

For a given pump of fixed dimensions, more *NPSH* would be required at higher flowrates. However, as the flow is reduced and approaches zero, the *NPSH* required begins to rise. This is illustrated in Fig. 4 for one pump. The increase in *NPSH* under these conditions is explained by the pump operating at "off design" conditions—resulting in large inefficiencies that may manifest themselves as noisy operation or excessive temperature rise at low or minimum flow.

When provided with insufficient *NPSH* available from

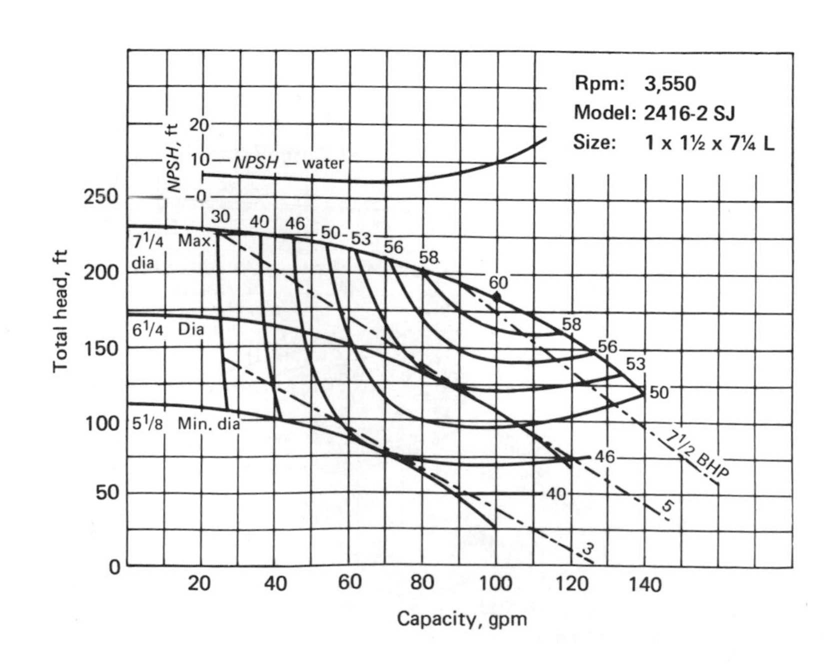


**NET** positive suction head in a given pump of fixed dimensions may increase as flow through pump is reduced and approaches zero—Fig. 4

the system in which it is to operate, a centrifugal pump will drop off in capacity from the full designed value. Expressed another way: at the required capacity point, the pump will develop less head than was planned. This phenomenon is known as cavitation. It is caused by excessive losses in *NPSH* in the entrance to the pump impeller. This results in the formation of vapor bubbles in the liquid, which then quickly collapse, releasing energy that attacks the impeller vanes or shrouds.

Although cavitation itself does not mean failure (that is, a pump may operate in cavitation occasionally with only minor damage), it should not be tolerated in the correct design of a pumping system.

An old definition for pump cavitation is illustrated in Fig. 5. When the difference between the developed head at rated capacity and flooded suction, and at the actual condition, is over 3%, the pump is said to be cavitating. Pumps handling pure liquids such as water may be at their worst during cavitation, inasmuch as the liquid is homogeneous and the vapor bubbles collapse at the same time. In addition, liquids such as water have a high vapor-to-liquid volume ratio. In the same pumping system, a mixed chemical or petroleum liquid, which are composed of many fractions that vaporize at somewhat different temperatures, will exhibit much less violent cavitation.

Much has been written on the subject of pump cavitation, and it is far beyond the scope of this article to discuss every aspect of this involved subject. In recent years, several good technical articles [*2, 3, 4, 5* and *6*] have appeared that cover the subject thoroughly. Suffice it to say that pumps must be selected that do not cavitate. How is this done?

In making the hydraulic study, sufficient *NPSH* must be made available to the pump, through a knowledge of the pumped liquid's characteristics and the physical location of the pump. Let us consider our first example again, in which the liquid is considered to be boiling. The height of the liquid above the pump can be controlled by designing the suction vessel with sufficient height abovegrade. However, if the liquid is at its boiling point in a storage tank on the ground with the pump located adjacent to it on the ground, we have a different pumping problem, as shown in Fig. 6.

The *NPSH* available, $NPSH_a$, is defined as:

$$NPSH_a = (P_{suct} - P_{frict} + P_{static} - P_{vapor})(2.31/sp\ gr)$$

where $P_{suct}$ is the absolute pressure of the liquid at its source, $P_{frict}$ is the pressure loss due to friction in the suction line, $P_{static}$ is net height of the liquid level above the centerline of the pump inlet (converted into psi for this equation), and $P_{vapor}$ is vapor pressure of the liquid at flowing temperature. A boiling liquid is considered to be at its vapor pressure in its source vessel, making *NPSH* available a function of static elevation and pipe friction only.

Normal practice in the chemical process industries is to obtain *NPSH* by elevation of the suction vessel. When this cannot be done, several procedures remain. One is to use a vertical pump, in which the pumping elements are in effect put beneath the ground, usually in a "can" or barrel. Thus, we obtain the necessary elevation below

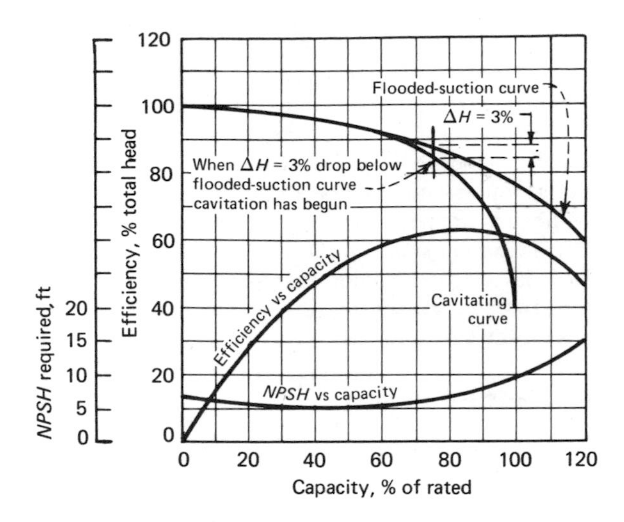


**CAVITATION** in a small centrifugal pump—Fig. 5

grade. This type pump is widely used for this purpose alone, although it can be used even where adequate *NPSH* exists abovegrade. Hot boiling liquids, volatile liquids, liquids such as butane or propane in storage spheres, hot-water condensate from vacuum steam condensers, etc., form some of the most popular applications for this type of pump.

With large process systems (especially above flowrates of 3,000 to 4,000 gpm), it is often uneconomical or completely unfeasible to elevate the suction vessel sufficiently to provide enough *NPSH* for a horizontal pump. Then, a booster pump may be used (a single-stage, often slower-speed pump) that takes the liquid from the suction vessel,

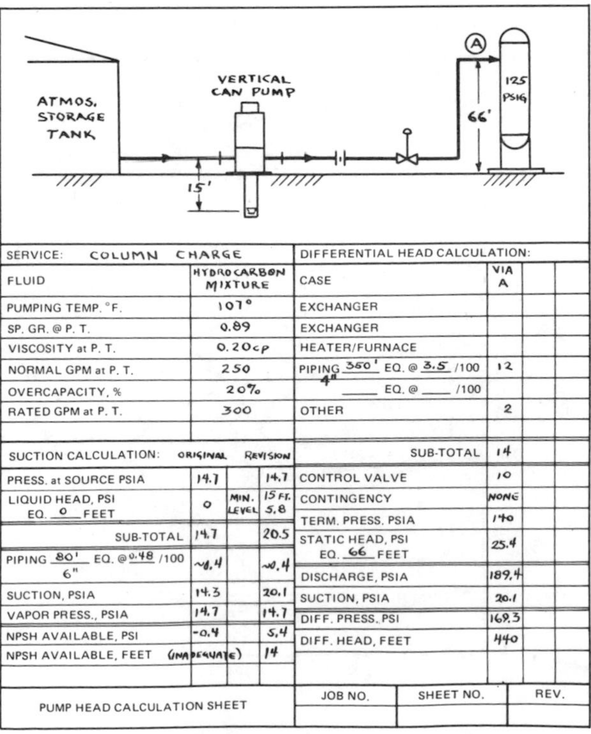


| SERVICE: COLUMN CHARGE | | | |
|---|---|---|---|
| FLUID | HYDROCARBON MIXTURE | | |
| PUMPING TEMP. °F. | 107° | | |
| SP. GR. @ P. T. | 0.89 | | |
| VISCOSITY at P. T. | 0.20cp | | |
| NORMAL GPM at P. T. | 250 | | |
| OVERCAPACITY, % | 20% | | |
| RATED GPM at P. T. | 300 | | |
| SUCTION CALCULATION: | ORIGINAL | | REVISION |
| PRESS. at SOURCE PSIA | 14.7 | | 14.7 |
| LIQUID HEAD, PSI EQ. 0 FEET | 0 | MIN. LEVEL | 15 FT. 5.8 |
| SUB-TOTAL | 14.7 | | 20.5 |
| PIPING 80' 6" EQ. @ 0.48 /100 | ~0.4 | | ~0.4 |
| SUCTION, PSIA | 14.3 | | 20.1 |
| VAPOR PRESS., PSIA | 14.7 | | 14.7 |
| NPSH AVAILABLE, PSI | -0.4 | | 5.4 |
| NPSH AVAILABLE, FEET | (INADEQUATE) | | 14 |

| DIFFERENTIAL HEAD CALCULATION: | | | |
|---|---|---|---|
| CASE | VIA A | | |
| EXCHANGER | | | |
| EXCHANGER | | | |
| HEATER/FURNACE | | | |
| PIPING 350' 4" EQ. @ 3.5 /100 | 12 | | |
| ____ EQ. @ ____ /100 | | | |
| OTHER | 2 | | |
| SUB-TOTAL | 14 | | |
| CONTROL VALVE | 10 | | |
| CONTINGENCY | NONE | | |
| TERM. PRESS. PSIA | 140 | | |
| STATIC HEAD, PSI EQ. 66 FEET | 25.4 | | |
| DISCHARGE, PSIA | 189.4 | | |
| SUCTION, PSIA | 20.1 | | |
| DIFF. PRESS., PSI | 169.3 | | |
| DIFF. HEAD, FEET | 440 | | |

| PUMP HEAD CALCULATION SHEET | JOB NO. | SHEET NO. | REV. |
|---|---|---|---|
| | | | |

**NPSH** calculations for boiling liquid in tank—Fig. 6

and boosts it into the main pump for delivery at the required pressure.

This concept of booster pumping is now being applied to smaller chemical-type pumps by use of a built-in "inducer," a form of impeller that acts in effect as a low-*NPSH* booster impeller. This impeller boosts the pumped liquid into the main impeller, and makes the total net requirement for *NPSH* very low (as low as 1 or 2 ft in certain small sizes).

Now let us consider our example from Fig. 3 again. From dozens of types of pumps, which one shall we choose to fit the data?

## AVS Standard Pumps

The American Voluntary Standard (AVS) type overhung chemical-process pump would fit very well in this application. [For more details, see the article in this issue on pp. 123–124.] This type is the current workhorse of the chemical process industries, having been developed by nearly all pump manufacturers in accordance with standardized coverage and mounting dimensions as set forth by the Manufacturing Chemists Assn., and proposed as the American Voluntary Standard for Pumps, in 1962 [7].

AVS pumps are now available in sizes from 1-in discharge, with best capacity at 3,550 rpm of 100 gpm, to 8-in discharge, peaking at about 3,000 gpm. Up to 700 ft of head is developed in certain smaller sizes at 3,550 rpm, and about 200 ft in the larger sizes running at 1,750 rpm.

These pumps incorporate the back pullout feature, so that the pumping element may be removed without breaking the piping connections. The AVS type have been designed for use of mechanical seals, and the majority of such pumps come with seals. However, for unusual applications, conventional packing can also be used.

Because of standardization, the pump manufacturers have many standard material combinations available for relatively quick shipment, including ductile iron, steel, several types of stainless steels, Alloy 20, and Hastelloy. Also, using the same bearing brackets, special liquid-ends have been developed for specialized pumping services—such as the self-priming liquid end; or that made of solid plastic, or lined, for example, with Teflon or ceramic material.

For the example shown in Fig. 3 (and assuming the AVS type was desired), let us proceed to see how a correct selection of size is made. For a given manufacturer, reference to a coverage chart (Fig. 7) will show the several sizes that might be considered. Looking at the individual curves for these several sizes (Fig. 8), we find that the smallest pump (2 × 3 —8), Fig. 8a, is too small for the given application because its peak-efficiency point is at 260 gpm—somewhat less than our requirement; also, that it is at its maximum impeller diameter, and requires about 15 ft *NPSH* at 300 gpm. Normally, the engineer attempts to select pumps having an impeller diameter not more than 95% of the maximum diameter that can be installed, in order to provide for future expansion, or correction in the event of minor revisions to pumping-system requirements.

The next larger size (2 × 3—10), Fig. 8b, might be an acceptable selection, although it is still past peak efficiency. Note that it has an efficiency of 55% at 300 gpm, and *NPSH* required is 11 ft. The third pump (3 × 4—8G), shown in Fig. 8c, would not cost much more. It is

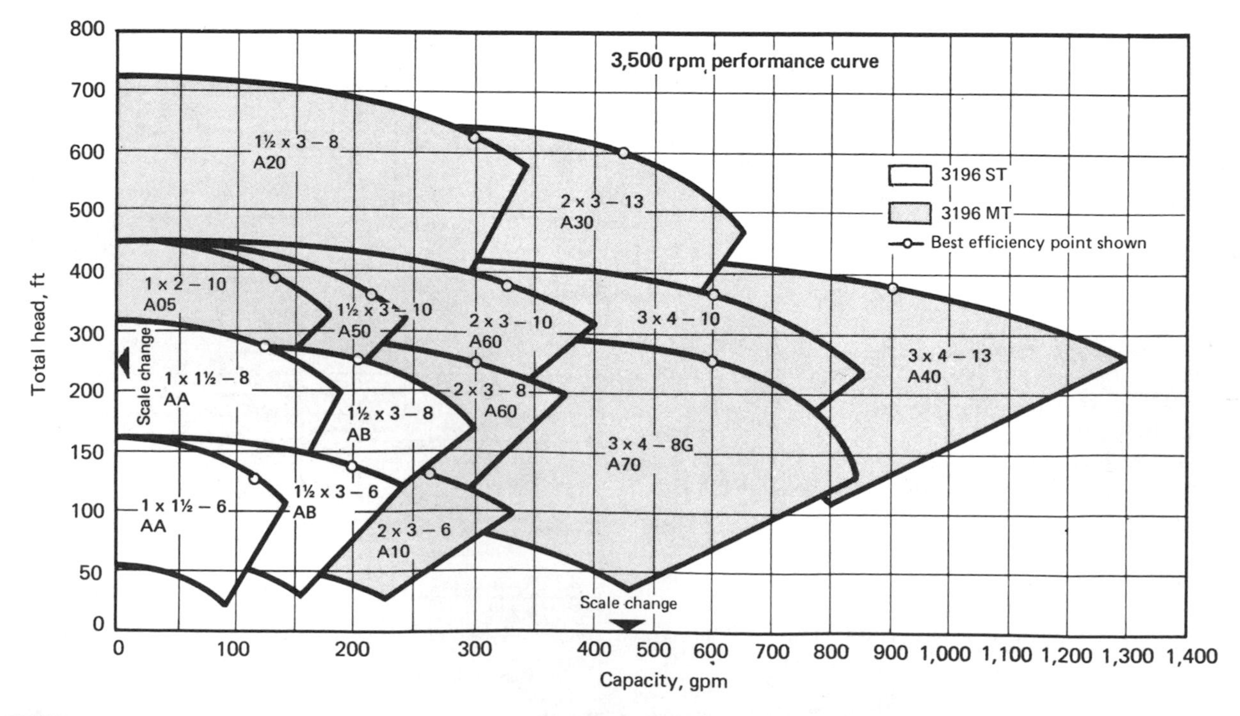


**CENTRIFUGAL**-pump coverage chart of one manufacturer shows available sizes that conform to AVS standards—Fig. 7

a larger pump, and more efficient (63%), requiring less *NPSH* (10 ft), and acceptable with regard to the impeller diameter.

Once the selection is made, we can record the efficiency at rated point; the rated and maximum impeller diameters; and the *NPSH* required, making sure this is lower than *NPSH* available. We then calculate pump horsepower according the the relationship:

$$BHP = \frac{(\text{gpm})(\text{ft head})(\text{sp gr})}{(3{,}960)(\text{pump efficiency})}$$

If the motor driver is required to cover the pump throughout the entire range of the curve, visual extension of the curve to the end of the chart is usually sufficient. From this step, another point of capacity, head and efficiency can be obtained, and the horsepower at the end of the curve calculated as shown. Unless otherwise noted, all pump curves are printed for liquids having a specific gravity of 1.0, and must be corrected as shown.

## Other Centrifugals

Older horizontal designs can be used for this same service. Overhung, tangential-discharge pumps are the predecessors of the AVS type, and many are still in use. They have gradually become less popular, due to the advantages, greater availability and lower cost of the AVS type.

The "close coupled" type could have been chosen, either with tangential discharge, or with AVS liquid-end dimensions. Here, the disadvantage in the chemical process industries is greater hazard, because the motor is closer to the pump seal and may be more vulnerable in the event of seal or packing failure. Also, the back pullout feature is not available, requiring either disconnecting of piping or of electric motor to remove the pump. Finally, flange-mounted special motors are required. These are often undesirable because of lack of interchangeability with the majority of electric motors in the plants. However, some plants have been designed to use this type of motor mount, and savings may accrue due to elimination of baseplate, and ease of alignment and installation.

Even the horizontal-split-case

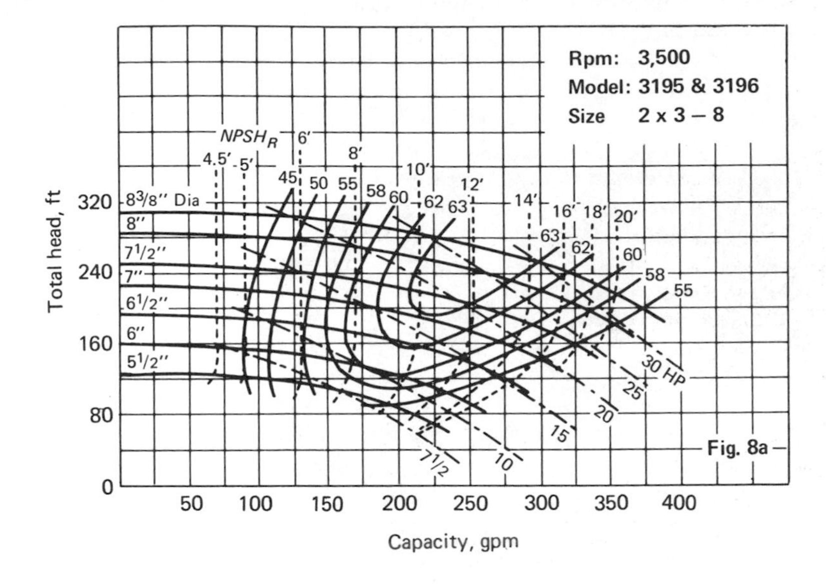


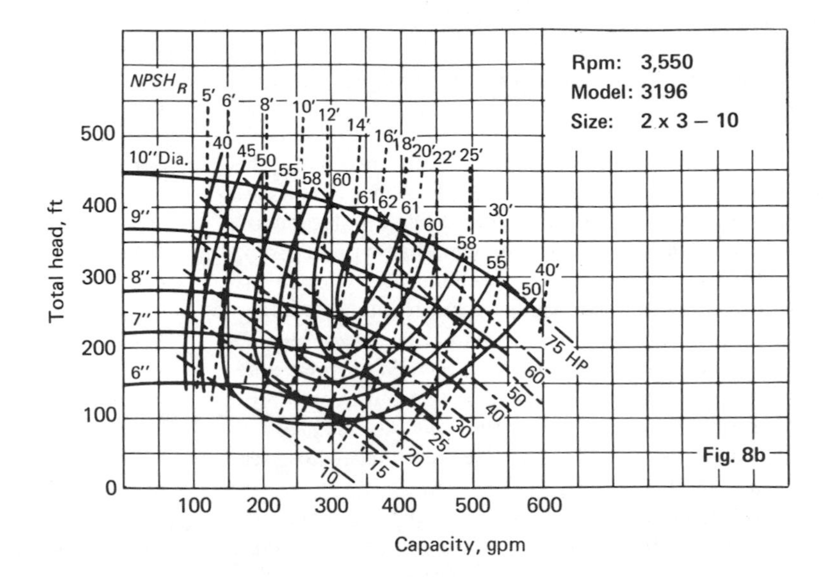


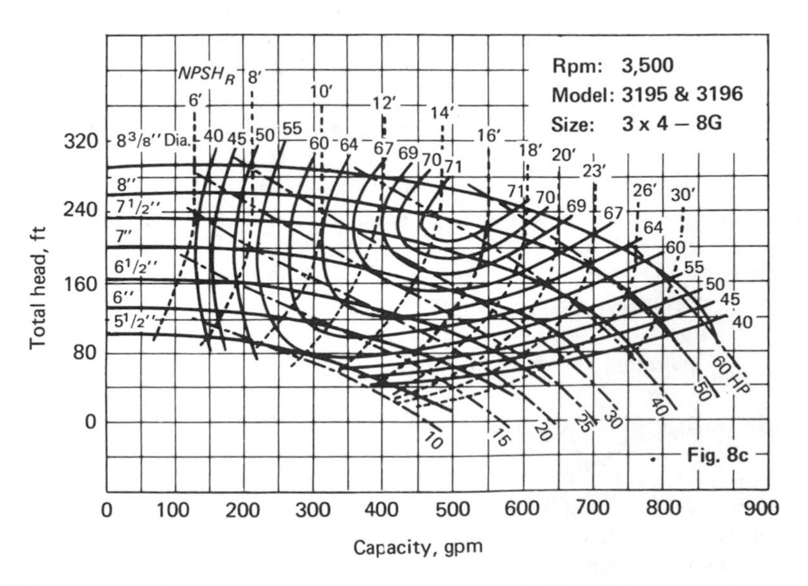


**PUMP** characteristics for some of the pumps shown in Fig. 7 chart—Fig. 8

**CENTERLINE**-supported pump for high temperature—Fig. 9

double-suction pump could have been chosen for this application. However, being generally higher in first cost, presenting no appreciable advantage in efficiency, having two stuffing boxes instead of one, this type of pump would not be chosen today for chemical service in this size. It still finds wide use as a utility pump, cooling-water or fire-water pump, or for larger flow requirements (above 3,000 gpm) for which the AVS type is not available.

Let us now consider the identical hydraulic conditions, but for a different liquid. This time, we have a corrosive, toxic, dangerous liquid. Even the best mechanical seal may fail.

In conventional pumps, double mechanical seals may be installed, with nontoxic liquid to buffer the leakage. In the event this system is not desired, consideration could be given to leakproof pumping. Several designs exist in overhung pumps up to 10-in size (about 4,000 gpm) that use a hydraulic, spring-loaded or air-loaded seal. This seal moves axially along the shaft upon startup, causing the stuffing box to remain sealed whether operating or at rest.

The canned-rotor pump is also available, in sizes from about 20 gpm to over 1,500 gpm, heads up to 500 ft in the smaller sizes. This type of pump permits the liquid to recirculate through motor bearings and return. A thin stainless-steel shell protects the motor windings from contact with the fluid. No shaft seal is required, and true leakproof pumping can be obtained.

## High-Temperature Pumps

We will now consider the same hydraulic problem but with the additional requirement that the pumped liquid is hot. The maximum temperature limit for the AVS-type, foot-supported casing is about 350 to 400°F. Most petroleum and petrochemical plants do not permit this type, even at this temperature, but require the use of the centerline-supported casing type (Fig. 9). Hydraulically similar, these types are heavier (maximum working pressure of casing is up to 600 psig, or higher, vs. a limit of about 300 psig for the AVS type), and suitable for temperatures to 800°F, and higher. They still retain the back pullout feature. In fact, it was this type design where the back pullout idea first originated. The centerline-supported casing provides for equal expansion above and below the pump centerline, essential in maintaining alignment when pumping hot liquids. For temperatures higher than about 250°F, the stuffing box and bearing housing may be water cooled.

These hot types are also available in larger sizes, using the double-bearing or between-bearing construction. The casing is split on the vertical axis, making it possible to use a confined gasket as in the overhung type, desirable when pumping hot, flammable or dangerous liquids.

Less-expensive versions of the centerline-supported overhung pumps are also available. These include all the features of the AVS types plus lugs to the pump casing, making centerline support possible. Such pumps should be applied with caution because they are not generally as heavy or as versatile as the standard overhung centerline types.

## Slurry Pumps

If the liquid to be pumped is a heavy slurry rather than hot, or corrosive, the lined slurry pump should be chosen. Also an overhung design, this type is usually much heavier and runs at slower speeds (often with V-belt or variable-speed driver). It may be hard-lined with renewable hard-metal plates, such as Ni-hard, which resist heavy slurry abrasion, or may be rubber-lined (preferred for certain materials). Slurry pumping is a specialized subject, and more details can be found in other articles [*8, 9*].

## Vertical Pumps: Inline or Sump

Hydraulic coverage similar to the AVS type can be obtained with the vertical inline chemical pump. This type is relatively new but has increased in popularity due to its compact design and the saving in space resulting from installation directly in the horizontal pipeline. The American National Standards Institute has published a tentative standard for vertical inline pumps for the chemical process industries [*10*]. Special alloy materials are available with this type as well. Larger-flow-size, higher-temperature, higher-pressure models are also available up to flows of approximately 2,500 gpm and heads of 150 ft.

If the pump meeting the requirements for the data in Fig. 3 must be in a tank or pit, a vertical pump will probably be chosen. (Self-priming horizontal pumps can be used where a suction lift exists, at the user's preference.) Vertical sump pumps are either single-stage, volute-type centrifugal, or single-stage or multistage vertical-turbine type. The latter is derived from the deep-well pump used for pumping water. It consists of one or more stages screwed or bolted together to form a multistage pump. Pumps of this type are available in flow ranges from 60 gpm to over 30,000 gpm, such as might be required for a cooling-water pump in a large plant.

On the other hand, the volute-type sump pump has many special variations for the chemical process industries, which cannot be obtained with vertical turbine pumps such as steam-jacketing for pumping molten liquids (Fig. 10), or vertical cantilever designs that eliminate any bearings along the shaft and the possibility of corrosion in the tank or pit. There are also vertical slurry pumps (again in special metals or rubber lined) with paddle-wheel-type impellers, which give up overall pumping efficiency in return for pumping reliability in very difficult, dirty services.

## Multistage Pumps

For heads above about 700 ft, pumps must either be operated at speeds above those available with 60-Hz electric motors, or multistage pumps must be used.

Dynamic-type pumps obey the affinity law:

$$\frac{N_1}{N_2} = \frac{Q_1}{Q_2} = \sqrt{\frac{H_1}{H_2}}$$

where $N$ is speed, $Q$ is flow and $H$ is head. From this relation, we see that by operating a pump at twice a given speed, the capacity is doubled, and the head developed will be four times as great. Operating at speeds above 3,550 rpm requires the use of speed-increasing gears, which may not be desirable.

Many pumps are built only with antifriction bearings, which are not used in pumps above about 4,000 to 4,500 rpm. Hence, only on specially designed pumps for higher speeds can higher heads be obtained in one stage. The more common practice is to use several stages in series in the same pump casing. The horizontal-split casing design is the most common, with impellers arranged either in one direction balanced by suitable balance drums and thrust bearings, or with opposed impellers where the thrust is partially balanced by the opposing forces on each impeller.

Higher pressures (above about 1,800 to 2,000 psi) are available with the vertically-split, barrel-type casing, often applied in higher-pressure boiler-feed service or in high-pressure charge pumps for certain chemical or petrochemical processes. Two-stage pumps are also available in horizontal-split case designs, and also as overhung or between-bearing centerline-supported casings, which can be applied in the range 50 gpm to 2,700 gpm at heads up to about 1,500 ft.

Vertical multistage pumps follow similar affinity laws, and can be built in many stages, as many as 20 or 25 for a small, vertical, turbine pump. However, before specifying a pump with this many parts, it is wise to investigate other types, such as the vertical inline type with speed-increaser gear, or possibly even a reciprocating triplex pump.

## Small-Volume Pumping

Often, many of the pumping requirements in the chemical process industries are for relatively small flows.

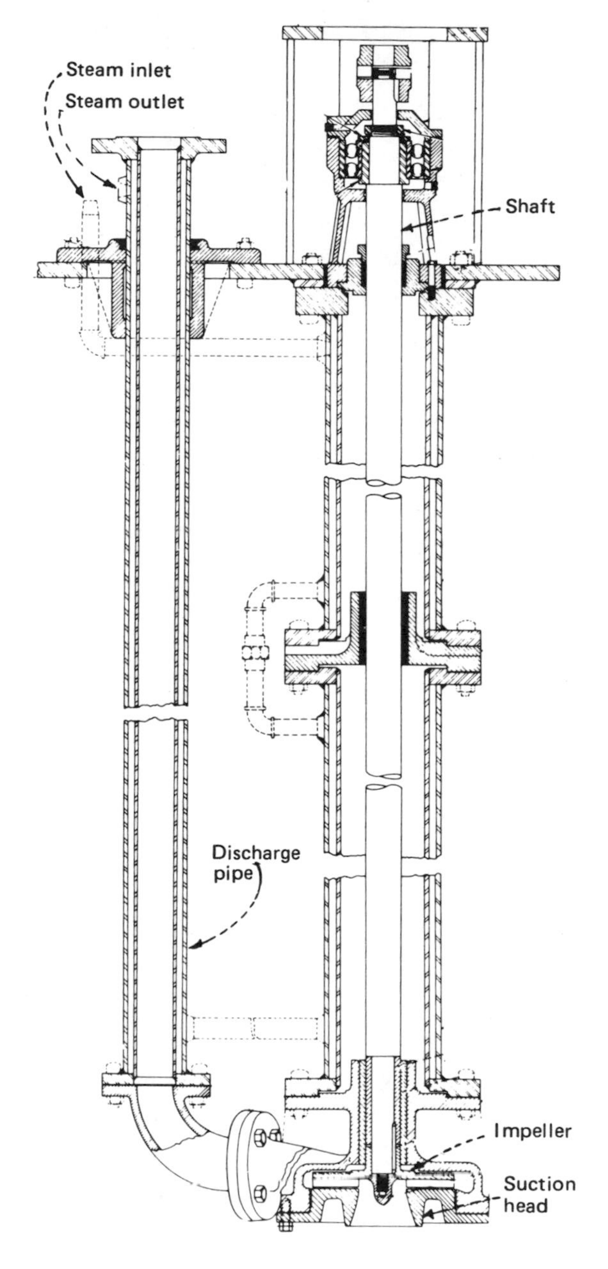
Steam inlet
Steam outlet
Shaft
Discharge pipe
Impeller
Suction head
Keep saturated steam on pump jacket at 35 psi. Any other pressure or temperature will change viscosity of sulfur and may foul submerged bearing.

**VERTICAL** steam-jacketed centrifugal pump—Fig. 10

Although we are constantly thinking about larger plants and larger outputs, it still remains a fact that the process plant of whatever size always requires many small pumping services. This brings us to the dilemma of not being able to find a pump small enough for efficient pumping, and compromises either in centrifugal-pump efficiency or pump type have to be made.

The most obvious compromise on small-volume pumping is to use a pump that is too large for the job. Let us consider the problem shown in Fig. 11. Visual inspection of this head-calculation sheet would at first

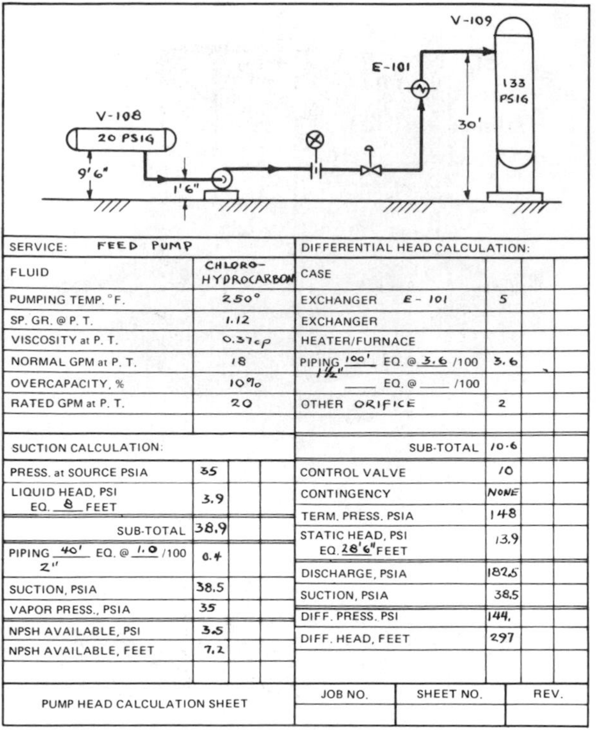


| SERVICE: FEED PUMP | |
|---|---|
| FLUID | CHLORO-HYDROCARBON |
| PUMPING TEMP. °F. | 250° |
| SP. GR. @ P. T. | 1.12 |
| VISCOSITY at P. T. | 0.37cp |
| NORMAL GPM at P. T. | 18 |
| OVERCAPACITY, % | 10% |
| RATED GPM at P. T. | 20 |

| SUCTION CALCULATION: | |
|---|---|
| PRESS. at SOURCE PSIA | 35 |
| LIQUID HEAD, PSI EQ. 8 FEET | 3.9 |
| SUB-TOTAL | 38.9 |
| PIPING 40' 2" EQ. @ 1.0 /100 | 0.4 |
| SUCTION, PSIA | 38.5 |
| VAPOR PRESS., PSIA | 35 |
| NPSH AVAILABLE, PSI | 3.5 |
| NPSH AVAILABLE, FEET | 7.2 |

| DIFFERENTIAL HEAD CALCULATION: | |
|---|---|
| CASE | |
| EXCHANGER E-101 | 5 |
| EXCHANGER | |
| HEATER/FURNACE | |
| PIPING 100' 1½" EQ. @ 3.6 /100 | 3.6 |
| ____ EQ. @ ____ /100 | |
| OTHER ORIFICE | 2 |
| SUB-TOTAL | 10.6 |
| CONTROL VALVE | 10 |
| CONTINGENCY | NONE |
| TERM. PRESS. PSIA | 148 |
| STATIC HEAD, PSI EQ. 28'6" FEET | 13.9 |
| DISCHARGE, PSIA | 182.5 |
| SUCTION, PSIA | 38.5 |
| DIFF. PRESS. PSI | 144. |
| DIFF. HEAD, FEET | 297 |

| PUMP HEAD CALCULATION SHEET | JOB NO. | SHEET NO. | REV. |
|---|---|---|---|
| | | | |

**PUMP-HEAD** requirement for small flows—Fig. 11

suggest the use of a conventional centrifugal pump. But the smallest size available—for instance, in the AVS type—is much too large (Fig. 12). The pump could be chosen anyway, operating back on the curve toward zero flow, by sacrificing efficiency. Reference to Fig. 2 would confirm that for this specific speed of 10.5 (with $Q$, ft$^3$/s, and at 3,550 rpm), no efficient conventional centrifugal pump can be found.

However, other types of pumps might be considered for this service. The partial-emission pump, built mostly in vertical inline types, has a single point of emission in the diffuser, as contrasted with a full-volute pump (Fig. 13). The partial-emission pump appears in Fig. 2 in a limited range. It is built both as a direct-coupled pump at 3,550 rpm for capacities between 5 gpm and 300 gpm, and heads to 700 ft; and as a gear-driven unit at rotating speeds up to 24,000 rpm, for capacities up to 400 gpm or heads to 6,000 ft.

The drag or regenerative-turbine pump is also available for small-volume pumping in either one or two stages. (Some multistage varieties also exist.) This type requires clean liquid, free of abrasive or finely divided solid materials, because the clearances between the rotating impeller and the stationary casing are very small. It is available in both horizontal-shaft and vertical-shaft models, and by some manufacturers in casings that meet the AVS outer dimensions. But both of these types are inherently inefficient. They are usually chosen because of their simplicity and relative ease of operation and control. If efficient low-volume pumping is required, the positive-displacement types must be used.

## Positive-Displacement Pumps

Positive-displacement pumps divide basically into rotary pumps and reciprocating pumps. R. W. Abraham recently presented a chart illustrating more than a dozen types of rotary pumps [*12*]. These are not only used for low-volume pumping but are made in larger sizes up to capacities of 2,500 gpm. They are also used for pumping viscous liquids that cannot be effectively handled by centrifugal pumps.

Reciprocating pumps are subdivided into crankshaft-driven piston and plunger pumps, and direct-acting steam pumps. This latter type, while not used much today, should not be discounted for certain specialized services when steam can be used for the motive power. It functions by the action caused by the steam exerted on the steam piston directly transmitted through piston rod and valve gear to the liquid piston. The relative pressure that

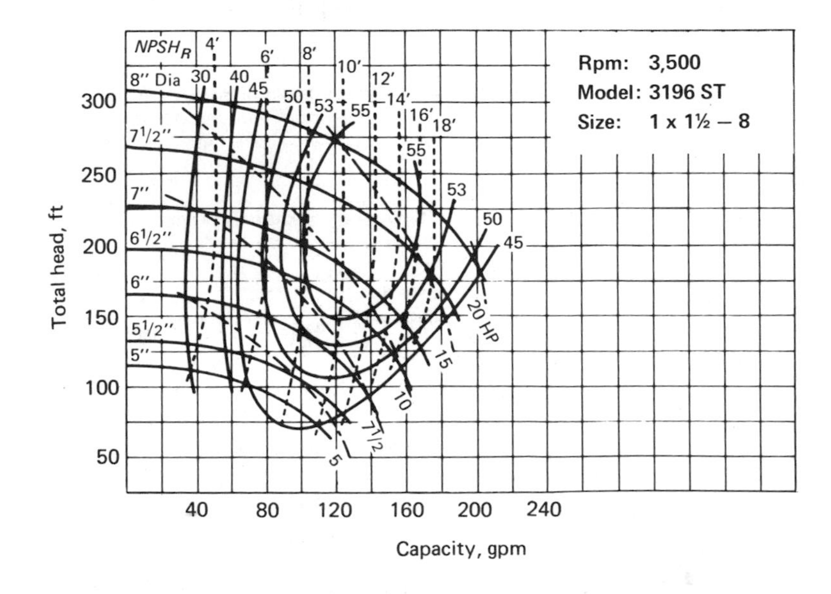


**CURVE** for smallest AVS standard centrifugal pump shows that pump can handle a small-volume flow by sacrificing efficiency—Fig. 12

**Maximum Speeds for Reciprocating Pumps in Process Service—Table I**

| Plunger Pumps | | |
|---|---|---|
| Stroke, In | Speed, Rpm | Plunger Speed, Ft/Min |
| 2 | 433 | 130 |
| 3 | 320 | 160 |
| 4 | 270 | 180 |
| 5 | 240 | 200 |
| 6 | 220 | 220 |
| **Direct-Acting Duplex Steam Pumps** | | |
| Stroke, In | Speed, Reversals/Min | Piston Speed, Ft/Min |
| 6 | 30 | 30 |
| 8 | 26.2 | 35 |
| 10 | 24 | 40 |
| 12 | 22.5 | 45 |

Values in this table are suggested for liquids below 200°F, and viscosities below 2,000 SSU (Saybolt Seconds Universal). For higher temperatures or higher viscosities, these values should be reduced.

the pump can develop is a ratio of the areas of steam piston to liquid piston.

Most reciprocating pumps used today are the crankshaft-driven types. These are built in combinations of two or more plungers (usually three, five or seven) to reduce the effect of pulsating flow, which is one of the disadvantages of positive-displacement pumping (Fig. 14). Although the pulsations may be relatively small with a triplex pump, nevertheless they do exist and must be accounted for in the process system, especially in relation to control. Pulsation dampers are often used to reduce pulsations to acceptable levels.

Inasmuch as the piston or plunger cannot displace every bit of the space in which it moves, a certain volume in the cylinder is left untouched, giving rise to volumetric efficiency. This is defined as the ratio of actual pumped capacity to theoretical displacement. Volumetric efficiency must not be confused with pumping efficiency, for a pump with a low volumetric efficiency may still have a high mechanical efficiency. This results in a very low power consumption, according to the relationship:

$$BHP = \frac{(\text{Flow, gpm})(\text{Pressure, psi})}{1{,}714\ (\text{Mechanical efficiency})}$$

Although the concept of pumping head is also true for positive-displacement pumps, it is seldom referred to because it is unnecessary in the determination of horsepower. Since the pumps are positive-displacement types, not dynamic types, they will develop the same pressure on any liquid that can flow into the pump chamber.

Reciprocating pumps are usually chosen with relation to crankshaft speed, rpm; and piston or plunger speed, ft/min. Piston speed is the average rate at which the piston or plunger travels while making one complete stroke, and can be calculated when the stroke and rotating speed are known:

$$S = 2LN/12$$

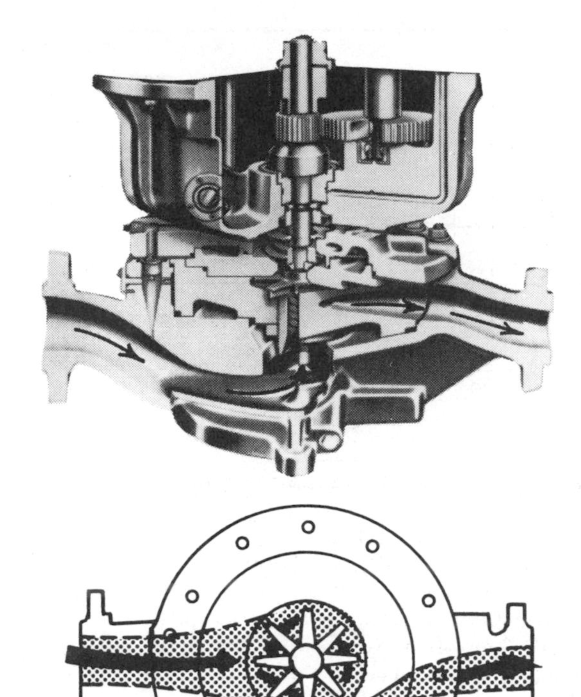

Single-emission-point pump

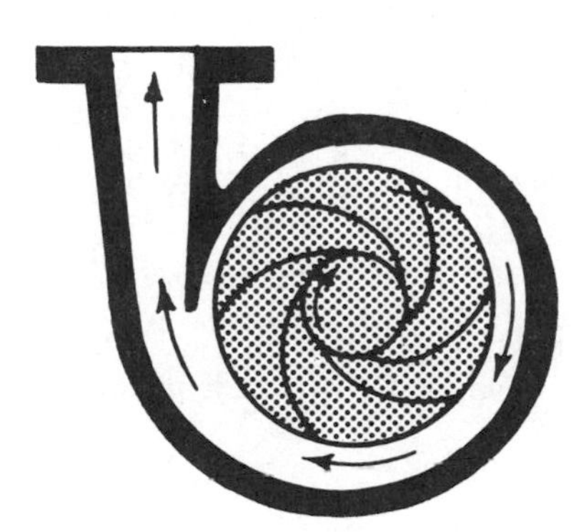

Full-emission volute pump

**FLOW** path through partial-emission pump—Fig. 13

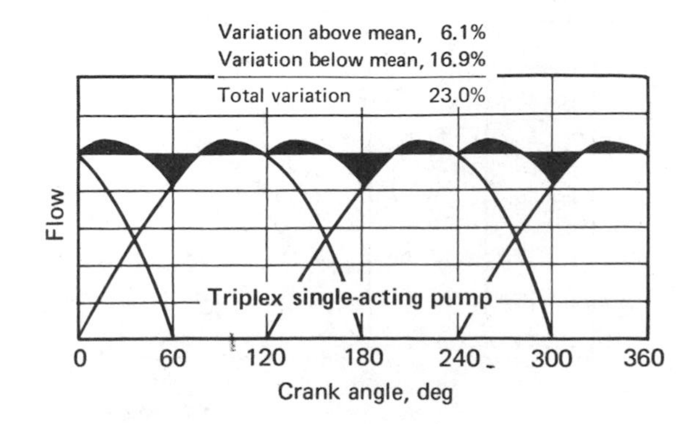


**PULSATIONS** in triplex reciprocating pump—Fig. 14

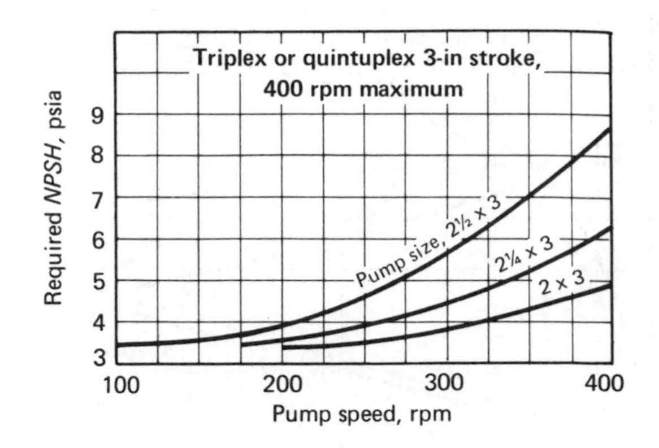


**NPSH** requirements for reciprocating pumps—Fig. 15

where $S$ is piston speed, ft/min; $L$ is stroke, in; and $N$ is rotating speed, rpm.

Permissible rotating speeds and piston or plunger speeds are often left to the manufacturer's recommendation. However, the limits shown in Table I are suggested as a guide wherever continuous uninterrupted pumping is required.

Reciprocating pumps also require *NPSH*. However, in the event that the *NPSH* available is insufficient, reciprocating pumps do not exhibit cavitation as do centrifugal pumps. Instead, the cylinder chamber will fail to become filled, with the liquid vaporizing as it flows through the inlet valves. This causes a reduction in outlet volume and possible short-stroking on direct-acting pumps, or vibration or noisy operation on crank-driven pumps. The *NPSH* required by a given size pump is likewise a function of speed (Fig. 15). Thus, the way to provide proper pump selection for low *NPSH* systems, requiring positive-displacement pumps, is to specify lower than maximum operating speeds.

Many types of valves exist for use in direct-acting and crank-driven pumps. These include plate, plug-and-disk, wing-guided, ball, and spherically-seated valves. Usually the correct selection of valves may be left to the pump manufacturer if complete details on liquid properties and operating conditions are given.

Like the rotary positive-displacement pump, the reciprocating pump finds application also for larger volumes (as high as 2,500 gpm), for high pressures (to 10,000 psig, and higher with special designs), or for special pipeline applications where centrifugal pumps cannot be built (for example, the coal-slurry pump in the Black Mesa project) [*11*]. Reciprocating pumps should not be considered as obsolete, as they very definitely have a place for certain special pumping applications.

## Pumping Viscous Liquids

The largest field of application for the rotary positive-displacement pump is for viscous materials. In a dynamic pump, as the viscosity of the liquid increases, the friction losses increase. This causes a drop in the head-capacity relationship for a given pump, and a drop in efficiency (Fig. 16). To combat this difficulty, the gear or screw pump can be used; here, the increasing viscosity actually helps the pump, as there is less liquid lost to slippage and more forward net flow. Of course, sufficient power must be provided to handle the viscous liquid. Fig. 17 shows a typical performance curve for a rotary pump at constant speed and varying liquid viscosity.

## Controlled-Volume Pumps

In a separate category are pumps (required for very small flows) that deliver an exact metered amount, such as in an injection system or a chemical feed system. For this duty, the pumps are basically crank-driven plunger types, with means for adjusting the stroke length. Usually, the adjustment is made while such a pump is in operation, and sometimes with an arrangement to make the adjustment automatically. The diaphragm liquid-end is also available for this type of pump. The diaphragm

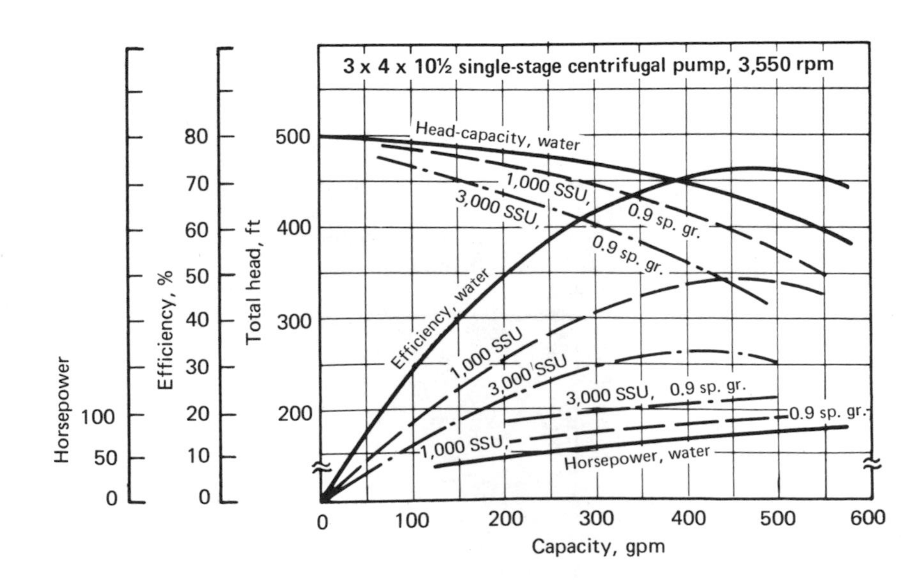


**CENTRIFUGAL**-pump performance changes when handling liquids of different viscosity—Fig. 16

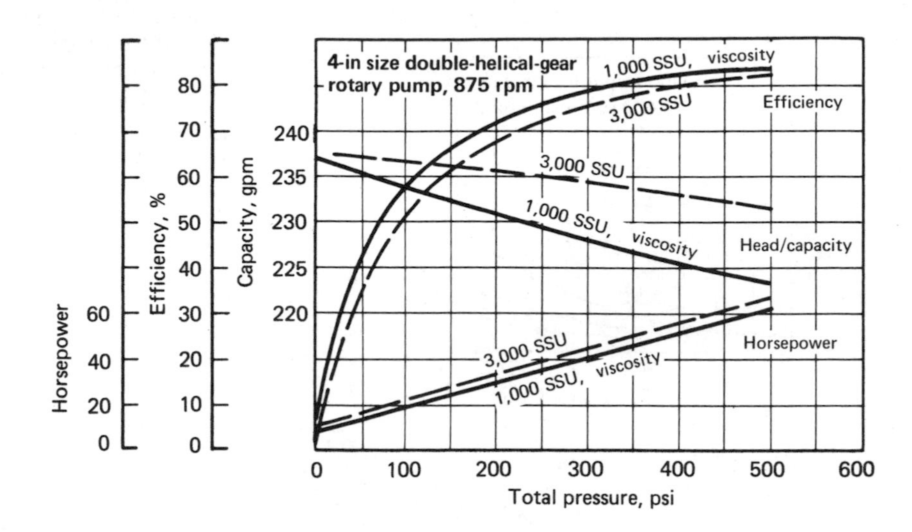


**ROTARY**-pump performance at constant speed and varying liquid viscosity—Fig. 17

separates the pumped liquid from the hydraulic oil handled by the plunger, providing leakproof pumping during the life of the diaphragm. Plunger or diaphragm pumps can be arranged as multiple units, either to handle larger flows or sometimes to handle several different liquid flows with the same drive-train.

## Guidelines for Pump Selection

Countless varieties of pumping equipment exist and confront the process engineer. In selecting a proper pump for a given application, we might reiterate the need for:

1. An accurate pump-head calculation.
2. A basic knowledge of the various types of pumps.
3. A decision about the type of pump desired (horizontal or vertical, inline, aboveground, belowground, etc.).

Assistance from pump manufacturers is often most helpful in determining the type to be used, and should precede the actual writing of specifications or inquiry for proposals. It is customary to send pump data-sheets with specifications for bid, and to make analysis of bids upon receipt. But this can only be done intelligently if the same types are being considered from each bidder.

Finally, there is still no substitute for common sense and experience in selecting a pump. Past history of successes and failures, refinery or plant personnel preferences, attempts at overall interchangeability within a given process unit or total plant complex—all these must be considered in making proper pump selections.

## Acknowledgements

The following sources have supplied background information and/or illustrative material: Aldrich Div., Ingersoll-Rand Co.; Allis-Chalmers Corp.; Aurora Pump Div., General Signal Corp.; Barrett, Haentjens Pump Co.; Buffalo Forge Co.; Byron-Jackson Pump Div., Borg-Warner Corp.; Chempump Div., Crane Co.; Dean Brothers Pumps Inc.; Deming Div., Crane Co.; Duriron Co., Inc.; Eco Pump Corp.; Envirotech Corp.; Fairbanks Morse Pump Div., Colt Industries Inc.; The Galigher Co.; Gorman-Rupp Co.; Goulds Pumps, Inc.; Hills-McCanna Div., I.U. International Corp.; Hydraulic Institute; Ingersoll-Rand Co.; Interpace Corp.; Johnston Pump Co.; La Bour Pump Co.; Chas. S. Lewis & Co., Inc.; Mission Mfg. Co., TRW Inc.; Nagle Pumps Inc.; Pacific Pumps Div., Dresser Industries Inc.; Peerless Pump Div., FMC Corp.; Roper Pump Co.; Roy E. Roth Co.; Milton Roy Co.; Sundstrand Corp.; Tuthill Pump Co.; Union Pump Co.; United Centrifugal Pumps; Wallace & Tiernan Inc.; A. R. Wilfley and Sons Inc.; Wilson-Snyder Pumps, Oilwell Div., U.S. Steel Corp.; Viking Pump Div., Houdaille Industries Inc.; Waukesha Foundry Co.; Worthington Corp.; Yarway Corp.

## References

1. Balje, O. E., A Study on Design Criteria and Matching of Turbomachines: Part B—Compressor and Pump Performance and Matching of Turbocomponents, *J. Eng. Power,* Jan. 1962.
2. Stepanoff, A. J., Cavitation in Centrifugal Pumps With Liquids Other Than Water, *J. Eng. Power,* Jan. 1961.
3. Salemann, V., Cavitation and NPSH Requirements of Various Liquids, *J. Basic Eng.,* **81,** 167-173 (1959).
4. Hefler, J. R., Figure NPSH for Proportioning Pumps, *Petrol. Refiner,* June 1956, pp. 161-169.
5. Miller, J. E., Effect of Valve Design on Plunger-Pump Net Positive Suction Head Requirements, presented at the ASME Petroleum Mechanical Engineering Conference, New Orleans, Sept. 18-21, 1966.
6. Hendrix, L. T., New Approach to NPSH, *Petrol. Refiner,* June 1958, pp. 191-194.
7. Centrifugal Pumps for Chemical Industry Use B73, proposed American Voluntary Standard, American National Standards Institute, New York.
8. McElvain, R. E. and Cave, I., Transportation of Tailings, Warman Equipment (International) Ltd., Madison, Wis., Nov. 1972.
9. Wick, K. E. and Lintner, R. E., Pumping Slurries With Rubber-Lined Pumps, Denver Equipment Div., Joy Mfg. Co., Denver, Colo., Nov. 1973.
10. Vertical Inline Pumps, working draft of Subcommittee #2, American National Standards Institute, New York, Apr. 25, 1972.
11. Love, F. H., The Black Mesa Story, *Pipeline Eng. Intern.,* Nov. 1969.
12. Abraham, R. W., Reliability of Rotating Equipment, *Chem Eng.,* Oct. 15, 1973, p. 105.

## Meet the Author

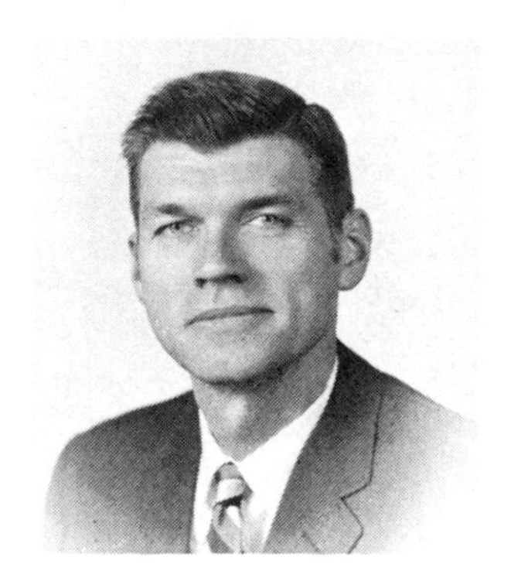

**Richard F. Neerken** is Chief Engineer, Rotating Equipment Group, for The Ralph M. Parsons Co., P. O. Box 54802, Los Angeles, CA 90054. He joined Parsons in 1957 and has worked continuously with rotating machinery such as pumps, turbines, compressors and engines on all projects for the company. Previously, he spent over 11 years as an application engineer for a major manufacturer of pumps, compressors and turbines. He has a B.S. in mechanical engineering from California Institute of Technology and is a member of the Contractors Subcommitte on Mechanical Equipment for the American Petroleum Institute.

# Saving energy and costs in pumping systems

For lowest cost in pumping systems, looking at first costs is not enough. You must evaluate the overall system, including flowrates and variable-capacity and material requirements.

*John A. Reynolds,* *Union Carbide Corp.*

☐ There are many ways to unknowingly squander energy in pumping systems. Some of the more prominent ones will be discussed in this article. But first, we should look at what energy is costing us, and find out how we can save operating horsepower by investing in a more efficient pump.

## Energy costs

Let us compare two pumps of different manufacture for water-booster service. Brand A requires 10 hp at the specified performance conditions; Brand B requires 9 hp. The electric utility rate is 2¢/kWh. The direct operating cost for a ten-year-life project would be (2¢/kWh)(8,760 h/yr)(0.746 kW/hp) (10 yr) ÷ 0.85 motor efficiency = $1,538/hp.

However, a dollar spent in the tenth year of the project is worth less than a dollar now, due to the time value of money. (Using a discounted-cash-flow method of analysis, one can enter the various elements of consideration, such as investment, direct costs, interest rates, project life, etc., and arrive at a precise number.) For simplicity in later comparisons, assume that $1,000 is the amount we can afford to spend in first cost to save one operating horsepower, which we will call the *investment equivalent* of the lifetime operating costs. Thus, if Brand A pump costs $1,500 and Brand B costs $1,700, Brand B would have the best overall economics, as shown by the comparison presented for Brand A and Brand B in Table I.

Once we have established an investment equivalent of utility usage, it is advisable to pass this value on to the supplier (even to having an entry on the equipment data sheet). This will help him to offer the pump with the best overall economics; it will encourage him to avoid limiting his bid to one that is lower in first cost only.

Originally published January 5, 1976

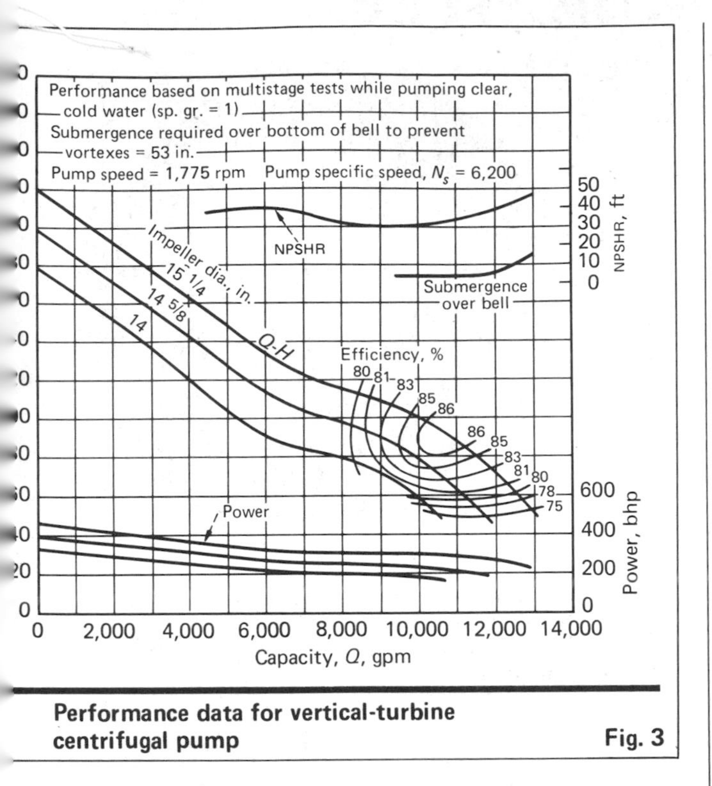


**Performance data for vertical-turbine centrifugal pump** **Fig. 3**

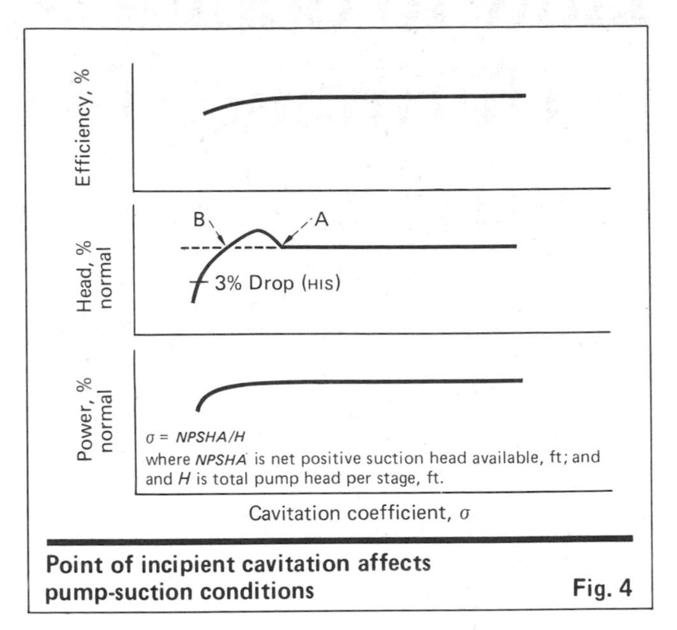


**Point of incipient cavitation affects pump-suction conditions** **Fig. 4**

ence of one does not have anything to do with the other.

Increasing the NPSHA usually has no effect on recirculation cavitation, because an addition to the suction head has little or no effect on the magnitude of the accelerative forces in the recirculation area. As an example, a mixed-flow pump was tested in a closed loop that incorporated a heat exchanger to remove the mechanical equivalent of heat. The pump speed was increased and the pump was throttled until cavitation could be heard. Then the pressure on the loop was increased 100 psi, which essentially increased the NPSHA by more than 230 ft. The pump continued to cavitate. Subsequent inspection of the impeller showed a band of cavitation on the rear of the vanes, with no evidence whatever of cavitation occurring on the leading edges.

Photographs have been taken during the operation of vertical turbine-type centrifugal pumps, such as those used for cooling-tower applications, where the pumped liquid passed out of the impeller eye and into the suction-bell area of the pump. Flow reversals of this type will cause an uneven pressure loading across the eye of the first-stage impeller, and give rise to severe radial deflection of the impeller assembly. This deflection will often cause the bottom bearings of such vertical pumps to show a "keyhole" wear pattern, resulting in pump failure.

Aggravated wear patterns also occur when pumping slurries at flows other than at the BEP. Due to the flow-phase-angle separation in this case, the solids do not follow the flow streamlines of the fluid and bounce off both the impeller and volute casing. This bouncing can cause abnormal wear patterns on the pump parts, and attrition of the solids when handling crystalline slurries—especially undesirable when pumping pharmaceutical materials. For this reason, variable-speed reducers or V-belts are used when handling solids, in order to keep the pump operating as near as possible to the BEP flow.

## Recommendations for process pumps

When a chemical-process centrifugal pump is purchased for operation far back on its performance curve, the minimum flow through the pump should be at least 25% of the BEP flow. This will reduce the amplitude of the hydraulic anomalies and lengthen the operating time before pump rebuild.

Minimum flow can be maintained, preferably, by installing a bypass line to the pump-suction pipe. Where a control system is set up so that there is a possibility of zero forward flow, the bypass line to the pump suction must be cooled. On small pumps, cooling can be accomplished by using finned tubing in an air-to-fluid heat exchanger. Occasionally, the bypass-tubing line is routed near the motor to make use of the air passing through the motor for cooling. Relatively inexpensive cooling can often be achieved by using the readily available tube-within-a-tube heat-exchanger "tees" supplied as tubing fittings. Cooling can also be done with simple heat exchangers, such as are often used to cool the fluid flow to mechanical seals.

Fig. 3 shows a typical performance curve for a vertical, diffuser-bowl, turbine-type centrifugal pump used for cooling-tower applications. Note that the illustration shows only a portion of the efficiency range for this pump. This is a very good indication that pump manufacturers are trying to show the limited rangeability for such a pump.

The minimum continuous flow for the pump whose hydraulic characteristics are shown in Fig. 3 is about 8,200 gpm, or 77% of the BEP flow. This is a normal, expected minimum flow for a pump having a design specific speed of 6,200. Recognizing the minimum and maximum flow limitations of pumps is extremely important when several pumps operate in parallel. If the pump has a flat performance curve, and one or more pumps are shut off, the remaining pumps run out on their curve until they intersect the system friction curve.

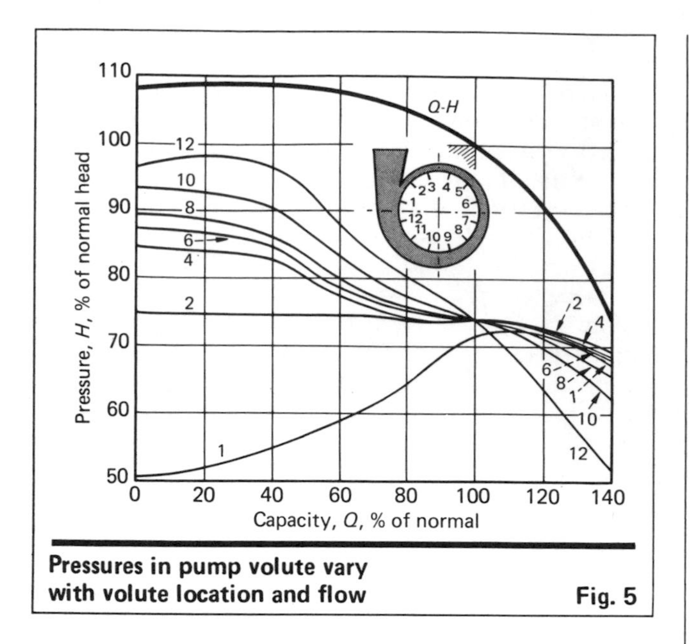


**Pressures in pump volute vary with volute location and flow** **Fig. 5**

As a result, pumps in many installations have been known to experience both internal recirculation and cavitation, and to destroy themselves.

In cooling-water circuits, heat exchangers often have throttling valves on the cooling-water flow, in order to regulate the temperature of the fluid being cooled. Reductions in plant production, combined with extremely low cooling-water temperatures in winter, often force the cooling-water pumps to operate dangerously far back on their performance curves. Usually, there is no automatic flow-control system to vary the number of pumps in operation in accordance with flow demand.

For example, a process operation was designed for a maximum cooling-water flow of 15,000 gpm. Three pumps, each having a capacity of 7,500 gpm, were purchased and installed. Production problems combined with winter operations caused the total flow through the system to be 3,000 gpm. Two of these pumps then shared the total flow of 3,000 gpm all winter, and major repairs were required in the spring.

## Net positive suction head required

What does the net positive suction head required (NPSHR) curve actually signify on a manufacturer's performance chart for a given centrifugal pump? This curve must be defined and interpreted before the information is of value. All major U.S. pump manufacturers test their pumps in accordance with the "Hydraulic Institute Standards" (HIS).* In complying with HIS, the manufacturer can show a series of points on an NPSHR curve where the head has dropped 3% as suction conditions are decreased. This curve then represents the level at which the pump is already cavitating.

Therefore, the levels of demarcation on an NPSHR curve between zero cavitation and the 3% head drop are relatively close in smaller, lower-specific-speed pumps. However, the difference between the stated NPSHR at a point of zero cavitation and the 3% head drop may be as much as 15 to 20 ft in a larger,

*"Hydraulic Institute Standards," 12th ed., Hydraulic Institute, 2130 Keith Blvd., Cleveland, OH 44115.

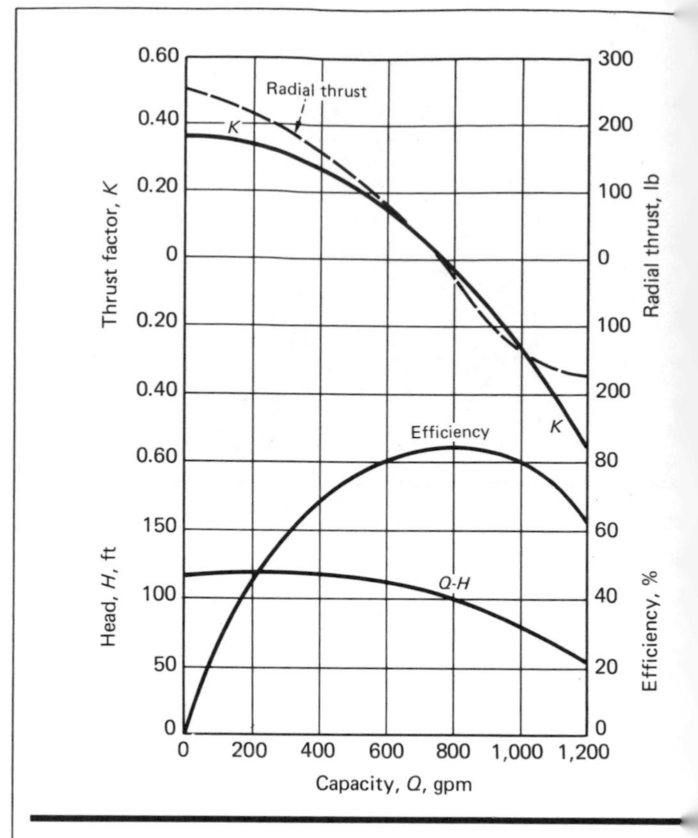


**Magnitude of radial thrust at various flows** **Fig. 6**

massflow, higher-specific-speed pump. In such a pump, the larger volume of liquid passing through the pump will often sweep the cavitating liquid through the casing, with a minimum drop in total head. This means that the pump can be suffering internal cavitation damage with no appreciable noise being noted. Performance will not deteriorate until ruinous damage has been done.

## Specifications to avoid cavitation

Fig. 4 is adapted from the "Hydraulic Institute Standards."* Point A has been added to show the point of "critical" NPSHR, and is defined by the author as that initial point at which no change or deterioration of performance occurs as the pump suction is suppressed. The "critical" NPSHR is further defined as that point at which the pump can operate indefinitely with no incipient cavitation occurring with changes in suction.

This is done to differentiate between Point A and Point B. In many centrifugal pumps, there is a hump between Points A and B, and the generated head is back to the original level at Point B as the suction conditions change.

Unless the pump specification spells out the permissible level of suppression for zero cavitation, the manufacturer may use Point B as the critical NPSH level. This would be incorrect because the pump is already cavitating at Point B.

The pump manufacturer should be required to show

*"Hydraulic Institute Standards," p. 44. See also pp. 42–44 for more information on NPSHR determination in centrifugal pumps.

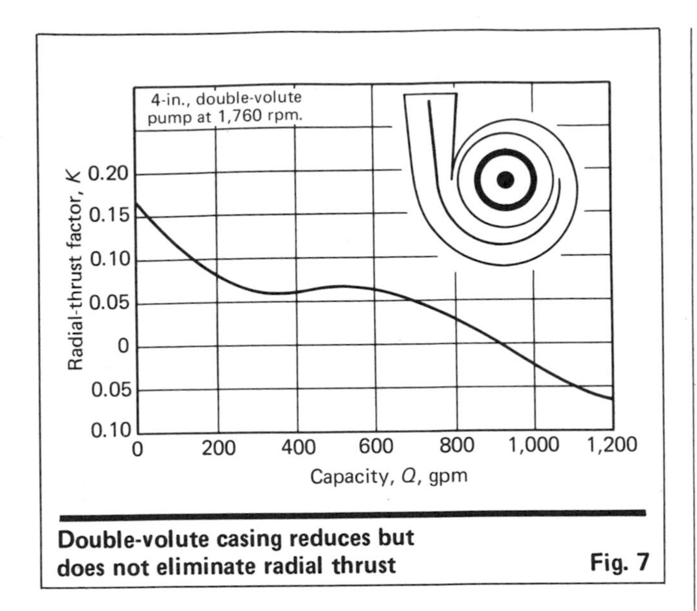


**Double-volute casing reduces but does not eliminate radial thrust** **Fig. 7**

both the critical NPSHR, and the 3% head-drop NPSHR on the pump's performance curve, as part of the specification. Then the engineer has a uniform basis on which to design the pumping system. Using the 3% level plus some rule-of-thumb increment only leads to overdesign of the pump, with its attendant increased cost, or leads to an inadequate level to compensate for pump wear.

However, the incipient-cavitation level is often exploited in the design of hydrocarbon and petrochemical pumps. Thus, many API-610* pumps are designed with very low NPSHR levels because hydrocarbon mixtures can partially cavitate, and the remainder of the liquid will carry the cavitating mixture through the pump to a higher pressure level. For corroboration, ask a pump manufacturer how much NPSHA should be added to such a pump if it is to operate on cold water.

Another characteristic of the NPSHR curve for a given pump is a curve that rises at both shutoff and cutoff, with a low NPSHR level between. This is one of the reasons why most published performance curves show only a portion of the NPSHR curve. Again, this is due to the desire of pump designers to show the area of safe, continuous operation.

## Other problems with centrifugal pumps

Another problem arising from cavitation, whether due to inadequate NPSHA or recirculation flow, is the effect on the pump's mechanical seal. Pressure surges and semiflashing that occur can load and unload the seal faces very rapidly. This can cause chatter, aggravated wear and, in many instances, breakage of the carbon-seal face, and failure of the seal.

An undesirable characteristic of off-BEP performance is the reaction of the impeller/volute assembly to radial thrust, which tends to impose a bending load at right angles to the pump shaft. Only at or near the BEP is the pressure the same at all volute sections around the impeller. Evidently, this is the most desirable condition for the liquid to discharge from the impeller. However, at either side of the BEP this equilibrium of pressure distribution is destroyed and the periphery of the impeller and its shroud is subjected to uneven pressure reactions.

Fig. 5 shows the typical pressure-distribution in the volute (based on tests made at Ingersoll-Rand). Fig. 6 shows the magnitude of radial-thrust measurements made on a 4-in. single-volute pump. Note that the thrust is maximum at zero flow, essentially zero at or near the BEP, and that it reverses direction but again increases in magnitude with flows to the right of (i.e., greater than) the BEP.

The immediate effect of these radial forces in a single-volute casing is excessive shaft deflection, which results in rapid wear at the close clearances in the pump (wearing rings, throat bushings, etc.), as well as the very large possibility of shaft breakage due to fatigue bending of the shaft material.

Pumps with double-volute casings are often used to reduce the net radial thrust. In such a casing, the flow is divided into two equal streams by two cut-waters, as shown in Fig. 7. Inequalities in the volute pressure are still present at off-BEP flow (the same as in a single volute). Owing to symmetry, there are two resultants of the radial forces opposing each other at all points around the periphery of the impeller. However, since the flow passage in one section of the volute is longer than the other, the radial forces are not in balance, due to the difference in flow through the passages.

A good rule-of-thumb is that the net imbalanced radial forces in a double-volute casing are about 30 to 40% those in a single-volute casing. In addition, due to the increase in internal friction from the twin-flow passages, BEP efficiency is about 1 to 2 percentage points lower than that in a single-volute pump.

## The author

**W. Stanley Tinney** is a project engineer for Stoner Associates, Inc., Box 629, Carlisle, PA 17013. He has more than 20 years experience with a number of firms—most recently with United Engineers & Constructors. He has also conducted seminars on pump design, specification, application and maintenance. He has a B.S. in mechanical engineering from Drexel University and is a registered professional engineer in N.J. and Pa.

*API-610, Standard of the American Petroleum Institute, Washington, D.C.

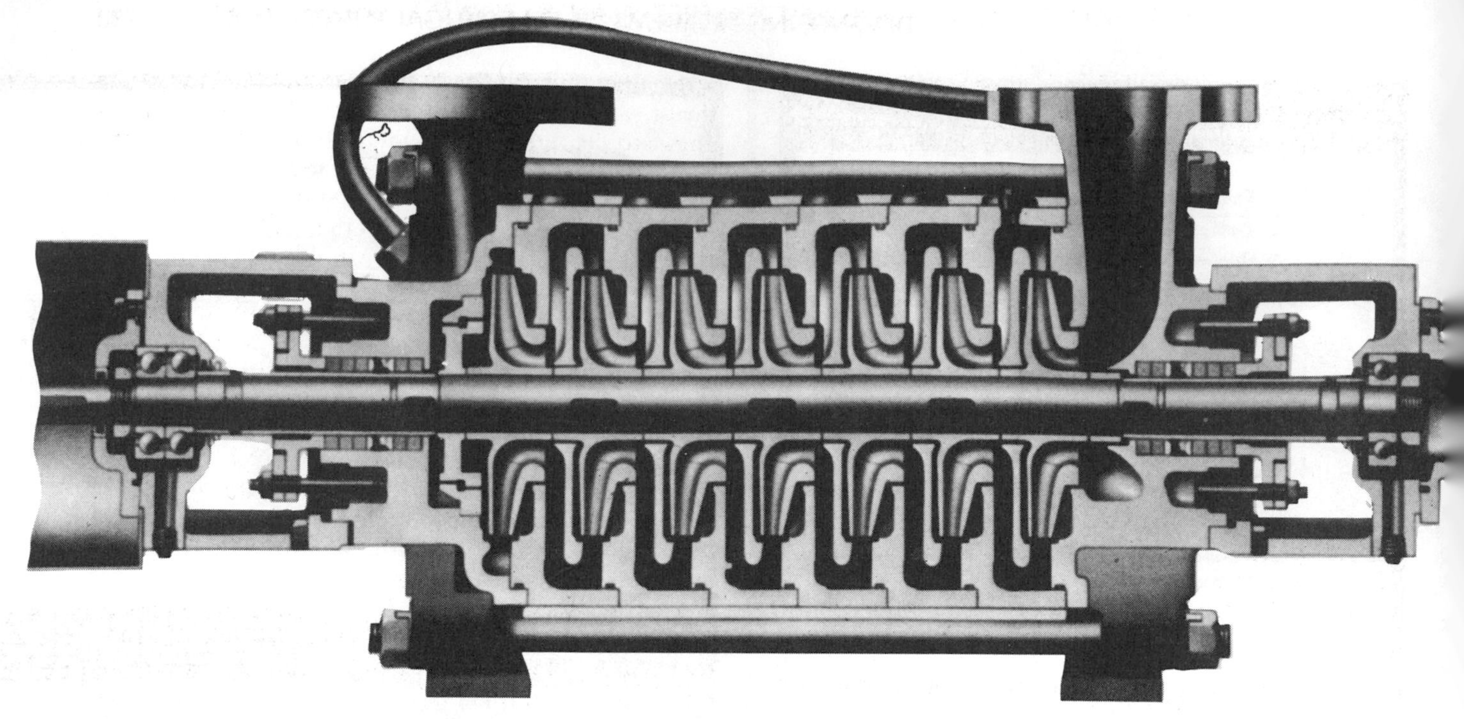

# Diagnosing problems of centrifugal pumps—Part I

*First of a three-part series, this article presents general principles by which engineers and operators can determine causes and cures for malfunctioning centrifugal pumps.*

**S. Yedidiah,** *Worthington Pump Corp.*

☐ The checklist (Table I) shows 14 different kinds of centrifugal pump malperformance that may confront a chemical engineer. All of these can be grouped under three classes: real hydraulic problems, real mechanical problems, and unreal hydraulic problems.

The real hydraulic problems are those in which the pump is unable to function according to specifications of capacity, head and efficiency. They may be due to faults in the pump or in the pump-driver assembly. Certain hydraulic problems, such as cavitation, can cause the second class of real problems, mechanical breakdown. Real mechanical problems display such symptoms as noise, vibrations, excessive heating, and can lead to hydraulic malfunction, through which the pump fails to meet its performance requirements.

The unreal problems are usually hydraulic in nature and generally result from faulty piping layout and faulty testing procedures. Unfortunately, correcting such problems is often more expensive than correcting those of the first two classes because of difficulty in diagnosis.

An itemized list of detailed causes for the 14 kinds of malperformance leads to 89 entries (Table II) and still leaves the subject not completely covered. Fortunately, 99 out of 100 cases of centrifugal pump malperformance can be solved with proper thinking. One good approach to such thinking is to question, first, if the pump has been damaged and if the malfunction is due to a blow to a pump part, and then second, to question the presence of gas in the system.

## Blow against a pump part

When smooth metal is struck by a hard object, as when dropped on a concrete floor, struck by a hammer, etc., a dent is created, with metal around the dent displaced and raised above the surface. If this surface is next fitted against another metal surface, the two will not assume a parallel position.

Such dents are not rare on pump parts. If they occur on surfaces like an impeller hub or the shoulder against which it bears, they will prevent the impeller from running square to its axis. A dent on shaft-sleeves covering a long shaft may cause the shaft to bend when the sleeves are tightened to each other. A dent on a shoulder

Originally published October 24, 1977

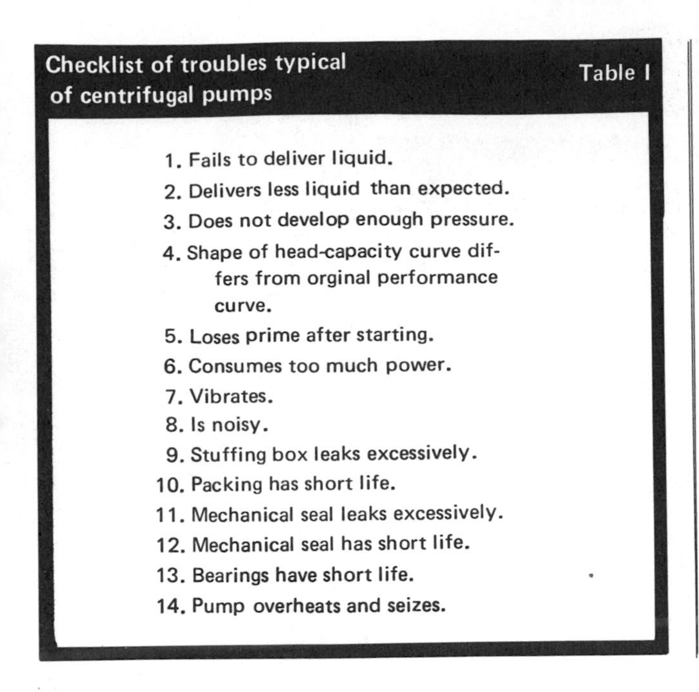
Checklist of troubles typical of centrifugal pumps — Table I

1. Fails to deliver liquid.
2. Delivers less liquid than expected.
3. Does not develop enough pressure.
4. Shape of head-capacity curve differs from orginal performance curve.
5. Loses prime after starting.
6. Consumes too much power.
7. Vibrates.
8. Is noisy.
9. Stuffing box leaks excessively.
10. Packing has short life.
11. Mechanical seal leaks excessively.
12. Mechanical seal has short life.
13. Bearings have short life.
14. Pump overheats and seizes.

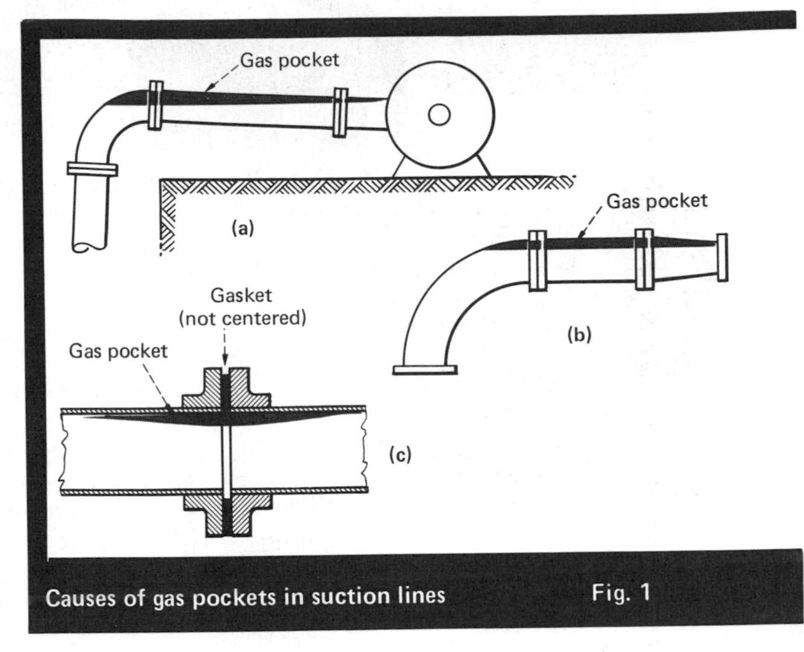

Causes of gas pockets in suction lines Fig. 1

against which a ball-bearing rides will usually cause the bearing to overheat.

Dirt between undamaged surfaces can cause similar effects. Also, when any pump part is allowed to fail, the mechanical breakdown is likely to damage related parts.

Among the more common effects of damaged parts are: overheated bearings, excessive wear on the sealing rings, noise and vibration, and/or excessive power consumption—all of which may result from a bent shaft. Similarly, a reduction in rate of flow or delivered pressure-head may result from a blow that has bent the impeller shrouds inward, thus reducing the areas of passage. When there is no practical way to straighten such a shroud, its effect can sometimes be compensated for by opening up the throat area through filing.

## Gas pockets

There are many instances in which pumps have functioned successfully with known stationary gas* pockets present in the suction line. This experience has misled pump users, encouraging them to believe that gas pockets are harmless.

The trouble starts when the gas pocket moves and enters the pump. The rotating impeller projects the heavier liquid outward, locking the gas within the eye of the impeller. Sometimes, this will completely block liquid flow through the pump, causing complete failure. At other times, liquid will continue to flow, but at a reduced rate.

*By "gas" we mean both vapor and air that have been trapped in the system, as distinguished from vapor formed in the system by liquid vaporization due to excessive pressure drop or cavitation.

This reduced flow can lead to either one of two results. First, depending on the higher velocities through the partially blocked areas and on pressure losses in the suction, the absolute pressure of the liquid passing through the gas lock may be less than at the suction inlet, causing more of the gas to leave the liquid. Second, because of the higher liquid velocities past the gas pocket, more of the gas is carried away. Depending on whether more of the gas is formed or carried away, the pocket will increase or vanish.

Since it is difficult if not impossible to predict which of these two events will take place, it is better to play safe and eliminate gas pockets, which can occur in the suction line, in the casing, or in the discharge line.

## Suction line gas pockets

Typical causes of vapor pockets in suction lines (see Fig. 1) are (a) high points in the line, (b) concentric reducers leading to the pump nozzle, and (c) gaskets smaller than, or eccentric with, the piping. These should be avoided by gradually sloping piping, eccentric reducers, and gaskets with inside diameters larger than those of the pipes.

When plant layout prevents installing suction lines with a gradual slope from tank to pump, and high points occur, all such points should be vented either to the atmosphere or to the vapor space in the supply tank. (When unvented high points do not cause problems, that is because the liquid velocity is so low that the gas is not carried out of the suction line into the pump.)

## Casing gas pockets

In single-stage pumps, gas pockets usually occur at the highest point of the volute. If such pockets are not

**Checklist of causes of centrifugal-pump problems** — Table II

1. Measuring instruments not correctly calibrated and/or incorrectly mounted.
2. Air entering the pump during operation, or the pumping system not completely deaerated before starting.
3. Insufficient speed.
4. Wrong direction of rotation.
5. Discharge pressure required by the system is greater than that for which the pump was designed.
6. Available NPSH too low (including suction lift too high).
7. Excessive amount of vapor entrained in liquid.
8. Excessive leakage through wearing surfaces.
9. Viscosity of liquid higher than viscosity of liquid for which the pump was originally designed.
10. Impeller and/or casing partially (or fully) clogged with solid matter.
11. Waterways of impeller and/or of casing very rough.
12. Fins, burrs, sharp edges, etc., in the path of the liquid.
13. Impeller damaged.
14. Outer diameter of impeller machined to a lower diameter than specified.
15. Faulty casting of impeller and/or casing.
16. Impeller incorrectly assembled in casing.
17. System requirements too far out on the head/capacity curve.
18. Obstruction in suction and/or discharge piping.
19. Foot valve jammed or clogged up.
20. Suction strainer filled with solid matter.
21. Suction strainer covered with fibrous matter.
22. Incorrect layout of suction and/or discharge piping.
23. Incorrect layout of suction sump.
24. The operation of one pump (in a system having two or more pumps operating either in parallel or in series, or in a combination of these) seriously affected by the operation of the other pumps.
25. Water level in suction sump (or tank, or well) below pump intake.
26. Speed too high.
27. Pumped liquid of higher specific gravity than anticipated.
28. Oversize impeller.
29. Total head of system either higher or lower than expected.
30. Misalignment between pump and driver.
31. Rotating parts rubbing on stationary parts.
32. Worn bearings.
33. Packing improperly installed.
34. Incorrect type of packing.
35. Mechanical seal exerts excessive pressure on seat.
36. Gland too tight.
37. Improper lubrication of bearings.
38. Piping imposing strain on pump.
39. Pump running at critical speed.
40. Rotating elements not balanced.
41. Excessive lateral forces on rotating parts.
42. Insufficient distance between outer diameter of impeller and volute tongue.
43. Faulty shape of volute tongue.
44. Undersize suction and/or discharge piping and fittings (sometimes causing cavitation).
45. Loose valve or disk in the system, causing premature cavitation in the pump.
46. Bent shaft.
47. Bore of impeller not concentric with its outer diameter and/or not square with its face.
48. Misalignment of parts.
49. Pump operates at a very low capacity.
50. Improperly designed baseplate and / or foundation.
51. Resonance between operating speed of pump and natural frequency of foundation and/or other structural elements of pumping station.
52. Rotating parts running off-center because of worn bearings or damaged parts.
53. Improper installation of bearings.
54. Damaged bearings.
55. Water-seal pipe plugged.
56. Seal cage improperly located in stuffing box, preventing sealing fluid from entering space to form seal.
57. Shaft or shaft sleeves worn or scored at the packing.
58. Failure to provide cooling liquid to water-cooled stuffing boxes.
59. Excessive clearance at bottom of stuffing box, between shaft and casing.
60. Dirt or grit in sealing liquid.
61. Stuffing box eccentric in relation to shaft.
62. Mechanical seal improperly installed.
63. Incorrect type of mechanical seal for given operating conditions.
64. Internal misalignment of parts, preventing seal washer and seal from mating properly.
65. Sealing face not perpendicular to shaft axis.
66. Mechanical seal that has been run dry.
67. Abrasive solids in liquids that come into contact with seal.
68. Leakage under sleeve due to gasket and O-ring failure.
69. Bearing-housing bores not concentric with water end.
70. Damaged or cracked bearing housing.
71. Excessive grease in bearings.
72. Faulty lubrication system.
73. Improper installation of bearings (damage during assembly, incorrect assembly, wrong type of bearings, etc.).
74. Bearings not lubricated.
75. Dirt getting into bearings.
76. Water entering the bearing housing.
77. Balancing holes clogged up.
78. Failure of balancing device.
79. Too-high suction pressure.
80. Tight fit between line bearing and its seats (which may prevent it from sliding under axial load).
81. Pump not primed and allowed to run dry.
82. Vapor or air pockets inside the pump.
83. Operation at too-low capacity.
84. Parallel operation of poorly matched pumps.
85. Internal misalignment due to too much pipe strain, poor foundations or improper repair.
86. Internal rubbing of rotating parts on stationary parts.
87. Worn bearings.
88. Lack of lubrication.
89. Rotating and stationary wearing rings made of identical materials with identical physical properties.

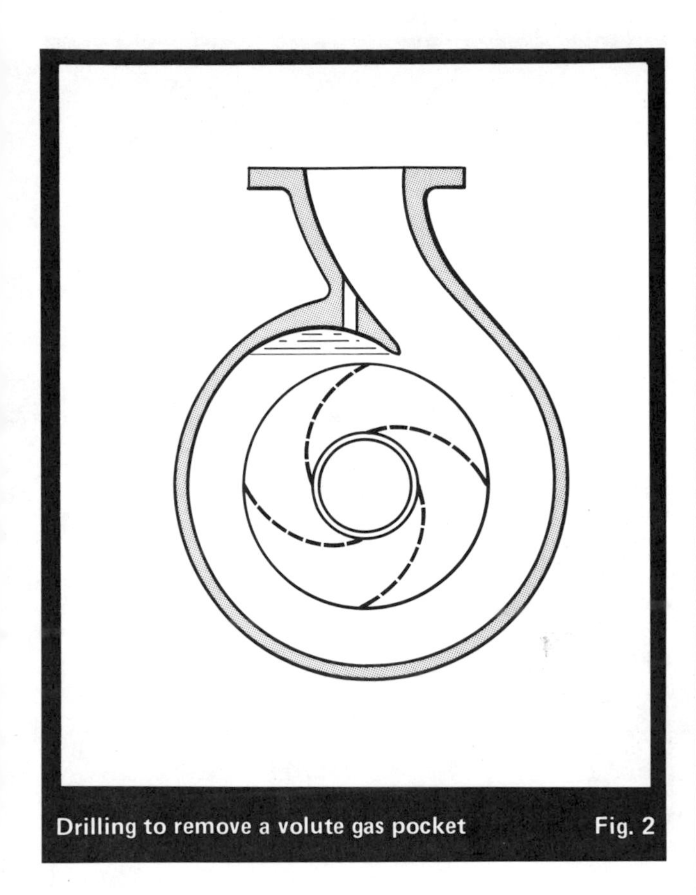

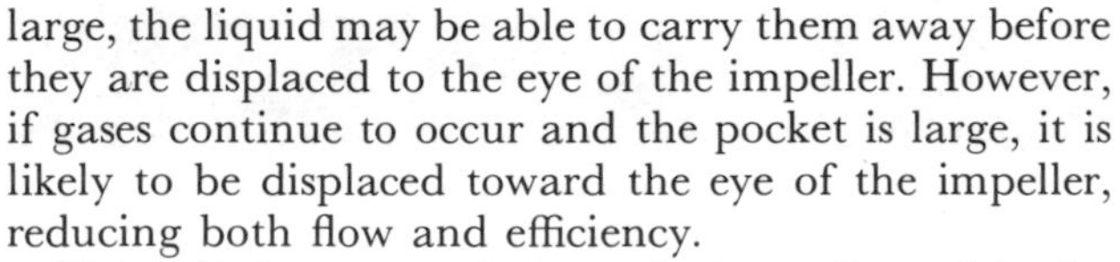

Drilling to remove a volute gas pocket Fig. 2

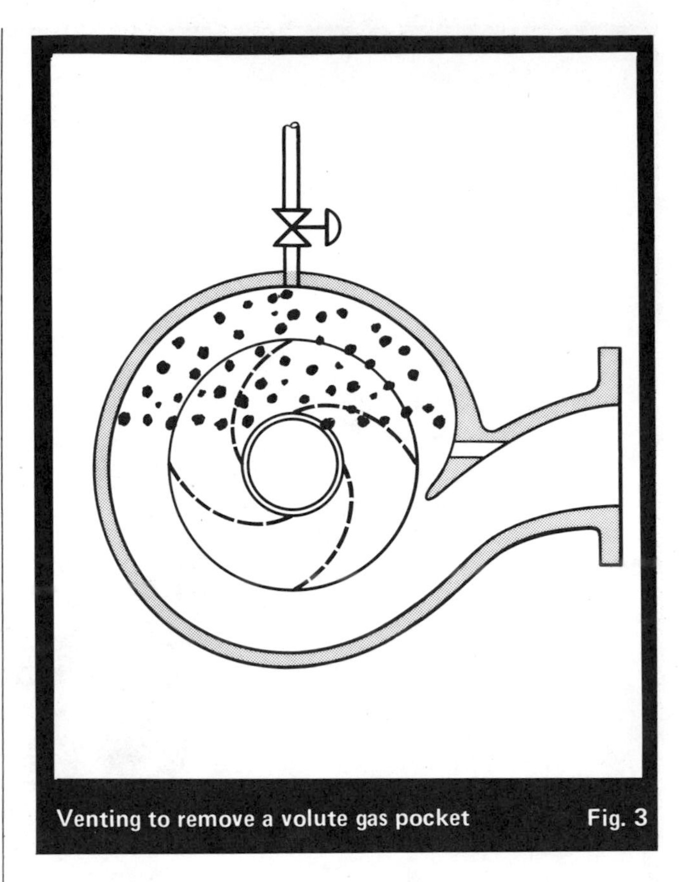

Venting to remove a volute gas pocket Fig. 3

large, the liquid may be able to carry them away before they are displaced to the eye of the impeller. However, if gases continue to occur and the pocket is large, it is likely to be displaced toward the eye of the impeller, reducing both flow and efficiency.

If the discharge nozzle is vertical, gas formed in the volute high-point may be drawn off through a hole drilled through the casing inside the nozzle (Fig. 2). If the discharge nozzle is horizontal, the gas must be let off through a vent (Fig. 3).

A gas pocket may also form in a horizontal suction nozzle above the eye of the impeller and may have to be vented.

Because of the diffusers present in multistage pumps (see photo p. 124), gas pockets cannot be displaced backward from casing high-points to the eye of the impeller. Instead, the liquid will usually carry trapped gas forward into the eye of the subsequent impeller—with damaging results.

Since the pressure increases with each consecutive stage of a multistage pump, there is a greater tendency for gases to be dissolved in the latter stages of such pumps. Thus it is rarely necessary to vent more than the first two stages.

## Gas pockets in discharge lines

Although it may seem remarkable, gas pockets in the discharge line can affect pump performance. They most frequently occur between the discharge block-valve and check-valve, when a pump has been shut down and the block valve shut off.

Sometimes, the discharge line is lower than the pump's centerline and the pump is primed with the discharge valve closed. In this case, when the discharge valve is opened before the pump is started (a common practice), gas trapped between the block valve and the check valve will escape backward into the pump casing and affect performance.

If the discharge line is higher than the pump's centerline, gas trapped between the block and the check valve may cause a sudden noise, as if the piping system had been hit with a hammer. This is the result of the check-valve disk swinging forward against its stop. Since the exposed downstream surface of the disk is larger than the upstream surface, the pressure head coming from the pump must be greater than the backpressure, in order to start the disk swinging open. Once the disk moves from its seat, however, the upstream surface is fully exposed, and the expanding pocket of gas throws the disk forward.

## Air leakage to pumps handling water

Air may leak into a pump that is handling water—either through the inlet (when the pump takes suction from a sump) or through holes in the suction line and joints and between the shaft and the shaft sleeve (Fig. 4). Pumps sometimes are equipped with a standpipe for priming them at startup, and this standpipe can be filled with the foot valve closed to test the suction line for leaks before startup.

Sometimes a vortex can form in the sump around the suction of a pump; and the spiralling channel of air at the tip of such a vortex can cause a troublesome sort of air leakage. Flowrates and inches of submergence to avoid such vortexes are shown in Fig. 5. Otherwise, the vortex may be suppressed by floating lumps of material

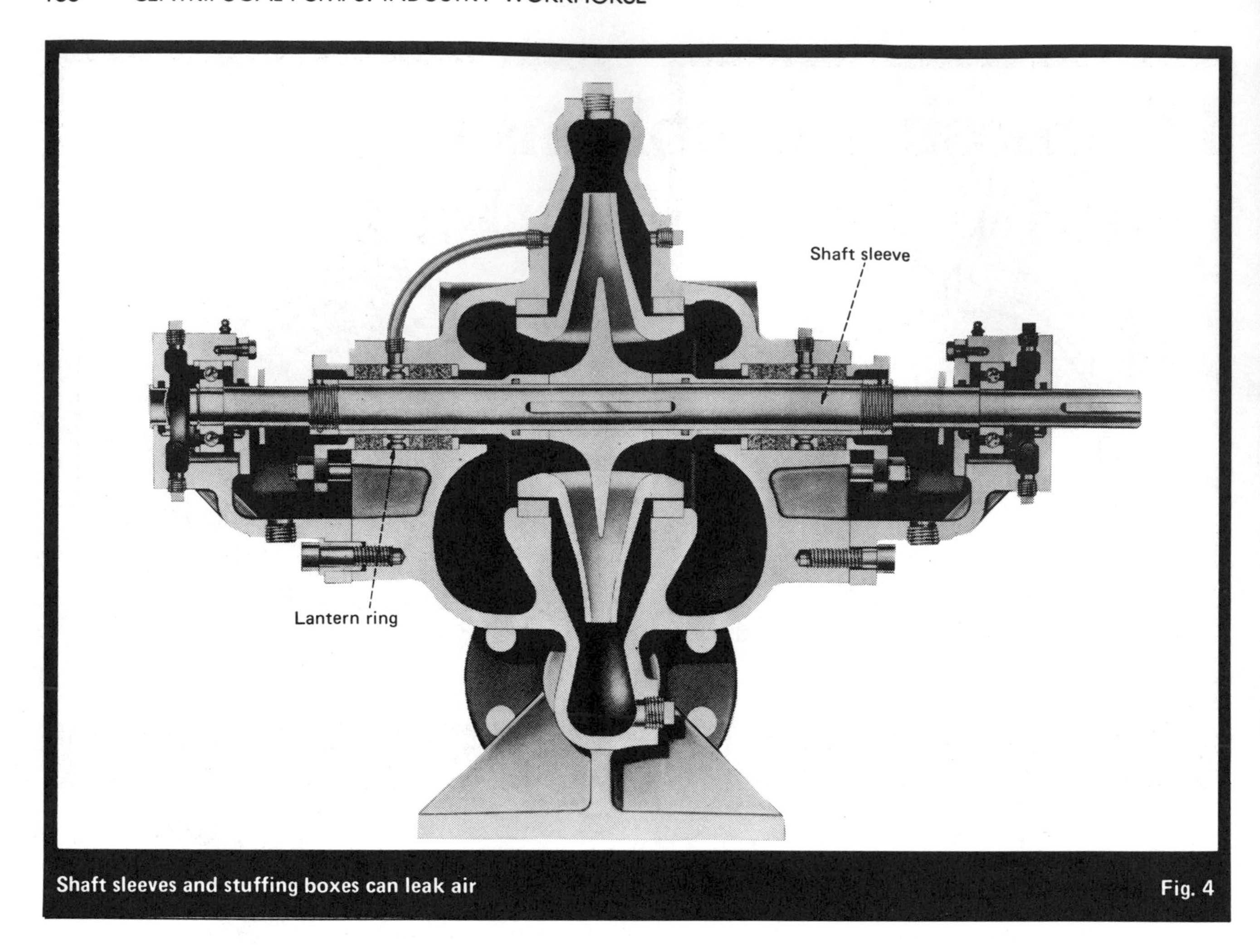


Shaft sleeves and stuffing boxes can leak air — Fig. 4

on the sump surface or equipping the suction pipe with a bell or strainer.

Air not entering the inlet will usually result from leaks at the suction-side stuffing box, flanges, bushings, nipples, drain plugs, vents, etc. These points can be checked either with an open flame (when the pump is in a nonhazardous area) or by pouring water over the suction piping. The leaks will cause a flame to flicker; and since water running over a leak will temporarily stop it, the presence of the leak may be indicated by a sudden jump on the pressure gage.

Suction-side stuffing boxes are usually equipped with lantern rings and a water-seal arrangement specifically to prevent air leaks. Consequently, if a stuffing box appears to be leaking air, the water seal should be checked to see that it is free-flowing. The lantern ring must be located correctly in the stuffing box for the water seal to function properly, and the inlet to the lantern ring should be free-flowing. Water leaking from the gland indicates that this seal is functioning.

Submergence, in.: 0, 10, 20, 30, 40, 50, 60, 70, 80
Flow, gpm: 100, 200, 400, 600, 800, 1,000, 2,000, 4,000, 6,000
4-in. I.D. pipe
6-in. I.D. pipe
8-in. I.D. pipe
10-in. I.D. pipe
$d$

Submergence required to prevent vortex in sumps — Fig. 5

## The author

**S. Yedidiah** is a hydraulic specialist at Worthington Pump Corp., 14 Fourth Ave., East Orange, NJ 07017, where his duties include serving all elements of the corporation as an expert in areas of hydraulic design. Active as a pump specialist since 1938, he has experience with all phases of pumps, from research and development through testing, production planning and management, and troubleshooting. A member of ASME, he has published many technical papers and holds a number of patents.

# Diagnosing problems of centrifugal pumps—Part II

The troubles analyzed here include those due to improper installation, assembly and machining, or testing.

*S. Yedidiah, Worthington Pump Corp.*

☐ Most cases of pump malperformance require no special analysis or explanation. However, because the possibilities are many, the few exceptional cases often defy analysis and cause delay and expense. Such malperformance problems should be approached, first, through a review of the possibilities for mechanical faults, and then through an analysis of performance curves.

The review of mechanical faults should include looking for effects of improper impeller mountings, and (for

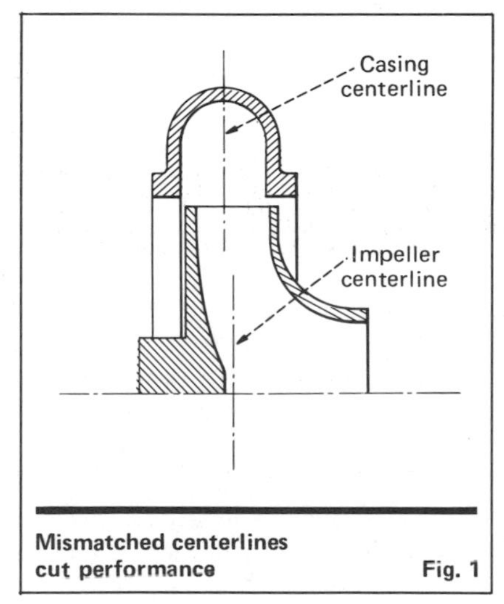


**Mismatched centerlines cut performance** **Fig. 1**

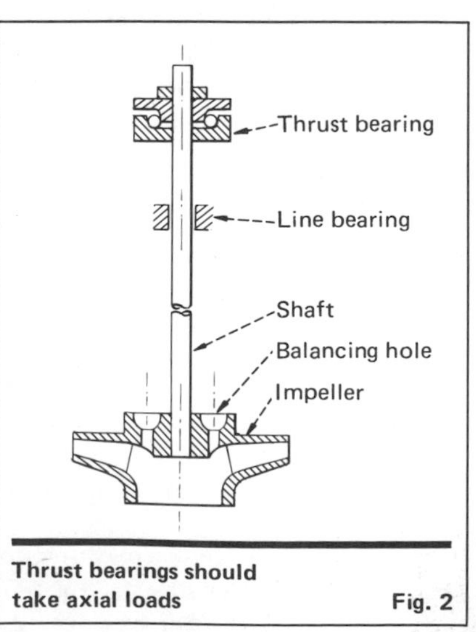


**Thrust bearings should take axial loads** **Fig. 2**

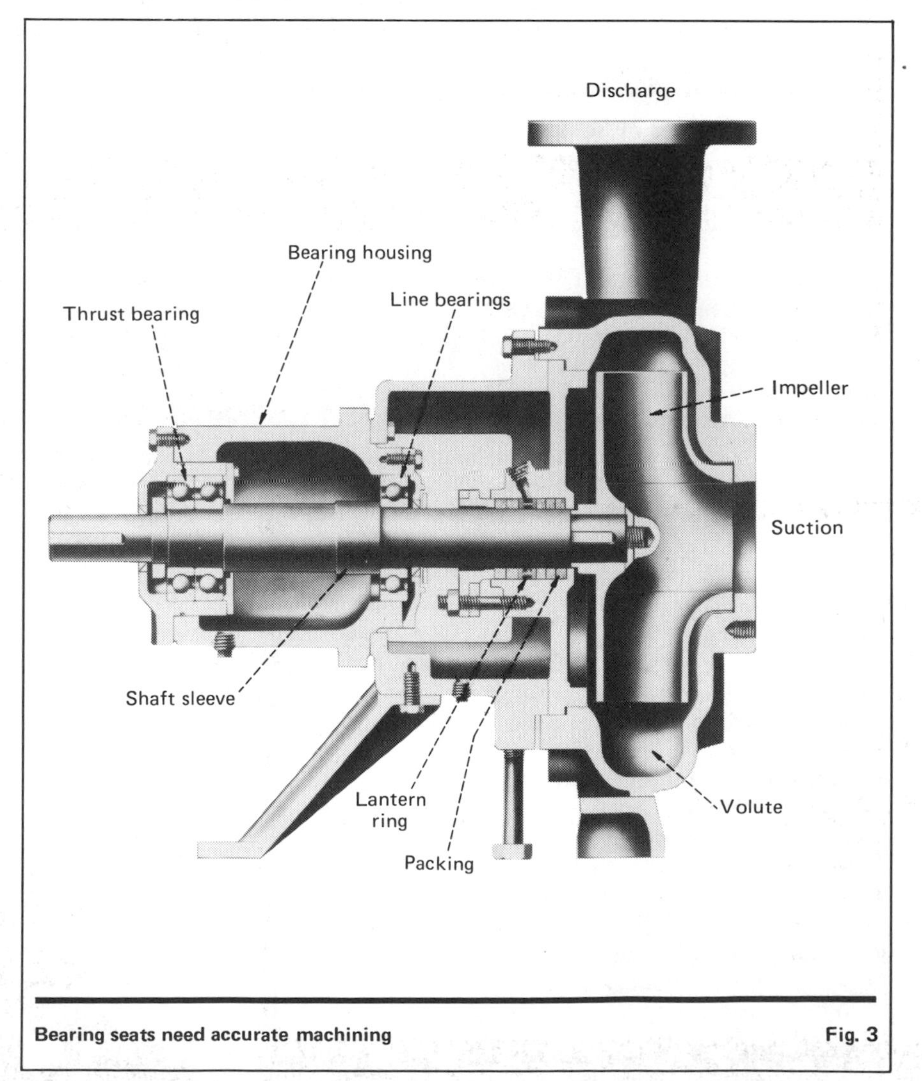


**Bearing seats need accurate machining** **Fig. 3**

Originally published November 21, 1977

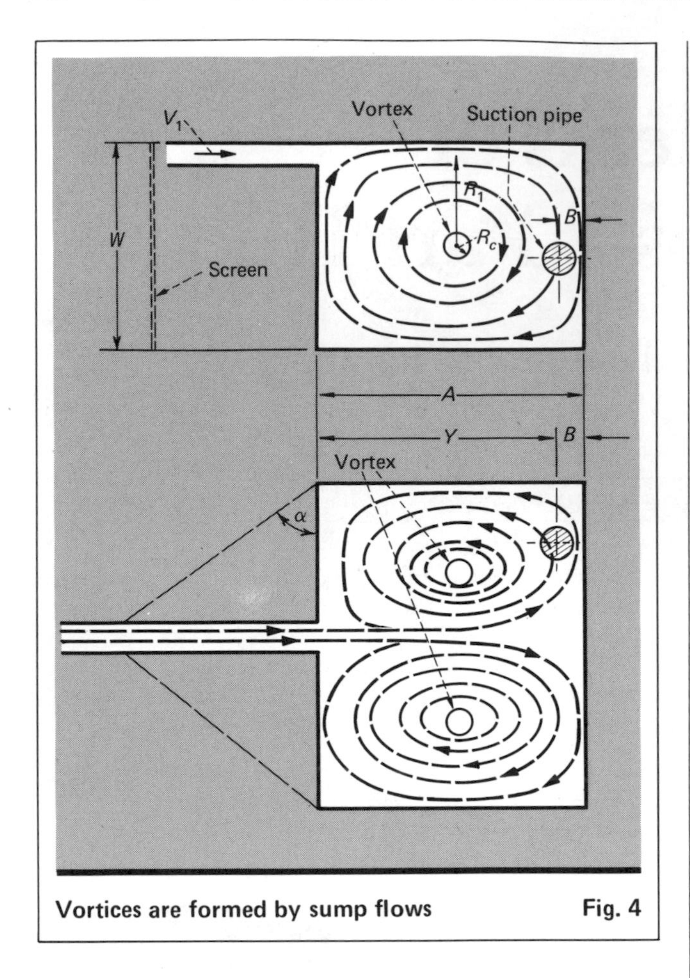


Vortices are formed by sump flows Fig. 4

sump pumps) the layout of the sump and suction line. Analysis of performance curves calls for a hydraulic test to provide data of head/capacity/efficiency/horsepower.

## Improper mountings

Two nuts located on the shafts of many pumps determine the axial position of the impeller. Such pumps should be assembled with care that the centerline of the impeller discharge coincides with the centerline of the casing. Any mismatch between these two centerlines (Fig. 1) may adversely affect performance, particularly if the clearance between casing and impeller is small.

Certain other mounting problems, which are related to the bearings, are less easy to detect. Sometimes, for example, an impeller is provided with balancing holes to allow liquid to escape under pressure from one side of the impeller to the other, and thus act to reduce axial thrust. However, when the shaft is light in a vertical pump (Fig. 2), its weight alone may not be enough to keep the races of the thrust-bearing in permanent contact with its balls. The results are noise, and vibrations that can ruin a pump in a short time. A simple remedy for such a pump is to plug off the balancing holes and thus increase the axial load on the bearing.

Whenever a pump impeller is subject to high axial loads, the thrust-bearing should be located on the outboard side of the frame (Fig. 3), with the line-bearing on the inboard.

In order for bearings to operate satisfactorily, their seats must be accurately machined, with no radial play between them and the casing. When the fit between an inboard line-bearing (Fig. 3) and its seat is too tight, however, the line-bearing may absorb the axial load, instead of letting this load ride on the thrust-bearing (Fig. 3). The consequent extra load may ruin a line-bearing in a short time.

A similar effect may also result when dirt is wedged between the outer race of the line-bearing and the bearing seat, even though the casing is machined to the correct tolerances (Fig. 3). To eliminate such overloads, care should be taken during assembly that the line-bearing is not too tight to be able to move in the axial direction.

On the other hand, no bearing should be so loose that its race can rotate in its seat. Generally, sliding friction is 10 to 15 times greater than rolling friction, so it is puzzling how the race of a ball bearing can rotate in its seat. However, this frequently happens in practice, and the result is always a breakdown of the bearings in a relatively short time.

## Sump design and suction-line layout

The flow of an ideal liquid obeys Bernoulli's well-known principle:

$$(V^2/2g) + (P/\gamma) + Z = \text{constant}$$

where: $V$ = velocity of the liquid
$g$ = acceleration due to gravity
$P$ = pressure
$\gamma$ = specific weight
$Z$ = static head

If Bernoulli's equation is applied to a liquid vortex, this vortex, known as the potential vortex, has a rotating velocity, $V$, which varies inversely with radius, $R$, of the vortex, or:

$$(V)(R) = \text{constant}$$

This equation has enormous implications for the effects of sump design on pump performance. It tells us that the velocity increases toward the center of the vortex, tending to become infinite as the radius approaches zero. Because of Bernoulli's principle, therefore, the absolute pressure near the center of a potential vortex becomes very low—considerably lower than atmospheric pressure.

Whenever such a low-pressure zone is in direct contact with the surrounding atmosphere, and the body of liquid is moving in the direction of the axis of the vortex, air enters the vortex. Depending on the shape and the size of the sump, this air is likely to reach the eye of the impeller, causing trouble.

Whenever air is prevented from entering the low-pressure zone of such a vortex, either a low-pressure core of liquid rotating as a solid vortex, or a vapor-filled cavity, is created. This latter occurs when the center of the vortex enters a suction pipe.

Fig. 4 shows schematically how these relationships affect the performance when liquid passes from a narrow channel into a wide sump. No matter how low the inlet velocity may be, it may cause one or more vortices;

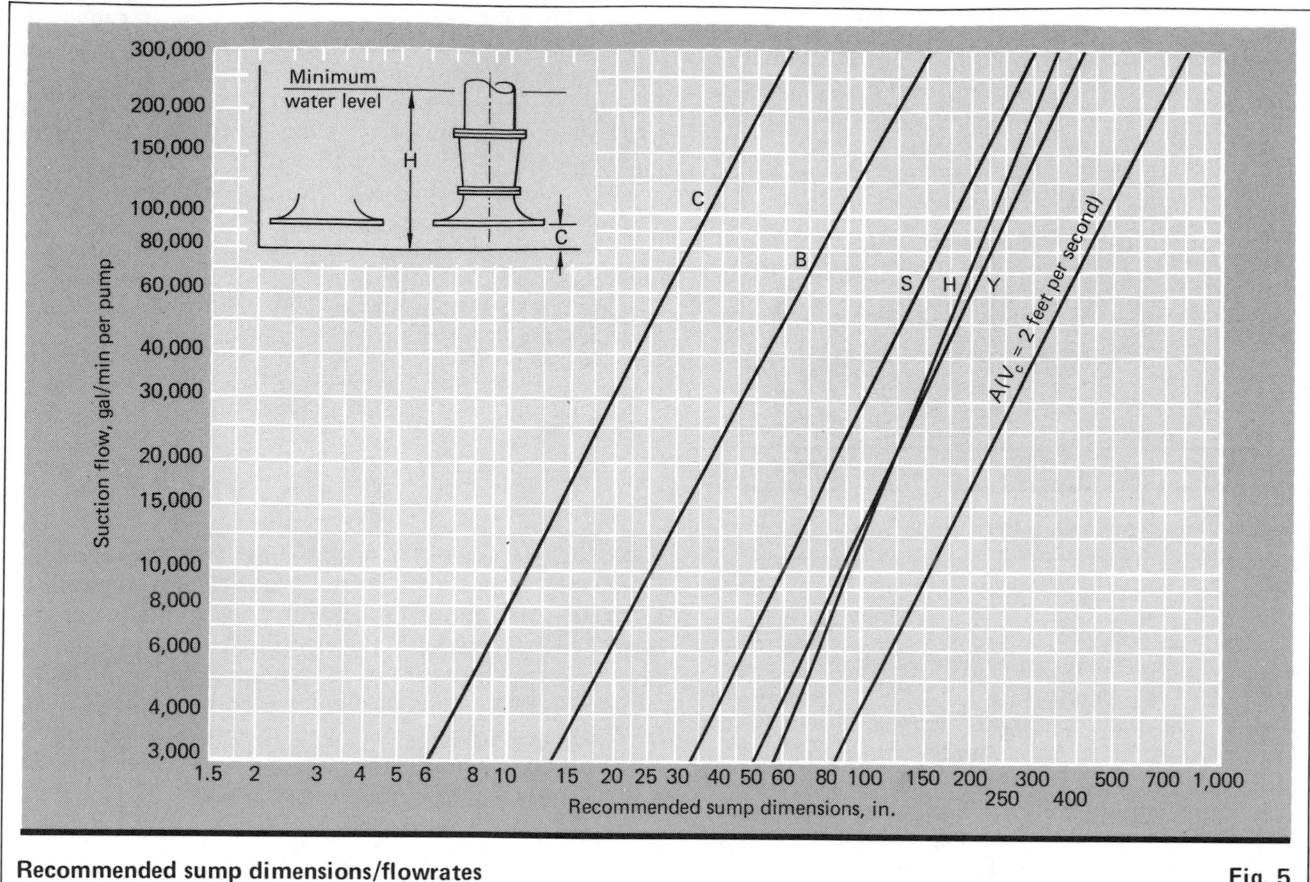


**Recommended sump dimensions/flowrates** **Fig. 5**

and the velocity of the liquid in these vortices will increase toward their centers, providing zones of low pressure. If the suction line to a pump is installed at one of the centers, and is large enough to cover the low-pressure zone, cavitation will develop, and the cavities will be carried into the pump.

There, when they reach a zone of high pressure, they will collapse vigorously, often causing serious damage. Also, they reduce the areas of passage at the inlet of the impeller blades, impeding the flow, or even stopping it completely.

On the other hand, if the suction pipe is either smaller than the low-pressure zone of the vortex or lies off-center from the vortex, an air funnel will be formed near the pump inlet, and air will be drawn through this funnel and into the pump (see Fig. 5, *Chem. Eng.* Oct. 24, 1977, p. 124).

Consequently, a sump should be large enough for the

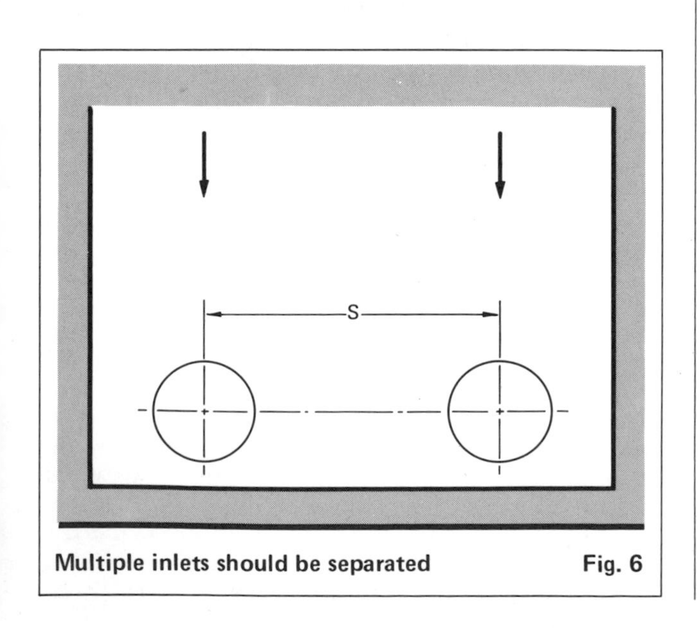


**Multiple inlets should be separated** **Fig. 6**

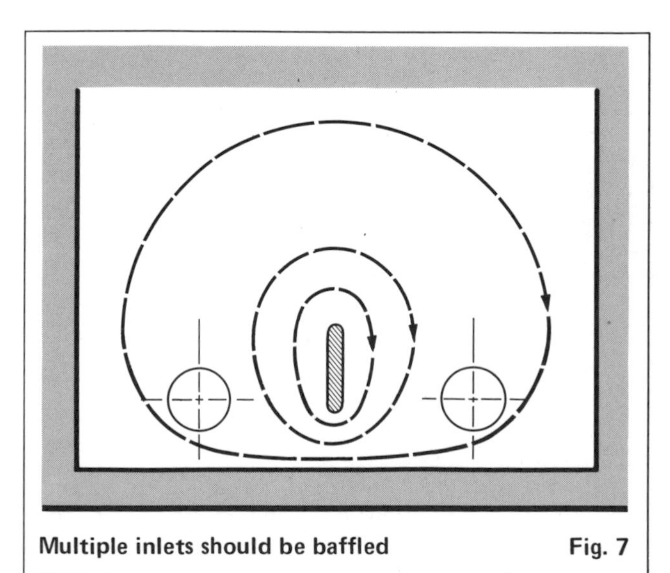

**Multiple inlets should be baffled** **Fig. 7**

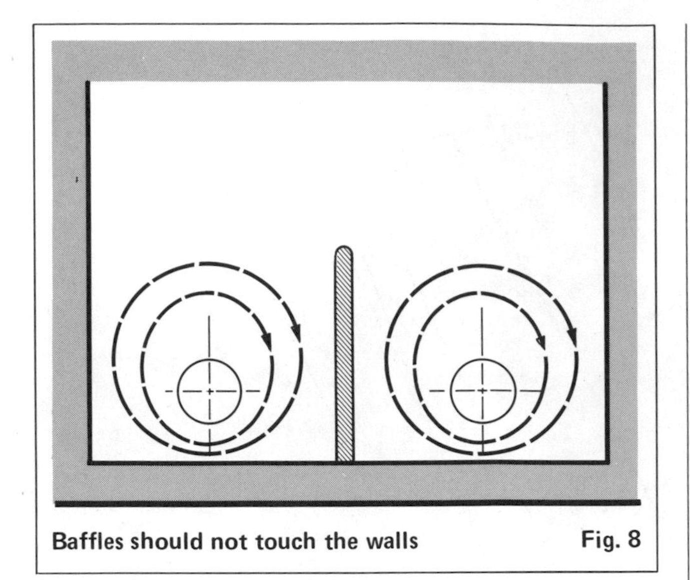

Baffles should not touch the walls Fig. 8

pump (or pumps) to be located far from the center of any vortex that may arise. The Hydraulic Institute (New York) has compiled a composite chart of suggested minimum sump-dimensions (Fig. 5). The dimensions to which curves A, B, S and Y apply are shown in Fig. 4, 6.

Fig. 4, in combination with Bernoulli's principle and the potential vortex, leads to a series of important guidelines relating to sump design. Since the pressure in a vortex increases with the distance from its center, the highest pressure in the sump will be near its walls. Also, the walls are at the greatest distance from the low-pressure zone in which an air funnel is likely to appear. The Hydraulic Institute's curve for the distance, B, of the suction pipe from the sump wall (Fig. 4) indicates that the suction pipe should be located as near to the wall as possible.

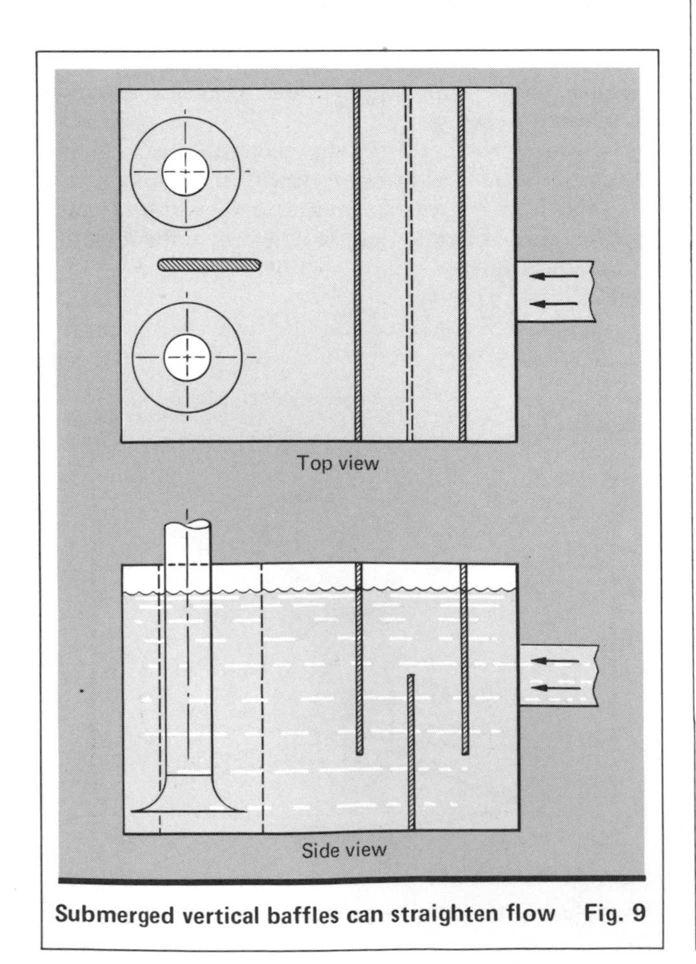


Submerged vertical baffles can straighten flow Fig. 9

If the liquid velocity in the channel of approach is $V_1$, and $R_1$ is the distance of the channel from the center of the vortex, then $V_c$ and $R_c$, the velocity and radius at which the vortex pressure falls below atmospheric, are given by:

$$V_c R_c = V_1 R_1$$

or

$$R_c = V_1 R_1 / V_c$$

Thus the size of the radius of the low-pressure zone is directly proportional to the inlet velocity, $V_1$, and the danger that an air funnel will reach a pump increases proportionately with velocity $V_1$.

The Hydraulic Institute recommends that $V_1$ should be kept as small as possible, preferably below 1.0 ft/s. Although pumps are known to operate satisfactorily with approach velocities up to 2.0 ft/s, the sumps in such cases are so large that the pump inlet can be far from the vortex.

Fig. 4 indicates that a vortex can be avoided, if the width, $W$, of the channel of approach is equal to the width of the sump. The same effect can be accomplished by means of a screen, to provide a uniform velocity distribution across the sump. Such screens are recommended as being part of one of the best sump arrangements.

When such an arrangement cannot be adopted, the next best thing is to make a conical approach (shown dotted in Fig. 4). The angle of such a cone should not be less than about 45 deg. Also, a flow straightener should be used at the end of the conical section (e.g., a bar screen), and a certain minimum distance, $Y$, should be maintained between the screen and the suction inlet.

When several pump inlets are installed in one sump, they should be parallel to each other along a line perpendicular to the direction of approach (Fig. 6) and not in series, because liquid sheds vortices as it passes a cylindrical object (e.g., a suction pipe) and such shed vortices may propagate toward a downstream inlet.

Even when the pump inlets are located in parallel, there may be harmful interaction between neighboring pumps, particularly when the velocity of the approaching liquid is not evenly distributed across the width of the sump. Consequently, it is advisable to install partitions between neighboring pumps (Fig. 7), but such partitions should never touch the end-walls (Fig. 8).

When air funnels appear in the vicinity of the pump inlets in an existing sump, an effective remedy is to cover the water surface with floating material (wood logs, etc.), which will usually break up the funnels and prevent new ones from forming. When the core of the vortex lies within the suction pipe, however, such floats are useless, and the only remedy that can sometimes correct the sump design consists of baffles or screens to

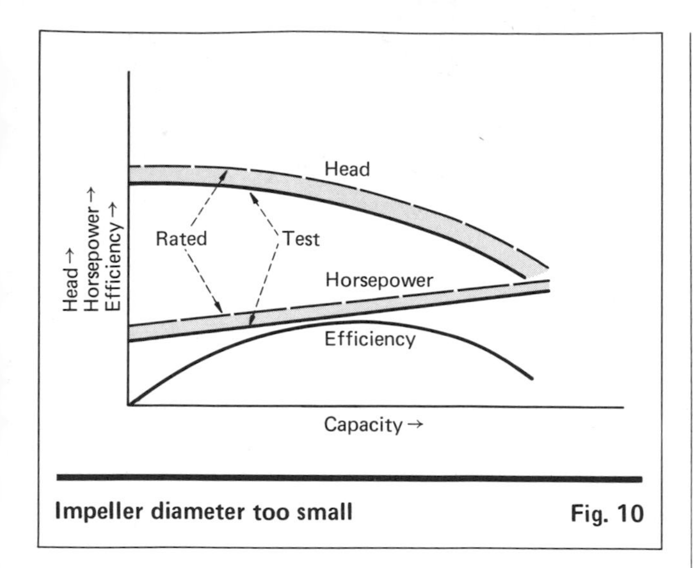


**Impeller diameter too small** **Fig. 10**

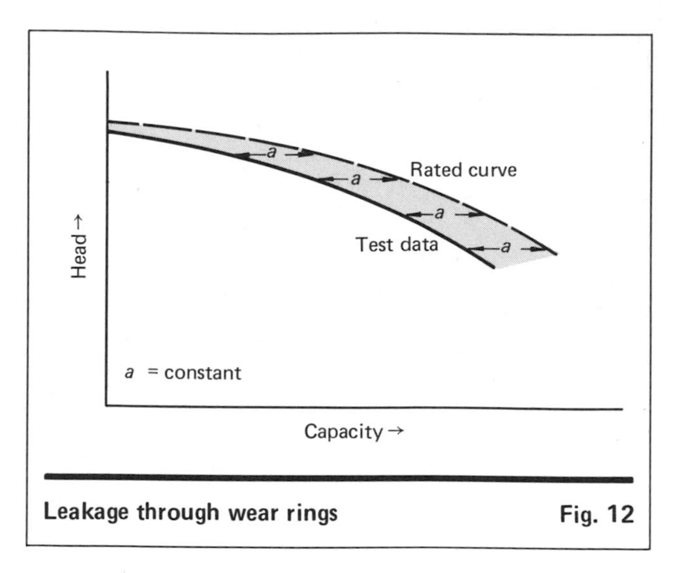


**Leakage through wear rings** **Fig. 12**

straighten the liquid flow. One successful system has been to pass the liquid over and under submerged baffles (Fig. 9).

When using submerged baffles, the liquid velocities above and below the baffles should be as small as possible, and no submerged baffle should ever be located so as to form a waterfall, which could entrain air bubbles. For the same reason, no entrance channel should be located above the lowest possible level of the sump.

The layout of a pump-suction line can present another form of trouble similar to troubles caused by poor sump design. Whenever there are several bends in the line, and all turn the liquid in the same direction, they may induce a vortex in the liquid entering the impeller.

## Interpretation of test results

Although the many possible causes for any given type of malfunction make diagnosis difficult and time-consuming, it is often possible to reduce the number through a careful study of a pump's performance curves. Some curves typical of malfunctioning pumps and the associated causes follow:

*Pump develops less head and consumes less power over its whole working range, while efficiency remains unaltered (Fig. 10)*: The most common cause for such behavior is a deformed impeller casting. Two other possibilities are rotational speed lower than specified, or an undersized impeller.

*Head falls off rapidly with an increase in flowrate while shutoff head is practically unchanged (Fig. 11)*: Such a curve indicates a reduced throat area of the volute, or a reduced area between diffuser vanes. Also, an obstruction at some location between the impeller outlet and the point of pressure readings may be the cause of such a curve.

*Test flowrate is lower than rated by a constant amount at any given head (Fig. 12)*: In a pump with closed impellers, this curve results from excessive (but not too excessive) leakage through the wearing rings. Usually, such a curve indicates that the rings are worn and need replacement. A worn wearplate and/or worn impeller vanes are indicated when this curve applies to a pump with semi-open impeller.

*Head, capacity, efficiency and horsepower are all lower over the entire curve (Fig. 13)*: Excessive clearances in the wearing rings or between the impeller vanes and wearplates

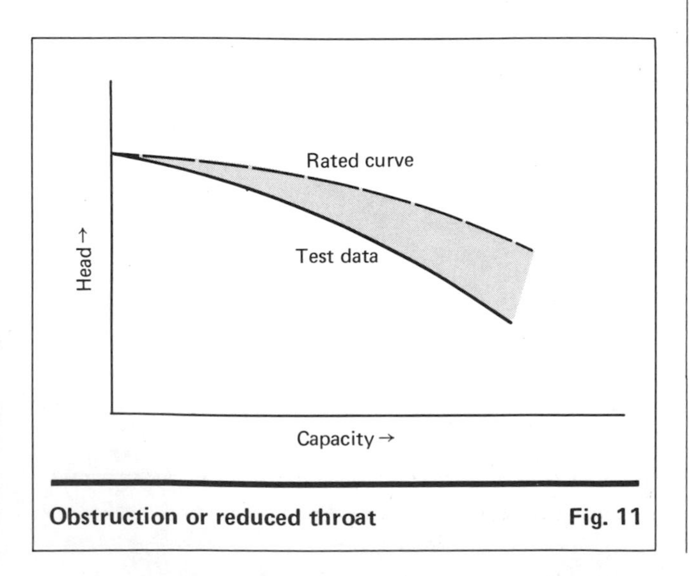


**Obstruction or reduced throat** **Fig. 11**

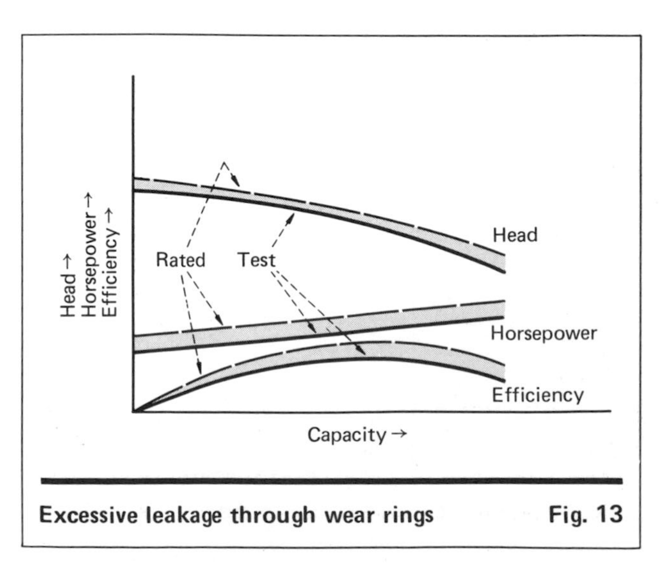


**Excessive leakage through wear rings** **Fig. 13**

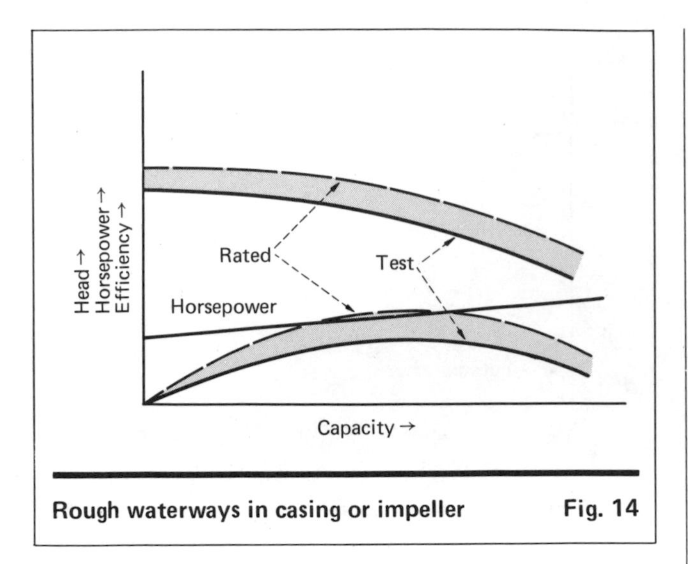


**Rough waterways in casing or impeller** **Fig. 14**

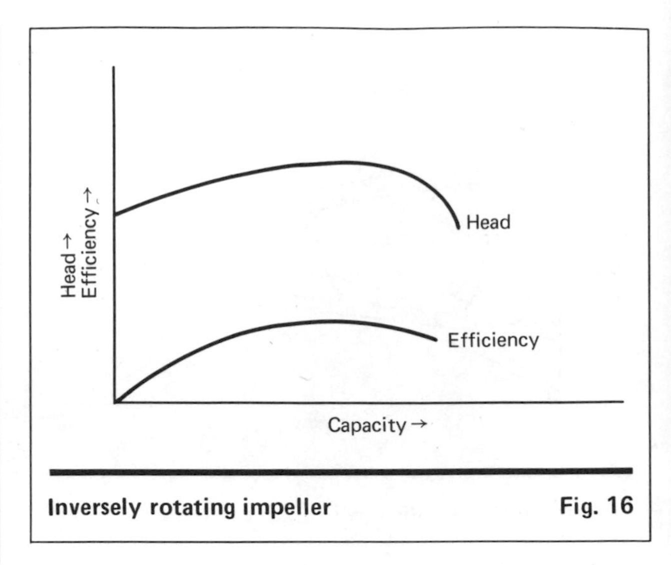


**Inversely rotating impeller** **Fig. 16**

(for semi-open impellers) usually produce this type of curve. If the wearplates are responsible, they may have become completely disintegrated. This type of curve can also occur when a mechanic forgets to reinstall a wearing ring or wearplate after a pump has been opened for repair or inspection.

*Reduced head and efficiency but horsepower unchanged* (*Fig. 14*): This is usually due to rough waterways (because of rust, scale, etc.) in the impeller and/or casing.

*Head/capacity curve correct but low efficiency causes increased horsepower:* This is usually due to mechanical losses, which may have resulted from: tight packing or mechanical seal, excessive hydraulic pressure on a seal or packing, faulty bearings, misaligned parts, misaligned pump and driver, bent shaft, operation near the critical speed, deformation of the casing by piping or baseplate stresses.

*Curve breaks off earlier than specified* (*Fig. 15*): This is due to insufficient net positive suction head (NPSH).

*Pressure head developed by the pump increases with an increase in flowrate* (*Fig. 16*): This occurs either when the pump is designed with forward blades (nowadays rare) or when the impeller is mounted inversely on the shaft. And it sometimes happens when the pump rotates in the wrong direction.

*During an NPSH test, head/NPSH line stops abruptly instead of continuing down to the breakpoint* (*Fig. 17*): During this test, the flowrate is kept constant and the NPSH is gradually reduced until the head breaks down, indicating the minimum required NPSH. Under certain circumstances, however, it may be impossible to maintain the required constant flowrate. This sometimes happens in a closed test-loop, indicating cavitation has evolved downstream of the measuring instruments, to block off the flow as the NPSH is reduced. The only remedy is to install a larger discharge line.

*Head starts to drop off gradually with reduced NPSH instead of falling off abruptly* (*Fig. 18*): In most cases, this curve indicates air leakage into the system. Also, such curves occur whenever a pump operates at a flowrate higher than design.

*NPSH requirements are higher at all flowrates* (*Fig. 19*): Whenever the difference between the required and rated NPSH is constant, leakage due to worn seal rings, etc., is indicated. Whenever the difference varies, rough waterways or protrusions are indicated.

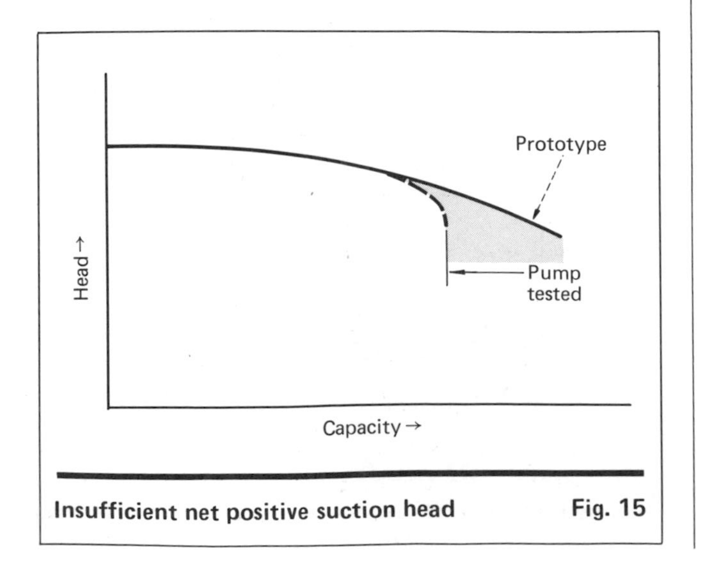


**Insufficient net positive suction head** **Fig. 15**

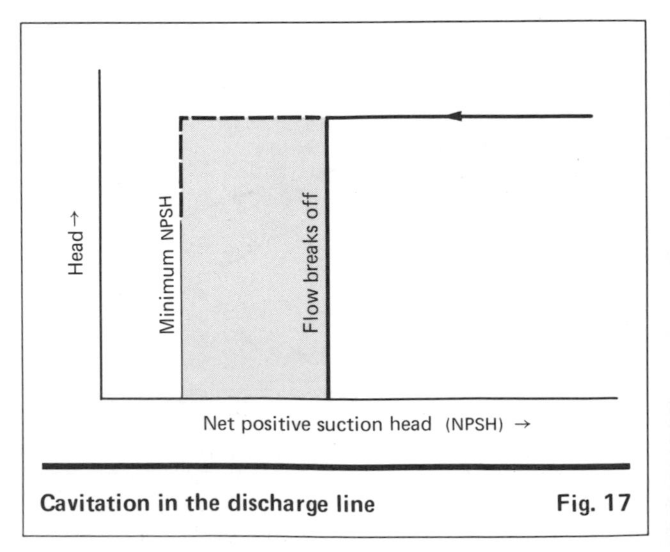


**Cavitation in the discharge line** **Fig. 17**

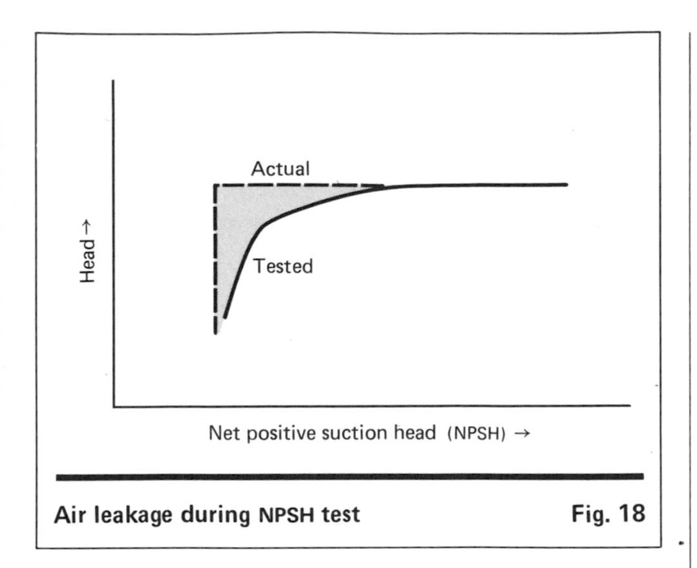


**Air leakage during NPSH test** **Fig. 18**

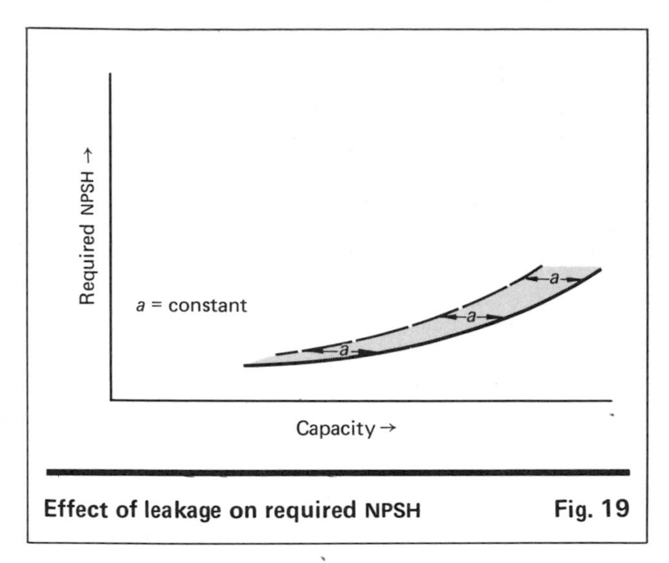


**Effect of leakage on required NPSH** **Fig. 19**

## Precautions for testing pumps

Armed with these typical curves, the engineer is prepared to diagnose a wide variety of pump problems. However, the curves must be compared to the actual performance curve for the pump. Unfortunately, such curves result only from tests carried out correctly, and this is often not the case.

A number of official and semiofficial institutions set forth procedures for testing centrifugal pumps. However, none of these provide a foolproof warranty that a test will be correctly performed, and it is worthwhile to describe some typical causes of faulty tests.

A very common practice that can lead to incorrect pressure-gage readings consists of leaving sharp edges or burrs of metal at the connections between the pump piping and the pressure-gage tubing. For correct readings, these edges should be carefully rounded off of all such connections.

The variation in pressure over the cross-section of a flowing liquid also frequently causes incorrect pressure readings. In fact, it is rarely possible to determine the true average pressure across a given pipe section. In order to reduce this error, a number of holes equally spaced around a given pipe section should be connected in parallel to the same pressure gage.

Air pockets and clogged tubes are other common sources of incorrect pressure readings.

Pressure readings in the suction line can be greatly affected by prerotation. At low flowrates, there occurs an interchange between the liquid that has already entered the impeller and that approaching the pump via the suction pipe. Part of the liquid that has been acted on by the impeller vanes returns to the suction pipe, where it mixes with the incoming stream and imparts a rotational motion derived from the vanes. This not only sets the incoming liquid into rotational motion but also increases its pressure near the outer diameter of the suction pipe.

The total head developed by a pump is defined as the total amount of useful energy transferred from the impeller blades to the liquid. In the absence of prerotation, this total head is equal to the total head measured at the pump discharge, minus the total head measured at the inlet to the pump. However, with prerotation, the impeller starts to transfer useful energy to the liquid still upstream of the pump. Since this liquid has thus absorbed a certain amount of energy, any pressure readings in the suction pipe will be a combination of the true suction pressure plus the added energy, and equivalent to taking these measurements somewhere in the pump itself.

The only correct method of determining true suction pressure in the presence of prerotation is to measure the pressure at the surface of the liquid in the suction tank or sump at a spot where the liquid is free from vortices or significant velocities. The true suction head will be equal to this pressure reading plus or minus the difference in the levels of the surface and the pump centerline, and minus the losses in the suction pipe.

Pressure readings in the discharge pipe may sometimes be affected by the pipe system, although such cases are less frequent. In a few instances, the small projection of a gasket into the waterways at the discharge flange has reduced the readings on the pressure gage by as much as 60 ft. (In other cases, however, even greater projections did not affect the pressure readings at all.)

Cavitation at the connections of the measuring instruments or in the instruments themselves makes for still another source of incorrect readings, even in the discharge line. This is apt to occur when the pump is part of a closed loop, and the pressures in the loop are relatively low. One of the most vulnerable spots is the throat of a venturi meter or an orifice. A second spot is the last valve or fitting that regulates the return of the liquid to the suction tank, since the pressure downstream of such a valve may be low.

Finally, methods for processing the data present another source of error in pump performance curves, particularly when the calculations involve velocity heads and similar losses. Ready-made charts are commonly used for such calculations, but such charts relate to pipes of a given schedule number, and the test piping may be fabricated from a different schedule. To avoid such errors, the actual pipe diameter should be carefully measured and accounted for in the calculations.

# Diagnosing problems of centrifugal pumps—Part III

## Here are some special types of pump malfunction that stem from the service and type of pump.

*S. Yedidiah, Worthington Pump Corp.*

☐ Since many specific causes of pump malfunction, such as gas pockets, cavitation, excessive clearance, etc., relate both to the use and the design of a pump, many special problems can be identified with: pumps operated in series, pumps operated in parallel, split-case double-suction pumps, pumps with semi-open impeller, or deep-well pumps.

### Pumps in series and pumps in parallel

One or more pumps in a series may fail due to loss of NPSH because of the failure of an upstream pump.

A reduced rate of flow may persist through pumps in series, even when one of them is not in operation. Such liquid flow through the idle pump will cause the impeller to rotate in the reverse direction, loosening nuts, etc., that secure the impeller and sleeves on the shaft. When the idle pump is then brought under power again, the loose parts may cause it to fail in a short time.

The suction manifold serving several pumps in parallel should receive special consideration for layout and sizing, because cavitation generated at the entrance to one suction line can be propagated along the manifold to other suction lines; or one pump might rob the others of suction pressure, thus reducing their available NPSH.

### Split-case double-suction pumps

This type of pump is subject to a number of special problems that occur because of its two-sided, symmetrical impeller.

A bend in the suction line (Fig. 1) may cause uneven distribution of flow to the impeller, with consequent lower head and efficiency, as well as noise and vibrations. When such an elbow cannot be avoided, it should be located at least 20 diameters upstream from the inlet, or special flow-straighteners should be installed.

The symmetrical nature of the double-suction impeller creates the possibility for mounting this type of impeller inversely on its shaft, after the pump has been dismantled for inspection or repair. However, the impeller vanes are shaped for rotation in only one direction, and the inverse rotation will result in noise, vibration, drastically reduced efficiency and a head-capacity curve completely different from the rated one.

Also, when a split-case pump has been dismantled for inspection or repair, misalignment may occur between the two casing halves (Fig. 2), impairing performance. A similar problem results when such a reassembled pump has the gasket between the casing halves protruding into the waterways.

The axial position of the double-suction impeller (Fig. 3) should be adjusted so that the centerline of the impeller coincides with the center of the volute. Significant separation of these centerlines may cause the shrouds of the impeller to rub against the inside of the casing. Less-drastic separation may cause excessive axial loads and a reduction in performance.

Corrosion or erosion in this type of pump can produce holes in the partitions between the suction and the volute (Fig. 3). Because of their location, such holes are often too small for easy detection but large enough to cause a significant reduction in performance.

### Pumps with semi-open impellers

A semi-open impeller has only one shroud, with the second shroud replaced by either a stationary wearplate

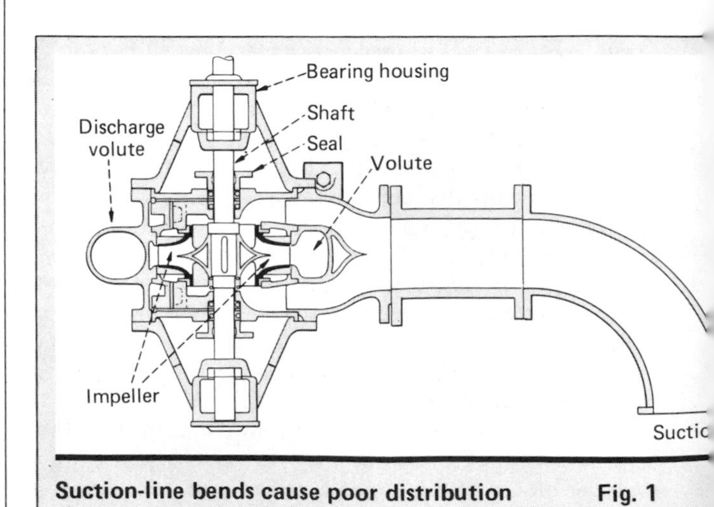


Suction-line bends cause poor distribution Fig. 1

*For the author's biography, see p. 186.

Originally published December 5, 1977

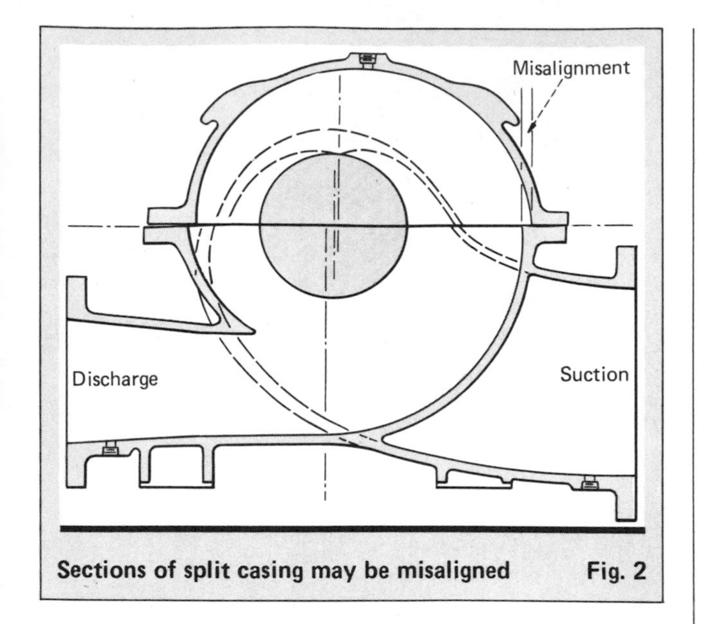


Sections of split casing may be misaligned Fig. 2

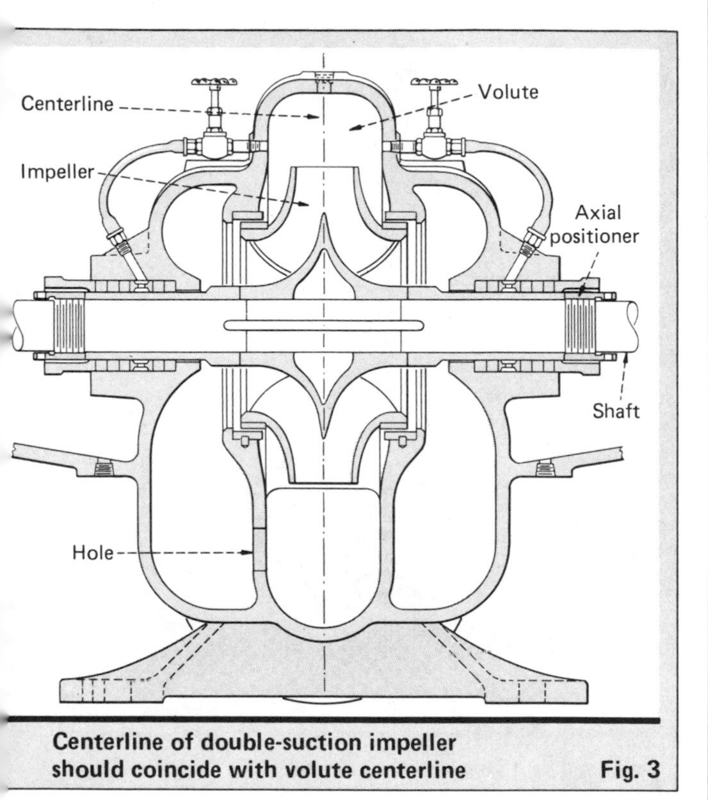


Centerline of double-suction impeller should coincide with volute centerline Fig. 3

or by a wear face cast integrally with the casing (Fig. 4). The clearance between the impeller vanes and the wear faces has a profound influence on the performance of the pump (see Buse, Fred, *Chem. Eng.,* September 26, 1977, p. 93.) Consequently, most pumps with semi-open impellers have some sort of arrangement for adjusting this clearance without dismantling the pump.

In horizontal pumps, this is usually done by mounting the thrust bearings in a cartridge and adjusting the axial position of the cartridge by means of shims. However, use of such shims is not necessarily foolproof, since the faces of an impeller's vanes may not be parallel to the wearing face of the casing. Whenever there is reason to believe that this is the case, the pump must be dismantled for accurate measurements.

Sometimes semi-open impellers are used in multistage pumps, and such impellers can be reassembled in the wrong order along the shaft, causing variation in their clearances.

An entirely different problem is encountered in these pumps when the pumped liquid contains stringy or fibrous solids. Such solids have a tendency to wedge between the impeller blades and the wearing face, breaking either vanes or shaft.

## Deep-well pumps

Deep-well pumps have small-diameter multistage impellers with diffusers, mounted at the end of a long shaft (Fig. 5), and are classified according to the type of impeller (whether closed or semi-open) and type of column structure, whether open (with water-lubricated rubber bearings) or closed (with oil-lubricated metal bearings). Some of the following problems occur in all types of deep-well pumps; others are characteristic of only one type.

*Variations in water level:* When a deep-well pump is not in operation, the level of the water is exactly the same as in the surrounding strata. When the pump is started, however, the water level falls until a state of equilibrium is established between rates of influx and removal. This equilibrium level is subject to seasonal and weather conditions. It may fall so low that air enters the pump, immediately reducing the flowrate and allowing the water level to rise, which in turn increases the flowrate to cause the water level to fall again in a cycling performance.

Such cycling can be temporarily remedied by partially closing the discharge valve. The only permanent remedy is to lengthen the column, which may require deepening the well.

*Effects of sand:* The immediate effect of sand in water is excessive erosion, and the remedies are either erosion-resistant materials or periodic replacement. Experience shows, however, that sand is most significant in newly drilled wells, and is usually reduced to tolerable levels after a period of time.

When the water does contain sand, the pump should never be stopped suddenly, but rather the discharge valve should be closed gradually and very slowly. If the pump is stopped abruptly, all the sand contained in the column will settle to clog the impellers and casings, rendering the pump inoperative. If the flow is reduced gradually, however, the amount of sand carried in is reduced proportionately, so that there is not so much sand to settle out on complete shutdown.

*Effects of air:* Deep-well pumps are usually provided with a check valve close to the discharge. However, this valve cannot prevent the liquid elevated in the column from flowing back into the well and creating a vacuum between the water level and the check valve. When the pump is started again, vaporization due to this vacuum may lead to cavitation shocks.

Also, in an open-column pump (Fig. 5) some air may leak through the stuffing box, draining lubrication from the packing.

To prevent such problems, a deep-well pump with a

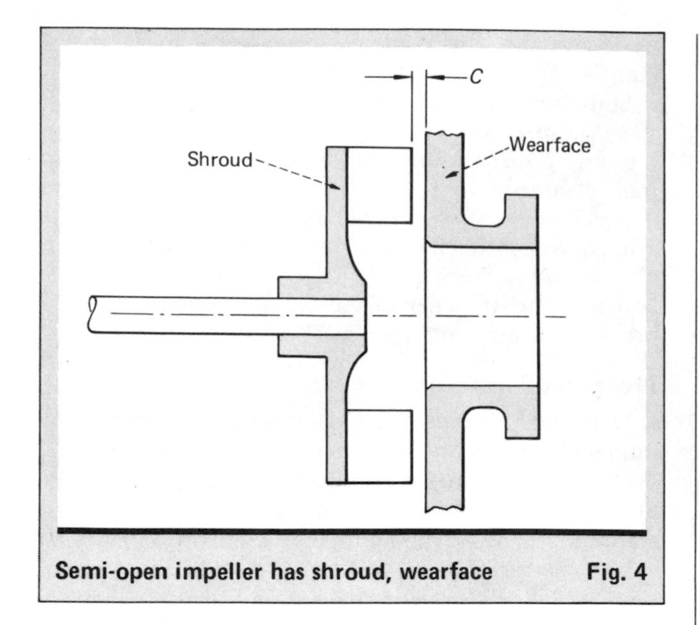


**Semi-open impeller has shroud, wearface** **Fig. 4**

long column should be provided with a vacuum-breaking valve.

Another problem arises when air is allowed to enter the column during shutdown but is unable to escape when the pump is started. This may cause severe blows in the check valve, as explained under "Gas pockets," p. 183.

*Faulty installation and/or faulty well:* Deep-well pumps are suspended from a head (Fig. 5), which rests on a solid foundation. Because of the column's length, any lateral force can easily misalign it, thus misaligning the bearings with resultant noise, vibrations and early deterioration of the pump. A deep-well pump will not operate satisfactorily if all line bearings and shafts are not properly aligned.

Accordingly, the head and motor must be installed with great care to avoid lateral forces from the well casing. When the well is straight and vertical, head assembly must be installed perfectly vertical and with its centerline in line with that of the well. When the well is straight but not vertical, satisfactory alignment may be achieved by inclining the head assembly to the same angle as the well, so that its center line forms an extension of the centerline of the well. When the well is not straight, only a pump having a sufficiently small outer diameter can be aligned satisfactorily.

On deep-well pumps, each shaft section and each column pipe is machined to perfect concentricity of threads and perfect squareness of faces. However, this machining can be ruined during assembly by dropping a column pipe, or striking a part, or allowing dirt to lodge between the faces.

*Water-lubricated columns:* When the pump is run dry, the rubber bearings of these columns possess a very high coefficient of friction and may be ruined in a few seconds. Accordingly, these bearings should not only be lubricated before starting the pump, but this prelubrication should be continued until the pumped water completely fills the column and starts to discharge from the pump head.

Most kinds of rubber have a tendency to swell when

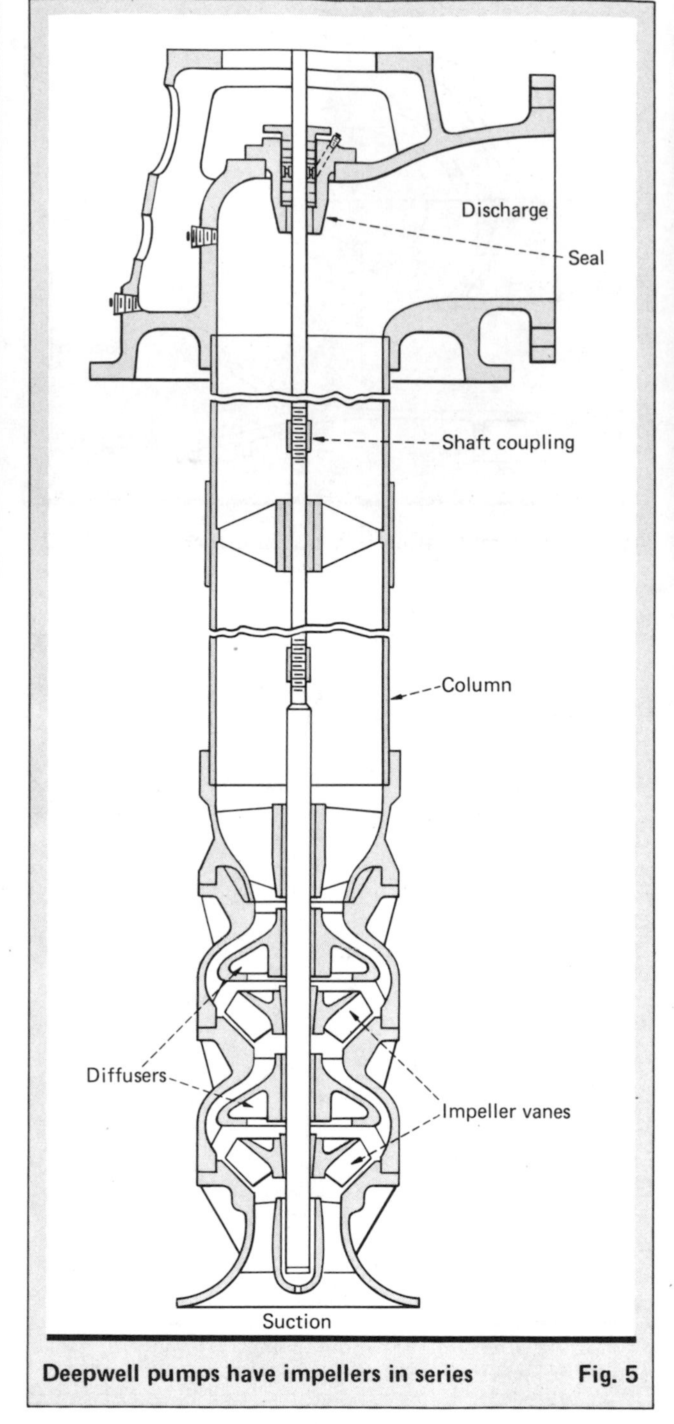


**Deepwell pumps have impellers in series** **Fig. 5**

exposed to water containing chemicals such as sulfides and/or sulfates. When this happens, the bearings may bind the shafts, causing breakdown.

In certain cases (when the rubber swells only so much and then stops), a good remedy is to run the pump without interruption for several weeks. This will sometimes wear away the excess rubber and allow the pump to operate trouble-free.

However, the best method is to make laboratory tests of different kinds of rubber in samples of water taken from the deep well. Then when a resistant rubber has been identified, this rubber can be included in the specifications for the pump.

# The effects of dimensional variations on centrifugal pumps

This article shows what to look for in single-stage end-suction centrifugal pumps with semi-open impellers, specific speeds of 600 to 1,000, 1- to 3-in.-dia. discharge nozzles, 6- to 11-in.-dia. impellers operating at 3,550 rpm, and axial adjustment for setting the running clearance between the impeller and casing.

*Fred Buse, Ingersoll-Rand Co.*

☐ A pump's hydraulic performance can differ drastically from its published performance expectation. When this happens, one should be able to find the cause of the discrepancy. Most times this will involve external aspects of the pump, and will be obvious. There may not be enough NPSH; vapor may be trapped in the suction lines because a high point is not vented back to the vapor space in the supply vessel; the motor may be connected for reverse rotation; discharge or suction lines may be blocked; the pump may not be properly primed, and so forth.

At other times, the cause of the malfunction may be faulty dimensions of internal parts. These are much more difficult to identify. The pump must be dismantled for a careful dimensional analysis of each major component.

This paper will point out the things to study during such an investigation—i.e., the effects that nonconforming dimensions of components have on pump performance.*

In general, the dimensions of three components are most critical to pump performance, in this order: the impeller, the casing's running surface, and the volute throat (Fig. 1).

*The original purpose of the paper was to establish commercial pump manufacturing-tolerances and quality-control standards that would result in hydraulic performance consistently within the Standards of the Hydraulic Institute (New York). Our study has also shown alterations that could be applied to substandard pumps to obtain standard performance.

Originally published September 26, 1977

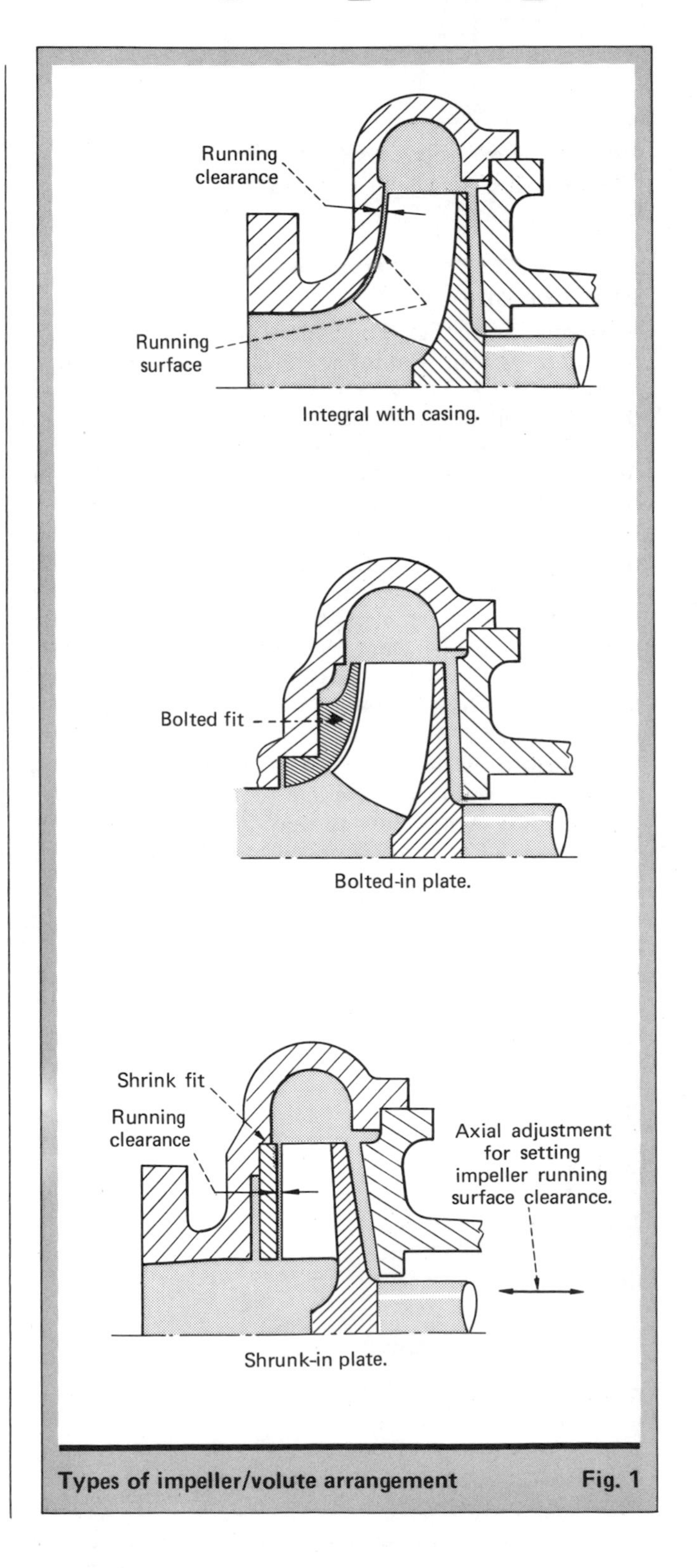


Types of impeller/volute arrangement Fig. 1

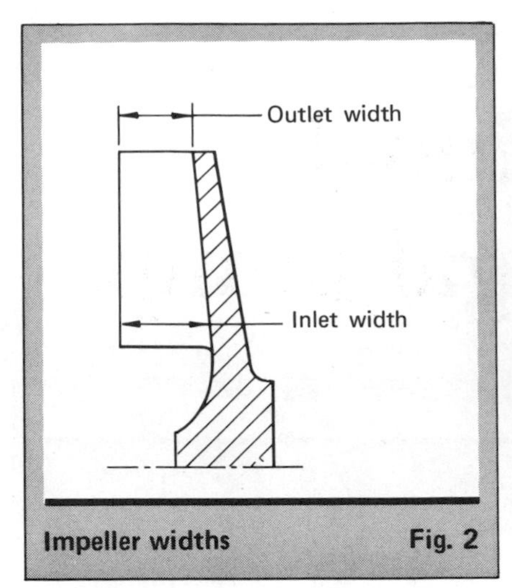


Impeller widths Fig. 2

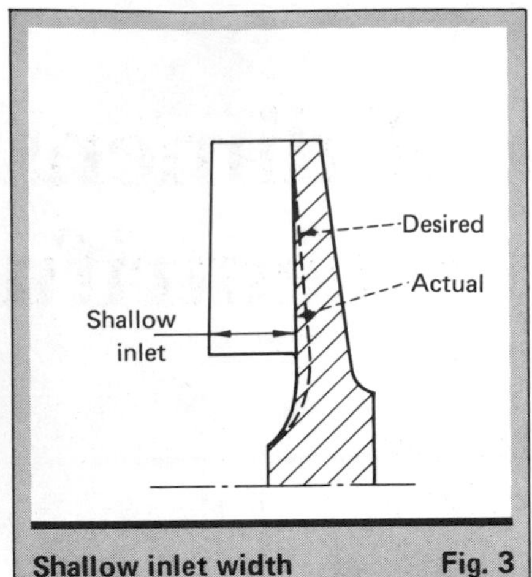


Shallow inlet width Fig. 3

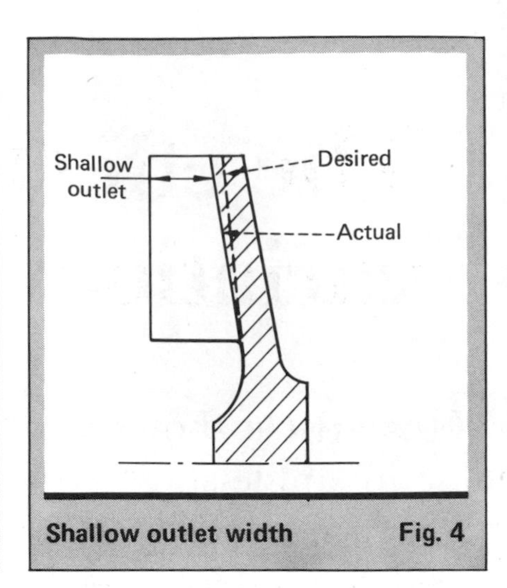


Shallow outlet width Fig. 4

## Impeller quality

Among the configurations most important to an impeller are: width, running surface, and attachment.

*Impeller width:* The width of the impeller (Fig. 2) has the biggest effect. Although an oversize width will cause little change, undersizing will cause a dramatic reduction in performance, whether it occurs on the thinner edge at the inlet or the outer circumference at the outlet.

Impellers having an outlet width equal to or wider than design, but a shallow inlet width (Fig. 3), will produce a performance curve parallel to, but lower than, the standard curve. Impellers having an inlet width equal to or wider than design, but a shallow outlet width (Fig. 4), will have the same type of characteristics. In both cases, heads 10 to 30 ft below standard will result (Fig. 5).

By measuring the hydraulic inlet and outlet widths, one can determine the average required area to meet performance. A new casting may be required if excessive material has to be removed.

Width reduction mainly comes from a concave or convex casting caused by poor casting heat-treating techniques. This results in reduced hydraulic area. Impeller inlet and outlet widths should be machined to within 0.000 to 0.015 in. of the design dimensions for consistent performance.

*Impeller-width vibration:* An impeller can be machined to the overall proper width and be dynamically balanced, and still result in a pump that vibrates excessively. This happens when one or more of the vanes is 0.125 in. shallower than others at the outlet area (Fig. 6). Because the vanes are not all loaded evenly, this causes a hydraulic imbalance and a shaft vibration.

*Running-surface clearance:* Performance with excessive running clearance (Fig. 7) will be parallel to the standard performance. Head will be reduced approximately 10 ft for every 0.010 inch of clearance for 6- to 11-in. impellers operating at 3,550 rpm (Fig. 8). This rate of reduction holds with wide clearances, as much as 0.120 in. having been tested. Also, efficiency decreases

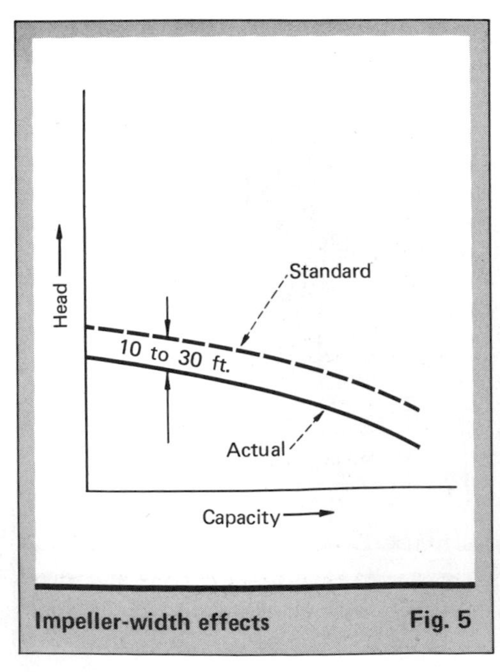


Impeller-width effects Fig. 5

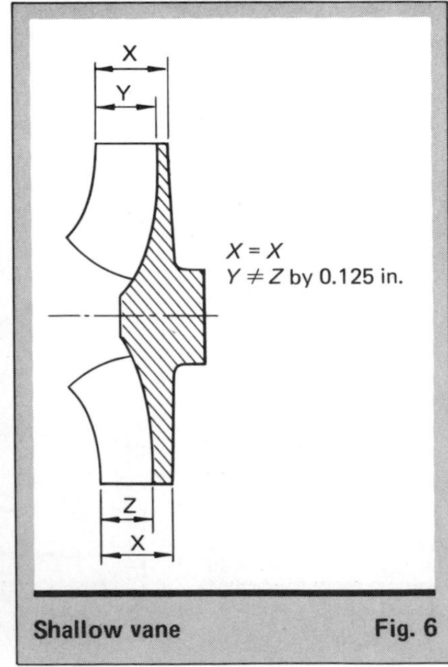


Shallow vane Fig. 6

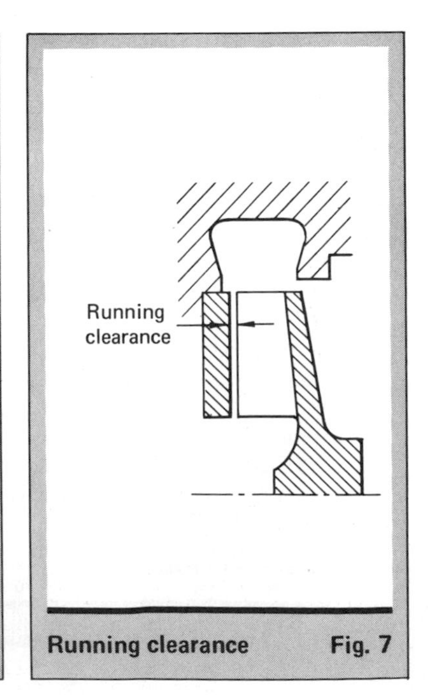


Running clearance Fig. 7

by two to three points for every increase of 0.010-in. in clearance.

*Concave running surface:* When machining the contour of the running surface of the impeller, care must be taken not to machine the casting too fast or to remove heavy metal without proper support. Otherwise the ends of the vanes will be pushed back, only to spring forward when the machining is complete. The result is a running surface that is concave by as much as 0.010 to 0.015 in. (Fig. 9). When the impeller is set for the normal running clearance, the absolute clearance will be 0.020 to 0.025 in. rather than 0.010 in. Performance will be similar to that for excessive running clearance.

*Impeller attachment:* When an impeller is screwed to its shaft, the male or female surfaces cannot produce more than 0.001 in. out of perpendicularity measured at the mating surfaces (Fig. 10) or be more than 0.001 in. out of line with the shaft centerline, measured at the impeller hub (Fig. 11). Otherwise only one vane will touch the running surface at the initial clearance setting.

**Head losses with excessive running clearances** **Fig. 8**

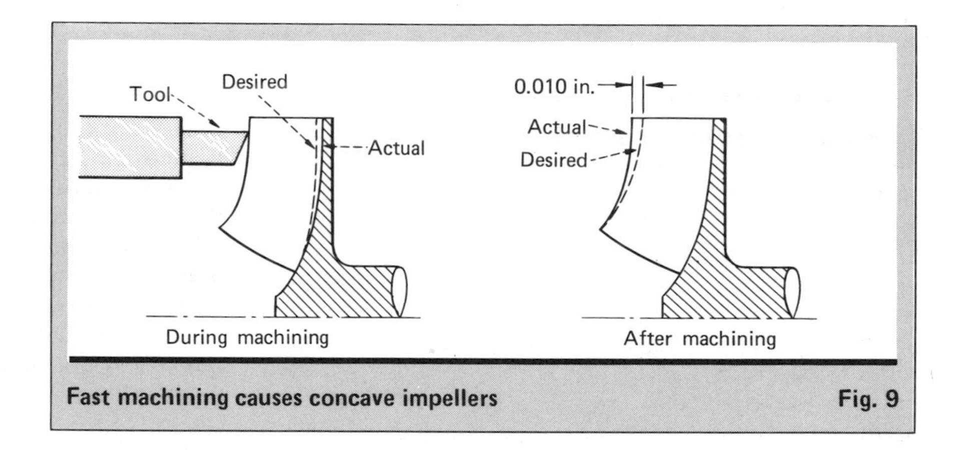


**Fast machining causes concave impellers** **Fig. 9**

Such cocked impellers will result in performance similar to that for impellers with excessive clearance.

Also, an impeller can be statically or dynamically balanced, and within all hydraulic dimensions, and there can still be excessive shaft vibration, as a result of looseness in the impeller/shaft fit, with consequent eccentricity of the two axes (Fig. 12). A 10-in.-dia. impeller with total indicator runout (TIR) of 0.004 in. will throw the mass of the impeller off center to cause shaft vibration.

*Underfiling adjustments:* One method for increasing pump performance is to underfile the exit vanes,

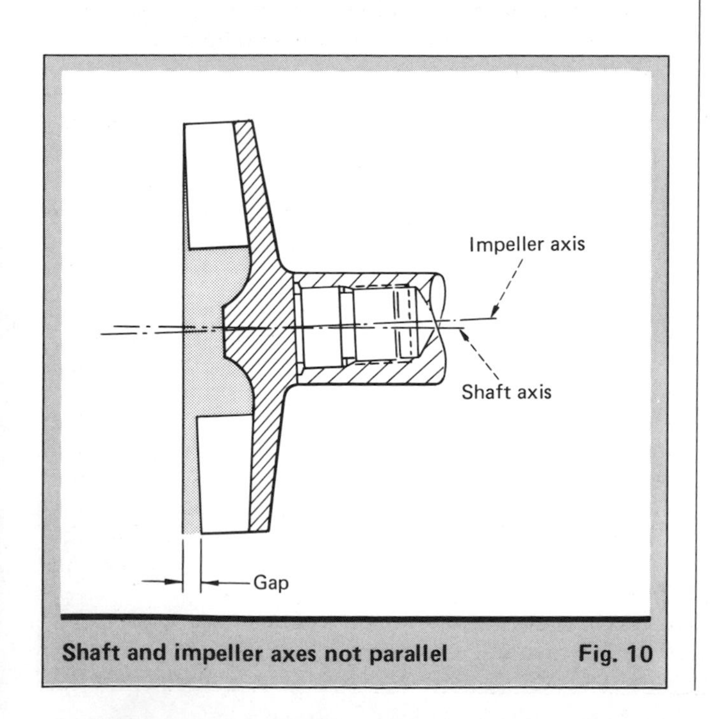


**Shaft and impeller axes not parallel** **Fig. 10**

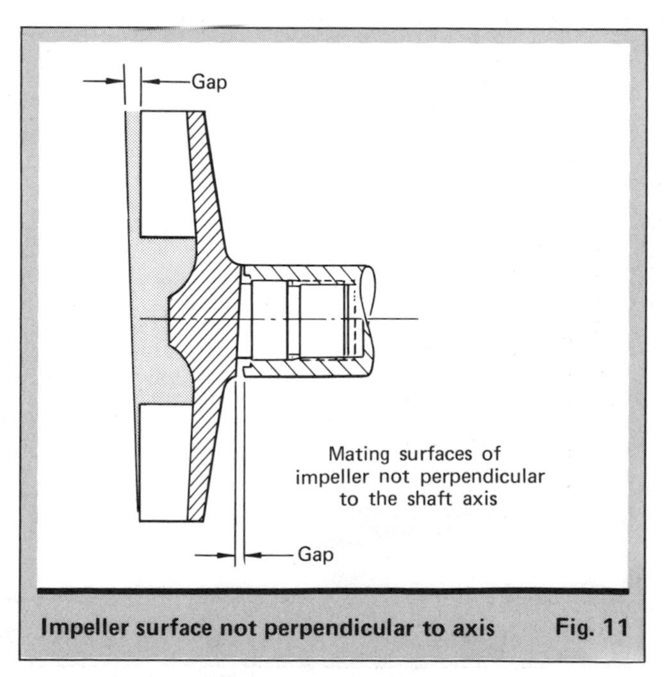


**Impeller surface not perpendicular to axis** **Fig. 11**

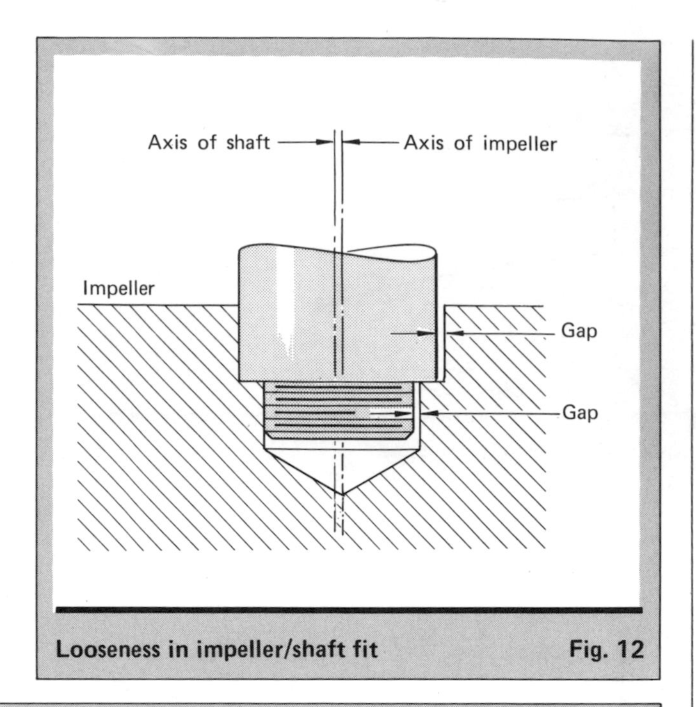


Looseness in impeller/shaft fit **Fig. 12**

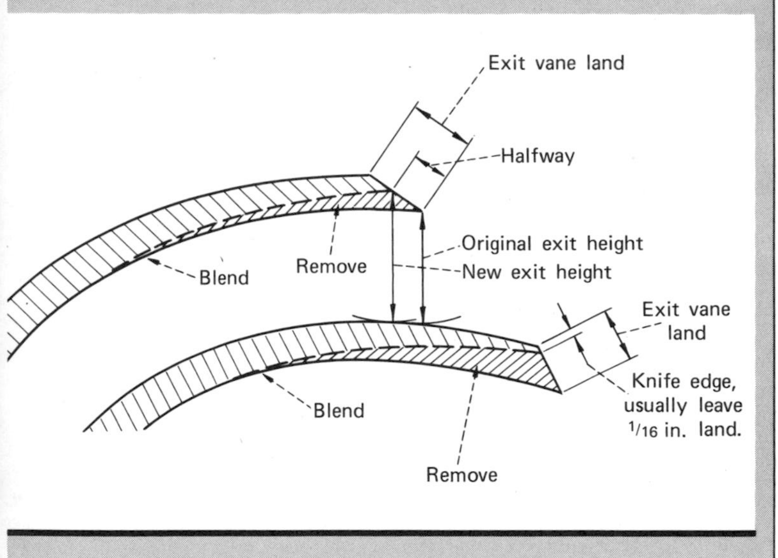


Underfiling adjustments to exit vanes **Fig. 13**

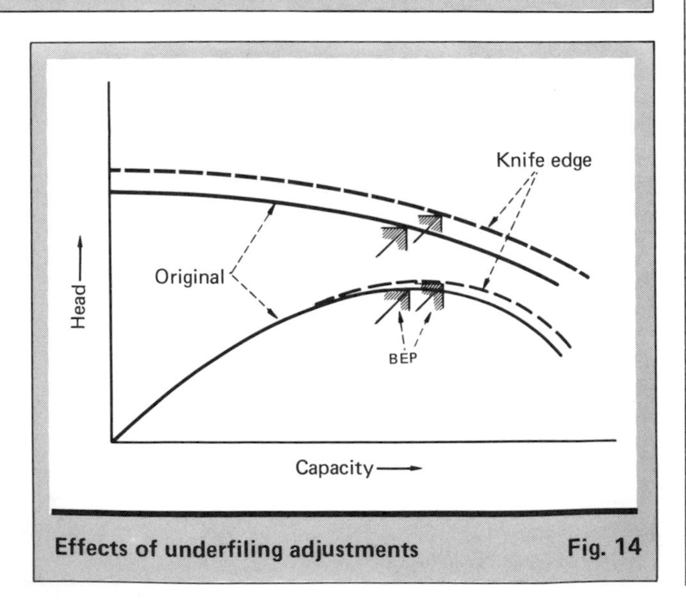


Effects of underfiling adjustments **Fig. 14**

**Effects of underfiling**

Gain in head, ft, and in efficiency, points

| Impeller dia., in. | Knife edge: Percent of BEP capacity 0 | 50 | 100 | Knife edge: Points at BEP | Halfway: Percent of BEP capacity 0 | 50 | 100 | Halfway: Points at BEP |
|---|---|---|---|---|---|---|---|---|
| 6 | 4 | 6 | 10 | 1 | 2 | 3 | 5 | 1 |
| 8 | 13 | 15 | 20 | 2 | 6 | 7 | 9 | 1.5 |
| 10 | 17 | 20 | 28 | 2 | 8 | 10 | 14 | 1.5 |

thereby increasing the outlet area. The extent of the underfiling is identified by the terms "halfway" and "knife edge" (Fig. 13). Halfway underfiling consists of removing half of the exit-vane land by underfiling and blending the filed surface into the lower part of the vane. Knife-edge underfiling consists of reducing the land to $\frac{1}{16}$ in., below which the tips of the vanes become too thin and brittle for practical use.

The approximate gains in head and efficiency by underfiling a 3,550-rpm impeller (Fig. 14) are as shown in the table.

## Casing-surface quality

The running surface of the casing can be machined integral with the casting; or it can be a separate plate that is mechanically attached to the casting by (1) a shrink fit, (2) internal bolting, (3) external bolting, or (4) tack welding.

The main disadvantage of machining the running surface integral with the casting is a lost casting in the event of improper machining or of surface porosity. The main advantage is more-liberal tolerances than a separate plate, since the machining precision of only one surface is involved.

The main disadvantages of a separate plate are the difficulties of proper installation and of properly maintaining the installed position under field operation. The main advantage is the savings in maintenance and replacement costs that usually result, because only the plate has to be replaced.

*Running surface not parallel to impeller surface:* When the casing running surface is not parallel to within 0.003 to 0.004 in. of the impeller running surface, the reduction in performance will be similar to that caused by excessive running clearance. When the surfaces are even more out of parallel, one of the impeller vanes will prematurely contact the casing surface on the initial adjustment of running clearance. If the casing running surface has a total indicator reading of 0.008 in., the running clearance would be 0.010 in. on one vane and 0.018 in. on the opposite vane (Fig.15).

This mismatch of running surfaces can come from:

1. Not machining (during the same tool setup) the running surface and either the mounting surface for the

bearing housing or the support-head of the casing. The result is surfaces nonparallel by as much as 0.010 in. (Fig.16).

2. Improperly mounting a separate plate in a casing, because (a) its mounting surface in the casing is not parallel within 0.001 in. either to the mounting surface for the bearing housing or to the support head (Fig. 17); (b) the mounting surface and running surface of the plate are not parallel to each other (Fig. 18); (c) the plate is not mounted properly because of burrs or dirt between it and the casing surface (Fig. 19); (d) the plate is mounted cocked in a shrink-fit, because the assembler did not bottom the plate to its fit (Fig. 20).

3. Permanently distorting an integral running surface, which was otherwise machined and inspected satisfactorily, by inadvertently exposing it to excessive pressure during hydrostatic testing.

4. Surfaces that are nonparallel by 0.002 in. within the support-head mounting surfaces or the bearing-housing mounting (Fig. 21), thus causing the running surfaces to have excessive total indicator runout.

*Concave or convex plate:* When a plate is shrunk into a casing with an excessive amount of fit, the plate can buckle up or down, creating nonparallel surfaces and resulting in excessive running clearances. Similarly, when the plate surface is not properly supported during machining, it will spring (Fig. 22).

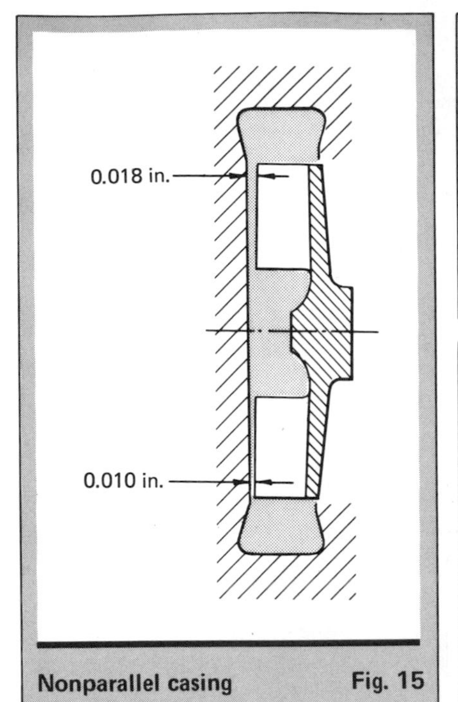

Nonparallel casing Fig. 15

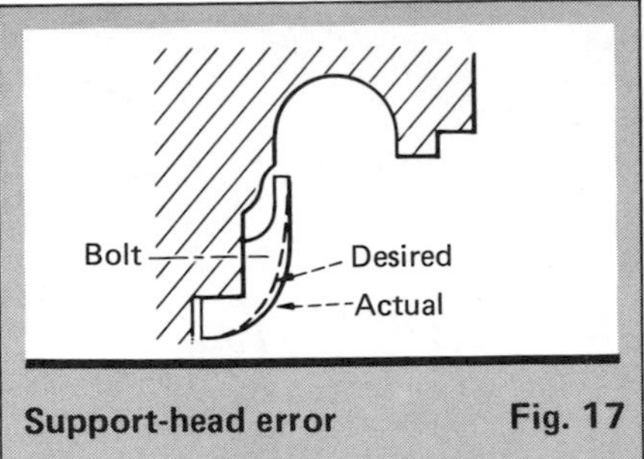

Support-head error Fig. 17

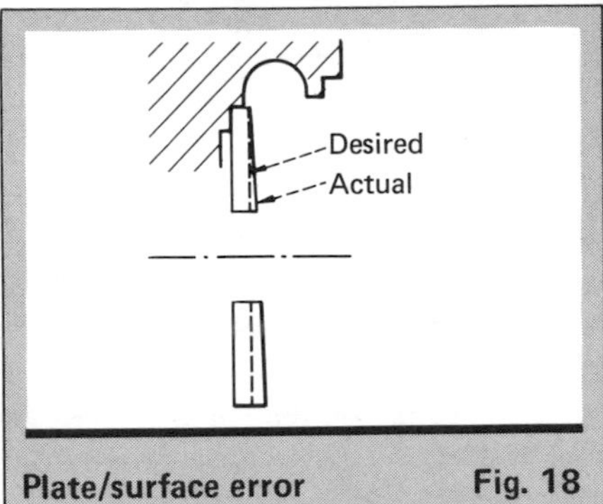

Plate/surface error Fig. 18

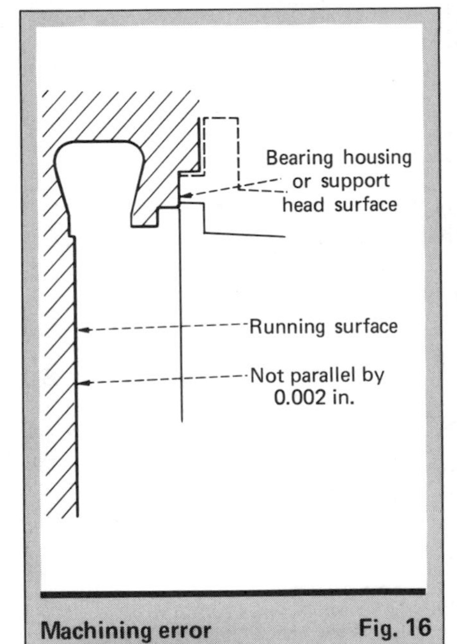

Machining error Fig. 16

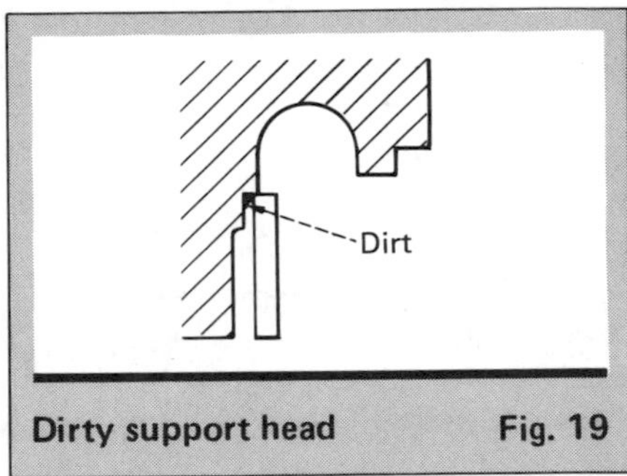

Dirty support head Fig. 19

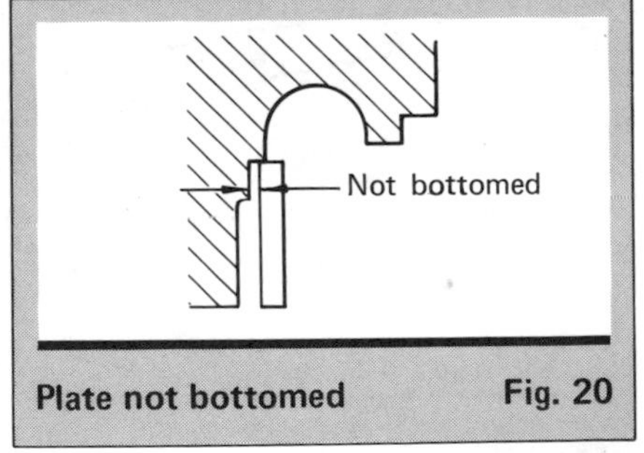

Plate not bottomed Fig. 20

## Volute quality

*Volute width:* The effects caused by varying from design can be significant for a semi-open impeller with 600 to 1,000 specific speed (Fig. 23). A series of tests on the effects of volute width have been made by raising and lowering the casing's running surface from the design datum (Fig. 24). Results are ratioed to a common impeller diameter of 10 in.

When the surface is lowered 0.010 in. below the design-datum line, there is a 15-ft reduction in head at the best efficiency point (BEP); when lowered 0.020 in., the reduction is 25 ft; when 0.030 in., the reduction is 35 ft—all from a design head of 400 ft (Fig. 25). The corresponding reductions at shutoff are 5, 10 and 15 ft, respectively.

Similarly, the capacity at BEP is reduced by approxi-

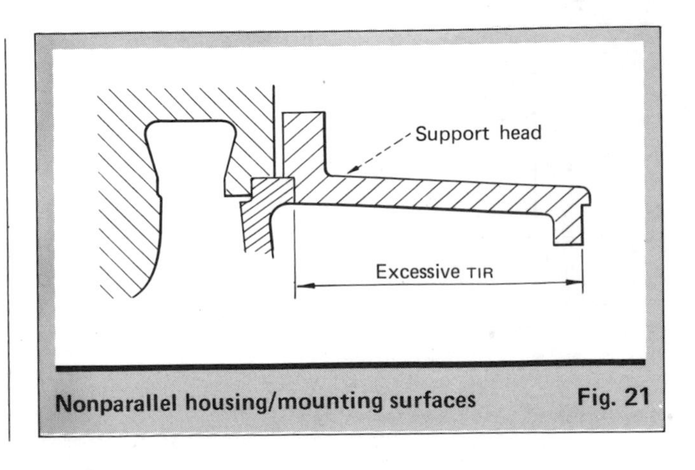

Nonparallel housing/mounting surfaces Fig. 21

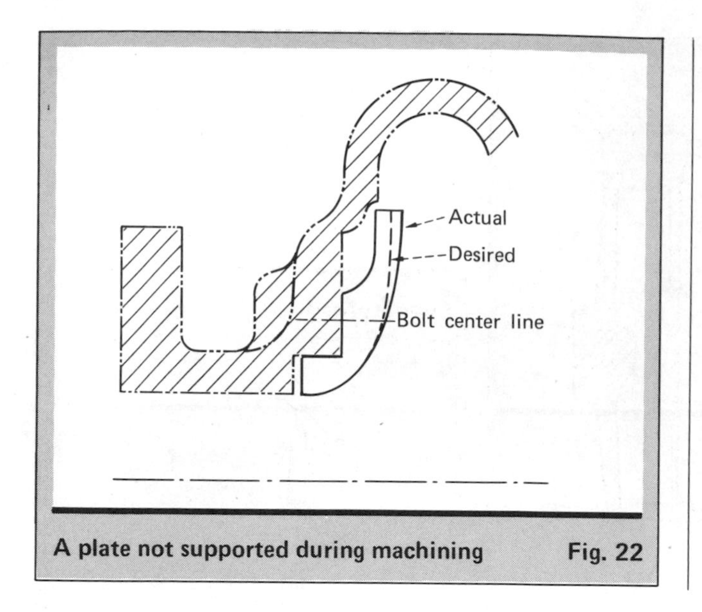


A plate not supported during machining Fig. 22

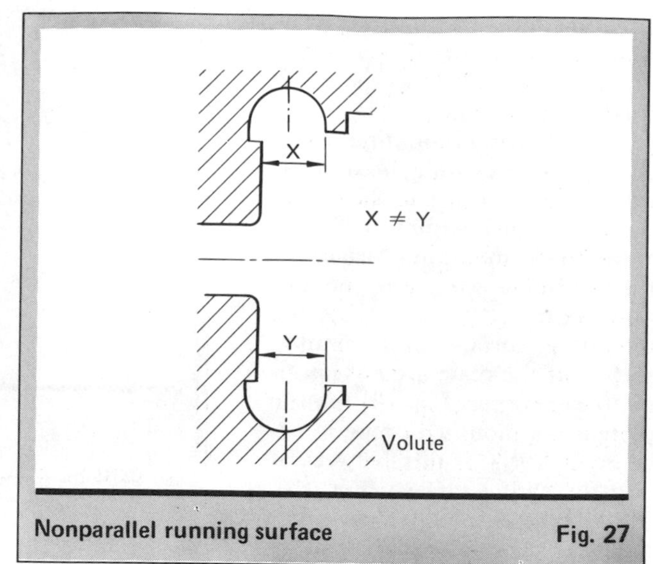


Nonparallel running surface Fig. 27

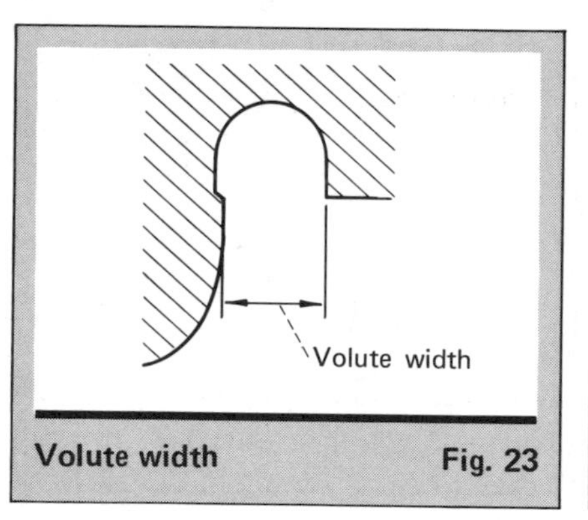


Volute width Fig. 23

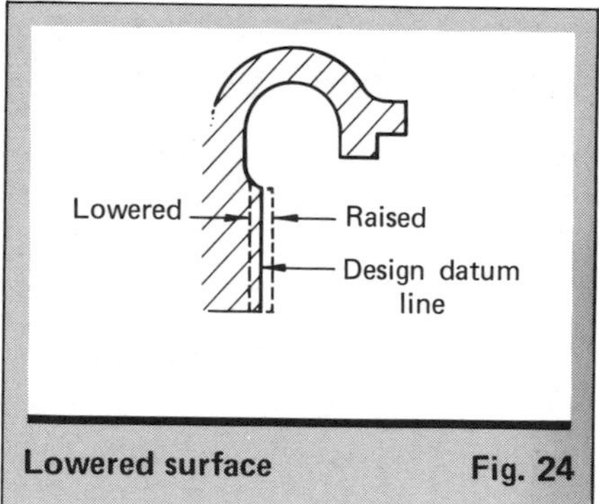


Lowered surface Fig. 24

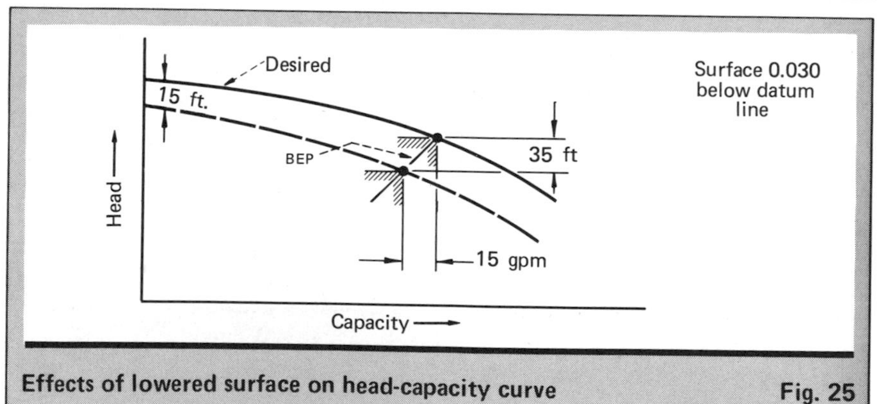


Effects of lowered surface on head-capacity curve Fig. 25

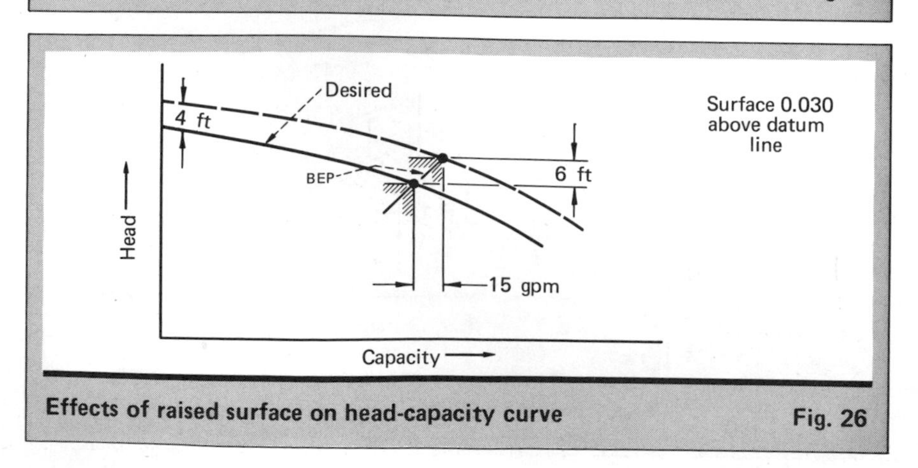


Effects of raised surface on head-capacity curve Fig. 26

mately 5, 10 and 15 gpm, respectively, and efficiency is reduced one to two points over the 0.010 to 0.030-in. range.

When the surface is raised above the design datum line, the head at BEP and shutoff increases, but not as rapidly, giving 2 ft at 0.010 in., 3 ft at 0.020 in. and 6 ft at 0.030 in. (Fig. 26). Similarly, the capacity at BEP will increase 5, 10 and 15 gpm, respectively, and efficiency will increase one to two points.

If the design is made with the datum line too low, there will be a large increase in head as the running surface is raised, and a smaller reduction in head when the surface is lowered. The large increase can overload the driver, increase radial and axial thrusts and impose improper conditions on the system.

*Running surface not parallel to centerline:* This is different from nonparallelism between the impeller and the casing running surfaces. In this instance, both the impeller and casing are cocked relative to the volute centerline, though parallel to each other. The result is a volute entrance area of varying width (Fig. 27). When this condition exists, there can be as much as a 60-ft drop in head across the entire performance curve, or a sharp drop in performance when starting up from shutoff.

The severity of malperformance depends on the plane on which the cocking occurs, relative to the location of the throat. The following are the maximum peripheral variances in volute width before performance is adversely affected. Above these tolerances, the pump can experience a significant loss in head.

| Impeller design diameter, in. | Maximum cocking, in. |
|---|---|
| 6 | 0.020 |
| 8 | 0.040 |
| 10 | 0.060 |
| 13 | 0.080 |

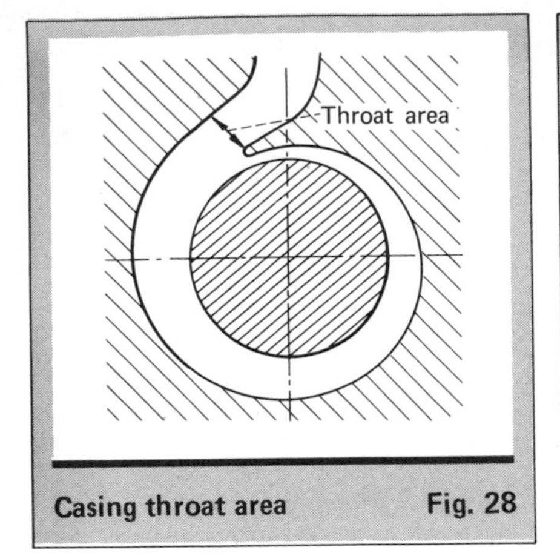


Casing throat area Fig. 28

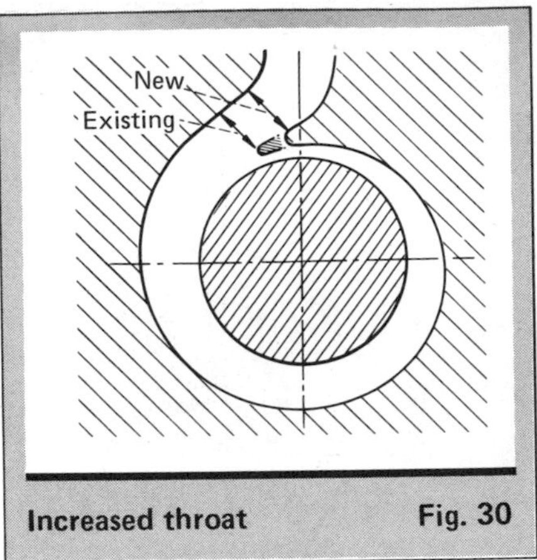


Increased throat Fig. 30

*Throat area:* The low-specific-speed (500 to 800) pumps are very sensitive to the area of the casing throat (Fig. 28). A slight reduction in this area, due either to a shift or break in core, or an obstruction, will have a throttling effect on performance (Fig. 29). Shutoff head will usually be the same as design; but the rest of the performance curve will be reduced proportionally to the throttling effect. If all other factors conform to design, the brake horsepower will be the same as design.

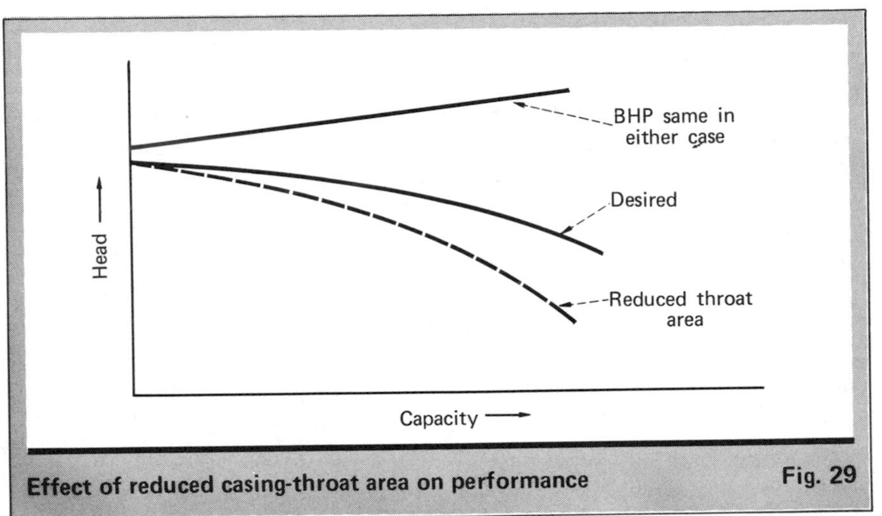


Effect of reduced casing-throat area on performance Fig. 29

It is sometimes necessary to increase the throat area above design to increase flow (Fig. 30). However, the short nozzle length typical of these pumps usually prevents a substantial gain in throat area. Also, such an increase sometimes results in a loss of head, because of the reduction of convergence in the nozzle. Although it is often relatively easy to increase the throat area of a pump operator's maintenance shop, the throat area should not be changed as a matter of course because of possible driver overload.

The following equation should be used to calculate a new throat area:

New gpm = (existing gpm)(new area/existing area)$^{1/2}$

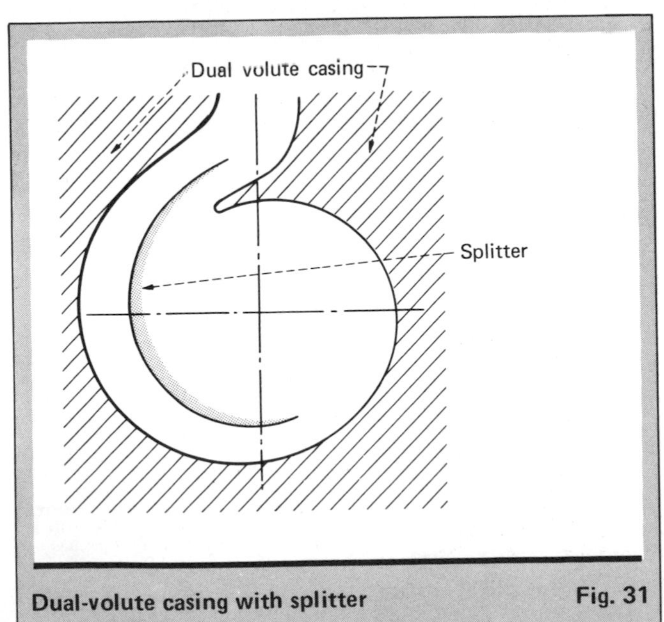


Dual-volute casing with splitter Fig. 31

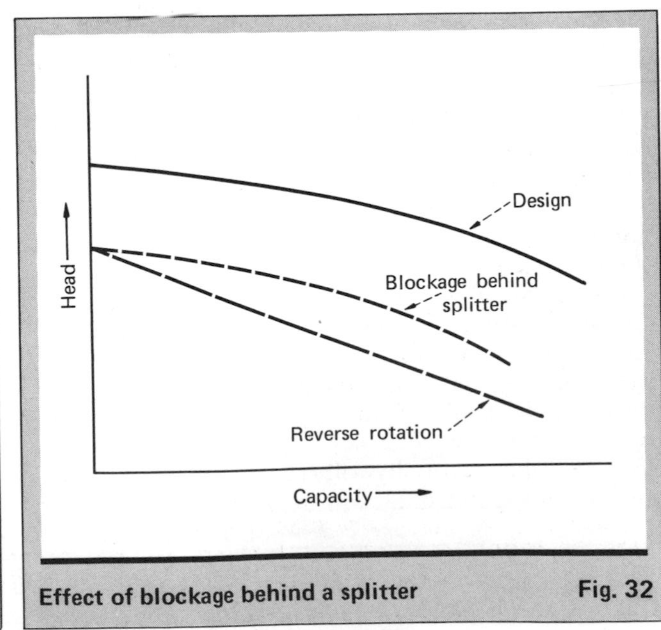


Effect of blockage behind a splitter Fig. 32

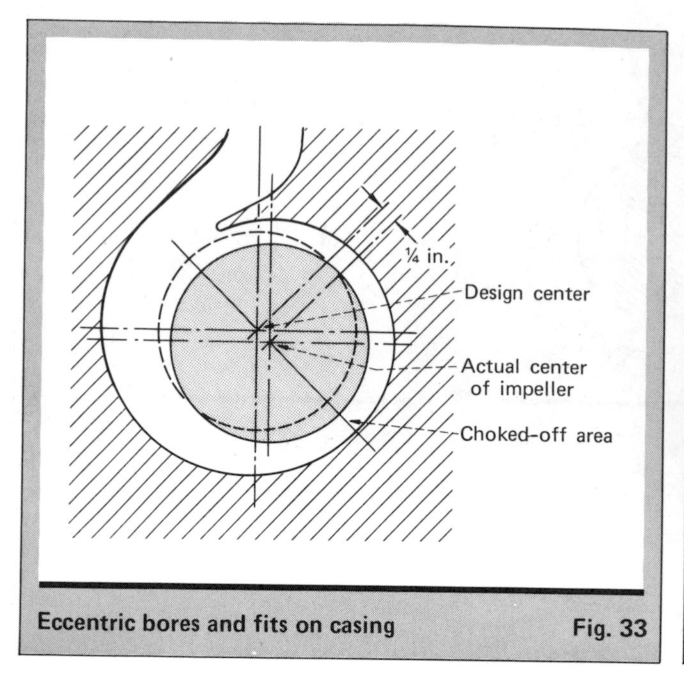


Eccentric bores and fits on casing Fig. 33

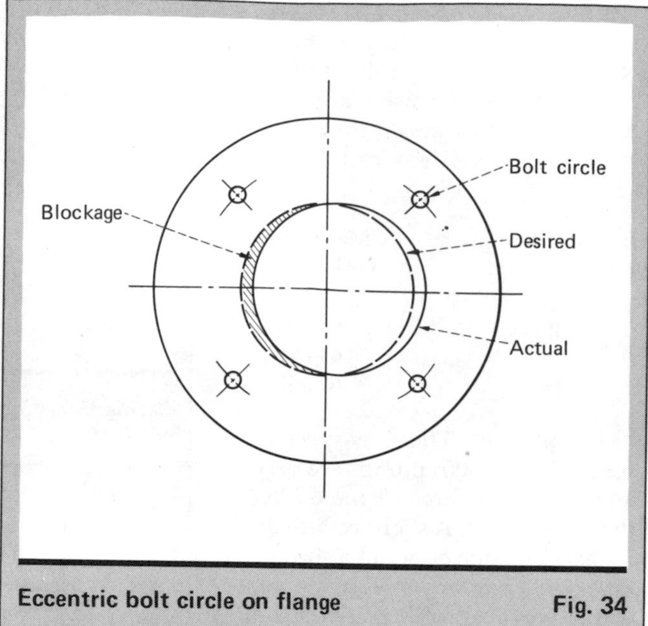


Eccentric bolt circle on flange Fig. 34

Both new and existing gpms are at BEP.

The relative gains in head and efficiency at BEP that are possible with an increase in throat area are:

| Specific speed | Change in head, % | Change in efficiency, points |
|---|---|---|
| 560 | −1 | +3 to +4 |
| 750 | 0 | +2 |
| 850 | +2 | +2 |
| 1,150 | +1 | +2 |

*Splitter blockage (dual-volute casing):* It is important to check that there are no obstructions behind the splitter in pumps having dual-volute casings (Fig. 31). When there are, the resulting head, including shutoff, is usually half the design value. At first glance, the performance resembles that of a pump operating in reverse rotation; but a closer examination will show that the reduced performance is not the straight-line characteristic of reverse rotation, but a line parallel to design performance (Fig. 32).

Performance will also be reduced if the area behind the splitter is smaller than design along its overall length, because of incorrect casting patterns or cores. A 5% reduction in area will result in a 20- to 30-ft reduction in head across the entire performance curve for a 10-in. impeller.

*Concentricity of volute and shaft:* When the aligning bores and fits of the casing are machined eccentric to the volute center, the impeller's center will be eccentric to the volute (Fig. 33). When the impeller is away from the throat, it will choke off the opposite side of the volute, reducing head and efficiency. With a 10-in.-dia. impeller, a 0.25-in. eccentricity will reduce the head 4 to 10 ft, with a three-point reduction in efficiency.

*Concentricity of discharge hole:* With a 1-in. discharge flange, there can be a reduction in head when the discharge flange is eccentric to the bolt circle. This comes from a core shift relative to the outside pattern of the casting. When the pump is installed, part of the discharge hole will be blocked by the mating pipe flange (Fig. 34).

However, when the liquid velocities in the nozzle and pipe are small, due to low capacity or low speed (1,750 vs. 3,550 rpm), this low may not be noticeable.

In conclusion, this paper has shown twenty-one ways in which nonconforming design dimensions affect pump performance. It is reasonable to expect that many of these will not normally contribute to malperformance. Parts whose machined surfaces are relative to other machined surfaces fall into this category—i.e., shaft, shaft-to-impeller attachment, support heads, and casing-support-head to running surface. The locations that have to be closely monitored are those having a cast surface in contact with a machined surface—i.e., the impeller width, volute width, volute concentricity and nozzle eccentricity. Generally, those locations are routinely checked as part of the quality control procedure of the pump supplier, before the assembled pump is packaged and shipped to its operating site. Results of the investigation point out the need for establishing manufacturing tolerances and maintaining quality control in order to obtain consistent hydraulic performance.

## The author

**Fred Buse** is Chief Engineer of the Standard Pump Aldrich Div. of Ingersoll-Rand Co. Holder of seven U.S. patents on centrifugal and reciprocating pumps, he was named 1976 Man of the Year for Hydraulic Institute, and is a member of the Hydraulic Institute, ANSI B.73 for centrifugal pumps, API Task Force for Reciprocating Pumps, and the ASME-Hydraulic Institute test code committee. A graduate of the New York State Maritime College (1958) with a B.S. in Marine Engineering, he has contributed to handbooks on pumps, and lectured for ASME.

# Inert gas in liquid mars pump performance

Here is a method for evaluating the effects of dissolved inert gas on the suction requirements of centrifugal pumps in order to obtain desired capacity and minimize or avoid mechanical damage.

*W. Roy Penney, Monsanto Co.*

☐ Suction requirements for centrifugal pumps are determined on the basis that the calculated net positive suction head available, $(NPSH_a)$, from Eq. (1) is greater than the experimental net positive suction head required, $(NPSH_r)$, from the pump performance curves.

To account for the effect of inert, dissolved gases on $(NPSH_a)$, two methods are common:

*Method 1*—The dissolved gases are ignored and the vapor pressure of the pure liquid is used. This is the textbook approach and is often used in industry.

*Method 2*—The vapor-pressure term in Eq. (1) is taken as the pressure at which the liquid is saturated with any inert gas. Some designers, notably engineering contractors, use this procedure because it prevents any dissolved gas from flashing. Consequently, it is very conservative and often very expensive.

A centrifugal pump will tolerate a reasonable amount of inert gas. In this article, we will attempt to develop a rational design method that is intermediate to the extremes of the two common methods. The overall approach for this rational method is to:

1. Predict the volume fraction of flashed inert gas within the lowest pressure regions of the pump as a function of system and operating parameters.
2. Recommend limits on the amount of flashed inert gas that a pump can handle without a significant decrease in performance.
3. Develop a design method that limits the amount of flashed inert gas to acceptable limits.

## Minimum pressure within pump

With reference to Fig. 1, the calculated $(NPSH_a)$ is defined as:

$$(NPSH_a) = \pi_{hs} + H - h_f - P_2^* \quad (1)$$

To prevent cavitation within the pump, the safe, minimum pressure (within the lowest pressure regions of the pump) should be equal to the vapor pressure, $P_2^*$, in Eq. (1) when $(NPSH_a)$ is equal to $(NPSH_r)$. Thus, it is reasonable to assume that:

$$\pi_m = P_2^* \quad (2)$$

when $$(NPSH_a) = (NPSH_r) \quad (3)$$

Originally published July 3, 1978

Substituting Eq. (2) and (3) into Eq. (1), we obtain a calculated minimum pressure that is safe and suitable for design purposes:

$$\pi_m = \pi_{hs} - (NPSH_r) + H - h_f \quad (4)$$

## Volume fraction of flashed gas

Considering the overall accuracy of the methods developed here, it is reasonable to make the following assumptions:

1. Perfect gas law applies.
2. Raoult's law applies for all components in the liquid phase, except inert gases:

$$P_i = P_i^* x_i \quad (5)$$

3. Dalton's law applies:

$$y_i = P_i/\pi \quad (6)$$

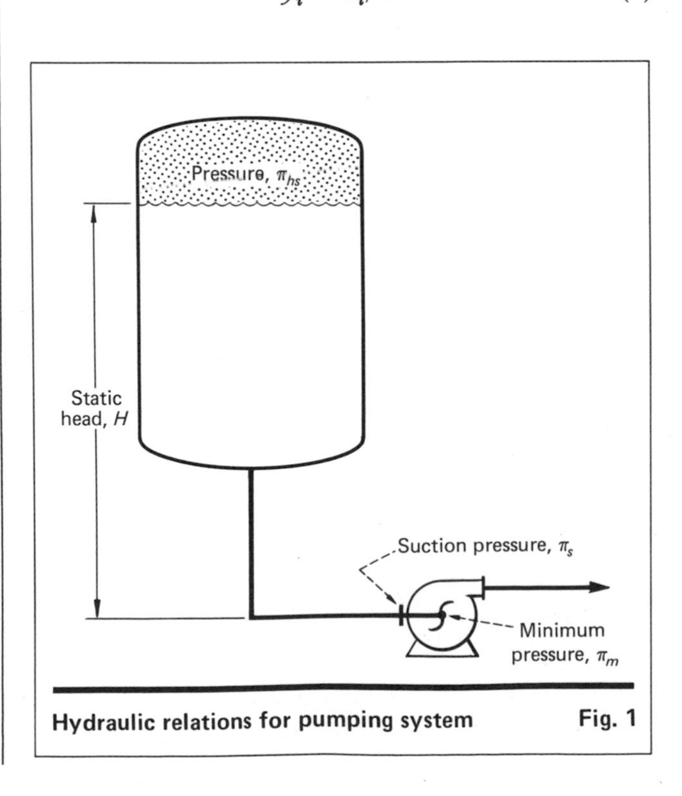


Hydraulic relations for pumping system — Fig. 1

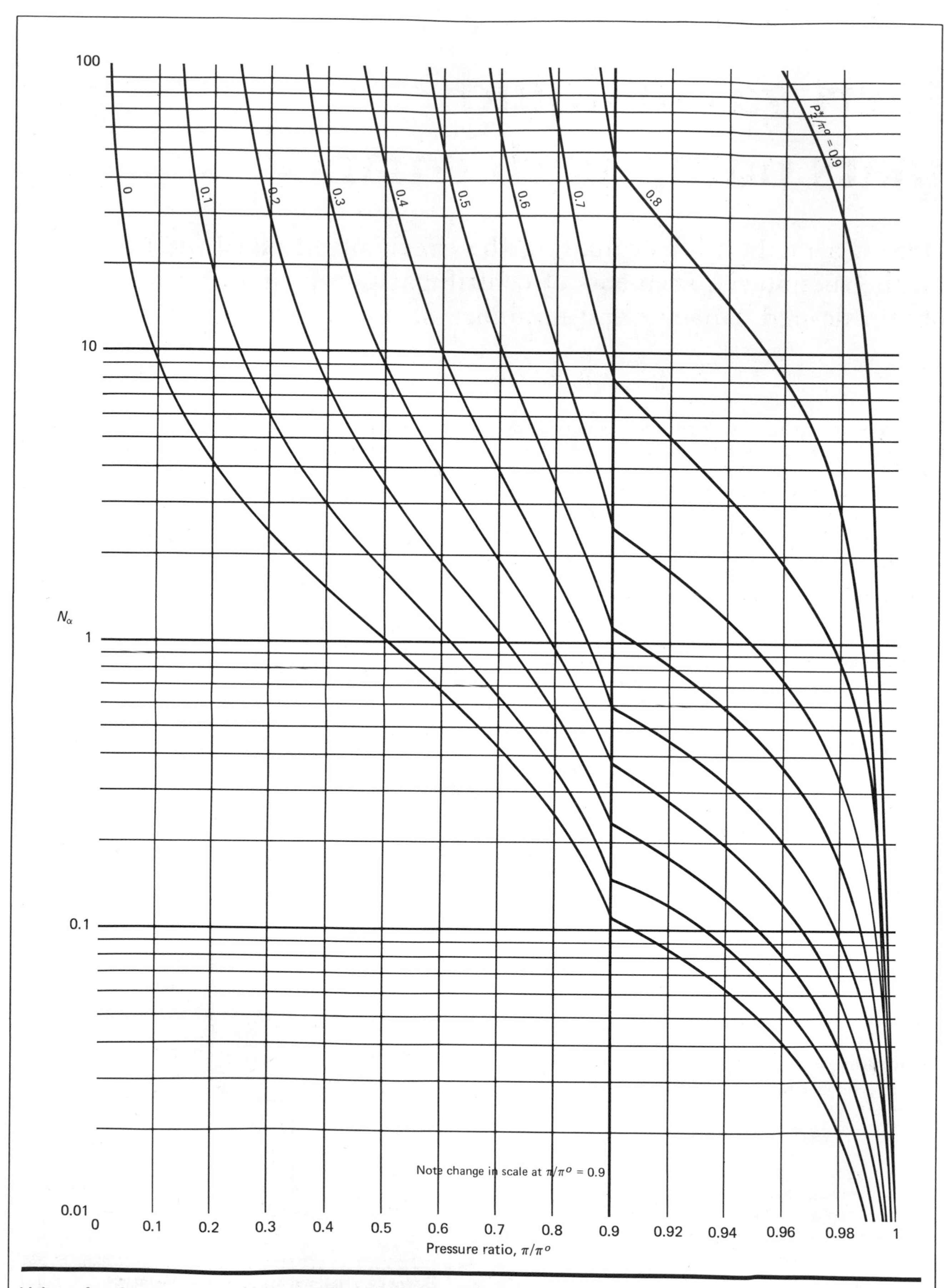


**Volume-fraction vapor in a liquid containing a dissolved, inert gas** **Fig. 2**

## Nomenclature

| | |
|---|---|
| $h_f$ | Friction loss in inlet piping to pump, ft-lb$_f$/lb$_m$ |
| $H$ | Static head of liquid from free liquid surface to centerline of pump suction, ft-lb$_f$/lb$_m$ |
| $H^*$ | Henry's law constant, Eq. (7) |
| $N_\alpha$ | $[\alpha/(1-\alpha)]/[W_1^o\rho_L/\rho_{1G}^o]$ |
| $(NPSH)$ | Net positive suction head, ft-lb$_f$/lb$_m$ |
| $(NPSH_a)$ | Calculated net positive suction head available, ft-lb$_f$/lb$_m$ |
| $(NPSH_r)$ | Experimental net positive suction head required from pump performance curve, ft-lb$_f$/lb$_m$ |
| $P_i$ | Partial pressure of Component $i$ at pressure $\pi$, ft-lb$_f$/lb$_m$ or atm |
| $P_1$ | Partial pressure of Component 1 at pressure $\pi$, ft-lb$_f$/lb$_m$ or atm |
| $P_1^o$ | Partial pressure of Component 1 at pressure $\pi^o$, ft-lb$_f$/lb$_m$ or atm |
| $P_i^*$, $P_2^*$ | Vapor pressure of Component $i$ and the liquid component, respectively, at operating temperature, ft-lb$_f$/lb$_m$ or atm |
| $V_{1G}$ | Ratio of volume occupied by the flashed inert gas to the liquid volume |
| $V_{2G}$ | Ratio of volume occupied by the vaporized liquid to the liquid volume |
| $V_L$ | Ratio of liquid volume after flashing to liquid volume before flashing (assumed $\cong 1$) |
| $V_T$ | Defined by Eq. (14) |
| $W_1^o$ | Weight fraction of dissolved inert gas at pressure $\pi^o$ |
| $W_1$ | Weight fraction of dissolved inert gas at pressure $\pi$ |
| $x_i$ | Mole fraction of Component $i$ in liquid phase |
| $y_i$ | Mole fraction of Component $i$ in vapor phase |
| $Z_1$ | Weight fraction of flashed inert gas at pressure $\pi$ |
| $\alpha$ | Volume fraction of flashed gas as pressure is lowered over a liquid saturated with a dissolved inert gas |
| $\pi$ | Pressure, ft-lb$_f$/lb$_m$ or atm |
| $\pi^o$ | Pressure at which the liquid is saturated with the inert gas (used as reference pressure), ft-lb$_f$/lb$_m$ or atm |
| $\pi_{hs}$ | Pressure in head space of suction reservoir, ft-lb$_f$/lb$_m$ or atm |
| $\pi_m$ | Minimum pressure in the lowest pressure regions of the pump, back-calculated from measured $(NPSH_r)$, ft-lb$_f$/lb$_m$ or atm |
| $\pi_s$ | Pressure at suction flange of pump, ft-lb$_f$/lb$_m$ or atm |
| $\rho_L$ | Liquid density, lb$_m$/ft$^3$ |
| $\rho_{1G}^o$ | Density of the inert gas at the reference pressure, $\pi^o$, and the operating temperature, lb$_m$/ft$^3$ |

4. Henry's law applies for the inert gas dissolved in the liquid:

$$P_i = H^* x_i \tag{7}$$

5. Isothermal conditions.

For highly nonideal gases, where large errors could occur by using these assumptions, the analysis can be done with more-accurate vapor-liquid equilibria.

Here, we shall consider only two components—an inert gas (Component 1) in a pure liquid (Component 2). The multicomponent case is an extension of the binary one.

To start the analysis, let us assume we know the weight fraction of dissolved gas entering the suction line to the pump. The saturation pressure, corresponding to this saturation condition, is the reference pressure, $\pi^o$. Partial pressure of the inert gas at the reference condition is:

$$P_1^o = \pi^o - P_2^* \tag{8}$$

Weight fraction of dissolved gas at any other partial pressure is:

$$W_1 = W_1^o(P_1/P_1^o) \tag{9}$$

Weight fraction of flashed gas at $P_1$ is:

$$Z_1 = W_1^o - W_1 = W_1^o[1 - (P_1/P_1^o)] \tag{10}$$

The ratio of volume occupied by the flashed inert gas to the liquid volume is given by:

$$V_{1G} = Z_1(\rho_L/\rho_{1G}) \tag{11}$$

In Eq. (11), we are assuming the liquid volume is not significantly decreased as the inert gas flashes.

Liquid density is assumed constant. Inert-gas density can be expressed in terms of the reference condition:

$$\rho_{1G} = \rho_{1G}^o(P_1/\pi^o) \tag{12}$$

Since the gas phase will be saturated with vaporized liquid, the volume fraction of the gas phase that is occupied by the vaporized liquid is given by:

$$V_{2G}/(V_{1G} + V_{2G}) = y_2 = P_2^*/\pi \tag{13}$$

Total volume occupied by the liquid, inert gas and vaporized liquid ratioed to the original liquid is:

$$V_T = V_{1G} + V_{2G} + V_L \approx V_{1G} + V_{2G} + 1 \tag{14}$$

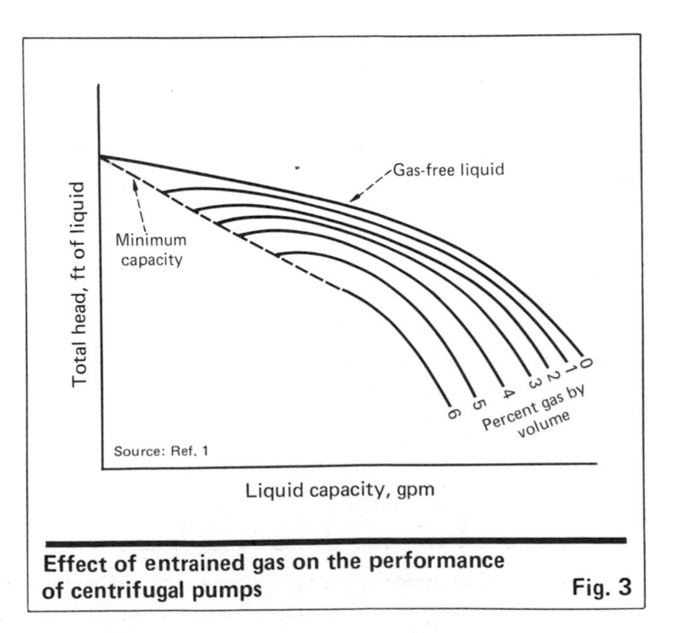


**Effect of entrained gas on the performance of centrifugal pumps** **Fig. 3**

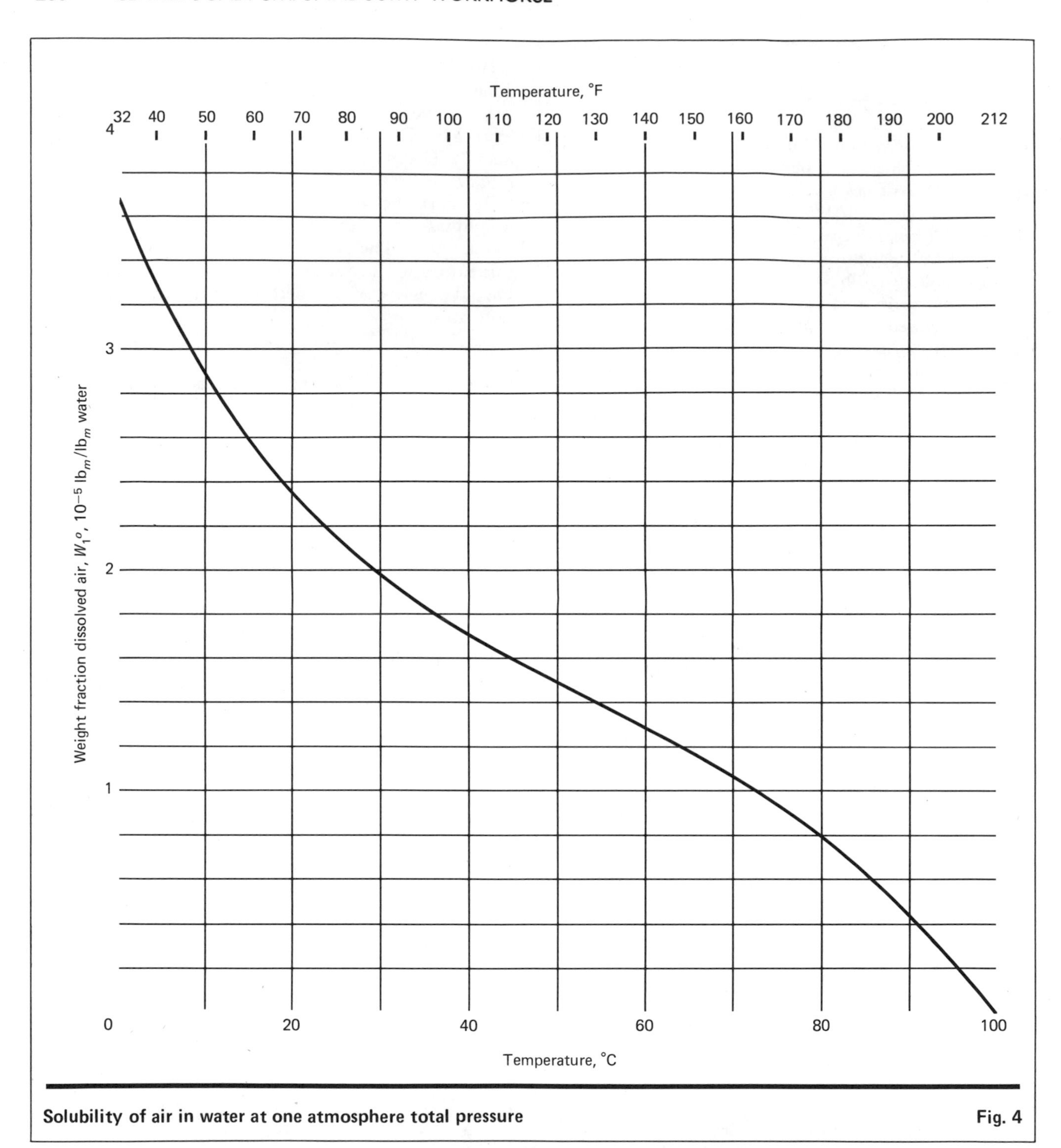


**Solubility of air in water at one atmosphere total pressure** **Fig. 4**

Eq. (8) through (14) can be solved for the volume fraction of the gas phase, $\alpha$, as the pressure, $\pi$, is reduced from the reference pressure $\pi^o$. After performing the several substitutions and rearranging, we find the volume fraction, $\alpha$, as:

$$\alpha = \frac{1}{\dfrac{\left(\dfrac{\pi}{\pi^o} - \dfrac{P_2{}^*}{\pi^o}\right)^2 \left(1 - \dfrac{P_2{}^*}{\pi^o}\right)}{\left(\dfrac{W_1{}^o \rho_L}{\rho_{1G}{}^o}\right)\left(\dfrac{\pi}{\pi^o}\right)\left(1 - \dfrac{\pi}{\pi^o}\right)} + 1} \quad (15)$$

Alternatively, we can express Eq. (15) in terms of $N_\alpha$:

$$N_\alpha = \frac{(\pi/\pi^o)[1 - (\pi/\pi^o)]}{[(\pi/\pi^o) - (P_2{}^*/\pi^o)]^2[1 - (P_2{}^*/\pi^o)]} \quad (16)$$

Eq. (16) is plotted parametrically in Fig. 2, which is very convenient to use for design purposes.

## Effect of entrained gas

The effect of entrained gases on centrifugal-pump performance has been reported [1,2], and is illustrated in Fig. 3 in relation to capacity and head. The maximum amount of inert gas should be 3% by volume.

Dissolved gases do not flash from solution instantaneously, and their volume fraction is not uniform throughout the pump. The recommended 3% by volume is for a constant fraction of inert gas entering and leaving the pump. Thus, the 3% by volume as a maximum for "flashed" dissolved gas should be more conservative than 3% by volume for entrained gas.

## Computations for "flashed" dissolved gas

Even though the 3% by volume as a maximum amount of flashed gas will not significantly affect the pump's performance initially, mechanical damage may gradually occur. This eventually will drastically affect the hydraulic performance.

*Example 1*: Cooling-tower pump—The pump in the system handles water at 80°F, which is saturated with air. Other pertinent data are:

| | |
|---|---|
| Flow, gpm | 25,000 |
| $\pi_s$, in. Hg | 28.5 |
| $\pi_s$, ft-lb$_f$/lb$_m$ | 32.2 |
| $(NPSH_r)$, ft-lb$_f$/lb$_m$ | 22.0 |
| $P_2^*$, ft-lb$_f$/lb$_m$ | 0.8 |
| $(NPSH_a) = \pi_s - P_2^*$, ft-lb$_f$/lb$_m$ | 31.4 |

Note that $\pi_s$ is the pressure measured at pump suction, and $(NPSH_r)$ is obtained from manufacturers' performance curves.

Initially, the performance of the pump was essentially as expected [$(NPSH_a)$ is 43% greater than $(NPSH_r)$]. However, the pump very distinctly produced the gravelly noise that is typical of cavitation. Mechanical damage occurred to the impellers, and impeller life was about two years.

In the absence of dissolved gas, the $(NPSH_a)$ seems more than adequate. Let us determine the maximum volume-fraction of flashed gas that could exist in the lower pressure regions of the pump.

In order to calculate the volume-fraction, $\alpha$, we will use Eq. (15). From Fig. 4, we find the solubility of air in water at 1 atm total pressure and various temperatures. We then evaluate as follows:

$P_2^* = 0.023$ atm
$\pi^o = 1$ atm
$P_2^*/\pi^o = 0.023$
$\pi = \pi_m = [\pi_s - (NPSH_r)]$, or
$\pi = 0.93 - 0.65 = 0.28$ atm
$\pi/\pi^o = 0.28$
$W_1^o = 0.000021$
$\rho_L = 61.2$ lb$_m$/ft$^3$
$\rho_{1G}^o = 0.072$ lb$_m$/ft$^3$
$(W_1^o \rho_L/\rho_G^o) = 0.018$

Substituting the various terms into Eq. (15), we find that $\alpha = 0.053$, or 5.3%.

The dissolved air, flashing in the pump, was very likely responsible for the mechanical damage. We do not know the amount of flashing that could be tolerated without significant mechanical damage. The 3% by volume, recommended here, is a reasonable design maximum. The suction pressure for these pumps would have to be increased about 3.5 ft of liquid in order to reduce $\alpha$ from 5% to 3%. We do not know how much the mechanical damage would have decreased because it cannot be predicted. To be absolutely sure that no such damage would occur, the pump must be designed for zero inert-gas flashing.

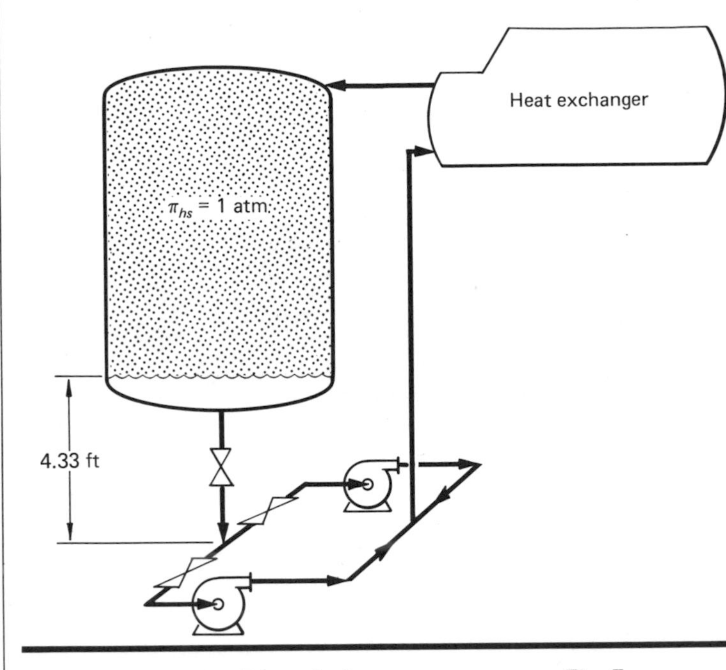

Pumping system for Example 2 Fig. 5

*Example 2*: Process pump—The pumping system (shown schematically in Fig. 5) handles a chlorinated hydrocarbon liquid of density 1.3 g/cm$^3$ and viscosity of 10 cP. The amount of dissolved inert gas (HCl) was measured at atmospheric pressure as 0.5% by weight. The vapor pressure of the liquid component is approximately zero, i.e., $P_2^* \cong 0$.

The pumps are Durco, Type AVS, size 4 × 3—10; and the performance curves for the pump are shown in Fig. 6. Suction piping is 4-in. Schedule 40 pipe, with a total length of 10 ft. Gate valves are used in the suction line. Other pertinent data are shown in Fig. 5.

This system was originally designed for a flow of 350 gpm, producing a velocity of 9 ft/s in the 4-in. suction line. From Fig. 6, we get $(NPSH_r)$ as 3.8 ft-lb$_f$/lb$_m$ for the 8.5-in. impeller at a flow of 350 gpm. The friction loss in the suction piping, $h_f$, was calculated as 3.9 ft-lb$_f$/lb$_m$. By substituting appropriate values into Eq. (4), we obtain $\pi_m$ as:

$$\pi_m = 24.2 - 3.8 + 4.33 - 3.9 = 20.8$$

To calculate $\alpha$, we enter Fig. 2 at $\pi/\pi^o = 20.8/24.2 = 0.86$ and at the parameter for $P_2^*/\pi^o = 0$ get $N_\alpha = 0.165$.

By definition:

$$N_\alpha = \frac{\alpha/(1-\alpha)}{W_1^o \rho_L/\rho_{1G}^o}$$

Substituting the appropriate values in the denominator yields:

$$\frac{W_1^o \rho_L}{\rho_G} = \frac{0.005(87)}{0.096} = 4.5$$

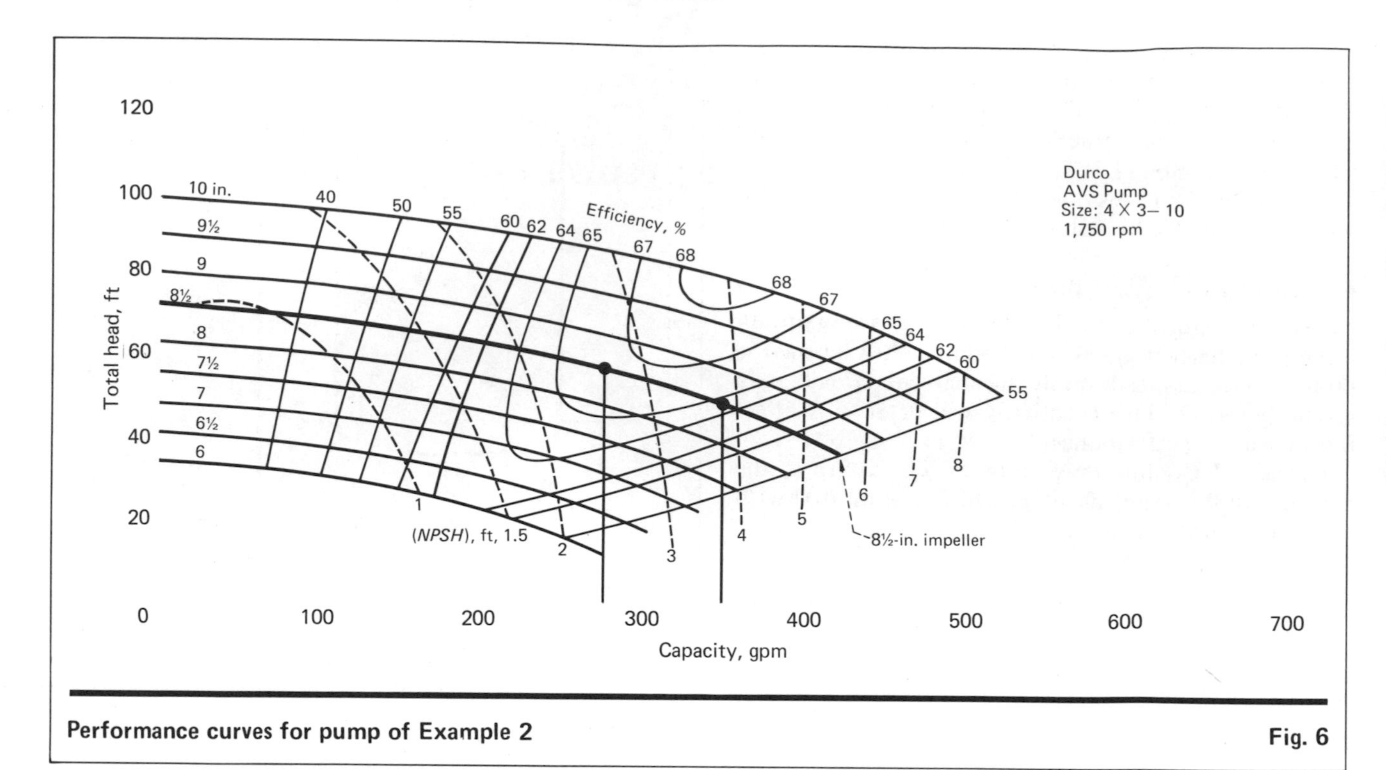


Performance curves for pump of Example 2

Fig. 6

Since $N_\alpha = 0.165$ for this problem:

$$\alpha/(1-\alpha) = 0.165(4.5) = 0.75$$

$$\alpha = 0.75/1.75 = 0.4, \text{ or } 40\%$$

This value of $\alpha$ is too high. We would expect the performance of the pump to suffer drastically with this amount of gas in the lowest pressure region of the pump.

Now, let us calculate the $\pi_m$ necessary to produce 3% by volume of gas in the lowest pressure region. With $\alpha = 0.03$:

$$N_\alpha = 0.03/(0.97 \times 4.5) = 0.007$$

From Fig. 2, we obtain $\pi/\pi^o = 0.992$. For practical purposes, we must design for $\pi_m$ not to drop below $\pi^o$. To accomplish this without decreasing the flow would have required increasing the liquid level in the vessel by 3.4 ft, i.e., (24.2 − 20.8). We could not increase the liquid level, so we actually designed the pumping system with the minimum $H = 5.33$ ft, and with a maximum flow of 275 gpm. With this flow, $\pi_m$ is 24.9 ft-$lb_f/lb_m$. Thus, the minimum pressure in the pump will exceed the saturation pressure by about one ft of liquid. We expect satisfactory performance from the pump.

This example illustrates the problems encountered when pumping liquids containing very-soluble inert gases. In such systems, either the amount of dissolved gas should be measured, or the conservative approach should be taken by not allowing $(NPSH_a)$ to drop below the saturation pressure of the inert gas.

## Recommended design methods

- If mechanical damage is of significant concern, or the solubility of the inert gas is unknown, and the conservative approach is desired, replace $P_2$* with $\pi^o$ in Eq. (1). That is, by not allowing the pressure in the lowest pressure region of the pump to go below the saturation pressure of the inert gas, no flashing will occur. Mechanical damage will certainly be of concern if corrosion-erosion is involved.
- If mechanical damage is of little concern, then calculate $(NPSH_a)$ from:

$$(NPSH_a) = \pi_{hs} + H - h_f - \pi \qquad (17)$$

- Determine $\pi$ from Fig. 2 by the following method:
  a. Calculate $P_2/\pi^o$.
  b. Obtain $W_1{}^o$, and calculate $W_1{}^o\rho_L/\rho_{1G}$.
  c. Calculate $N_\alpha$ at $\alpha = 0.03$ (or any $\alpha$ deemed suitable), i.e., 3% by volume as the maximum flashed gas.
  d. Determine $\pi/\pi^o$ from Fig. 2, and calculate $\pi$.
  e. Select a pump such that $(NPSH_a)$ from Eq. (17) is greater than $(NPSH_r)$ from manufacturers' performance curves.

## References


1. Doolin, J. H., *Chem. Eng.*, Jan. 7, 1963, p. 103.
2. Stepanoff, A. J., "Centrifugal and Axial Flow Pumps," 2nd ed., p. 230, Wiley, New York, 1957.


## The author

**W. Roy Penney** is an engineering manager in the Corporate Engineering Dept., Monsanto Co., 800 N. Lindbergh Blvd., St. Louis, MO 63166. His industrial experience includes service with Phillips Petroleum Co. as a heat-transfer specialist and several years at Monsanto as a fluid-mechanics specialist. He has a B.S. and an M.S. in mechanical engineering from the University of Arkansas and a Ph.D. in chemical engineering from Oklahoma State University.

# ESTIMATING MINIMUM REQUIRED FLOWS THROUGH PUMPS

F. CAPLAN, P.E.
Oakland, Calif.

Near the shutoff point, the efficiency of a centrifugal pump is almost zero, and almost all of the pump energy goes into heating the liquid in the pump. Unless a minimum flow passes through the pump to carry away the heat, overheating will cause liquids with low vapor pressure to boil, and the pump can seize. One common method of ensuring a minimum flow is to put a restriction orifice in a line leading from the pump discharge through some heat-dissipation device to the suction.

If the vendor's pump curve is available, with pump efficiency and required NPSH, one can accurately determine the pump minimum flow required to prevent overheating. In the absence of such data, one can estimate the required flow, and square-edged orifice size, in terms of the specified design head and flow, as follows:

$$Q_m = 6.1 \times 10^{-5}(Q_d H_d) \quad (1)$$

$$d_o = 0.282(Q_m^{1/2}/H_d^{1/4}) \quad (2)$$

where: $Q$ = flow, gpm
$H$ = head, ft
$d_o$ = square-edged orifice diameter, in.
$m, d$ = subscripts indicating minimum and design, respectively.

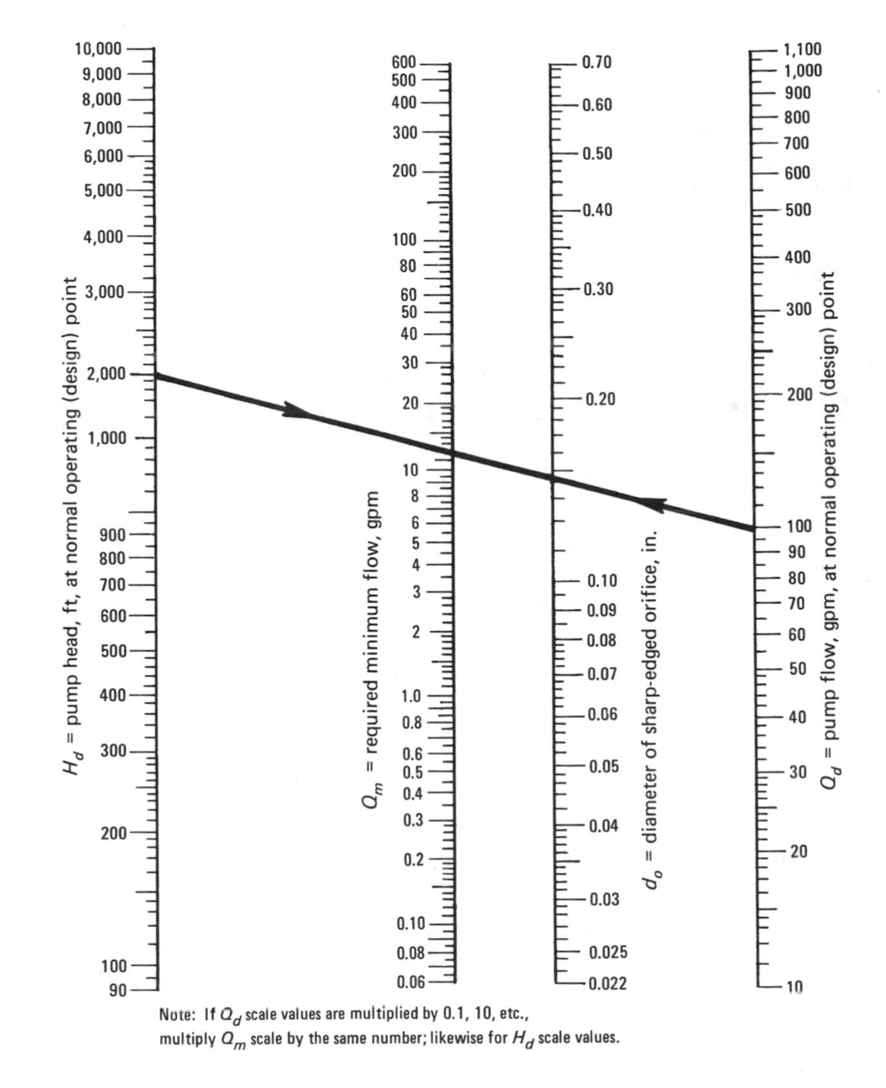


Eq. (1) is derived from the following assumptions: (1) Near shutoff, the pump brake horsepower is one-half that at the design point. (2) The loss in efficiency heats the pumped liquid no more than 15°F. Thus:

$$0.5(\text{bhp})_d(2545 \text{ Btu/hp}) = Q_m(\text{sp gr})(\text{gpm } 500 \text{ lb/h})C_p(15°\text{F}) \quad (3)$$

Also:

$$(\text{bhp})_d = (Q_d)(H_d)(\text{sp gr})/3960(e_d) \quad (4)$$

Where: sp gr = liquid specific gravity
$C_p$ = heat capacity, btu/lb (assumed = 1.0)
$e_d$ = pump efficiency at design point, expressed as a decimal and assumed as 0.705

Combining Eq. (3) and (4) and solving for $Q_m$ gives Eq. (1).

Eq. (2) is derived from the following assumptions: (1) Flow through a sharp-edged orifice is $Q_m = 19.64 (d_o^2 C_d H_s^{1/2})$. (2) The coefficient of discharge, $C_d = 0.61$. (3) The head at shutoff, $H_s = 1.1H_d$. Substituting these assumed values and solving for $d_o$ gives Eq. (2).

The nomograph permits a rapid, simultaneous solution of both equations. *Example:* What minimum flow and orifice diameter are required if the design conditions are 100 gpm and 2,000 ft of head? Align $H_d = 2{,}000$ with $Q_d = 100$, then read $Q_m = 12.2$ gpm and $d_o = 0.147$.

This indicates that a line from the pump discharge through a 0.147-in orifice to the suction would automatically protect it against minimum flow. This line, which would be about 1 in diameter, should also have a small exchanger in it, either field fabricated or purchased for the purpose.

Originally published March 17, 1975

# Specifying Centrifugal and Reciprocating Pumps

Maximum suction pressure for centrifugal pumps, minimum differential pressure and maximum discharge pressure for reciprocating pumps, and available net positive suction head for either, require different procedures for selecting the appropriate design parameters of each type.

UDAY S. HATTIANGADI, Continental Oil Co.

Most articles are notably silent on procedures to be followed once a pump is selected for a particular application. Very often a pump performs satisfactorily even when incorrectly specified; and the design engineer concludes that the procedure used for the specification is correct. Only in those instances when the pump does not perform as expected is the specification procedure reexamined and reevaluated in order to find the probable cause of difficulty.

In this article, I will review the principles of operation of two types of pumps commonly used in the chemical process industries: the centrifugal and crank-driven reciprocating pumps. I will also illustrate how problems in specifying these pumps may be handled. The tendency to treat both types in the same manner (when analyzing their operating criteria) can lead to serious errors.

## SPECIFYING CENTRIFUGAL PUMPS

The following are normally specified for a centrifugal pump:

1. Temperature of fluid to be pumped.
2. Fluid properties such as specific gravity and viscosity.
3. Flowrate at which the pump is expected to operate.
4. Differential pressure at which the pump is expected to operate.
5. Maximum suction pressure.
6. Net positive suction head available.

The reasons for the first four items and the way they are specified are straightforward and will not be covered here. Only the last two will be considered.

### Maximum Suction Pressure

The maximum suction pressure of a centrifugal pump is given to enable the vendor to design the pump casing. Depending on flowrate, a centrifugal pump creates a certain differential between its suction and discharge ends. This differential is maximum at no flow. The maximum pressure to which the casing will be subjected can be determined by adding the differential at no flow to the maximum suction pressure. It is for this reason that this pressure has to be specified for every centrifugal pump.

The maximum suction pressure should be the maximum pressure that the pump may be subjected to at any time. For example, if the pump is used to transfer the contents of a tank, the maximum suction pressure will be the pressure at which the safety relief valve on the tank is set plus the pressure due to the head of liquid in the tank.

Originally published February 23, 1970

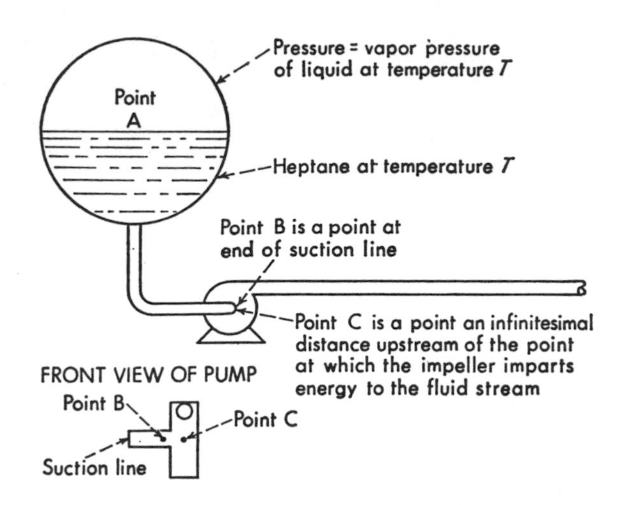


**FLOW** relations in liquid system and in centrifugal pump establish available and required NPSH—Fig. 1

## Net Positive Suction Head

The concept of net positive suction head (NPSH) is based on the thermodynamic principle that if the static pressure on a liquid is higher than its vapor pressure at the temperature, no vapor can exist. However, if the pressure drops below the vapor pressure of the liquid at the pumping temperature, some flashing occurs. If this happens in the suction of a centrifugal pump, the vapors are returned to the liquid state when the pump increases the pressure of the stream, and cavitation occurs.

Consider the case of a centrifugal pump being used to transfer a pure liquid (heptane) from a closed storage vessel (Fig. 1).

The pressure in the vessel will be the vapor pressure of heptane at the vessel temperature. Consider three points in this system: Point A on the liquid surface, Point B in the suction line just before it enters the pump, and Point C in the pump. We know that the total head at B is equal to the total head at A less the frictional drop in the suction line. The total head at any point is the sum of the elevation, static pressure and velocity heads.

Now let us see what happens inside the pump. The fluid enters with the total head corresponding to the head at B. Some of this total head is lost to friction in going into the pump. If C is a point in the pump an infinitesimal distance upstream of the point at which the impeller first starts imparting energy to the fluid stream, then the total head at C is equal to the total head at B less the frictional

## Basic Operating Principles of Centrifugal Pumps

A centrifugal pump consists of an impeller rotating at fairly high speed in a casing. Liquid is fed through the pump-suction line to the eye of the impeller, and is forced outward by centrifugal action. Capacity of the pump depends on the differential head against which it has to act. A typical pump characteristic (head-capacity) curve is:

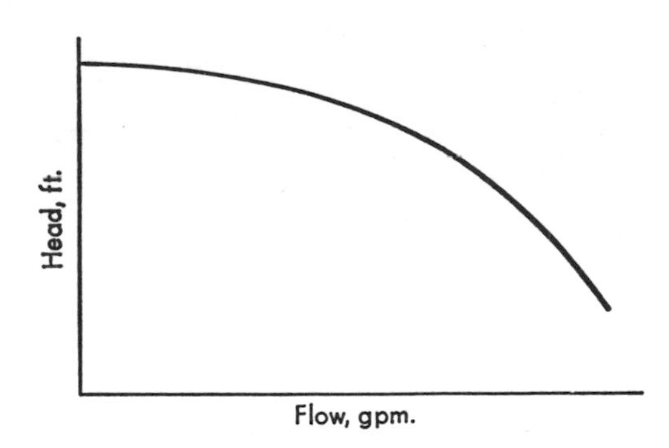


The centrifugal pump can operate at any point along its characteristic. The exact point is determined by the system in which the pump is installed.

Consider a simple system where a centrifugal pump is used to pump the contents of a tank to the atmosphere through a long line and a throttling valve. Assume also that either the cross-section of the tank is fairly large so that the decrease in liquid level at any time is small, or that the tank is continually filled to maintain a constant liquid level.

In this case, the system consists merely of a line with a valve, and it is possible to construct a system head-curve for it. If velocity heads for the system can be neglected, the only significant factor is the frictional drop in the line. As flowrate increases, frictional drop increases. The system head-curve would be a plot of the flowrate vs. the pressure (converted to ft. of liquid). If the valve is throttled, another system head-curve is obtained. Further throttling yields yet another. At total shutoff, the system head-curve coincides with the vertical axis. These relations for pump and system are shown in the following diagram:

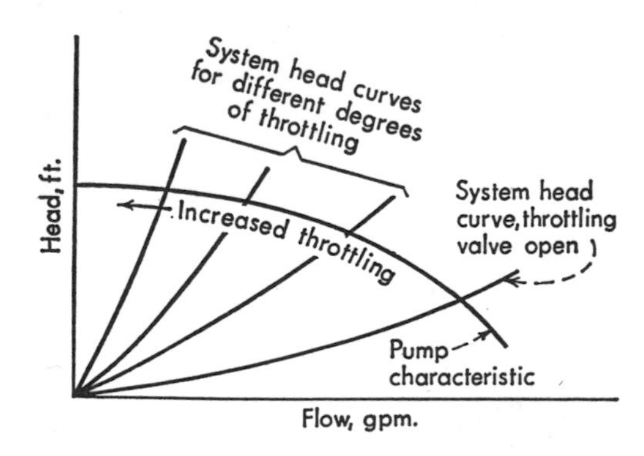


The operating point of the centrifugal is determined by the point of intersection of the system head-curve with the pump's characteristic curve. Thus, the flowrate of the centrifugal pump can be varied by a procedure as simple as throttling its discharge. The pump also reaches its maximum head at shutoff conditions.

head loss from B to C. In general, the velocity at C will also be much higher than at Point B. Since the total head at C is fixed, an increase in the velocity head at C means a corresponding decrease in pressure head at that point. If this decrease is enough to cause the pressure at Point C to fall below the vapor pressure of heptane, vaporization will occur.

The total head at B minus the vapor pressure of the liquid at the pump temperature is called the NPSH available. The frictional head loss from B to C plus the velocity head at Point C is referred to as the NPSH required. The NPSH available has to be calculated and specified by the design engineer. The NPSH required depends on the pump design and has to be specified by the pump vendor.

As long as the NPSH available is greater than the NPSH required, the pressure at Point C will be greater than the vapor pressure of heptane at the pumping temperature, and no vaporization will occur in the pump. The following analysis summarizes these relationships:

$$H_C = H_B - h_{BC} \quad (1)$$

$$p_C = H_C - V_C \quad (2)$$

Substituting Eq. (1) into Eq. (2) gives:

$$p_C = H_B - h_{BC} - V_C \quad (3)$$

By definition, NPSH available and required are:

$$s_a = H_B - p_v$$

$$s_r = h_{BC} + V_C$$

Therefore, Eq. (3) can be written as:

$$p_C = s_a - s_r + p_v \quad (4)$$

In the above equations, $H_C$ is total head at $C$, $H_B$ is total head at $B$, $h_{BC}$ is frictional head loss from B to C, $p_C$ is pressure head at C, $V_C$ is velocity head at C, $s_a$ is NPSH available, $s_r$ is NPSH required, and $p_v$ is vapor pressure of heptane at pumping temperature corrected to feet of heptane.

Unfortunately, the required NPSH cannot be calculated directly. It is determined by the pump vendor performing tests using pure deaerated water. The tests normally consist of (a) operating the pump on water and throttling its suction until vaporization in the pump (as evidenced by a drop in capacity) is observed, (b) calculating the available NPSH at this point, and (c) using the relation that required NPSH equals available NPSH at the point of incipient vaporization.

Since pump capacity does not drop sharply until a considerable amount of vaporization takes place, it is not possible to determine the point of incipient vaporization and, hence, the exact required NPSH.

Therefore, for the rest of the discussion, I shall assume that when the available NPSH is equal to the required NPSH (as determined by the water test), the pump is at the point of incipient cavitation. I shall also assume that the required NPSH to prevent any vaporization does not change with the liquid. This second assumption is reasonable because it is the velocity-head term that is the major contributor to the NPSH required by the pump.

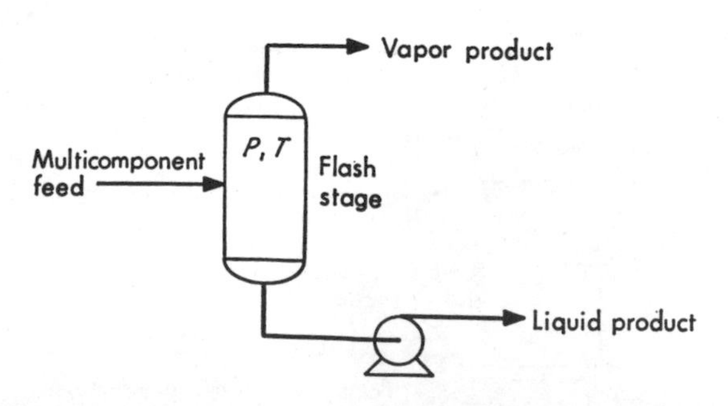


**MULITCOMPONENT** feedstream makes the specification of available NPSH more difficult—Fig. 2

If these assumptions are made, it is possible to make a rough calculation on the volume percent vaporization that would occur in a pump when the required NPSH is greater than the available NPSH. To do this, the pressure head at C is found by subtracting the required NPSH from the total head at Point B. Once the pressure at Point C is known, an isenthalpic flash calculation can be made to determine the volume percent vaporized at Point C.

## Available NPSH for Mixtures

Specifying available NPSH for mixtures becomes a little more complicated. Consider the multicomponent feedstream entering a flash stage, as shown in Fig. 2.

Suppose that the pressure in the flash stage is $P$ and the temperature $T$, and suppose that a pump is used to remove the liquid from this flash stage. The question arises as to what is the correct vapor pressure for calculating the available NPSH.

At first glance, this seems like a straightforward problem. Since the flash stage is assumed to be an equilibrium flash stage, the vapor pressure of the liquid at the flash-stage temperature is equal to the pressure in the flash stage. If the components in the feed are close-boiling, it would seem logical to use the pressure of the flash stage as the vapor pressure of the liquid.

But what if the feed contains one component that is much lighter than the others? The amount of this light component in the liquid phase will be very small, yet enough to raise the vapor pressure of the liquid stream to the flash-stage pressure. Certainly it would be conservative to use the pressure of the flash stage as the vapor pressure of the liquid because any decrease in pressure below this value would result in some vaporization. However, most of the material vaporizing would be the light component; and since there is only a small amount of it in the liquid phase, excessive vaporization cannot occur.

An extreme condition arises when one of the components is a dissolved gas. Its presence in the liquid phase is very small; yet this amount can raise the "vapor pressure" of the liquid to an extremely high value.

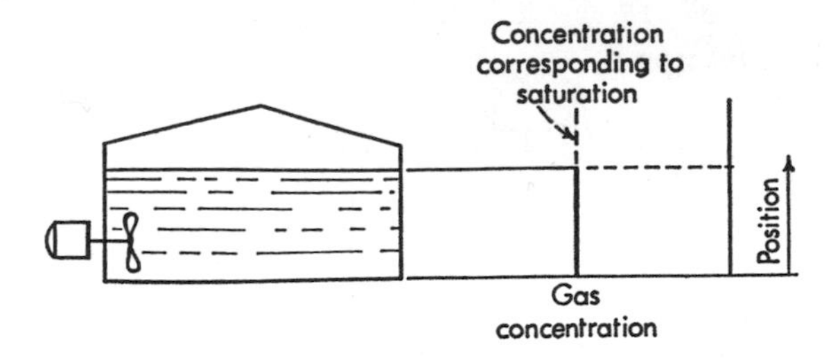


**DISSOLVED** gases add to the problem of specifying available NPSH when liquid in tank is agitated—Fig. 3

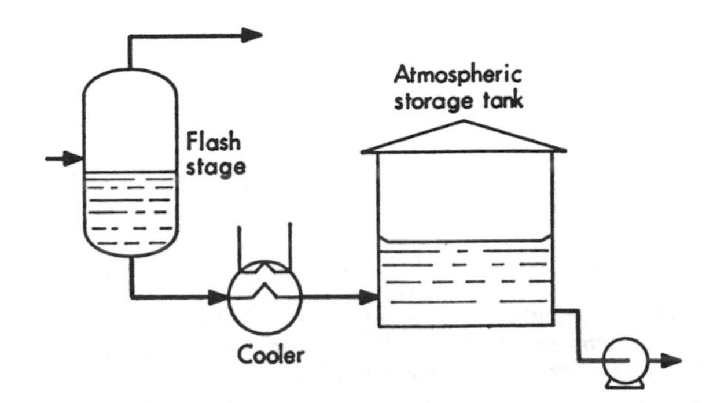


**LIQUID** stream may pick up light hydrocarbons or gases during its passage through processing steps—Fig. 4

The problem of dissolved gas occurs more often than might be expected. Consider an agitated storage tank that is used to store a hydrocarbon whose vapor pressure is less than atmospheric (Fig. 3).

Since the liquid contents of this tank are agitated, the liquid will eventually come to equilibrium with the vapor space above it. To do this, the liquid must absorb enough air to bring its vapor pressure up to 14.7 psia. Therefore, the design engineer is faced with the following dilemma:

Should the vapor pressure of the hydrocarbon for calculating the available NPSH be the vapor pressure of the pure hydrocarbon at the pumping temperature; or should the dissolved gas be included, and the vapor pressure assumed to be 14.7 psia.? To be sure, some of the air will be liberated when the pressure in the pumping system drops below 14.7 psia. Yet this amount liberated may be so small that it will not create a problem. In instances like this, it is clear that specifying "no vaporization will occur in the pumping system" may be a little too stringent, and may result in such a low value of specified available NPSH that an expensive pump will have to be purchased.

Whistler[1] states that most centrifugal pumps can handle about 2% by volume of vapor. This is probably as good a criterion as one is likely to get. On this basis, I suggest the following methods of specifying available NPSH for two types of mixtures.

For mixtures containing very light components or dissolved gases:

1. Calculate the available NPSH using the conservative method. For example: If the liquid is in an atmospheric tank and agitated, use 14.7 psia. as the effective vapor pressure.
2. Check with pump manufacturers' bulletins to see if it is possible to find a pump with a required NPSH lower than the available NPSH calculated in Step 1. If it is possible, then specify the available NPSH as calculated in Step1. If not, proceed to Step 3.
3. Calculate the amount of dissolved gas (or light components) that would be present in the liquid. To calculate the amount of dissolved gas, assume saturation.
4. Calculate the pressure at which the light components or dissolved gases would occupy a volume equal to 2% of the liquid stream at the known stream temperature. Assume that all light components (or dissolved gases) are in the vapor phase at this pressure. Convert pressure to feet of liquid.
5. Use this pressure rather than the vapor pressure of the liquid to compute and specify the available NPSH.

Note that saturation with air only becomes a problem when the tank is well agitated. In unagitated tanks, the top layer of liquid may become saturated. Since filling and emptying of such tanks is done from the bottom, the liquid is rarely saturated.

Whistler's criterion[1] of using a vapor pressure halfway between 14.7 psia. and the vapor pressure of the liquid is probably conservative enough. However, be sure to include any gases or light hydrocarbons that the stream may have picked up in the process when computing the vapor pressure of the liquid. For example, consider the system shown in Fig. 4 where liquid from a flash stage is cooled, and then sent to an atmospheric storage tank.

The vapor pressure of the liquid at storage-tank conditions can be obtained by assuming that its vapor pressure at flash-drum temperature is equal to the flash-drum pressure. Then extrapolate to the storage-tank temperature on a Cox chart to obtain the vapor pressure.

For mixtures containing a wide range of close-boiling components, the following procedure is perhaps warranted:

1. Calculate the available NPSH, as before, by the conservative method. If it is not too stringent, specify this value. If too stringent, proceed to Step 2.
2. Perform isenthalpic flash calculations on the liquid being pumped at different pressures. Select that pressure that yields 2% by volume of vapor.
3. Convert the pressure selected in Step 2 to feet of liquid. Use this pressure instead of the vapor pressure for specifying available NPSH.

## Comments on Suggested Methods

The procedures outlined are not thermodynamically exact. For example: whenever vaporization occurs,

## Operation of Reciprocating Pumps

A reciprocating pump consists of a piston moving up and down in a cylinder. Check valves are provided for suction and discharge. As the piston moves down, the suction valve opens and allows fluid to enter the cylinder. On each upward motion of the piston, the suction valve closes, the discharge valve opens, and fluid is ejected into the discharge line as shown:

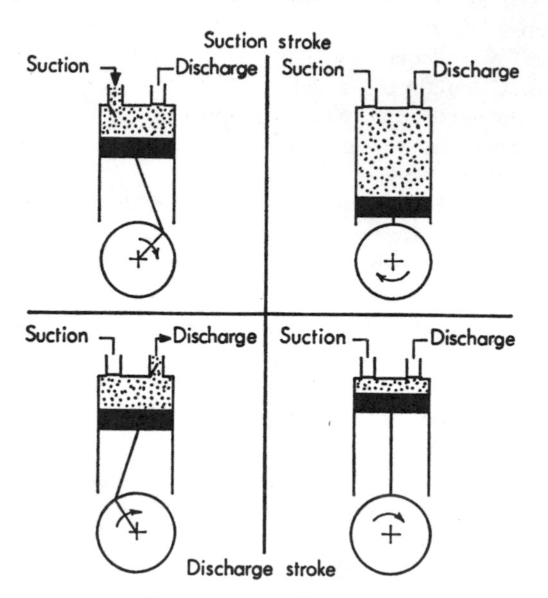


Since the cylinder fills on each suction stroke and empties on each discharge stroke, the volume of liquid moved by a reciprocating pump depends only on the pump characteristics: cylinder diameter, stroke length, and strokes per minute. A reciprocating pump puts out as much head as discharge conditions require until mechanical or other failure occurs.

Once a centrifugal pump stabilizes at its operating point, the flow through it is constant. In most instances, the piston of a reciprocating pump is moved by a crank operating at a constant angular velocity. Hence, the piston moves up and down in simple harmonic motion, and flow through this pump is pulsating.

The pulsating nature of the flow from reciprocating pumps is extremely important and must not be ignored when these pumps are specified.

### Flow Variation in Simplex Pumps

The most elementary reciprocating pump is a simplex type that consists of a single piston moving in a single cylinder. Such a pump is shown in the following diagram:

In the diagram, $L$ is stroke length, $A$ is cross-sectional area of cylinder, and $N$ is constant speed of rotation of crank. The position of the piston is shown a little after the pump begins its discharge stroke.

It is assumed that the length of the connecting rod is much larger than the radius of the crank circle, so that the change in piston position is not affected by the change in the length of the projection of the connecting rod on the axis of the cylinder.

The variation of flow with time for this type of pump may be derived from the preceding diagram, and is given by:

$$V = A(L/2)(1 + \cos 2\pi Nt) + V_c$$

where $V$ is volume of liquid in cylinder, $t$ is time, and $V_c$ is clearance volume.

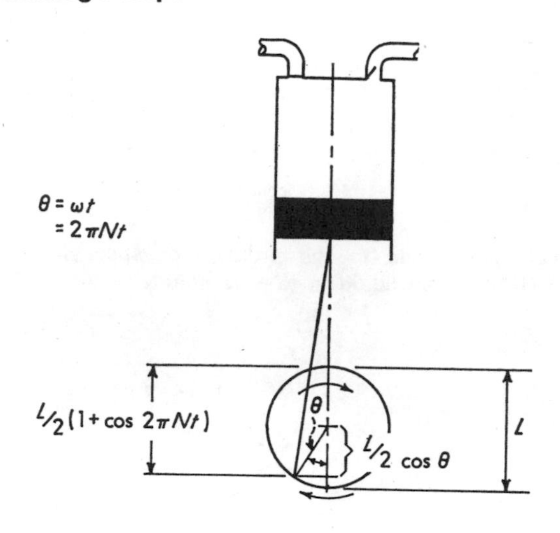


The instantaneous volumetric flowrate in the discharge line is:

$$dV/dt = -ALN \sin 2\pi Nt \tag{A}$$

Therefore, the volumetric flowrate in the discharge line during the discharge stroke of the simplex pump varies sinusoidally. Of course, during the suction stroke, there is no flow in the discharge line. Hence, flow from a simplex pump can be represented as shown by the first flow pattern in the accompanying diagram.

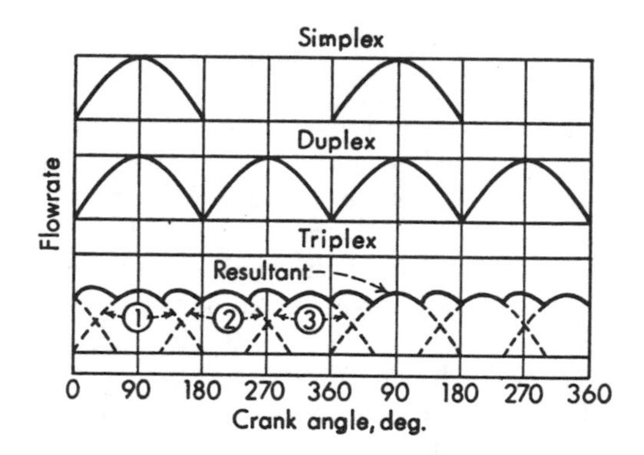


### Flow Variation in Multiplex Pumps

A duplex pump consists of two cylinder-and-piston arrangements driven by a common driver. The two pistons and two cylinders are so placed that when one combination is in its suction stroke, the other is in the discharge stroke (i.e., they are 180° out of phase). Therefore the flowrate in the discharge line of a duplex pump conforms to the second pattern shown in the diagram.

A triplex pump consists of three piston-and-cylinder combinations arranged so that they are 120° out of phase with each other. The variation in flowrate in the discharge line for each individual piston-and-cylinder combination is shown by the dotted lines in the diagram. The solid line shows the resultant total flow.

This analysis can be extended to pumps that contain four or more piston-and-cylinder combinations.

In the above discussion, the flow variations are for the discharge line. The suction line will also be subject to the same flow variations but these will be 180° out of phase from variations in the discharge line.

## Pulsating Flow Affects Specifications for Reciprocating Pumps

**The pulsating nature of flow from reciprocating pumps requires that they be specified in a different way than centrifugal pumps. Engineers often tend to overlook this difference and make calculations based on average velocity. This practice, while having no basis in theory, is usually nonconservative and leads to serious errors.**

**Using the average flowrate for calculating items such as available NPSH and maximum discharge pressure of reciprocating pumps is nonconservative because:**

**1. Flow to and from a reciprocating pump is always fluctuating. Hence, the liquid in the suction and discharge lines has to be continually accelerated and declerated. A pressure drop is required to provide this acceleration. This pressure drop, converted to feet of liquid, is often referred to as "acceleration head." The acceleration-head loss is extremely important because it is often the governing factor in determining items such as the available NPSH.**

**2. Peak flowrate is much greater than average flowrate. For example, in a simplex pump, the maximum flowrate is $\pi$ times the average. For a duplex pump, the maximum flowrate is $\pi/2$ times the average.**

**Fortunately, the maximum acceleration-head loss occurs at the point of minimum frictional loss. That this is true for a simplex pump can be seen easily from the following argument:**

**Flowrate for a simplex pump varies as the sine function. The acceleration at any point can be obtained by differentiating the instantaneous flowrate. Since the differential of the sine function is a cosine function, the acceleration will vary as the cosine function. Hence, the maximum value of the cosine function will occur at the same time as the minimum value of the sine function. Therefore, the acceleration-head loss will be maximum when the instantaneous volumetric flowrate is minimum. Since the frictional drop depends on the instantaneous flowrate, the acceleration head will be maximum when the frictional drop is minimum, and vice versa.**

**Except for very low-speed pumps, the maximum pressure drop in lines serving reciprocating pumps will be determined by the maximum acceleration-head loss rather than the maximum frictional drop. [See box on next page for development of acceleration-head relations.]**

there is an increase in the average stream velocity because the density of the vapor is less than that of the liquid. The increase in velocity has been neglected.

When engineers specify the available NPSH for a pump transferring the contents of an atmospheric storage tank, and base it on the vapor pressure of the pure liquid, they are allowing some evolution of dissolved gas in the pump. The procedures given here help the engineer check that this evolution is not excessive. I am not suggesting that pumps be designed to allow 2% vaporization. It is best to have none. Only when the available NPSH calculated by the conservative method yields such a low value that it is impossible to find a standard pump to fit is it necessary to allow some vaporization to occur.

Whenever vaporization or gas evolution is allowed to occur, there is the possibility of cavitation and resulting impeller damage. Cavitation is more pronounced where there is vaporization because vapor bubbles collapse readily. Evolution of dissolved gas may not create much of a cavitation problem because the dissolved gases cannot go back into solution as fast as they were evolved. However, the slow rate of dissolving can cause the pump to be "gas locked" when excessive gas is evolved.[3]

## SPECIFYING RECIPROCATING PUMPS

The following should be specified for a reciprocating pump:

1. Type of pump simplex, duplex or triplex.
2. Temperature of fluid to be pumped.
3. Fluid properties such as specific gravity and viscosity.
4. Flowrate at which the pump is expected to operate.
5. Differential pressure at which the pump is expected to operate.
6. Minimum differential pressure.
7. Maximum discharge pressure.
8. Net positive suction head available.

The first five items are self-explanatory. Only the last three will be discussed. [See material on preceding page for a review of operating principles.]

### Minimum Differential Pressure

For a reciprocating pump to operate satisfactorily (particularly in metering operations), the discharge pressure must exceed the suction pressure by a certain minimum amount. We can see why this is so by considering the extreme case when the suction pressure exceeds the discharge pressure. In this case, the liquid will not only enter the cylinder of the reciprocating pump on the suction stroke but will continue to flow through the pump to the discharge end. Thus, flow out of the pump is not limited to the discharge stroke as it must be for metering applications. In such cases, a backpressure valve must be provided on the discharge line. Backpressure valves for frequent open-close service should be recommended only in unavoidable cases because of possible problems at the valve.

Specifying a minimum differential pressure allows the vendor to select a suitable backpressure valve for a given pump.

The minimum differential pressure is obtained by subtracting the maximum pressure at the pump suction from the minimum pressure at its discharge.

## Maximum Discharge Pressure

The maximum discharge pressure is used by the vendor to design the pump casing. A common error is to ignore the difference between reciprocating and centrifugal pumps. The design differential pressure is added to the maximum suction pressure. This value is specified as the maximum discharge pressure.

Of course, this is wrong, because a reciprocating pump will always put out as much head as discharge conditions require until some kind of failure occurs. The maximum suction pressure for a reciprocating pump has no bearing on the maximum discharge pressure. Unfortunately, due to the pulsating nature of the flow, an accurate determination of the maximum discharge pressure is not possible until the pump has been selected.

The following procedure may be used:

1. Calculate the operating discharge pressure by adding: (a) pressure in discharge tank, (b) pressure corresponding to the static head in discharge tank, and (c) frictional line loss. The frictional line loss is to be calculated at peak flow, which is 3.2 times the mean flow for a simplex pump, 1.6 times for duplex, and about 1.1 times for triplex.

2. Increase the operating discharge pressure by 10 psi., or 10%, whichever is greater. Specify this as maximum discharge pressure.

3. Specify that a safety valve be installed from the discharge line to the suction line, and that the

## Acceleration Head Determines Maximum Pressure Drop in Pipes

**Liquid in the suction and discharge lines of reciprocating pumps has to be accelerated because flow varies with time. This requires alternate pressure drops and surges. The instantaneous pressure drop required to accelerate the mass of fluid in the suction line, or the instantaneous pressure rise required to accelerate the mass of fluid in the discharge (converted to feet of liquid), is referred to as acceleration head.**

**The concept of acceleration head and how it may be calculated is best understood by considering a simplex pump. In the suction line of this pump, let:**

$A$ = cross-sectional area of cylinder, sq. ft.
$A_p$ = Cross-sectional area of pipe, sq. ft.
$D$ = Cylinder dia., in.
$D_p$ = Pipe dia., in.
$L$ = Stroke length, ft.
$l$ = Stroke length, in.
$L_p$ = Length of suction line, ft.
$M$ = Mass of liquid in suction line, lb.
$N$ = Crank speed, rpm.
$n$ = Crank speed, rps.
$P_A$ = Presssure at Point A, lb./sq. ft.
$P_B$ = Pressure at Point B, lb./sq. ft.
$X$ = Displacement of liquid in line, ft.
$\rho$ = Density of liquid, lb./cu. ft.

**Consider the time instant just after the pump has started the suction stroke as shown by the relations in the following diagram:**

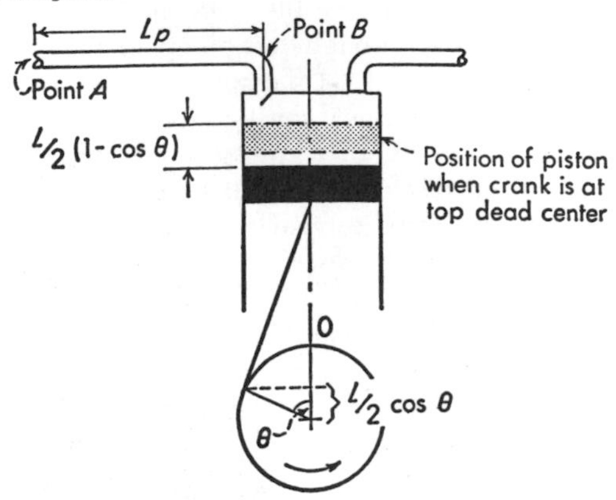


**Volume of liquid $V$ in cylinder:**

$$V = A(L/2)(1 - \cos 2\pi nt)$$

**Volumetric liquid flowrate $dV/dt$ into cylinder at time $t$ is:**

$$dV/dt = \pi A L n \sin 2\pi nt$$

**Instantaneous liquid velocity $dX/dt$ in suction line at time $t$:**

$$\frac{dX}{dt} = \frac{\pi A L n \sin 2\pi nt}{A_p}$$

**Instantaneous acceleration $d^2X/dt^2$ in suction line is:**

$$\frac{d^2X}{dt^2} = \frac{2\pi^2 A L n^2}{A_p} \cos 2\pi nt$$

**The maximum acceleration becomes:**

$$\left(\frac{d^2X}{dt^2}\right)_{max.} = \frac{2\pi^2 A L n^2}{A_p}$$

**The force $F$ required to accelerate the mass of liquid in the suction line is provided by the differential pressure between Points A and B, acting on the cross-sectional area $A_p$ of the pipe, or:**

$$F = \left(\frac{M}{g_c}\right)\frac{d^2X}{dt^2} = (P_A - P_B)A_p$$

**The mass of liquid is:**

$$M = A_p L_p \rho$$

**Using the above relations, find the maximum pressure drop in order to evaluate the acceleration head, $h_a$:**

$$(P_A - P_B)_{max.} = \frac{L_p\rho}{g_c}\left(\frac{d^2X}{dt^2}\right)_{max.}$$

$$= \frac{L_p\rho}{g_c}\left(\frac{2\pi^2 A L n^2}{A_p}\right)$$

$$= \frac{2L_p\rho\pi^2}{g_c}\left(Ln^2\right)\frac{D^2}{(D_p)^2}$$

**Converting $(P_A - P_B)_{max.}$ to feet of liquid yields the acceleration head, $h_a$, in feet. Hence:**

$$h_a = \frac{2L_p\pi^2 l N^2 D^2}{g_c(12)(3{,}600)(D_p)^2}$$

$$h_a = \frac{L_p l N^2 D^2}{(7.06 \times 10^4)(D_p)^2} \qquad \text{(B)}$$

set pressure on the valve be the maximum discharge pressure. Also, specify that the driver be designed to handle the maximum discharge pressure.

4. Check to see that the maximum discharge pressure orignally specified is not exceeded when acceleration head is considered for the recommended pump.

For simplex and duplex pumps, the maximum discharge pressure equals pressure in discharge vessel plus static head (converted to psi.) plus acceleration head (converted to psi.). (Note that frictional drop is not included because acceleration head is maximum at start of discharge stroke when flow, and hence frictional drop, is zero.) The acceleration head may be computed by using the same expression derived elsewhere as Eq. (B) but applying it to the discharge line rather than the suction line.

For triplex pumps, the maximum discharge pressure equals pressure in discharge vessel plus static head (converted to psi.) plus acceleration head (converted to psi.) plus frictional drop. The acceleration head for a triplex pump equals the values found from Eq. (B) divided by 2.7.

5. If the maximum discharge pressure for the acceleration-head limiting condition is higher than originally specified, apply one of the following:

a. Respecify maximum discharge pressure and relief-valve set pressure.

b. Select lower pump speed.

c. Increase size of discharge line.

d. Install a pulsation dampener in discharge line.

## Net Positive Suction Head

Selecting the best value to be used for the vapor-pressure term in the calculation for the available NPSH is the same as for centrifugal pumps.

Evolution of dissolved gases reduces volumetric efficiency but does not cause excessive hammering, perhaps because the rate at which these gases redissolve is slower than the rate at which they are evolved.[3] Vaporization can cause serious hammering and must be avoided.[2] Conservative values must be used for vapor pressure.

Again, the acceleration-head loss in the suction line cannot be determined unless the pump is selected. This makes specifying the available NPSH at the pump suction difficult. Close coordination with the pump vendor is required.

One way of avoiding the problem is specifying the available NPSH at the entrance to the suction line and requesting the vendor to include the acceleration-head loss in the suction line when determining the required NPSH for the pump. Some vendors do this as a matter of course; others do not. This is an important point to remember when comparing vendors. The vendor who specifies the required NPSH at the suction of his pump may appear to have a product superior to the vendor who includes the acceleration-head requirement in the suction line and specifies the required NPSH at entrance to the suction line.

If it is not possible to have the vendor calculate the acceleration head in the line, proceed as follows:

1. Specify the available NPSH as is done for centrifugal pumps, assuming frictional loss in the suction line is the governing factor. (This will generally not be true. It is true only for pumps with a very-low speed.)

Available NPSH (ft.) equals vessel pressure (ft. of liquid) plus static head minus frictional drop in suction line (ft.) minus vapor pressure.

Frictional drop is computed at peak flow, which is 3.2 times the mean flowrate for simplex pumps, 1.6 times for duplex pumps, and 1.1 times for triplex pumps.

2. When the vendor returns with a pump recommendation, check to see that the required NPSH for the pump at its suction is less than the available NPSH when the acceleration-head loss in the suction line is considered.

For simplex and duplex pumps, when acceleration head is the governing factor: Available NPSH equals vessel pressure plus static head minus acceleration head minus vapor pressure. All values to be expressed in ft. of liquid. The acceleration head is obtained from Eq. (B).

For a triplex pump when acceleration head is the governing factor: Available NPSH equals vessel pressure plus static head minus frictional drop minus acceleration head minus vapor pressure. All values expressed in ft. of liquid. The acceleration head is the value obtained from Eq. (B) divided by 2.7. Frictional drop is to be computed at 90% of mean flow.

3. If the available NPSH so calculated is less than the required NPSH:

a. Choose a lower speed pump, or one with more cylinders.

b. Increase size of the suction line.

c. Install a pulsation dampener in the suction line so that available NPSH exceeds required NPSH.

## References

1. Whistler, A. M., A Yardstick for NPSH Requirements, *Petrol. Refiner,* Jan. 1960, pp. 175-180.
2. Whistler, A. M., personal communication, May 5, 1969.
3. Taylor, I., What NPSH for Dissolved Gas?, *Hydrocarbon Process.,* Aug. 1967, pp. 133-134.

## Meet the Author

**Uday S. Hattiangadi** is a process engineer in the process engineering department of Continental Oil Co., Ponca City, Okla. 74601. He has a B. Tech. (Hons) degree from the Indian Institute of Technology (Bombay) and an M.S. from Cornell University. He is an associate member of AIChE.

# Section V
# Pumps for Special Applications

# Plastic centrifugal pumps for corrosive service

**In many cases, plastic pumps have advantages over conventional metal types when handling corrosive fluids. Here is how to select them.**

***Ed Margus,*** *Vanton Pump & Equipment Corp.*

In a process plant, the centrifugal pump is one of the most costly items to maintain. Its high operating speed, susceptibility to the fluids being handled, and inherent weak points, such as bearings, tend to adversely affect its life. For this reason, special construction materials as well as innovative design features are paramount in making the pump perform efficiently with a minimum of maintenance.

Over the years, metallurgists have developed a variety of alloys that resist most corrosives, but even the most advanced metals or alloys frequently are unable to provide satisfactory performance. Among the reasons are the high cost of special alloys needed to withstand the rigors of the service, and the limitations of these metals for use in the variety of fluids commonly encountered in the chemical industry.

So, more and more firms are switching to specially engineered plastic centrifugal pumps. The combined design and material of these pumps offers the greatest life expectancy with a minimum of maintenance problems. In many cases, the cost of plastic pumps is considerably less than for their metal counterparts. Pumps made out of these less-exotic and lighter materials are more readily available and, thanks to numerous design advances, maintenance is minimal and parts replacement is simple.

## First uses of plastics in pumps

Plastic centrifugal pumps have come a long way since they were introduced more than 25 years ago. Initially, the plastics used were not ideally suited for the applications. Early attempts to protect metal components against corrosive and erosive liquids involved coatings of epoxy or polyester resins. Although these materials worked for reasonable periods of time, they needed renewal at regular intervals—a costly and time-consuming procedure.

Originally published February 28, 1977

In addition, plastic-coated pump parts had several other major drawbacks, including a tendency to wear quickly as well as to crack and peel under normal use. To resolve these problems, pump manufacturers began producing metal-based pumps having key parts made of various plastics—such as polyvinyl chloride (PVC), chlorinated PVC, polypropylene, and proprietary formulations such as Teflon, Hypalon, Kynar, Nordel and Viton. Yet these plastic-component pumps also had serious drawbacks, most notably with wear and lack of strength. Since plastics were merely substituted for metal parts (and made to metal-part designs), they were unable to stand up to the constant rigors of pump operation.

## Enter all-plastic pumps

Pump manufacturers were slow to adopt plastics for use throughout centrifugal pumps because the early plastics lacked high structural strength and were difficult to machine or form to required tolerance, and because performance under various conditions of chemical and physical atmospheres could be verified only by extensive and costly laboratory and field tests.

This is no longer the case. Today, pumps are being constructed with plastic parts throughout, so that no metal parts come into contact with the solutions being handled. Most important, plastic pumps have proven their durability in years of use. They now provide a dependable and economical way to handle corrosive chemicals such as hydrofluoric and sulfuric acids, nitric acid and other oxidizing chemicals, as well as caustics, solvents and other problem materials, at temperatures to 325°F, pressures to 150 psi and flows up to 550 gpm.

Design of these new all-plastic pumps takes into consideration the plastics' resiliency. (It should be noted that when the term "all plastic" is used, the pump is made entirely of plastic parts with the exception of the motor and the stainless-steel shaft—some manufacturers offer a shaft with wetted areas encapsulated with a heavy sectioned plastic sleeve to isolate the shaft from the fluids being handled.)

As a result of plastics' resiliency, impeller blades can bend or flex to absorb possible shocks due to hydraulic hammer, or shocks during startup or shutdown that result from the inertia of fluid in a piping system. Impeller blades are therefore designed thicker at the root for greater strength, and thinner at the tips for flexibility

**Suitability of plastics for various corrodents** **Table I**

| Fluid being handled | Polyethylene | Polypropylene | Teflon | Stainless steel | Polyvinylidene fluoride (Kynar) | Buna N | Viton |
|---|---|---|---|---|---|---|---|
| Acetic acid, glacial, 100% | C | B | A | — | B | C | C |
| Aluminum chloride | A | A | A | B | | A | A |
| Aluminum fluoride | A | A | A | C | | A | — |
| Aluminum nitrate | A | A | A | A | | A | A |
| Aluminum potassium sulfate | A | A | A | C | | — | A |
| Aluminum sulfate | A | A | A | A | | A | A |
| Ammonium hydroxide | A | A | A | A | A | C | A |
| Ammonium nitrate | A | A | A | A | | C | — |
| Ammonium persulfate | A | A | A | A | | — | A |
| Ammonium sulfate | A | A | A | B | | A | A |
| Amyl acetate | C | C | A | A | | C | C |
| Aniline | C | C | A | A | B | C | A |
| Aqua regia | C | C | A | C | A | C | A |
| Arsenic acid | B | A | A | A | | — | A |
| Barium hydroxide | A | A | A | B | | A | A |
| Barium nitrate | A | A | A | A | | — | A |
| Barium sulfide | A | A | A | A | | — | A |
| Benzaldehyde | C | C | A | A | B | C | C |
| Benzene | C | C | A | A | B | C | A |
| Benzyl chloride | — | A | A | A | | C | C |
| Butyl alcohol | B | A | A | A | | A | A |
| Calcium bisulfide | C | C | A | C | | C | A |
| Calcium bisulfite | A | A | A | A | | A | A |
| Calcium hypochlorite | A | A | A | C | A | B | — |
| Carbolic acid | B | B | A | B | | C | A |
| Carbon disulfide | C | C | A | A | | C | A |
| Carbon tetrachloride | C | C | A | A | B | C | A |
| Caustic potash | A | A | A | A | | A | A |
| Caustic soda | A | A | A | A | | A | A |
| Chlorine water | C | A | A | C | | C | A |
| Chrome-plating solution | B | B | A | — | | C | A |
| Chromic acid | B | B | A | A | A | C | A |
| Cupric carbonate | A | A | A | C | | A | A |
| Cupric chloride | A | A | A | C | | A | A |
| Cyclohexane | C | C | A | A | | C | A |
| Cyclohexanol | C | C | A | A | | C | A |
| Diethylamine | — | A | A | A | | C | A |
| Dibutyl phthalate | C | C | A | A | | C | C |
| Diethyl benzene | B | B | A | A | | C | A |
| Dimethyl amine | C | C | A | A | | C | A |
| Ethyl acrylate | A | A | A | C | | C | C |
| Ethylene oxide | C | C | A | A | A | | |
| Ethylene trichloride | — | — | A | A | | | |
| Ferric chloride | A | A | A | C | | A | A |
| Formic acid | A | A | A | A | A | C | C |
| Gasoline | C | C | A | A | | A | A |
| Gold-plating solution | A | A | A | C | | — | A |
| Hydrobromic acid | B | A | A | C | | C | A |
| Hydrochloric acid | A | A | A | C | | — | A |
| Isopropyl ether | A | A | A | A | | C | C |
| Magnesium ammonium sulfate | A | A | A | A | | A | A |
| Mercuric chloride, 20% | A | A | A | C | | C | B |
| Mercuric nitrate | A | A | A | A | | A | A |
| Methyl acetate | — | A | A | A | | A | C |
| Methyl chloride | — | C | A | A | | C | A |
| Methyl ethyl ketone | C | C | A | A | | C | C |
| Methyl salicylate | — | — | A | A | | C | A |
| Methylene chloride | — | C | A | A | | C | B |
| Naphtha | C | C | A | A | A | B | A |
| Naphthalene | C | C | A | A | A | C | B |
| Nitric acid, to 30% | C | A | A | C | A | C | A |
| Nitric acid, 30-60% | C | C | A | C | A | C | A |
| Nitric acid, 70% | — | C | A | C | A | C | A |
| Perchloric acid, 10% | — | C | A | C | A | C | A |
| Phenol | C | C | A | A | B | C | A |
| Potassium hydroxide | A | A | A | A | A | B | A |
| Radium chloride | A | A | A | C | | C | A |
| Resorcinol | C | C | A | A | | C | A |
| Sodium acetate | A | A | A | A | | C | C |
| Sodium bromide | A | A | A | C | | C | A |
| Sodium hypochlorite (all concentrations of bleach) | A | A | A | C | A | A | — |
| Sodium peroxide | — | A | A | A | | C | A |
| Sodium sulfide | A | A | A | A | | C | A |
| Sodium uranate | A | A | A | A | | A | A |
| Stannous chloride | A | A | A | C | | C | A |
| Sulfamic acid | A | A | A | A | | C | A |
| Sulfuric acid, 0-50% | A | A | A | C | B | C | A |
| Sulfuric acid, 50-85% | C | A | A | C | B | C | A |
| Sulfuric acid, 85-96% | C | C | A | C | B | C | A |
| Sulfurous acid | A | A | A | C | | C | A |
| Toluene (toluol) | C | C | A | A | | C | C |
| Tributyl phosphate | C | A | A | A | | C | C |
| Trichloracetic acid | C | A | A | C | | C | C |
| Trichloroethylene | C | C | A | A | A | C | A |
| Tricresyl phosphate | C | A | A | A | | C | A |
| Triethanolamine | C | C | A | A | | C | C |
| Trimethyl amine | C | C | A | A | | C | C |
| Zinc chloride | A | A | A | C | | A | A |
| Zinc sulfate | A | A | A | A | | A | A |

Key:

- A Excellent
- B Fair, consult pump manufacturer
- C Not applicable
- — No results available

and shock resistance. In addition, vanes are set at a greater pitch to move more fluid. Pumps themselves are being offered in ever-larger sizes, permitting them to operate at lower speeds to minimize problems of cavitation (a major cause of vane erosion in metal pumps' impellers). Plastic pumps can be expected to offer ever-increasing efficiencies as low-speed versions become more widely available.

All-plastic pumps can be substantially lighter than metal pumps of equal capacity. And, even after years of service, absence of troublesome pit corrosion with all-plastic pumps permits rapid disassembly, whereas it might be completely impossible to disassemble metal. A varied and ever-growing number of nonmetallic materials are used to provide durable and long-lasting bearing performance. These include ultrapure ceramics, Teflon, ultrapure carbon, Rulon, Ryton, etc.

A more recent design development, the hollow-shaft pump, completely eliminates sleeve bearings operating within the liquid area, permitting the pump to operate dry without burnout. In both horizontal and vertical configurations, the shaft is supported entirely by the remotely positioned motor ball-bearings. Should the shaft ever need replacing or resleeving, it can be easily removed.

Engineers have wide latitude in choosing the best plastic centrifugal pump for each application. Selection is based on space requirements, initial cost, installation cost, spare-parts purchase, service, availability and maintenance.

It must be noted that the selection of plastics is usually dictated by considerations of corrosion, erosion, personnel safety and liquid contamination. If solids are present in a highly corrosive liquid to be handled, the choice of pumps and materials becomes very limited. Fortunately, newer plastic pumps easily accommodate these fluids.

## Determining service parameters

The major features to be considered in selecting a plastic centrifugal pump for corrosive services are:

- Nature of liquid to be handled.
- Head of liquid.
- Flowrate or capacity.
- Piping material and sizes.
- Operating parameters—effects of temperature, flows, transport velocities.
- Wear on pump from hydraulic or thermal shock.
- Anchorage and support.
- Ease of installation.

The conditions to which the pump will be subjected—including the temperature, suction conditions and the lubricating properties of the liquid being pumped—are necessary to determine the type of mechanical seals used. Seal failure is perhaps the greatest cause of pump failure. Supplying the necessary information so that the pump manufacturer can recommend an appropriate seal will contribute immeasurably to the operating life of the pump.

Once the specifying parameters have been determined, it is necessary to select the pump manufacturer.

First, since most pumps are made in standard sizes, does the manufacturer being considered offer both the size and type of pump that most nearly fits the service in question?

Following that, does the manufacturer offer in-field service or, should the occasion warrant, can the pump be taken to the central facilities of the pump maker for repair, and can the engineer bringing the pump actually watch and learn from the repairs being made? (Many pump manufacturers consider their manufacturing/repair operations proprietary.)

Another important point to consider before a final selection is: How extensive is the maker's recommended spare-parts kit? While most pump manufacturers recommend purchase of such a kit, those kits with numerous replacement parts may suggest trouble. By contrast, new hollow-motor-shaft design pumps are likely to be quite reliable when one considers that the only spare parts are the replaceable shaft and a casing.

One of the major advantages of centrifugal pumps is their simplicity of operation and design. Centrifugals are the ideal choice for handling liquids of all types, including those that contain solids—such as abrasives and other foreign matter. However, many pump users will often specify a need for filters or special strainers to eliminate substances in the fluids being handled. In 99% of all applications, filters are not needed, and in many cases strainers are not the answer. A properly engineered plastic centrifugal pump can handle solids without problem, and a strainer will only impede flow and require frequent cleaning.

When solids are present, the pump manufacturer will provide the units with internal passages having adequate dimensions to handle them. He will work with engineers to help them avoid having pockets or dead spots in the system where solids might accumulate, and ensure that internal clearances between stationary and moving parts in the pump are kept to a minimum. He can also help design a system flush to periodically purge the system of any solids buildup.

Before selecting a pump, work closely with the manufacturer to ensure that the overall system into which the pump is fitted is designed to facilitate maintenance and inspection and to accommodate possible future changes. Flow into the pump should be as uniform and straight as possible, with a minimum of abrupt upstream changes in cross-section. Without careful installation plans, noise, vibration, loss of output, driver overload and excessive bearing and shaft failures can result. Most important, the pump should be highly accessible to permit ease of removal for onsite repairs.

## The author

**Ed Margus** is Vice-President, Engineering, of Vanton Pump & Equipment Corp., 201 Sweetland Ave., Hillside, NJ 07025, where he is responsible for product development and manufacturing techniques on all-plastic centrifugal, rotary and sump pumps. He has B.S. and M.S. degrees from Newark College of Engineering and has authored many papers on all-plastic fluid-handling systems.

# Polyvinylidene fluoride for corrosion-resistant pumps

**Advances in fabricating polyvinylidene fluoride has made possible corrosion-resistant pumps in which no metal parts contact the fluid.**

*Edward Margus,* *Vanton Pump & Equipment Corp.*

☐ Improvements in engineered plastics—better physical, thermal and chemical properties combined with excellent molding and machining properties—have extended their use to the most demanding pumping applications.

Precision-molded and -machined pump components of polyvinyl chloride, chlorinated polyvinyl chloride, polypropylene, polyethylene and polyvinylidene fluoride ($PVF_2$), for example, are giving long, low-cost performance now in pumping solutions where only stainless steel or other special alloys were used previously (often with less than satisfactory results). In handling many solutions, any metal part exposed to the fluid or its fumes is almost certainly an area of future trouble.

The obvious advantage of plastics over metals in resisting attacks by corrosive solutions was not immediately available to pump users. Many pump manufacturers have been slow to adopt plastics because they once lacked high structural strength and were difficult to machine or form to tolerance. Also, their performance under various conditions of chemical and physical atmospheres could be predicted only after extensive and costly laboratory and field trials. Probably nowhere is the stamina of a new plastic more thoroughly tested than in pumps for the process industries. They are simultaneously subjected to corrosion, heat and abrasion as well as the stresses and pressures of high output.

Slowly, however, plastics have lived up to their potential in process pumps, through the developments by the plastics manufacturers in engineering tougher, stronger products and through the efforts of a few pioneering pump manufacturers who have taken advantage of these new qualities in the polyvinyls and polypropylenes to improve the service life of pumps.

Today, pumps can be constructed with no metal touching the solution being pumped—including rotary, horizontal centrifugal, vertical centrifugal and completely submersible pumps. This extends the use of plastic pumps not only into the super-clean areas such as handling foods, pharmaceuticals, pure water and other products where metallic contamination cannot be tolerated, but also into the areas where combined chemical and physical conditions would heretofore rapidly damage pumps.

## Latest and toughest

The latest development in engineered plastics for use in critical pump components is the fluoroplastic, $PVF_2$. It withstands unfavorable conditions beyond the parameters of polyvinyl chloride, polyethylene or polypropylene and is responsible for new levels in pump performance in applications that preclude the use of most other materials—plastics or metal alloys.

Polyvinylidene fluoride (marketed under Pennwalt Corp.'s registered trademark of "Kynar") is mechanically strong and tough. It can be molded and machined to the close tolerances and complex configurations of pump components better than most of the engineered plastics.

Its useful temperature range (below −80 to +300°F) is broader than polyvinyl chloride, polypropylene or chlorinated polyvinyl chloride, and it is highly resistant to chemicals including halogens, oxidants, and most bases and acids (except fuming sulfuric). It shows al-

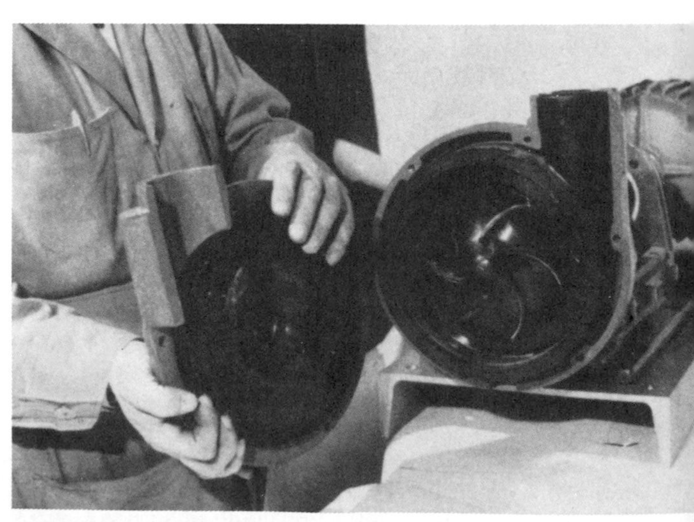

**Centrifugal pump** with all wetted parts of $PVF_2$

Originally published October 27, 1975

**Performance of $PVF_2$ pumps under various conditions**

| Acids | 70°F | 150°F | 212°F | 275°F |
|---|---|---|---|---|
| Acetic, glacial | 1 | 2 | 3 | NR |
| Aqua regia | 1 | 1 | 1 | NR |
| Formic | 1 | 1 | 1 | 1 |
| Hydrochloric, 20% sol. | 1 | 1 | 1 | 1 |
| Hydrofluoric, 35% sol. | 1 | 1 | 1 | 1 |
| Nitric | 1 | 1 | 1 | ? |
| Nitric/sulfuric, 50/50 | 1 | 1 | 2 | ? |
| Phenol, 10% | 1 | 1 | 1 | ? |
| Phenol, 100% | 1 | 1 | 2 | ? |
| Sulfuric, conc. | 1 | 1 | 2 | |
| **Bases** | | | | |
| Ammonium hydroxide | 1 | 1 | 1 | 1 |
| Aniline | 1 | 2 | 3 | NR |
| *sec*-Butylamine | 1 | 3 | NR | NR |
| *tert*-Butylamine | 1 | 2 | 3 | NR |
| Sodium hydroxide, 10% | 1 | 1 | 1 | ? |
| Sodium hydroxide, 50% | 1 | 1 | 1 | ? |
| UDMH/hydrazine, 50/50 | 1 | 2 | ? | ? |
| **Solvents** | | | | |
| Acetone | 3 | NR | NR | NR |
| Benzaldehyde | 1 | 2 | ? | ? |
| Benzene | 1 | 2 | ? | ? |
| Carbon tetrachloride | 1 | 1 | 1 | 1 |
| Diacetone alcohol | 1 | 3 | NR | NR |
| Diethyl cellosolve | 1 | 1 | 1 | 1 |
| Diisobutylene | 1 | 1 | 1 | 1 |
| Ethanol | 1 | 1 | 1 | 1 |
| Ethyl acetate | 1 | 2 | ? | ? |
| Ethyl ether | 1 | 2 | ? | ? |
| Perchloroethylene | 1 | 1 | 1 | 1 |
| Trichloroethylene | 1 | 1 | 1 | 1 |
| Turpentine | 1 | 1 | 1 | 1 |
| **Other Chemicals** | | | | |
| Bromine (dry) | 1 | 1 | | |
| Bromine water | 1 | 1 | 1 | |
| Chlorine (wet or dry) | 1 | 1 | 1 | |
| Refrigerant 113 | 1 | 1 | 1 | |
| Ethylene oxide | 1 | 1 | 1 | |
| Fluorine | 1 | | | |
| Hydrogen peroxide, 30% | 1 | 1 | 1 | |
| Hydrogen peroxide, 90% | 1 | | | |
| Nitrogen dioxide | 1 | 1 | 1 | |
| Nitrogen tetroxide | 1 | 1 | | |

1 - Little or no effect.
2 - Some effect, but not indicative of impaired serviceability.
3 - Noticeable effect, although possible serviceability.
NR - Not recommended.
? - Conclusive data not available.

most complete resistance to aliphatics, aromatics, alcohols, and chlorinated solvents.

While casings and internal pump parts of $PVF_2$ are up to 15 times more expensive than the same parts in polypropylene, $PVF_2$ pumps are considered economical for applications requiring long service life, abrasion resistance, and high temperature performance:

- A large national chemical company, after 3,000 hours of tests with a Kynar centrifugal pump, pumping 65% nitric acid at 180 to 200°F, flooded suction, 80 gpm, at 80 ft total discharge head, adopted Kynar pumps for company-wide use in the demanding application. No other type pump tested by the company withstood the tests.
- One entire industry, the bromine industry, owes its progress to Kynar components. Bromine attacks pumps of any other material. Kynar is impervious, not only to the bromine, but also to the even more corrosive bromine vapors as well as several corrosive chemicals used in related processing. A number of major chemical companies are using Kynar pumps to make the processing of bromines feasible.
- A Pennsylvania company, specializing in recycling of concentrated acid and salt pickling solutions from metals-producing plants at temperatures of 180°F and above, turned to Kynar pumps after repeated failures of other pumps due to impeller erosion.

Pumps made of Kynar are capable of performing under such rugged conditions as indicated in Table I. Tests and actual experience with Kynar performance in pumping a number of chemicals at elevated temperatures, indicated that polyvinyl chloride should be used only to 145°F; chlorinated polyvinyl chloride to 190°F, polypropylene to 185°F and Kynar to 250°F.

Tests in which Kynar pumps performed successfully where polypropylene pumps were less than adequate included the chemicals in Table II:

It should be emphasized that in the same series of tests, polypropylene performed equally as well as Kynar in pumping many other chemicals. Among some 200 chemical solutions tested, however, only in the pumping of hot benzene sulfonic acid did polypropylene components show better performance than Kynar. Neither polypropylene nor Kynar gave top test results in pumping hot cyclohexanol.

## Wide range of problems solved

The impervious nature of polyvinylidene fluoride ($PVF_2$) makes it applicable to a wide range of hot, abrasive, corrosive chemicals and solvents. At the other end of the scale of applications, it is nontoxic and has been approved by the Food and Drug Administration for use in contact with food and food products.

The scope of applicability of this new engineering plastic, combined with advances in pump design and better understanding of the requirements of the process industries, allows pump manufacturers to provide all-plastic centrifugal pumps to solve many fluid handling problems that previously plagued the chemical process industries.

## The author

**Edward Margus** is chief engineer for Vanton Pump & Equipment Corp, 201 Sweetland Ave., Hillside, NJ. He is now engaged in helping Vanton set up its second manufacturing plant in Europe. He formerly worked for a machine tool builder (Gould & Eberhardt, Inc.) and later as a consulting engineer to plants in the sheet metal industry. He has a bachelor's degree in mechanical engineering from Newark College of Engineering (now New Jersey Institute of Technology), from which he also received his master's degree.

# Slurry pump selection and application

*Slurry handling is becoming more common in process plants (e.g., many pollution-control projects involve pumping slurries). This article will help in selecting a slurry pump to meet your requirements.*

*J. Ingemar Dalstad*, *Morris Pumps, Inc.*

☐ Most engineers seldom have the need to select slurry pumps. Unfortunately, the improper selection of a pump for handling slurries can have costly and dramatic results.

This article will discuss when a slurry pump should be used (it is not always obvious), the design features that are most desirable, and the other criteria and pitfalls involved in selection and application processes.

## When to use a slurry pump

Much of the confusion in deciding when to specify a slurry pump arises from the lack of agreement on the meaning of the word "slurry." This is due in large part to the nearly infinite number of solid-liquid mixes. In place of the many academic slurry definitions, this broader, more functional definition if offered: A slurry is any mixture of liquid and solids capable of causing significant pump abrasion, clogging, or mechanical failure due to high loads or impact shocks.

Under some circumstances, it may seem superfluous to consider the "when" of slurry pump selection. Obviously a pump employed to move "deliberate" slurries such as mine tailings or chemical concentrates must be designed and constructed with exceptional strength, abrasion resistance, and solids-passage ability.

But what about a pump employed to supply large quantities of water from a sandy river for cooling purposes? In such "accidental" slurries, the transport of liquid is the prime purpose—the presence of solid materials is not intended (or, sometimes, even recognized). Nevertheless, failure to use a slurry pump for this type of application can frequently result in excessive maintenance, parts, and down-time costs.

The "when" of slurry pump selection might best be answered by a rule of thumb which says that whenever the fluid to be pumped contains more solids than are found in potable water, at least consider the use of a slurry pump.

## Typical slurry pump features

What are the features of a slurry pump (Fig. 1) that set it apart from a standard centrifugal? Outwardly there are few differences, although the slurry pump is usually larger in size. Internally, however, there are many characteristics that make the slurry pump a very specialized breed.

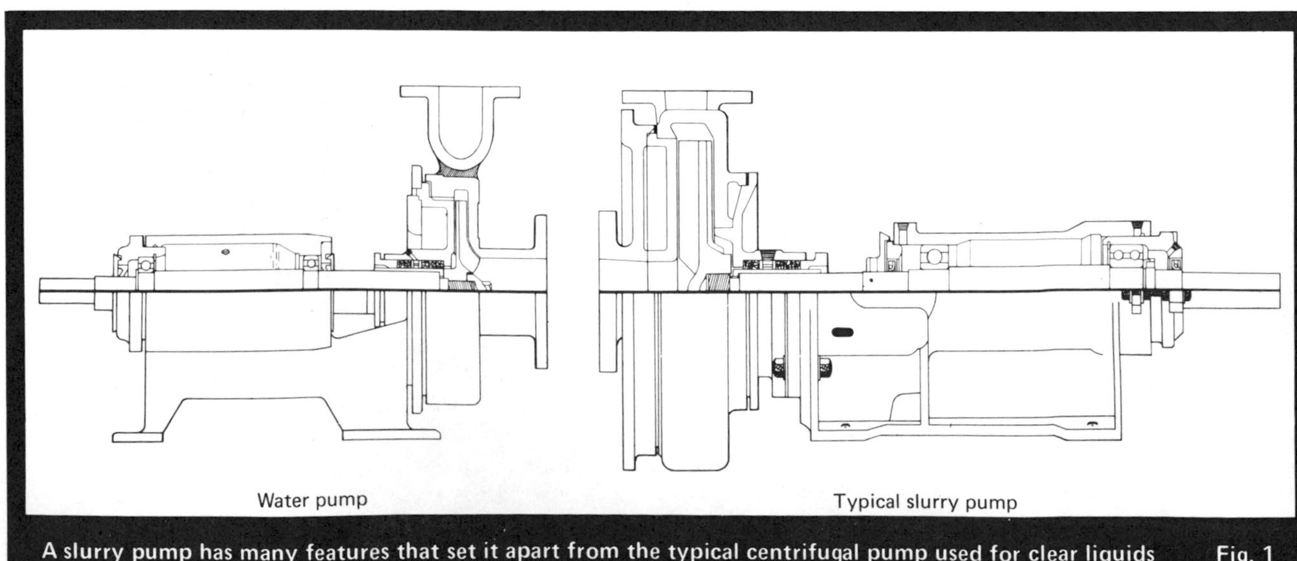


A slurry pump has many features that set it apart from the typical centrifugal pump used for clear liquids — Fig. 1

Originally published April 25, 1977

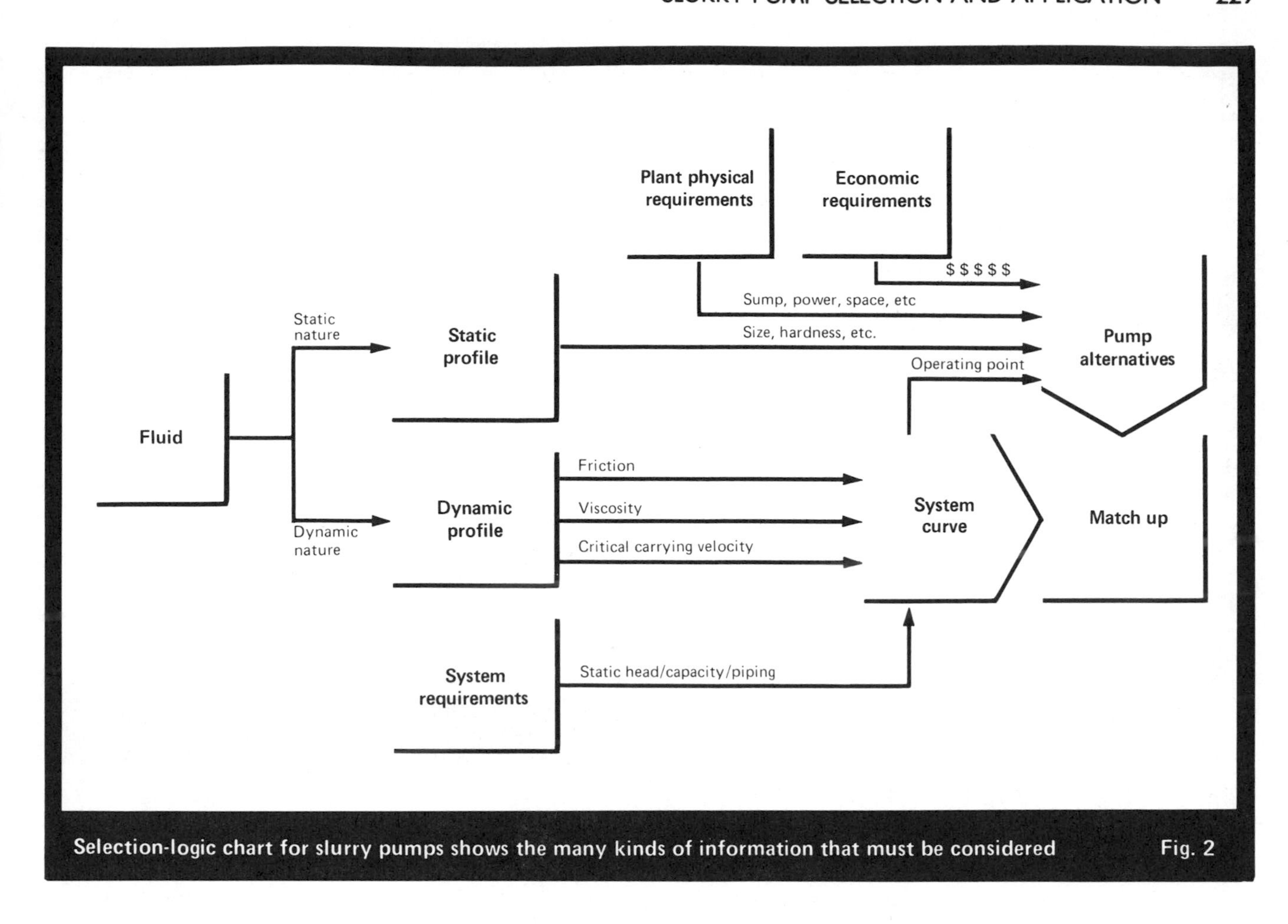


Selection-logic chart for slurry pumps shows the many kinds of information that must be considered Fig. 2

Wall thicknesses of wetted-end parts (casing, impeller, etc.) are greater than what is used in conventional centrifugals. The cutwater, or volute tongue (the point on the casing at which the discharge nozzle diverges from the casing) in the casing is generally less pronounced in order to minimize the effects of abrasion. Flow passages through both the casing and impeller are large enough to permit solids to pass without clogging the pump. Slurry pumps are available in a variety of materials of construction, to best handle the abrasion, corrosion, and impact requirements of nearly any solids-handling application.

Replaceable liners are used in critical areas of wear to reduce the cost of parts replacement. On many light-to-medium-duty pumps, the liners are used only on the suction side. A slurry pump for severely abrasive duty will use an additional liner on the hub side of the casing and, frequently, a liner in the suction nozzle as well. Another approach is the use of a complete shell lining—a sort of casing within a casing. Rubber, ceramic and other synthetic linings are used in many specialized applications.

Because the gap between the impeller face and suction liner will increase as wear occurs, the rotating assembly of the slurry pump must be capable of axial adjustment to maintain the manufacturer's recommended clearance. This is critical if design heads, capacities, and efficiencies are to continue. Other specialized features include extra-large stuffing boxes, replaceable shaft sleeves, and impeller back vanes that act to keep solids away from the pump's stuffing box.

Both radial and axial-thrust bearings on the slurry pump are generally heavier than for standard centrifugals, owing to the demands imposed by slurries of high specific-gravity. Although impeller back-vanes (used to lower stuffing-box pressures) do actually reduce axial thrust, these vanes can wear considerably in abrasive services. Consequently, the bearings must be of ample capacity to handle thrust loads by themselves. Balancing holes through the impeller should not be used to reduce axial thrust since they can either clog or initiate excessive localized impeller wear.

Nearly all slurry pumps have larger-diameter impellers than units for pumping clear liquids, to enable heads and capacities to be met at reduced rotational speed. Low-speed operation is one of the most important wear-reducing features of a slurry pump. In fact, experience shows that abrasive wear on any given pump rises at least with the third power of rpm increase.

## Which pump should be selected?

Once having determined that a pump specifically designed for solids handling is required, the next steps are to select and properly apply the pump. Although selection and application are complex and interrelated processes, the task can be simplified as follows:

- Establish a group of possible pump alternatives from the many units on the market.
- Plot a system curve depicting required pump heads at various capacity levels.
- Match pump-performance curves with the system

curve in order to determine the final pump selection.

The slurry-pump-selection logic chart (Fig. 2) depicts the many sources of information that must be analyzed to arrive at a group of pump alternatives, the system curve, and the final match-up. (If the intended application will employ a standard centrifugal pump for handling water, virtually all that must be considered are head and capacity requirements, which will dictate the initial list of pump alternatives.) When selecting a slurry or solids-handling pump, on the other hand, the first consideration must be the material to be pumped, both in a static state and as a dynamic entity.

## Static profile of slurry

An analysis of the static profile of the slurry will help determine the solids-passage ability, abrasion resistance, and mechanical strength required of the pump. The most important elements in the static profile can be assigned to four categories:

1. *Size of the solids*—what are the largest particles the pump must handle? Are these solids similar, or random, in size?
2. *Nature of the solids*—Are they pulpy or hard, light or dense, round or jagged, or abrasive, or corrosive?
3. *Nature of the liquid*—How corrosive is the liquid? Will it lubricate the solids and reduce abrasion?
4. *Concentration of the solids*—It is the ratio of solids to liquid that determines how the characteristics of the solids will influence the slurry as a whole.

These four static characteristics create unique demands, requiring specific pump design and construction features.

For example, Fig. 3 shows a pump designed to handle wastes, light slurries, and random large solids. Unlike the slurry pump shown in Fig. 1, this unit does not use wear liners. The emphasis here is on very large flow-passages through the casing and impeller. Because such units are generally used for pumping sewage and relatively non-abrasive industrial wastes, certain wear-reducing design features can be compromised to increase hydraulic efficiency.

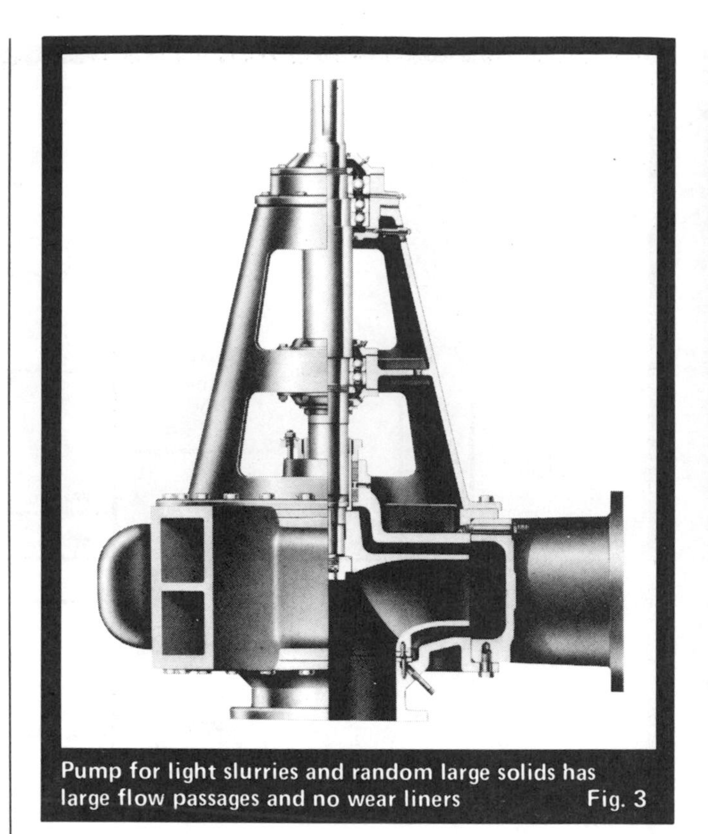

Pump for light slurries and random large solids has large flow passages and no wear liners Fig. 3

Other applications involve not only large solids, but high levels of abrasion and impact shock as well. Fig. 4 shows how added protection is supplied by the extensive use of replaceable liners. Units such as these are used in the pumping of furnace slag, high-concentration ash slurries, and extremely abrasive chemical wastes.

When chemical sludges or wastes containing large solids must be pumped, the vortex pump (Fig. 5) is often the best answer. Because its impeller is fully recessed into the rear of the casing, a relatively small pump can be used to handle liquids containing very large solids.

Still other slurries may exchange the problems of large solids for the equally difficult pumping idiosyncracies associated with high concentrations of small solids. More often than not, such slurries present extreme abrasion problems. Typical are those associated with lime-slurry pumping, the handling of ore concentrates, kaolin clay, or cement slurries. Fig. 6 shows an extremely heavy-duty slurry pump ideal for such applications.

## Materials of construction

In addition to pump design itself, the many available materials of construction present a wide spectrum of choice to meet the challenges found in slurry pumping. Cast iron is the basic pump material, being both low in cost and well suited for mildly abrasive applications. Cast iron with 3% nickel is sometimes used when protection against mild corrosion is needed. Ductile-iron and cast-steel offer extra strength in high-pressure applciations or when the pump is exposed to high impact shock from large, hard solids.

As abrasion increases, materials such as Ni-Hard or heat-treated high-chromium iron, with Brinnel hardness ratings in the 550 to 600 range, offer extended parts life. Chromium iron is recommended when mild corrosion will

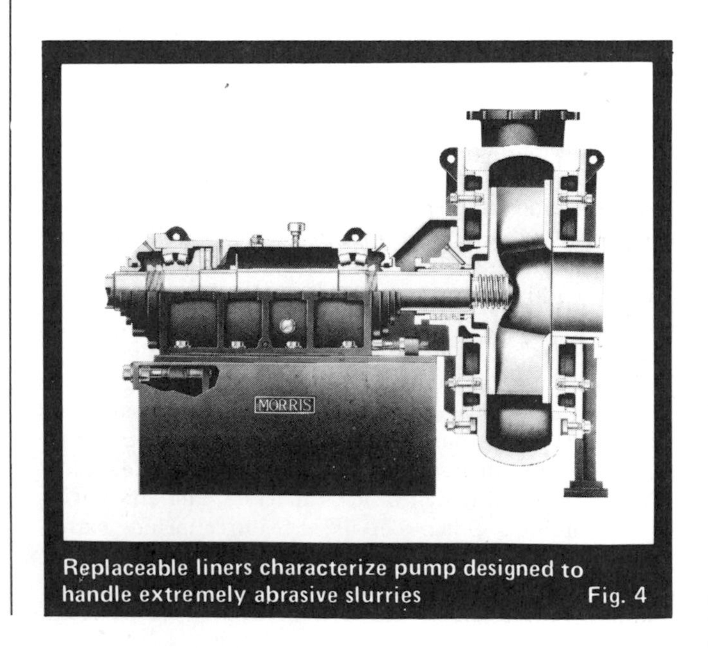


Replaceable liners characterize pump designed to handle extremely abrasive slurries Fig. 4

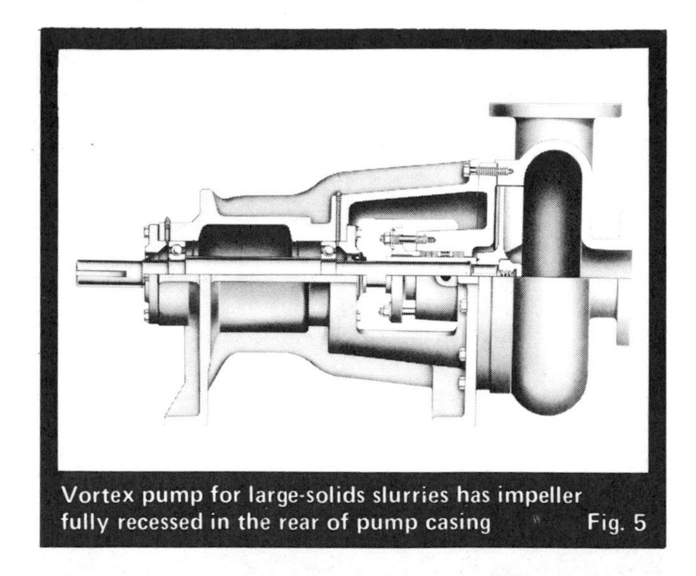

Vortex pump for large-solids slurries has impeller fully recessed in the rear of pump casing Fig. 5

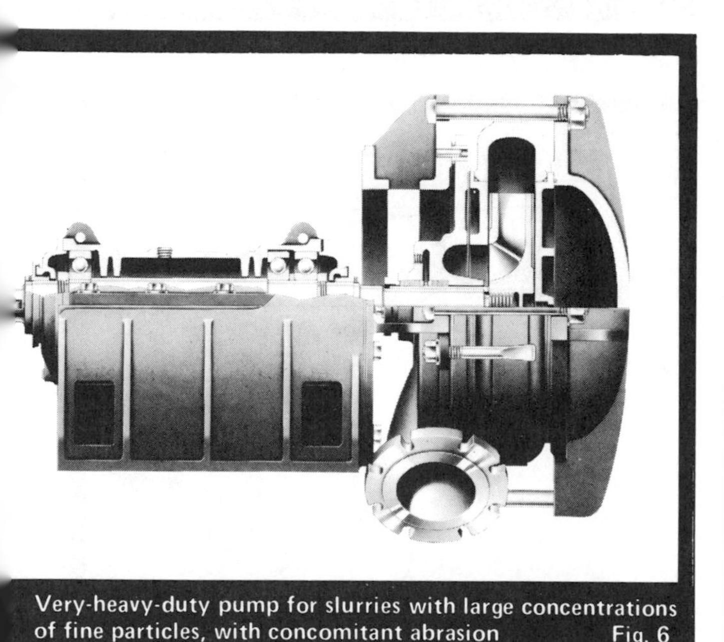

Very-heavy-duty pump for slurries with large concentrations of fine particles, with concomitant abrasion Fig. 6

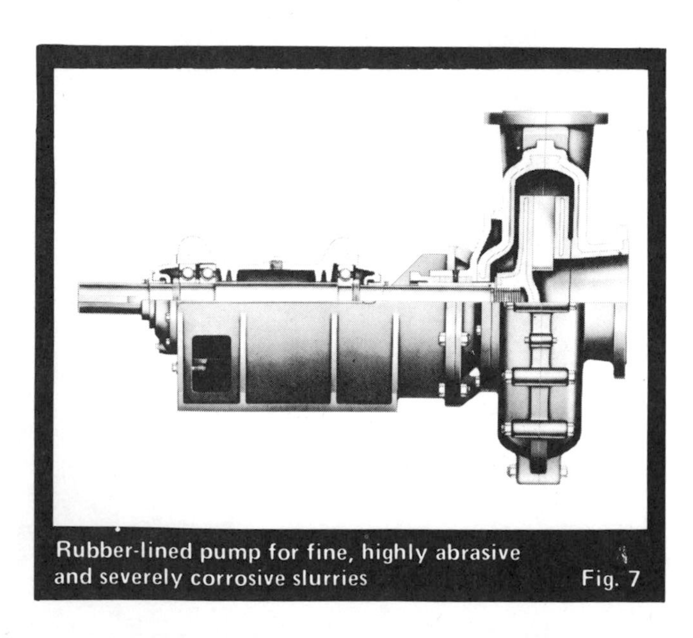

Rubber-lined pump for fine, highly abrasive and severely corrosive slurries Fig. 7

accompany abrasion. Rubber lining is frequently used in pumps for handling fine, highly abrasive, and more severly corrosive slurries (Fig. 7).

Finally, in highly corrosive situations, a number of alloys of the stainless steel and high-nickel groups are available. Ceramics and a number of plastic materials also offer protection from severe corrosion and are being used with increasing frequency to reduce costs.

Referring back to the pump-selection logic diagram (Fig. 2), it can be seen that the many static characteristics of the pumpage, combined with the physical and economic requirements of the pumping job, allow the choice of a number of pump alternatives from the many units available. This information alone, however, is not enough to enable the design engineer to make a good slurry-pump selection.

## Dynamic profile, and system curves

Still to be examined are the system requirements and the dynamic profile of the material to be pumped. These two information sources will yield the system curve, the working basis for final pump selection and application. Following is a brief review of the system curve and some common errors that lead to its misuse.

The system curve (Fig. 8) is a plot of the head that must be met by the pump at any given capacity. This is entirely a function of the piping and in no way determined by the pump itself. The vertical coordinate, expressed in this example in feet of liquid, is the sum of two factors: (1) *static head*, or the actual difference in liquid levels which the pump must overcome, and (2) *friction head*, the expression of energy needed to overcome friction in the piping system. Pipe friction is a function of flowrate, increasing as flowrate increases. Friction values for pipe and fittings can be taken from tables, slurry laboratory tests, or field tests, once the slurry and the type, length, and diameter of the pipe are known. The curve in Fig. 8 describes the behavior of a Newtonian slurry. As shown, pipe friction increases by a constant percentage as capacity is increased, thereby producing a curve with a parabolic shape.

Not all slurries will exhibit Newtonian behavior. This is illustrated by Fig. 9, which shows curves for two sand slurries with different solid concentrations. At a concentration of 60%, the sand slurry is Newtonian and will yield a predictable, parabolic curve. But, when the concentration is stepped up to 70%, the curve is drastically different. The slurry becomes non-Newtonian (in this case, plastic) and the system curve cannot be plotted using normally available friction tables. Instead, pipe friction must be arrived at by actual testing in the laboratory or plant.

Fig. 10 demonstrates what will happen if the non-Newtonian nature of the 70% slurry is not recognized. The system curve (A) is plotted as if the slurry were Newtonian, and a pump with a performance curve that will intersect the system curve at a desired 600 gpm is selected. It can be seen, however, from the actual curve (B) that the intersection will never take place. If applied this way, the pump would not pump.

Another factor that will become important is demonstrated in Fig. 9—the concept of critical carrying velocity. This is the velocity at which the material suspended in the

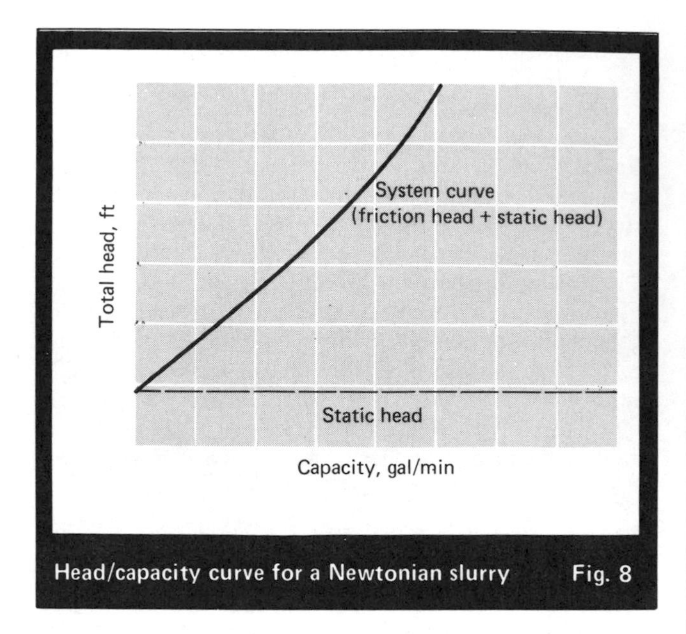


Head/capacity curve for a Newtonian slurry Fig. 8

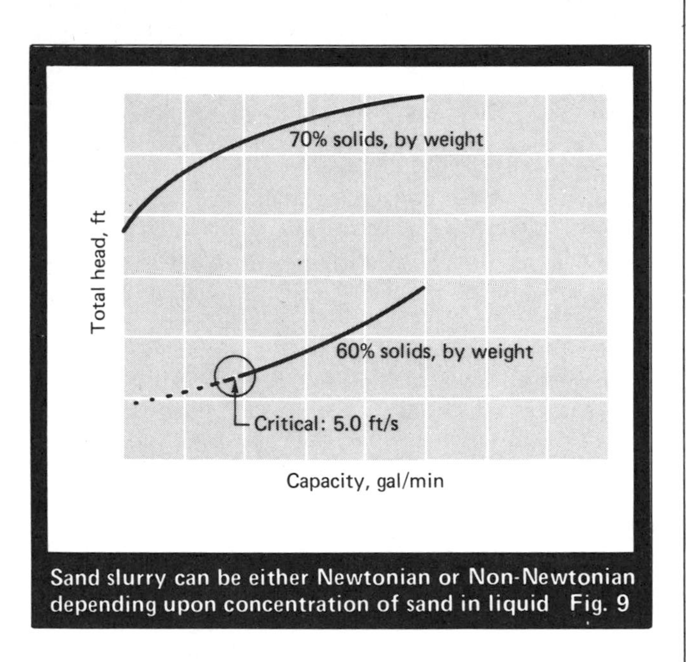


Sand slurry can be either Newtonian or Non-Newtonian depending upon concentration of sand in liquid Fig. 9

liquid will begin to settle out in the piping. The operating point of the pump must be selected so that pipeline velocity exceeds this critical carrying velocity. If this results in too great a flowrate, a smaller pipe size must be used. Critical carrying velocity can only be determined with any degree of accuracy by actual testing, although results from a test with a single pipe diameter may be extrapolated to other pipe sizes.

There are other ways to go astray in the development and use of a system curve. One relates to the overestimation of pipe friction. In Fig. 11, two system curves are plotted. The curve with the steeper slope intersects the pump performance curve at a desired capacity of 800 gpm (Point A). If, however, pipe friction has been overestimated, the true system curve might intersect the pump curve at, for example, 1,100 gpm (Point B). The effect of this error is overcapacity, which results in excess velocity in the system. This mistake will cause increased pump wear and the expenditure of excess horsepower.

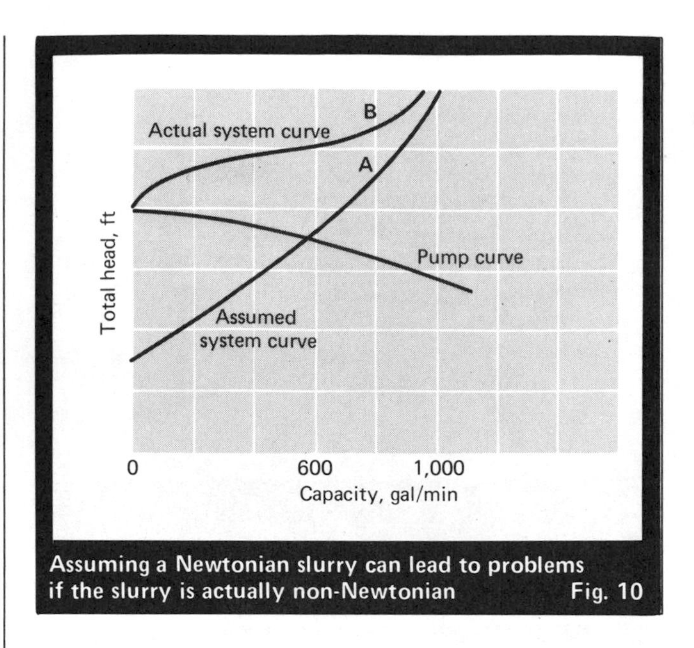


Assuming a Newtonian slurry can lead to problems if the slurry is actually non-Newtonian Fig. 10

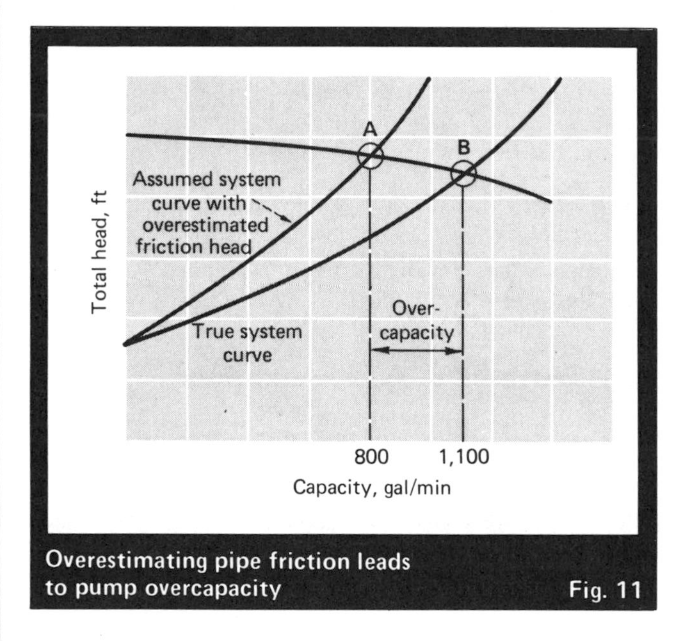


Overestimating pipe friction leads to pump overcapacity Fig. 11

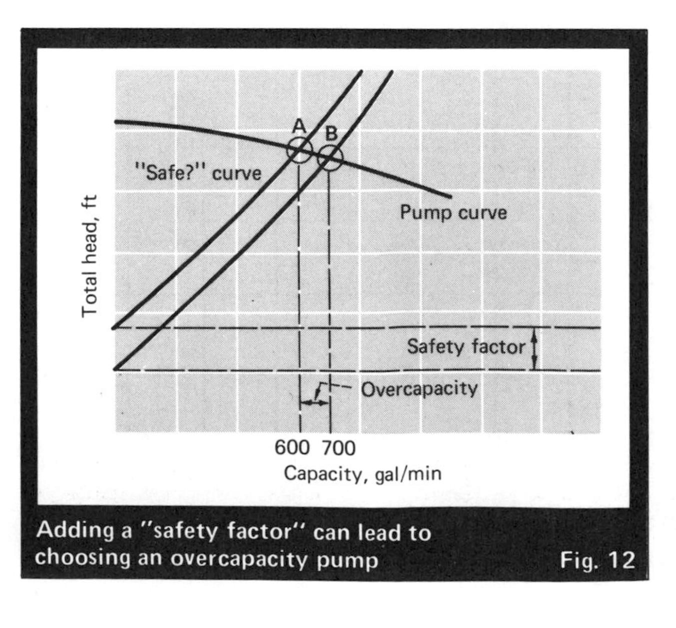


Adding a "safety factor" can lead to choosing an overcapacity pump Fig. 12

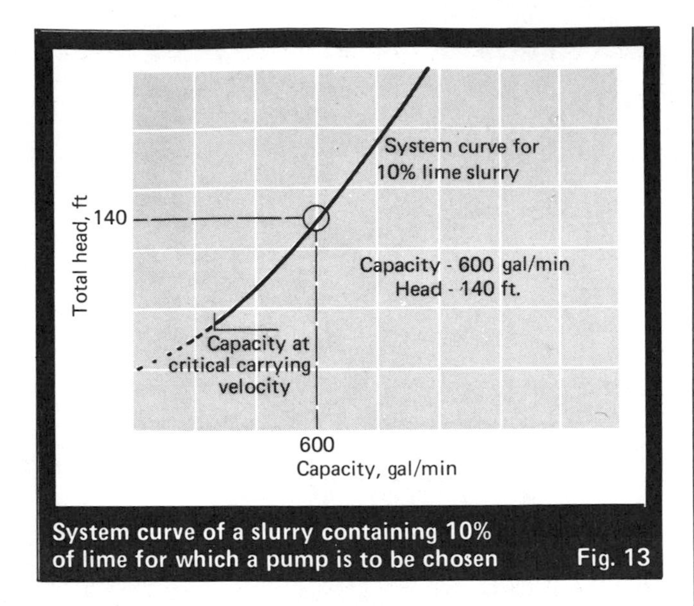


System curve of a slurry containing 10% of lime for which a pump is to be chosen Fig. 13

Depending upon just how far out on the curve the pump runs, cavitation and even mechanical failure can also result.

Possibly an even more common error is made in the name of "safety." In Fig. 12, a correct assessment of the friction has been made, and the slurry has been properly identified as Newtonian. This time, the concern is about the ability of the pump to do the job, so a few feet are added to the static head ". . . just to be sure." As a result the slope is true but the intersection planned for Point A actually takes place at Point B. Overcapacity with all its problems enters the picture again.

## The match-up of system and pump

Fig. 13 and 14 demonstrate how a correctly executed system curve can help the engineer to select and properly apply a pump. In this example, the material to be pumped is a 10%-concentration lime slurry. The required capacity is 600 gpm and the critical carrying velocity is determined

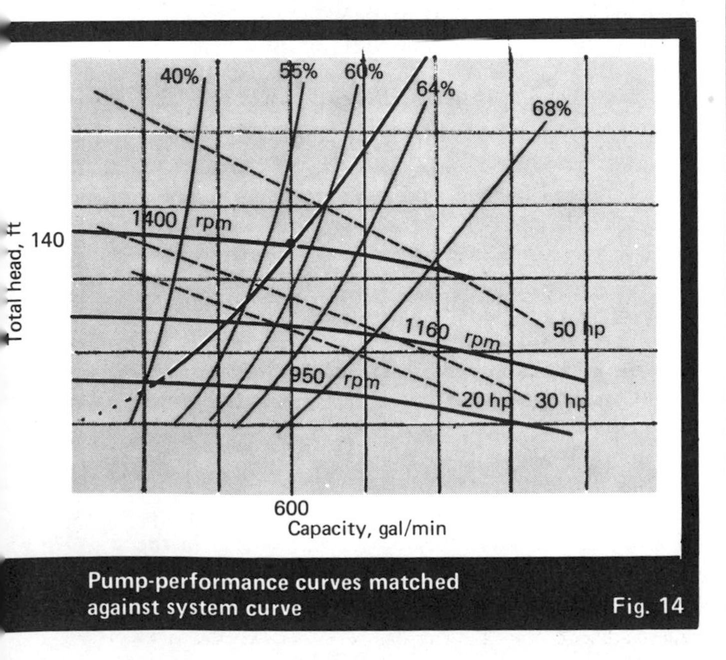


Pump-performance curves matched against system curve Fig. 14

to be well below the fluid velocity at the planned operating point. The system curve shows that, at 600 gpm, a total head of 140 ft is required.

By taking the data compiled in the examination of the static nature of the slurry and of the economic and physical requirements of the plant, and comparing them to various manufacturers' specifications, one pump in particular is generally found to have the required design and also to be available in materials of construction to meet the requirements of the job. When the pump's performance curves are matched to the system curve, an intersection occurs at a pump speed low enough to minimize wear. The curves shown in Fig. 12 are for a 4-in. horizontal slurry pump with a 14-in. impeller. Intersecting the system curve at the prescribed point is a pump curve for 1,450 rpm. In this case, 1,450 rpm is low enough to minimize wear. The selected operating point falls between 30 and 50 hp, and efficiency is just over 55%, which is quite acceptable for a relatively small slurry pump.

## Driving the slurry pump

It has been pointed out that low-rpm operation is of the utmost importance to wear reduction. For highly abrasive applications, then, operating points should *not* be met by varying the impeller diameter. Instead, pump speed should be altered and a maximum-size impeller used. In addition to reduced parts wear, money will be saved by the slightly higher pump efficiency gained. Yet another advantage of using only full-diameter impellers is that the pump user does not need to carry as many replacement impellers in his inventory, because of the fact that slurry pump impellers made of Ni-Hard or chromium iron are most readily available from pump manufacturers in full sizes.

The most common way of driving a slurry pump is by using a V-belt drive with an electric motor. However, for applications above 300 hp, gear reducers are more practical. By using V-belts drives or gear reducers, high-speed motors can be used, with resulting cost savings over more-expensive, lower-speed units.

Because of the difficulty in determining friction values—and thus total pump head—for certain slurries for which test data are not readily available, it is sometimes advisable to use V-belt sheaves with adjustable pitch-diameters. Without increasing the initial purchase cost to a great extent, these sheaves simplify balancing the system at startup and allow the pump to meet future changes in operating conditions.

## The author

**J. Ingemar Dalstad** is Sales Manager-International, for Morris Pumps, Inc., Baldwinsville, NY 13027, where he is responsible for the establishment and management of the company's many overseas representatives and licensed manufacturers. He was formerly Chief Application Engineer for the company. He holds a B.S. in mechanical engineering from the Technical Institute of Malmo, Malmo, Sweden, and has done graduate work at the Royal Institute of Technology in Stockholm.

## Engineers concerned with pumping abrasive slurries should be familiar with the . . .

# Miller Number

**TEST MACHINE** for determining Miller numbers—Fig. 1

Weight-loss data from running wear-block tests in slurries form the basis for the Miller number, a relative measure of a slurry's abrasivity and attrition. The number is useful for estimating the service life of pumps and valves.

JOHN E. MILLER, Consultant

Even though slurry pumping is a common operation in the chemical process industries, there is a lack of information on slurry abrasivity, which may be the controlling factor in service life of pumps, valves and pipelines. The Miller number, which we have developed over the last six years, is a useful tool for determining the relative abrasivity of slurries as well as the rate at which abrasivity changes with time (attrition).

The Miller number consists of two values. The first represents abrasivity and is obtained from computer analysis of weight loss in a wear-block test. The second value is a measure of attrition—that is, the loss or gain in abrasivity as particles break down during the test period. A typical Miller number might be 74 – 7.

We include a constant in the derivation of the abrasivity value, so that 1,000 is approximately equal to a slurry of 220-mesh corundum, which has reproducible abrasivity. Slurries of most industrial materials have wide variation in their abrasivity, depending on composition and particle size and shape. This can be easily seen from the typical Miller numbers in Table I.

### Developed Out of Necessity

Our work on the Miller number originated through necessity during a full-scale loop test of a 560-hp reciprocating pump handling a magnetite slurry. To evaluate the life of expendable fluid-end parts, we had to consider the amount of attrition in a recirculated slurry. We expected that attrition would cause the slurry to become less abrasive during the latter part of the test and thus give erroneous data on parts life.

We investigated several methods for measuring the slurry's change in abrasivity, but none seemed to fully duplicate the type of back-and-forth rubbing action found in a reciprocating pump. Then we recalled a "rubbing block" device used some years ago to determine the quality of hard-facing materials. This was an attempt to suspend a small amount of abrasive particles in a viscous, colloidal mud. Results were erratic because it was difficult to maintain a consistent slurry concentration under the wear block, and the reciprocating speed was too slow to provide good agitation.

We built a crude machine designed to overcome the shortcomings of the earlier device. We solved the problem of maintaining a consistent slurry by having the wear block run at a higher speed in a V-shaped tray that confined the slurry to the path of the block.

This test enabled us to arrive at a replenishment program in which the abrasivity of the slurry used in the pump study was held constant by adding "fresh" magnetite on a regular basis. Results of several years operation

Originally published July 22, 1974

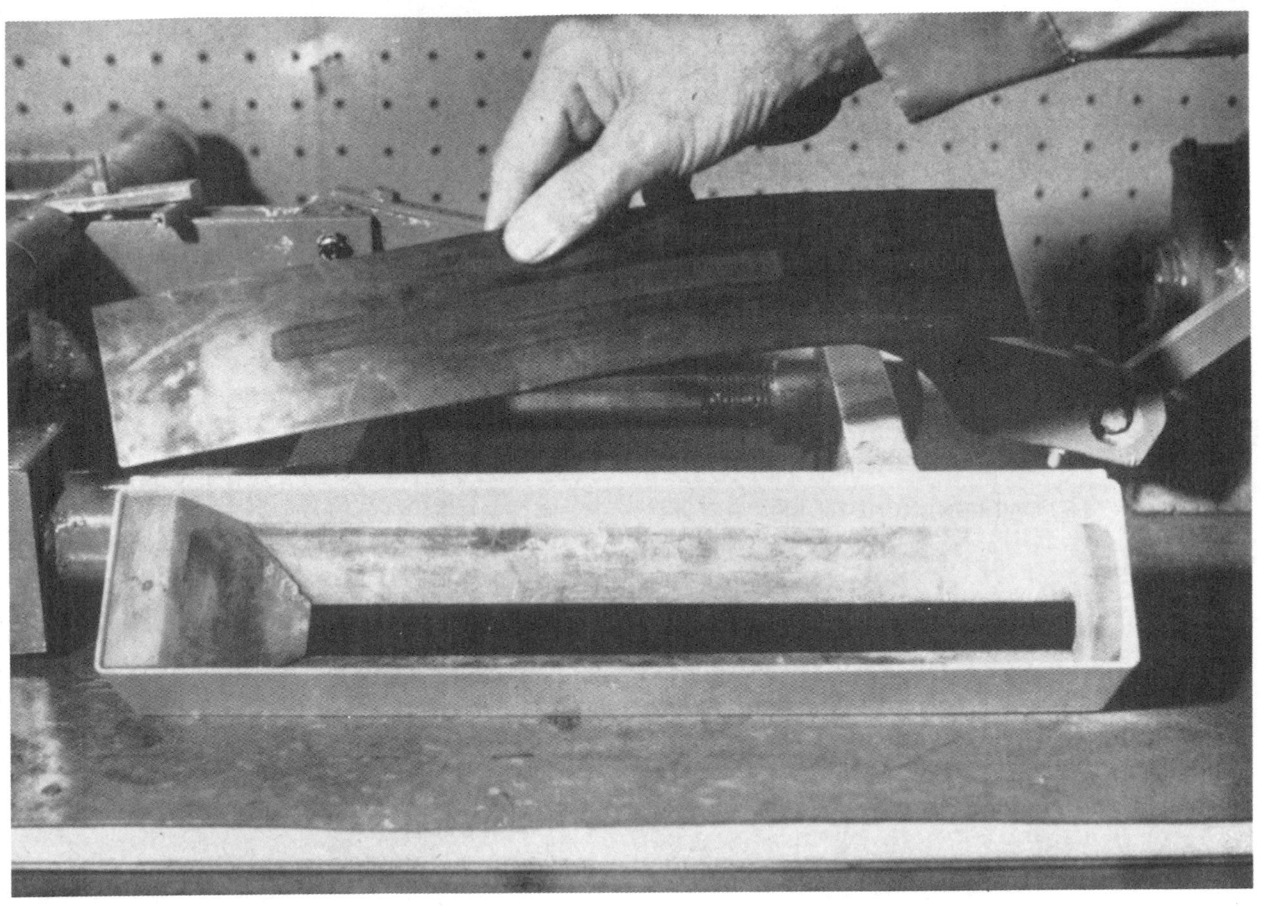

**TRAY ASSEMBLY** in test machine is designed to confine slurry particles in the path of the wear block—Fig. 2

of the pump confirm that our test procedure accurately predicted the wear rate of fluid-end parts.

Our success in measuring slurry abrasivity in the pump evaluation suggested that the procedure could have much wider application in handling slurries. For example, the relative abrasivity of a slurry dictates whether a piston or plunger type pump is required and whether flushing the packing is necessary. Valve selection is also influenced by the abrasivity and particle size of the solids in the slurry.

## More-Elaborate Machine

To extend the usefulness of the method, we constructed a more-elaborate testing machine (Fig. 1). This consists of a standard metal wear block (1.27 x 2.54 cm) that is driven at 48 strokes/min over a distance of 20 cm. As shown in Fig. 2, the V-shaped tray that holds the wear block confines the slurry particles to the block's path.

For each test, the bottom of the tray is fitted with a new piece of neoprene rubber. Then a 50% by weight slurry of the solids to be tested is placed in the tray. (Some tests are run with actual slurries as supplied.)

A 2.3-kg weight is added to the wear block, and the machine is set in motion. At the end of each stroke, a cam lifts the block 0.32 cm to allow a flow of fresh slurry under the block. Both the block holder and the tray are made of plastic to prevent electrolysis.

Our procedure calls for weighing the wear block to 0.1 mg, after it has been scrubbed in detergent and dried for 15 min under a heat lamp. After 4 h, the block is removed from the machine, washed and weighed again. This is repeated four times, while duplicate samples are run simultaneously to verify the results.

Originally, we selected a wear block made of Colmonoy No. 6 alloy because the plunger in the test pump was constructed of this material. When using replacement blocks, we found that the method of applying the surface coating yielded variable results. The popularity of 27% chrome iron in pump liners for drilling-mud service led to picking this reproducible and readily available material as a standard. Many of the early tests had been rerun with 27% chrome iron blocks in an attempt to correlate the wear resistance of the two metals. We had enough data on hand to retain the same Miller number for all of the older tests.

## Analyzing Results

After running a series of tests on available slurries, and ones made from pure minerals, we noted, as expected, that the abrasivity was not the same for each incremental run. This meant there had been a change in the characteristics of the particles, which apparently was a function of the materials' friability.

We reasoned that the rate of metal loss early in the

```
                    ABRASIVITY TEST RESULTS
              FØR DETERMINATIØN ØF MILLER NUMBER


TEST NUMBER - M126
TEST DATE   -  6/14/72
ABRASIVE MATERIAL-                  IRØN ØRE 50 PCT BY
WEAR BLØCK MTL.  - 27 PCT CHRØME IRØN        WT SLURRY

                            WEAR    BLØCK         A= 6.8577E+00

                             1        2           B= 9.7193E-01

                         ---------  ---------   * WL=(A)(HØURS)**B
IDENTIFICATIØN NUMBER       3          4
                                                AVE MILLIGRAMS LØSS
WEIGHT IN GRAMS                                 -------------------
                                                ACTUAL   *PREDICTED
     INITIAL               15.3535    12.9685   --------  ---------

     AFTER  4 HØURS        15.3258    12.9472     24.5       26.4

     AFTER  8 HØURS        15.2962    12.9192     53.3       51.8

     AFTER 12 HØURS        15.2716    12.8957     77.4       76.7

     AFTER 16 HØURS        15.2477    12.8728    100.8      101.5

                         ABRASIVITY  ATTRITIØN   LAP MATERIAL WEAR

MILLER NUMBER              121.7       -1.4          TRACE


                       100-------------------------------------1
                          -        -        -        -       **
  PERCENT               90-        -        -        -     ** 2
                          -        -        -        -   **   -
    ØF                  80-        -        -        - **     -
                          ------------------------------1**--------
   105.8                70-        -        -      **2        -
                          -        -        -    **  -        -
MILLIGRAMS              60-        -        -  **    -        -
                          -        -        1 **     -        -
   LØSS                 50----------------**------------------
                          -        -    **2          -        -
                        40-        -  **    -        -        -
1-WEAR BLØCK 1            -        - **     -        -        -
                        30-        - **     -        -        -
2-WEAR BLØCK 2            ---------*1*--------------------------
                        20-     ** 2        -        -        -
3-WEAR BLØCK 1&2          -   **   -        -        -        -
                        10- **     -        -        -        -
*-CURVE FIT LØSS          -**      -        -        -        -
                        03--------------------------------------
                                   4        8        12       16
                                           HØURS
```

**TYPICAL** printout of abrasivity-test results shows a curve of the fitted test data and gives calculated Miller number—Fig. 3

### Typical Miller Numbers for Slurries — Table I

| | |
|---|---|
| Alundum 400 mesh | 241+21 |
| Alundum 200 mesh | 1058–15 |
| Aragonite | 7+8 |
| Bauxite 16 mesh | 9–35, 22+6, 33–23 |
| Calcium carbonate | 14–2 |
| Carborundum 220 mesh | 1284–15 |
| Clay | 34–1, 36–9 |
| Clay residue | 66–17, 184–18, 226–17 |
| Coal | 6–26, 9+11, 12–17, 21–7, 28–17, 47–0, 57–3 |
| Copper conc. | 19–8, 37–11, 58–23, 68–3, 111–13, 128–5 |
| Detergent | 0–0, 5–22 |
| Gilsonite | 10+6 |
| Iron ore | 28–7, 37–13, 64+1, 79–4, 122–1, 157–11 |
| Lignite | 14–6 |
| Limestone | 22–2, 27+5, 29–6, 30+11, 33–5, 39–0, 43+6, 46–1 |
| Limonite | 113–3 |
| Microsphorite | 76–13 |
| Magnetite | 64–1, 67–4, 71–3, 134–1 |
| Mud, drilling | 10–14, 10–4 |
| Nickel | 31–7 |
| Phosphate | 74–7, 84+1, 134–12 |
| Potash | 0–0, 10+1, 11+2 |
| Pyrite | 194–4 |
| Rutile | 10–11 |
| Sand & sand fill | 51–10, 59–6, 68,9 85,16, 93–2, 246–9 |
| Sea bottom | 11–0 |
| Shale | 53–0, 59–3 |
| Sodium sulfate | 4–38 |
| Sewage (digested) | 15–7 |
| Sewage (raw) | 25–8 |
| Serpentine | 134+5 |
| Sulphur | 1–38, 1–12 |
| Tailings | 24–8, 58–8, 76–10, 91–5, 159–11, 217–15, 480–21, 644–14 |

test would be more indicative of what occurs when a slurry passes through a pump or valve for the first time. We decided to calculate abrasivity at a point two hours into the test to override the initial effect of wear on the block's finish and obtain a wear rate before attrition started. First, the weight-loss data are fitted to the equation:

$$W = At^B \qquad (1)$$

where $W$ is weight loss in g, and $t$ is time in h. Using the least-squares linear-taylor differential-correction technique, the values of $A$ and $B$ are determined for a particular series of test data. Abrasivity is then defined as the partial derivative of Eq. (1) at $t$ equal to 2 h. As mentioned earlier, we add a constant so that the abrasivity of 220-mesh corundum is about 1,000. Thus we have:

$$\text{Abrasivity} = C(dW/dt) = (C)(A)(B)(t)^{B-1}$$

Finally, attrition is calculated by taking the second derivative of Eq. (1) at $t$ equal to 2 h.

$$\text{Attrition} = \frac{d^2W/dt^2}{dW/dt} \times 100$$

where $\quad d^2W/dt^2 = (A)(B)(B-1)(t)^{B-2}$

We have prepared a computer program so that we can easily obtain a complete report (Fig. 3) showing the calculated Miller number and a curve of the fitted test data. The closeness of the actual weight loss to that predicted by Eq. (1) demonstrates that the wear-block weight loss is a true mathematical function.

## Applications

Experience shows that slurries with a Miller number below 50 are not abrasive in the sense they can be pumped with double-acting piston-type units. Materials such as coal, limestone and sulfur are usually in this range.

When slurry abrasivity is above 50, one should consider pumps other than piston types. The first choice would be a plunger pump in which the packing and plunger can be protected by a barrier of clean fluid that is injected into the stuffing box. Also, remember that in any slurry pump the valves cannot be isolated from the fluid, and they degrade at a rate proportional to the Miller number.

We should emphasize that the number should only be used to compare the life of parts in equipment of similar design. For example, a duplex piston pump handling coal with an abrasivity of 40 would have half the parts life of the same pump moving coal of number 20. However, this experience could not be used to estimate the life of a plunger pump that can be flushed continuously.

In general, the finer the material, the less abrasive it is. Corundum at 220 mesh is about four times more abrasive than 400 mesh. Although a considerable reduction in particle size is necessary to get a significant difference in abrasion, the savings in expendable pump parts could sometimes warrant the expense of milling to a finer particle size.

At present, the usefulness of the attrition factor in the Miller number is limited. As more data are accumulated, it may become important in predicting rheological changes such as those that occur in a long pipeline.

## Meet the Author

**John E. Miller** is a consulting engineer at 9850 Mercer Drive, Dallas, TX 75228. Formerly he was a development engineer with the Oilwell Div. of U.S. Steel, where he had worked since 1933 except for a five-year leave during which he was self-employed. Mr. Miller has a degree in petroleum and natural-gas engineering from Pennsylvania State University and is a registered professional engineer in Texas.

# Positive-Displacement Pumps

## Types—Principles—Application—Systems Design

*Types of positive-displacement pumps are so numerous that a user interested in their unique advantages may be badly confused. This report not only will aid his choice but will also show him how to design a complete pumping system.*

F. A. HOLLAND and F. S. CHAPMAN, *Lever Bros. Co.*

Pumps are perhaps the most common kind of equipment used in the chemical process industries. Although there are doubtless more centrifugal pumps in operation than positive-displacement pumps, the latter are numerous, particularly for higher-head applications, and extremely diverse in principle and design. Since there are so many types of positive-displacement pumps, it is worthwhile to classify them and to examine the principles, advantages, disadvantages and applicable ranges of each type.

Whereas the delivery of a centrifugal pump will vary with the pressure differential under which it is operating, a positive-displacement pump's delivery will be substantially invariable (except for steam pumps and for variable slip) regardless of the differential. It is for this reason that positive pumps are used for metering in many cases; it explains too why such pumps can be built for use against high pressures.

All positive pumps operate by trapping a quantity of liquid and then forcing it out against the pressure of the discharge. They can be classified under three main headings: (1) rotary pumps; (2) reciprocating pumps; and (3) miscellaneous pumps that are neither true rotary nor reciprocating. As a general rule, they make use of some form of rotating piston, plunger or eccentric.

The first class, rotary pumps, forcibly transfers liquid from suction to discharge through the action of such members as rotating gears, lobes, vanes or screws, operating inside a rigid container. These pumps do not require valves. Normally their pumping rates are varied by changing the speed of the rotor. They may be classified under the headings of (1) external gear; (2) internal gear; (3) lobe; (4) vane; (5) screw; and (6) flexible impeller.

Pumps of the second class, reciprocating pumps, forcibly discharge liquid against a pressure by changing the internal volume of the pump. These pumps require valves on both their suction and discharge sides. Their pumping rates are varied by changing either the frequency or the stroke length. They may be classified as either: (1) piston or plunger; or (2) diaphragm.

There are also a number of miscellaneous positive-displacement pumps that have some of the characteristics of both rotary and reciprocating pumps.

Positive-displacement pumps are self-priming. This is a great advantage in many applications. In general, they are capable of much higher pressures than even multistaged centrifugal pumps.

## Rotary Pumps

### External Gear Pumps

Fig. 1 shows a typical external gear pump. The fixed casing contains two meshing gears of equal size. These gears may be of the spur type, helical or double-helical (herringbone). One of the gears is coupled to the drive shaft that transmits the power from the

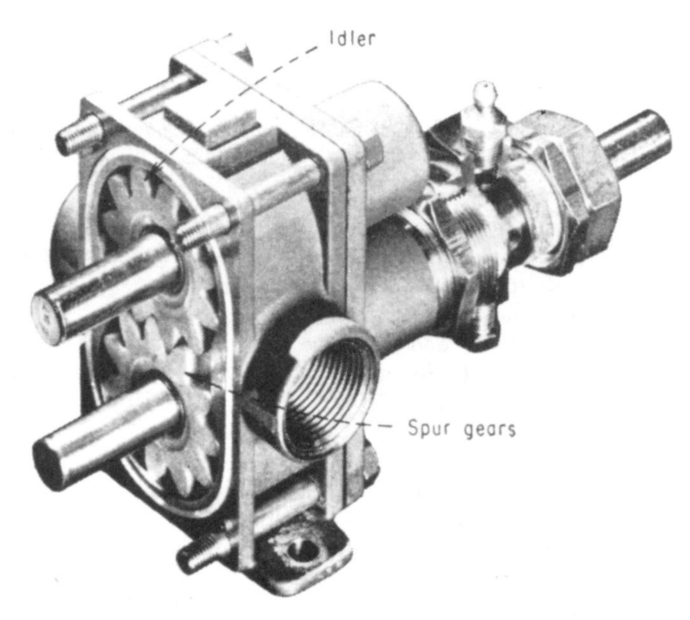


**Cutaway view of an external spur-gear pump—Fig. 1**

Originally published February 14, 1966

motor. Usually the other gear runs free, i.e., it is an idler gear. For severe service, however, both pumping gears are driven by timing gears.

In-line gear pumps are now available that do not require a foundation, as shown in Fig. 2.

A partial vacuum is formed by the unmeshing of the rotating gears. This causes liquid to flow into the pump. The liquid is then carried to the other side of the pump between the rotating gear teeth and the fixed casing, where the meshing of the rotating gears prevents return of liquid to the suction side.

The direction of gear rotation determines the discharge side of the pump. A basic external gear pump can discharge liquid either way, depending on the direction of the gear rotation. However, most external gear pumps are equipped with built-in relief valves to limit the discharge pressures generated. Thus, in these pumps, the direction of gear rotation is fixed and is usually clearly marked on the pump. The rotating gears are capable of pumping out any air present in the pipeline.

*Advantages*—These pumps are therefore self-priming. They give constant delivery for a set rotor speed, uniform discharge with negligible pulsations, and do not require check valves. They can if necessary pump in either direction, have small space requirements, are light in weight, and can handle liquids containing vapors and gases. Changes in capacity are small with variations in viscosity and discharge pressure, while volumetric efficiencies are high.

*Disadvantages*—The liquid pumped must be comparatively clean. The pump cannot be operated against a closed discharge without damage—hence relief valves are required. Close clearances are essential between moving parts, so alignment is critical. Shaft seals are required. Liquids containing vapors and gases tend to cause erosion, while the pumps are prone to damage by foreign bodies. They depend on the liquid pumped to lubricate the internal moving parts and can be damaged if run dry. Variable-speed drives are required to provide changes in pumping rate.

*Ranges*—External gear pumps containing spur gears are normally low-capacity pumps. The upper limit is about 200 gpm. For higher capacities, the noise level and thrust on the bearings become too great. Helical and herringbone gears are therefore used in higher-capacity pumps. Herringbone-gear pumps are available with capacities up to 5,000 gpm. Furthermore, helical and herringbone-gear pumps can, in general, handle higher-viscosity liquids. The latter has been known to handle liquids with viscosities up to 5 million SSU.

Spur gears are always used in very small pumps because of the difficulty of cutting small helical and herringbone gears.

Rotor speeds up to 1,800 rpm. are used, although the most common speed is 1,150 rpm. Mechanical efficiencies depend on the liquid pumped but can be as high as 90%. Line sizes range from $\frac{3}{8}$ in. NPS for small spur-gear pumps to 16 in. NPS for large herringbone pumps. External gear pumps have been made to operate at temperatures up to 800 F. Normally the upper limit of pressure for external gear pumps is 500 psig., although pumps have been made to operate against pressures up to 5,000 psig. The cost of these pumps, in steel, without motors, ranges from $50 to $6,000, depending on the size and construction.

**Some** positive pumps, such as this gear type, are available for in-line mounting, without a foundation —Fig. 2

## Internal Gear Pumps

Fig. 3 shows a typical internal gear pump. The drive shaft from the motor is coupled to a rotor that has internally cut gear teeth. These mesh with the teeth of an externally cut idler gear. The idler gear is set off-center from the rotor; the resulting space on the side opposite the meshing point is filled with a stationary crescent-shaped filler piece attached to the casing.

As the rotor turns, a partial vacuum is formed by the unmeshing of the internal teeth of the rotor and the external teeth of the idler. This causes liquid to flow into the pump. Liquid is then carried to the other side of the pump between the teeth of both the rotor and idler, and the fixed casing. The crescent part of the fixed casing divides the liquid flow between the rotor and idler gears.

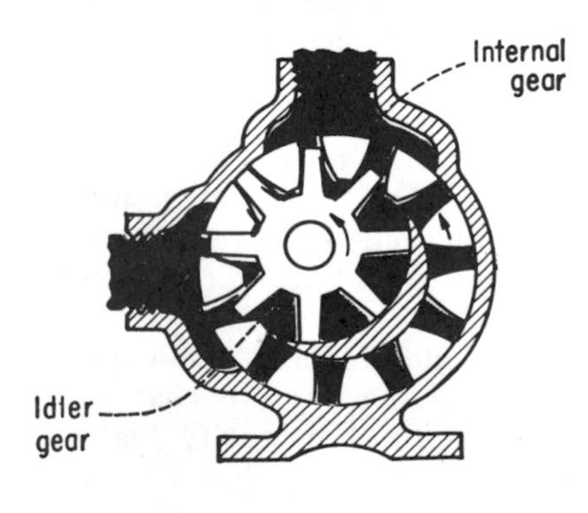


**Both** internal and external gears trap liquid and move it through this internal-gear type of pump —Fig. 3

*Advantages and Disadvantages*—These are the same as listed for external gear pumps. However, in contrast, the gears in an internal gear pump rotate in the same direction. The slow relative rotation and the rolling contact between the gears reduces friction, wear and turbulence, and also permits pumping of liquids that are more shear sensitive.

*Ranges*—Line sizes range from ¼ to 8 in. NPS. Internal gear pumps are available with capacities up to 1,100 gpm. In general, discharge pressures are limited to 100 psig.

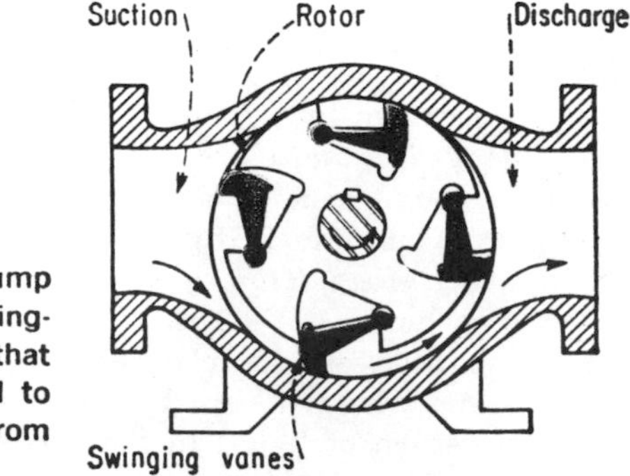


**Swinging-vane** pump resembles sliding-vane type except that vanes are pivoted to swing outward from the rotor—Fig. 6

## Lobe Pumps

These operate like external gear pumps. In lobe pumps, the gear wheels are replaced by impellers of two, three or four lobes. The principle of a three-lobe pump is illustrated in Fig. 4. The two impellers are driven independently and usually a small clearance is maintained between them.

*Advantages and Disadvantages*—These are the same as for external gear pumps, with the following exceptions. The output from lobe pumps pulsates more than that from external gear pumps. Since the lobes operate as gears, the greater the number of lobes, the less the pulsation.

Lobe pumps tend to wear less than external gear pumps. In some pumps, the lobe ends are fitted with replaceable packing strips to protect the lobes from wear. The alignment is not quite as critical as for external and internal gear pumps.

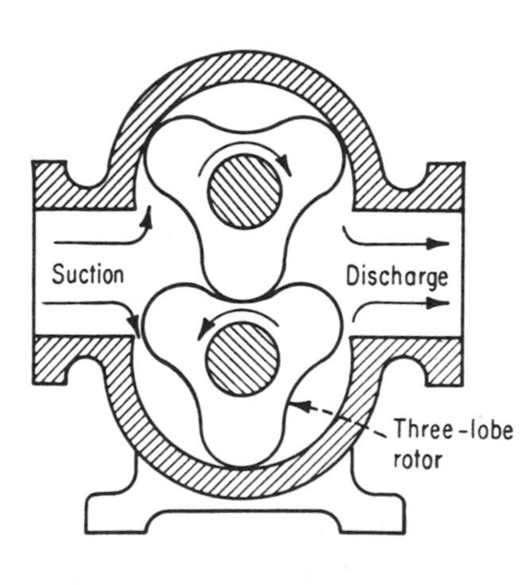


**Lobe pumps,** generally in three- and four-lobe types, are similar to gear pumps in principle—Fig. 4

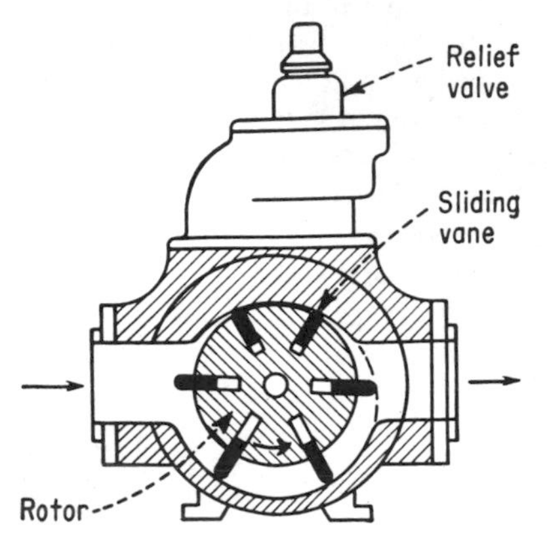


**Several** sliding vanes in slots in an eccentric rotor carry liquid through pumps of the sliding-vane type — Fig. 5

*Ranges*—Lobe pumps are readily available in line sizes from ¾ to 6 in. NPS, and in capacities up to 600 gpm. However, lobe pumps up to 2,000-gpm. capacity have been made. The commonly available pumps operate at temperatures up to 600 F. and against discharge pressures up to 400 psig. Costs of common cast-iron pumps, without motors, range from $300 to $2,000. Commonly available stainless-steel pumps, without motors, range in cost from $600 to $5,000, depending on capacity.

## Vane Pumps

Fig. 5 shows a typical sliding-vane pump. In this pump, the rotor is mounted off-center. At regular intervals around the curved surface of the rotor are rectangular vanes that are free to move in a radial slot. As the rotor revolves, the vanes are thrown outwards by centrifugal force to form a seal against the fixed casing. The eccentricity of the revolving rotor produces a partial vacuum at the suction side of the pump, causing an inflow of liquid. This is carried to the other side of the pump in the space between the rotor and the fixed casing.

Normally, pumping rates of rotary pumps are varied by changing the speed of the rotor. In certain vane pumps, however, the pumping rate can also be varied by changing the degree of eccentricity of the rotor, since this determines the amount of liquid carried through per cycle. This is usually done by turning a graduated dial to operate a pinion and rack, which in turn rotates an eccentric cylindrical liner in the fixed casing. The arrangement lends itself well to automatic control.[13]

Another type of vane pump is the swinging-vane pump shown in Fig. 6.

*Advantages*—Vane pumps are self-priming, robust, and give constant delivery for a set rotor speed. They produce uniform discharge with negligible pulsations. Their vanes are self-compensating for wear, and the original capacity is not affected until the vanes are critically worn. Vanes and liners are easily replaced. These pumps do not require check valves, they can pump in either direction and, in addition, they require little space.

Also they are light in weight, and can handle liquids containing vapors and gases. Only small changes in capacity occur with variations in viscosity and discharge pressure, while volumetric efficiencies are relatively high. In some vane pumps, the eccentricity of the rotor can be adjusted to provide changes in pumping rate for a fixed rotor speed.

*Disadvantages*—Vane pumps cannot be operated against a closed discharge without damage to the pump—hence relief valves are required. They cannot handle abrasive liquids; seals are required; and foreign bodies can damage the pump.

*Ranges*—Vane pumps are available in line sizes from 1 to 10 in. NPS, and in capacities up to 2,000 gpm. They operate at temperatures up to 450 F. and against discharge pressures up to 150 psig. Such pumps can also be made to operate up to 2,000 psig. The commonly available pumps handle liquids with viscosities up to about 500,000 SSU. Rötor speeds up to 960 rpm. are used. Costs of the commonly available basic cast-iron pumps, without motors, range from $100 to $4,000, depending on capacity.

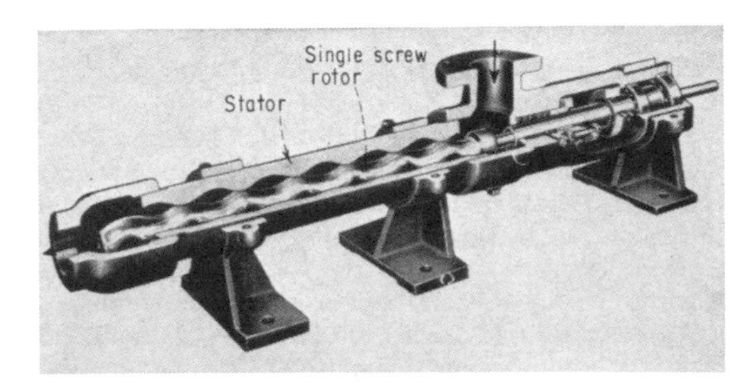


**Single-screw** pump makes use of a "progressing cavity" to carry liquid from suction to discharge—Fig. 7

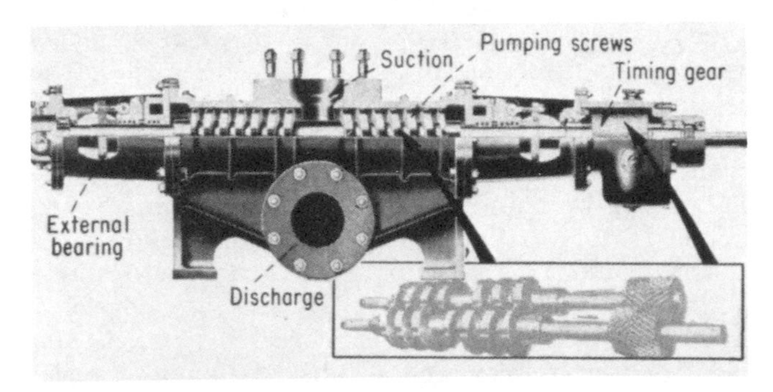


**Action** of a twin-screw pump resembles gear pump—Fig. 8

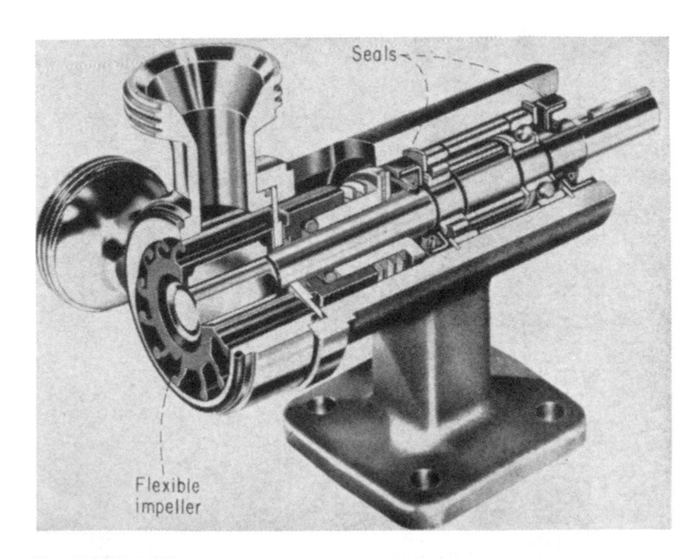


**Flexible-impeller** pump has rubber-like rotor with teeth giving action somewhat like that of a vane pump—Fig. 9

## Screw Pumps

Fig. 7 shows a typical single-screw pump. A helical-screw rotor revolves in a fixed casing. The latter is so shaped that cavities formed at the suction move towards the discharge side of the pump as the helical screw rotates. The creation of a cavity produces a partial vacuum that causes liquid to flow into the pump, after which it is carried to the other side of the pump inside the progressing cavity. At the discharge side, the shape of the fixed casing causes the cavity to close. This generates an increase in pressure, forcing the liquid into the outlet line. The discharge pressure required determines the length and pitch of the helical-screw rotor.

Another screw pump illustrated in Fig. 8 employs twin screws with involute helicoid surfaces in place of a single helical-screw rotor. On entering the suction chamber, liquid is divided and flows to opposite ends of the pump body. At these points, liquid enters the twin rotors and is conveyed to the center of the pump where it is discharged into the outlet line. Timing gears lubricated by the liquid being pumped prevent contact between the twin-screw rotors.

Although screws will convey liquid either way, depending on the direction of rotation, the types shown in Fig. 7 and 8 are designed to operate in one direction only.

*Advantages*—Screw pumps are self-priming, and give uniform discharge with negligible pulsations. They do not require check valves, produce relatively low wear on moving parts, and can handle liquids containing vapors and gases. Liquids containing a substantial amount of solids can be pumped. Exceptionally long life and low shear are features.

*Disadvantages*—These pumps are bulky and heavy, and subject to relatively large changes in capacity with variations in viscosity and discharge pressure. Seals are needed. Pumps cannot be operated against a closed discharge without damage—hence relief valves are essential. Variable-speed drives are required to provide changes in pumping rate.

*Ranges*—Screw pumps are available in line sizes from ¾ to 12 in. NPS and in capacities up to 3,000 gpm.

The commonly available pumps operate at temperatures up to 500 F. and against discharge pressures up to 1,000 psig. Rotor speeds as high as 3,600 rpm. are used with single double-screw pumps. With twin double-screw pumps, rotor speeds are normally limited to 1,150 rpm. Screw pumps can pump liquids with viscosities as high as 5 million SSU. Costs of single double-screw pumps range from about $600 to $4,000, depending on size, for the basic cast-iron pump, without motor. Twin double-screw pumps, without motors, range in cost from $1,000 to $12,000 for cast-iron pumps having internal timing gears. The corresponding price range for twin double-screw pumps with external timing gears is $1,300 to $14,000.

### Flexible-Impeller Pumps

These are eccentric-rotor pumps, illustrated in Fig. 9. The rotor consists of an impeller with flexible blades that unfold as they pass the suction port. This unfolding creates a partial vacuum that draws in liquid behind each successive blade. Rotation then carries the liquid from the suction to the discharge port. The eccentricity of the impeller causes the flexible impeller blades to bend against the fixed casing at the discharge port. This results in a squeezing action on the liquid to force it into the outlet line.

*Advantages*—These pumps are self-priming and produce uniform discharge with negligible pulsation. They do not require check valves, can pump in either direction, and are of simple construction, easily disassembled and light in weight. They have small space requirements, can handle liquids containing vapors and gases, produce low shear, and can pump liquids containing solids. They can be operated against a closed discharge for short periods without damage to the pump. The flexible impellers are easily replaced at low cost.

*Disadvantages*—These pumps have limited range of application, cannot pump against high pressures, and cannot be used to pump abrasive liquids. They are not suitable for heavy-duty applications. Variable-speed drives are required to provide changes in pumping rate. Shaft seals are necessary.

*Ranges*—Flexible-impeller rotary pumps are available in line sizes from ¼ to 3 in. NPS, and in capacities up to 100 gpm. These pumps normally operate at temperatures up to 180 F. and against discharge pressures up to 30 psig. Pumps are also available to pump up to 60-psig. discharge pressure. Rotor speeds are in the range of 75 to 1,750 rpm. Only liquids with viscosities less than 5,000 SSU can be handled satisfactorily with these pumps. Cost of the basic steel pump, without motor, ranges from $20 to $400, depending on the size.

## Reciprocating Pumps

### Piston and Plunger Pumps

In piston pumps, the pumping action is effected by a piston that moves in a reciprocating cycle through a cylinder. The piston is attached to a rod passing through a packed seal at one end of the cylinder. Some form of crank transforms the rotary motion of the drive to a reciprocating motion on the piston rod.

As the piston moves to increase the available volume in the cylinder, liquid enters through the one-way suction valve. Then the piston moves in the reverse direction to decrease the available volume in the cylinder, and liquid is forced into the outlet line through the one-way discharge valve. The pumping rate varies as a sine curve from zero at the point when the piston reverses its direction, to a maximum when the piston is approximately half way through its stroke. In addition, the pumping rate remains zero for the time it takes to fill the cylinder with liquid. This delay can be avoided by using each side of the piston to pump liquid. In this case, the pump is double-acting.

The fluctuations in pumping rate for both single-acting and double-acting pumps can be reduced if more than one cylinder is employed in parallel. Simplex, duplex, triplex and quadruplex pumps employ one, two, three and four cylinders, respectively. These

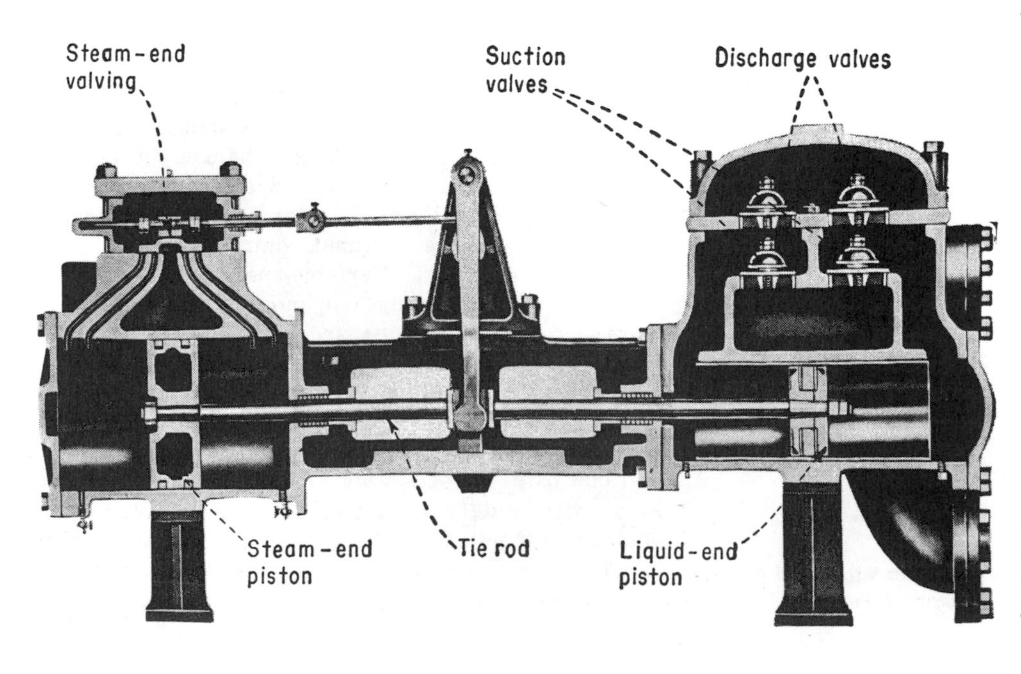


**Steam pumps** are an old standby. Steam cylinder at left drives pump cylinder at right —Fig. 10

may either be single- or double-acting. Air chambers are also used on piston pumps to dampen the pulsations from the pump's discharge.

In steam pumps, a double-acting piston is directly driven by steam. The pump piston is connected to the steam piston by a tie rod, without any intermediate crank or flywheel, as shown in Fig. 10. In so-called power pumps, the pistons are driven by a motor-driven crankshaft. In both types of piston pumps, the pumping rates may be varied either by changing the reciprocating speed or the stroke of the piston. The speed of a steam pump will vary with discharge pressure. This is not the case with power pumps.

A plunger is used in place of a piston for operation at higher pressures. Plunger pumps operate in almost the same manner as piston pumps. They differ in that the plunger moves through a stationary packed seal, whereas in piston pumps the packing seal is carried on the piston. Plungers are necessarily single-acting devices. A duplex pump illustrating the plunger principle is shown in Fig. 11.

Plunger pumps are commonly used not only for higher pressures, as in multiplex power pumps, but also as metering or proportioning pumps. The latter are either of the plunger type or employ diaphragms. (See Fig. 12.) Proportioning pumps commonly have some method of varying the length of stroke for adjusting the delivery. Often this adjustment, as in Fig. 12, is possible while the pump is operating.

Another class of variable-delivery plunger pump is the multiple-plunger type with swash-plate drive. Such pumps were formerly used for accurate delivery as rayon spinning pumps although they have largely been replaced today in fiber-spinning applications by gear pumps. Pumps of the same general type are extensively used for hydraulic drives where an easy method of speed variation is needed, as in certain mixers, cooling-tower-fan drives, and the like.

In these pumps, a rotating cylinder carries a number of bores parallel to the axis of rotation. The cylinder rides in a casing having circumferential inlet and discharge passages near one end. Plungers lapped to fit the bores accurately are attached flexibly at their outer ends to a rotating plate (swash plate) that is tiltable with respect to the axis of pump rotation. If the plate is perpendicular to this axis, rotation of the cylinder does not move the plungers and no delivery results. As the disk is tilted, the amount of motion of the plungers—and hence the delivery—increases. Thus the delivery can be adjusted at any value from zero to some maximum.

The same general principle, but using only a single

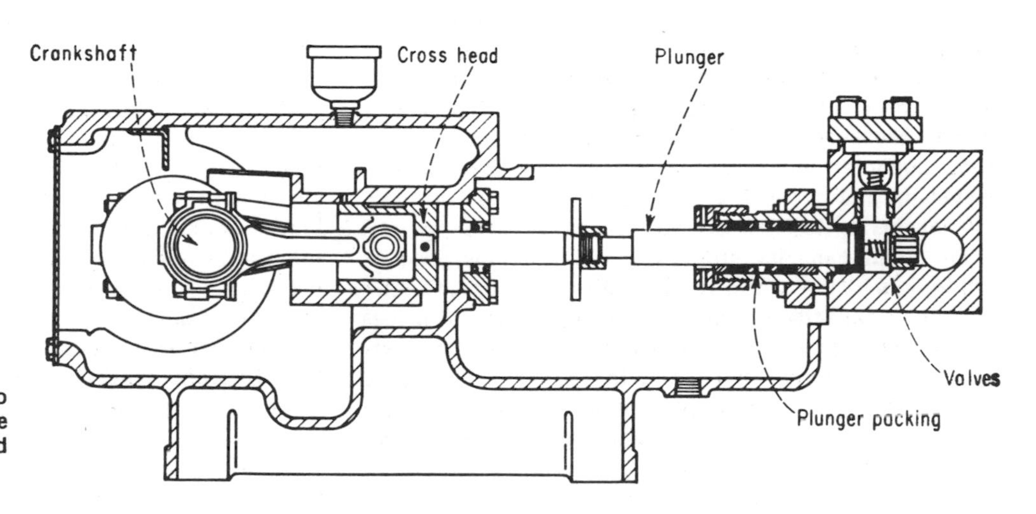


**Plunger pumps,** in single- to several-cylinder types, are sealed with a packing around the plunger—Fig. 11

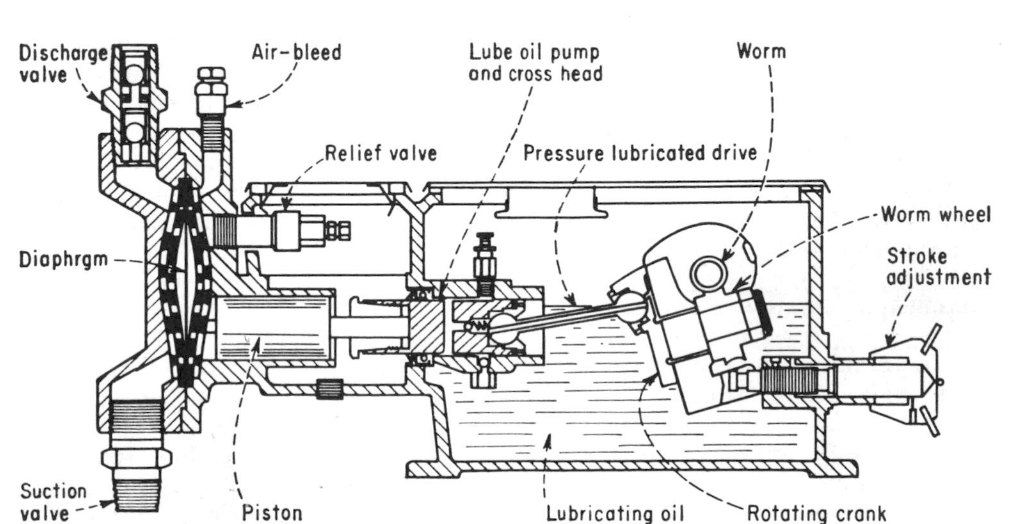


**Liquid-drive** diaphragm pump has variable-stroke device, is used for metering precise quantities—Fig. 12

plunger, appears in the variable-stroke mechanism shown at the right of Fig. 12.

*Advantages*—Piston and plunger pumps are self-priming. If the stroke is variable, they do not require variable-speed drives to alter pumping rates. They can handle liquids containing vapors and gases, and have exceptionally long life.

*Disadvantages*—They tend to be heavy and bulky and have a pulsating discharge. In the case of steam pumps, a steam source must be available. Power pumps cannot be operated against a closed discharge (hence relief valves are required). Most types require inlet and discharge valves.

*Ranges*—Steam pumps range in capacity from 20 gpm. to 2,000 gpm., and are available to operate against discharge pressures up to 20,000 psig. Typical prices range from $450 for a cast-iron duplex pump of 20-gpm. capacity, for discharge pressures to 300 psig., to $75,000 for a specially designed carbon-steel duplex pump delivering 800 gpm. against an 800-psig. discharge head.

The expense of these pumps is offset by their complete reliability and long life. At the Seager Evans & Co. factory in England, which makes "Coates' Plymouth Gin," a steam pump has been running without repair or replacement since 1855—111 years!

Power pumps range in capacity from 10 gpm. to 1,000 gpm. and can operate against discharge pressures up to 1,000 psig. Typical prices for power pumps range from $700 for a duplex 10-gpm. plunger pump with a 5-hp. electric motor, operating against a discharge pressure of 500 psig., to $125,000 for a 1,000-gpm. nonoplex pump with a 1,500-hp. electric motor, operating against a discharge pressure of 1,000 psig. Efficiencies up to 95% are possible with power pumps.

The small-capacity simplex plunger pumps used as single or multiple units for metering or proportioning operate at capacities as low as 0.15 gph. Typical prices for metering pumps are $170 for a 6 to 10 gph. cast-iron simplex pump with a $\frac{1}{4}$-hp. electric motor, to $9,000 for a 1,200-gph. duplex pump with a 5-hp. electric motor. In general, the smaller the plunger diameter, the greater the discharge pressure possible. A 2-gph. pump is available with a $\frac{5}{16}$-in.-dia. plunger that can operate against a discharge pressure of 50,000 psig.

Slip leakage in reciprocating piston and plunger pumps is normally less than 5% and is sometimes less than 1%. Piston and plunger pumps operate at temperatures up to 550 F. Power and steam pumps operate with speeds up to 500 and 150 strokes/min., respectively.[10] The upper speed limit for metering pumps is about 350 strokes/min.[10]

## Diaphragm Pumps

Diaphragm pumps can operate either mechanically or through fluid pressure on the drive side of the diaphragm.

A typical diaphragm pump of the second type (using a liquid drive) is shown in Fig. 12. The diaphragm takes the place of a piston. The only moving parts in contact with the process liquid are the flexible diaphragm and the inlet and outlet one-way valves, which may be of either the ball-check or the flapper-check type.

The pulsating motion of an air-driven diaphragm pump sometimes used for handling slurries is obtained by alternately admitting air to, and venting air from, the drive side of the diaphragm. This can be accomplished by using time-controlled solenoid valves. For high-suction applications, vacuum is required to restore the suction position of the diaphragm.

In liquid-drive pumps (Fig. 12), liquid is used to transmit motion from a reciprocating piston to the diaphragm. The liquid must be recirculated to avoid overheating. This type of actuation is used in many proportioning pumps.

In mechanical pumps, the motion is directly transmitted to the diaphragm as in the case of piston pumps. For higher pressures, as in filter pumps, it is normal practice to use concentric rings to support the flexible diaphragm.

Liquid- and mechanical-drive pumps give a liquid pumping rate that is substantially independent of the liquid-discharge pressure. In contrast, air-operated pumps (like steam pumps, which are also pressure exchangers) decrease in liquid-pumping capacity with an increase in liquid-discharge pressure. The pumping capacity falls to zero when the liquid-discharge pressure becomes equal to that of the actuating air.

*Advantages*—Diaphragm pumps are usually self-priming. If provided with a variable-stroke mechanism (as in Fig. 12), they do not require variable-speed drives to alter pump speeds. They can handle liquids containing vapors and gases, or liquids containing a substantial amount of solids. Process liquid does not come into contact with the prime moving parts, and the pump can be run dry for an extended period of time. No packings or seals are required. Some types are adjustable while in operation. Air and hydraulically driven diaphragm pumps can be operated against a closed discharge without damage to the pump.

*Disadvantages*—Diaphragm pumps are bulky, and the air-driven type suffers large variations in capacity with changes in discharge pressure. They have a pulsating discharge and, with mechanically driven diaphragms, cannot be operated against a closed discharge without damage to the pump. Like piston pumps, they require check valves.

*Ranges*—Diaphragm pumps normally range in capacity from 4 to 100 gpm. However, diaphragm metering pumps are available to operate at capacities as low as 1 gph. Some of them can operate against discharge pressures up to 3,500 psig. Normally, diaphragm pumps are not used to handle liquids with viscosities greater than 3,500 SSU.

Typical costs for air-driven diaphragm pumps range from $300 for an unlined, cast-iron pump of 5 gpm. capacity, to $2,000 for a cast-iron pump of 90 gpm. capacity, lined with Hypalon. Typical costs for mechanically driven proportioning pumps, without motors, are $150 for a 1-gph. polyvinyl chloride dia-

**phragm pump, and $3,000 for a 10-gpm. pump with a Teflon diaphragm.**

# Miscellaneous Pumps

## Eccentric-Cam Pumps

These pumps are sometimes called rotating-piston or -plunger pumps. In a true rotary pump, the liquid being pumped is in direct contact with a rotating surface. In addition, the total volume available in the pump for liquid remains constant. Eccentric-cam pumps can be divided into those that maintain a constant volume of liquid in the pump, and those that have a variable liquid volume in the pump.

*Constant-Volume Eccentric-Cam Pumps*—These resemble true rotary pumps in operating with a constant volume of liquid in the pump—but differ from them in not having a rotating surface in contact with the pumped liquid. A typical constant-volume eccentric-cam pump is illustrated in Fig. 13. The rotating eccentric cam moves inside a cylindrical plunger in direct contact with the pumped liquid. This creates a cavity at the suction side of the pump that progressively increases as the cam rotates. The partial vacuum created draws in liquid and fills the pump. At the end of the cam cycle, the liquid in the pump is open to the discharge side and sealed from the suction side. Then, during the next cam cycle, the volume available for this liquid progressively decreases and the liquid is forced into the outlet line.

These pumps do not require check valves. A loose septum fitting prevents backflow from the discharge to the suction side of the pump. This fitting also prevents the cylindrical plunger from rotating. The only rotating component of the pump is the eccentric cam that rotates inside the cylindrical plunger. This in turn causes the cylindrical plunger to change position relative to the fixed casing. A somewhat different idea, the rotary plunger pump, is illustrated in Fig. 14.

Still another type of constant-volume eccentric-cam pump is the flexible-liner or squeegee type shown in Fig. 15. The eccentric cam rotates within a flexible liner. A progressive squeegee action is exerted on

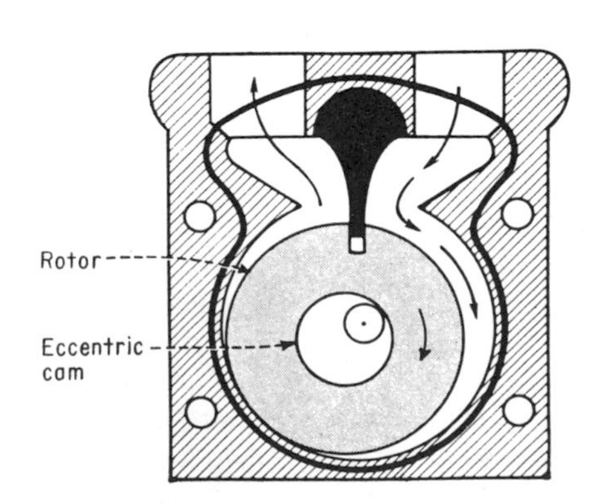


**Several** types of eccentric-cam pump are built. In this one, a tongue at the top seals the rotor—Fig. 13

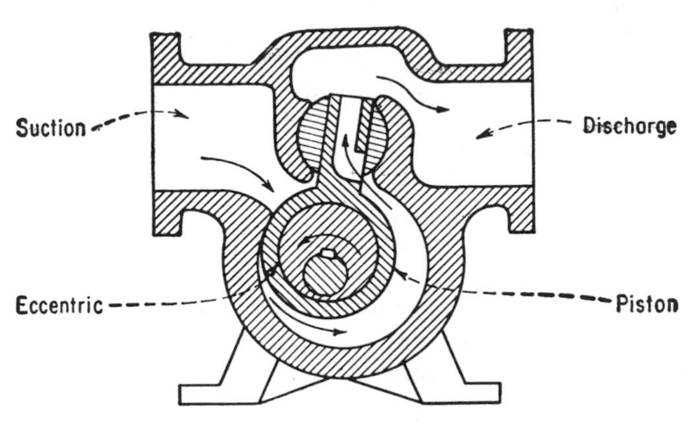


**Another** eccentric-cam type, the rotary plunger, discharges through a sliding extension of the plunger—Fig. 14

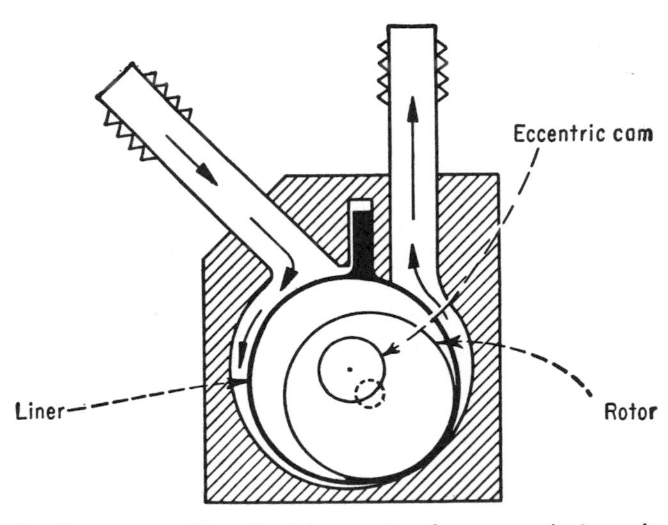


**Another** pump with a flexible member forces an elastomeric liner against the body by an eccentric rotor—Fig. 15

the liquid trapped between the flexible liner and the fixed casing. No packing seals are required because the flexible liner isolates the process liquid from direct contact with any primary moving parts. The pumping action is analogous to the squeezing of a soft tube containing liquid.

A fourth type of constant-volume eccentric-cam pump is the eccentric-cam diaphragm pump. The eccentric cam rotates within a circular housing inside the circular diaphragm. This rotation causes the circular diaphragm to change position relative to the fixed casing. The movement produces a cavity adjacent to the suction port. This draws liquid in. Further movement of the diaphragm forces liquid through the pump and out through the discharge port.

*Variable-Volume Eccentric-Cam Pumps*—These resemble true rotary pumps in having a rotating surface in direct contact with the pumped liquid. They differ from them in not maintaining a constant volume of liquid throughout the pumping cycle.

An eccentric cam fits closely into a shoe, as shown in Fig. 16. (Actually there are three such elements on the same shaft, with cams 120° apart.) As the eccentric cam rotates, the volume available for liquid inside the shoe successively increases and decreases. The cam also produces an up-and-down movement of the shoe and functions as a slide valve, alternately connecting the interior of the shoe to the suction and discharge sides of the pump. Since the available volume in the shoe is increasing when the suction side is connected—and decreasing when the discharge side is connected—a liquid-pumping action results. The three overlapping cam-shoe systems provide a more uniform pumping rate than a single system. This type is known as the sliding-shoe pump.

### "Peristaltic" Pumps

A typical peristaltic pump is shown in Fig. 17. At each end of the rotor is a roller. A flexible tube passes through the fixed casing of the pump. As the rotor revolves, the rollers press and run on the flexible tube, exerting a squeegee action on the liquid in the tube and producing an even flow of liquid. The tubing must be sufficiently flexible to allow the impeller rollers to squeeze it until it is completely closed.

Such pumps require no check valves or shaft seals. The process liquid is completely isolated from the prime moving parts.

All the pumps discussed under the "miscellaneous" category have the following advantages: they are self-priming, do not require check valves, can pump in either direction, have small space requirements, are light in weight, can handle liquids containing vapors and gases, and produce low shear.

In addition, flexible-liner and peristaltic pumps have the additional advantage that the process liquid does not come into contact with the prime moving parts of the pump. Thus no seals are required. These two pumps can also be run dry for an extended period of time.

All the miscellaneous pumps discussed are relatively low in capacity (40 gpm. or below). In addition, they can operate only against a relatively low discharge pressure (100 psig. or below). In general, they can only handle liquids with viscosities below 5,000 SSU.

Typical costs are $100 to $250 for peristaltic pumps, without motors, and $150 to $600 for flexible-liner pumps, without motors.

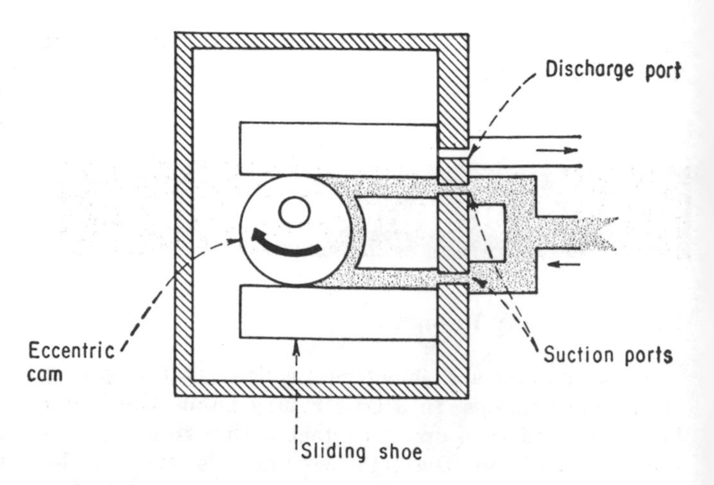


**"Sliding shoe" encloses an eccentric rotor, acts like a slide valve; three such "cylinders" overlap—Fig. 16**

## Pump Auxiliaries

### Pump Drives

The total power consumed by a pumping system is the sum of the powers required to (1) raise the liquid level; (2) overcome friction in the suction line, discharge line, pump internals, bearings, shaft seals and packing glands; and (3) compensate for power losses in the drive train between the motor and pump. Normally, pump drives should be overpowered by about 10%.

Although internal combustion engines, air and steam are used to provide power, by far the most common drive for pumps is the electric motor. Alternating-current electric motors are widely used, with direct-current motors much less common. Synchronous and induction motors are the most readily available types of a.c. motor.

*Synchronous Motors*—These operate at a constant speed under widely varying load conditions. The designed motor speed is a function only of the frequency of the power supply. Synchronous motors are equipped with a d.c. generator that supplies the rotor current. The d.c. generator is driven either by the synchronous motor itself or by a small auxiliary a.c. motor. The stator uses a.c. power to produce the required magnetic field. This is synchronized with the d.c. magnetic field in the rotor, hence the term "synchronous" motor.

The efficiencies of synchronous motors are high, falling between 92 and 97%. These motors require complex rotor windings and control devices for the rotor's d.c. field. The expense of the auxiliary equipment precludes the use of this type of motor below 250 hp. While operating speeds range between 300 and 3,600 rpm., the most common speeds are 900 and 1,200 rpm. Standard synchronous motors are available in sizes up to 10,000 hp.

*Induction Motors*—In these motors, a.c. power in the stator is used to induce a current in the rotor. Induction motors are available in two types: wound rotor and squirrel-cage rotor. In the former, the rotor is wound with wires that cut the magnetic field of the stator when the armature is rotated. This action induces a current and magnetic field in the rotor.

The squirrel-cage rotor is not wound with wires but consists of numerous parallel bars. Squirrel-cage motors provide a relatively constant speed. However, the variation in speed is substantially greater than with synchronous motors. These motors operate best in the 600 to 3,600-rpm. range. They are the simplest and usually the least expensive of the three types dis-

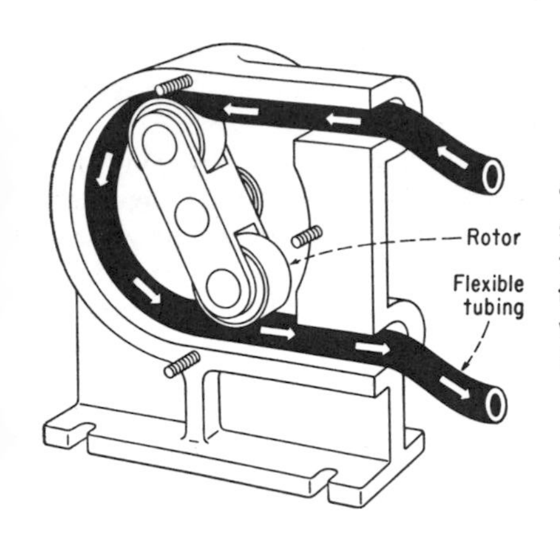


**"Peristaltic" pump simulates a body function, squeezing tube progressively toward the discharge— Fig. 17**

cussed. Efficiency is about 3% less than for comparable synchronous motors. Squirrel-cage induction motors are available in sizes from fractional to 10,000 hp. However, above 500 hp., synchronous motors with their higher efficiency become more economical.

*Motor Enclosures*—If the atmosphere in the immediate vicinity of a pump contains potentially explosive vapors, then a totally enclosed "explosion-proof" motor should be used. This type of enclosure prevents sparks produced at the rotor from coming into contact with the hazardous vapors. Since pumps may leak, especially at the seals, totally enclosed, explosion-proof motors should be used when pumping hazardous liquids. Explosion-proof motors are relatively expensive and bulky.

For general use in nonhazardous areas, the drip-proof or splash-proof types of motor enclosure are adequate. In the drip-proof design, the working parts are shielded from falling liquids and particles. However, protection is not afforded against fumes, dust, etc., which may be carried into the motor by the air. Dust screens may be used to prevent solid particles from entering the motor.

In most open and semi-open electric motors, the natural circulation of the atmospheric air is sufficient to dissipate the heat produced during operation. For totally enclosed motors, and some large open and semi-open motors, additional means of heat removal must be provided. Totally enclosed motors may be fan-cooled (TEFC); air is circulated over an internal heat-transfer area. Open motors may also be equipped with fans to provide forced air-circulation cooling.

## Power-Transmission Units

Most electric motors are constant-speed devices. Although the speed of a synchronous motor depends on the current frequency, the latter is normally constant. The speed of a d.c. motor can be changed by incorporating a rheostat into the electrical system.

If the required pump speed is the same as the speed of the electric motor, a direct coupling can be used. However, most electric motors operate most efficiently at or near their rated maximum speed. Frequently this is greater than the required pump speed, so in this case it is normal practice to use a constant-speed gear-reduction unit.

*Constant-Speed Gear-Reduction Units*—These are available in a wide range of speed-reduction ratios. For example, a 5:1 unit will enable a 1,750-rpm. motor to drive a 350-rpm. pump. Three types of constant-speed gear-reduction units are available: integral motor-coupled, shaft-mounted and integral pump-mounted. Essentially all of them operate in the same manner but differ in their position in the drive train. Each type has its own advantages. The integral motor-coupled and pump-mounted units are compact and require only one flexible coupling. Although the shaft-mounted unit requires two flexible couplings, it can be removed from the drive train without disturbing the pump or motor. However, these units usually require belt drives, which may not always be desirable. Furthermore, shaft-mounted units may occupy more space than the integral units.

*Variable-Speed Units*—Pumping rates should not be changed by throttling either the discharge or suction lines. The latter may cause cavitation and the former may produce a sufficient pressure to fracture the pump casing. Both methods can appreciably shorten the life of a positive-displacement pump.

Variable-speed units are readily available to enable the pumping rate of a positive-displacement pump to be adjusted without change in the drive-motor rpm. These units utilize gears, cones, belts and/or clutches to effect speed changes. Some types give an infinitely variable speed output between zero and the speed of the drive motor; others yield a series of specific speed ratios such as 10:1. Speed changes can be made while the unit is in operation.

Variable-speed drives are bulky, heavy and have a high initial cost. Nevertheless they are a good investment because of their flexibility. Their use normally increases the life of a pump. Manufacturers have tried to make variable-speed units as compact as possible in both horizontal and vertical types. The former are used when head space is limited but floor area is available. Both horizontal and vertical units are available as either "Z" or "C" units. In the Z type, the motor and output shaft are on opposite sides of the unit. In the C type, they are on the same side. The latter are usually employed when the unit is to drive a belt. In this case, it is an advantage to have the motor on the opposite side, to avoid interference with the belt.

## Flexible Couplings

The alignment between a pump and its drive train is extremely critical. The same is true of the component parts of the drive train. Misalignment results in excessive vibration, wasted power and ultimate damage to the pump and drive. It also leads to excessive wear at shaft seals and packing glands. Damage to gearing and bearings in reduction units, pumps and drives is a common result of misalignment, which does not need to be large to cause serious damage to precision parts. Two types of misalignment are possible: angular and parallel. These are illustrated in Fig. 18. Both types may be present in the same system.

Angular misalignment occurs when the center-line

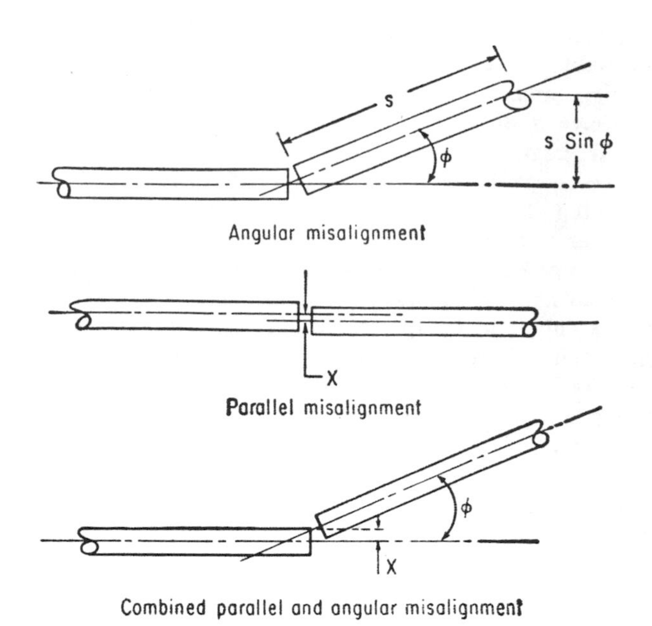


**Pump and driver must be carefully aligned to avoid angular or parallel misalignment shown here—Fig. 18**

of one shaft is at an angle to the connecting shaft. Parallel misalignment occurs when the centerlines of two shafts are parallel but in different planes displaced by a distance $x$. The allowable limits for maximum alignment error depend on such factors as shaft speed, shaft length, and construction of the component parts. In practice, the angular misalignment should be less than 1°, although most flexible couplings can generally correct for this amount of misalignment.

The adverse effect of angular misalignment increases with shaft length. For a shaft length $s$ and a misalignment angle $\phi$, the displacement at the end of the shaft is $s \sin \phi$. Therefore, if the driving shaft is rigid and the pump shaft absorbs all the angular misalignment, the circle swept out by the end of the pump shaft will have a radius of $s \sin \phi$. This is true only if the pump shaft is free to adjust. Hence small angular misalignments in long shafts can exert serious strains on pumping equipment.

To reduce the effects of misalignment, flexible couplings may be used between shafts. Two general types of flexible couplings are available: single-engagement and double-engagement.

Single-engagement couplings are capable of transmitting power between angularly misaligned shafts. They have only a single degree of freedom, since they will not handle parallel misalignment. A typical example of a flexible coupling in this group is the common universal joint.

Double-engagement couplings are capable of transmitting power between shafts that suffer from both angular and parallel misalignments. A number of double-engagement couplings are available. For example, jaw couplings consists of two meshing jaw-flanges separated by a floating shock-absorbent spacer block. They are used primarily for medium-to-high speed and low-to-moderate power transmission. Since they consist of only three parts, they are easily installed and maintained.

Chain couplings are of two general types. One design consists of two sprocketed flanges held together by a standard type roller or multi-link silent chain. Frequently a cover or bonnet surrounds the coupling. Another type of chain coupling utilizes a nylon chain, which forms its own safety cover since the outer surface is smooth. Silent-chain couplings are claimed to provide increased strength and less play.

Biscuit couplings are an extremely flexible type, made in two designs: single and double biscuit. The first of these consists of two flanges. One flange is coupled to a divider plate by flexible "biscuits" of an elastomeric compound. The other flange is directly coupled by a floating tie-link. This coupling provides excellent angular and parallel adjustment for low-to-moderate speed and moderate power loads. The double-biscuit design has two sets of biscuit assemblies, one on each side of the divider plate, an arrangement that gives extra flexibility.

In another type of coupling, the shafts terminate in geared hubs that fit into an internally cut gear sleeve. For low-to-moderate service, unlubricated nylon sleeves are used. For heavy-duty service, metallic sleeves with self-contained lubrication are used.

Other types of flexible couplings incorporate bellows, flexible central disks, cushioned pins or radial-slip clutches of various designs.

Costs of flexible couplings range from $15 for an open roller-chain for a ½-in. shaft, without safety cover, to $5,000 for an 8-in.-shaft radial coupling with cover.

## Shaft Packing and Seals

Any liquid in a container will flow to conform to its shape. Most pumps are not completely closed containers since power must be transmitted to the pump mechanism by a reciprocating or rotating shaft passing through the pump casing. In many pumps, the process liquid is in direct contact with the prime moving parts. In these pumps, the point at which the shaft passes through the casing is a potential source of leakage for the high-pressure process liquid. Although the clearance between the shaft and the pump casing can be made extremely small, it can never be completely eliminated. Therefore, packings and mechanical seals are used to minimize the leakage of process liquid from the pump.

*Shaft Packings*—Although these do not usually stop leakage entirely, they can reduce it to a very low level.[2] Packing consists of pliable or semipliable material that is forced against the shaft as it passes through a stuffing box. A movable collar around the shaft, the gland, is usually secured to the stuffing box with bolts. It can be tightened to compress the packing around the shaft.

Some slight leakage of the process liquid is necessary for the packing to operate successfully. If the seal is too tight to allow a small amount of leakage,

**Operating temperature ranges for shaft packings—Table I**

| 250–525 F. Range | Maximum Temperature, F. |
|---|---|
| Duck-and-rubber-base packing | 250 |
| Plant-fiber-base packing | 250 |
| White asbestos, braided, with light-oil-and-graphite lubricant | 350 |
| White asbestos, braided, with solvent-resistant-and-graphite lubricant | 350 |
| Lead-base packing | 400 |
| Teflon | 500 |
| **525–700 F. Range** | |
| White asbestos, lattice braid | 550 |
| White asbestos over Teflon core | 550 |
| Blue African asbestos braid, asbestos over Teflon core, 35% Teflon impregnated throughout | 550 |
| Blue African asbestos, square-plaited braid | 550 |
| 100% Teflon fiber yarn, lattice braid | 550 |
| 25% glass-filled Teflon U-rings | 550 |
| Commercial-grade packings made of asbestos fiber | 600 |
| White asbestos, braid over braid, mineral-oil-and-graphite lubricant | 650 |
| **700–1,000 F. Range** | |
| Aluminum foil over asbestos core, castor-oil-and-graphite lubricant | 750 |
| White asbestos braid over brass-wire core | 750 |
| "Junk rings" of braided copper wire and aluminum packings | 900 |
| Aluminum-foil laminations, castor-oil-and-graphite lubricant | 1,000 |

the packing will overheat and burn or scorch, and its surface will become hard and smooth, no longer providing a good shaft seal. When hazardous liquids are pumped, the use of packings poses a safety problem.

The choice of a packing material depends on the liquid to be pumped, its temperature and pressure, shaft speed, and the materials of construction of the pump. Shaft packing is produced in two forms: preformed rings and flexible spirals that can be cut to size. For heavy-duty applications, laminated packings may be used. Examples of these are asbestos-wrapped copper wire, aluminum foil over an asbestos core, and asbestos over a Teflon core. Table I gives the approximate operating temperature ranges for some common packings.

Specially preformed seal rings are available for pressures up to approximately 8,000 psig. In general, high-pressure packing is weakened by a rotating shaft. Packings such as braided asbestos and Monel metal over a mica-neoprene core are generally used with non-rotating or infrequently rotated shafts, e.g., on valves.

Some of the advantages and disadvantages of shaft packings are as follows:

*Advantages*—Packings are relatively inexpensive, easily replaced in the field, and are available in a wide variety of materials. There is an extensive range of applications for most types.

*Disadvantages*—Packings require some leakage to provide lubrication. A packing gland needs frequent adjustment, and the break-in period is critical. Packing failure may result in shaft damage. A packing offers a short life if used on abrasive materials, together with increased power consumption resulting from frictional drag. Packings may corrode the gland and shaft; for example, graphite packings corrode stainless-steel shafts.

The basic source of difficulty with packed seals is that the shaft moves within a stationary packing. Mechanical seals have been designed to overcome this.

*Mechanical Seals*—In these, the sealing surface rotates with the shaft.[6, 12] Mechanical seals consist of two basic parts: a seal ring that fits tightly on the shaft, and a stationary insert secured to the pump casing. The rotating seal ring contacts the stationary insert at a face that is precisely lapped to provide a leak-proof sealing surface. A thin film of liquid must separate the two surfaces. The film may flow to the area either under the pressure of the process liquid, by capillary action, or through an auxiliary feeding device. In most mechanical seals, compression springs or bellows are used to sustain good contact between the two surfaces.

Two general categories of mechanical seals are available: hydraulically unbalanced and balanced. The former provides no way of counteracting the hydraulic forces tending to push the seal ring against the stationary mating insert. In high-pressure pumps, the force between the two faces can result in an excessive generation of heat and damage to the liquid film. To avoid this effect of hydraulic unbalance, means are available to balance the forces at the mating contact faces. The most common method is to provide a step in the shaft (see Fig. 19) that decreases the direct load on the contact surfaces. There are also more elaborate devices. Some mechanical seals are totally balanced, i.e., the internal pressure does not contribute to the forces mating the contact faces. More commonly, balance is partial.

In some seals, the process liquid is recirculated through the seal and a temperature control unit, thus providing lubricating liquid at the optimum temperature. This is useful when the process liquid tends either to vaporize or crystallize. The first of these may lead to dry running of the sealing surfaces.

Double mechanical seals are used with valuable or hazardous liquids or under extremely high pressures.

For general service, mechanical seals are commonly equipped with a ceramic seal ring and a carbon insert. With highly corrosive liquids, the seal ring may be made of tungsten carbide and the insert of highly inert carbon.

Some of the advantages and disadvantages of mechanical seals are as follows:

*Advantages*—The power consumption resulting from frictional drag is only about one-sixth of that in a packing. Since there will be relatively little wear on shafts and shaft sleeves, the unit can run for relatively long periods without repair if the seals were properly installed. Seals are good for vacuum service, and for corrosive service with suitable materials of construction. Mechanical seals do not usually contaminate the process liquid, although balanced seals operating with an additional feed of lubrication liquid are an exception.

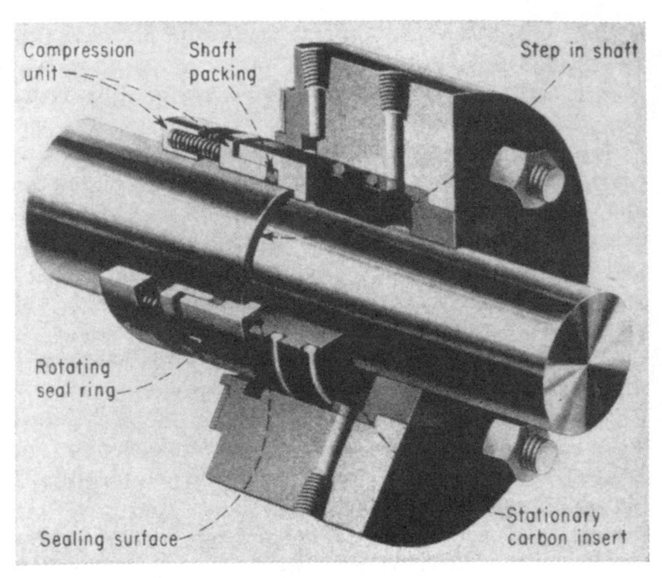


**Mechanical seals are of many types. This one uses a step to achieve hydraulic balance—Fig. 19**

*Disadvantages*—Mechanical seals are relatively expensive. Failure of the sealing surfaces can lead to sudden and heavy leakage. They are difficult to repair, are sensitive to mechanical abuse or maladjustment, have relatively complicated construction, and are subject to damage by liquids containing solids. Some mechanical seals require expensive auxiliary equipment.

Typical costs for mechanical seals are $25 for a 30-psi. unbalanced single ⅜-in.-dia. shaft seal with a ceramic seal ring, and $1,200 for a 600-psig. balanced, single 5-in.-dia. shaft seal with a tungsten-carbide seal ring. The corresponding double-seal for a 5-in.-dia. shaft costs $2,000. The pressure balancing equipment for this seal could cost an additional $2,000.

### Pump Protection Equipment

Most positive-displacement pumps require protection against overpressure. If these pumps are operated against a closed or clogged discharge, a dangerous pressure buildup will take place. As the discharge pressure increases, it will eventually exceed the design limits of either the pump casing, the discharge pipeline or the shaft seal. Fracture of the weakest component can result. The danger is particularly great with mechanically driven pumps. Pressure relief or safety valves are commonly used to protect positive-displacement pumps from damage by overpressure. Automatic overload cutoffs on the drive motor are also used.

*Pressure-Relief Valves*—These consist of four basic components: (1) the main body of the valve encasing the internal parts; (2) a movable spindle that seats into the entrance port of the valve; (3) a spring to force the spindle against the valve seat; and (4) an adjusting screw to regulate the compression of the spring.

Relief valves are located in the discharge line of the system as close to the pump discharge port as possible, or they may be built into the pump casing to bypass liquid back to the suction. The compression in the spring is adjusted to correspond to the maximum allowable pressure in the pump. When the predetermined pressure is reached, the spindle lifts from the valve seat, venting liquid from the discharge to the suction side and preventing the pump discharge pressure from rising further. The opening of a relief valve is a function of the discharge pressure, only sufficient liquid being released to allow the spindle to return to the valve seat.

Use of a built-in relief valve eases maintenance and repair work since the process piping need not be disturbed.

*Safety Valves*—These differ from relief valves in that a diaphragm or packed stem separates the liquid passages from the valve internals. In relief valves, the discharged liquid floods the area enclosing the compression spring. Two common types of safety valves are the balanced-pressure or bellows valve and the unbalanced-pressure type. Safety valves are installed in the same way as relief valves.

*Strainers and Screens*—Damage can result when foreign materials enter the pump internals—close-tolerance pumps such as gear pumps being particularly prone to damage. Therefore, pipeline strainers and screens are frequently used to protect pumps. These are placed in the suction line ahead of the pump intake port. Suitable strainers are designed to permit rapid cleaning and replacement.

It is particularly important to use screens during the startup of new pumping systems, since a substantial amount of dirt, scale and welding slag is usually present in new pipelines.

The area of the strainer should be about three to four times greater than the cross sectional area of the pipeline. Coarse strainers are used with slurries. Oversized strainers are used with liquids of high viscosity.

## Performance Factors

### Properties of the Liquid

*Viscosity*—This property causes the liquid to resist flow, however small. For a Newtonian liquid, the coefficient of viscosity $\mu$ is the factor of proportionality between the shear stress $R$ and the shear rate $dv/dy$ or $R = \mu\ (dv/dy)$. Viscosity coefficient data are most frequently given in centipoises where 1 cp. = 0.000672 lb. mass/(ft.)(sec.).

The coefficient of kinematic viscosity $\mu/\rho$ takes into account the liquid density. If the coefficient of viscosity $\mu$ is in centipoises and the liquid density $\rho$ is in grams/cc., then the coefficient of kinematic viscosity $\mu/\rho$ is in centistokes. For a liquid density $\rho = 1$ g./cc., centipoises and centistokes have the same numerical value.

The viscosity of liquids is also expressed as the time

required for a given liquid volume to flow through a particular orifice under a specified pressure head. This method is particularly popular with pump suppliers and manufacturers. Seconds Saybolt Universal (SSU) units are the most frequently used units of this type. Conversion tables for viscosity units are available in many handbooks, e.g., Ref. 7.

The higher the viscosity, the greater the head losses due to friction in the pipeline and the pump. Suction head and the available net positive suction head both decrease with an increase in liquid viscosity for the same pumping rate. At the same time, the discharge and total heads both increase with an increase in liquid viscosity for the same pumping rate. Thus the power required for pumping increases with liquid viscosity.

Sometimes the viscosity of a liquid must be reduced to permit efficient pumping, which can be accomplished by using heated process lines and pumps.

*Shear Characteristics*—For many liquids, the factor of proportionality between the shear stress $R$ and the shear rate $dv/dy$ does not remain constant. These are the so-called non-Newtonian liquids. The ratio of the shear stress $R$ to the shear rate $dv/dy$ is the apparent viscosity $\mu_a$ corresponding to that particular shear rate. For pseudoplastic liquids, an increase in shear rate leads to a decrease in apparent viscosity $\mu_a$. When the apparent viscosity continues to decrease during the time for which the shear is applied, the liquid is said to be thixotropic. For dilatant liquids, an increase in $dv/dy$ leads to an increase in $\mu_a$. When the apparent viscosity continues to increase with the time for which the shear is applied, the liquid is said to be rheopectic.

Pseudoplastic and dilatant liquids immediately return to their original state when the shear is removed. Thixotropic and rheopectic liquids also return to their original state when the shear is removed, but in this case there may be a time delay that varies over wide ranges for different liquids. Shear has no permanent effect on any of these liquids.

In the case of some liquids, shear can lead to a permanent change in viscosity. Other liquids can be damaged by shear in a number of ways. For example, certain liquid detergents can be permanently broken down into two phases if they should be subjected to too much shear.

*Volatility*—Vapor pressure is a measure of the volatility of a liquid at a given temperature. An increase in vapor pressure leads to a decrease in available NPSH. As the temperature of the process liquid is increased, the vapor pressure may eventually reach a point where the NPSH becomes zero so that there is no head available to get the liquid through the suction piping.

The presence of vapor in the process liquid reduces the volume available for liquid in the pump. This in turn results in a falloff in pumping rate.

*Corrosiveness*—Corrosion produced by a liquid may be either localized or general over-all. Localized corrosion can be either of the intergranular, electrolytic, stress or pitting type. In austenitic stainless steels such as 304 or 316, intergranular corrosion takes place at the grain boundaries as a result of damage that causes carbide precipitation. Concentration differences occurring in pits or recesses can cause electrolytic corrosion. Stress corrosion, the acceleration of the corrosion rate by stress, has been observed in almost all metals and their alloys. Pitting corrosion is caused by local concentrations of the corrosive liquid, by carbide precipitation, or by stresses.

The corrosion rate of alloys increases as the flow rate of the corrosive liquid increases. The smoother the surface of the metal, the greater the resistance to corrosion.

The corrosiveness of the liquid to be pumped determines the materials that can be used in the pump construction. Corrosion is more serious for intermittent than for continuous pump operation.

*Lubricating Qualities*—Zisman[16] has shown that a relationship does not necessarily exist between the wettability or spreadability of a liquid on a solid surface and its ability to act as a boundary lubricant. The contact angle $\theta$ between a drop of liquid and a plane solid surface is a measure of the wettability of the liquid on that surface. When $\theta = 0$, the liquid completely wets the surface; when $\theta \neq 0$, the liquid is nonspreading.

Zisman[16] has also shown that the ability of a liquid to wet a solid surface is a function only of the liquid surface tension and the nature of the surface atoms of the solid.

Paraffin hydrocarbons and methyl silicones are liquids that spread freely over all metals. Paraffin hydrocarbons are among the best, and methyl silicones are among the poorest, of the boundary lubricants for common metals. Liquid fatty alcohols and acids are nonspreading, yet they are excellent boundary lubricants for many metals.

The free-surface energies of the common liquids are less than 100 erg/sq. cm. at ordinary temperatures. Ordinary metals have free-surface energies in the range of 1,000 to 5,000 erg/sq. cm. Most organic solids have a free-surface energy of less than 100 erg/sq. cm. Zisman points out that one would expect a liquid to spread freely over a high-energy metal surface, since this would decrease the free energy of the whole system. He explains the fact that this does not always occur by assuming that in this case the liquid coats the metal in such a way that it behaves like a low-energy surface.

With liquids having poor lubricating qualities, the wear on the moving parts of a pump is likely to be relatively rapid. For this reason, piston pumps are usually provided with replaceable liners.

*Abrasiveness*—An abrasive liquid also produces wear in the pump, particularly on the moving parts. Mechanical seals should not be used with abrasive liquids, packed seals being recommended for these. Pump speeds should be kept low with abrasive liquids to prolong pump life.

*Suspended Solids*—In addition to making the liquid abrasive, suspended solids can lead to clogging in pump valves and moving parts.

**Nomenclature**

| | |
|---|---|
| $d_i$ | Inside pipe diameter, ft. |
| $D_i$ | Inside pipe diameter, in. |
| $D_T$ | Inside tank diameter, ft. |
| $D$ | Plunger diameter, in. |
| $E$ | Mechanical efficiency, %. |
| $f$ | Fanning friction factor (dimensionless) |
| $g$ | Gravitational acceleration, 32.2 ft./sec./sec. |
| bhp | Brake horsepower |
| gpm | Gallons per minute |
| hp | Horsepower |
| fhp | Friction horsepower |
| $h$ | Head, ft. |
| $\Delta h$ | Total head, ft. |
| $h_f$ | Head loss due to friction, ft. |
| $h_s$ | Suction head, ft. |
| $h_d$ | Discharge head, ft. |
| $h_c$ | Head loss due to contraction, ft. |
| $h_e$ | Head loss due to expansion, ft. |
| $j_f$ | Basic friction factor, $R/\rho u^2$ (dimensionless) |
| $l$ | Actual pipe length, ft. |
| $l_e$ | Equivalent pipe length, ft. |
| $L$ | Stroke length of pump, in. |
| $m$ | Factor in Eq. (14), hp./SSU |
| $m'$ | Factor in Eq. (16), gpm./SSU |
| $N_{Re}$ | Reynolds number, $\rho ud/\mu$ (dimensionless) |
| NPSH | Net positive suction head |
| $N$ | Pump speed, rpm. or strokes/min. |
| $P$ | Pressure, psi. |
| $P_v$ | Vapor pressure of liquid, psi. |
| $R$ | Shear stress, poundals/sq. ft. |
| $s$ | Shaft length, ft. |
| $u$ | Mean linear velocity of liquid inside a pipe, ft./sec. |
| $v$ | Point velocity, ft./sec. |
| $W$ | Mass flow rate of liquid, lb./min. |
| $y$ | Distance between two parallel shear planes, ft. |
| $z$ | Liquid height, ft. |
| $\alpha$ | Factor-in Eq. (1) (dimensionless) |
| $\mu$ | Viscosity of liquid, lb./(ft.)(sec.), cp. or SSU |
| $\rho$ | Density of liquid, lb./cu.ft. |
| $\theta$ | Liquid contact angle, deg. |
| $\phi$ | Angular misalignment, deg. |

## Mechanical Factors

*Slip*—This is the leakage that occurs between the discharge and suction sides of a pump through the pump clearances. The extent of this leakage depends on the width, length and shape of these clearances, the viscosity of the pumped liquid, and the pressure difference between the discharge and suction sides of the pump. Pump speed does not influence slip.

Generally the bulk of the slip occurs at the shortest sealing surfaces, e.g., at the gears in a gear pump. Slip decreases with increasing liquid viscosity. The heating of the liquid in the pump clearances leads to some decrease in viscosity and hence to an increase in slip.

No slip occurs with zero pressure difference between the discharge and suction sides of the pump. Provided the internal parts of the pump do not distort with increasing pressure, the slip increases linearly with the pressure difference across the pump if the pumped liquid is in laminar flow. For turbulent flow, slip increases with the square root of the pressure difference across the pump.

Since slip is independent of pump speed, it is an advantage to pump low-viscosity liquids at high speeds. With very viscous liquids, slip becomes negligible.

*Condition of the Pump*—Wear in a pump will increase clearances and thus lead to a greater amount of slip. In rotary pumps, incorrect alignment of the rotating parts with the fixed casing can produce excessive wear and can cause the gears to seize. The suction and discharge pipes should be supported independently of the pump to avoid putting a strain on the pump casing that might cause distortion.

*Seals*—Air leaks through the shaft seals reduce the liquid pumping rate. Also, if packed seals are too tight, the pump will consume an excessive amount of power. In addition, the shaft may become scored and the packing burnt. The packing gland should be adjusted to allow a very small amount of the pumped liquid to leak, and so provide lubrication and cooling.

*Temperature of Operation*—This determines the material of construction of the pump and also the clearances of the moving parts. Frequently the packing determines the upper temperature limit of operation. Plastic component parts can be used only in pumps operating at relatively low temperatures.

*Pump Speed*—The higher the speed of the moving parts in a pump, the greater the wear. This is particularly true when pumping abrasive liquids. An increase in pump speed produces a decrease in apparent viscosity for pseudoplastic liquids and an increase in apparent viscosity for thixotropic liquids. This leads to a decrease in power consumption for pseudoplastic and an increase for thixotropic liquids. Liquids that are prone to permanent deformation or damage by shear require low pump speeds.

There are two general classes of rotary pumps: large-liquid-cavity, low-speed pumps and small-liquid-cavity, high-speed pumps. The latter require more power when pumping highly viscous liquids.

*Bearings*—Heat is the main destructive factor in bearing wear. This can be reduced by a continual flow of cooling liquid through the bearing.

In positive-displacement pumps, the pressure differential across the pump is balanced by the reactive force of the bearings. This so-called hydraulic unbalance increases the wear on the bearings. Pumps can be equipped with channels to carry high-pressure liquid that is fed to various bearing points to reduce the hydraulic unbalance.

## Cavitation and Priming

For smooth operation, pumps should be completely filled with liquid. When liquid does not completely fill the pump chamber, a loss of capacity results. This may occur through the vaporization of some of the process liquid in the suction line or the pump chamber. The vapor bubbles are carried into higher-pressure regions of the pump where they collapse, resulting in noise and vibration. This phenomenon is called cavitation; it can occur in all types of pumps.

Cavitation can also occur from air leaks through the pump seals. The repeated blows of the collapsing bubbles cause rapid wear in the pump and may also accelerate the corrosion rate. Pitting of pump parts is a common result of cavitation. The greater the slip in a pump, the smaller the effect of cavitation. Cavitation is usually more serious with viscous liquids than with thin liquids. High-speed pumps are more prone to cavitation than low-speed pumps.

In a leak-free pump, cavitation can be avoided by ensuring that the available NPSH in the suction line is always greater than that required by the pump. Short suction lines of large diameter will reduce the frictional head loss and increase the available NPSH.

When not in operation, pumps and suction lines may be full of air. This air must be removed (by priming) before the pump can start to move liquid. Normally, positive-displacement pumps are self-priming, although for suction lifts greater than 28 ft. such pumps may require priming. This may be accomplished either by admitting liquid to the suction line to displace the air or by exhausting the air from the suction line and pump by a vacuum system.

## System Design Equations

The function of a pump is to overcome friction losses in a piping system in which a liquid is flowing, and also to raise the liquid to a higher level if necessary. The various elements of energy imparted by a pump are called "heads," and a consideration of these heads in the light of the Bernoulli equation will lead to a method for flow-system design.

A liquid of density $\rho$ lb./cu. ft. is flowing at a mean linear velocity of $u$ ft./sec. in a circular pipe of inside diameter $d_i$ ft., as shown in Fig. 20. Consider two points 1 and 2 in the pipe, at a distance $l$ ft. apart.

The combined head of the liquid at any point is made up of the following three components:

- The static head or head of position $z$ measured in feet above an arbitrarily chosen base line. (See $z_1$ and $z_2$.)

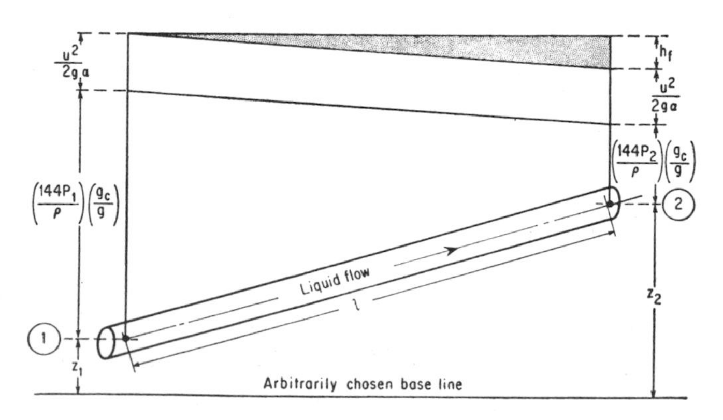


Diagram portrays "heads" in Bernoulli equation—Fig. 20

- The pressure head as given by a pressure gage placed at the point. If the reading of the pressure gage is $P$ psi., then the pressure head in feet is $(144\ P/\rho)(g_c/g)$ ft. (Since the ratio $g_c/g$ is substantially unity, it is omitted in the rest of the article.)
- The velocity head $u^2/2g\ \alpha$ is in feet if $u$ is in ft./sec. and $g$ the gravitational acceleration is in ft./sec./sec. Here $\alpha$ is a dimensionless factor that for fluid flow in a circular pipe can be shown to be 0.5 for laminar flow and approximately 1 for turbulent flow.[3] It serves as a correction factor to account for the velocity distribution across the pipe. The velocity head remains constant if $u$ is constant.

Thus, at point 1, the combined head is:

$$h_1 = z_1 + \frac{144\ P_1}{\rho} + \frac{u^2}{2g\ \alpha} \tag{1}$$

At point 2, the head is:

$$h_2 = z_2 + \frac{144\ P_2}{\rho} + \frac{u^2}{2g\ \alpha} \tag{2}$$

For a constant pipe diameter, $u_1 = u_2$, so that the velocity head remains constant as the liquid flows from point 1 to point 2. The change in static and pressure heads with position in the pipe is illustrated in Fig. 20.

The combined head $h_2$ at point 2 is less than the head $h_1$ at point 1 by the head loss $h_f$ due to friction. This last head is:

$$h_f = 8\left(\frac{R}{\rho u^2}\right)\left(\frac{l}{d_i}\right)\left(\frac{u^2}{2g}\right) \tag{3}$$

where $R/\rho u^2$ is the dimensionless basic friction factor that is obtained from plots of $R/\rho u^2$ versus the dimensionless Reynolds number $N_{Re} = \rho u d_i/\mu$ as shown in Fig. 21.*

Alternatively, the head loss due to friction $h_f$ is given by the equation:

$$h_f = 4f\left(\frac{l}{d_i}\right)\left(\frac{u^2}{2g}\right) \tag{4}$$

where $f$ is the long-used Fanning friction factor. The $l/d_i$ ratio in Eqs. (3) and (4) is a dimensionless shape factor defined as the length-to-diameter ratio of the pipe section considered. In the Reynolds number $N_{Re}$, $\mu$ is the viscosity of the liquid in appropriate units, in this case lb./(ft.)(sec.), where 1 cp. = 0.000672 lb./(ft.)(sec.).

In pipelines containing valves and fittings, the equivalent pipe length $l_e$ should be used in Eqs. (3) and (4).

Clearly, a pump is required to supply head in order to make good the head loss $h_f$ due to friction. Equating heads using the Bernoulli equation is equivalent to making an energy balance on 1 lb. of material.

* There is considerable confusion among the three types of friction factor used in pipe-flow calculations. Designating the older, conventional Fanning friction factor as $f$, a newer, so-called basic friction factor as $R/(\rho u^2)$, and a third one widely used among pump manufacturers as $F$, we have the following relations: $f = 2R/(\rho u^2) = F/4$. The Reynolds number-friction factor chart of Fig. 21 is based on the second of these, and consequently shows at every point an ordinate value of half that of the older and more familiar form of the chart based on $f$, which until recently was taught most frequently to engineering students. Now, however, the chart based on $R/(\rho u^2)$ is rapidly becoming more popular, owing to its use in modern teaching of transport processes.

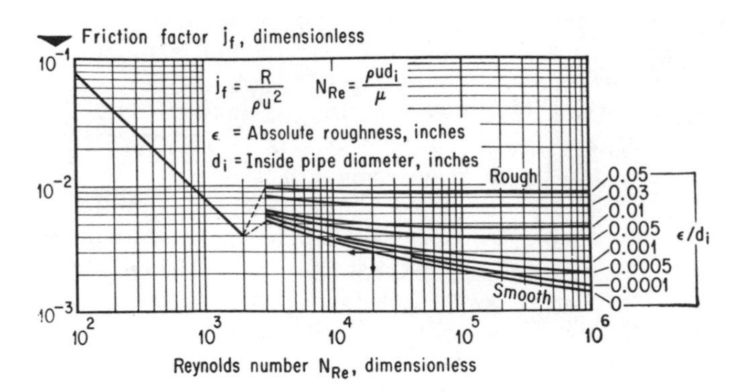


**Basic** pipe-flow, friction-factor chart—Fig. 21

## System Heads

The following definitions are given in reference to the typical pumping system shown in Fig. 22, where the arbitrarily chosen base line is the centerline of the pump. Here:
Suction head, in feet, is:

$$h_s = z_s + 144\,(P_s/\rho) - h_{fs} \tag{5}$$

Discharge head, in feet, is:

$$h_d = z_d + 144\,(P_d/\rho) + h_{fd} \tag{6}$$

Here $h_{fs}$ and $h_{fd}$ are the head losses due to friction in the suction and discharge lines, respectively, and $z_s$ and $z_d$ are the static heads on the suction and discharge sides of the pump, respectively, measured above the centerline of the pump.

If the liquid level on the suction side is below the centerline of the pump, $z_s$ is negative. $P_s$ is the gas pressure in psi. above the liquid on the suction side of the pump, and $P_d$ is the gas pressure in psi. at the end of the discharge line.

The head that the pump must impart is called the total head and is the difference between the discharge and suction heads, or:

$$h_d - h_s = (z_d - z_s) + \left[\frac{144\,(P_d - P_s)}{\rho}\right] + (h_{fd} + h_{fs}) \tag{7}$$

The pressures $P_d$ and $P_s$ can either be gage or absolute, but they must be consistent with each other.

The suction head decreases and the discharge head increases with increasing liquid velocity because of the increasing value of the frictional-head-loss terms $h_{fs}$ and $h_{fd}$. Thus the total head increases with pumping rate.

The head available to get the liquid through the suction piping, the net positive suction head (NPSH), in feet, is defined as:

$$\text{NPSH} = z_s + [144\,(P_s - P_{vp})/\rho] - h_{fs} \tag{8}$$

where $P_{vp}$ is the vapor pressure in psi. of the liquid being pumped at the particular temperature in question.

NPSH is seen to be the suction head minus the vapor pressure in feet of the liquid. Since $P_{vp}$ would normally be in psia. units, the gas pressure Γ above the suction side liquid should in this case also be written in psia. units. The available NPSH in a system should always be positive, i.e., the suction head must always be capable of overcoming the vapor pressure of the liquid.

Since the frictional head loss $h_{fs}$ increases with increasing liquid flow, the available NPSH decreases with increasing flow rate. When the liquid is at its boiling temperature, $P_s$ and $P_{vp}$ are equal and the available NPSH becomes $z_s - h_{fs}$. In this case, $z_s$ must be positive and no suction lift is possible, Normally, liquid can be lifted from below the centerline of the pump, i.e., when $z_s$ is negative, provided the term $(P_s - P_{vp})$ is sufficiently large.

## Pumping Efficiencies

*Volumetric Efficiency*—This expresses the delivered capacity per cycle as a percentage of the true displacement per cycle. The lower the internal slip losses, the higher the volumetric efficiency. For zero slip, the volumetric efficiency is 100%. The true displacement is the delivered capacity for zero pressure difference across the pump. A knowledge of the volumetric efficiency for the required range of application is very important when the positive-displacement pump is to be used for metering. Entrained air or gas in the pumped liquid reduces volumetric efficiency.

*Mechanical Efficiency*—This is the liquid horsepower expressed as a percentage of the brake horsepower. Liquid horsepower is the rate of useful work done on the liquid and can be defined as volumetric liquid flow rate times the pressure difference across the pump, divided by an appropriate conversion factor, or:

$$\text{Liquid hp.} = \text{gpm.} \times \text{psi.}/1{,}714 \tag{9}$$

Alternatively, liquid horsepower can be expressed in

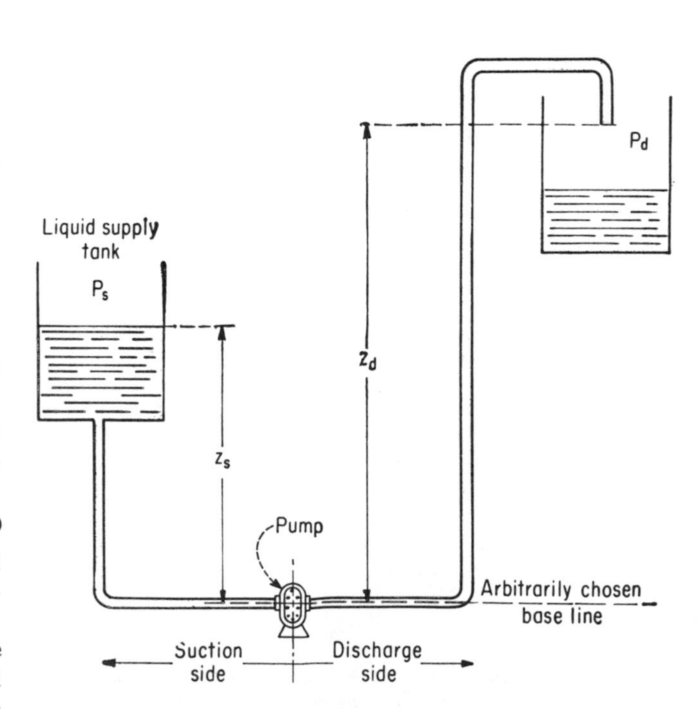


**Heads** in a typical pumping system—Fig. 22

**terms of the weight rate of flow in lb./min., times the head difference across the pump, as follows:**

$$\text{Liquid hp.} = W\Delta h/33{,}000 \qquad (10)$$

Brake horsepower is the actual power delivered to the pump and is the sum of liquid horsepower and friction horsepower. In terms of the mechanical efficiency $E$, it can be expressed, in terms of Eq. (10), as:

$$\text{Brake hp.} = (W\Delta h/33{,}000)\,(100/E) \qquad (11)$$

Since the amount of power required to overcome friction rises with increased liquid viscosity, mechanical efficiency decreases as liquid viscosity increases. Some improvement in mechanical efficiency can be obtained with viscous liquids by using larger clearances in the pump. However, with thin liquids, this leads to an increase in slip and lower mechanical efficiency.

Slip is independent of pump speed. Thus the power used in recirculating liquid through slip back from the discharge to the suction side of the pump does not depend on pump speed. Since the ratio of liquid horsepower to slip horsepower increases with pump speed, thin liquids are pumped more efficiently at high pump speeds.

Power losses in timing gears, bearings and seals reduce mechanical efficiencies. A tightly packed seal can be a large consumer of power while contact between the rotor and the fixed casing also increases the power losses in a rotary pump. These losses are not proportional to pump size, small pumps tending to have lower efficiencies. The best efficiencies are usually found in relatively large pumps. Pumps handling tacky liquids require more power to start than those pumping noncohesive liquids.

Expansion

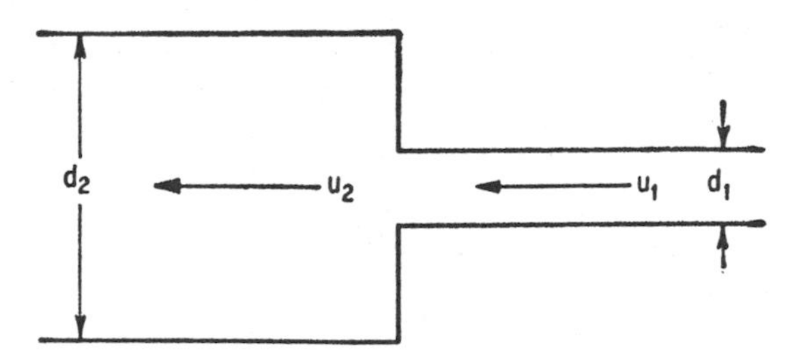


The loss in head due to sudden expansion is:

$$h_e = \frac{(u_1 - u_2)^2}{2g} = \frac{u_1^2}{2g}\left(1 - \frac{d_1^2}{d_2^2}\right)^2$$

Contraction

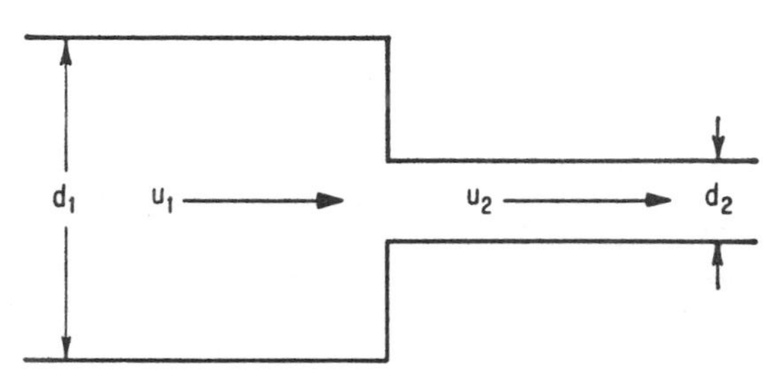


The loss in head due to sudden contraction is:

$$h_c = K\left(\frac{u_2^2}{2g}\right)$$

When:

$$\frac{d_2^2}{d_1^2} \begin{cases} < 0.715, & K = 0.4(1.25 - d_2^2/d_1^2) \\ > 0.715, & K = 0.75(1 - d_2^2/d_1^2) \end{cases}$$

**Expansion** and contraction head losses in pipes—Fig. 24

# Design Calculations

The important factors to consider is the design of a pumping system are: (1) the physical and chemical properties of the liquid to be transferred; (2) the required liquid-transfer rate; and (3) the pipe dimensions and system heads, which are also a function of the liquid physical properties. This information enables the horsepower requirements of the system to be determined and the appropriate pump to be selected.

*Liquid Properties*—The rotational or reciprocating speed of a positive-displacement pump should be reduced as the viscosity of the pumped liquid increases. This minimizes noise, vibration and the loss of pumping capacity. Fig. 23 gives recommended operating speeds[14] for a typical internal gear pump as a function of rated speed and liquid viscosity. Similar plots are available for other pump types. Thus larger pumps are required to handle the same capacity of viscous liquids. Viscosity affects the Reynolds number and hence the friction head, as in Eq. (3).

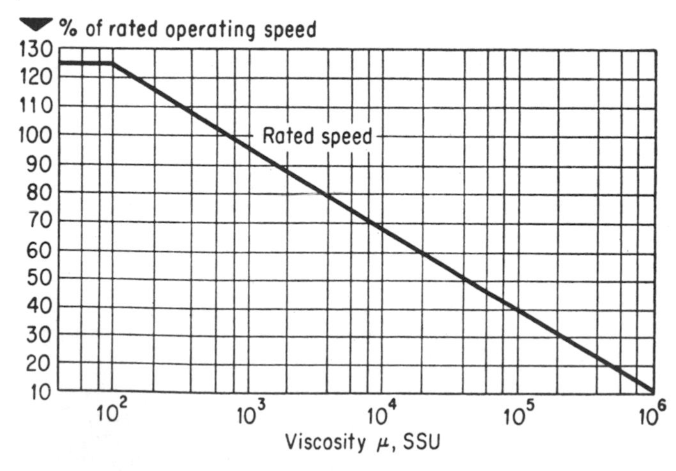


**How viscosity** affects allowable pump speed for one type of positive pump. Others are affected similarly—Fig. 23

Liquid density effects show up in both the suction and discharge heads, and hence in the total head of the system, as in Eqs. (5), (6) and (7). Finally, the volatility of the pumped liquid at the operating temperature of the system determines the available NPSH, as in Eq. (8).

The temperature and corrosiveness of the pumped liquid determine the materials of construction used for the pump and the pipelines. The maximum allowable contamination of the pumped liquid is also a factor in materials-of-construction selection. In general, materials with very low corrosion rates are expensive. But cheaper materials involve higher maintenance costs. The choice will depend on the application.

*Pumping Rates*—Positive-displacement pumps provide a constant average (though not necessarily uniform) flow rate for a given pump speed. If other flow rates are required, the pump must be equipped for either variable speed or variable displacement. The system heads and horsepower requirements need to be calculated not only for each flow rate but also for each liquid viscosity that may be encountered if viscosity changes during the process.

*System Heads*—As we have seen, the system heads are a function of the liquid properties, pumping rate, pipe diameter, pipe roughness, pipe lengths and heights. Fittings in the suction and discharge pipelines also contribute to friction losses and must be accounted for.

A design problem may be one of the following: (1) to design a new pipeline and select a pump; (2) to select a pump for an existing pipeline; or (3) to design a new pipeline for use with an existing pump.

All these problems require the calculation of head losses in the pipeline plus the fittings. The frictional head loss $h_f$ is a function of the equivalent length $l_e$ ft. of pipe of internal diameter $d_i$ ft., plus fittings, as given by Eq. (3). For use in this equation, it is necessary to determine a friction factor $j_f = R/(\rho\, u^2)$ from Fig. 21. For this purpose, one needs a Reynolds number $N_{Re} = \rho\, u\, d_i/u$ and a roughness factor for the pipe material $\epsilon/d_i$ from Table II. The equivalent lengths of various pipeline fittings are given in Table III, and the entrance and exit losses for the line that must be added are given in Fig. 24.

*Example*—Calculate the system heads in the pumping system shown in Fig. 25, using the following data:

- The liquid has a viscosity of 1,000 cp., specific gravity of 1.2 and vapor pressure of 1.0 psia. at the pumping temperature.
- The 6-ft. suction and 172-ft. discharge lines are of 2-in. Schedule 40 steel pipe.
- There is one 90° standard-radius elbow and one gate valve in the suction line.
- There are three 90° standard-radius elbows and one globe valve in the discharge line.
- The liquid transfer rate is 40 gpm.
- The tanks connected to the suction and discharge lines are both cylindrical and 12 ft. in diameter.
- Initially the liquid depths in the suction and discharge tanks are 6 ft. and 1 ft., respectively.
- The pump is stopped when the liquid depth in the suction-side tank has fallen to 1 ft.
- Pressure above both the suction and discharge tanks is 14.7 psia.

The calculation procedure is as follows:

- Calculate the linear velocity of liquid in the pipelines. A mean linear velocity of liquid of 1 ft./sec. in a 2-in. Schedule 40 steel pipe gives (from standard piping tables such as Ref. 8) a volumetric flow rate of 10.45 gpm. Hence a volumetric flow rate of 40 gpm. gives a mean linear velocity of 3.83 ft./sec. in a 2-in. Schedule 40 steel pipe.
- Calculate the velocity head to be $u^2/2g = 3.83^2/(2 \times 32.2) = 0.228$ ft.
- Calculate the liquid density to be $\rho = 62.4 \times 1.2 = 74.9$ lb./cu. ft.
- The inside diameter of a 2-in. Schedule 40 steel pipe is 2.067 in., so $d_i = 2.067/12 = 0.1722$ ft.
- Calculate the liquid viscosity in pounds-foot-second units. Since 1 cp. = 0.000672 lb./(ft.)(sec.), then 1,000 cp. = 0.672 lb./(ft.)(sec.).
- Calculate the Reynolds number $N_{Re}$ in the pipeline to be $N_{Re} = \rho\, u\, d_i/\mu = 74.9 \times 3.83 \times 0.1722/0.672 = 73.6$. Thus the liquid in the pipeline is in laminar flow.
- From Fig. 21, read the value of the basic friction factor $j_f = R/(\rho u^2) = 0.1086$, corresponding to a Reynolds number $N_{Re} = 73.6$. The basic friction factor can also be calculated in this case, since for laminar flow in pipelines:

$$j_f = R/(\rho u^2) = 8/N_{Re} \qquad (12)$$

- Estimate the ratio of equivalent length to pipe diameter $l_e/d_i$ for the suction pipeline. This is made up of the following component parts:

(1) The actual length of the suction pipe, which contributes $6/0.1722 = 34.8$ to the $l_e/d_i$ ratio. (2) The 90° elbow and gate valve which, from Table III, con-

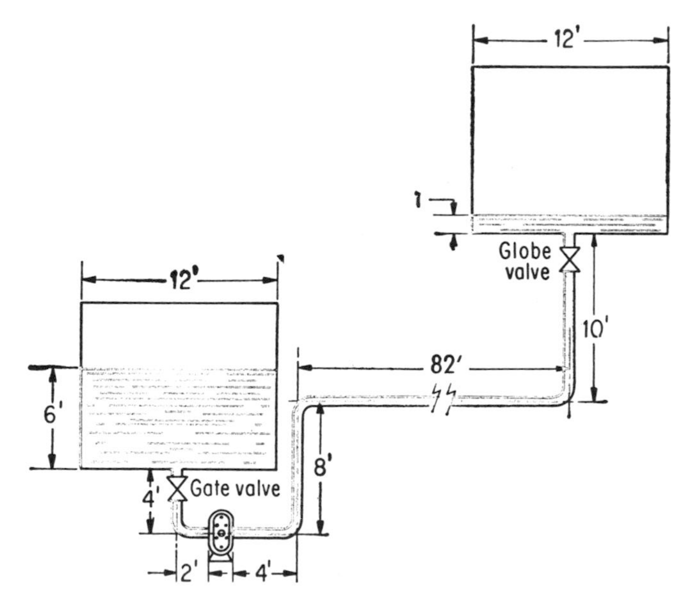


**Dimensioned** pumping system shown in example—Fig. 25

**Values of absolute roughness $\epsilon$, feet[4]—Table II**

| Material | $\epsilon$, ft. |
|---|---|
| Drawn tubing | 0.000005 |
| Commercial steel and wrought iron | 0.00015 |
| Asphalted cast iron | 0.0004 |
| Galvanized iron | 0.0005 |
| Cast iron | 0.00085 |
| Wood stave | 0.0006–0.003 |
| Concrete | 0.001 –0.01 |
| Riveted steel | 0.003 –0.03 |

**Equivalent lengths of screwed fittings, valves, etc.[7] —Table III**

| | Equivalent Length in Pipe Diameters |
|---|---|
| 45° elbows | 15 |
| 90° elbows, standard radius | 32 |
| 90° elbows, medium radius | 26 |
| 90° elbows, long sweep | 20 |
| 90° square elbows | 60 |
| 180° close return bends | 75 |
| 180° medium-radius bends | 50 |
| Tee (used as elbow, entering run) | 60 |
| Tee (used as elbow, entering branch) | 90 |
| Couplings, unions | negligible |
| Gate valves, open | 7 |
| Globe valves, open | 300 |
| Angle valves, open | 170 |
| Water meters, disk | 400 |
| Water meters, piston | 600 |
| Water meters, impulse wheel | 300 |

tribute ratios of equivalent length to pipe diameter of 32 and 7, respectively.

Thus, for the suction pipeline, the total equivalent-length-to-pipe diameter ratio $l_e/d_i = 34.8 + 32 + 7 = 74.8$.

• Similarly, estimate the equivalent-length-to-pipe-diameter ratio $l_e/d_i$ for the discharge pipeline. This is made up of the following component parts: (1) The actual length of the pipe, which contributes 104/0.1725 = 601.1. (2) Three 90° standard-radius elbows, each from Table III, having an equivalent-length-to-pipe-diameter ratio of 32. (3) The globe valve which, from Table III, has an equivalent-length-to-pipe-diameter ratio of 300.

Thus, for the discharge pipeline, the total equivalent-length-to-pipe-diameter ratio $l_e/d_i = 601.1 + 96 + 300 = 997.1$.

• Calculate the head loss $h_{fs}$ due to friction in the suction line.

From earlier steps, the values of the terms $R/\rho u^2 = 0.1086$, $l_e/d_i = 73.8$ and $u^2/2g = 0.228$ ft. for the suction line are substituted into Eq. (3) to give $8 \times 0.1086 \times 73.8 \times 0.228 = 14.7$ ft., the head loss due to pipe friction in the suction line. Add the contraction head loss due to liquid entering the suction line. From Fig. 24, $K = 0.5$. Therefore $h_c = 0.1$ and $h_{fs} = 14.7$ ft.

• Calculate the head loss $h_{fd}$ due to friction in the discharge line. Similarly to the above, for the discharge line, $R/\rho u^2 = 0.1086$, $l_e/d_i = 997.1$ and $u^2/2g = 0.228$ ft. These are substituted into Eq. (3) to give $8 \times 0.1086 \times 997.1 \times 0.228 = 197.3$ ft., the friction head loss due to pipe friction in the discharge line. Add the expansion head loss due to liquid leaving the discharge line. From Fig. 24, $h_e = 0.228$ ft. Therefore $h_{fd} = 276.2$.

• Calculate the initial suction head from Eq. (5). From Fig. 25, $z_s = 10$ ft. Also, $144\ P_s/\rho = 144 \times 14.7/74.9 = 28.3$ ft.

From the $h_{fs}$ calculation above, $h_{fs} = 14.7$ ft. Therefore, from Eq. (5), $h_s = 10 + 28.3 - 14.7 = 23.6$ ft., the suction head at the start of the operation.

• Similarly, calculate the initial discharge head from Eq. (6). From Fig. 25, $z_d = 19$ ft. Also $144\ P_d/\rho = 144 \times 14.7/74.9 = 28.3$ ft. From the $h_{fd}$ calculation above, $h_{fd} = 197.5$ ft. Therefore $h_d = 19 + 28.3 + 197.5 = 244.8$ ft., the discharge head at the start of the operation.

• Calculate the initial total head $\Delta h = h_d - h_s$. As above, $h_d = 244.8$ ft. and $h_s = 23.6$ ft. Therefore $\Delta h = h_d - h_s = 244.8 - 23.6 = 221.2$ ft., the total head at the start of the operation.

• Calculate the initial available net positive suction head from Eq. (8). From Fig. 25, $z_s = 10$ ft. Since $P_{vp} = 1.0$ psia., $144\ (P_s - P_{vp})/\rho = 144 \times (13.7/74.9) = 26.4$ ft., and $h_{fs} = 14.7$ ft., therefore NPSH = 10 + 26.4 − 14.7 = 21.7 ft. Hence there is + 197.5 = 244.8 ft., the discharge head at the start.

• Proceeding as in the calculation of the initial suction head, calculate the final suction head to be $h_s = 18.6$ ft.

• And, as in the calculation of the initial discharge head, calculate the final discharge head to be $h_d = 249.8$ ft.

• Then, as in the initial total-head calculation, calculate the final total head to be $\Delta h = h_d - h_s = 231.2$ ft.

• Concluding the head calculations as in the initial available NPSH calculations, calculate the final available net positive suction head to be NPSH = 16.7 ft. Thus there is still sufficient available head to get the liquid into the pump at the end of the operation.

## Horsepower Calculations

Eq. (10) gives the liquid horsepower in terms of the mass flow rate of the liquid. In terms of the volumetric flow rate, Eq. (10) can be rewritten as:

$$\text{Liquid horsepower} = \text{gpm.} \times \Delta h \times \text{sp. gr.}/3{,}960$$

where sp.gr. = specific gravity of the liquid referred to water at 68 F., weighing 62.3 lb./cu. ft.

*Example*—Calculate the liquid horsepowers at the start and finish of the pumping operation described in the previous example, where the volumetric liquid-flow rate is 40 gpm. and the specific gravity of the liquid is 1.2. Assume that a rotary gear pump is available with a mechanical efficiency of 48% under these conditions. Estimate the size of the drive unit.

• Calculate the initial liquid horsepower. Since the initial total head $\Delta h = 221.2$ ft., the initial liquid horsepower, from Eq. (12), is $40 \times 221.2 \times 1.2/3{,}960 = 2.64$ hp.

• Calculate the final liquid horsepower. Since the final total head $\Delta h = 231.2$ ft., the final liquid horsepower, also from Eq. (12), is $40 \times 231.2 \times 1.2/3{,}960 = 2.80$ hp.

• Calculate the brake horsepower on the basis of a final liquid horsepower of 2.80.

$$\text{Brake hp.} = (\text{liquid hp.} \div \%\ \text{mech. efficiency}) \times 100$$

Therefore the brake horsepower = 2.80/0.48 = 5.83 hp. A 7.5-hp. electric motor can be used to drive the rotary gear pump. If the liquid is tacky, the additional power available from the motor will be required to start the pump.

Usually brake horsepower (bhp) is given in manufacturers' bulletins plotted against pump discharge pressure in psig. for a particular liquid viscosity. Frequently plots for two different viscosities $\mu_1$ and $\mu_2$ are given.

Zalis[15] presents a method for calculating the brake horsepower of rotary pumps at viscosities $\mu$ other than those specified for a particular pump-discharge pressure. The method is based on the fact that brake horsepower = liquid horsepower + friction horsepower. The liquid horsepower is calculated as already described. However in this case the volumetric flow rate into the pump is used in the calculation, i.e., the pumping capacity assuming zero slip. The friction horsepower (fhp) is the brake horsepower value at zero discharge pressure. This can be read from the brake-horsepower versus pump discharge-pressure plots for each of the two viscosities $\mu_1$ and $\mu_2$ mentioned above. A plot is then made of friction horsepower versus liquid viscosity on logârithmic coordinates by joining the two points to form a straight line. The friction horsepower values at viscosities other than $\mu_1$ and $\mu_2$ can be read from this plot.

Alternatively, the equation:

$$\text{fhp}/\text{fhp}_1 = (\mu/\mu_1)^m \tag{13}$$

can be used to obtain the friction horsepower at a particular liquid viscosity, where

$$m = \frac{\log_{10} \text{fhp}_2 - \log_{10} \text{fhp}_1}{\log_{10} \mu_2 - \log_{10} \mu_1} \tag{14}$$

In Eq. (14), $\text{fhp}_1$ and $\text{fhp}_2$ are the known friction horsepowers, respectively, at the given liquid viscosities $\mu_1$ and $\mu_2$.

## Finding Brake Horsepower

The brake horsepower is then obtained by adding the calculated liquid horsepower to the friction horsepower as calculated from Eq. (13).

*Example*—Consider the previous example. Head $h$ ft. is related to pressure $P$ psi. by the expression $h = 144\ P/\rho$, where $\rho$ is the liquid density in lb./cu. ft. The final discharge head $h_d$ was calculated in our first example to be 249.8 ft. Since the liquid density $\rho$ = 74.9 lb./cu. ft., the discharge head is calculated to be 130.0 psia., or 115.3 psig. The liquid has a viscosity of 1,000 cp. and a specific gravity of 1.2. Thus the kinematic viscosity is 833 centistokes, which is equivalent to a viscosity $\mu$ of 3,790 SSU.

Now, calculate the brake horsepower under the above condition if the friction horsepowers at $\mu_2$ = 10,000 SSU and $\mu_1$ = 1,000 SSU viscosity are 4 and 2 hp., respectively, at a pump discharge pressure of 115.3 psig.

First, find the value of $m$ in Eq. (14).

$$m = \frac{\log_{10} 4 - \log_{10} 2}{\log_{10} 10{,}000 - \log_{10} 1{,}000} = 0.302$$

Then calculate fhp at $\mu$ = 3,790 SSU from Eq. (13), when $\text{fhp}_1 = 2$ and $\mu_1$ = 1,000 SSU. Therefore, fhp = $2\ (3.79)^{0.302} = 2.99$. Finally, to obtain the brake horsepower, add fhp = 2.99 to the liquid horsepower value of 2.80 calculated in the last example. This totals bhp = 5.79.

This method of obtaining brake horsepower is less accurate than the method of direct calculation already described.

## Slip Calculations

Zalis[15] also presented a method for calculating the liquid bypassed from the discharge to the suction side of a rotary pump, due to slip at a particular pressure and at viscosities $\mu$ other than those specified.

Usually the capacity of a rotary pump is given in manufacturers' bulletins in the form of a plot of gpm. versus pump discharge pressure in psig. for a particular liquid viscosity. Frequently, plots for two different viscosities $\mu_1$ and $\mu_2$ are given. The basis of the Zalis method is as follows:

The pump capacity with no slip is the capacity at 0 psig. discharge pressure. This can be read from the appropriate gpm. versus discharge-pressure plot, together with the capacity at the particular discharge pressure in question. The difference between the two values is the slip capacity. This can be calculated for each viscosity $\mu_1$ and $\mu_2$ at the discharge pressure required. A plot is then made of slip gpm. versus liquid viscosity on logarithmic coordinates, by joining the two points to form a straight line. The slip gpm. values at viscosities other than $\mu_1$ and $\mu_2$ can be read from this plot.

Alternatively, the equation:

$$\text{Slip gpm.}/\text{slip gpm.}_1 = (\mu/\mu_1)^{m'} \tag{15}$$

can be used to obtain the slip gpm. at a particular viscosity $\mu$, where

$$m' = \frac{\log_{10} \text{slip gpm.}_2 - \log_{10} \text{slip gpm.}_1}{\log_{10} \mu_2 - \log_{10} \mu_1} \tag{16}$$

The value of $m'$ is negative since slip capacity decreases with increasing viscosity. In Eq. (16), slip $\text{gpm}_1$ and slip $\text{gpm}_2$ are the known slip capacities, respectively, at the given liquid viscosities $\mu_1$ and $\mu_2$.

The pump capacity at the discharge pressure and liquid viscosity $\mu$ in question is then obtained by subtracting the slip capacity calculated from Eq. (15) from the capacity at 0 psig. discharge pressure.

*Example*—Consider the previous example. The liquid viscosity $\mu$ = 3,790 SSU and the final discharge pressure is calculated to be 115.3 psig. Calculate the slip and discharge capacities under the above conditions if the slip capacities $\mu_2$ = 10,000 SSU and $\mu_1$ = 1,000 SSU viscosity are 0.3 and 0.7 gpm., respectively, at a pump discharge pressure of 115.3 psig.

First find the value of $m'$ in Eq. (16):

$$m' = \frac{\log_{10} 0.3 - \log_{10} 0.7}{\log_{10} 10{,}000 - \log_{10} 1{,}000} = 0.368$$

Now calculate the slip gpm. at $\mu$ = 3,790 SSU from Eq. (15), when slip gpm. = 0.7 and $\mu_1$ = 1,000 SSU. Therefore slip gpm. = $0.7\ (3.79)^{-0.368} = 0.43$ gpm. Finally, calculate the pump capacity at a discharge pressure of 115.3 psig. and a liquid viscosity $\mu$ of 3,790 SSU. If the pump capacity at 0 psig. discharge pressure is 40 gpm., then the capacity with 0.43 gpm. of slip will be 40 − 0.43 = 39.57 gpm.

### Piston and Plunger-Pump Calculations

At the peak of the suction stroke in a piston or plunger pump, the available net positive suction head can be calculated from Eq. (8) as discussed earlier. The following equation gives minimum NPSH required by the pump at this point to avoid cavitation:

$$NPSH_{min} = 25\, D_i\, h_{fs}/l_e \qquad (17)$$

where these quantities apply to the suction line.

At the start of the suction stroke, the liquid has no velocity. The available net positive suction head at this point is not reduced by a friction head and is found to be:

$$NPSH = z_s + 144\,(P_s - P_{vp})/\rho \qquad (18)$$

The following equation gives the minimum net positive suction head required by the pump to avoid cavitation at the start of the suction stroke[1]

$$NPSH_{min} = lLD^2 N^2/(5.19 \times 10^4 \times D_i^2) \qquad (19)$$

where $l$ = actual (not equivalent) length of the suction pipeline (including fittings) in ft. and pump speed $N$ is in strokes/min.

*Example*—Calculate the maximum allowable pump speed $N$ in strokes/min. for a plunger pump on the basis of the following data: Pipeline system data: $z_s = 10$ ft.; $P_s = 14.7$ psia.; $P_{vp} = 1.0$ psia; $\rho = 72.0$ lb./cu. ft.; $l = 20$ ft.; $l_e = 50$ ft.; $h_{fs} = 1.0$ ft.; and $D_i = 1.049$ in. (for 1-in. Schedule 40 steel pipe). Pump data: $L = 4$ in. and $D = 1.5$ in.

First, calculate the available NPSH from Eq. (18): NPSH = 10 + (144/72) × (14.7 − 1.0) = 37.4 ft. Then, for the NPSH value just calculated, determine the stroke speed $N$ from Eq. (19):

$$N = \left[\frac{5.19\, D_i^2\, NPSH}{lLD^2}\right]^{1/2} 10^2 = \left[\frac{5.19 \times 1.049^2 \times 37.4}{20 \times 4 \times 1.5^2}\right]^{1/2} 10^2 = 109 \text{ strokes/min.}$$

This is the maximum allowable pump speed. At greater speeds, cavitation will occur.

Now calculate the available NPSH from Eq. (8), as NPSH = 10 + [(144/72) (14.7 − 1.0)] − 1.0 = 36.4 ft., and calculate the minimum required NPSH from Eq. (17) as $NPSH_{min} = 25\,(1.049)\,(1.0/50) \cong 0.5$ ft. Thus there is adequate NPSH available at the peak of the suction stroke.

*Example*—Recalculate the maximum allowable pump speed $N$ for a vapor pressure of 10.0 psia., with all the other data remaining the same as in the previous example. (In practice, the frictional-head loss changes with liquid flow rate. The same value is taken here only for illustration purposes.)

First, calculate the available NPSH from Eq. (18) as NPSH = 10 + [(144/72) (14.7 − 10.0)] = 19.4 ft. Then, for this NPSH value, calculate the stroke speed $N$ from Eq. (19):

$$N = \left[\frac{5.19\, D_i^2\, NPSH}{lLD^2}\right]^{1/2} 10^2 = \left[\frac{5.19 \times 1.049^2 \times 19.4}{20 \times 4 \times 1.5^2}\right]^{1/2} 10^2 = 78 \text{ strokes/min.}$$

This is the maximum allowable pump speed.

## Installation and Startup

A pump must be properly installed in order to operate efficiently and have a long life. Improper installation frequently results in rapid and excessive wear. This also increases the downtime required for repairs and replacements. The installation and startup of a pumping system involves the following steps:

- Selection of the pump location.
- Layout of the suction and discharge pipelines.
- Provision of a suitable foundation for the pump and drive train.
- Installation of auxiliary equipment.
- Alignment of the pump, drive train and piping.
- Testing of the system, and startup.

### Location and Layout

The pump location should be such that repair and maintenance work can be carried out easily. For example, it should not be necessary to dismantle pipelines and other equipment to get at the pump.

If possible, the pump should be positioned below the level of the liquid supply. The suction piping should be kept as short and free from bends as possible. The availability of utilities should not determine pump location. If necessary, utilities should be brought to the best available pump site.

Suction-side pipelines should be sloped to avoid air pockets. For the same reason, obstructions should be bypassed in the horizontal plane rather than the vertical. In pumping systems with suction lifts, the suction line should contain a check valve.[11] This keeps the prime on the pump when not in operation.

The suction and discharge pipelines should be supported independently of the pump, to avoid any strain on the pump casing. For the same reason, the suction and discharge pipelines should be perfectly aligned with the pump flanges.

Since the pump is rigidly secured to a foundation, taking up misalignment in the pipe connections will put a strain on the pump casing. Some strain is unavoidable because of flexures in the pipelines, which tends to increase with pipe length. Strain on the pump casing can also result from thermal expansion of the pipes. If the pipeline expansion is likely to be greater than 0.01 − 0.02 in., expansion joints should be used.[11] Strain due to piping misalignment can be avoided by incorporating dampener sections in the pipelines. Dampeners are lengths of rubber pipe reinforced with multiple plies of high-strength fabric and helical steel-wire reinforcement. These have the additional advantage of reducing vibration and line noises.

Excessive strain can distort the pump casing and lead to rapid wear and damage to the pump internals. This is especially true in close-clearance pumps such as rotary gear pumps. Mechanical seals and shaft packings are very sensitive to slight distortions in pump shape. They wear excessively when strain is present.

The pipelines to the pump should not have diameters that are less than those of the suction and discharge ports.

## Pump Foundations and Alignment

Most pump foundations are constructed of concrete. It is usual to bolt the pump and drive train to a cast-iron or steel base plate. This is then secured to the concrete foundation.

Foundations for small pumps need only be large enough to accommodate the base-plate assembly. Large pumps require foundations that are three to four times the weight of the pump and drive-train assembly; they should preferably consist of concrete reinforced with steel rods.

Anchor bolts are used to secure the base-plate to the foundation. In a common method each bolt is fitted with a washer and passed through a pipe sleeve that has a diameter three to four times greater than the bolt. The bolt-sleeve unit is set into the concrete foundation at the carefully predetermined base-place hole positions. The sturdiest base plate will flex to some degree if strained when the bolts are tightened. However, the flexibility in the sleeve-washer unit allows minor adjustments to be made in the bolt position prior to final tightening. This is the case even after the concrete foundation has set.

Metal shims are used to position the pump on the foundation. Adjustments should be made until the pump shaft and port flanges are completely level. The alignment between the pump and drive train is then adjusted before connecting the pump to the suction and discharge lines. The latter should have previously been aligned during the initial positioning of the base plate.

After the pipelines have been securely bolted, the entire pump assembly should be rechecked for flexure due to pipe strain. If the drive-train alignment has not been changed by bolting the pipelines, the space between the base plate and the concrete foundation is filled with grouting, which should be sufficiently fluid to fill all the available space under the base plate.

It is essential that the alignment between the piping, pump and drive train should not change. Preferably, alignments should be made at the operating temperature of the pumping system. Although this is not easily accomplished, it eliminates any alignment changes due to thermal expansion.

In-line positive-displacement pumps are now available that do not require any foundation. An example shown in Fig. 2 delivers 104 gpm. of liquid at viscosities up to 100,000 SSU and pressures up to 150 psig.[9] Centrifugal pumps of this type are also available. In-line pumps save valuable floor space. A further advantage is that the straight runs of pipe into and out of the pump reduce frictional head losses.

Drives for in-line pumps are usually part of an integral frame that provides ready alignment. As in other types of pump, however, alignment between the pump and piping must be precise to avoid pipe strain on the pump casing.

**Probable causes for pump operating troubles—Table IV**

| No Liquid Delivered | Pump Delivers Less Than Rated Capacity | Pump Losses Prime While Operating | Pump Is Noisy | Pump Wears Rapidly | Pump Takes Too Much Power |
|---|---|---|---|---|---|
| Pump not primed. | Air leaks in suction line or pump seal. | Liquid level falls below the suction-line intake. | Cavitation. | Pipe strain on pump casing. | Speed too high. |
| Insufficient available NPSH. | Insufficient available NPSH. | Air leak develops in pump or seal. | Misalignment. | Grit or abrasive material in liquid. | Shaft packing too tight. |
| Suction-line strainer clogged. | Suction-line strainer partially clogged or of insufficient area. | Air leak develops in suction line. | Foreign material inside pump. | Pump running dry. | Liquid more viscous than specified. |
| End of suction line not in liquid. | Wear on pump leads to increased clearances and slip. | Liquid vaporizes in suction line. | Bent rotor shaft (for rotary pumps) | Corrosion. | Misalignment. |
| Relief valve set lower than minimum required discharge pressure. | Relief valve wrongly set. | | Relief valve chattering. | | Obstruction in discharge line raises operating pressure. |
| Relief valve jammed open. | Relief valve jammed open. | | | | Discharge line too small. |
| Pump rotates in wrong direction. | Speed too low | | | | Discharge valve partially closed. |
| Suction or discharge valves closed. | Suction or discharge valves partially closed. | | | | |
| Bypass valve open. | Bypass valve open. | | | | |
| | Liquid viscosity differs from that specified. | | | | |

## Testing and Startup

Positive-displacement pumps are particularly prone to damage at startup. Thurlow[11] suggests the use of a temporary stainless-steel conical screen in the suction line. This provides protection against damage from bolts, welding slag, scale, etc., left in the pipeline during construction. A vacuum gage in the suction line and a pressure gage in the discharge line are useful aids in diagnosing system malfunctions.

The procedure for startup should be:

- Recheck alignment of pipeline connections, pump, and drive train.
- Rotate the pump by hand. If it does not move freely, check for internal obstructions.
- Momentarily "kick" in the motor to ensure that the pump is rotating in the proper direction.
- Check the operation of the relief valve.
- Prime the pump if it is not self-priming.
- Open all valves in suction and discharge lines.
- Ensure the tightness of all joints.
- Check the lubrication of packings, seals, etc.
- Start the pump.
- If the pump fails to deliver liquid within 1 min., shut it down. A positive-displacement pump should never run dry for any longer than is necessary.

The pump should also be stopped for any of the following reasons:

- Excessive vibration indicating misalignment.
- Noisy operation due to cavitation. This may be caused by an air leak into the suction line.
- Noisy operation caused by a foreign object.
- Excessive leakage at shaft seals and packings.
- Overheating of the pump or drive motor. This may arise from the binding of close-tolerance moving parts, from too great a discharge pressure, or from inadequate cooling facilities.

Packing glands should not be taken up completely during startup. A break-in period should be allowed to enable the packing to seat itself. If this is not done, scorching and damage to the packing and shaft can result. After the break-in period, the packing gland should be tightened to allow only sufficient leakage to provide proper lubrication.

When the pump is in operation, its performance should be checked against the manufacturer's specifications. Periodic checks on performance will provide an estimate of pump wear.

Table IV lists six types of pump troubles and their possible causes during pump operation.

## Acknowledgments and Credits

The following companies supplied information and/or illustrative material for this article. Numbers in parentheses are figure numbers of illustrations supplied by the companies: Barish Pump Co., New York; Blackmer Pump Co., Grand Rapids, Mich. (5, 6); Dorr-Oliver Inc., Stamford, Conn.; Durametallic Corp., Kalamazoo, Mich. (19); Eco Engineering Co., Newark, N. J. (1, 13); Jabsco Pump Co., Costa Mesa, Calif. (9); Kunkle Valve Co., Fort Wayne, Ind.; Milton Roy Co., Philadelphia, Pa. (12); Morse Chain Co., Ithaca, N. Y.; The Randolph Co., Houston, Tex. (17); Robbins & Myers, Inc., Springfield, Ohio (7); Roper Pump Co., Commerce, Ga.; Sier-Bath Gear Co., North Bergen, N. J.; Union Pump Co., Battle Creek, Mich. (11); Vanton Pump & Equipment Corp., Hillside, N. J. (15); U. S. Electrical Motors, Los Angeles, Calif.; Ulrich Mfg. Co., Roanoke, Ill.; Viking Pump Co., Cedar Falls, Iowa. (3, 23); Warren Pumps Inc., Warren, Mass. (2, 8, 10).

## References

1. "Controlled Volume Pumps," Bulletin 553-1, Milton Roy Co., Philadelphia.
2. Coopey, W., *Chem. Eng.*, **65**, 131, Sept. 27, 1958.
3. Coulson, J. M., and Richardson, J. F. "Chemical Engineering," Vol. 1, p. 33, Macmillan, New York, 1964.
4. *Ibid.*, p. 51
5. "Flow of Fluids Through Valves, Fittings and Pipe," Tech. Paper No. 410, p. B-4, Crane Co., Chicago, 1957.
6. Norton, R. D., *Chem. Eng.*, **63**, 197, Sept., 1956.
7. Perry, J. H., Editor, "Chemical Engineers Handbook," 3rd Ed., p. 390, McGraw-Hill, New York, 1950.
8. *Ibid.*, p. 415.
9. "Roline Gear-Type Rotary In-Line Pumps," Warren Pumps Inc., Warren, Mass.
10. Thurlow, C., *Chem. Eng.*, **72**, p. 117, May 24, 1965.
11. Thurlow, C., *Chem. Eng.*, **72**, p. 213, June 7, 1965.
12. Tracey, H. E., *Chem. Eng.*, **64**, 239, Apr., 1957.
13. "Vari-Flow Controlled Volume Pumps," Bulletin 600, Blackmer Pump Co., Grand Rapids, Mich.
14. "Viking Rotary Pumps," Viking Pump Co., Cedar Falls, Iowa.
15. Zalis, A. A., *Pet. Refiner*, 40, Sept. 1961.
16. Zisman, W. A., Naval Research Lab Report 4932, Washington, D.C., 1957.

## Meet the Authors

F. A. Holland

F. S. Chapman

F. Anthony Holland is chief of the engineering development section in the Research and Development Div. of Lever Brothers Co., Edgewater, N. J. He was formerly a production manager with Imperial Chemical Industries, Ltd., England. He has a B.S. and an M.S. in physics from the University of Durham, a B.S. in physics from the University of Oxford, and a diploma in chemical engineering and Ph.D. in chemical technology from the University of London.

Frederick S. Chapman specializes in process design and development in the fluid flow, heat transfer and liquid mixing in the engineering development section at the Research and Development Div. of Lever Brothers Co., Edgewater, N. J. He is currently completing his course work for a degree in chemical engineering at Newark College of Engineering.

# Positive Displacement Pumps

Reciprocating and rotary positive displacement pumps find wide applications in the CPI, especially in demanding situations and for metering and proportioning.

MYRON GLICKMAN, Dynamic Technology, Inc.

The primary characteristic of positive displacement pumps is that they exhibit a direct relationship between the motion of the pumping elements and the quantity of liquid moved. This is contrasted with the operation of centrifugal pumps, in which the flow rate for any given speed is determined by the resistance to flow in the discharge line.

Positive displacement pumps have a dynamic seal (sliding or rotating) that separates the discharge fluid from the inlet or suction side. Liquid displacement is theoretically equal to the swept volume of the pumping element. Flow rate is determined by the speed of the pumping element, or the number of cycles per time unit. In other words, the theoretical or ideal flow from a positive displacement device is fixed solely by its size and speed, and is independent of pressure.

In actual practice, however, pressure does influence capacity because of leakage through the dynamic seals, which is proportional to the pressure gradient. The difference between the ideal and actual flow rates is sometimes called "slip" and is illustrated in Fig. 1. Actual flow divided by ideal flow is volumetric efficiency, a basic characteristic of positive displacement pumps.

Volumetric efficiency can range from less than 50% up to 98%. Units in good condition, operating within their design speeds and pressures, should have efficiencies between 85 and 95%. The figure will be lower at very high pressures or very low speeds.

## RECIPROCATING PUMPS

The piston or plunger pump is probably the simplest form of a reciprocating positive displacement device. In the direct-drive pump, a common rod connects the drive piston to the pumping piston. The drive is powered by any convenient medium such as steam or air. A simplex pump has one drive and one pumping piston; a duplex unit has two of each.

The drive piston is controlled by a directional valve that is connected by a linkage to the piston rod. The linkage shifts the direction valve at the end of the stroke and causes the drive piston to reverse.

Displacement is determined by the pumping piston's area and stroke, while flow is fixed by the stroking rate. This, in turn, is controlled by the flow rate of the driving fluid into the drive piston chamber. Discharge pressure is equal to the drive pressure multiplied by the drive piston area divided by the pump piston area. Actual pressure is less than ideal because of friction losses.

Direct acting pumps usually use double-acting cylinders; that is, liquid is pumped on both the forward and reverse strokes. A single-acting cylinder discharges liquid only during the forward stroke.

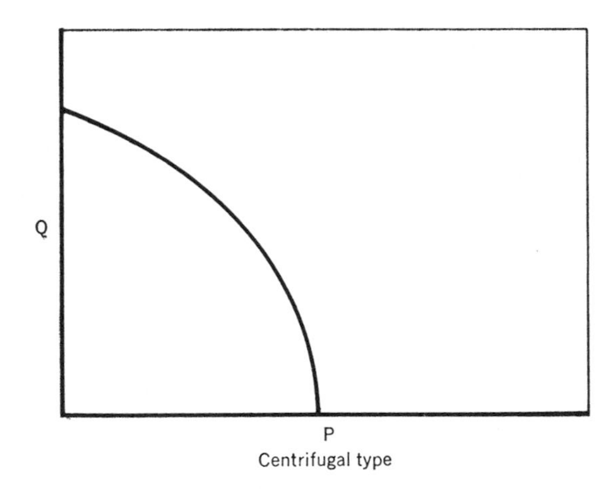


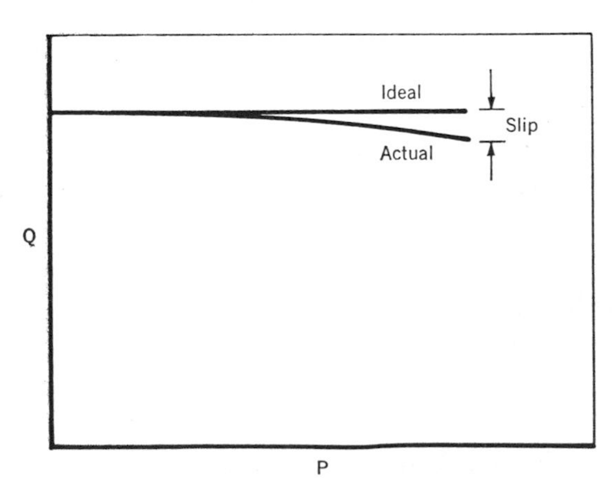


**FLOW VS. DISCHARGE** characteristics of the two principle pump classes—Fig. 1

Originally published October 11, 1971

## Power Pumps

In power pumps, the piston is driven by an external power source operating through a crankshaft. These pumps are commonly driven by electric motors or internal combustion engines. When running at constant speed, these units have a nearly constant flow rate over a wide range of discharge pressures.

Because of the relatively high speed and inertia of the moving parts in power pumps, very high pressure peaks develop if the discharge line is blocked or if flow resistance suddenly increases. To protect the pump and piping from damage, a relief valve is placed on the discharge line.

Power pumps usually have single-acting pistons. They are available with three, five and seven pistons, with three (triplex) being the most common. This type can be used for applications requiring up to several thousand psi.

## Flow Characteristics

Reciprocating pumps have a pulsating discharge, as contrasted to centrifugal and most rotary pumps, which produce relatively steady flow. At the end of the stroke, where the piston stops and reverses, the flow theoretically drops to zero. With double-acting cylinders, however, this can be prevented by cushion chambers or accumulators.

Discharge curves for simplex pumps with single and double-acting cylinders are shown in Fig. 2. Duplex pumps show twice as many dips in their discharge curve, but the dips are not as severe because the cycle of one cylinder is usually displaced half a stroke from the other.

In power pumps, where the piston is driven by a crankshaft, the piston moves with approximately sinusoidal motion (exact sinusoid action would require an infinitely long connecting rod). The approximate sine-wave motion of the piston results in flow curves such as those shown in Fig. 3.

Interior design varies widely in reciprocating piston pumps. As a general rule, flat-faced stem guided valves are used with low-pressure pumps. A wing-guided flat or bevel-faced valve is used for intermediate pressures, while high-pressure applications generally rely on only bevel-faced valves. Depending on operating conditions, the valve material may be iron, bronze, steel, stainless steel or plastic.

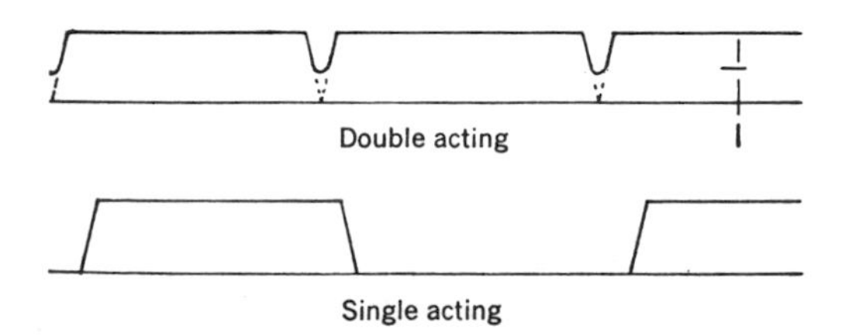


**RECIPROCATING** pumps have pulsating flow—Fig. 2

Packing material also varies widely, but is usually made of a soft, deformable material. In many designs, the piston rod packing gland is adjustable and may be tightened to maintain proper compression. Piston packings take many forms, ranging from the familiar metallic piston ring to preformed, molded rings. Piston packing is subject to the same environmental restraints as the rod packing, except that it usually cannot be tightened in service.

## Diaphragm Pumps

The diaphragm pump uses a flexible membrane as the displacement element. It may be moved directly by an eccentric mechanism or by a secondary pumping liquid. No packings or dynamic seals are required because the drive mechanism is completely isolated from the pumped fluid by the diaphragm. For this reason, the diaphragm pump is applied where leakage or contamination of the process fluid cannot be tolerated.

The diaphragm in a mechanically driven pump cannot be supported over its entire area. Stresses place a limit on the pumping pressure, which is usually held to less than 125 psig. The fluid-driven type, which does not have this drawback, is used for higher pressures.

Diaphragms are fabricated from elastomers, plastics and metals. Fluid-driven pumps with metal diaphragms may be used up to 45,000 psi. With plastics such as Teflon, operation is usually restricted to below 1,500 psi. and 280 F. Elastomeric materials are usually limited to 750 psi. and 212 F.

## ROTARY PUMPS

Gear pumps, which are probably the most widely used type of rotary pumping unit, have three basic configurations: the spur gear, internal gear and gerotor.

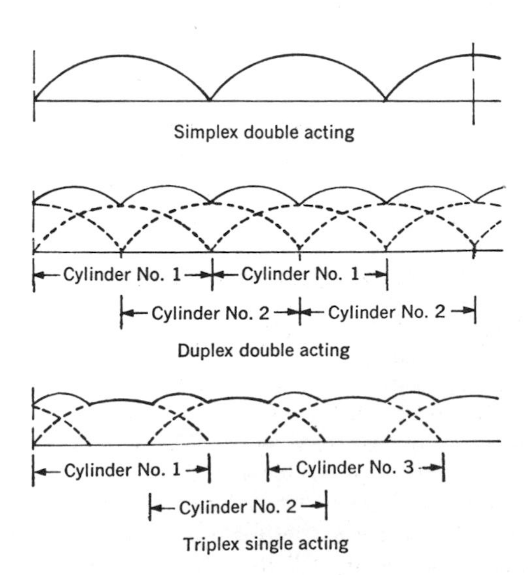


**CHARACTERISTICS** of power pumps—Fig. 3

These are suitable for pressures up to 3,000 psi. and capacities to 100 gpm.

A typical spur gear pump is shown in Fig. 4. The drive gear and driven gear are encased in a housing with minimum clearance at the gear tips. Liquid is carried around the periphery of the revolving gears from the suction to the discharge side.

The quantity of liquid displaced per revolution of the drive shaft is constant and is determined by the pitch and face width of the gears. Volumetric losses in gear pumps are caused by leakage in the form of viscous flow between the mating surfaces and by jet losses at the line contact of the teeth. Efficiencies are usually in the range of 80 to 95%.

A certain amount of clearance is required between the gears and side plates and between the gear tips and the housing. This is necessary to prevent seizure at high temperatures, but also results in slippage when pressure is increased. The fall-off in delivery is not serious at typical operating pressures (less than 1,000 psi.).

Moveable pressure-loaded side plates are used for high-pressure applications. Operating pressure is directed into compartments on the back sides of the plates to equalize pressure on both sides. Manufacturers of high-pressure gear pumps employ various designs for pressure-loading of the side plates. The objective is to attain an optimum clearance that prevents metal-to-metal contact and reduces leakage to a minimum.

## Other Design Factors

The spur gear pump is hydraulically unbalanced because of the opposed location of the inlet and outlet ports. This unbalanced pressure, which tends to crowd the gears toward the inlet side of the housing, requires heavy shafts and bearings. Some gear pumps use journal bearings that are lubricated by the pumped fluid, but most have heavy-duty roller or needle bearings.

Operating a gear pump continuously at higher than design pressure may cause scoring of the housing on the inlet side. This usually results in increased leakage and a substantial reduction in volumetric efficiency.

Another consideration is that as the gears rotate some liquid is trapped between the meshing teeth (Fig. 5). This is returned to the suction side and in the process generates extra pressure in the tooth space, causing additional loads on the shaft and bearings. Relief grooves milled in the side plates and end housings are used to transmit the trapped liquid to the discharge side of the pump.

Two types of internal gear pumps are shown in Fig. 6. These pumps have one gear with internally-cut teeth that meshes with a smaller, externally-cut gear. Those with standard involute gears (Fig. 6a) have a crescent-shaped divider that separates the inlet and outlet sides.

In the gerotor type (Fig. 6b), a divider is not required. The internal gear has one less tooth than the external gear. The gear contour is such that line contact is always maintained between each external tooth and the internal gear. As they pass the suction port, the space between the gears increases, pulling liquid into the pump. This space is reduced at the discharge, forcing liquid out of the pump. An interesting feature of this pump is that it has low sliding velocity at the gear contact lines.

Gear pumps can operate at speeds ranging from a few hundred to several thousand rpm. Practical speed is only limited by gear accuracy, noise level, bearing life and the pressure loss that results from the fluid's centrifugal action at high speeds.

Typical applications for gear pumps include process fluid transfer, fuel pumps for oil-fired furnaces and supplying hydraulic power for industrial and mobile machinery. Modifications of the basic design, often involving helical rather than spur gears, are used for pumping high viscosity fluids such as paint, varnish, ink, heavy oil, etc.

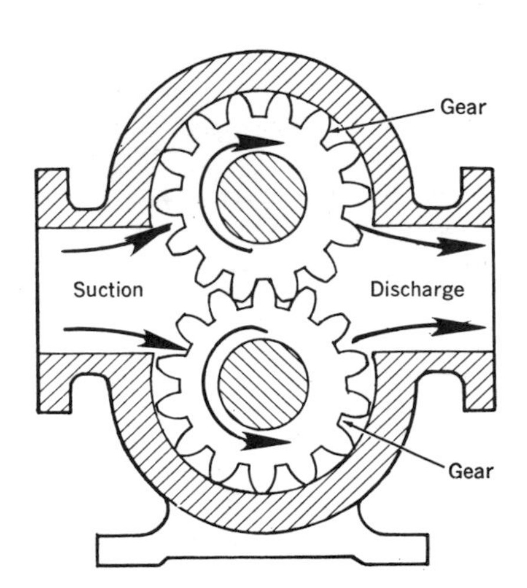


**SPUR GEAR** pump configuration—Fig. 4

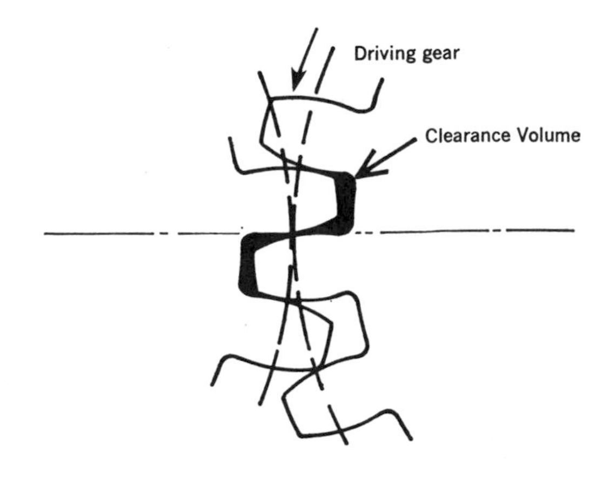


**LIQUID** trapped in teeth affects efficiency—Fig. 5

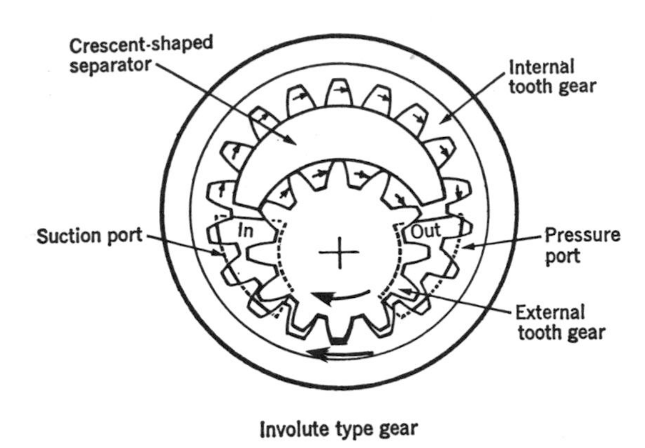


Involute type gear

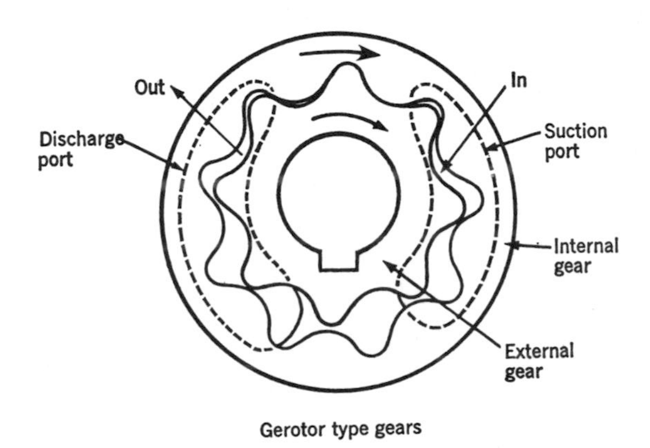


Gerotor type gears

**TWO TYPES** of internal gear pumps—Fig. 6

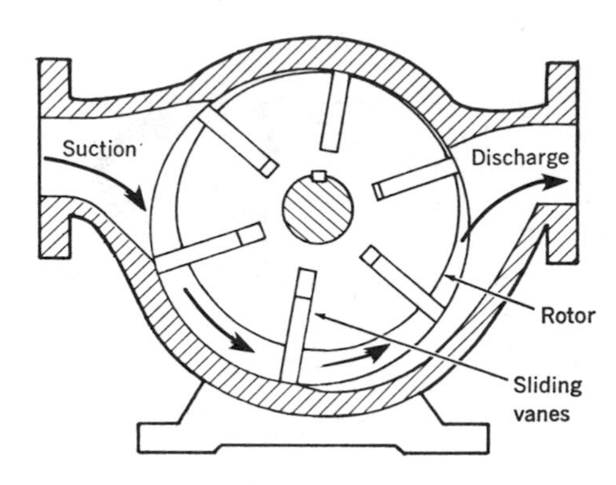


**VANE PUMPS** have uses similar to gear pumps—Fig. 7

## Vane Pumps

The vane pump in Fig. 7 is another type of rotary device. It has a rotor mounted in a housing, generally between closely fitted end plates. Vanes are inserted in slots in the rotor, either radially or in a slanted arrangement. The rotor shaft is located eccentrically in relation to the housing, and the vanes slide on a hardened surface inside the housing.

As vanes proceed from the point of shortest distance between the rotor and housing, the space between the two increases and fills with liquid from the suction port. As the cycle continues, the spaces between the vanes become smaller and liquid is discharged through the opposite port.

Vane pumps may be either fixed or variable displacement. The variable displacement design has a mechanism for changing the eccentricity between the cam ring that contacts the vane tips and rotor. Decreasing the eccentricity reduces the amount of fluid displaced during each revolution of the rotor. The cam ring may be moved manually or in reaction to the hydraulic pressure.

Variable displacement pumps are used, obviously, where flow rate requirements vary considerably. It is often more practical to change the pump's displacement rather than alter the speed of the driver, which is usually an electric motor or governed engine.

In the pressure-compensated variable displacement pump, increased pressure reduces the cam ring's eccentricity with respect to the rotor. This produces a characteristic in which flow decreases as pressure increases. The shape of the curve is determined by a spring loading on the cam ring.

To minimize leakage, the vanes must be kept in contact with the cam ring. Centrifugal force is not enough, and various arrangements involving spring and pressure loadings are employed. The problem is one of optimization: the vanes must be supported on a liquid film that is thick enough to prevent metal-to-metal contact, yet thin enough to minimize leakage. In many designs, pressurized liquid is ported under the vanes to keep their tips in contact with the cam ring.

In a simple vane pump, the inlet and outlet ports are opposed, which causes the same type of hydraulic unbalance as is present in gear pumps. This can be compensated by providing two sets of diametrically opposed inlet and outlet ports around the cam ring's periphery. This feature can only be incorporated in fixed displacement pumps, and there is no equivalent method for balancing variable vane pumps.

Volumetric efficiency of vane pumps is generally above 90%. They are used over roughly the same pressure range as gear pumps, but are found most frequently in high-pressure applications (1,500 to 2,500 psi.). Vane pumps are common as the power source on industrial hydraulic machinery.

## Other Rotary Pumps

The screw pump, represented in Fig. 8, has two or three pumping elements with helical threads. The elements mesh so that their contact points progress axially down the screws, creating the pumping action. Screws are accurately machined and do not require timing or spacing gears. They produce continuous flow with relatively little pulsation or agitation of the fluid.

Screw pumps are quiet and efficient. They are available in speed ranges up to 3,500 rpm., pressures up to 2,500 psi. and flow rates greater than 100 gpm.

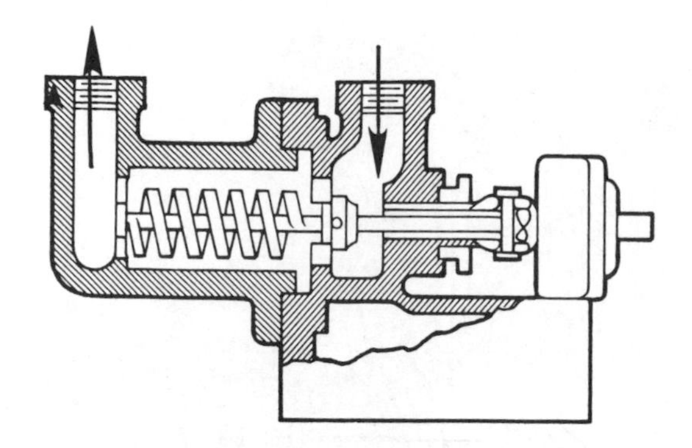

**SCREW PUMPS** are quiet and efficient—Fig. 8

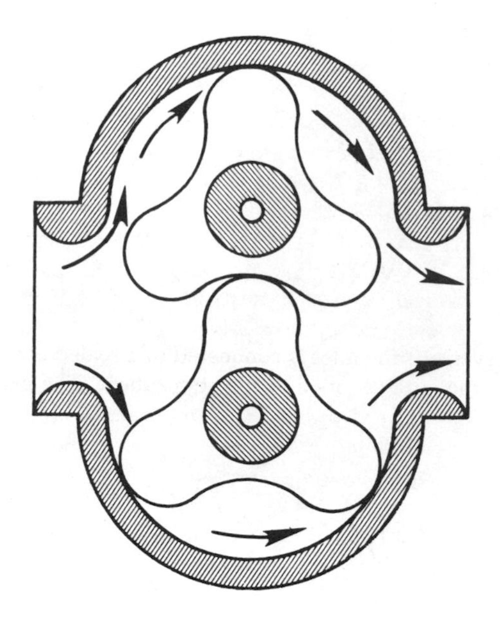

**LOBE PUMP** design is similar to Roots blower—Fig. 9

Applications range from process transfer pumps to hydraulic power pumps for submarine systems.

The lobe pump has a configuration similar to that of a Roots blower. A three-lobe design is shown in Fig. 9 and they are also available with two or four lobes. Timing gears hold the lobes in place, and the lobes are positioned so that there is a small clearance between themselves and the housing at any mesh position. Sometimes helical lobes are used in lower pressure applications. The cam pump, a related design, employs two cam-shaped rotors that mesh together to create pumping action.

Cam, lobe and screw pump configurations find important applications in food processing. These are usually low-speed types constructed of stainless steel. They are designed so that relatively large pockets are formed by the rotation of the rotors. This way the suspended solids often present in food products may be pumped without being crushed.

The radial-piston and axial-piston pumps have the highest volumetric efficiencies of all positive displacement pumps. They are used mostly in high-pressure (1,500 to 3,000 psi.) services for hydraulic power supply. Some axial-piston pumps develop pressures up to 8,000 psi.

In the radial type, spring-loaded pistons are housed in a cylinder block that is eccentric to the pump housing (Fig. 10). As the block rotates, the pistons move radially in and out of their respective cylinders, producing the pumping action. A variation has pistons contained in a stationary housing and spring loaded against a rotating cam.

The axial-piston pump has pistons placed parallel to the pump drive shaft. They are pressed by springs against a plate that is at an angle to the shaft axis. When operating, either the drive shaft rotates the cylinder block against a non-rotating angle plate, or rotates an angle plate against a non-rotating cylinder block. The result causes reciprocating axial motion of the pumping piston.

Both radial and axial-piston pumps are available in fixed and variable displacement versions. To achieve high volumetric efficiencies (over 95%) these pumps require close clearances and very accurate machining. Piston clearances are usually held to less than 0.001 in.

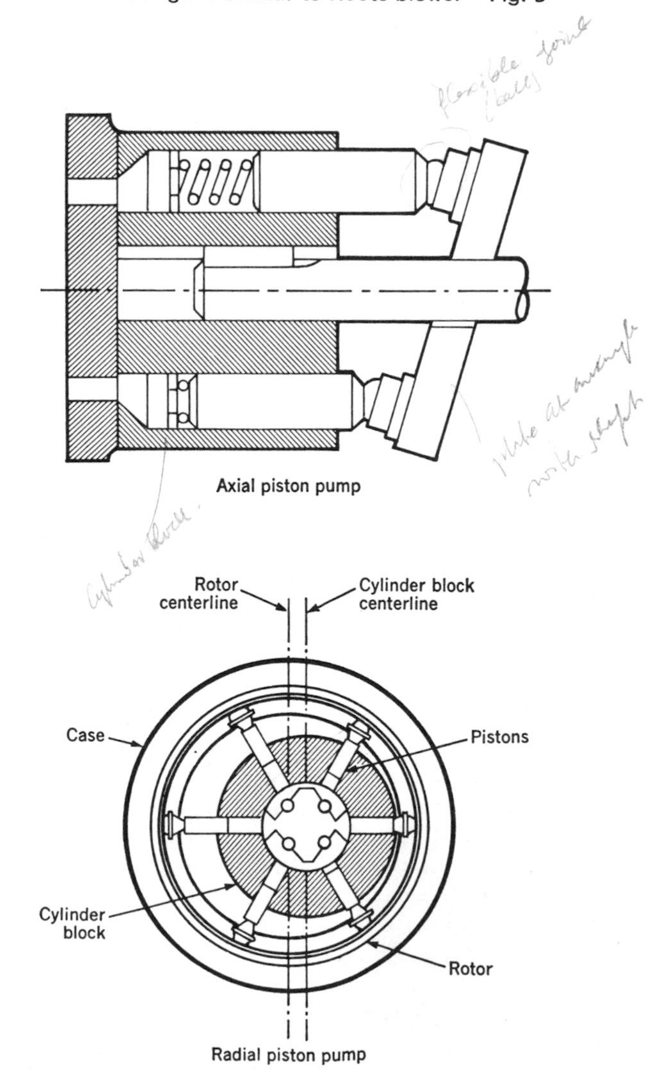


**PISTON PUMPS** for high pressure applications—Fig. 10

## Vacuum Pumps

The region below atmospheric pressure can be roughly divided into low and high vacuum. Low vacuum can be produced by mechanical pumps, while high vacuum refers to extremely low pressures, which are achieved by molecular or ionic action. The lowest pressures ordinarily reached through mechanical pumps are in the range of 0.1 to 0.0001 Torr (1 Torr = 1 mm. hg. = 0.0193 psi.).

Mechanical vacuum pumps are essentially positive displacement compressors, except that the connections are reversed: the inlet is connected to a sealed chamber while the outlet is open to the atmosphere. The general types are the reciprocating piston, vane, lobe (Roots blower) and single lobe.

A good seal between the inlet and outlet sides is essential in all mechanical vacuum pumps. To reach pressures of 0.1 Torr and below, the pump is designed so that a film of oil is maintained between the fixed and moving surfaces. The oil is usually silicon based and has an extremely low vapor pressure.

## Metering Pumps

Metering or controlled volume pumps are simply very accurate positive displacement devices, usually of the plunger or diaphragm type. In the plunger variety, the theoretical amount of fluid discharged equals the area of the plunger multiplied by the stroke length. In practice, however, some liquid slips back into the cylinder during the time required for the check valves to seat. Worn valves will further reduce the actual output volume. Properly-serviced plunger pumps should have a volumetric efficiency of 95% or better depending on speed, fluid viscosity and discharge pressure.

Metering pumps are almost always mechanically driven, usually by standard electric motors. A reduction gear is used to accommodate standard motor speeds. Variable-speed motors make it possible to change feed rate in some proportioning pumps. Practically every manufacturer has its own method for adjusting stroke length. This usually involves an electric, pneumatic or hydraulic control that changes the crank radius.

In hazardous locations or other situations where electric power cannot be used, metering pumps may be driven pneumatically. With these drives it is difficult to maintain constant speed and as a result accuracy suffers.

Glandless pumps in plunger and diaphragm designs are applied where the liquid is toxic, corrosive, volatile, abrasive or where contamination cannot be tolerated. One design, a mechanically-actuated bellows pump, is shown in Fig. 11. This has a larger capacity than the diaphragm-type head, and is particularly suited for metering into and from a vacuum.

There are several configurations of the hydraulic diaphragm pump. Displacement is controlled by adjusting the stroke of the plunger on the oil side.

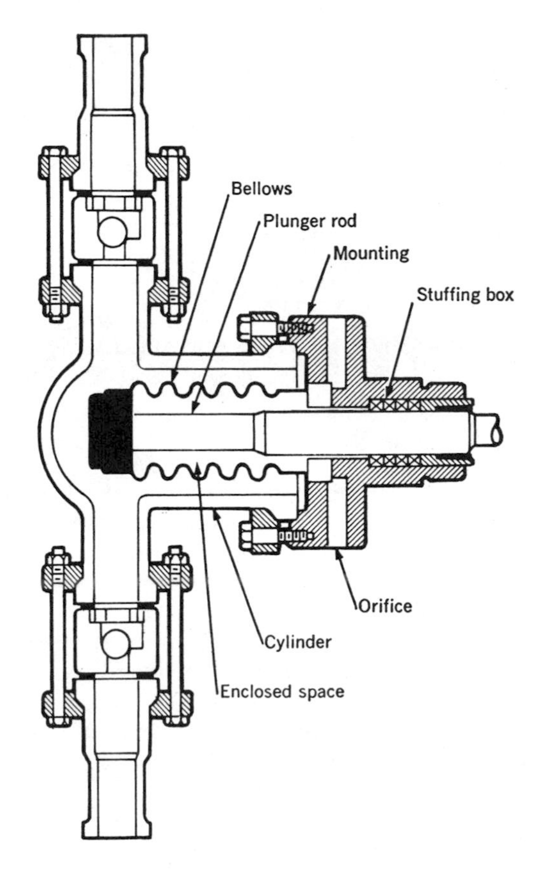


**GLANDLESS** plunger pump for metering—Fig. 11

Accuracy depends on minimizing leakage past the plunger packing and no air in the oil. Entrained air could cause an error in diaphragm displacement due to increased compressibility of the oil.

Metering pumps find extensive applications in the chemical process industries. They handle all types of fluids including those that are highly corrosive, flammable or toxic. Temperatures may be in the cryogenic range or as high as 1,000 F., and flow rates can vary from a trickle to thousands of gallons per hour.

## Bibliography

1. Hicks, Tyler G. and Edwards, Theodore W., "Pump Application Engineering," McGraw-Hill, New York, 1971.
2. Ernst, Walter, "Oil Hydraulic Power and Its Applications," 2nd. Edition, McGraw-Hill, New York, 1960.
3. Elonka, Steve and Johnson, Orville H., "Standard Industrial Hydraulics Questions and Answers," McGraw-Hill, New York, 1967.
4. Vetter, G. and Bohm, O., Design Characteristics and Uses of Glandless Metering Pumps, *British Chemical Engineering*, Sept., 1968.
5. Hernandez, L. A., Controlled Volume Pumping, *Chem. Eng.*, Oct. 21, 1968.
6. Arlidge, Dean B., The Basics of Fluid Power, *Mechanical Engineering*, Sept., 1968.
7. Thomas, G. M. and Henke, R. W., Pumps for Fluid Power, *Mechanical Engineering*, Sept., 1968.

## Meet the Author

**Myron Glickman** is vice president of Dynamic Technology Inc., a consulting engineering firm specializing in fluid power. Previously, he has done fluid power research and development at IIT Research Institute, TRW Corp. and Martin Marietta. Mr. Glickman holds B.S. and M.S. degrees from Illinois Institute of Technology.

# Metering With Gear Pumps

While they have limitations, gear pumps with a bypass flow arrangement provide a low-cost, simple answer to many process metering problems.

ARTHUR W. TRENT, Eco Pump Corp.

When an engineer specifies or selects a chemical-metering system, the choice usually centers around a piston or diaphragm-type metering pump.

While these pumps are the preferred choice for many applications, an available alternative merits consideration, particularly for moderate-pressure, open- or closed-looped systems: the relatively simple, low-cost, rotary gear pump.

Piston or diaphragm pumps handle quantities in units of gal/hr; rotary gear pumps are designed for gal/min delivery—but a simple bypass arrangement on the gear-pump system produces the desired lower flowrates.

Because of the pump design, backpressure is a limitation on gear-pump metering systems: overall system backpressure cannot exceed 75 to 100 psi. And abrasive solids in the metered liquid must be avoided, since they will damage the pump. Mindful of these advantages and limitations, let us describe some typical metering systems.

## Basic Metering System

The main parts of a basic system (Fig. 1) include chemical-storage facilities; positive-displacement gear pump; bypass piping loop; bypass (control) valve; and flow indicator (rotameter) to show rate of chemical flow.

In essence, the gear pump, driven by a low-cost, constant-speed motor, takes fluid from the storage tank, pressurizes it, and sends it to the point of feed. Rate of feed is clearly indicated on a visual-flow indicator or rotameter, since the gear pump output is steady and linear, with no pulsations.

To vary the feedrate, a bypass line must be installed after the pump discharge and returned to the storage tank. Depending on the relative pressure drops in the bypass line and the feed line, flowrate to process can be metered from 0 to 100% of pump capacity.

Pressure drop in the bypass or recirculating line is controlled by a manual or automatic bypass valve. Piping size and length control the pressure drop in the feed line, which in itself is often enough to provide necessary backpressure. Backpressure can also result from system backpressure at the point of feed, or it can be controlled by a manual or automatic valve in the feed line, as shown in Fig. 1.

Note, in the examples given later, that the systems are designed so that all backpressures are minimal.

For all its simplicity, the system shown in Fig. 1 can produce accurate and reproducible process flowrates, ranging from 0.004 to 20 gal/min, all from the same pump.

## Key Element: Storage Facilities

The storage tank will probably be the biggest capital outlay for the entire metering system. Such a tank takes many forms. For highly corrosive chemicals, there are

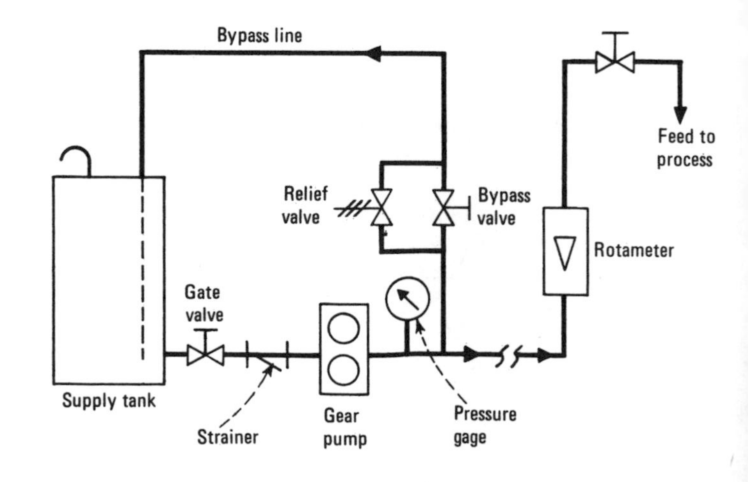


**BASIC METERING** system contains bypass line—Fig. 1

Originally published January 20, 1975

stainless-steel and glass-lined tanks; for handling large quantities, carbon-steel tanks may be custom-erected and lined with a sprayed-on plastisol. But the best choice is probably a filament-wound fiber-glass reinforced plastic tank with an approximately 60/40 glass-to-resin ratio for good fire resistance. Fiber-glass tanks are available in many standard sizes and shapes, and with capacities ranging from a few gallons to 30,000 gal or more.

A word of warning: While filament-wound fiber-glass tanks will be adequate for a great range of chemicals, they are not intended to handle everything. For example, mineral acids are best stored in a carbon-steel or plastisol-lined steel tank. Heat buildup due to condensation dripping into the acid could cause a violent exothermic reaction, seriously weakening a fiber-glass reinforced plastic tank. A desiccant basket placed in the tank vent can minimize these problems.

The least-expensive fiber-glass tank is the simple, noninsulated type, usually installed indoors. But for various reasons—to save space or comply with safety regulations—outdoor chemical storage might be best. In this case, thermal insulation and heat-tracing for the tanks, associated pumps and transfer lines may be needed (also, recirculating flow from the bypass loop will help stabilize temperature by keeping the liquid moving).

Some tank fabricators offer factory-installed heat-tracing and weather-proof insulation, and this prefabricated arrangement can represent a savings up to 40% over field installation.

Regardless of inside or outside storage, the main holding area should always be located for quick, convenient delivery by tank truck or railroad tank car.

## Piping, Valves and Fittings

Fiber-glass pipe is suitable for the tank-truck unloading line, but since this pipe is chemical resistant mainly on its inside surface, it should not be used for a dip pipe, recirculation line, or any lines wholly or partially immersed in the solution.

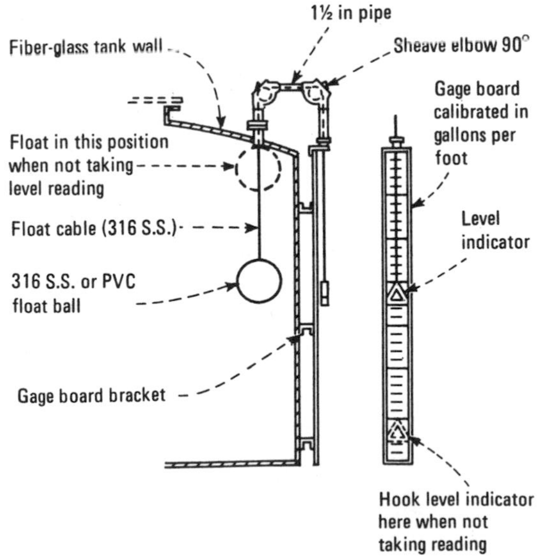

For these applications, schedule 10 316 stainless steel or schedule 80 polyvinvyl dichloride pipe, secured to the tank wall with brackets, would be a good choice. The dip and recirculation, or bypass, pipes should extend to within a foot to 18 in of the bottom of the tank (this prevents foaming and formation of solids that might occur if there were no agitation at the bottom of the tank).

A syphon breaker (½ in) between unloading line and tank vent prevents syphoning back through the line.

One nice point about the indoor, noninsulated fiberglass tank: it is translucent, so checking liquid level is a very simple task. In outdoor installations, however, thermal insulation removes the visibility advantage, and a tank requires a liquid-level gage (the tank needs an extra nozzle to accommodate the gage pipe and cable).

A simple manual "float-ball and board" liquid-level gage, as shown in Fig. 2, does an adequate job for chemical storage, particularly for viscous chemicals. Lowering the ball until it floats produces a reading on the gage board. Be sure to fasten the indicator to a hook at the bottom of the gage board when not in use: this hoists the float ball to the top of the tank, above liquid level. If the ball stays on the surface between readings, solids buildup may destroy the accuracy of the gage.

The unloading-hose connection is an ASA standard fitting, which will connect to the reducer carried on tank trucks as standard equipment.

The tank outlet valve should be Type 20 stainless or similar material, since it must function as the main shutoff valve during maintenance and leakage cannot be tolerated. Other line valves can be stainless or, in some cases, plastic. Optimum body and trim for the stainless bypass valve would be high-nickel Hastelloy C.* A lower-cost 316 stainless valve will do the job, but the trim has to be watched and replaced as required.

Nonimmersed piping may be fiber-glass plastic, stainless, polyvinyl chloride or polyvinyl dichloride. A stainless-steel line strainer should be installed in the pump-suction line and equipped with a Monel or Hastelloy C* screen to protect the system from abrasive solids. To provide unimpeded flow, strainer size must have a total open area in the mesh at least 1.5 times the area of the suction line I.D. Of course, the bypass line will be the same pipe size as the pump discharge line. The valves should be capable of handling average flow conditions.

**LIQUID-LEVEL** indicator: simple, low cost—Fig. 2

## Rotary Gear Pump

The heart of this metering system is the rotary gear pump. Equipped with suitable bypass control, a positive-displacement gear pump can provide accurate unattended metering—on the order of ±2% in open-loop systems, ±0.5% or better in closed-loop systems.

Advantages the gear pump brings to this system include moderate cost, easy maintenance, and nonpulsating flow—a particular advantage in metering, because there is no slug feeding, and flowrates can be easily checked with a simple rotameter.

Gear pumps feature a simple, straight-through de-

* Two "C" alloys are available: Hastelloy C-276 and Hastelloy C-4.

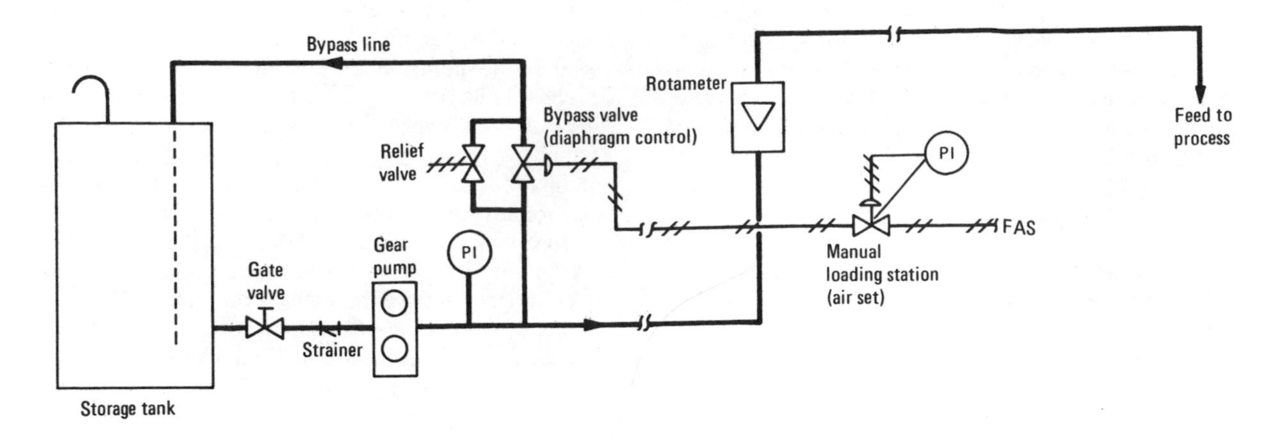


**PNEUMATIC** system permits feed control at some distance from the equipment—Fig. 3

sign—there are no check valves, balls, springs or diaphragms. For this reason, gear pumps can be expected to operate troublefree in many difficult situations. Of course, these pumps cannot achieve the high-discharge pressures of plunger pumps.

Because the unmeshing of the gears evacuates any air between pump inlet and fluid level, atmospheric pressure forces a continuous column of fluid to the inlet. At the same time, continuous meshing of the gears forcibly expels fluid from the housing while preventing return of fluid to the suction side. The result is self-priming operation, resulting in excellent suction capabilities.

## Bypass Control Provides Flexibility

It is important to have a good grasp of the possibilities of bypass or recirculating control, because this is the key to the flexibility of any gear-pump chemical metering system.

Fig. 1 shows the details of the simplest form of such a system. It is possible to control flow with a variable-speed motor, but this is more costly, and flowrate variation is limited by speed rangeability of the drive. Greatest economy and flexibility comes from running the pump at constant speed, pumping its full capacity, and using a bypass valve (manual or automatic) and recirculation line to return part of the flow to storage.

Note the auxiliary relief-valve loop around the bypass valve. While not strictly essential, this is an excellent safeguard, protecting the pump and system from excess pressure buildup in case a line clogs or a valve accidentally closes.

The system shown in Fig. 1 is used for unattended delivery of chemicals to metering points at a fixed or predetermined flowrate. The operator need only adjust the bypass valve as flow parameters or other conditions change. This system, of course, can be developed for fully automatic control, but at the cost of added instrumentation and increased system complexity.

One type and size of pump to handle maximum capacity needed can be specified for more than one metering system. Different feedrates can be obtained by bypassing any part of the flow up to 100%, with no excess heat buildup or risk of damage. This standardization gives excellent flexibility, with minimal parts stocking.

## More Advantages of Loop

A further important advantage of the bypass loop: As wear takes place within the pump, tending to reduce output, bypass flow can be readjusted to compensate for the change, keeping the pump in operation until a scheduled maintenance outage.

When not metering, bypass flow very effectively mixes and agitates chemicals in the storage tank. Even with the system in normal metering operation, enough mixing and agitation will be created by the bypass flow to keep solids from forming, helping prevent buildup on the bottom of the tank. For this reason, the recirculating pipe should always be located on the opposite side of the storage tank from the pump suction. Locating the pipe outlet near the bottom of the tank prevents aeration problems.

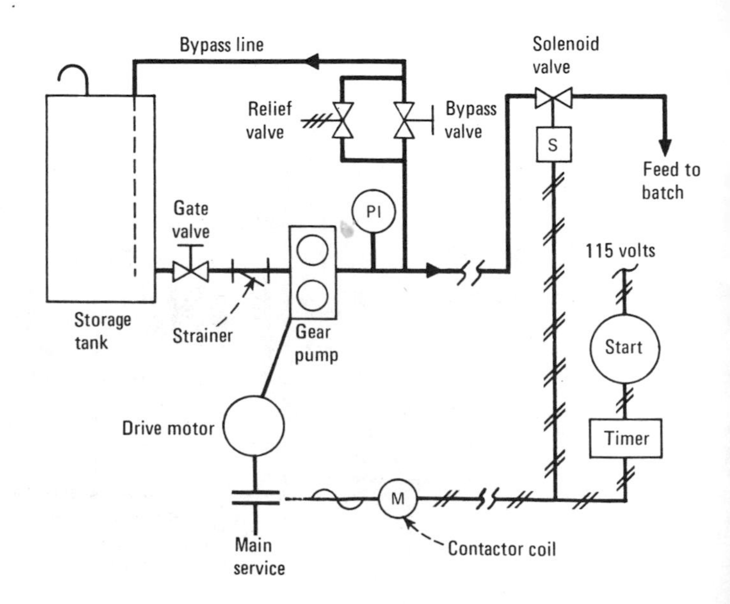


**TIMED CYCLE** meters predetermined quantities—Fig. 4

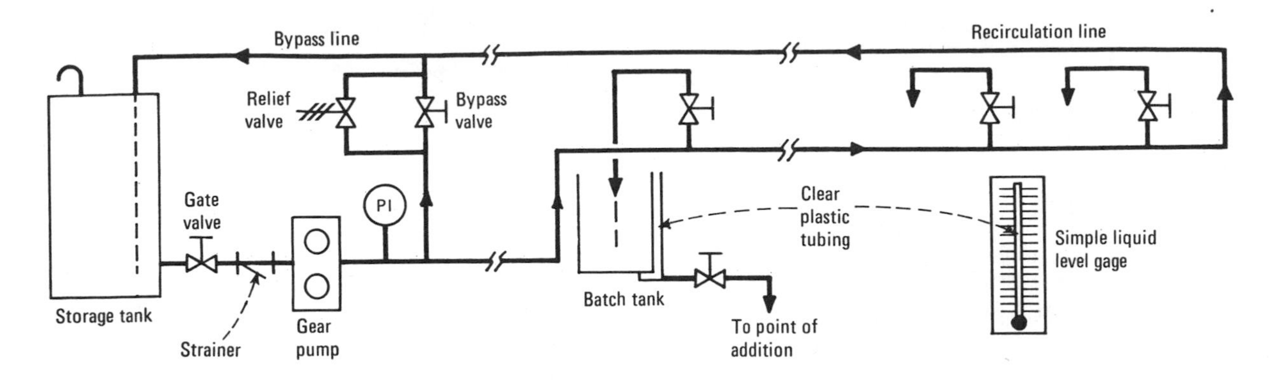


**BATCH METERING** system uses intermediate holdup tanks for feeding liquid into the process—Fig. 5

## Visual Indicators for Flow

Because of the nonpulsating flow of a gear pump system, it is possible to accurately measure flow with a simple rotameter.

Locating the point of feed above the highest liquid level in the storage tank will eliminate the need for a valve in the feed line. When this cannot be done, a needle valve is necessary in the discharge line after the rotameter for good control. Then by setting the bypass valve to maintain a head of 10 to 20 psig on the system, flow is controlled by adjusting the needle valve. (Close the valve when shutting down the system to prevent gravity flow of chemicals from the storage tank through the feed line.)

For many applications, it is best to specify rotameter body and fittings of 316 stainless steel, and a Hastelloy C float. Rotameters have a 10:1 range, so the calibrated tube should have average flow midrange of the tube.

## Manual With Pneumatic Control

The manual metering system with pneumatic control shown in Fig. 3 involves a slightly more complex arrangement than is shown in Fig. 1. This system is particularly useful when locating point of feed at some distance from the tank, pump and bypass line. Rate of flow is still determined by the operator, based on rotameter reading, but the operator sends his command long-distance, by increasing or decreasing air supply to the bypass valve, via an air filter-regulator in the pneumatic supply line.

Many processing operations are conducted on a batch basis rather than continuously, and the system shown in Fig. 4 is a simple, inexpensive way to meter a predetermined amount of additive to a particular batch.

Here, a normally closed solenoid valve located in the feed line, and the contactor coil in the pump-motor starter, are both triggered simultaneously by an elapsed-time-cycle timer. The operator merely pushes the start button, the pump starts and the solenoid valve opens. After the preset timed interval, the pump stops, the valve closes, and the timer resets itself for another batch.

A variation of this system has the pump running continuously, feeding several controlled-time-cycle solenoid valves, each with its own pushbutton timer. Calibration of this arrangement is more complex than the single system, since pressure balances in the bypass line and feed lines depend on number of valves open at any time.

A typical application of the system would be addition of alum to a dump chest in a paper mill. Assuming a 3,000-lb dump chest, and a requirement of 21 lb of dry

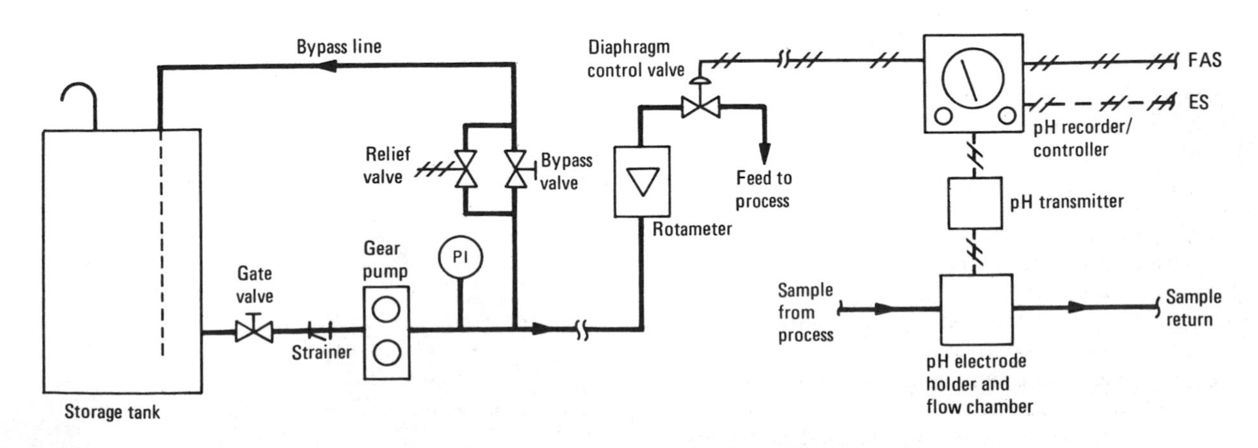


**CLOSED-LOOP** arrangement, with a controller modulating a diaphragm valve, controls process pH—Fig. 6

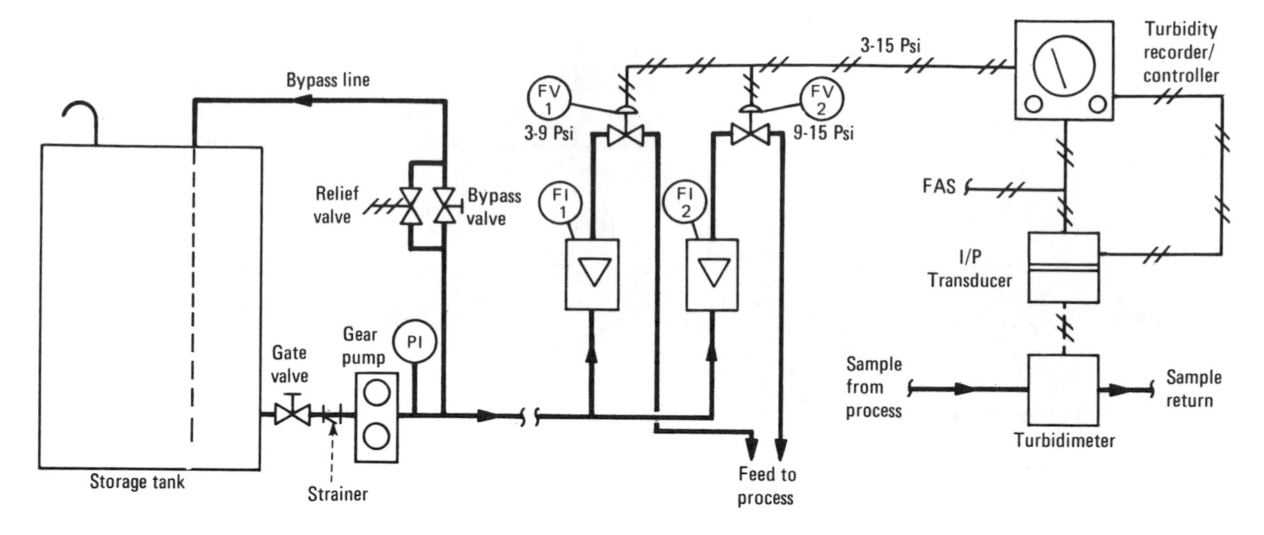


**AUTOMATIC** metering for turbidity control has two-level design for handling usual situations—Fig. 7

alum per batch, a flowrate of 2 gpm requires a time cycle of just under 2 min.

## Multiple Batch

Another batching system is shown in Fig. 5. Here, multiple-batch, or head tanks are filled to individually predetermined levels from a recirculation line serving all tanks. The valves feeding each tank may be manually operated as shown, or could be solenoid or air-operated valves.

Each tank comes fitted with a very simple sight gage consisting of a clear plastic tube connected to a tee in the tank's outlet, with the open end clamped at the top of a calibrated board. The operator fills the head tank to the required level, and then allows the contents to feed by gravity to the point of addition.

## Automatic pH Control

One of the most common applications of rotary gear pumps in metering systems is closed-loop pH control. Fig. 6 shows a typical system for feeding commercial sulfuric acid to maintain process-stream pH. Variation in pH not exceeding ±0.1 unit can be regularly achieved with such a system.

A sample of the process stream continuously feeds to a pH sensor assembly, connected to a pH transmitter. When the pH sensor and pH transmitter come from different manufacturers, an interfacing module between the two is generally required.

Output from the pH transmitter, usually a milliamp signal, goes to the pH recorder/controller. Output of the controller, a 3 to 15-psi air signal, modulates a reverse-acting-diaphragm control valve in the acid-feed line.

## Automatic Turbidity Control

Fig. 7 illustrates a two-level demand system for turbidity control and other applications. For a process parameter that can vary over a wide range, a second feed handles occasional demand bursts.

In the illustration, a sample stream of raw water is continuously monitored by a turbidimeter calibrated in Formazin turbidity units (normal range of variation in the raw-water supply, for example, could be 25 to 100 FTU). The signal from the turbidimeter, fed to the recorder controller, is such that pneumatic output from the controller varies from 3 to 9 psi over this range of FTU. Hence, only flow-control valve No. 1 operates to feed alum to the raw water.

However, high turbidity can occur from time to time due to some water-supply disturbance, perhaps by a rainstorm. In this event, turbidity may rise to 500 FTU and signal from the controller calls for both valves to open, with perhaps some modulation of flow-control valve No. 2.

Typical feedrates under normal conditions might be 3 grains of alum (equivalent dry weight)/gal of water, while after a storm, feedrates could range to 10 grains/gal.

Note that both valves in the feed line are reverse acting, since the control-system arrangement calls for more alum feed with increasing air signal. Note also that both valves are designed to close in the event of air failure, assuring treatment until the trouble can be located and corrected.

## Meet the Author

**Arthur W. Trent** is an engineering and technical services advisor to Eco Pump Corp., 2387 So. Clinton Ave., South Plainfield, N.J. 07080. He was previously employed for 23 years in the Engineering and Technical Services Dept. of American Cyanamid at Wayne, N.J. Prior to that, Mr. Trent worked for Fritzsche Brothers, Inc., Clifton, N.J., and Western Electric, Kearny, N.J. Mr. Trent holds a B.S. from Wilson School of Engineering, and attended Newark College of Engineering.

# Basics of reciprocating metering pumps

The author gives tips on the specification and operation of this important class of metering pumps used to inject a liquid stream into a process at a controlled flowrate.

*James P. Poynton,* *Milton Roy Co.*

☐ Metering pumps are widely used to control the rate at which a volume of fluid is injected into a process. They are capable of high accuracy, and may be adjusted while in operation to vary the flowrate.

Also referred to as "controlled-volume" and "proportioning" pumps, these devices include a variety of positive-displacement-type designs. In this article, the term metering pump will be used to denote any chemical pump whose steady-state delivery is accurate and whose output is adjustable while in operation.

Metering pumps operate by various pumping mechanisms. The two largest categories are reciprocating pumps, which include packed-plunger and diaphragm (both disk and tubular) designs, and rotary pumps, which include screw, vane, lobe and peristaltic designs.

In many chemical processing applications, it is desirable to select a metering pump whose output can be varied through manual or automatic adjustment of the displacement mechanism. Ideally, the pump should be capable of accommodating a wide range of fluids, from abrasive slurries to corrosives. If the fluid is to be injected into a process, the pump should be able to generate high discharge pressures.

## Reciprocating metering pumps

This article will focus on the operating characteristics of the first category of metering pumps, reciprocating.

In general, this category features four basic components: (1) a prime mover, either electrically or pneumatically driven, (2) a drive mechanism, (3) a device for capacity adjustment, and (4) a displacement chamber or liquid end.

Reciprocating units commonly employ one of two types of liquid ends—packed-plunger (or piston) and diaphragm. In the packed-plunger liquid end, a piston contacts the pumped fluid directly. In the diaphragm or bellows-type liquid end, some type of positive barrier arrangement prevents the fluid from contacting the plunger.

These pumps may be driven by an electric motor (the most common drive), or by a pneumatic cylinder in which the cylinder rod is directly coupled to the plunger. Pneumatically driven pumps are particularly appropriate for locations where electric power would be unsafe, or for systems that use digital techniques to control pump delivery.

## Packed-plunger designs

The packed plunger is the basic design from which all types of liquid ends originate. Fluid motion is induced by one or more sealed plungers that draw in and expel fluid. One-way check valves on the inlet and outlet ends of the pump operate 180 deg out of phase so as to permit the filling of the displacement chamber during the suction (or "vacuum") stroke, and to prevent backflow to the supply system during the discharge stroke (Fig. 1).

The plunger displaces a set volume of fluid during each cycle. Assuming that it is properly fitted to the displacement chamber, the plunger displaces a volume of fluid equal to the volume of a cylinder having the same diameter as the plunger, and a length corresponding to the stroke distance. However, the amount of liquid actually pumped may differ from this theoretical value due to:

1. Excessive time associated with check valve operation, resulting in "fall-back" loss.
2. Fluid compression caused by high discharge pressure.
3. Entrapped gases.

Originally published May 21, 1979

4. Elastic deformation of the pump head.
5. Leakage through defective piping and seals.

The actual amount of liquid displaced by a properly operating pump is about 95% of theoretical, if the liquid being handled is not compressible. A simplified equation for determining volumetric efficiency is:

$$\text{Vol. eff.} = \frac{\text{Delivered volume}}{\text{Swept volume}}$$

If the pump is not operating at rated capacity, its volumetric efficiency may be calculated by modifying this equation slightly:

$$\text{Vol. eff.} = \frac{\text{Actual flow}}{\begin{matrix}\text{Theoretical delivery} \\ \text{at maximum stroke}\end{matrix} \times \begin{matrix}\text{Percentage} \\ \text{of stroke}\end{matrix} / 100}$$

The packed-plunger design has several advantages over other configurations: low initial cost, design simplicity and high discharge pressures.

In addition, only the vapor pressure of the fluid really limits the ability of the sealed piston to lift fluids to the pump inlet. This pump, in its basic form, may be operated at pressures up to 30,000 psi, with repetitive accuracy to within $\pm 1\%$ of rated capacity.

However, the plunger-type metering pump has several inherent disadvantages that make it a poor choice for certain applications. One drawback is in the design of the plunger mechanism, which requires packing to effect a tight seal between the plunger and plunger bore. A small amount of controlled leakage of process fluids must be tolerated in most instances, in order to cool and lubricate the plunger.

Also, as the plunger reciprocates, the unavoidable friction that results wears the packing and increases the leakage of process fluid past the plunger. Periodic adjustment becomes necessary, because such wear decreases the volumetric efficiency of the pump. Replacement of worn packing contributes to maintenance costs.

What kind of packing material is used depends on a number of factors, such as temperature, pressure, liquid composition and pump speed. If abrasive slurries are to be pumped, "lantern" rings should be provided as a means of flushing the packing. Often specified for plunger pumps are chevron, lip-seal and square-ring packings, which come in a variety of materials.

The instantaneous pumping rate of a reciprocating pump usually varies sinusoidally, with the maximum rate occurring when the plunger is approximately halfway through its stroke (Fig. 2). At the point where the plunger reverses its direction, the displacement chamber is filling with liquid and the rate of discharge is zero.

These fluctuations in pumping rate are sometimes reduced by combining the discharges from several cylinders operating out of phase. A manifold collects fluid from several simplex (single-cylinder) pumps operating off the same drive-shaft. Such multiple-cylinder units are known as duplex, triplex, or quadruplex pumps, depending upon the number of cylinders employed.

In summary, packed-plunger pumps are employed in applications that require a low-cost pump, and that can tolerate a certain amount of leakage and maintenance. Applications typically involve injection of a nontoxic, noncorrosive chemical into either a batch or continuous operation.

## Diaphragm pumps

For moving corrosive acids or pollutants whose leakage must be avoided, diaphragm metering pumps, though more costly, offer an alternative.

In these units, the diaphragm acts as an interface

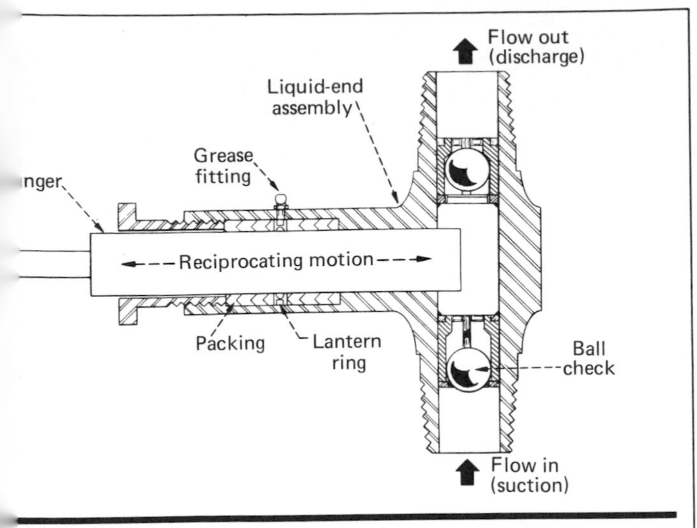


**Operating principle of a metering pump having a packed-plunger liquid end** **Fig. 1**

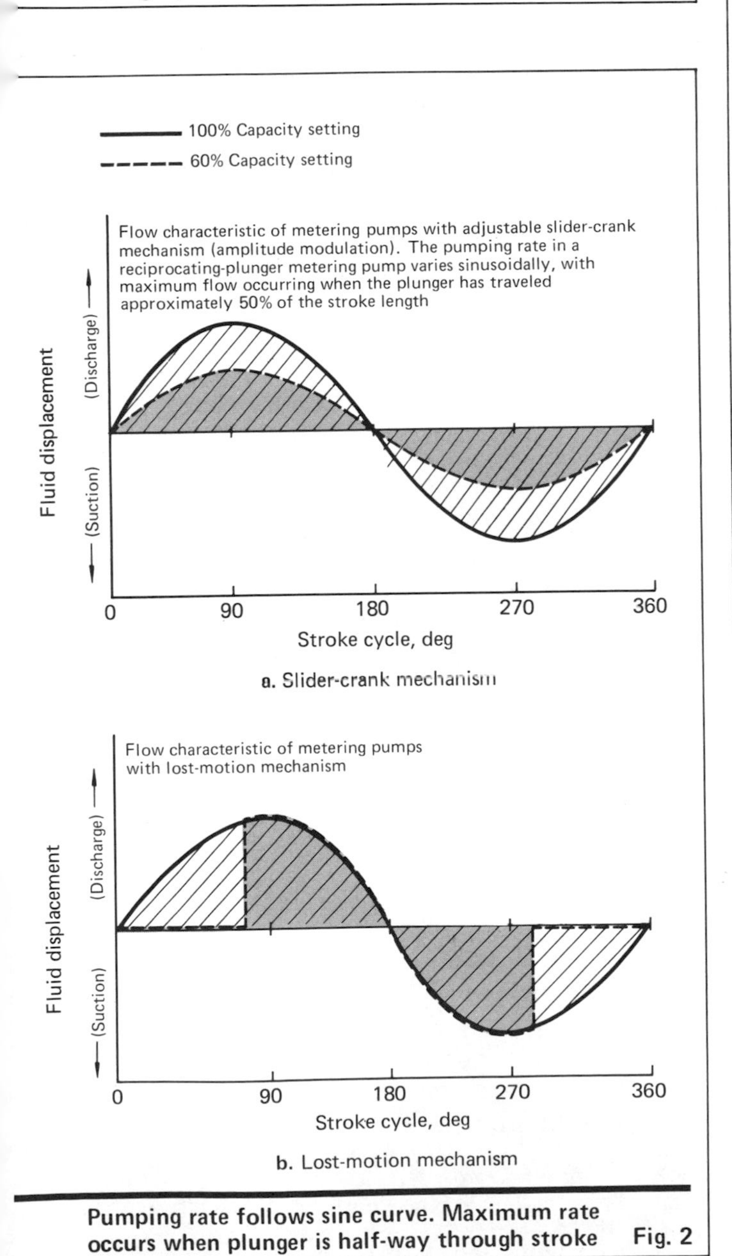


**Pumping rate follows sine curve. Maximum rate occurs when plunger is half-way through stroke** **Fig. 2**

between the plunger and the process fluid. The plunger still travels through a bore in this design, but instead of directly displacing the process fluid, it actuates a diaphragm across which energy is transmitted. On the opposite side of the diaphragm, a proportional amount of process fluid is drawn into the displacement chamber and is discharged with each cycle of the plunger.

Diaphragms are either mechanically actuated—i.e., directly coupled to the plunger—or hydraulically actuated. The hydraulically actuated diaphragm is balanced between two fluids, so as to lessen diaphragm fatigue and permit higher discharge pressures.

To ensure the accuracy of the hydraulically actuated liquid-end, the proper volume of hydraulic fluid must be kept in the hydraulic chamber. A three-valve system is often installed in order to keep the hydraulic fluid relatively free of air or entrained gases, and to protect against overpressure in both the hydraulic and process fluids.

A *refill valve,* which also acts as a vacuum breaker, maintains a constant volume of hydraulic fluid by replenishing any that leaks out of the chamber.

A *gas bleed valve* removes air or gas bubbles from the hydraulic fluid.

A *relief valve* in the hydraulic chamber provides overpressure protection for both the hydraulic and process fluids.

Controlled porting of the hydraulic fluid may also be used to achieve fluid-refilling and gas-bleeding. With this design, check valves located on the suction and discharge sides of the pump admit and expel the process fluid in response to the negative and positive pressure exerted by the flexing diaphragm. Double check valves increase the accuracy of the pump and provide redundant sealing action.

## Disk diaphragm pumps

In these pumps, the diaphragm—generally made of TFE (tetrafluoroethylene)—flexes between two dish-shaped supporting plates that have flow-through holes. The plates (often called contour plates) provide containment for the diaphragm and prevent its rupturing under high pressure (Fig. 3).

This arrangement solves the problem of process fluid leaking past the plunger seals and is effective for pumping corrosive fluids such as acids. It is mainly used to handle liquids in applications where even minimal leakage is not acceptable.

This design can cause problems when viscous fluids or slurries are pumped. With viscous fluids, a very noticeable pressure drop results across the small flow-through holes in the contour plates, often causing suction cavitation. The resulting formation of random gas bubbles in the fluid usually reduces the accuracy of the pump below acceptable levels.

## Tubular diaphragm pumps

In this design, a plunger reciprocates as described above, but a tube-shaped elastomeric diaphragm expands or contracts with the pressure exerted by the hydraulic fluid. Constriction or expansion of the tube, combined with one-way action of suction and discharge check valves, drives a pulse of metered liquid through the pump. Cavitation problems are reduced since there is no contour plate used on the process side of the tube.

In the tubular diaphragm pump shown in Fig. 4a, the plunger is located within the elastomeric tube itself. During the suction stroke, a vacuum is created within the tube as the plunger is withdrawn. This causes the tube to contract, drawing a predetermined amount of process fluid through the suction check valve and into the displacement chamber located along the exterior of the diaphragm. The opposite occurs during the discharge stroke: pressurization of the hydraulic fluid within the tube causes the tube to expand and displace a proportionate amount of process fluid through the discharge valve.

Other tubular-diaphragm variants include the double tube (Fig. 4b) and combination disk/tube (not shown). In these twin-diaphragm pumps, the plunger exerts hydraulic pressure to flex a primary diaphragm which, in turn, flexes a secondary diaphragm via an intermediate liquid.

The secondary diaphragm (tube), which handles the process fluid, isolates it entirely from the pump head, precluding the need for costly materials to resist liquid-end corrosion by the process fluid.

Because the secondary diaphragm may be considered to be an extension of the suction and discharge piping, it offers little resistance to flow. Such pumps are ideally suited for metering viscous liquids and abrasive slurries and are not prone to sludge accumulation.

Twin-diaphragm pumps offer the safety of a backup diaphragm should the process diaphragm fail, and usually employ an inert hydraulic fluid between the diaphragms. This latter feature is useful when handling products that may react violently with hydraulic oils, or when contamination of the process chemical must be prevented in the event of secondary diaphragm failure.

The intermediate chamber between the two diaphragms is sometimes equipped with a sight glass to allow monitoring. If the chamber is filled with a liquid whose color is pH-sensitive, the change of color warns of diaphragm failure. Electrodes may also be provided in the intermediate chamber to monitor conductivity of the intermediate fluid. A change in conductivity can be detected by the electrodes to provide a warning of diaphragm failure.

### Other diaphragm configurations

Another liquid-end configuration is the double TFE disk diaphragm (Fig. 5). This design, with the inherent safety that twin diaphragms offer, is well suited for handling solvents and chemicals that attack elastomeric tube materials. Primary disadvantage of such a design lies in the requirements for potentially expensive metallurgical construction.

Both double- and single-diaphragm heads may be remotely located to handle hazardous products such as high-temperature, cryogenic and radioactive fluids. Remote assemblies may also be employed where space does not permit installation of a complete unit, or where corrosive fluids must be lifted from top-opening tanks.

All of these diaphragm designs are in use in various services throughout the chemical process industries. Selection of the most suitable design for any application

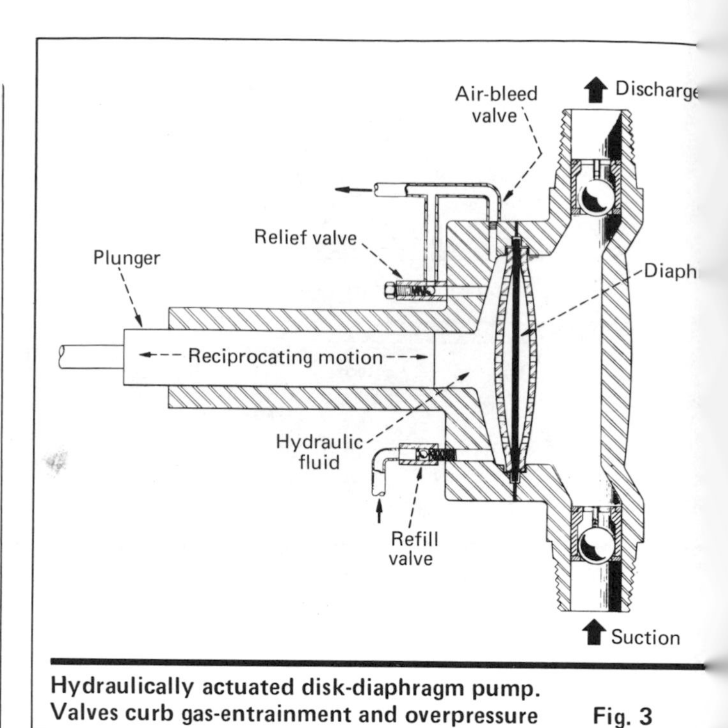


**Hydraulically actuated disk-diaphragm pump. Valves curb gas-entrainment and overpressure** **Fig. 3**

depends on the characteristics of the fluid being handled, the amount of maintenance permissible, safety aspects and desired system cost. In addition to the various pumping mechanisms, many drive mechanisms have also been evolved to transmit motive power as efficiently as an application will permit.

## Drives and prime movers

The capacity of a metering pump is a function of the diameter of the plunger, the effective length of the plunger stroke, and the rate or speed of stroking. Since the plunger diameter is constant for a prescribed pump, stroke length and pump speed are the variables available to adjust the output capacity while the pump is operating. Speed and/or stroke adjustments can be effected manually or automatically, in accordance with the process demand, to vary the volume of liquid delivered by the pumps.

Reciprocating pumps are so named because their operation depends upon the back-and-forth action of a plunger to displace a process fluid either directly or indirectly as a result of diaphragm flexure. A necessary characteristic of metering pumps is adjustable output. This is made possible by varying the actual or effective length of the displacement stroke.

There are two main types of stroke-length adjustments. One alters the radius of eccentricity of the plunger drive mechanism—sometimes referred to as amplitude modulation. This method is employed by various two- and three-dimensional slider-crank configurations (Fig. 6) and the shift-ring drive.

The second type varies the amount of fixed-crank travel that is transmitted to the plunger. Many common designs use some sort of "lost-motion" arrangement. These designs include both mechanical (eccentric cam) and hydraulic-bypass configurations (Fig. 7a and b).

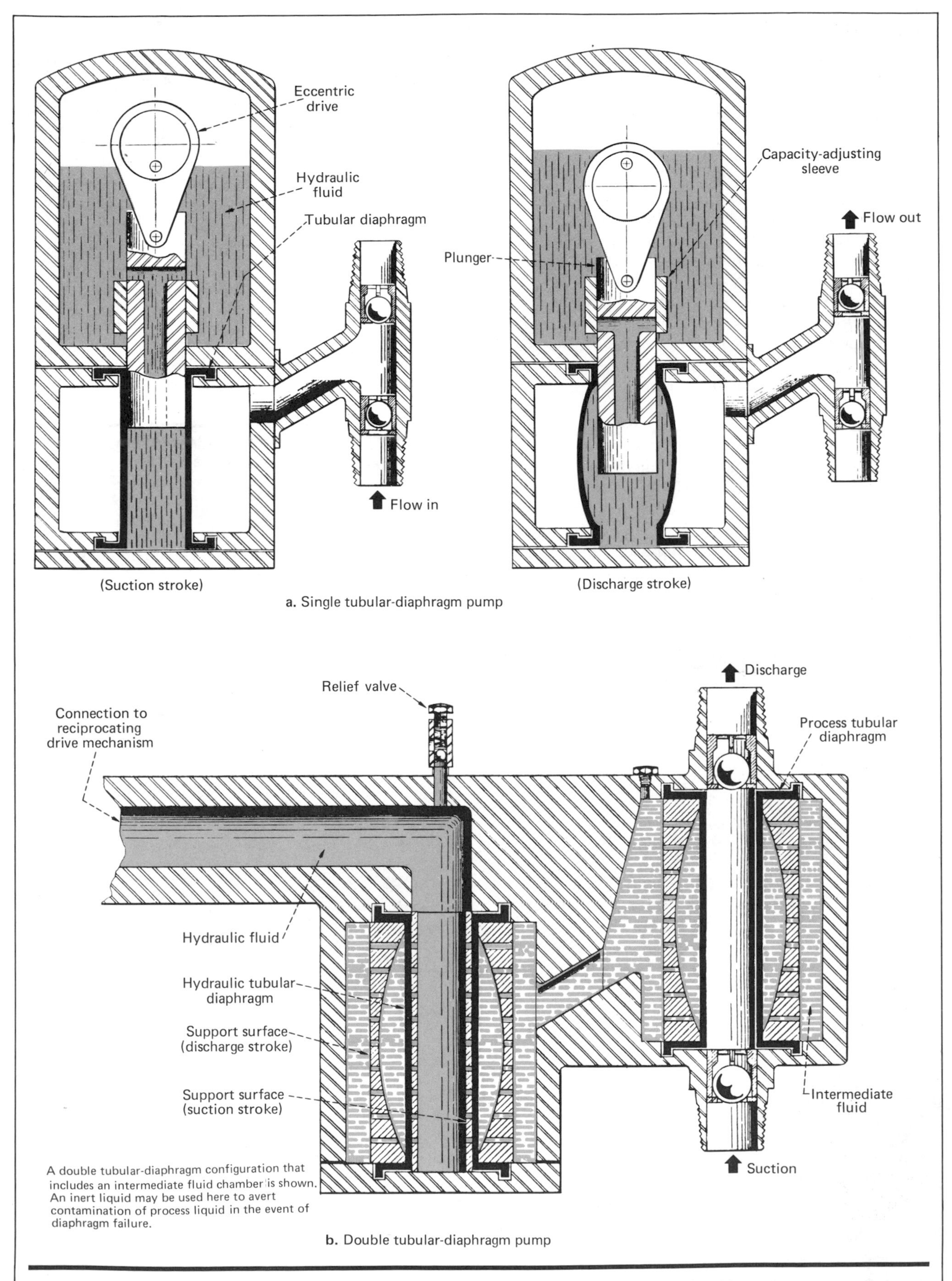


**Design of single and double tubular-diaphragm pumps reduces cavitation** **Fig. 4**

## Amplitude modulation

The slider-crank mechanism, a forerunner of most modern drive designs, features a turnbuckle-type adjustment that permits one to alter stroke length by changing the length of a pivot arm. Many variations of this drive have evolved, and while they are associated with different names and trademarks, the same basic operating principle is readily recognizable. All slider-crank mechanisms are attached to the plunger, although the various methods of linkage may differ.

The slider-crank drive is generally chosen for higher-capacity pumps that operate under greater pressures.

The shift-ring drive is intended to minimize mechanical vibration and shock loads in the drive train. A spring-loaded plunger rides within the main drive shaft, and reciprocation is achieved by adjusting the position of the ring within which the plunger rotates.

## Lost-motion drives

As the name implies, lost-motion drives provide a method of altering the stroke length by somehow "losing" or not utilizing the drive thrust available over a full pumping cycle. Lost-motion drives can be categorized as either hydraulic or mechanical.

Hydraulic units change the *effective* stroke length of the pump, rather than its actual stroke length. The plunger continues to reciprocate the full length of the stroke at all times, but some of the hydraulic fluid that deflects the diaphragm escapes through a bypass valve. This unused fluid returns to the sump or hydraulic fluid reservoir. The bypass valve can be externally adjusted while the pump is in operation.

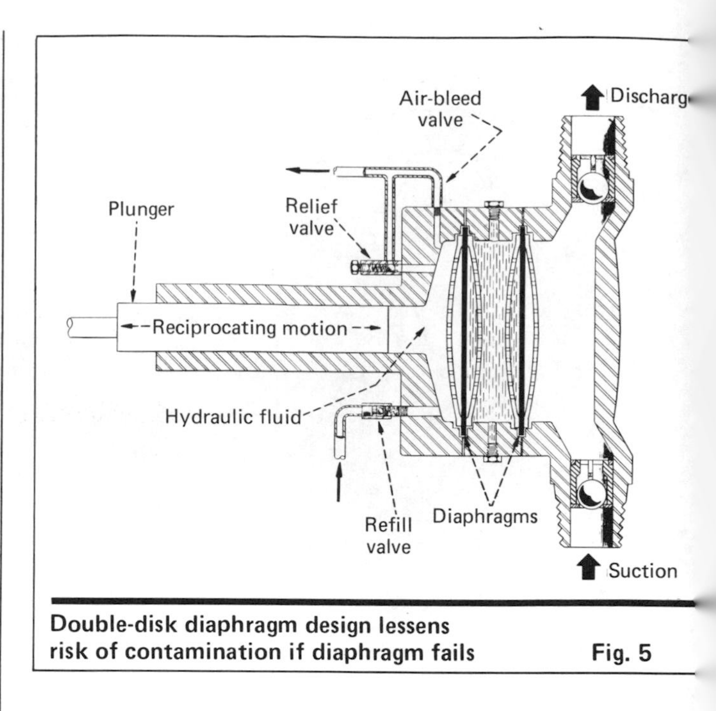


**Double-disk diaphragm design lessens risk of contamination if diaphragm fails** **Fig. 5**

The lost-motion drive mechanism can be built at lower cost than the more complex slider crank, but it is not intended for the higher capacities (in excess of 150 gal/h).

One advantage of hydraulic over mechanical lost motion is that the diaphragm is hydraulically balanced, making for longer fatigue life and higher discharge pressures.

In mechanical lost-motion drives, the diaphragm is mechanically attached to the plunger, or deflecting

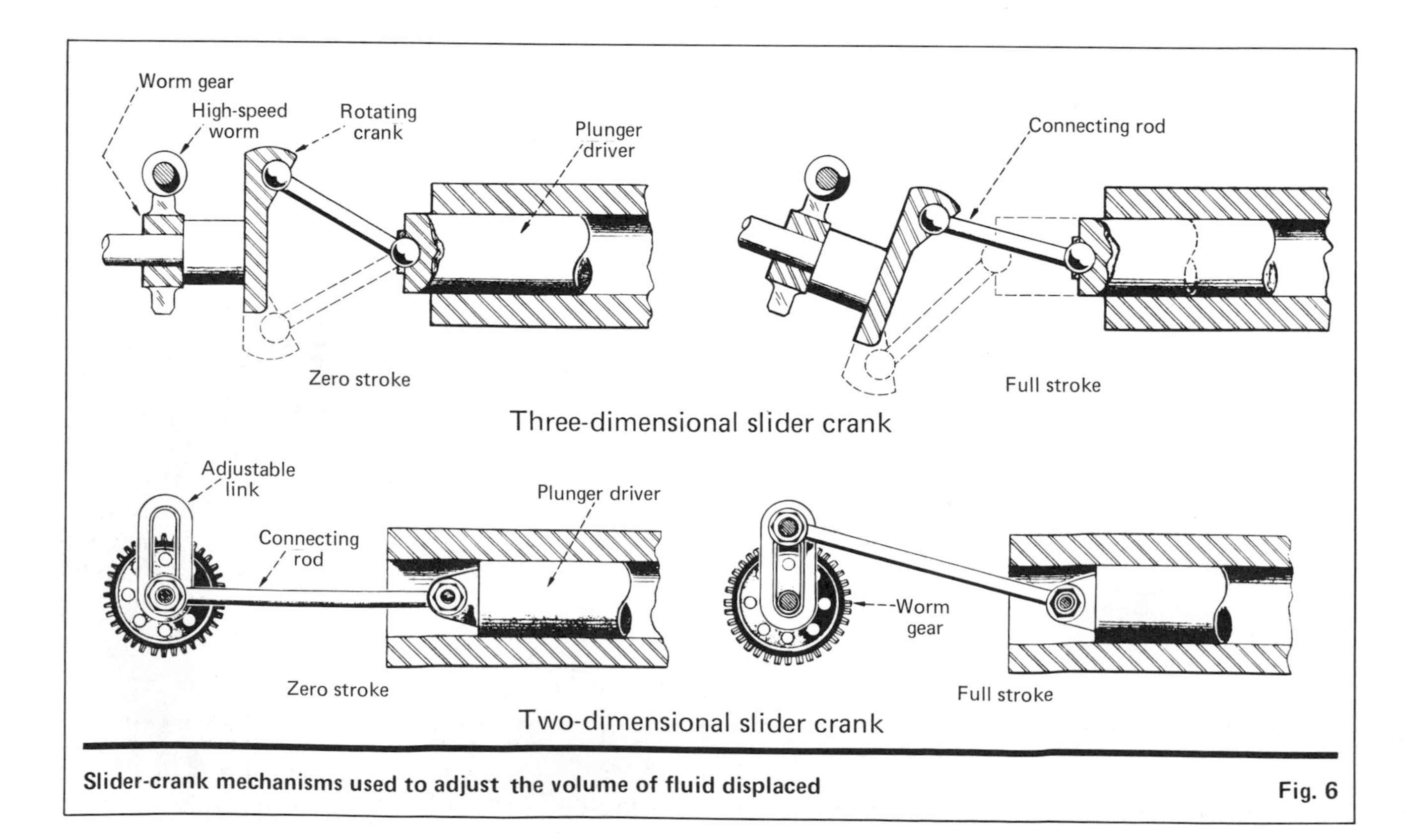


**Slider-crank mechanisms used to adjust the volume of fluid displaced** **Fig. 6**

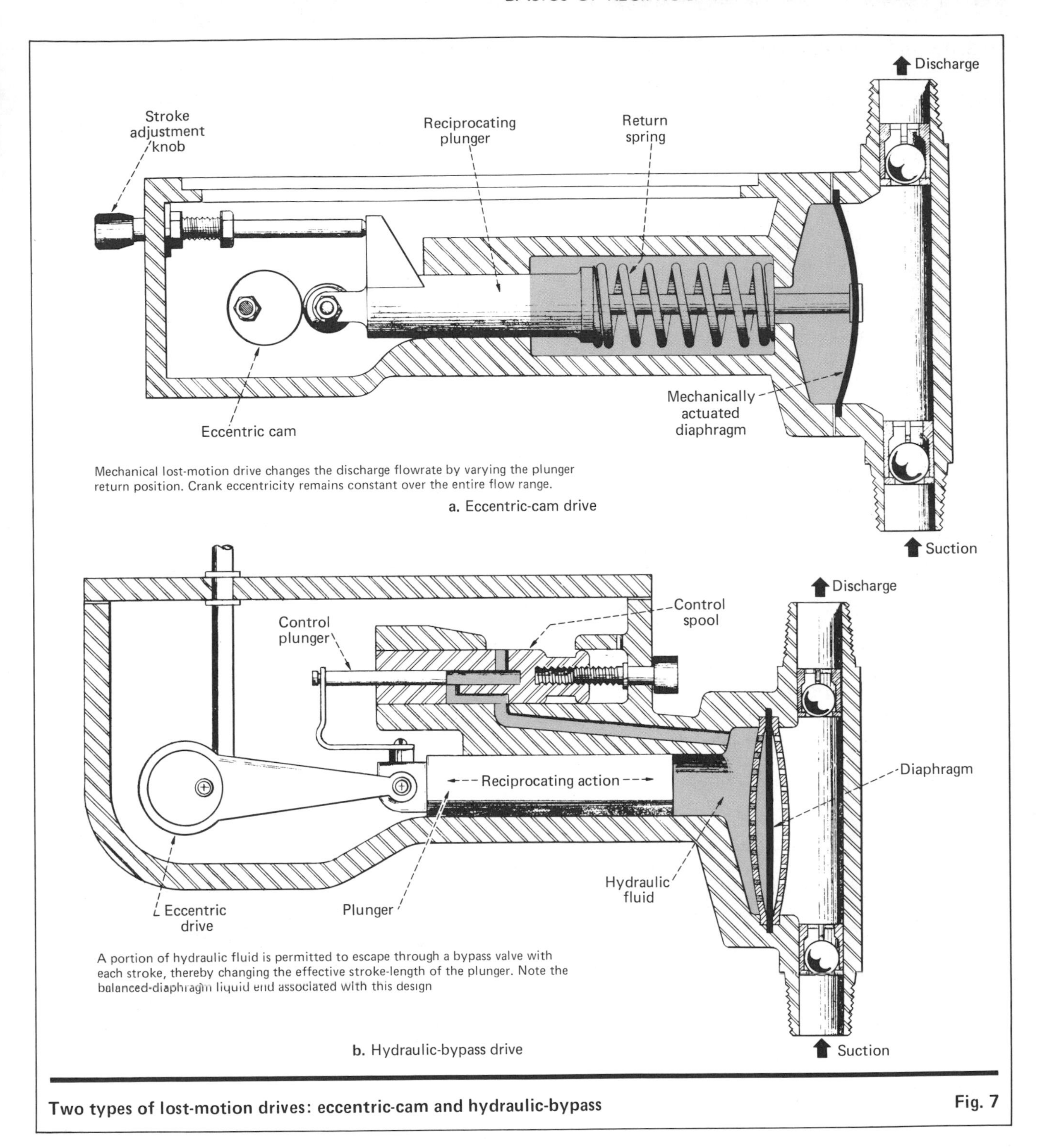


**Two types of lost-motion drives: eccentric-cam and hydraulic-bypass** **Fig. 7**

mechanism. The mechanism is driven forward by an offset cam, causing the diaphragm to deflect a predetermined amount of fluid.

The flowrate is varied by limiting the return of the mechanism, as shown in Fig. 7a. A spring generally causes the driving mechanism to return to its original position after each cycle of the offset cam. If the plunger is allowed to return only half-way to the original position, the stroke and accompanying deflection will be decreased by 50%.

Mechanically actuated diaphragms are not generally used at pressures over 250 psi, or in applications where capacity requirements exceed 25 gal/h. Typical applications include water treatment systems for cooling towers, boilers, water and sewage treatment plants, and laboratory installations.

## The prime mover

The constant-speed a.c. electric motor is the most commonly used driver for metering pumps. The motors are either flange-mounted and coupled to the pump input drive shaft, or foot-mounted and supported on a bracket attached to the casing of the metering pump.

These motors are often matched with mechanical variable-speed drives, which are either flange-mounted or supported on a common base with the pump drive. Speed variation is accomplished by changing the ratio of conical disks, spring-loaded to provide belt tension. Belts should be of the non-slip type to maintain constant speed at the output shaft of the driver.

An alternative to this approach is to specify variable-speed d.c. motors, which are mounted in a similar manner. Silicon-controlled rectifier circuits control the motor speed, which can be paced automatically by a remote analog signal.

Pneumatic power is sometimes used to drive plunger and diaphragm liquid-end pumps. This type of drive uses a reciprocating pneumatic cylinder, sized so that it provides the necessary thrust-differential between driving force and generated pressure.

Pressure is admitted to the cylinder by a four-way valve that alternately opens to opposite sides of the piston. The valve may be actuated by electrically energizing a solenoid, which changes the direction of the pressure, or by porting the pressure internally so as to cycle the piston automatically.

Pneumatic power facilitates the use of digital control, since the electrical on-off pulses to the solenoid are directly analogous to plunger reciprocation.

Metering pumps with solenoid drives are available for applications that require low thrust. Although these drives are ideally suited for electronic capacity control, the limited cycle life of the solenoid should be taken into consideration.

# Metering systems

In the CPI, metering pumps are used most often for continuous processing, batching and wastewater treatment. The pumps are installed singly, or in multiples. An installation may contain one, two, three or more liquid ends, all driven by the same power source, but with separate, adjustable capacities.

Metering pumps are found in both open- and closed-loop systems. In a fairly stable process that does not dictate frequent capacity changes, a simple open loop in which flowrate is changed manually may be adequate.

Open-loop systems, as shown in Fig. 8a, are characterized by the fact that there is no feedback or corrective action by the controls, and system operation is strictly proportional. Good sensitivity and a high degree of linearity in the controls (including the stroke control on the pump) are necessary features.

Open-loop systems require some way of adjusting the dosage rate to maintain the proper ratio between the measured stream flow and the flow from the metering pump. As it is usually not necessary to attain zero flow conditions on these systems, it is practical to use motor speed controls. This frees the manual stroke control for easy ratio, or dosage, adjustment.

Closed-loop systems (Fig. 8b) are characterized by feedback, or corrective action from the controller. Such controllers are quite sophisticated, and read the process variable after the chemical reagent has been injected by

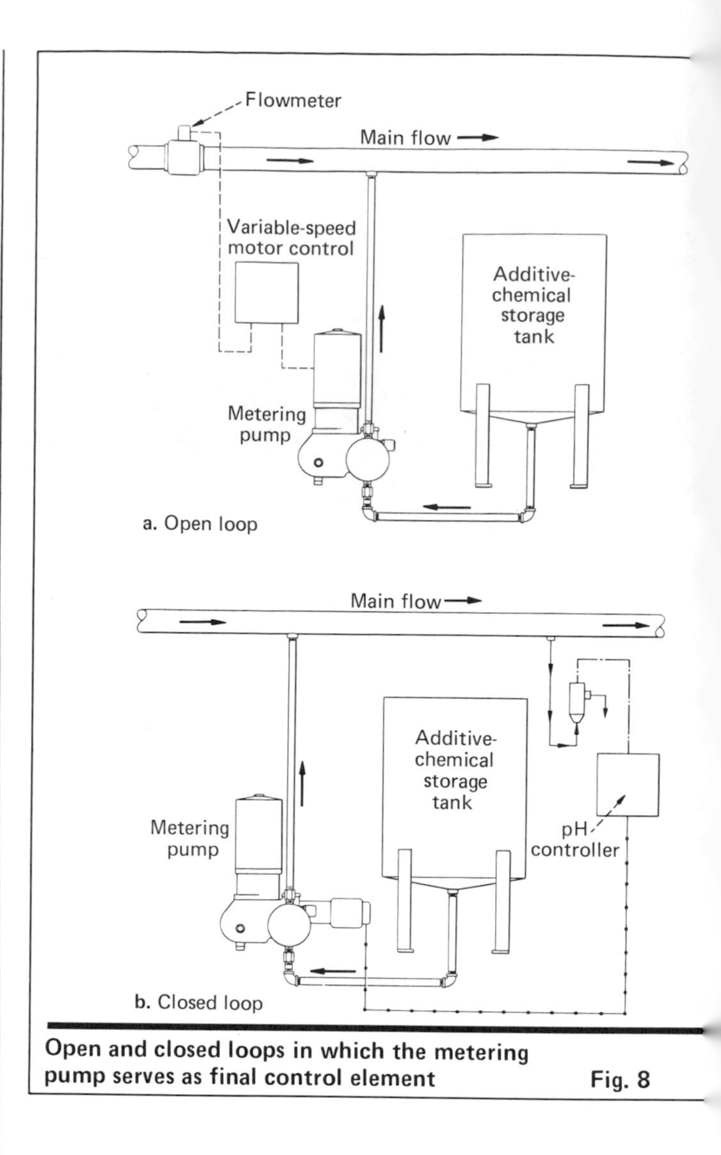


**Open and closed loops in which the metering pump serves as final control element** **Fig. 8**

the pump. The accuracy of the pump and its mode of control (pneumatic or electric) determine how close the process variable can be kept to the setpoint without considerable modulation by the process controller.

In closed-loop systems, it is essential that the pump capacity be able to go to zero. It is more practical to attain zero output by controlling the stroke than it is by varying the motor speed, since many variable-speed drives cannot be satisfactorily operated at zero speed. Capacity controls employed in closed-loop systems are usually pneumatic or electric analog devices.

A metering pump can also be chosen as the final control element for two-process-variable controllers. For example, if the process-liquid flowrate varies considerably, a closed-loop feedback system is not adequate for following large swings in flowrate, as well as changes in the process variable.

However, a combination of motor-speed and capacity control can be used to achieve the desired range. One type of combined system, an open-loop proportional-feed system cascaded to a closed-loop process-variable control system, is shown in Fig. 9.

Turndown ratios must be considered. If a pump has a 10 : 1 turndown ratio, the output can be very accurately adjusted between 100 and 10% of rated flow. Below

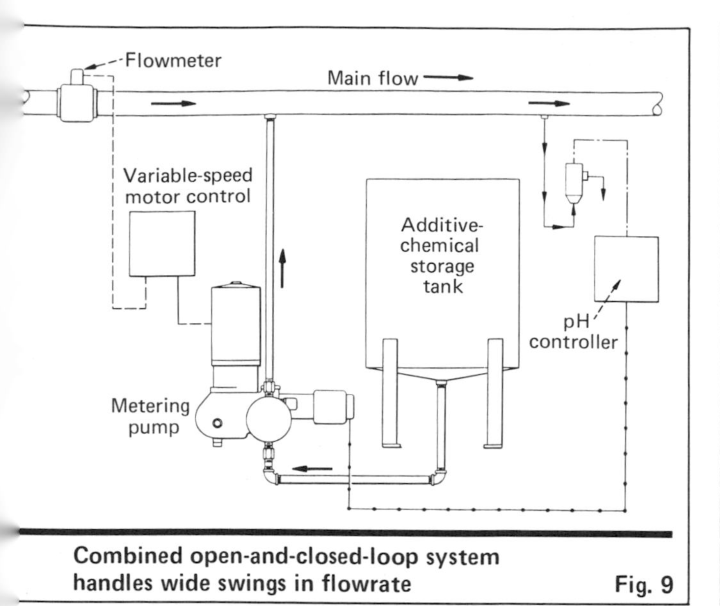


**Combined open-and-closed-loop system handles wide swings in flowrate** **Fig. 9**

10%, however, it is difficult to achieve reliable delivery without further refining the accuracy. This can be accomplished by combining stroke control with motor-speed control. With the motor running at one sixth of rated speed, and the capacity adjustment set at 10%, the actual output would be 0.1666 × 10% of total pump capacity.

Variable-speed d.c. motors generally cost less than variable-speed a.c. motors, and are a good choice if motor speed-control is elected. In hazardous atmospheres, however, brushless a.c. motors may be required. In extremely hazardous locations, pneumatic drives and controls are recommended.

Use of direct-coupled air-cylinder drives also affords control versatility. Direct digital control is available, using accessory electronic modules such as a stroke pacer for speed control, a stroke counter for pumping a predetermined volume (based on stroke count), an analog-to-digital signal converter, and a fail-safe device that monitors pump operation.

Independent adjustment of plunger traverse speeds can be specified for pumping high-viscosity fluids and for improving suction lift. A typical pneumatically powered pump system is shown in Fig. 10.

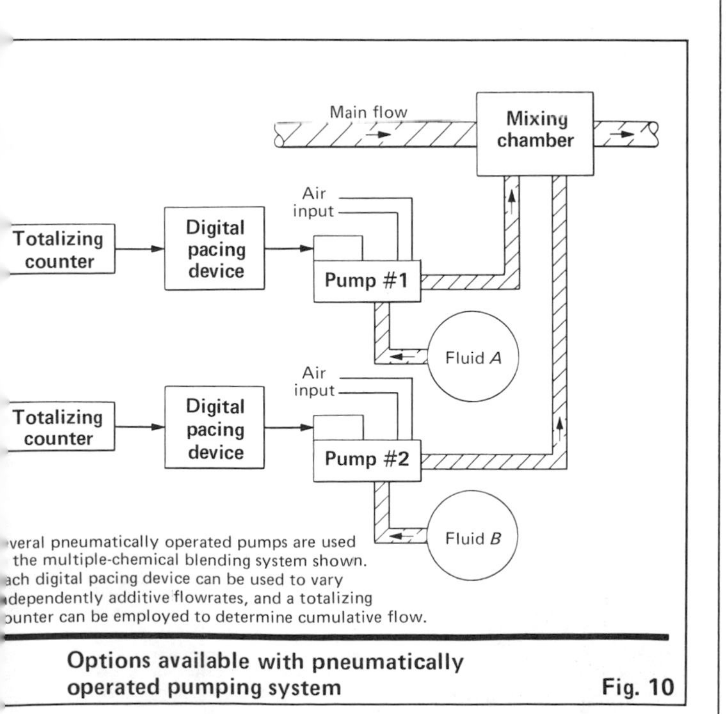


**Options available with pneumatically operated pumping system** **Fig. 10**

# Installation and maintenance

With so many models and materials available, it is difficult to provide a broad-brush guide to installation and maintenance. There are, however, certain general guidelines that will enable optimum performance and service life.

Unless the pump is a portable unit designed for operation without mounting, it should be firmly supported on a concrete foundation, preferably with its base above the floor level to protect it from washdowns and to provide service access.

Suction and discharge piping must be adequate to handle instantaneous peak flow. In motor-operated pumps, the flow follows a sine curve, due to the reciprocating motion of the plunger. Peak instantaneous flow is approximately $\pi$ times the average flow.

Piping must therefore be designed to handle a flow considerably greater than the rated pump capacity. Many operating problems are caused by undersized piping. One way of checking whether the piping has been adequately sized is to unhook the suction tubing and measure the free flow of liquid from the pipe. If the flow is $\pi$ times the rated capacity of the pump, then the size of the piping is usually adequate.

In the selection of materials, care should be exercised that galvanic corrosion is avoided at pump connection points. Inside diameters should match perfectly between connections. Burrs and sharp edges should be removed, and weld splatter should be cleared from the lines. All lines must be blown or flushed before final connections to the pump are made. If fluid temperatures are extreme, allowance for expansion and contraction of the piping must be made. Piping should be supported so that strain is not placed on the pump, and piping should never be laterally loaded or sprung when making connections.

Both suction and discharge runs should be as straight and short as possible. Avoid unnecessary elbows in piping. Where possible, 45-deg or long-sweep 90-deg fittings should be utilized. Piping should be sloped to eliminate vapor pockets.

## Suction piping

Suction piping must be absolutely air-tight to ensure accurate pumping. After installation, piping should be tested for leaks.

The tendency toward vapor formation, or flashing, during pump suction is proportional to increases in pump speed, fluid viscosity and density, plunger size, length of piping and lift height. Net positive suction head (the amount of fluid pressure in excess of the limiting vapor pressure available to move liquid into the pump) must be sufficient to ensure satisfactory operation as process conditions become more severe.

Practical approaches to improving NPSH are:

■ Increasing the inlet pipe diameter.

■ Decreasing the pump speed while increasing the plunger diameter to maintain flowrate.

■ Increasing the head pressure by raising the level of the supply tank relative to the pump inlet.

■ Pressurizing the supply tank, if the fluid has an especially high vapor pressure.

## Discharge piping

Excessive pressure losses during the discharge stroke of the pump can be prevented by installing adequately sized discharge piping. The pipe pressure rating must be in excess of the safety-valve setting or rupture-device rating.

"Flow-on" or overfeeding may be troublesome at low discharge pressures. This condition can be prevented by decreasing the suction or discharge pipe diameters, or by placing a back-pressure valve within the discharge line. Back-pressure valves are mandatory when fluids are pumped into a vacuum process.

When back-pressure valves are installed to provide an artificial discharge head, one should call for an accumulator or pulsation dampener in the discharge line. The accumulator smooths out the flow from the pump to the back-pressure valve. This extends valve life and provides a more uniform flow to the process.

## Safety valves

Because motor-driven metering pumps can build up high discharge pressures in one or two strokes, thermal-overload detectors and similar safety devices in the motor starter circuit may act too late to provide protection. Damage to the pump, discharge line or process equipment can occur if the discharge line is obstructed. To prevent such occurrences, equip the discharge line with a suitable safety valve or rupture device, sized to handle system pressures and to discharge the maximum pump flowrate safely.

The safety valve should be located in the discharge line between the pump and the first downstream shutoff or back-pressure valve. The safety-valve outlet is then piped back to the suction supply tank, or to a suitable disposal point. The open end of the return piping should be located where it is visible, so that any safety-valve leakage can be detected.

## Pump maintenance

Metering pumps can provide years of dependable service with a little routine maintenance. Individual manufacturer's recommendations should be consulted for specific maintenance procedures, but a few basic steps are necessary with all models.

Pumps with packed-plunger-type liquid ends require periodic replacement of their packing. Generally, packings depend for their proper functioning on compression in the stuffing box. If too tightly compressed, packing will overheat and may disintegrate rapidly. Excessively tight packing may score the plunger, necessitating replacement of the plunger itself. Packing should also be regularly lubricated according to the manufacturer's recommendations. Lubrication schedules may range from daily to more infrequent intervals. Pumps that have plastic liquid ends may not require periodic lubrication of their packing.

The level of hydraulic oil in the pump housing should be monitored so that the pump does not run dry. The hydraulic oil lubricates all moving parts, and often provides the force necessary to actuate the pump. Pumps having bronze worm gears should be filled with a lubricant such as AGMA (American Gear Manufacturers Assn.) 8 Compounded, which contains inactive additives for extreme pressures.

Motors should also be lubricated yearly, or more often if specifications dictate.

## Packaged chemical-feed systems

Chemical-feed systems are available as complete package units from many pump makers. These systems usually consist of a supply tank ranging in capacity from 50 to 1,500 gal, an agitator, valves, piping and instrumentation, all mounted on a common frame or skid. The systems are generally provided ready for installation on the job site, and single- or multiple-feed systems may be specified.

The packages are routinely used for a wide variety of chemical processing applications and are generally custom-designed. Water-treatment feed systems are usually compact, and specifically designed for each application as well.

## Accessories and controls

A wide variety of accessories may be used with metering pumps. Those most commonly used are:

■ Automatic controls to open valves and flush the pumping assembly.

■ Anti-siphon or back-pressure valves, which prevent liquid from flowing through the pump without proper metering action.

■ Metering-rate timers and counters that totalize pump pulses. These may be set to shut off the flow when a specified limit has been reached.

■ pH analyzers/controllers, which are frequently used in closed-loop systems to adjust the pump output.

■ Pulse dampeners to overcome acceleration losses in the suction line and reduce pressure surging in the discharge line.

■ Safety relief valves, which prevent system overpressure.

■ Steam jackets to maintain or raise the temperature of the process fluid.

## The author

**James P. Poynton** is engineering manager for the Flow Control Div., Milton Roy Co., 201 Ivyland Rd., Ivyland, PA 18974. Responsible for a number of new hydraulic products with biomedical and industrial applications, he has written various articles on metering pumps, and has served on an international committee of the Scientific Apparatus Makers Assn. A member of the pump standards task force of the API, and a senior member of the ISA, he was formerly propulsion-systems project engineer for Grumman Aerospace. He holds a B.S. in aeronautical engineering from Notre Dame University.

# Industrial wastewater pumps

This is an overview of the available types of industrial wastewater-treatment pumps, with emphasis on process considerations of effluent transfer, chemical addition, slurry and sludge transport, and sampling systems.

*Jacoby A. Scher, Fluor Engineers and Constructors, Inc.*

□ Within the last decade, industrial wastewater treatment has taken a much more important role in overall plant facilities. Increasingly stringent effluent requirements promulgated by all agencies have resulted in a surge of more sophisticated waste-treatment methods. Now, even contaminated storm-water must be treated, and more emphasis must be placed upon intricate gathering and treatment networks.

The problem of gathering, transporting and treating waste liquids in existing plants is much more complex than for a grassroots facility. Frequently, the wastewater-treatment system is relegated to a distant corner of available plot space, where ease of hydraulic transport is not necessarily a prime consideration. Liquids must be moved, chemicals supplied, slurries processed and other liquid or semiliquid services provided.

More often than not, pumps of many types and functions are used to ensure that the process operates. When examining a particular pump application, it is useful to check with others who have used pumps in similar situations. Also, pump manufacturers can be of great help in matching the right pump for the service involved.

Processing units generally dispose of wastewater by gravity into underground sewers or into open channels. Surface runoff of contaminated stormwater flows by gravity to a centralized collection sump. Cooling water and boiler blowdowns, aqueous water-treatment and/or slurried-waste streams—as well as other utility-type wastes—must be segregated or combined, and moved to the treatment area.

Various types of chemical solutions, whether acids or bases for neutralization, coagulants for clarification, conditioning chemicals for sludge dewatering, etc., must be provided at the various unit waste-treatment operations. Slurries such as lime, biological mixed liquor, or similar suspensions must be transported. Sludges of varying viscosities and composition must be mechanically dewatered or moved to thickeners, pits, etc., for concentration or drying. Finally, continuous samples for process and effluent quality control must be obtained. All of these manipulations usually require pumps of various types and construction, that must be manufactured for a wide variety of services.

## Types of pumps

Of the many different types of pumps, the ones listed below are the most generally used for industrial wastewater-treatment systems. They are all listed for general information. Some are so common that they will not be discussed here, but others will be described in detail because of the uniqueness of their design and application. The various kinds are:

*Centrifugal*—This is a very general type of pump. In this article, the pump referred to is the American Voluntary Standard (AVS) type of overhung pump. It is so commonly used that it will not be dealt with here.

*Sump*—This is essentially a vertical centrifugal pump, which can be either the single-stage, volute type, or the single or multistage turbine type. The latter is an adaptation of the deep-well kind that is used for pumping water. Sump pumps, as shown in Fig. 1 (F/1), are designed to be installed in a pit or sump where an established water level exists.

*Submersible*—This is a centrifugal pump especially constructed so that the entire pump is submerged, including the motor (F/2). They are frequently designed to pass large particles, and can run dry for reasonable periods without damage.

*Piston (positive-displacement)*—This pump utilizes a cam-driven piston to directly move the fluid. It is available in a wide range of capacities and is capable of

Originally published October 6, 1975

attaining discharge pressures as high as 10,000 psig.

*Diaphragm*—This is an especially constructed positive-displacement pump. The diaphragm acts as a piston when it is motivated into reciprocal motion by either mechanical linkage, compressed air, or a fluid from an external pulsating force. The diaphragm acts as a seal between the motivating force and the liquid being pumped. F/3 shows a cutaway view of the pump, which indicates how the liquid is draw in and expelled by the diaphragm.

*Rotary*—Common pumps in this category are the gear and the progressive-cavity or screw pump. Defined by the Hydraulic Institute, a rotary pump consists of a fixed casing containing gears, cams, screws, plungers or similar elements actuated by rotation of the drive shaft. The rotary pump of primary interest is the progressive-cavity or screw pump. It consists of a rotor turning within a stator whereby cavities are formed moving toward the discharge that transports the slurry. These pumps (F/4) can be accompanied by units* that grind certain materials to convert them to slurries.

*Called MAZ-O-RATOR grinders, made by Robbins & Myers, Inc., Springfield, Ohio.

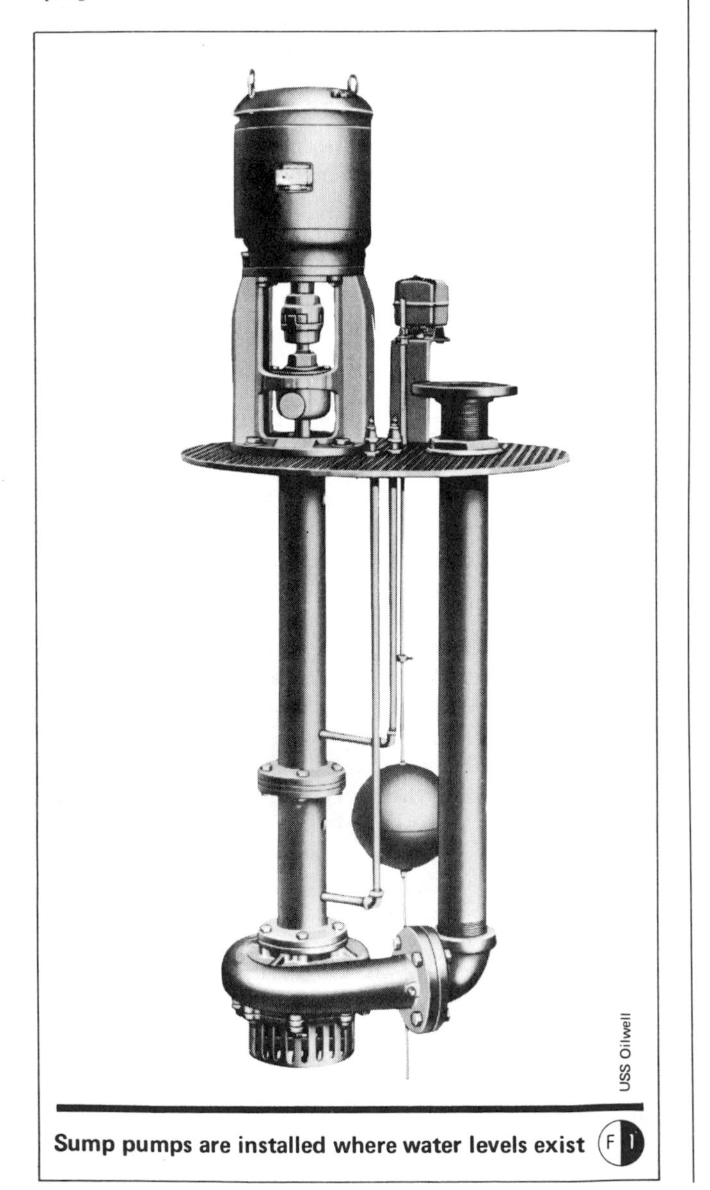


**Sump pumps are installed where water levels exist** F 1

*Peristaltic*—This is a special type of rotary pump that utilizes a piece of tubing wrapped around a cam moving eccentrically. As the cam rotates, the liquid is mechanically squeezed through the tubing (F/5).

All rotating equipment require some kind of seal between the shaft and the housing. Two types are common: the packed stuffing box and the mechanical seal. The packed gland has multilayers of packing placed inside a compression ring, which is gradually tightened as the packing wears away.

The lubrication for this seal is the fluid being pumped, which must be allowed to constantly drip from the stuffing box. Since the compression ring must never be tightened to the point where dripping ceases, this continuous drip prevents stuffing damage. If the pumped liquid contains suspended material, the packing can be quickly eroded and a severe leak may develop. A packed stuffing seal is fine for clear, rather innocuous fluids. It should not be used for corrosive, hazardous or abrasive solutions.

A mechanical seal has one stationary face and one rotating member held together by springs, a backup plate and other appurtenances. The close coupling provides a true mechanical seal that prevents leakage of fluid from the housing. If the solution being pumped is particularly hazardous or toxic, a double mechanical seal may be chosen, which has a nontoxic liquid between the seals to buffer any leakage. Also, to prevent scoring or seal crystallization, a mechanical seal with a flushed face may be selected. Water is circulated between the two seal faces to keep them clean and to extend their service life. Mechanical seals are very efficient in their operation if applied properly. However, they are more expensive than stuffed packing; constant replacement can add to maintenance costs.

A pump without a seal is also available. Called "canned rotor pump," it permits liquid to circulate through the motor bearings and then return to the general flow. A thin plate prevents fluid from contacting the motor windings; no shaft seal is required as

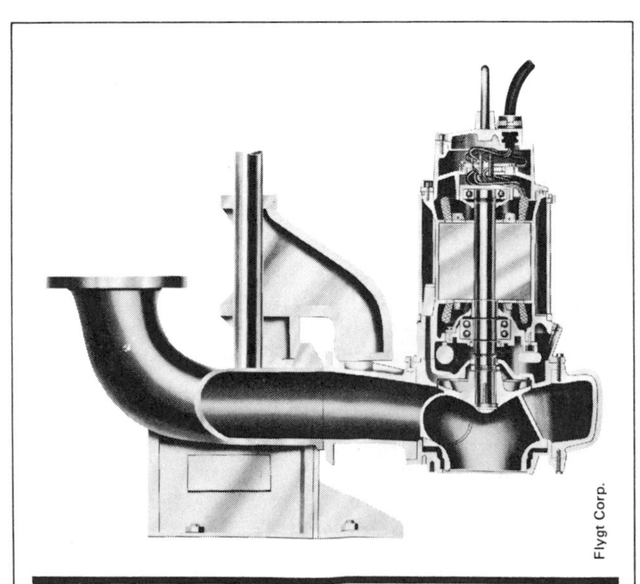


**Entire pump assembly, and the motor, are submersible**

there is no leakage whatever. This pump is more compact than the conventional ones. For pumps smaller than 30 hp, no foundations are needed because shaft-alignment problems are not encountered (F/6).

Since most wastewater-pumping applications—such as that of stormwater, cooling tower, boiler blowdowns, clarifier effluent, etc.—are within a pH range of 6–9, the liquid generally does not contain hazardous or corrosive chemicals in high concentration. In such instances, pumps can be made of carbon steel, cast iron or other mild materials of construction.

In services where chemicals are added, or when skimmed oil, ballast-water treatment, or other corrosive fluids are transported, other materials of construction are needed that are more resistant to various chemicals. The accompanying table indicates typical materials of construction for some liquids frequently encountered in industrial wastewater treatment.

## Effluent transfer

Collection systems for wastewater treatment are unlike process transport systems. Instead of pressurized enclosed-pump networks, wastewater effluents are usually drained into gravity sewers that flow to a common low-point collection area. Wastewater flows are also much more hydraulically irregular (e.g., stormwater runoff) than process flows. Equalization of flows prior to treatment is necessary because the various waste-treatment unit processes are rate dependent and sensitive to shock mass or hydraulic loadings.

Collection sumps and equalization ponds are utilized to smooth out fluctuations in gravity flows. Sump design and pump-capacity selection must be closely investigated so that the entire anticipated flow range can be adequately handled. Pump-capacity selection is important because a pump that continually cycles on/off/on may burn out a motor; and a pump that runs very infrequently may not function at all when it is needed. Often, if a collection sump experiences a flow range over several orders of magnitude, various different pumps with different capacities (set to actuate by sequential level control) may be the optimum configuration.

A typical installation recently encountered had a sump that collected a continuous effluent stream of 250 gpm, with additional stormwater flows up to 12,500 gpm. Four sump pumps were provided: one of 250 gpm to pump the continuous stream; two of 3,000 gpm each, to discharge low-intensity storm runoffs; and one of 6,500 gpm that would operate only under high-intensity storm flows.

Vertical sump pumps, whether of the centrifugal volute type or turbine type, are ideally suited to pump liquids from an open or closed collecting sump. Capacities for the volute type range from a few to several thousand gpm, with typical heads of 150–200 ft. Vertical turbine pumps with multistages have typical capacities of 100,000+ gpm and discharge pressures up to 1,000 psig or even higher.

These pumps can be operated by a float-type level controller, which prevents the pump from running if the level drops too low in the sump, but activates the pump when the liquid level rises. Vertical sump pumps require a minimum submergence above the intake bell to prevent vortex formation. Also, an intake screen is usually provided to prevent trash from being sucked into the pump. Sumps should be designed with a well, or boot, below the main sump floor into which the pump will fit. When this is done, the sump can be completely emptied, except for the capacity of the boot.

A submersible-type pump can perform several services in effluent transfer. It can also operate over a wide range of liquid levels, and can be obtained in capacities up to several thousand gpm and heads up to 100+ psig. This unit is somewhat capacity-limited, because the entire pump and motor are housed in a casing and placed on the bottom.

The advantages of the submersible type are that it can pump completely dry for a period of time, and can also pump large suspended solids if equipped with an open type of impeller. Maintenance is somewhat difficult because the entire pump assembly must be lifted from the bottom. It pumps very well from ponds that have widely fluctuating liquid levels; it can even pump a pit or pond dry.

## Chemical feed

Many of the unit processes associated with waste-treatment systems require addition of chemicals for optimization. Typical additions include:

*Acid and/or base for pH adjustment*—These chemicals are usually sulfuric or hydrochloric acid, sodium hydroxide, sodium carbonate or lime solution.

*Nutrients for biological treatment*—To satisfy biological phosphorus and nitrogen needs, phosphoric acid, forms of sodium phosphate, gaseous or aqueous ammonia, or ammonium-salt solutions are added.

*Coagulants for suspended solids control*—Typical of these

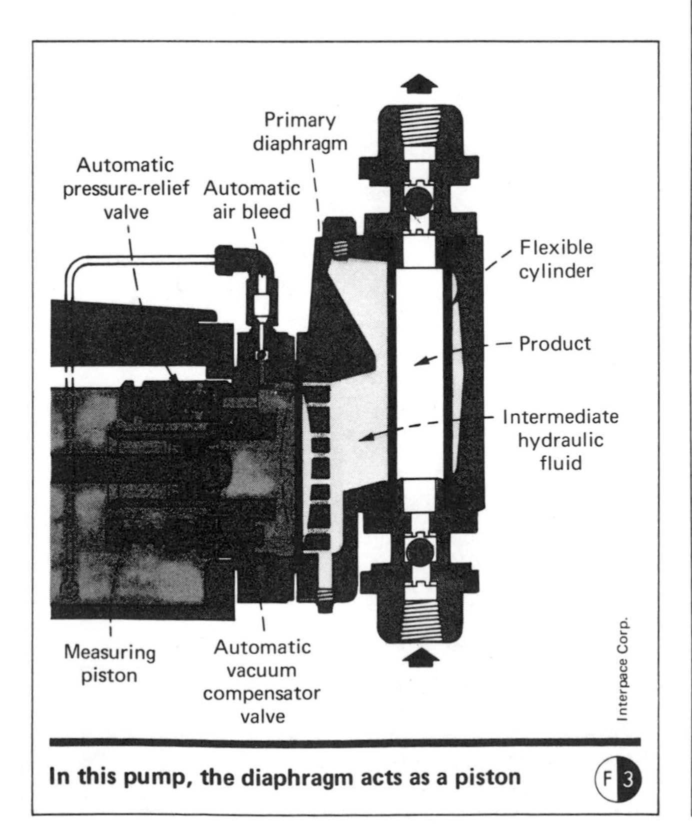


**In this pump, the diaphragm acts as a piston** F3

**Chemical resistance data for pumps***

| Material | Epoxy | Polypropylene | Polyvinyl chloride Type 1 | Nylon | Stainless steel 316 | Stainless steel 304 | Titanium | Buna N | Viton |
|---|---|---|---|---|---|---|---|---|---|
| Acetaldehyde | A | A | D | A | A | A | A | D | A |
| Acetic acid, glacial[1] | — | A | D | A | A | B | B | D | D |
| Acetic acid | A | A | A | D | A | B | B | C | C |
| Acetone | A | A | D | A | A | A | A | D | D |
| Alcohols | | | | | | | | | |
| Butyl | A | B | A | A | A | A | — | A | A |
| Ethyl | A | A | A | A | A | A | A | A | A |
| Hexyl | A | — | — | A | A | A | — | A | A |
| Isopropyl | A | A | — | A | A | A | — | C | A |
| Methyl | A | A | A | A | A | A | B | A | A |
| Aluminum chloride, 20% solution | A | A | A | A | C | D | B | A | A |
| Aluminum fluoride | A | A | A | A | C | — | A | A | — |
| Aluminum hydroxide | A | — | A | A | A | A | — | A | A |
| Aluminum sulfate | A | A | A | A | C | C | A | A | A |
| Ammonia, anhydrous | A | A | A | A | A | B | B | B | D |
| Ammonia, liquid | A | A | A | — | A | A | — | B | A |
| Ammonium nitrate | A | A | — | — | A | A | — | C | — |
| Ammonium phosphate | | | | | | | | | |
| Dibasic | A | — | A | A | A | A | — | A | A |
| Monobasic | A | — | A | A | A | A | — | A | A |
| Tribasic | A | — | A | A | A | A | — | A | A |
| Ammonium sulfate | A | A | A | A | B | B | A | A | A |
| Aromatic hydrocarbons | A | — | D | — | A | — | — | D | A |
| Benzene[2] | A | D | D | A | A | A | A | D | A |
| Bromine[2] | D | D | D | D | D | D | A | D | A |
| Calcium carbonate | — | — | A | — | A | A | A | A | A |
| Calcium chloride | A | A | A | A | A | A | A | A | A |
| Calcium hydroxide | A | A | A | A | A | A | A | B | A |
| Calcium hypochlorite | A | A | A | D | C | D | A | B | — |
| Calcium sulfate | A | — | A | A | A | A | — | B | A |
| Chlorine, anhydrous, liquid | C | D | D | D | D | D | B | C | A |
| Chlorox (bleach) | A | D | A | D | A | A | — | C | A |
| Citric acid | A | A | A | A | A | A | A | B | A |
| Copper chloride | A | A | A | A | D | D | A | A | A |
| Copper sulfate, 5% solution | A | A | A | D | A | A | A | A | A |
| Cresols[2] | A | D | D | — | A | A | — | C | A |
| Cresylic acid | A | — | A | D | A | A | — | C | A |
| Ethylene glycol[4] | A | D | A | A | A | A | — | A | A |
| Ferric chloride | A | A | A | D | D | D | A | A | A |
| Ferric nitrate | A | A | A | — | A | A | — | A | A |
| Ferric sulfate | A | A | A | A | C | C | A | B | A |
| Ferrous chloride | A | A | A | D | D | D | A | B | A |
| Ferrous sulfate | A | A | A | D | A | A | A | B | A |
| Formaldehyde | A | A | A | — | A | A | B | C | A |
| Fuel oils | A | A | A | A | A | C | — | A | A |
| Gasoline[1,4] | A | D | D | A | A | A | — | A | A |
| Hydraulic oils | | | | | | | | | |
| Petroleum[1] | A | D | — | A | A | A | — | A | A |
| Synthetic[1] | A | D | — | A | A | A | — | C | A |
| Hydrochloric acid | | | | | | | | | |
| 20% solution[4] | A | C | A | D | D | D | C | B | A |
| 37% solution[4] | A | C | A | D | D | D | C | A | A |
| Hydrofluoric acid | | | | | | | | | |
| 20% solution[1] | A | A | D | D | D | D | D | C | A |
| 50% solution[1,2] | A | D | D | D | D | D | D | C | A |
| 75% solution[1,2] | A | D | D | D | D | D | D | D | B |
| Hydrogen peroxide | C | A | A | A | C | C | B | A | A |
| Hydrogen sulfide, aqueous solution | A | A | A | A | C | B | A | B | A |
| Kerosene[2] | A | D | A | A | A | A | — | A | A |
| Lime | A | — | A | — | A | A | A | — | — |
| Lubricants | A | A | A | A | A | A | — | A | A |
| Magnesium hydroxide | A | A | A | A | A | A | A | A | A |

**(table continued)**

| Material | Epoxy | Polypropylene | Polyvinyl chloride Type 1 | Nylon | Stainless steel 316 | Stainless steel 304 | Titanium | Buna N | Viton |
|---|---|---|---|---|---|---|---|---|---|
| Magnesium sulfate | A | A | A | A | B | B | A | A | A |
| Nitric acid | | | | | | | | | |
| 5-10% solution | A | A | A | A | A | A | A | D | A |
| 20% solution | B | A | A | A | A | A | A | D | A |
| 50% solution[2] | D | D | A | A | B | B | A | D | A |
| concentrated | D | D | D | A | D | D | A | D | D |
| Phenol (carbolic acid) | C | A | A | D | A | A | B | D | A |
| Phosphoric acid | | | | | | | | | |
| 10-40% solution | A | A | A | D | A | B | B | C | A |
| 40-100% solution | A | A | A | C | B | C | C | C | A |
| Potassium cyanide solutions | A | — | A | A | A | A | — | A | A |
| Potassium hydroxide | A | A | A | A | B | B | B | B | A |
| Potassium permanganate | A | A | A | D | B | B | A | A | A |
| Sea water | A | A | A | A | A | A | A | A | A |
| Sodium aluminate | A | — | — | — | A | — | B | A | A |
| Sodium bisulfite | A | A | A | D | A | A | A | A | A |
| Sodium carbonate | A | A | A | A | B | B | A | A | A |
| Sodium chloride | A | A | A | A | A | A | A | A | A |
| Sodium hydroxide | | | | | | | | | |
| 20% solution | A | A | A | A | B | B | A | A | A |
| 50% solution | A | A | A | A | D | D | A | D | A |
| 80% solution | A | A | A | A | D | D | A | D | B |
| Sodium hypochlorite, up to 20% solution[3] | C | D | A | A | C | C | A | C | A |
| Sodium metaphosphate[2] | A | D | — | A | A | — | — | A | — |
| Sodium metasilicate | A | — | — | — | A | — | — | — | A |
| Sodium polyphosphate, mono, di, tribasic | A | — | — | — | A | A | — | A | A |
| Sodium silicate | A | A | A | A | A | A | A | A | A |
| Sodium sulfate | A | A | A | A | A | A | A | A | A |
| Sulfate liquors | A | A | — | — | C | C | — | — | — |
| Sulfur dioxide[2] | A | D | D | D | A | A | A | C | A |
| Sulfur trioxide, dry | A | — | — | — | C | — | — | C | A |
| Sulfuric acid | | | | | | | | | |
| up to 10% solution | A | A | A | D | C | D | A | C | A |
| 10-75% solution[2] | C | D | A | D | D | D | D | D | A |
| Sulfurous acid | A | A | A | D | A | C | A | C | A |
| Water, acidic from mines | A | A | A | A | A | A | — | A | A |
| White liquor (pulp mill) | A | A | A | A | A | A | — | — | — |
| White water (paper mill) | A | A | — | A | A | A | — | — | — |
| Zinc sulfate | A | A | A | A | B | B | A | A | A |

*The ratings for these materials are based upon chemical resistance only. Added consideration must be given to pump selection when the chemical is abrasive or viscous. Pressure should also be considered; maximum internal pressure or pumping head is 12-15 lb. The chemical-effect rating is as follows:

A = no effect, excellent; B = minor effect, good; C = moderate effect, fair; D = severe effect, not recommended.

[1]Polyvinyl chloride, satisfactory to 72°F.

[2]Polypropylene, satisfactory to 72°F.

[3]Polypropylene, satisfactory to 120°F.

[4]Buna N, satisfactory for "O" rings.

are sodium aluminate, aluminum chloride or sulfate, and ferric or ferrous chlorides or sulfates. These chemicals form flocs under controlled water-chemistry conditions and assist gravity sedimentation, dissolved-air flotation, and sludge thickening. They are also used in raw-water treatment and sludge conditioning.

*Coagulant aids*—To supplement the primary coagulants in floc formation, various aids may be added, such as natural or artificial polymers, or sodium metasilicate. Aids are also used in raw-water treatment and sludge conditioning, either in conjunction with primary coagulants or by themselves, under proper conditions.

*Miscellaneous*—Aqueous chlorine solution, or sodium or calcium hypochlorite for bacterial disinfection; methanol for denitrification; and potassium permanganate for chemical oxidation are a few of the other types of controlled chemical additions used in waste treatment.

Chemical addition—used also in process and utilities control—is a highly developed science, for which the diaphragm and piston-type pumps are commonly used.

Since most chemical-addition quantities are in the gallons per hour range, the diaphragm pump is ideal for this service. A plunger or cam, pushing against an internal fluid (usually a petroleum-based oil), generally actuates the diaphragm.

Check valves on the suction and discharge ends ensure consistent addition. Discharge-capacity control is

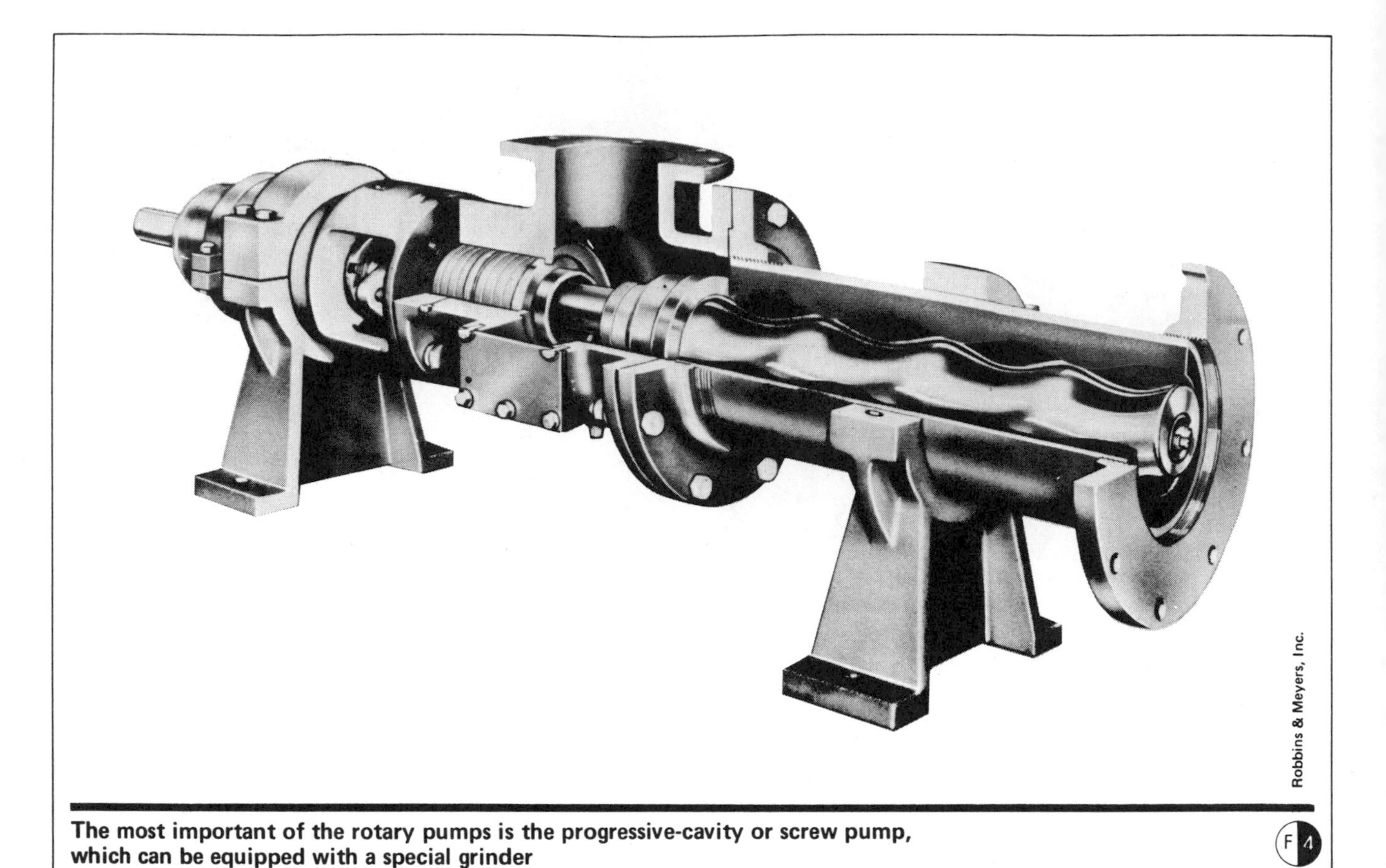


**The most important of the rotary pumps is the progressive-cavity or screw pump, which can be equipped with a special grinder**

established by adjusting the column of hydraulic fluid that bypasses the diaphragm cavity. The stroke adjustment—usually from zero to 100%—can be performed when the pump is operating by manual, electronic or pneumatic control. Capacities of these pumps range from zero to several thousand gallons per hour, with discharge pressures available up to 1,000+ psig.

Since the flow is pulsed because of the motion of the plunger, chemical addition is not uniform. One way to circumvent this phenomenon is to use two or three heads on the same pump to smooth out the flow. Multiple heads also raise the addition volume.

Diaphragm pumps are able to move highly viscous polymer solutions and are available in a wide range of metallic or plastic construction. Pumps of this type are quite reliable if suspended solids are not present in the liquid. Any maintenance required is fairly easily performed. This pump is truly the workhorse of industrial-waste treatment, with respect to chemical addition.

If larger capacities or higher discharge pressures are needed, a piston pump may be used. This pump has essentially the same pulsed-delivery characteristics as the diaphragm pump, except that the plunger interacts directly with the fluid. Multiple heads and pulsation dampers can be had to smooth out the flow. This pump is common where large volumes of chemicals (e.g., acid or caustic for neutralization) must be delivered.

Other pumps occasionally used for chemical addition are the so-called "canned" pump and the peristaltic pump. The former is a sealed, leak-proof centrifugal pump with delivery capacity in the gallons per minute range, while the latter is a very low-capacity pump that has flexible tubing wrapped around a rotating cam. Peristaltic pumps are primarily for laboratory or pilot-scale facilities, while the canned pump delivers larger volumes of hazardous or toxic chemicals.

## Slurry and sludge transport

There are many sources of slurries and sludges in wastewater treatment. Stringent effluent requirements dictate that suspended solids be removed to a level of 10–20 mg/l, and sometimes even lower. Biological treatment produces a mixed liquor that must be pumped for solids recycling. Excess biomass must be digested, thickened and dewatered. Typical solids concentrations of these streams range from 0.5% to 15%.

Chemical-physical treatment generates slurry streams of varying chemical constituency, solids content, particle size and abrasiveness. Sludges from the petroleum and organic chemical industry are usually processed in the waste-treatment area for oil removal.

The entire metals industry produces sludges that vary widely in composition. Even the fossil-fueled power industry, with its air-pollution-control scrubbers, must handle and dispose of large quantities of slurries and sludge. The list is endless, with each sludge or slurry being a little bit different from the others.

Misapplication of pumps for slurry or sludge transport has been common because the chemistry, particle size, abrasion characteristics, and other parameters have not been investigated in detail. Frequently, the liquids are non-Newtonian.

Rotary pumps, particularly the progressive-cavity or screw pump, have found wide application in slurry or sludge transport. The pumping action of all rotary pumps is similar. As the pumping elements are rotated, they open on the inlet side to create a void, which is filled by the fluid as it is forced by atmospheric pressure. The continued turning of the rotors encloses the fluid between the rotating elements themselves, or between the elements and the pump casing. At this point, the rotors are under inlet pressure until the enclosed portion opens into the outlet chamber.

Particles in suspension do not affect the pump operation. The general rule is that if the fluid is able to move into the pumping elements, it can also be forced out. These pumps are self-priming up to 28 ft of suction lift, they provide a uniform rate of flow, and they are generally nonfouling. Pumping capacities are available up to several hundred gpm, with discharge pressures up to 300–400 psig.

**In a peristaltic pump, the liquid is mechanically squeezed through the tubing** F5

**This "canned" rotor pump operates without a seal** F6

Although it is common to pump slurries with 5–15% solids content, sludges with 70% solids have also been moved. The operating speed of the pump is influenced mostly by the properties of the suspension. The greater the solids content, the greater the required horsepower and the lower the operating speed. For any material, however, the throughput per motor revolution is constant. The abrasive properties of a suspension affect the pump speed; capacities are lower with highly abrasive materials. Maintenance can be expected to be greater than for the other pumps already discussed.

Other pumps for slurry movement include the rotary-gear, piston and diaphragm pumps. The rotary-gear pump is similar to the progressive-cavity pump, except that intermeshing gears are used instead of a screw. The rotary-gear kind can pump a wide range of materials such as oils, tar, polymers, or any liquid that does not contain hard solids.

## Sampling systems

An often overlooked factor in wastewater treatment is the continuous or semicontinuous sampling of the various streams for operational control or effluent-quality reporting. Some samplers continuously bleed liquid from a pumped stream, while others obtain a sample at a preselected time interval. If at all possible, the easiest way to obtain a representative sample is from a small slipstream off a pump discharge. Frequently, samples must be obtained from open-channel or sewer flows, or from pits, ponds or lagoons. In these instances, pumping to the sampling device is ordinarily required.

The type of pump selected generally depends upon the sampling-apparatus configuration, as well as on the characteristics of the stream to be sampled. Flowrate to the sampler, suspended solids, entrained gases, chemical ions in solution, and other considerations may affect the pump selection. Typical installations use centrifugal pumps, various rotary pumps (including the peristaltic kind), and diaphragm pumps. Each case and its pump selection must be investigated individually.

## References


1. "Controlled Volume Pumps and Systems," Milton Roy Co. brochures.
2. Edwards, James A., Pumps for Pollution Control, *Pollution Eng.*, Nov. 1974, p. 26.
3. Aieks, Tyler G., "Pump Selection and Application, BME," McGraw-Hill, New York (1957).
4. Karassik, Igor J., and Carter, Roy, "Centrifugal Pumps," McGraw-Hill, New York (1960).
5. Neerken, Richard F., Pump Selection for the Chemical Process Industries, *Chem. Eng.*, Feb. 18, 1974, p. 104.
6. "Progressive Cavity Pumps," Robbins & Myers, Inc., brochures.
7. "Standards of the Hydraulic Institute," Hydraulic Institute, New York (1947).


## The author

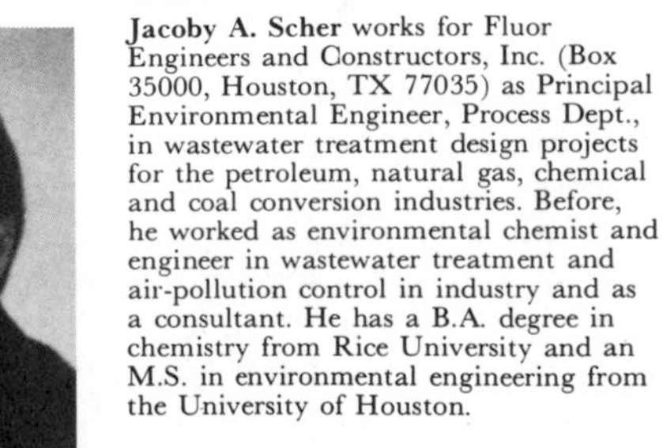

**Jacoby A. Scher** works for Fluor Engineers and Constructors, Inc. (Box 35000, Houston, TX 77035) as Principal Environmental Engineer, Process Dept., in wastewater treatment design projects for the petroleum, natural gas, chemical and coal conversion industries. Before, he worked as environmental chemist and engineer in wastewater treatment and air-pollution control in industry and as a consultant. He has a B.A. degree in chemistry from Rice University and an M.S. in environmental engineering from the University of Houston.

# How gear pumps perform in polymer-

*These two types of pumps are compared in an attempt to acquaint polymer-process engineers with the special problems involved in pumping highly viscous liquids.*

☐ Many polymer-processing operations involve the pumping of highly viscous liquids, as for example, in moving molten polymers out of polymerization-reaction vessels, and polymer melts and solutions through spinnerettes in fiber manufacturing. These high viscosities create special problems that preclude the use of conventional pumps.

Two types of pumps—gear pumps and screw pumps—are generally used in polymer processing. The first type is essentially a positive-displacement device, while the second depends upon viscous drag for fluid transport.

This article discusses in some detail the capacity relationships and energy efficiencies of both types of pumps. Screw pumps designed for maximum output per revolution have an energy efficiency of about 20%, independent of operating conditions and pump size. In contrast, the energy efficiency of a gear pump is a function of liquid viscosity, pump size and operating parameters. In general, it decreases with increasing viscosity and rises with increasing pressure-head. And, depending on specific conditions, it may be greater or less than that of a screw pump of comparable capacity.

The capacity-limiting factor for both types of pumps may be the size of the feedport, and the effectiveness with which the liquid fills cavities in the gear wheels or the channel of the screw.

Originally published September 27, 1976

The technical literature [*1*] contains much data, both theoretical and experimental, on the performance of screw pumps. But available information on gear pumps deals primarily with fluid power applications and aspects of mechanical design—none of which are particularly relevant to the polymer-process engineer.

The authors aim to draw comparisons between gear and screw pumps for polymer-processing applications, which will appeal mainly to the polymer-process engineer, not to a designer of pumps or fluid-power systems. Therefore, this article stresses the following aspects of pump performance:

■ Energy efficiency, which determines the amount of heating that the pumped material undergoes as it passes through the pumps, hence the extent to which the pump wastes energy. Because of the high viscosities involved and the fact that many polymeric compounds are thermally sensitive, energy efficiency is an important process consideration.

■ Discharge characteristics—i.e., the relationship between pumping rate and discharge pressure, and how this is affected by fluid viscosity, pump speed, and pump dimensions.

■ Surge resistance. In many polymer-processing applications, it is essential that the polymer melt or solution be delivered to a forming device at an extremely uniform rate, despite variations in melt temperature,

# and screw pumps processing applications

*James M. McKelvey, Washington University, St. Louis, Mo.*
*Urs Maire, Luwa Corp., Charlotte, N.C.*
*Fritz Haupt, Maag Gear Wheel Co., Zurich, Switzerland*

pressure, and viscosity of the fluid entering the pump, or changes in the flow resistance downstream from the pump.

## Fluid-transport mechanisms

Fig. 1 is a schematic representation of the two mechanisms of fluid transport. The positive-displacement one is illustrated with a piston-cylinder device; volumetric flowrate is obtained by multiplying the cross-sectional area of the cylinder by the speed at which the piston moves forward. The pressure at the end of the cylinder, designated $P^*$, depends upon piston speed, fluid viscosity, and the dimensions of the pipe through which the fluid flows.

The viscous-drag mechanism of fluid transport is illustrated with a device consisting of two parallel plates separated by the distance $H$. The lower plate is stationary, and the upper one moves at a constant speed $V$ to the right. The space between the plates is filled with a liquid that adheres to both surfaces. At the far right, the liquid flows through a pipe.

In the viscous-drag apparatus, liquid in contact with the upper surface moves to the right at speed $V$, while liquid in contact with the lower surface is stationary. Flow is assumed laminar. Because of the viscosity, the moving layers exert a dragging force on adjacent layers. The result is transport of the liquid to the right.

Flow is more complicated in this device because pressure $P^*$ at the entry to the pipeline not only causes the liquid to flow to the right through the pipe but also causes a back flow to the left. In the idealized isothermal Newtonian case, the drag flow to the right results in the linear velocity profile 1, while the pressure flow to the left results in the parabolic profile 2. The net result of these two flows produces profile 3.

A brief study of the nature of the two devices' discharge characteristics should first consider the positive-displacement device. As shown in Fig. 2, the volumetric pumping rate $Q$ is independent of $P^*$ and directly proportional to the speed $V$. Two pump characteristics are shown as horizontal lines: one at speed $V_1$, and the other at $V_2$.

Pump performance also depends upon the characteristic of the flow through the pipe from which it is discharging. A non-Newtonian fluid generates a characteristic curve that is nonlinear and convex to the pressure axis—i.e., a given increase in pressure causes a proportionally larger increase in flowrate. These curves are also very sensitive to fluid viscosity. Two characteristics are shown: one for a relatively low viscosity, $\eta_1$, and the other for a higher viscosity, $\eta_2$.

Operating point of the pump is the intersection of the pump characteristic line with the pipe characteristic line. There are four points shown in the diagram. They indicate graphically how changes in speed and fluid viscosity affect the performance of the positive-displacement device.

Fig. 2 also shows a similar diagram for the viscous-drag device. The pipe characteristics are, of course, the same, but pump characteristics differ markedly. Note that for a given speed, the pump characteristic, instead of being horizontal, is concave to the pressure axis. In addition, it is a function of the fluid viscosity. Therefore, two characteristics must be shown for each speed, one for viscosity $\eta_1$ and the other for viscosity $\eta_2$. The four operating points indicate how pump performance is affected by pump speed and fluid viscosity.

## Capacity of pumps and feedpipes

Fig. 3 is a schematic representation of a screw pump. It consists of a hollow cylinder (barrel), with a rotating inner cylinder (screw) that has a helical flight. The previously mentioned viscous-drag mechanism occurs in the fluid contained in the annular space.

Most screw pumps, as shown in Fig. 3, have a single screw. However, there are some twin-screw pumps in operation. Depending upon their design (primarily on the way in which the screw flights intermesh), their characteristics may resemble either those of positive-displacement devices or viscous-drag devices. In general, twin-screw pumps require a more-complicated drive mechanism. Hence, they are often uneconomical for straight pumping applications and are more likely to be found in uses involving liquid-solid mixtures.

### Nomenclature

| | |
|---|---|
| $D$ | Diameter of screw or gear wheel, in |
| $H$ | Depth of channel in screw, in |
| $L$ | Axial length of screw, in |
| $N$ | Frequency of rotation, rpm |
| $P_*$ | Pump discharge pressure, psi |
| $Q$ | Volumetric flowrate, in$^3$/s |
| $W$ | Width of gear wheel, in |
| $X$ | Scale factor, dimensionless |
| $c$ | Specific heat, Btu/(lb)(°F) |
| $n$ | Flow index of power law fluid, dimensionless |
| $p$ | Total power supplied to pump, hp |
| $p'$ | Power wasted by pump, hp |
| $\alpha$ | Capacity of gear pump, in$^3$/r |
| $\gamma$ | Shear rate, s$^{-1}$ |
| $\epsilon$ | Energy efficiency of pump, dimensionless |
| $\eta$ | Non-Newtonian viscosity, $(lb_f)(s)/in^2$ |
| $\theta$ | Helix angle of screw, deg |
| $\mu$ | Newtonian viscosity, $(lb_f)(s)/in^2$ |

Some of the principal design parameters for screw pumps are also shown in the figure. These include the length, $L$, diameter, $D$, channel depth, $H$, and helix angle, $\theta$. Pump performance depends on the design parameters, operating variables (primarily the frequency of rotation $N$ of the screw) and flow properties of the material being pumped.

Extensive hydrodynamic analyses of screw pumps

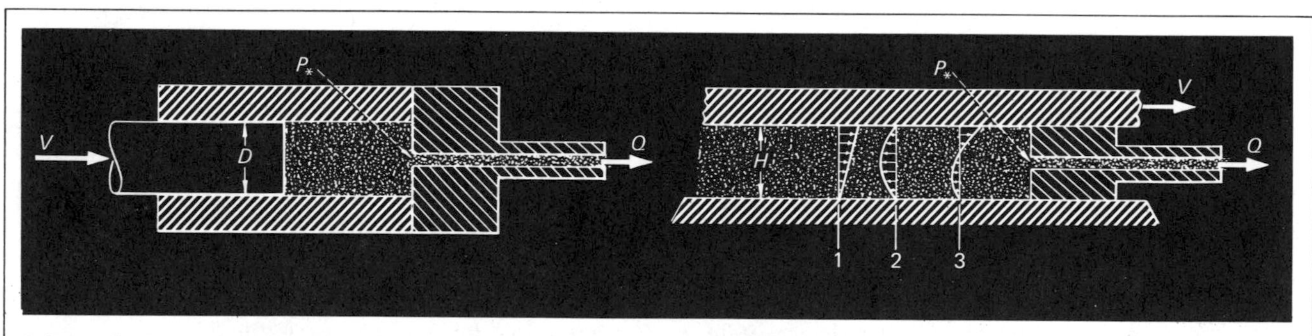


**Schematic drawing of a positive-displacement device (left) and a viscous-drag device (right)** **Fig. 1**

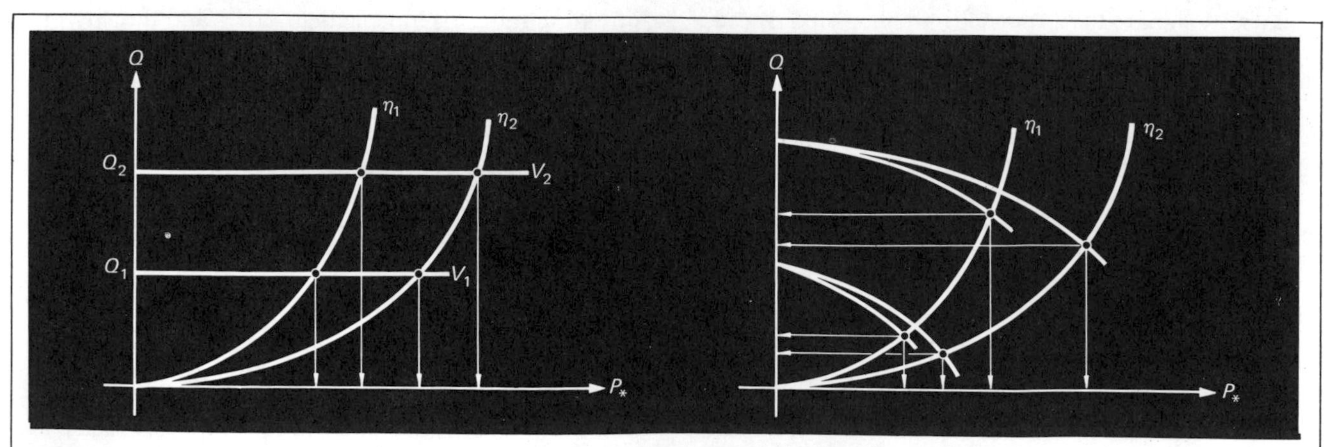


**Discharge characteristics of the positive-displacement (left) and viscous-drag devices (right)** **Fig. 2**

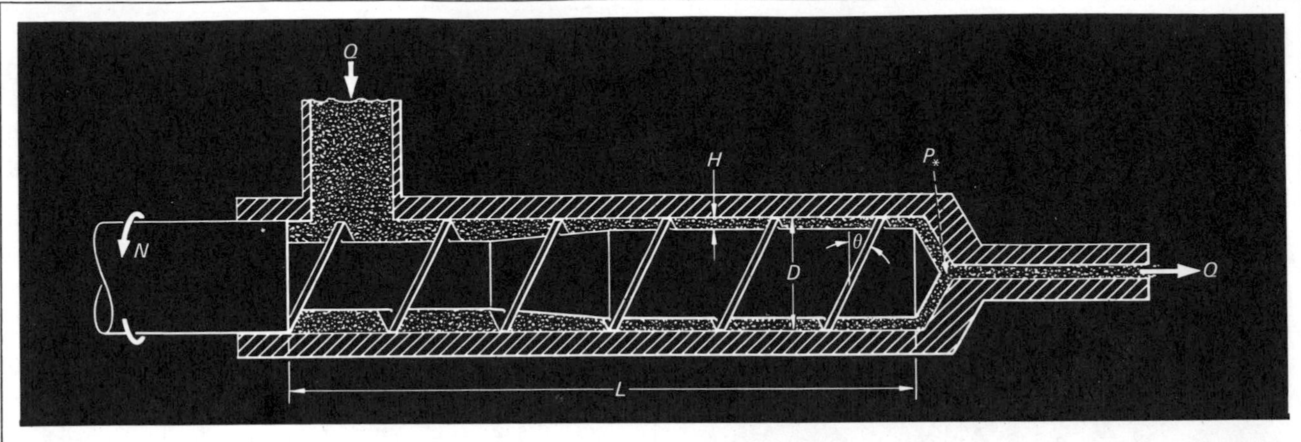


**Helical flight is typical of screw pump** **Fig. 3**

have appeared in the technical literature. Reference [2] provides an introduction to this subject. Although a complete analysis would be extremely complex, involving a three-dimensional, nonisothermal, non-Newtonian flow, it is possible to make a number of simplifying assumptions and arrive at a very simple set of equations that provide a fair approximation to the performance of a screw pump.

The simplified flow theory shows, given a pump of length $L$ and diameter $D$, that there is a set of values for $H$ and $\theta$ that will maximize output per revolution. This optimization requires the helix angle, $\theta$, to be 30° and the channel depth, $H$, to have the value given by:

$$H = \left[\frac{24\mu QL}{\pi DP*}\right]^{1/3} \quad (1)$$

Under these conditions, the output per revolution is given by:

$$\frac{Q}{N} = \frac{\sqrt{3}\pi^2 D^2 H}{12} \quad (2)$$

For the special case where the aspect ratio $(L/D)$ of the screw is equal to 10 (which is reasonable design practice), the pumping rate, obtained by combining Eq. (1) and (2), is given by:

$$\frac{Q}{N} = 14.86D^3\sqrt{\frac{\mu N}{P*}} \quad (3)$$

In Eq. (3), the quantity $(\mu N/P*)$ is dimensionless, and if the diameter, $D$, is expressed in inches, then the output per revolution, $(Q/N)$, has units of $in^3$.

Eq. (3) reveals some interesting information about screw pumps. First, it shows that the pumping capacity per revolution increases with increasing viscosity and decreases with increasing pressure. This is a direct consequence of the viscous-drag mechanism of such pumps, and of the fact that there is a backward-directed pressure-flow component. Furthermore, Eq. (3) shows that with increasing speed, the output per revolution increases. Note, finally, that the capacity is proportional to the third power of the diameter, which provides a useful clue to scaleup.

Gear pumps (Fig. 5) consist of intermeshing gears enclosed in a housing. When one gear is driven, it in turn drives the other. The meshing of the gears isolates the high-pressure discharge side from the low-pressure inlet side, and displaces the liquid contained in the opposite cavity, thus providing the positive-displacement mechanism.

If frequency of rotation of the gears is $N$, and if $v$ represents the total volume of liquid displaced from each gear wheel during one revolution, then capacity is given by:

$$Q = 2vN \quad (4)$$

From geometric considerations, it is seen that the displaced volume, $v$, is given approximately by:

$$v = \frac{\pi}{4}\left[D^2 - D_0^2\right]W = \frac{\pi D^2 W}{4}\left[1 - \left(\frac{D_0}{D}\right)^2\right] \quad (5)$$

where $W$ = width of the gear wheel. Eq. (4) and (5) yield:

$$Q = \alpha N \quad (6)$$

where: $$\alpha = \frac{\pi D^2 W}{2}\left[1 - \left(\frac{D_0}{D}\right)^2\right]$$

Eq. (6) shows that pump capacity is theoretically independent of factors such as viscosity, pressure, and pump speed.

This equation also allows the development of capacity scaleup relationships for gear pumps. Consider, for example, a geometric scaleup where, if $X$ represents the scale factor, all linear dimensions of the small pump are increased by the factor $X$. Hence:

$$D = XD' \\ W = XW' \\ D_0 = XD_0^* \quad (7)$$

where the primed quantities refer to the small pump, and the unprimed to the large pump. If these relationships are introduced into Eq. (6), we have:

$$\alpha = x^3\alpha^1 \quad (8)$$

which shows that the capacity of gear pumps scaled

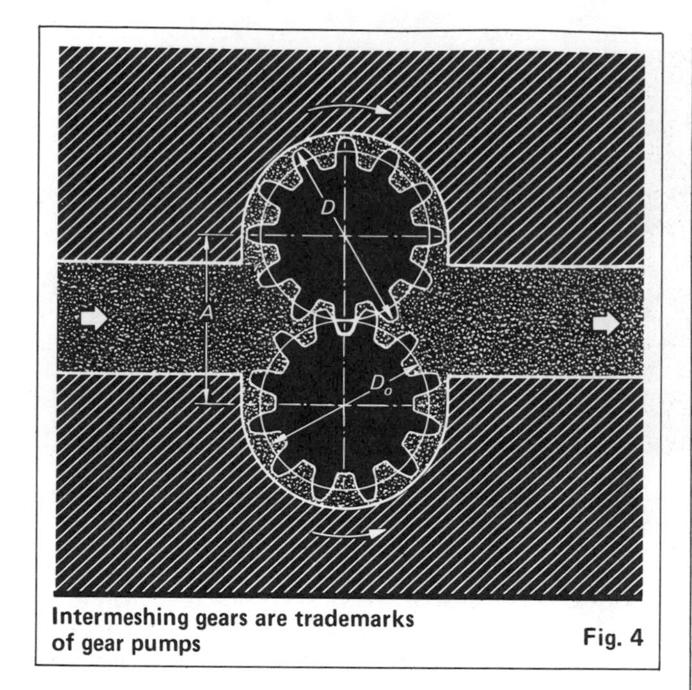

**Intermeshing gears are trademarks of gear pumps** **Fig. 4**

geometrically should increase roughly in proportion to the third power of the scale factor.

*Example 1*—A pumping application requires the delivery of 20,000 lb/h of a liquid having a density of 60 lb/ft$^3$ and a viscosity of 5,000 p against a head of 1,500 psi. Make an estimate of (a) the size of a screw pump and (b) the size of a gear pump, which would meet the stated capacity requirement, assuming that either type of pump would operate at a speed in the range of 60–180 rpm.

If $G$ is the pumping rate in lb/h, then the volumetric rate, $Q$, expressed in in$^3$/s is given by:

$$Q = \frac{1728G}{3600p} = \frac{(1728)(20{,}000)}{(3600)(60)} = 160 \text{ in}^3\text{/s} \quad (9)$$

For the screw pump, a relationship is sought between diameter, $D$, and speed, $N$, which will satisfy the requirement of a flow of 160 in$^3$/s. Eq. (3) is applicable, written as:

$$D = \left[\sqrt{\frac{P*}{\mu}}\left(\frac{Q}{14.86}\right)\right]^{1/3} N^{-0.5} \quad (10)$$

$$D = 11.57N^{-0.5} \quad (11)$$

Eq. (11) indicates that the $D$ vs. $N$ relationship will plot as a straight line with a slope of $-0.5$ on logarithmic paper, as shown in Fig. 5.

For the gear pump, the $D$ vs. $N$ relationship is again employed. Design parameters of the gear pump are as follows:

$$W = D$$

$$(D_0/D) = 0.90$$

Eq. (6) is the applicable one, written as

$$D = \sqrt[3]{\frac{320}{0.19\pi N}} \quad (12)$$

hence:

$$D = \frac{8.19}{\sqrt[3]{N}} \quad (13)$$

Eq. (13) indicates that the $D$ vs. $N$ relationship will produce a straight line on logarithmic paper. In this case, the slope is $-0.33$.

Fig. 5 also shows the $D$ vs. $N$ curve for the gear pump. In the range of speeds from 60 to 180 rpm, the required pumping capacity can be more adequately satisfied with a gear pump of smaller diameter, rather than with a screw pump.

The choice of an operating speed for either type of pump is not entirely arbitrary. While a small pump operating at high speed may satisfy the capacity requirement, it is possible that the feedport of the pump may be too small to permit the required rate of liquid inflow. In fact, in the case of highly viscous liquids, the rate of flow through the feedport may become the capacity-limiting factor for the pump.

In analyzing liquid flow through the feedport, the non-Newtonian nature of polymer melts and solutions should be taken into consideration. Under the assumption that the liquid obeys the power law, then its viscosity, $\eta$, is related to the shear rate, $\gamma$, by:

$$\eta = \eta^0\left(\frac{\gamma}{\gamma^0}\right)^{n-1} \quad (14)$$

where $\eta^0$ is the viscosity in the standard state, characterized by the standard-state shear rate $\gamma^0$; the exponent $n$ is the flow index of the fluid; the degree of departure of $n$ from unity is a measure of how non-Newtonian the fluid is.

The volumetric flowrate, $Q$, of a power-law fluid through a conical pipe of length, $L$, under a pressure head, $\Delta P$, is given approximately by:

$$Q = \frac{n\pi R_0^3\gamma^0}{3n+1}\left[\frac{R_0F}{2\eta^0\gamma^0}\left(\frac{\Delta P}{L} + \rho\right)\right]^{1/n} \quad (15)$$

where:

$$F = \frac{3n(\beta - 1)}{\beta(\beta^{3n} - 1)} \quad (16)$$

$$\beta = \frac{R_0}{R*} \quad (17)$$

$R_0$ = radius of pipe at top

$R*$ = radius of pipe at bottom

The density of the fluid is $\rho$.

Note that for the special case where the pipe is not tapered, that is, where $\beta = 1$, the function $F = 1$; and Eq. (15) is reduced to the familiar relationship for the flow of power-law fluids through straight tubes.

*Example 2*—The liquid feed to a pump must flow through a feedpipe having a height of 19.7 in. at a rate of 5,000 lb/h (38.46 in.$^3$/s). The liquid has the following properties:

$\eta^0 = 10{,}000$ p (at $\dot{\gamma}^0 = 1$ s$^{-1}$) $= 0.145$ (lb$_f$)(s)/in$^2$

$n \doteq 0.60$

$\rho = 1.00$ g/cm$^3$ $= 62.4$ lb/ft$^3$

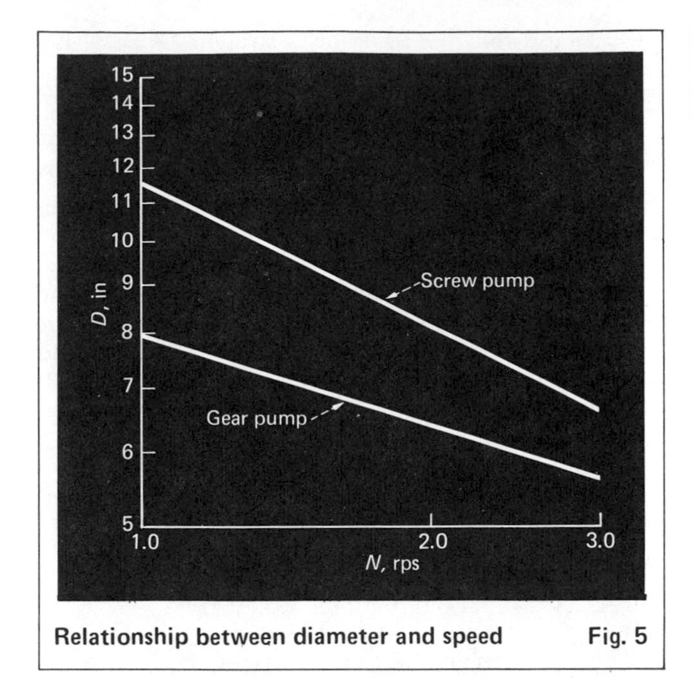


**Relationship between diameter and speed** **Fig. 5**

Compute the required feedpipe diameter for the following cases:

(a) straight pipe with gravity feed
(b) straight pipe with $\Delta P = 7.35$ psi
(c) conical pipe ($\beta = 8$) with gravity feed
(d) conical pipe ($\beta = 8$) with $\Delta P = 7.35$ psi

The calculations are based on Eq. (15). Note that for the straight pipe, $F = 1.00$. For the tapered pipe with $\beta = 8$, Eq. (11) yields $F = 0.0382$. The following results are obtained:

| Case | $\beta$ | $F$ | $\Delta P$ (psi) | $R_0$ (in) | $R^*$ (in) |
|---|---|---|---|---|---|
| a | 1 | 1 | 0 | 5.01 | 5.01 |
| b | 1 | 1 | 7.35 | 2.10 | 2.10 |
| c | 8 | 0.0382 | 0 | 16.09 | 2.01 |
| d | 8 | 0.0382 | 7.35 | 6.76 | 0.84 |

This example illustrates the beneficial effects of having a tapered pipe and a positive-pressure driving force. Note that in case (d), where both conditions prevail, a feedport of 0.87 in radius would accommodate the 5,000-lb/h flowrate, while in case (a), a port of 4.96 in. radius is required.

## Power requirements

The rate at which useful work (flow work) is done by any pump is the product of the volumetric flowrate, $Q$, and the pressure head, $P*$, against which it is working. Hence, if $p$ represents the total mechanical power input to the pump, then the pump energy efficiency, $\varepsilon$, is:

$$\varepsilon = \frac{QP*}{p} \tag{18}$$

The wasted power is $(1 - \varepsilon)p$. It is accounted for as heat imparted to the liquid as it passes through the pump.

*Example 3*—A pump with an energy efficiency, $\varepsilon$, of 0.2 is pumping a liquid having a density of 0.0361 lb/in$^3$ and a specific heat of 0.5 Btu/(lb)(F°) against a pressure of 1,500 psi. What is the temperature rise of the liquid, if the heat losses from the pump are negligible?

If $E_1$ represents the rate of energy input to the liquid from the pump, then:

$$E_1 = (1 - \varepsilon)p = \left(\frac{1-\varepsilon}{\varepsilon}\right)QP*$$

When $Q$ is expressed in in.$^3$/s and $P*$ in psi, then $E_1$ has units of (lb$_f$)(in.)/s. The conversion factor $1.07 \times 10^{-4}$ Btu/(in.)(lb$_f$) is used to convert $E_1$ to thermal units:

$$E_1 = 1.07 \times 10^{-4}\left(\frac{1-\varepsilon}{\varepsilon}\right)QP*$$

If $E_2$ represents the rate of absorption of heat by the liquid, then:

$$E_2 = \rho CQ\Delta T$$

For the case where there are no heat losses in the pump (the case of maximum temperature rise):

$$E_1 = E_2$$

and:

$$\Delta T = 1.07 \times 10^{-4}\left(\frac{1-\varepsilon}{\varepsilon}\right)\left(\frac{P*}{\rho C}\right)$$

$$\Delta T = 35.6°\text{F}$$

It has been shown [3] that for a screw pump designed with channel depth, $H$, given by Eq. (1), and with helix angle $\theta = 30$, the output per revolution will be a maximum, and such a pump will have an energy efficiency of 0.20, independent of the viscosity of the liquid and the speed of the screw. This result is based upon a simplified analysis that assumes Newtonian flow behavior, and must be taken only as a first-order approximation.

The magnitude of the heating effect with screw pumps is readily seen by referring to Example 3. Clearly, if the pressure head is much above a few thousand psi, the heating effect should not be ignored. Unless sufficient heat transfer is provided, the temperature rise will be significant.

In the analysis of energy requirements for gear pumps, it is convenient to break the total power requirement into two parts, as shown by:

$$p = p' + QP* \tag{19}$$

where $p'$ represents the power dissipated as heat. Combining Eq. (18) and (19), the energy efficiency of any gear pump can be written as:

$$\varepsilon = \frac{QP*}{p' + QP*} \tag{20}$$

It is not possible to derive a theoretical expression for $p'$ that would be useful for this analysis. One knows, in general, that for a given pump design the dissipated power will increase with increasing pump speed ($N$), viscosity ($\mu$), and size ($D$) of the pump. Hence, an empirical expression for $p'$ can be written as:

$$p' = KN^{A}\mu^{B}D_0^{C} \tag{21}$$

where the coefficient $K$ and the exponents $A$, $B$ and $C$ would have to be experimentally determined.

Experimental power measurements have been carried out on a series of commercially available gear pumps*

*The Vacorex® pumps used in these tests were developed by Maag Gear Wheel Co. and Luwa AG, both of Zurich, Switzerland. Luwa Corp. (Charlotte, N.C.) markets the units in North America.

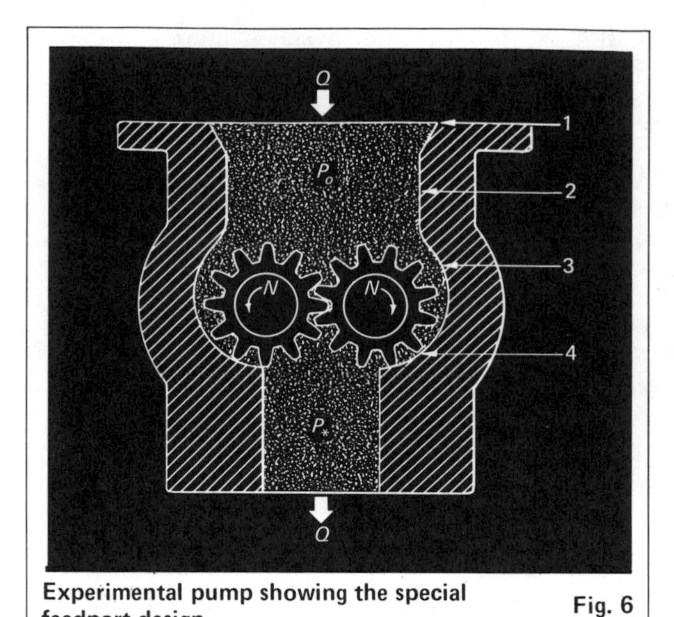

**Experimental pump showing the special feedport design** **Fig. 6**

encompassing a wide range of sizes, operating speeds, and viscosities. From the data, the following values for the coefficient and exponents in Eq. (18) have been established:

$$K = 38.6 \times 10^{-4}$$
$$A = 1.2$$
$$B = 0.62$$
$$C = 2.67$$

In Eq. (21), if $N$ is expressed in rpm, $D_0$ in inches, and $\mu$ in $(lb_f)(s)/in^2$, then the units of $p'$ will be given in horsepower.

Note from Eq. (20) and (21) that the efficiency of a gear pump will decrease with increasing viscosity. Since the efficiency of a screw pump is independent of viscosity, gear pumps could be expected to be the preferred devices for relatively low-viscosity liquids, and screw pumps for higher-viscosity liquids. However, it is not possible to make any such sweeping generalizations. It is known that the energy efficiency of a well-designed screw pump will be on the order of 20%, and that this is independent of size and operating conditions.

*Example 4*—This example considers an application in which a liquid having a density of 62.4 lb/ft$^3$ and a viscosity of 10,000 p (0.145 $(lb_f)(s)/in^2$) is pumped against a head of 1,500 psi at a rate of 6,500 lb/h. The problem is to establish the sizes of the gear pump and the screw pump that would meet the performance requirements summarized below:

| | |
|---|---|
| Required pumping capacity | 6,500 lb/h ($Q = 50$ in$^3$/s) |
| Pressure head | 1,500 psi |
| Viscosity | 10,000 p |
| Pressure at feedport | 1.5 psi |
| Density of liquid | 62.4 lb/ft$^3$ ($\rho = 0.0361$ lb/in$^3$) |
| Height of feedpipe | 4 in. |

It is first necessary to determine the diameter, $D^*$, of the feedport that has sufficient capacity to handle the required flowrate. Assuming that the fluid is Newtonian ($n = 1$) and that the feedpipe is straight ($F = 1$), Eq. (15) is reduced to:

$$D^* = 2R^* = 2\left[\frac{8\mu Q}{\pi(\Delta P/L + \rho)}\right]^{1/4} \quad (22)$$

From this equation, the required feedport diameter is found to be 3.1 in.

The size and power requirements are now established if a screw pump is considered for the application. The possibility of using a 4-in-dia. pump is first studied. The question is, for this diameter, what screw speed would be needed to meet the capacity requirements? Eq. (3) can be rearranged to give the following explicit expression for speed $N$:

$$N = \frac{1}{6.04D^2}\left[\frac{Q^2P^*}{\mu}\right]^{1/3} \quad (23)$$

Eq. (23) indicates that for a 4-in-dia. pump, the required operating speed would be 46 rpm, and for a 5-in-dia. pump, it would be 120 rpm.

The power needed for the screw pump is obtained by first calculating the power, $QP^*$, needed for the actual flow work:

$$QP^* = (50)(1{,}500) = 75{,}000 \text{ (in)(lb)/s (11.4 hp)}$$

Noting that the energy efficiency is 0.2, the total power required for the screw pump is given by:

$$p = \frac{QP^*}{\varepsilon} = \frac{11.4}{0.20} = 57 \text{ hp}$$

It should be noted that the calculated power, 57 hp, does not take into account mechanical losses in the gear reducers and other mechanisms associated with the pump. It is the actual power that must be delivered to the shaft.

The next step is to establish the size and power for a gear pump that will meet the performance requirements. It is assumed that the pump will have the following design characteristics:

$$W = D; \qquad \left(\frac{D_0}{D}\right) = 0.9$$

From Eq. (6):

$$N = \frac{2Q}{\pi D^3[1 - (D_0/D)^2]}$$

For a 4-in-dia. gear wheel, it is seen that a speed of 157 rpm is required. For a 5-in gear wheel, the required speed is 80 rpm. This is the better choice. If $D = 5$ in, then $D_0 = 4.5$ in, and the power requirement is estimated from Eq. (21), using the parameters given previously for the Vacorex pumps. Hence:

$$p' = 38.6 \times 10^{-4}(0.145)^{0.62}(80)^{1.2}(4.5)^{2.67}$$
$$= 12.4 \text{ hp}$$

The dissipated power is 12.4 hp, so the energy efficiency for the pump is given by:

$$\varepsilon = \frac{QP^*}{QP^* + p'} = \frac{11.4}{11.4 + 12.4} = 0.48$$

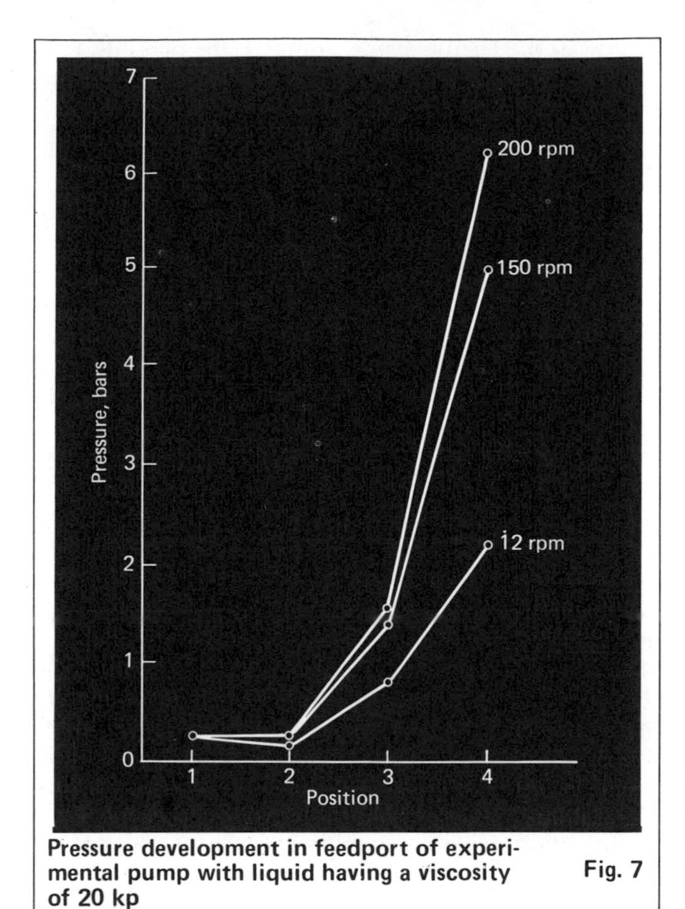


**Pressure development in feedport of experimental pump with liquid having a viscosity of 20 kp** **Fig. 7**

## Feedport design and performance

In polymer-processing applications, gravitational forces are often inadequate to ensure that the cavities in the gear wheels of conventional gear pumps are completely filled with liquid. When this happens, the pump operates in a "starved" condition. This is particularly probable at higher speeds and when the inlet-side pressures are low.

A somewhat analogous problem is encountered with screw pumps. As shown in Fig. 3, it is customary to make the screw channel deeper in the region of the feedport than in the forward part of the screw. It is possible to incorporate a somewhat similar design principle in gear pumps, as shown in Fig. 6. Here the pump housing on the inlet side is designed so that a compression effect is obtained. The four arrows indicate the location of pressure taps.

Fig. 7 shows results of pressure measurements at four locations made on the experimental pump. Measurements were made at three speeds with a liquid of 20,000-p viscosity. The suction-side pressure was 200 mm Hg absolute for all measurements.

Note first the slight decrease in pressure between positions 1 and 2. This slight pressure head provides a driving force that assists the gravitational flow of the liquid to the gear wheels. Between positions 2 and 3, the effects of the compression become apparent, and between positions 3 and 4 the compression effect is quite pronounced. At a speed of 200 rpm, a pressure of about 5 atm is developed at position 4, where the effective

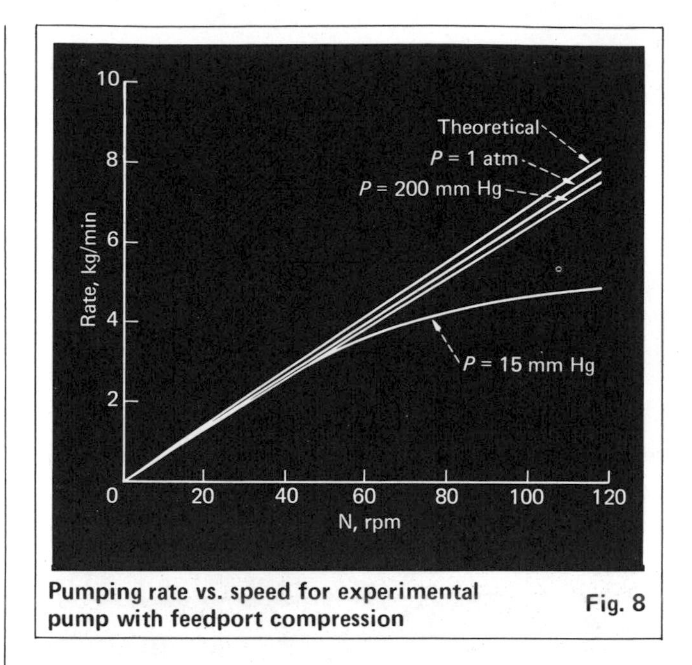


**Pumping rate vs. speed for experimental pump with feedport compression** **Fig. 8**

clearance between the tip of the gear teeth and the housing becomes effectively zero. From this point on, the pump operates essentially as a constant-displacement device.

Fig. 8 shows the discharge rate vs. speed for the experimental Vacorex pump operating with a liquid having a viscosity of 25–30 kp. The upper line shows the theoretical discharge rate. Note that when the inlet is at atmospheric pressure, there is only a very small deviation from the theoretical discharge rate over the entire speed range to 120 rpm. Reduction of the inlet pressure to 200 mm Hg causes only a small reduction in discharge rate. It is only when the pressure is reduced further, to the point where the liquid boils and foams (15 mm Hg) that significant losses in the discharge occur.

The importance of the inlet compression is illustrated by the curves of Fig. 9. Here the experiments shown in Fig. 8 were repeated, using the inlet feedport configuration in Fig. 4. Note that at inlet-side pressures below atmospheric, the volumetric efficiency of the pump not having a compression capability on the suction side becomes extremely low.

## Other design considerations

It should be noted that screw pumps, when operating on highly viscous liquids, have an almost unlimited pressure-building capability. Pressures in excess of 10,000 psi do not present particularly serious mechanical design problems. Aside from having a barrel and head of sufficient strength, the main limitation is in the capacity and life of the thrust bearing.

With gear pumps, the situation is somewhat different. At high discharge pressures, the axis of the gear wheels can be bowed from a straight-line configuration, and thus cause the pump to malfunction. The discharge pressure is further limited by the admissible load of the friction bearings. These are the primary factors that limit the width of the gear wheels.

A mechanical feature specific to gear pumps is the

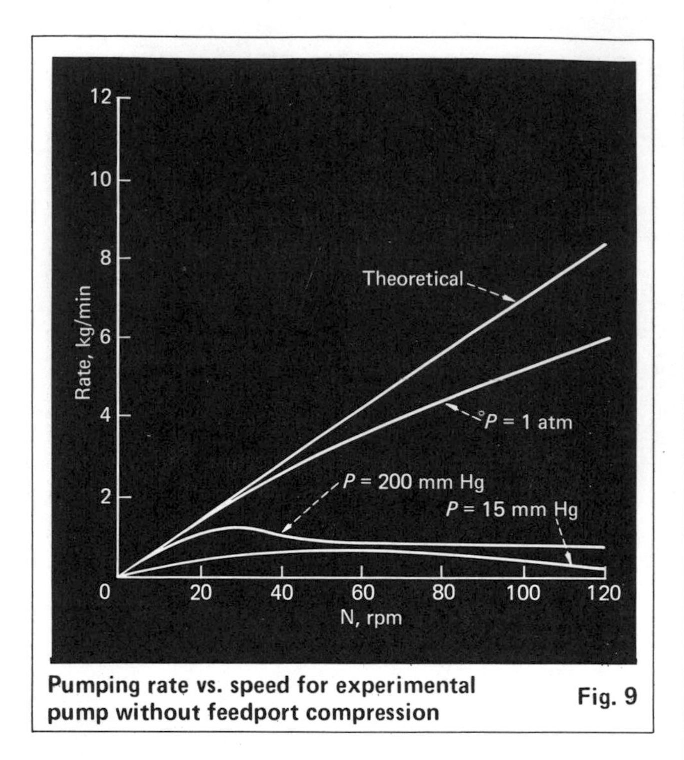


**Pumping rate vs. speed for experimental pump without feedport compression** **Fig. 9**

close tolerance with which the gear faces must fit to the sides of the housing. Otherwise, leakage through this space would seriously reduce the volumetric efficiency. However, given today's manufacturing standards these tolerances pose no particular problem, and gear pumps operate with high volumetric efficiencies. In fact, even higher precision is obtained in the manufacture of high-pressure, high-speed oil pumps.

Another important matter to be considered in gear pumps is the lubrication of the bearings without contaminating the liquid being pumped.

In comparing gear pumps with screw pumps, size should also be considered. A gear pump is a very compact device, which is often an advantage in equipment layout. On the other hand, the barrel surface area of a screw pump provides an opportunity for extensive heating and cooling by heat transfer. In fact, a screw pump can be thought of as a heat-transfer device that is especially suitable for highly viscous liquids. While some heating and cooling is admittedly possible with a jacketed gear pump as well, the short residence times and the limited surface area that is in contact with the fluid definitely limit the amount of heating or cooling that takes place.

Another mechanical feature of gear pumps that should be considered is that of pulsations. Because these pumps are essentially positive-displacement devices, they are surgefree so long as the volumetric efficiency (which depends upon conditions occurring in the feedport, as previously discussed) does not change. However, the meshing of the gears results in a periodic change in the volume of the outlet-side cavity of the pump. If the gear wheel has 12 teeth, and if it is rotating at $N$ rpm, then there will be 24 $N$ pulsations/min. At normal pump speeds, this means that there may be as many as 10 or 12 pulses/s. Because of the large capacity of the downstream system and the viscoelastic nature of the liquids being pumped, the magnitude of the pressure pulses diminishes, and this is not ordinarily a serious factor in polymer processing.

Pulses also can be observed in screw pumps. With a single-flighted screw, the frequency of the pulsations is equal to the frequency of rotation of the screw. Such pulsations are not normally a matter of serious concern. In gear pumps, their magnitude can be reduced by the use of helical gears instead of spur gears. Another advantage of helical gearing is the self-cleaning effect in the tooth root, thanks to a minimum bottom clearance. There are, however, some disadvantages to this option. For example, the helix imparts an axial thrust to the gear wheels.

Finally, it should be noted that for some materials (those undergoing chemical reaction, for example) the residence time in the pump can be an important factor. In both gear pumps and screw pumps, there will be a distribution of residence times, with some "particles" of the fluid taking a much longer time to advance through the pump than others.

Considering the superficial residence time that can be calculated by dividing the total free volume of the pump, from feedport to discharge head, by the volumetric flowrate, it can be seen that the residence time in screw pumps is much greater than that in gear pumps, by about an order of magnitude. This difference can be either an asset or a liability, depending upon the pump's specific application.

## References


1. Tadmor, Z., and Klein, I., "Plasticating Extrusion," Van Nostrand Reinhold Co., 1970.
2. McKelvey, J. M., "Polymer Processing," Chap. 10, John Wiley & Sons, 1962.
3. Ibid, p. 254.

## Meet the Authors


**James M. McKelvey** is professor of chemical engineering and Dean of the School of Engineering and Applied Science at Washington University, St. Louis, Mo. He has wide consulting experience in the field of polymer technology, and is the author of "Polymer Processing," published by John Wiley & Sons in 1962.

**Urs Maire** is a sales engineer with Luwa Corp. Formerly a project/process engineer specializing in polymer applications at Luwa Corp. and Luwa AG, he earned a B.S. in mechanical engineering at the University of Lucerne, Switzerland.

**Fritz Haupt** is technical manager of the pump division of Maag Gear Wheel Co. (Zurich, Switzerland). A mechanical engineer, he has had experience in the design of heavy-duty gears and couplings in 1964, he became head of the research and development department of Maag pumps.

## Section VI

# Seals, Packing, Piping and Layout

# How to choose and install mechanical seals

Leak-free operation, low maintenance and meeting environmental standards are major advantages when using seals on shafts.

*John H. Ramsden, Badger America, Inc.*

☐ Mechanical seals prevent the escape of all types of fluids, either gases or liquids, along a rotating shaft that extends through a casing or housing. Extensive applications of such seals in the chemical process industries (CPI) range from the containment of cryogenic fluids to high-temperature heat-transfer fluids.

The mechanical seal has certain advantages over packing because it:

- Offers a more positive seal.
- Eliminates periodic manual adjustment.
- Requires replacement of the seal only, and not the pump shaft or sleeve.

The types of equipment using mechanical seals include centrifugal and rotary pumps; centrifugal, axial-flow and rotary compressors; and agitators. In this article, we will deal primarily with the sealing of liquids in rotating pumps because this is the most common application.

Mechanical seals for compressors are of highly sophisticated design, generally much larger and normally tailor-made by the compressor manufacturer. Also, these seals are used to contain a gas or compressible fluid rather than a liquid, which presents unique design and operating problems. For additional information on mechanical seals for compressors, consult Ref. *1*, and for agitators, Ref. *2*.

## Characteristics of mechanical seals

A mechanical seal is a device used to prevent shaft leakage by means of two sealing surfaces, one stationary and the other rotating in close contact with it. These sealing surfaces, or faces, are perpendicular to the shaft rather than parallel with it. A mechanical seal is similar to a bearing in that it involves a close-running clearance with a liquid film between the faces [*3*].

The two sealing surfaces are called the primary ring and the mating ring, as shown in Fig. 1. Either ring may be stationary. However, most designs use a rotating primary ring and stationary mating ring. The faces of the two rings are lapped to a flatness that is measured in millionths of an inch. Therefore, the faces remain in contact over their entire contact-surface area to provide a nearly complete seal. The primary ring is flexibly mounted in order to allow axial and radial movement of the ring so that contact can be maintained with the mating ring.

Secondary seals allow flexible mounting of the primary ring. These are U cups, chevron rings, bellows, wedge-shaped rings, and O rings. Springs, metal bellows or magnetism supply the closing force necessary to maintain contact with the mating ring. The mating ring uses either flexible mounting with O rings or gaskets, or a pressed-in mounting.

## Classification of mechanical seals

Mechanical seals are classified by type of mounting (i.e., inside or outside mounted), and whether they are balanced or unbalanced.

If the primary ring is mounted inside the container holding the fluid, it is referred to as an "inside seal." When the primary ring is mounted external to the container, it is referred to as an "outside seal." Inside and outside seals are shown in Fig. 1.

Outside seals are preferred for ease of maintenance. They also permit the metal parts to be isolated from a corrosive environment. Some disadvantages:

1. Hydraulic force tends to open the seal faces.
2. Lubrication and flushing of the faces is restricted.
3. Abrasive-dirt particles in the fluid can pack in the annular opening. These are then pushed between the seal faces by centrifugal force and cause rapid wear.

Inside seals are generally preferred for better operation. The entire primary ring is surrounded by the fluid. Hydraulic forces are acting with the springs to keep the seal faces in contact. Flushing and lubrication can be designed to be more positive for better cooling at the seal faces.

The forces acting on the area of the primary face of

Originally published October 9, 1978

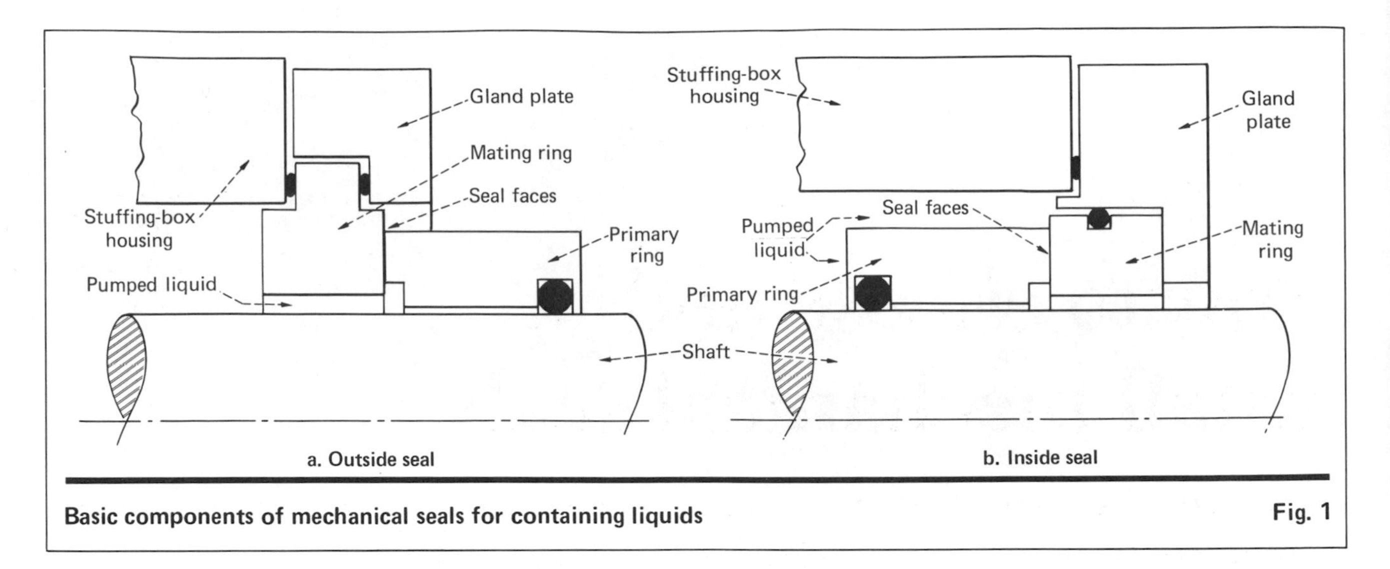


**Basic components of mechanical seals for containing liquids** **Fig. 1**

an inside seal that is exposed to the hydraulic pressure in the stuffing box may create an unbalanced condition. Fig. 2a depicts an unbalanced inside seal. Pressure, acting on the back of the primary ring, results in a force that pushes the seal faces together. With a seal operating at high stuffing-box pressure, the forces can become excessive and cause rapid wear of the seal faces. Manufacturers of mechanical seals use pressure-velocity relationships to determine the pressure limitations for unbalanced seals. Generally, the use of unbalanced seals is limited to a stuffing-box pressure of 200 psig (1,380 kPa), depending on shaft size and speed. API Standard 610 specifies a lower, more conservative limit, as shown in Table I [*4*].

The force acting on the seal faces can be reduced by changing the relationship of the closing area to the face area. Reducing the area on which the pressure acts, while holding the face area constant, will reduce the force on the face. This is called balancing the seal. A step in the shaft, sleeve or seal hardware (Fig. 2b) is used to fill in the area.

## Application of mechanical seals

Most mechanical seals installed are single mechanical. Such seals are suitable if the pumped liquid is clean, free from grit or solids, and is not toxic or lethal. A typical, single, balanced, inside seal is shown in Fig. 3.

To lengthen operating life, the seal must be kept cool. Therefore, the stuffing box is flushed with a liquid. If the pumped liquid is clean and cool, a bypass stream from the pump's discharge can be used for the seal flush to remove the heat of friction generated by the rubbing seal faces.

If the pumped fluid cannot be used, an external fluid compatible with the pumped fluid is supplied. The external flush liquid must be clean, cool and at a pressure greater than the maximum stuffing-box pressure. Stuffing-box pressure varies with pump type and manufacturer. For centrifugal pumps, it ranges from a few psi above suction to the full discharge pressure.

The quantity of external flush fluid can be reduced by means of a restriction between the stuffing box and pump cavity. This is done to reduce contamination or

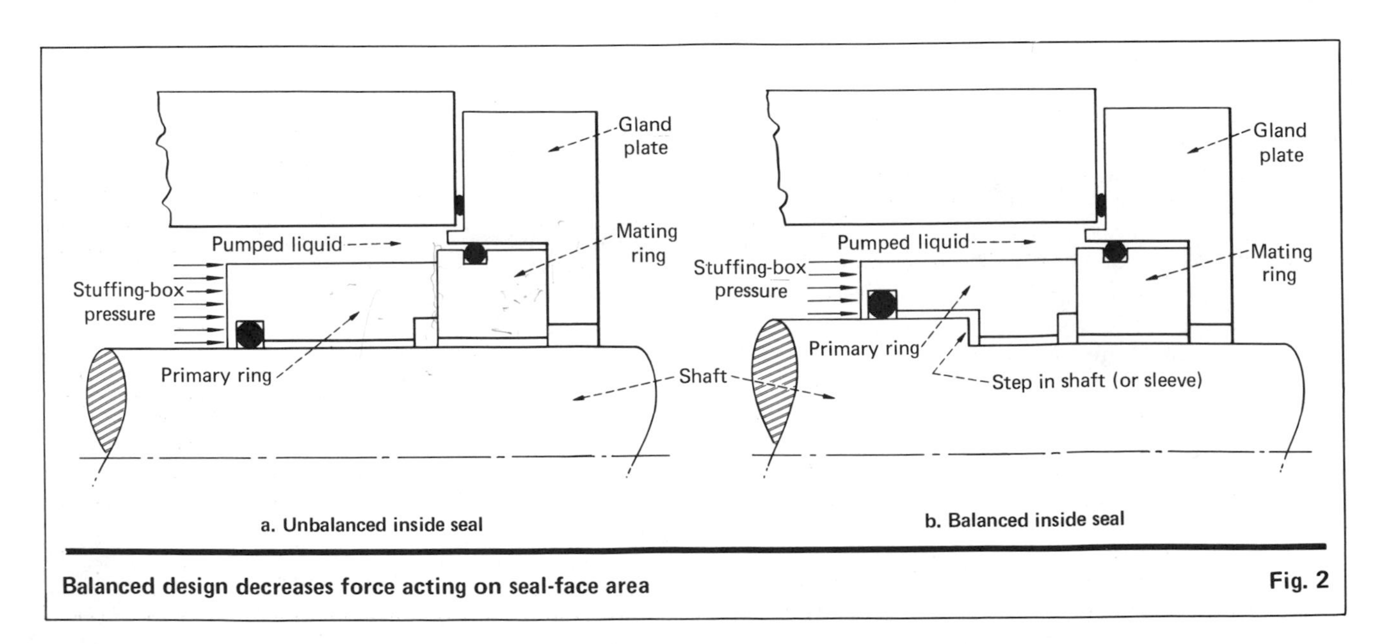


**Balanced design decreases force acting on seal-face area** **Fig. 2**

dilution of the pumped liquid and to reduce operating cost. Several types of restrictions are available, such as a lip seal or throat bushing.

The lip seal is used primarily to prevent pumped liquid from entering the stuffing box. It also provides some restriction to the flush liquid entering the pump.

The throat bushing is a close-clearance bushing to restrict flow. The clearance between the throat bushing and shaft must be large enough to prevent rubbing, and depends on the shaft eccentricity and deflection. The larger the clearance and shorter the bushing, the higher the dilution rate of the pumped liquid. A free-floating throat bushing is used to further reduce the dilution rate. This type of bushing is mounted so that it can follow the movement of the shaft and maintain a close clearance without rubbing.

In order to prevent the pumped liquid from entering the stuffing box, one seal manufacturer recommends that the flush-liquid velocity at the throat should be 10 to 15 ft/s (3 to 4.6 m/s). On a pump having a 2-in. (50 mm)-dia. shaft with a fixed throat bushing having a radial clearance of 0.007 in. (0.18 mm), a flushing rate of 2.1 gpm (8 L/min) is required to maintain a 15-ft/s (4.6 m/s) velocity. For the same pump with a floating throat bushing having a radial clearance of 0.003 in. (0.08 mm), 0.9 gpm (3.4 L/min) is required.

A mechanical seal is not a zero-leakage device. The seal operates on the principle of providing a thin film of liquid between the sealing faces to lubricate and cool them. This is the reason why the flush liquid must be clean and cool. Depending on the condition and flatness of the seal faces, the leakage rate is quite low (as little as one drop per minute) and the flow often is invisible.

If the seal leakage vaporizes or condenses when exposed to atmospheric conditions, it will be necessary to provide an auxiliary sealing device such as packing, or a throttle bushing outboard of the seal faces, in the gland plate. Vent and/or drain connections are provided to either vent the vapor to atmosphere or a safe location, drain the condensate, or quench the seepage with a quench liquid (Fig. 6).

In circumstances where zero leakage is required because of toxicity, emission controls, etc., a single seal is not suitable, and a double mechanical seal (Fig. 4) is generally selected. This is the most common multiple-seal arrangement. Two seals are oriented back to back, providing a closed cavity between them. A seal liquid whose temperature and pressure are controlled is circulated through this cavity to provide good seal life.

The seal liquid must be at a pressure above the operating pressure of the stuffing box in order for the seal to function. Thus, some leakage of seal liquid across the seal faces is necessary. The liquid going across the inboard face will enter the stuffing box and mix with the pumped fluid. The liquid going across the outboard face will go to atmosphere. Therefore, the seal liquid must be compatible with the pumped fluid and must not cause environmental problems. Either a clean, cool liquid from another pump system in the plant can be used as a seal liquid, or a closed sealing system servicing one or several pumps must be available.

Another seal arrangement frequently used to solve environmental and safety-related problems incorporates tandem seals (Fig. 5). Tandem seals differ from double seals in three major respects:

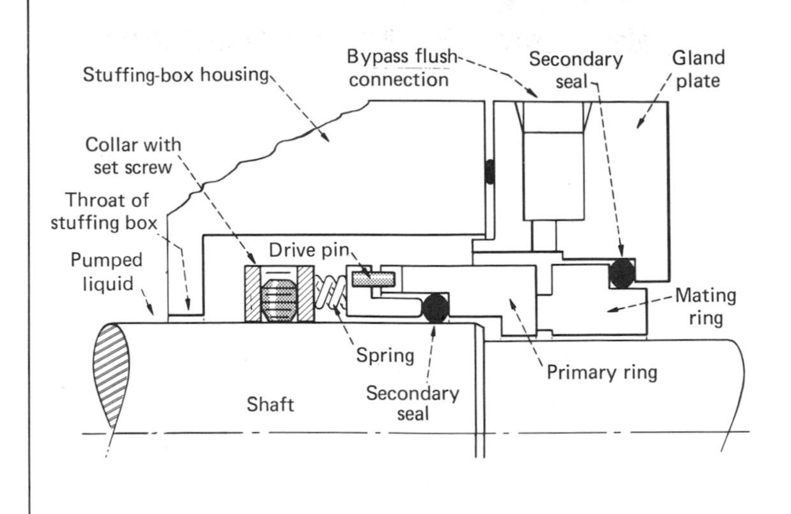


**Component relationships for a single, balanced, inside mechanical seal** **Fig. 3**

First, both seals are faced in the same direction rather than back to back.

Second, the fluid in the seal cavity is used as a barrier fluid, and is at a pressure lower than that in the stuffing box. Therefore, leakage will be from the stuffing box into the seal cavity containing the barrier fluid.

Third, a seal flush is used in the stuffing box to remove the heat generated by friction. The secondary seal (outboard) provides a backup for the primary or inboard seal.

## Accessories for mechanical seals

A requirement for the fluid used to flush the seal is that it be clean and free from dirt. Suspended particles can get between the seal faces and damage them.

*Filters*—One method for assuring clean liquid is to install a filter in the bypass line or feed line to the seal. Two questions must be considered before selecting such a filter to clean the fluid:

1. What amount of solids must be filtered out? If the liquid system is extremely dirty, the filters will rapidly fill and clog, and cause high maintenance costs.

2. Does the pump operate in a closed loop or not? If the pump is in a once-through system, the filter will periodically fill and require frequent changing. If this situation exists, a different flush system should be chosen. However, if the pump is installed in a closed-loop system, the filter will eventually clean the entire system,

**Pressure limits for unbalanced seals (American Petroleum Institute Standard 610)** **Table I**

| Inside diameter of seal, in. | mm | Shaft speed, rpm | Sealing pressure, psig | kPa |
|---|---|---|---|---|
| ½ to 2 | 13 to 50 | Up to 1,800 | 100 | 690 |
| | | 1,801 to 3,600 | 50 | 345 |
| Over 2 to 4 | Over 50 to 100 | Up to 1,800 | 50 | 345 |
| | | 1,801 to 3,600 | 25 | 172 |

Source: Ref. 4

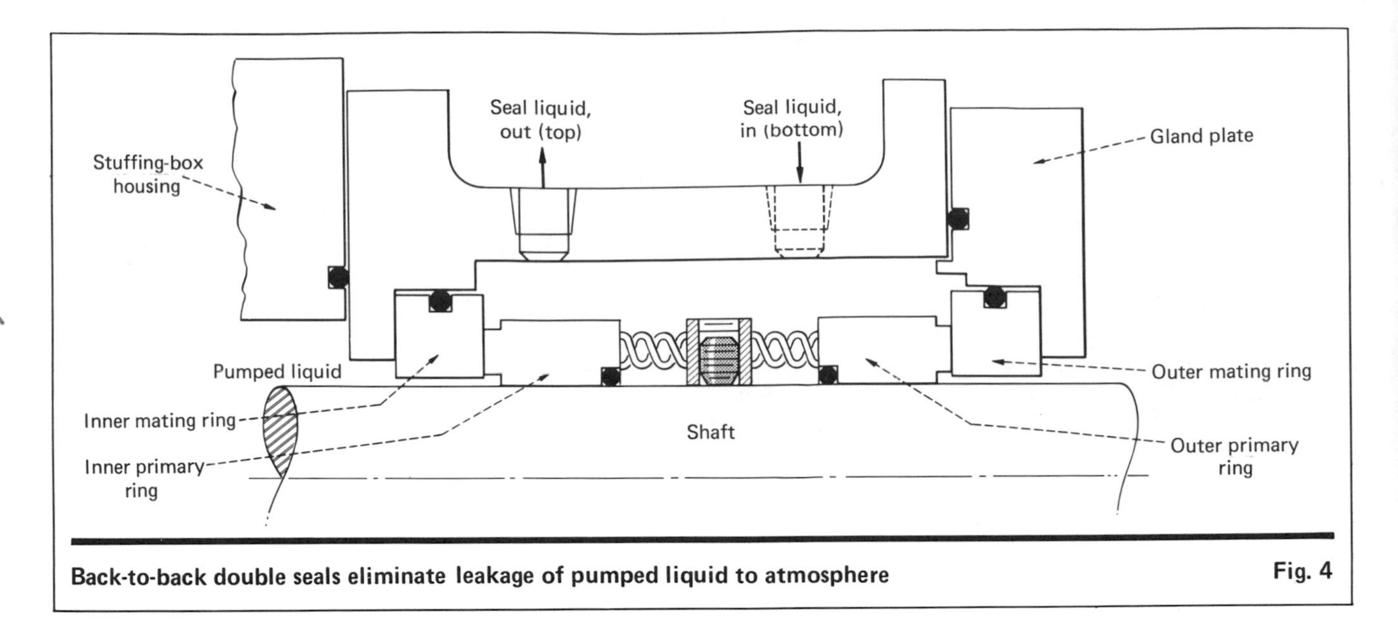


**Back-to-back double seals eliminate leakage of pumped liquid to atmosphere** **Fig. 4**

and the frequency of changing will become much lower.

Filters should be installed in pairs so that one filter can be in operation while the second is being cleaned and then put in standby service.

Filter elements can be made from various materials. Care must be used in selecting an element that is compatible with the liquid stream to be filtered.

*Cyclone separators*—These devices are particularly suited for a once-through pumping system, in which a filter would rapidly become clogged while removing abrasives and solids from a bypass flush stream. Bypass liquid enters the cyclone tangentially, near the top. The heavier particles are thrown to the outside of the cyclone and drain down to the bottom. The lighter, clear liquid moves inward and upward toward the top and flows to the seal's flush connection on the stuffing box. Solids and some liquid return to the pump suction from the bottom of the separator.

The efficiency of the cyclone separator depends on the particle size of the solids, solids concentration, relative specific gravity of the solids and liquids, and pressure drop across the separator. Efficiency of the unit increases with larger particle size, higher solution concentration, larger difference in specific gravity, and higher pressure drop across the cyclone.

## Temperature control

Mechanical seals are designed for operation up to 750°F (400°C). Seals are also available for higher temperatures. However, the cooler the flush liquid can be maintained, the longer will be the life of the seal, with fewer maintenance problems. Several methods are available to control the temperature in the stuffing-box.

Most pumps are furnished with, or can be supplied with, stuffing-box jackets or a region surrounding the stuffing box in which cooling water can circulate. This method provides some reduction of temperature. Also, the stationary face of the seal can be drilled to allow the circulation of cooling water. This is more effective for removing the heat generated by the rubbing faces of the seal. However, if the stationary seal ring is made of carbon, heat removal is small and this method is not very effective.

The best method is to use a heat exchanger in the bypass flush system. With the heat exchanger, the flush stream can be cooled directly before injection into the stuffing box. A temperature of less than 200°F is preferred for the flush liquid.

A closed system, consisting of a pumping ring and a heat exchanger in a closed piping loop, can be used on single or double mechanical seals. The pumping ring is a slotted ring (mounted on the shaft between the seals) that rotates and serves as a low-capacity, low-head pump. These rings develop enough head to circulate the seal liquid from the stuffing box through the heat exchanger and back to the stuffing box. The heat exchanger can be air cooled or liquid cooled. The pumping ring is often preferred over a bypass flush system having a heat exchanger, because it may allow the use of a small exchanger.

The flush line must be as short as possible, and unnecessary turns eliminated, to keep friction losses to a minimum. The capacity and head of the pumping ring are proportional to the peripheral speed of the ring. A minimum speed of 800 ft/min (4 m/s) is usual.

To use a closed system with a double mechanical seal, a means of maintaining the loop pressure above the stuffing-box pressure is required in order to prevent leakage of the pumped fluid out of the inner seal. One way of maintaining the seal-loop pressure is to sense the pump-suction or pump-discharge pressure and maintain a fixed differential above this pressure by means of a spring or static head.

Not all temperature control is for cooling. When pumping materials such as heat-transfer liquids, heavy oils, etc., that have melting points well above ambient temperatures, a means of heating the stuffing box is required to prevent the material from crystallizing and/or solidifying. Stuffing-box jackets can be used as steam jackets. Steam-heated gland plates are available. If steam is not available or not hot enough, electric heat-tracing can be applied.

## Function of the gland plate

The gland plate is an important part of the mechanical seal because the stationary ring is mounted to it. It is also the part that is bolted to the stuffing box and forms a section of the pressure-containing housing through which the shaft protrudes.

In addition to providing a means of mounting the stationary ring, the gland plate also allows the installation of some safety features. A mechanical seal is just that—a mechanical device to prevent leakage. As such, it is subject to failure, and when it fails, leakage occurs. We must recognize this risk and determine whether the leakage is a danger to equipment, personnel or the environment. When such danger exists, a means of controlling the leakage must be provided.

We previously discussed two options (double seals and tandem seals) for containing the leakage in the event of seal failure. The third option is to provide a means of containing and collecting the leakage and directing it to a safe location. This is done by using a gland plate that maintains a close clearance at the shaft and provides a means of venting, draining or quenching.

Fig. 6a shows a gland plate with a vent or quench connection, a drain connection, and a fixed throttle bushing. The fixed throttle bushing is similar to a fixed throat bushing and requires a relatively large clearance to prevent rubbing on the shaft. If leakage must be further reduced, a floating throttle bushing (Fig. 6b) can be installed. Leakage rates across either of these devices are determined as those for throat bushings.

A more-positive means of preventing leakage along the shaft is an auxiliary stuffing box, as shown in Fig. 6c. The auxiliary stuffing box usually consists of one or more rings of packing with a packing gland for a retainer. It is necessary to circulate a fluid such as water through the vent and drain connections to provide lubrication for the packing.

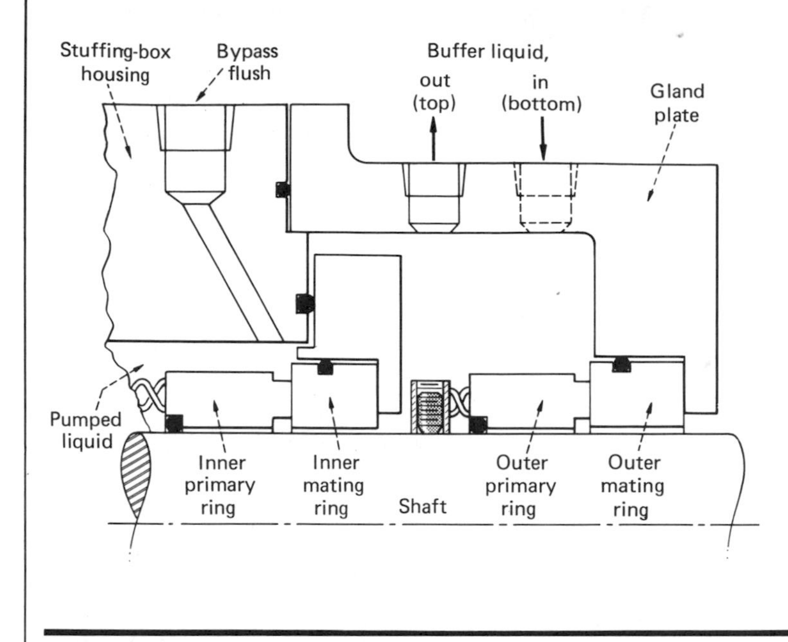


**Buffer liquid is at lower pressure than stuffing box in tandem-seal arrangement** **Fig. 5**

The gland plate can also provide a flushing connection close to the primary seal faces. Most pumps, especially those designed to use packing in the stuffing box, have a connection in the stuffing box that lubricates the packing. This connection can be used as the flushing connection for a mechanical seal. However, because the connection is quite often between the seal faces and the

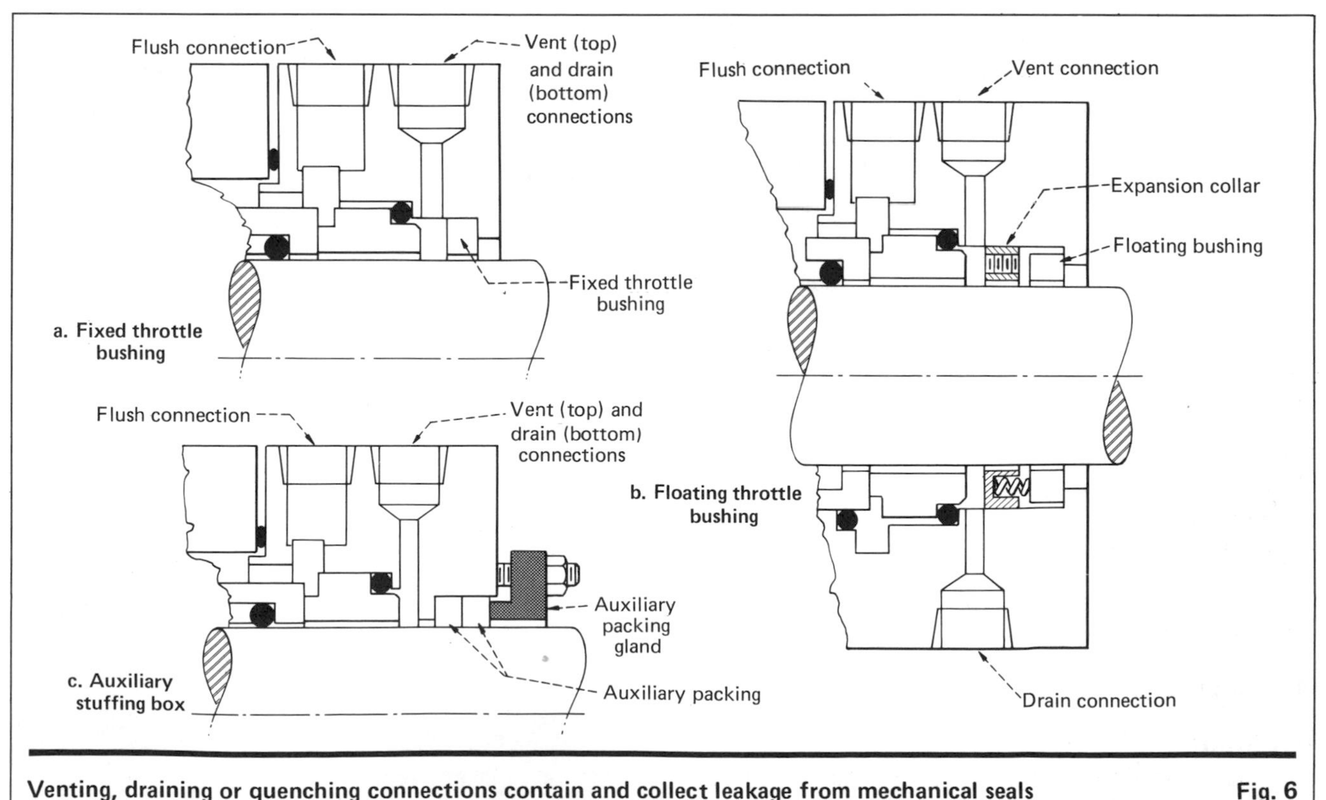


**Venting, draining or quenching connections contain and collect leakage from mechanical seals** **Fig. 6**

pump itself, it allows a stagnant region in the stuffing box (outboard of the connection) that can decrease heat removal from the seal faces and allow the accumulation of solids or dirt.

The preferred injection point for the flush liquid is just outboard of the primary seal faces, so that the liquid will flow past the seal faces and back into the pump through the throat of the stuffing box. Fig. 3 shows a gland plate with flushing connection that will allow the flush liquid to flow past the seal faces.

## Materials of construction

Selection of the proper materials for the operating environment is extremely important in order to assure a long operating life for the seal. Seal design, operating environment and lubricity must be considered.

The type of seal chosen will be affected by the fluid being contained. For instance, if fluorocarbon secondary seals are required or preferred for resistance to the pumped liquid, a seal design using V rings or U cups is necessary to overcome the cold-flow property of a pure fluorocarbon resin. If a filled fluorocarbon is used, e.g., glass-filled Teflon, then a seal design using O rings is suitable.

Temperature and pressure also influence seal design. Materials for the primary and secondary seals, springs and gland plate are determined by temperature and corrosiveness as well as compatibility with the liquid.

Earlier, we stated that a mechanical seal is similar to a bearing with a liquid film between the faces. If the liquid is not a lubricant such as a light hydrocarbon, then self-lubricating seal faces are required.

The majority of material combinations for seal rings will use carbon/graphite as one of the seal faces. This material is chosen because it has good wearing characteristics, is softer than other materials, and is compatible with a wide range of temperatures and corrosive environments. Some other materials commonly used for seal faces are Stellite, tungsten carbide, stainless steel, ceramics and Ni-resist. These have maximum temperature limits ranging from 350°F (177°C) to 750°F (400°C).

To satisfy corrosive environments, gland rings, springs and bellows are available in a variety of materials such as stainless steels, Monel and Hastelloys.

Secondary seal materials range from buna N and neoprene through the fluorocarbon resins to graphite. Each material has its own operating temperature limitations that range from −320°F (−196°C) to 800°F (427°C).

Each manufacturer of mechanical seals has developed selection charts for common liquids. These charts recommend the type of seal and materials for the majority of liquids being handled.

Because of the wide variation in materials and designs, seals are available for services that range in temperature from −350°F (−212°C) to 750°F (400°C), and in pressure from subatmospheric to 2,500 psig (17,238 kPa).

## Installation and operation

Proper installation of mechanical seals is critical. Axial movement of the shaft should be less than 0.004 in. (0.1 mm). Excessive axial movement can cause wear or fretting of the shaft or sleeve at the point of contact with the secondary seal. It can also cause overloading or underloading of springs, as well as chattering, with resulting seal failure. Shaft deflection of more than 0.003 in. (0.08 mm) can cause wear of the seal faces and wear on the shaft at the point of secondary-seal contact. Face squareness of the stuffing box and concentricity of the bore should also be checked. Installation instructions for the mechanical seals should be followed carefully to assure trouble-free operation.

The most critical time for a seal is when the pump is first started. Generally, the pump is flooded but the seal faces can run dry for a short period until stable operation is attained. Startup is the time when solids may settle out and cause damage to the seals. This is also the time when a pump is likely to run near shutoff, causing excessive heating of the pumped liquid, and when unstable operating conditions can occur. Under these conditions, damage to the seals can take place.

If a seal has been properly selected and installed, its operating life can be up to two years after the pump has been started, assuming that the seal is operating satisfactorily. Some of the problems that can occur with mechanical seals:

1. Loss of film between the seal faces, which can cause heat-checking of the harder seal face, or explosion of the carbon ring.

2. Wear of seal faces, caused by abrasive materials in the liquid or a liquid that tends to crystallize between the faces.

3. Distortion of the stationary seal ring, caused by overtightening of the gland bolts.

The mechanical seal allows the pumping of liquids that, for environmental considerations, cannot be handled by a pump whose shaft is sealed with packing. While the investment for mechanical seals is higher than for packing, they require less attention and have lower maintenance costs, which more than offsets the additional investment.

## References

1. Boyce, M. P., How to Achieve Online Availability of Centrifugal Compressors, *Chem. Eng.*, June 5, 1978, pp. 122–125.
2. Ramsey, W. D., and Zoller, G. C., How the Design of Shafts, Seals and Impellers Affects Agitator Performance, *Chem. Eng.*, Aug. 30, 1976, pp. 105–108.
3. Karassik, I. J., Krutzsch, W. C., and Messina, J. P., eds., "Pump Handbook," pp. 2-82 to 2-89, McGraw-Hill, New York, 1976.
4. API Standard 610, "Centrifugal Pumps for General Refinery Services," 5th ed., American Petroleum Institute, Washington, D.C., March 1971.

## The author

**John H. Ramsden** is chief engineer for rotating equipment at Badger America, Inc., One Broadway, Cambridge, MA 02142. For over 20 years, he has been responsible for the selection and application of pumps, compressors, expanders and drivers for a variety of projects for the chemical process industries. He has a B.S. in chemical engineering from Tufts University and an M.B.A. from Northeastern University, and is a registered professional engineer in Massachusetts.

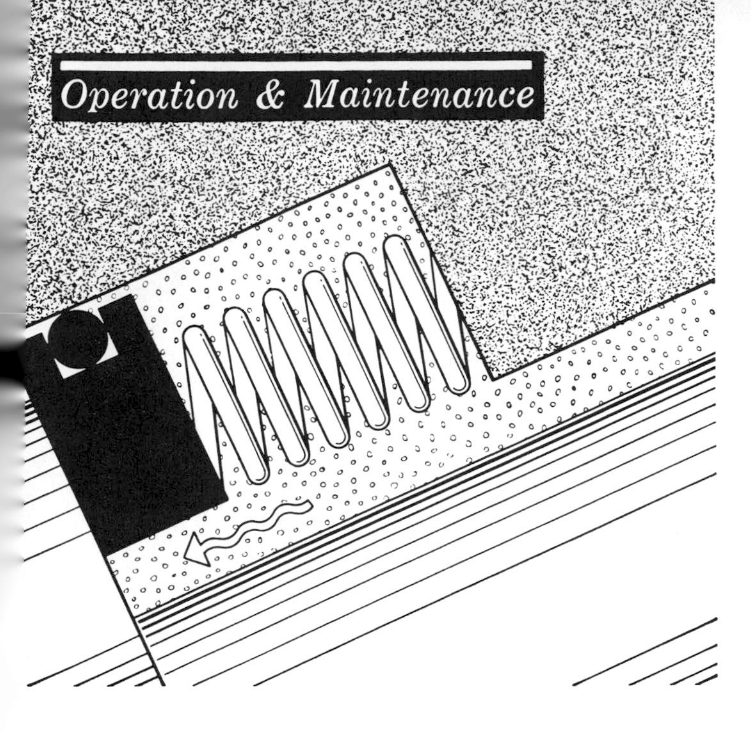


# Mechanical Seals: Longer Runs, Less Maintenance

A thin, clean, liquid film between the mating faces of mechanical seals can be the dividing line between long, trouble-free operations and high maintenance costs. Here are sealing-liquid arrangements for a variety of pump conditions.

ALEXANDER A. SAMOILOFF, *The Badger Co.*

A mechanical-seal assembly consists of three seals, generally classified as primary and secondary. The primary is a single seal; the secondary, two seals (Fig. 1).

Two basic components of a mechanical seal are the stationary and rotating rings. The first is fastened to the stuffing box and does not move; the second to the shaft or shaft sleeve, with which it rotates. The contacting surfaces of these two rings are called "mating surfaces," and these comprise the primary seal. One of the rings is commonly carbon.

Mating surfaces, flat and highly polished, are compressed against each other by the hydraulic pressure of the sealing liquid plus the force of mechanical springs or bellows.

## Primary Seal's Dependence on Sealing Liquid

The primary seal is the most vital part of the seal assembly. For it to operate satisfactorily, its mating faces must at all times be parallel to each other, and between these faces there must always be a film of sealing liquid, which must also be clean.

To keep mating faces parallel, flexibility must be provided between the rotating ring and the shaft and between the stationary ring and the stuffing box to absorb shaft movements relative to the box. This function is performed by the secondary seals, which in addition prevent leakage from the box (Fig. 1).

Secondary seals are made of a variety of flexible materials (such as synthetic rubbers, Teflon and neoprene), which are selected on the basis of resistance to heat or to corrosive liquids, or both, and are shaped into O, V and wedge rings, and U cups. Bellows constructed of thin metal, being extremely flexible, are sometimes used as secondary seals at the shaft.

The maximum temperature that many secondary-seal materials can withstand is often considerably lower than operating temperatures in chemical processes. For this reason, seal materials should be chosen with care. If necessary, the temperature in the stuffing box at the secondary seals should be suitably lowered by means of cooling jackets, external heat exchangers or other devices.

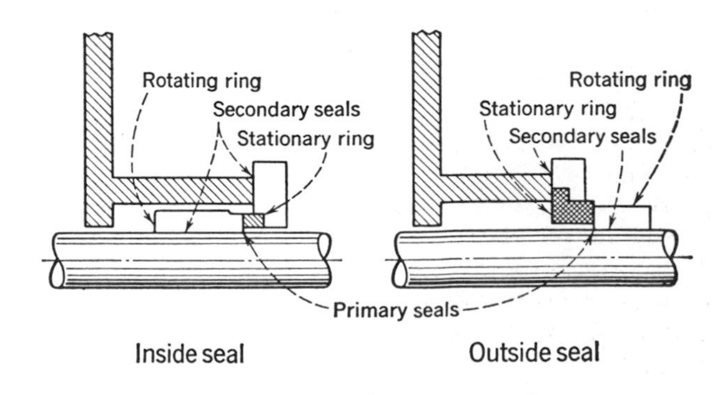


**THREE** sealing points of mechanical seals—Fig. 1

Originally published January 29, 1968

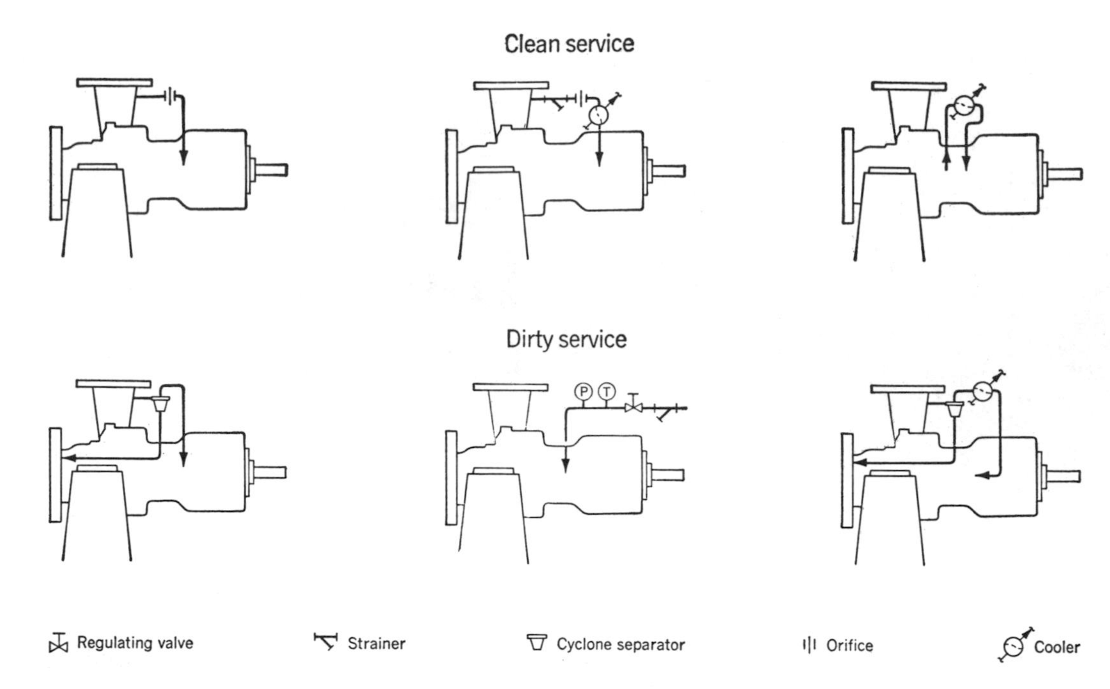


**SEALING LIQUID,** upper right, is circulated by pumping ring; lower middle, lip seal keeps sealing-liquid pressure up—Fig. 2

## Properties of Sealing Liquid Must Be Suitable

How well a clean film can be maintained between the mating surfaces of a mechanical seal depends on the physical properties of the sealing liquid.

If the properties of the liquid being pumped are suitable, it can serve as the sealing liquid. It can be delivered to the stuffing box directly from the pump casing through the annular space around the shaft or through an outside line from a point of high pressure in the casing (Fig. 2). If the pumped liquid cannot be used as the sealing liquid, a liquid from an outside source has to be introduced at a proper pressure into the stuffing box.

When analyzing pump operating conditions for seal application, the vapor pressure of the liquid at its operating temperature must be considered. Net positive suction head (NPSH) requirements for the satisfactory operation of a centrifugal pump are similar to those for a mechanical seal because neither can tolerate cavitation. To assure pump operation free of cavitation, the current practice (based mostly on field experience and experimental research, supplemented by limited theory) is to select a pump whose required NPSH is lower than, or equal to, that available for specific operating conditions.

Liquid in a stuffing box vaporizes when the heat generated by mating-surface friction is not adequately removed and the stuffing box pressure is not sufficiently high. Such a condition damages mating surfaces because the liquid film between the rotating faces is replaced partially by vapor.

Unfortunately, no rules—such as NPSH requirements for pumps—have been formulated for mechanical seals. However, it can be stated that, in general, when a pump's available NPSH is close to that which is required, its stuffing box pressure under operating conditions should be checked. If this pressure is not considerably higher than the suction pressure, corrective measures should be applied.

When a pump operates under vacuum suction, it is essential that its stuffing box pressure be checked, because a seal cannot operate when the pressure is below atmospheric. The most common method of improving a seal's operating conditions is to flush (circulate) liquid from the pump casing at high pressure through the stuffing box and back into the casing. To maintain the higher pressure in the box, a restrictive throat bushing should be provided at the outlet.

Seal types, modifications and accessories, it should be emphasized, are intended primarily to maintain the liquid film between mating faces and to keep the film clean. Water, of course, is one of the most difficult liquids to handle with centrifugal pumps. At low temperatures, NPSH requirements are extremely critical, because then water vapor has a very large specific volume, which condition leads to severe cavitation. It is for this reason that water has been adopted as a standard medium for testing NPSH. At temperatures above 160 F., water as a sealant presents great difficulties because of the crystallization of salts and minerals, which cause mating faces to wear rapidly.

A change in the operating conditions of a centrifugal pump should prompt an evaluation of the pump's characteristics and the physical properties of the liquid being pumped. A consultation with the seal manufacturer is also recommended.

Inside seals (which are most common) have the stationary ring installed with the mating surface facing the pump impeller and the rotating seal ring mounted on the shaft inside the stuffing box. Springs and other components are submerged in liquid (Fig. 1).

Outside seals have the stationary ring installed with the mating surface facing away from the impeller and the rotating ring outside the box. Springs and other components are exposed to the atmosphere (Fig. 1).

## Balancing Inside and Outside Seals

Both inside and outside seals may be balanced or unbalanced. Unbalanced seals are for low and moderate stuffing-box operating pressures, balanced seals for high pressures. API Standard 610 for centrifugal pumps recommends balanced seals for pressures above 75 psig. However, unbalanced seals are available, and are commonly used, for pressures much higher than this limit.

*Inside-seal balancing*—The liquid film between the mating surfaces of the inside seal is subjected to two opposing forces, a closing and an opening force. The closing force, which compresses the film, consists of two components: a spring force and a hydraulic force. The hydraulic force is the stuffing box pressure of uniform intensity applied to the rotating ring. The opening force, a hydraulic force that separates the mating surfaces, is variable, changing from a maximum at the outer circumference of the surfaces to atmospheric pressure at the inside.

Experiments and field experience have established that when stuffing box pressure is raised, a pressure is reached at which the liquid film between the mating surfaces cannot be maintained, and the seal ceases to function. The rotational speed of the mating faces, the face materials and the sealing liquid's physical properties (viscosity and lubricity) determine this limiting pressure for unbalanced seals.

When an inside seal is adapted for high-pressure service, the magnitude of the hydraulic component of the closing force is lessened by reducing the area exposed to hydraulic pressure.

*Outside-seal balancing*—In an outside, unbalanced seal, the forces exerted by stuffing box pressure on the mating faces are different from those in an inside seal. Only a spring supplies closing force; the hydraulic component is absent.

These seals are used only for low operating pressures. When adapted for high-pressure service, mechanical modifications create the hydraulic component of the closing force. The operating pressure limit of a balanced, outside seal is considerably lower than that of an inside seal.

## Double Seals and Lip Seals

A single seal provides sealing between the liquid in the stuffing box and the atmosphere. It can be inside or outside, balanced or unbalanced.

A double seal consists of two mechanical seals. Double seals are used when it is necessary or advantageous to lubricate the seal faces with a liquid from an outside source. Such an arrangement is necessary when the pumped liquid contains abrasive solids, when even the slightest quantity of it cannot be allowed to escape into the atmosphere, or when an outside liquid cannot be tolerated in the process stream. One seal functions as a barrier between the liquid in the casing and the sealing liquid in the stuffing box; the other, as a barrier between the liquid in the stuffing box and the atmosphere.

Sealing liquid from an outside source is introduced into the stuffing box at a proper temperature and pressure and returned to its source for recirculation, or it is run to a drain. Whether recirculated or wasted, it is important that a restrictive orifice or valve, at the sealing-liquid's outlet, control the pressure in the stuffing box above that of the pumped liquid. However, the sealing liquid's temperature must be low enough and its flow high enough to adequately remove the heat generated by the two seals.

A lip seal is used in conjunction with a single mechanical seal when the liquid being pumped is not suitable for flushing, and flushing is done from outside.

Lip seals, usually made of a flexible material such as Teflon, act like check valves. They permit liquid to flow only from the stuffing box into the pump casing. Because they are not primarily flow controllers, a rotometer is usually installed in the flushing stream.

The rate of flushing must be adjusted between two limits: (1) the maximum dilution rate, i.e., the maximum injection rate of outside liquid that the process stream can tolerate; and (2) the minimum acceptable flow for flushing and cooling. Because the rate of flow is usually small, it should be directed at the mating surfaces (Fig. 3).

Minimum flow-rates to remove only the heat generated by the seal faces are presented in Fig. 4 (courtesy of Crane Co.). Graphs I and II are based on water as the cooling liquid, a 2-in. shaft rotating at 1,750 rpm., a cooling liquid inlet temperature of 100 F., and a cooling liquid temperature rise of 20 F. To find cooling flow-rates for other than 2-in. dia. shafts, multiply the flow-rate from Graph I or II by the factor from Graph III; for shaft speeds other than 1,750 rpm., multiply the flow-rate by the ratio: new speed rpm./1,750 rpm.; and for cooling liquids other than water, multiply the flow-rate by the ratio: 1/(new specific heat) (new specific gravity).

## Circulating Ring for Seal Cooling

A circulating ring is used in high-temperature applications, when the pumped liquid is clean and suitable as a sealing liquid. Placed on the pump shaft, it forms a pump inside the stuffing box. It circulates the liquid from the stuffing box through a heat exchanger and back to the box. A close-fitting throat bushing in the stuffing box restricts the mixing of hot liquid in the casing and cooled liquid in the box. (See third pump for clean service in Fig. 2.)

A minimum shaft peripheral speed of 800 ft./min. for delivery of 0.5 gpm. is recommended by one manufacturer of circulating rings. Specially-designed heat exchangers, in which water is circulated under pressure, are available for this application.

Exchanger capacity should be selected with care. If required by operating conditions, circulating-ring cooling can be supplemented by circulating cooling water through the stuffing-box jacket. Water-circulating rates should be specified by the pump manufacturer. A new heat exchanger that operates under atmospheric pressure on the water side is shown in Fig. 5.

## Water Jacketing and Cyclone Separators

Circulating cooling water through a pump water jacket is an effective and economical method of lowering stuffing box temperature. It is used when the pumped liquid is clean and suitable for sealing, and flushing is not necessary. It is mandatory (as in circulating-ring applications) that the stuffing box be dead-ended to maintain effective cooling. (Dead-ending is accomplished by installing a close-fitting throat bushing at the bottom of the box.)

The application of cyclone separators to remove abrasive particles from seal flushing streams is comparatively new. Liquid from the pump discharge enters the cyclone tangentially. Clean liquid from the center of the cone is piped to the stuffing box for flushing. The liquid containing the abrasive particles is returned to the pump's suction. Flows from the two outlet streams are about equal. Separation efficiency for particles as small as 2.5 microns is about 87%, and higher for larger particles. Capacities, based on the clean stream, range from 1 to 3½ gpm., at pressure drops from 20 to 110 psi.

## Functions for Throat and Throttle Bushings

A throat bushing, placed at the bottom of the stuffing box, restricts the flow of liquid between the pump

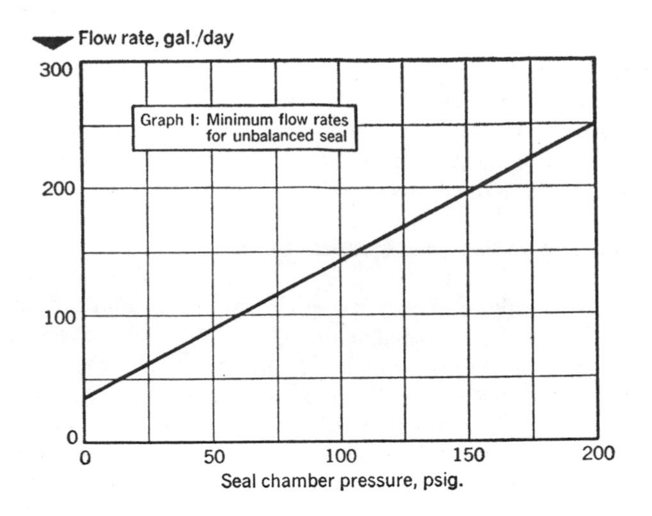


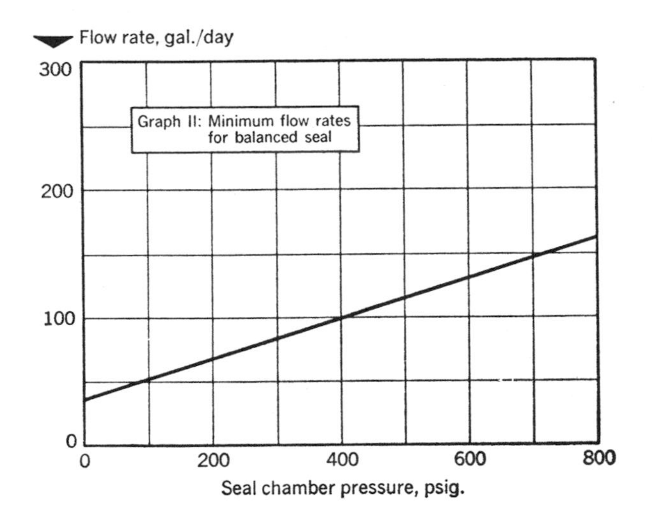


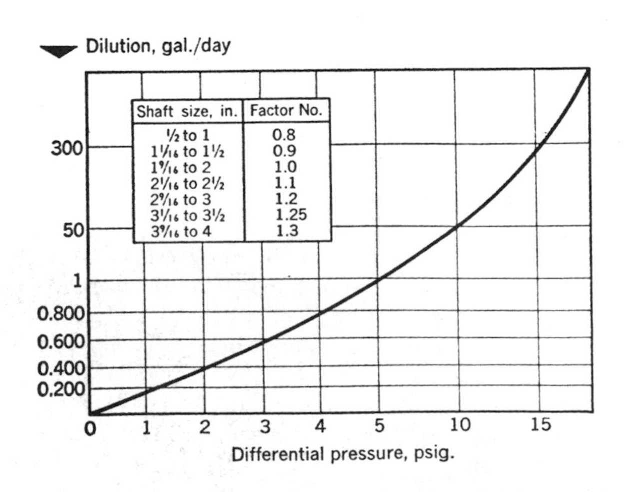


| Shaft size, in. | Factor No. |
|---|---|
| ½ to 1 | 0.8 |
| 1 1/16 to 1½ | 0.9 |
| 1 9/16 to 2 | 1.0 |
| 2 1/16 to 2½ | 1.1 |
| 2 9/16 to 3 | 1.2 |
| 3 1/16 to 3½ | 1.25 |
| 3 9/16 to 4 | 1.3 |

**DILUTION RATES** for 2-in. shaft: for other shafts multiply gal./day by the Factor No. (Durametallic Corp.)—Fig. 3

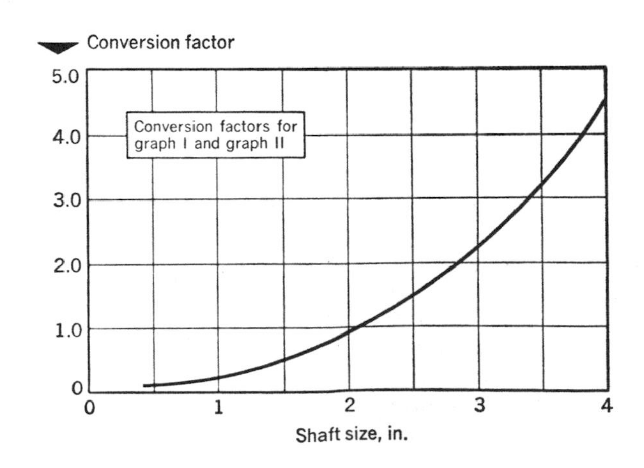


**FLOW RATES** estimated for cooling seals: Graphs I and II are based on water as the cooling fluid, entering at a temperature of 100 F. and exiting at 120 F. (20° rise), and on a 2-in. shaft rotating at 1,750 rpm. (Crane Co.)—Fig. 4

casing and the stuffing box. It keeps the liquid in the stuffing box from mixing with that in the pump casing when the stuffing box is cooled or heated by cold or hot liquids circulating through the water jacket. It also maintains the required pressure in the stuffing box when pumped liquid is discharged from the casing and circulated through the stuffing box for flushing and cooling the mechanical seal.

A throttle bushing, placed at the outside end of a gland plate (also called a follower plate), restricts the escape of liquid from the stuffing box when the seal leaks excessively or fails completely. This application is limited to inside seals.

## Vacuum and Crystallization Problems

When the pressure in the stuffing box is below atmospheric pressure, it cannot force the sealing liquid to form a film between the seal's mating surfaces. One way to overcome this condition is to bring in a flushing stream from the pump discharge. Another is to supply a lubricating liquid at an adequate pressure from an outside source to the mating surfaces, through a series of specially drilled holes in the stationary ring, which are hydraulically connected to an annular groove in the gland. Grease or circulating oil, under pressure, are used as lubricants.

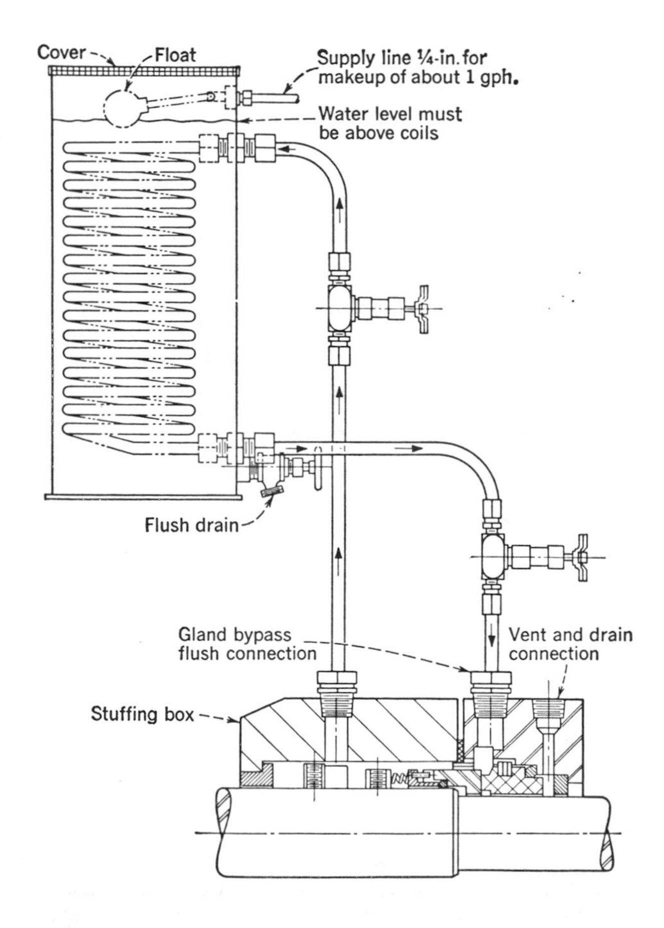


**WATER** moved by circulating ring (Durametallic)—Fig. 5

When handling liquids whose crystallization temperatures are above ambient, a means should be provided for preventing crystal formation at sensitive parts of the seal, mating surfaces and springs. This can be done one of the following ways: (1) a double mechanical seal, with an outside flushing liquid, would preclude contact by the pumped liquid with all parts of the seal, except for limited contact with the mating surfaces of the inner seal; (2) with a single seal, steam could be supplied to the pump's water jacket when the pump was shut down; and (3) a single seal could be flushed with an outside liquid if dilution could be tolerated, in which case a restrictive throat bushing should be provided at the bottom of the box.

## Thermal Aspects of Mechanical Seals

Vaporization (also called boiling or flashing) occurs at the seal faces when the heat generated is not removed adequately. It results in leakage that cannot be stopped, or in popping or puffing at the seal faces, with intermittent leakage. Popping is caused by local vaporization, which parts the seal faces slightly. After leakage cools the faces, they close again. This cycling tends to wear the carbon face; it pits it and chips its edges.

Heat is dissipated in mechanical seals by a combination of conduction and convection. Heat generated at the faces is carried by conduction through the seal body, and then by convection to the surrounding medium.

Because natural convection removes heat inadequately from stationary seals, forced convection is introduced by flowing a fluid around the stationary ring. The rate of heat removal by convection from rotary seals depends mainly on the speed of rotation, and not on the flow of liquid around the seal ring. The rotary ring's heat-transfer coefficient is higher than that of the stationary ring, even with forced convection. Heat removal by conduction through the seal rings depends on the thermal conductivity of the seal materials.

In mechanical seals, the rings are made of different materials, with the rotary seal ring having a higher convection heat-transfer coefficient than the stationary ring. To obtain a higher over-all heat-transfer coefficient for the entire assembly, the material having the higher coefficient of thermal conductivity should be used for the rotary seal, in order to satisfy heat removal requirements.

## Meet the Author

Alexander A. Samoiloff is a pump and compressor consultant with The Badger Co. (363 Third St., Cambridge, Mass. 02142). Formerly, he was a mechanical engineer with E. B. Badger & Sons Co.

He holds a degree from Harvard Engineering School, and is a member of ASME and Soc. of Harvard Engineers and Scientists. He has taught technical Russian at M.I.T. A registered professional engineer in Massachusetts, he is the author of several technical articles in the field of mechanical engineering.

# How to select and use mechanical packings

When properly specified and installed on rotating shafts, packings can isolate a fluid contained in process equipment from the environment.

***Richard Hoyle,*** *A. W. Chesterton Co.*

☐ New mechanical packings enable the production of a wide range of products that can seal better, last longer and reduce wear on equipment.

Though the current trend is to a totally sealed plant using mechanical or end-face seals, mechanical packings provide a viable alternative to seals for a wide variety of services. Let us review the technology of the mechanical packings and, as appropriate, discuss the relative merits of packings and seals.

The main value of packings is flexibility of application and ease of installation into a stuffing box, as shown in Fig. 1a. Packings also avoid the catastrophic failure possible with mechanical seals.

Packings operate on the principle of controlled leakage in dynamic applications. They never attempt to totally prevent fluid from leaking from the equipment. Rather, they allow a controlled amount of leakage, and we will discuss this in detail. On the other hand, mechanical seals attempt to completely stop leakage. In order to be quite accurate, however, we must define leakage.

## Leakage at packings and seals

A mechanical seal (Fig. 1b) is a device that transfers the wear from the shaft or sleeve of the equipment to the integral parts of the seal called seal faces or wear faces. If these faces are flat enough and smooth enough, they will essentially stop leakage from entering the atmosphere. If we define leakage as being a visible fluid, then we can state that mechanical seals stop leakage completely. Furthermore, they generally leak only trace amounts of vapor during the entire operation.

From a technical viewpoint, mechanical seals continually leak, but a year's leakage of a successful mechanical seal in water service, for example, would not yield a full cup. On the other hand, packing, leaking at 60 drops/min, would produce 15 cups/d. This seems like an excessive amount of leakage. However, for a pump handling a modest capacity of 300 gpm, the leakage rate is 0.00026%. In essence, the control, rather than elimination, of leakage becomes the basic purpose for using mechanical packings.

Mechanical seals can be said to prevent leakage because their leakage is considered to be inconsequential. However, seal leakage can occasionally be substantial and, just as important, can be uncontrollable in the event of a seal failure, requiring the equipment to be removed from service at an inopportune or unscheduled time.

## Types of mechanical packings

The terms soft packings, jam packings, compression packings and braided packings are used to describe portions or all of the ranges for mechanical packings. Other descriptions, such as metallic or plastic packings, refer to specific products within the general category of such packings.

Most mechanical packings are designed for rotating

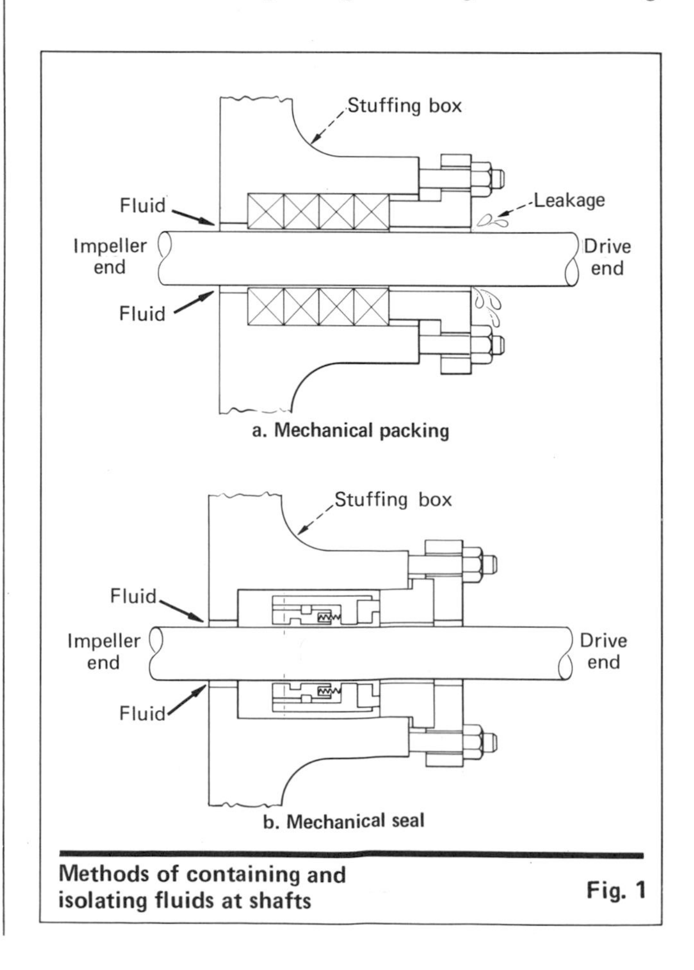


**Methods of containing and isolating fluids at shafts** **Fig. 1**

Originally published October 9, 1978

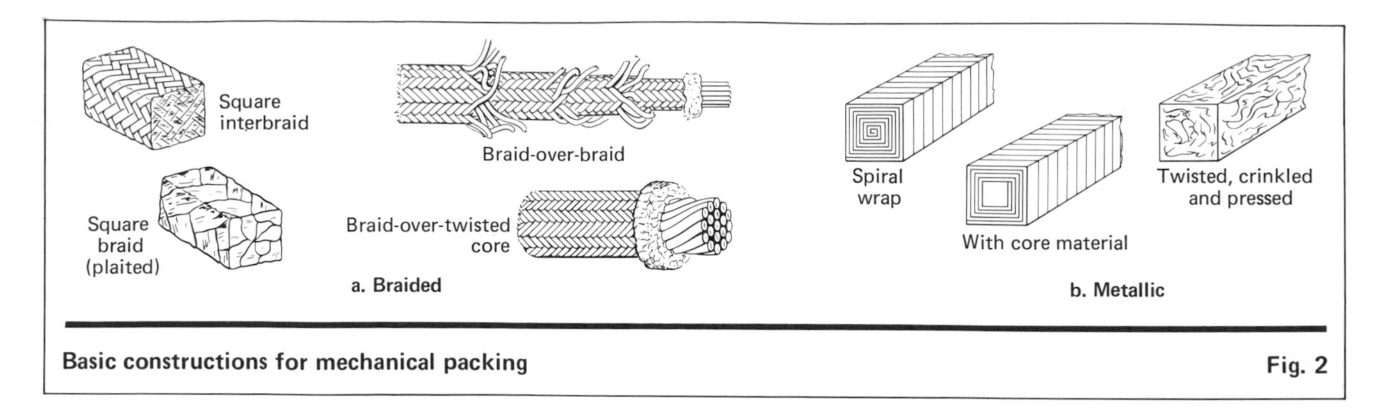


**Basic constructions for mechanical packing** **Fig. 2**

equipment. These same packings also find general use in valves and in other applications, including door gaskets, mixers, expansion joints and reciprocating pumps. When a packing is used on pumps, it must leak. When used for valves, expansion joints and door gaskets, it generally shows no leakage.

In this article, we will limit our discussion to mechanical packings as used in pumps, with an occasional reference made to valve services. Gaskets and packings of the automatic type will not be discussed nor will hydraulic packings.

The four types of braided packings are the square-interbraid, square-plaited, braid-over-braid, and braid-over-core types. Of these, the square-interbraid and the square-plaited are most popular. The definition of the braid represents the different types of machines on which the packing is manufactured (Fig. 2a).

Interbraid packing is made on a machine commonly called a lattice braider. The yarns are formed diagonally through the packing, as shown in Fig. 2a. This is the best packing for retaining a square shape and for controlling manufacturing tolerances. The square braid or plaited braid also generally retains its square shape but is usually a soft and absorbent braided structure that can carry a large amount of lubricant. The third type, braid-over-braid, is braided in a round form, and then put through a squaring press or calender to form it into the square shape. The fourth type of braided packing is braid-over-core, which is also braided in the round form and converted to the square shape by calendering. The basic materials of construction for these four packings are animal, vegetable, mineral and various synthetic fibers. These will be discussed in greater detail.

Metallic packings are made of lead or babbitt, copper or aluminum, and are either spirally wrapped or made in a folded, twisted construction. Other materials can be used but the preceding are in the majority. Metallic packings frequently have a core of a resilient or compressible absorbent material and have other lubricants present (Fig. 2b). The core is usually a synthetic rubber cord or asbestos roving. Metallic packings are made for their strength, nonabsorbency, heat resistance, or any combination of these.

Plastic packings are available in a homogeneous construction or occasionally formed over a core. Frequently, they have an asbestos or other jacket braided over them to help retain their shape. These packings are generally made from base materials of asbestos fiber, graphite and mica; and oil or grease; and with a number of other products possibly added to yield a finished product with the desired properties.

Rubber-and-duck and rubber-and-asbestos fabric packings are two other types. Rubber-and-duck packings are laminated plies of a fabric duck (usually cotton) that is treated with an uncured rubber compound. The curing produces a finished fabric packing of desired shape, size and strength. The resulting product is impregnated with lubricants (both dry or solid, and wet types). Asbestos-fabric packings are very similar to rubber-and-duck packings. Both types are also used as "bull" rings and for slow-speed pumps handling high-viscosity fluids. In such services, these packings are often wire-reinforced to add strength.

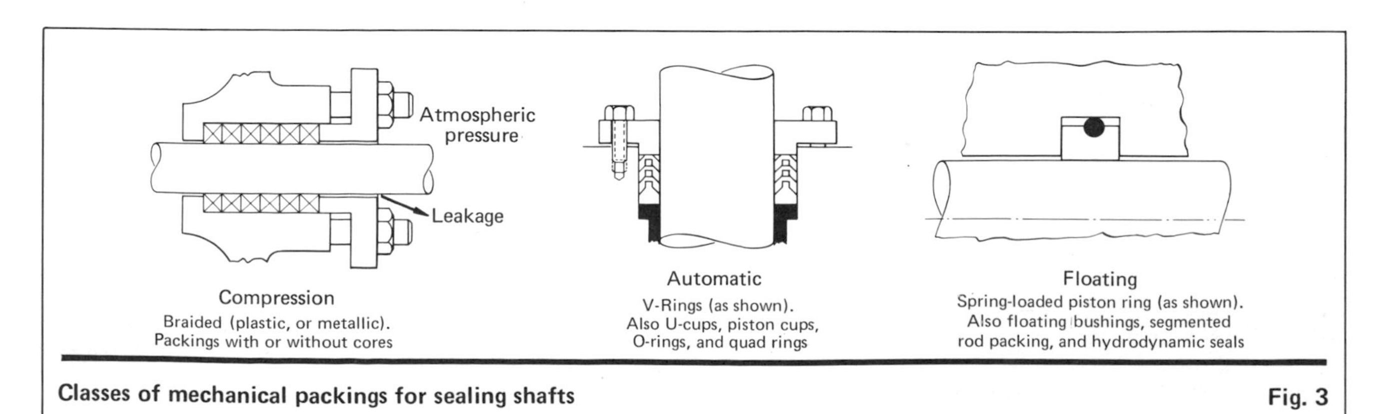


**Classes of mechanical packings for sealing shafts** **Fig. 3**

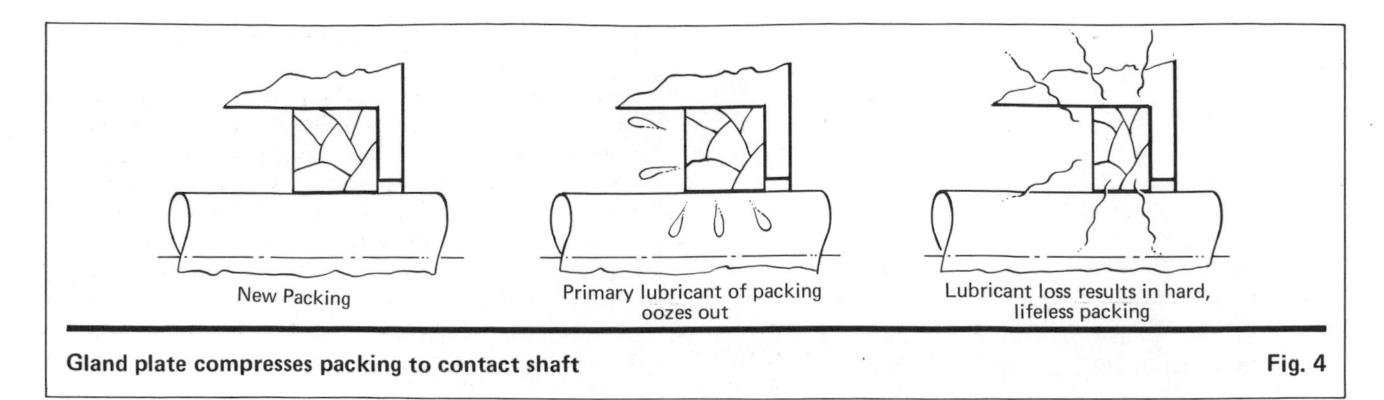
New Packing
Primary lubricant of packing oozes out
Lubricant loss results in hard, lifeless packing

Gland plate compresses packing to contact shaft — Fig. 4

The lubricants for mechanical packings are solid or dry types and liquid ones. Examples of the solid or dry lubricants are tetrafluoroethylene (TFE), graphite, mica and molybdenum disulfide. Examples of liquid lubricants include oils (both refined and synthetic), greases, animal fats and various waxes. Some packings contain their own lubricant, the best example being the graphitic-type packing.

## Classes of packings

Mechanical packings can be divided into three broad classes, which are compression or jam-type, automatic and floating packings. Examples of each are shown in Fig. 3.

Compression packing uses the force created by the gland plate or end plate to contact the shaft. Under these conditions, the lubricant eases the shaft contact and is gradually lost over a period of time. With total loss of the lubricant, the packing becomes lifeless and is ready for replacement (Fig. 4).

Automatic packing has a construction so that shaft contact is either not dependent on gland compression or is only dependent on an initial gland compression. Such packing is generally installed so that pressure helps to activate the sealing forces. Any automatic packing of lip construction generally seals in only one direction, and is most often used in reciprocating services.

An example of a floating packing is a piston ring. Any form of segmented packing that operates in a limited space and is usually held together by springs would be a floating packing. Neither floating packings nor automatic packings will be discussed in this article.

## Properties for each packing

The desirable properties of a mechanical packing are resilience, chemical resistance, and strength:

- Resilience will give the packing the ability to be inserted into a stuffing box and be slightly deformed to take the shape of the container. This will also provide for deformation under shaft-deflection conditions during operation.
- Chemical resistance will prevent attack by the fluid that the packing is called upon to seal. This chemical resistance will extend to the lubricant. Lubricant loss due to chemical attack or washout is often tolerated by the end-user of packings. For example, a solvent could dissolve a petroleum lubricant in a packing. Therefore a different packing might be in order in this case. With the loss of lubricant, the braided material is lifeless, often abrasive, and replacement must soon be made or definite shaft or sleeve damage will occur.
- Strength will protect the packing from mechanical destruction, particularly by shaft gyration (shaft whip), or by any mechanical action created by the fluid. An example of the latter is when a liquid forms crystals in the packing—creating mechanical wear between the packing and the shaft or sleeve. A lantern ring and a flush should be used for this condition.

To be desirable, mechanical packings should:

- Contain a sacrificial lubricant so that during initial break-in, or in the event of some minor overtightening, lubricant is lost rather than the packing destroyed.
- Maintain physical volume and not lose it quickly. This can be accomplished by (a) using no lubricants or (b) using a combination of lubricants so that volume loss is slow and controllable. For example, the use of

### What is packing and why is it needed?

If we add wax or petrolatum to a twisted hemp rope, we have a crude mechanical packing. This would answer the need of preventing water from entering a boat at the point that a drive shaft protruded into the water from the boat's interior. The cavity where the packing is stuffed is called a stuffing box. With this example, we have introduced the primary elements of a mechanical packing—namely, a fibrous material to which is added a lubricant.

With use, the wax may be washed out and the rope may rot from prolonged immersion in the water. In order to make this mechanical packing work successfully in the boat, we would want to select materials that are resistant to rot, and a lubricant that would not wash out readily in either fresh or salt water, and that would not bind the shaft when the boat was not used for a period of time.

In industry, many applications are similar to the example of the boat shaft, but compounded many times over. Packing is used on every conceivable type of liquid product, as well as with all types of equipment and under various service conditions. For example, packings are required to seal at temperatures from −300°F to over 2,000°F, and pressures from vacuum conditions to over 1,000 psig. Hydraulic packings are now used for pressures in excess of 15,000 psig.

lubricants that melt at different temperatures can control volume loss.

- Minimize scoring of the shaft or sleeve.
- Have the widest possible usage range for an individual design. This is usually possible with only the more expensive packings, such as graphite-filament or graphite-ribbon types, and certain TFE packings.

## Packing materials

Due to increasing service demands, packings manufactured from animal and vegetable fibers, and leather, are falling into limited or special uses. The mineral fibers (asbestos, glass, ceramic and metal) and synthetic fibers (Teflon and carbon) are today's common materials. Cotton, flax and leather still find some uses. Leather is found in cup applications, and cotton is the low-cost selection for simpler applications. Flax is in common usage for marine packings due to its resistance to rot, its compressibility, and its tensile strength.

Since asbestos is a proven carcinogen, let us summarize the regulations covering its use. Asbestos as a restricted material requires strict processing methods, and until it is put into a particular finished form must meet the exposure requirements of the Occupational Safety and Health Act (OSHA). As the finished forms of the majority of asbestos packings contain a lubricant or other binder, it is no longer considered to be under OSHA's control. The reference cited is that:

> "The asbestos fibers have been modified by a bonding agent, coating, binder, or other materials so that during any reasonably foreseeable use, handling, storage, disposal, process or transportation, no airborne concentrations of asbestos fibers in excess of exposure limits as defined by OSHA would be released. There is no foreseeable use of these products which would create any measurable amount of asbestos airborne particulate matter. If it is necessary to alter these materials in your plant, they should never be sawed or abrasive cut in any way but should, in fact, be cut by a knife-type action."

Asbestos is a material having exceptionally high chemical and heat resistance as well as being an excellent reservoir for lubricants. The most important asbestos used in mechanical packing is white crocidolite, due to its length, strength and flexibility. The temperature range for various asbestos packings is shown in Table I.

Some companies are banning the use of asbestos in their plants. If OSHA and industry ban asbestos products, then other materials will have to be substituted. A user converting to packings containing TFE, graphite or ceramic fibers would certainly pay more, whereas a user who downgrades to the less-expensive cotton fibers will get poorer results.

**Temperature range for asbestos packing** — **Table I**

| Grade | Asbestos content, % | Approximate service temperature, °F |
|---|---|---|
| Commercial | 75 – 80 | Up to 400 |
| Underwriters' | 80 – 85 | 450 |
| A | 85 – 90 | 550 |
| AA | 90 – 95 | 600 |
| AAA | 95 – 99 | 750 |
| AAAA | 99 – 100 | 900 |

Fiberglass has been applied to several mechanical packings. It is chemically resistant and can be readily braided into packing. Unfortunately, it has several drawbacks. The primary one is its tendency to disintegrate, and substantially wear the equipment on which it has been installed. While more work is being carried out with fiberglass, use of ceramic fibers (while considerably more expensive) appears to be a logical step to the eventual replacement of asbestos. Ceramic, with its high-temperature resistance and chemical inertness, polishes rather than abrades a shaft or sleeve. Therefore, it has the potential to become a widely used mechanical packing, with the one major drawback being very high cost. Fiberglass may eventually be the low-cost substitute for ceramic packing.

Graphite yarn has met wide acceptance in the past few years but its brittleness and high cost have been drawbacks. It tends to be porous but this has been overcome by carbon fillers that are dispersed between the fibers, acting as a liquid-blocking agent and at the same time reducing fiber breakage. This has essentially resulted in a product whose only drawback may be cost.

One of the considerations of many new products having high heat resistance is that the weak link may no longer be the packing. Historically, when a packing has been overtightened or similarly misused, the packing has failed but the lubricant present would substantially protect the equipment. Ceramic or graphite packings, on the other hand, may not fail for the same reasons and could through misuse generate enough heat to melt a shaft or sleeve. Therefore, special care must be exercised in installing and breaking-in graphite packings.

## Lubricants for packings

Mica is a hydrated silica, and as a lubricant is similar to talc. Both mica and talc are still used in valve packings but seldom for rotary applications due to their higher friction values. They are also still employed where the discoloration caused by graphite or molybdenum disulfide presents a problem.

Graphite is the most common lubricant for mechanical packing and is inert to most chemicals. Its lubricating value is attributed to the very thin wafers that adhere to the packing and to other surfaces that it comes in contact with. One of the problems associated with graphite is that it contributes to electrolytic or galvanic corrosion. The most notable example is the pitting of valve stems in high-pressure steam service.

Molybdenum disulfide is a dry lubricant similar in appearance, form and feel to graphite, but it does not cause electrolytic corrosion. Its main value is in preventing galling on metal surfaces. It inherently adheres to metal shafts, which enhances its lubricating qualities for packing usage, but it has the disadvantage of oxidizing at approximately 650°F (resulting in its no longer being a lubricant).

Various other lubricants such as grease, tallow and petroleum oils have limited temperature resistance and limited chemical resistance. Petroleum oils can carbonize at elevated temperatures and thereby have their value as lubricants reduced or even eliminated.

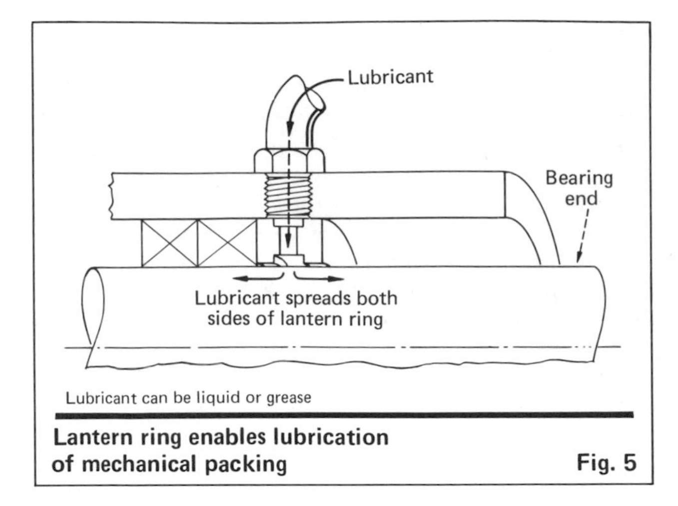


**Lantern ring enables lubrication of mechanical packing** **Fig. 5**

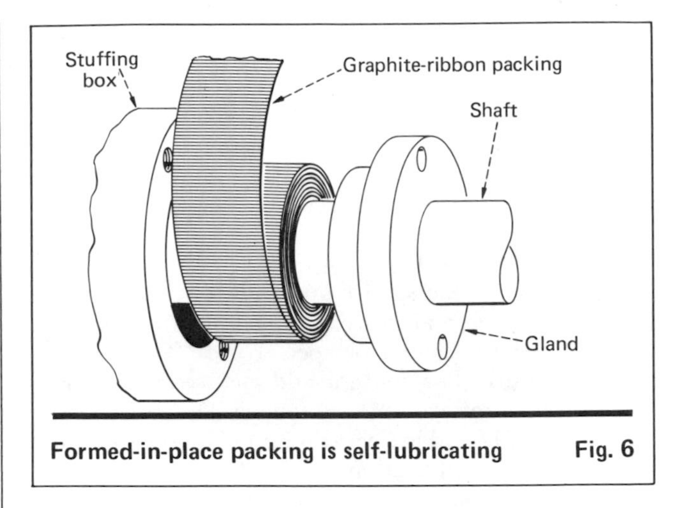


**Formed-in-place packing is self-lubricating** **Fig. 6**

Tungsten disulfide is another lubricant that can operate at very high temperatures (approximately 2,400°F), and is very highly corrosion-resistant. While it does not have the lubricating qualities of molybdenum disulfide or graphite, it does have the high-temperature capabilities and is a normal component of high-temperature packings—primarily for steam valves and expansion joints.

Undoubtedly, TFE has been the most significant lubricant developed for packing and is a component of most packings. Such packings may have a TFE content of up to 35%, depending upon the type of construction and absorbency characteristics of the base yarn. TFE has a temperature limit of 500°F and is virtually inert to all chemicals. Exceptions are molten alkali metals and some rare halogenated compounds.

Various high-temperature silicone oils find use as packing lubricants. Silicone oils have been developed with increased corrosion-resistance and higher operating-temperature capabilities. Silicone lubricants most often are added to an installed packing at the lantern ring during installation or during operation.

The ideal packing lubricant should:

1. Provide lubrication between the packing and shaft in order to prevent excess wear, galling, or seizing. A low coefficient of friction is essential.
2. Act as an interfiber blocking agent in order to prevent excess fluid from escaping directly through the stitches of the packing.
3. Be insoluble in the fluid being pumped.
4. Be capable of handling the recommended operating temperature of the base packing, except in the case of a "sacrificial" lubricant, which aids in the break-in process.
5. Have a prolonged shelf life without hardening or losing the base characteristics.
6. Be compatible with the fluid being pumped and not contaminate it.

**Maximum service limits for various mechanical packings** **Table II**

| Packing | Break-in leakage,[1] drops/min | Running leakage,[1] drops/min | Maximum temperature,[4] °F | Pressure at maximum temperature,[4] psig | Maximum pressure,[4] psig | Temperature at maximum pressure,[4] °F |
|---|---|---|---|---|---|---|
| Asbestos/PTFE | 120 | 60 | 500 | 50 | 200 | 100 |
| PTFE (lubed) | 120 | 60 | 500 | 50 | 200 | 100 |
| Asbestos/graphite | | 60 | 400 | 50 | 250 | 100 |
| Graphite-fiber | | 60 | 1,000 (600)[2] | 50 | 350 | 300 |
| Graphite-ribbon | | 60 | 1,000 (600)[2] | 50 | 350 | 300 |
| Lead | | 60 | 350 | 50 | 400[3] | 100 |
| Aluminum | | 60 | 800 (500)[2] | 50 | 400[3] | 200 |
| Flax | | 60 | 200 | 50 | 200 | 200 |
| Plastic | | 60 | 350 | 50 | 200 | 200 |

1. Leakage rate: 1 mL/min = 10 to 20 drops/min.
2. Larger number is nonoxidizing environment; smaller number is oxidizing environment.
3. Assumes rings are die-formed.
4. Temperature is product temperature; pressure is stuffing-box pressure.

**Basic data:** 2-in. shaft, 3,550 rpm. Controlled leakage for 720 h. Pumped liquid is water. Assumes maximum $\Delta T$ of 100°F (50°F for flax) due to shaft friction. Satisfactory results can be expected by using these maximum limits and following FSA (Fluid Sealing Assn.) Test Procedure #1.

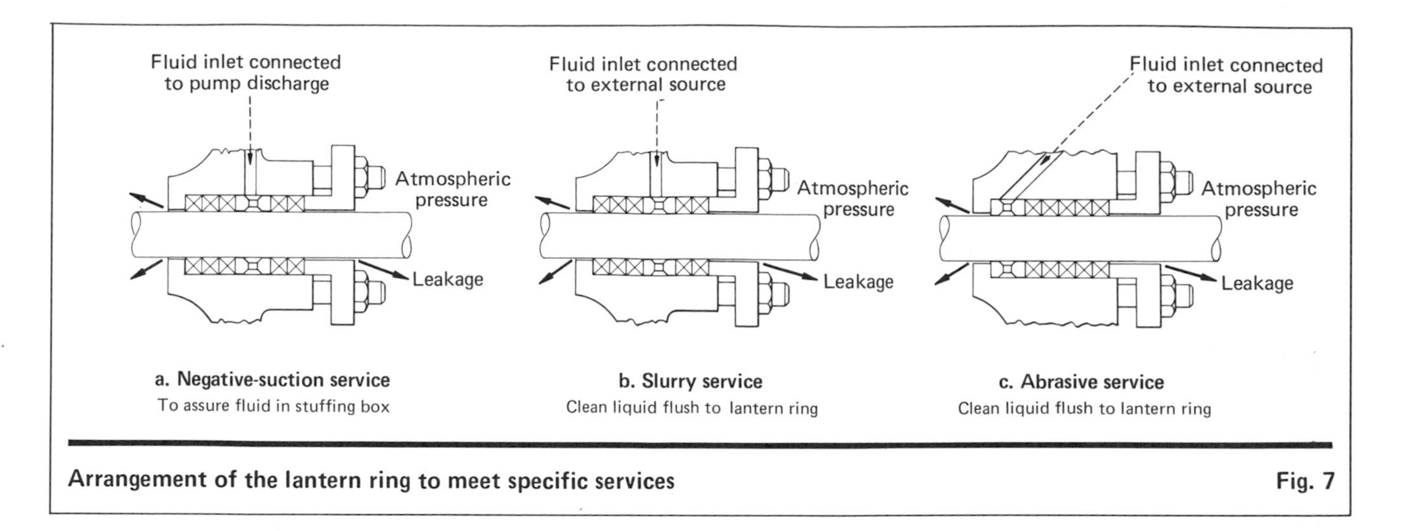


**a. Negative-suction service**
To assure fluid in stuffing box

**b. Slurry service**
Clean liquid flush to lantern ring

**c. Abrasive service**
Clean liquid flush to lantern ring

**Arrangement of the lantern ring to meet specific services** **Fig. 7**

7. Not promote galvanic or electrolytic corrosion.

The service recommendation chart (Table II) summarizes material and lubricant limits.

## Adding lubricant to packings

The lantern ring is a device made from a rigid material such as bronze, stainless steel, nylon or TFE, and is of open construction to allow free passage of lubricant. Normally, the lubricant enters the outside of the ring, and flows to the shaft or sleeve. The lantern ring usually has packing rings on either side (Fig. 5).

## Other forms of packing

In recent years, several new forms of packing have been introduced, such as the cord forms of TFE, and the graphitic-ribbon packings.

TFE cord packing is available on spools and is not unlike a slightly hard bead of toothpaste. When put into a stuffing box, it takes the shape of the container, and has all of the advantages of TFE. It primarily serves as a gasket and as a valve packing. A major advantage of this product is that it can substantially reduce the inventory of gaskets and valve packings.

Graphitic-ribbon packings are formed in place on the shaft, as shown in Fig. 6. The coiled ribbon is then inserted into the stuffing box and pressed into rings. Advantages of this packing are its self-lubricating characteristics, flexibility, high heat-conductivity, high-temperature resistance, superior corrosion resistance, and ability to fit any-size stuffing box. It is somewhat awkward to install, but the results are outstanding. Inventory requirements are also reduced.

Some extruded-TFE and graphite-TFE products are used for easy formability into a stuffing box. Such packings provide good seal life on a shaft or sleeve that may be scored. Of necessity, the scores must be smooth. However, such scoring is a characteristic of packing wear, and this packing will conform to some irregularities on rotating shafts. For use on reciprocating shafts, the scores must be axial and smooth.

## Selecting the packing

There are as many guides to packing selection as there are packing manufacturers. Packing selection is more art than science. Factors that must be considered in selecting a packing involve all of the fluid's conditions, such as temperature, lubricity and pressure; all equipment parameters, such as speed, equipment condition, material of shaft or sleeve; and miscellaneous factors, such as dimensions, space available, continuous or intermittent service; and any combination of these conditions. For these reasons, inplant training of personnel is generally required.

The two most common factors for packing selection are *PV* and pH. The *PV* factor is the pressure (*P*, psig) in the stuffing box multiplied by the velocity (*V*, ft/min) of the shaft surface. This is a number that is a measure of the relative difficulty of a packing application. The higher the number, the more difficult is the application. For example, a $1\frac{7}{8}$-in. shaft rotating at 1,800 rpm and operating at 50 psig has a *PV* factor calculated as follows:

$$PV = 50(1.875\ \pi/12)(1{,}800) = 44{,}178$$

A 4-in. shaft at 1,200 rpm and 40 psig has a *PV* factor of 50,265. This would be the more difficult application, all other conditions being equal.

The pH factor is a measure of the acidity or alkalinity of a solution. The scale is from 0 to 14, with 0 representing a strong acid, 7 being neutral (distilled water), and 14 being a strong caustic. Packing guides refer to pH values for purposes of selection.

Many other factors must also be considered. For example, flushing to a lantern ring may be required, or adding a quench and drain system to a gland, or heating or cooling of the packed shaft.

## Bull rings or end rings

From the early days of mechanical packing, rings have been inserted in the bottom of the stuffing box or in the top of the stuffing box at the gland. These are termed "bull" rings. Their primary purpose is to prevent extrusion of the adjacent rings of packing into excess clearances either at the bottom of the stuffing box or at the inside and outside diameters of the gland. Bull rings, normally made from a dense packing, are frequently metallic, or are sometimes woven if operating conditions permit. In recent years, the function of the bull ring has been expanded, it now acting as an initial blocking ring to prevent solids from entering the stuffing-box area and destroying the packing.

Bull rings are made from babbitt, aluminum, and various woven fabrics often vulcanized to a high degree of hardness. The rings are usually cut from flat stock, and custom-fitted to the dimensions of a stuffing box. A more recent construction is to make them from solid materials such as TFE or carbon graphite. These self-lubricating materials enable the packing user to obtain very close tolerances between shaft and stuffing box—thus preventing any measurable extrusion. This is particularly important when using some of the more extrudable packings such as graphite-ribbon and pliable TFE products.

## Alternating rings

By using packing rings of different materials and alternating them in a stuffing box, we can achieve a specific set of results or obtain a combination of features that is not available in any individual packing. For example, alternating a very soft packing that has excellent gland response with a harder packing will help in resisting deformation under pressure. Or, alternating a soft all-graphite packing with a TFE packing will help in controlling the rapid expansion of TFE with temperature changes. The softness of the carbon packing will protect the TFE packing during expansion. Alternate-ring usage is generally custom-selected by the user and the manufacturer of the packing. Packing being, again, more art than science, it is frequently difficult to prove results from a special design.

When a user exercises the same care in installing packing as he does in installing mechanical seals, significantly better results can be obtained by applying alternate rings of packing. With TFE packing, the results can be significantly improved if the alternate ring allows greater abuse in gland tightening. Here, the alternate ring helps to prevent charring of the TFE packing by providing for the expansion of the TFE when heat is generated. In addition, the alternative material may continue to function to some degree even when the TFE packing is glazed.

One problem with alternate rings is that it is often difficult to apply effective packing sets when the stuffing box is shallow and a lantern ring must be used.

## Die-formed rings

A die-formed ring is a particular packing material of the user's selection that is put into a mold, with pressure exerted on it so that the voids in the packing are essentially removed. The mold is made to a specific size so that the resultant ring matches the shaft or sleeve diameter and the stuffing-box bore diameter. The die-formed rings are then inserted into a stuffing box, and the requirement for gland takeup during the break-in period is minimized. Die-formed rings enhance extrusion resistance, help exclude abrasive materials, and are capable of sealing at higher pressures. These factors may frequently outweigh the higher cost of such rings for use with packings.

The primary use of die-formed rings is for higher-pressure applications where lengthy periods of break-in

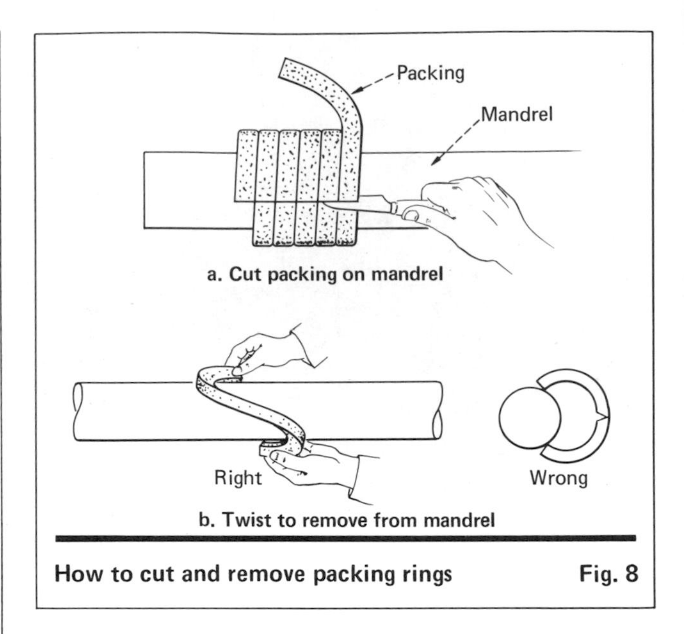


**How to cut and remove packing rings** **Fig. 8**

would be required for non-die-formed packings. Thus, break-in time is substantially reduced.

Die-formed rings can be of considerable help in maintaining a lantern ring in its proper position. In this case, the rings between the lantern ring and the bottom of the stuffing box would be die-formed, and the rings between the lantern ring and the gland would not be die-formed. However, there is the likelihood that the leakage from the lantern-ring inlet to the gland would be greater than from the lantern ring into the product fluid. It is recommended that the full set be die-formed.

## How to pack a centrifugal pump

It is estimated that up to 75% of all packing problems arise from poor installation. The installation method is critical, and is so frequently assumed to be routine that the problems created are overlooked. The damage is frequently very slight and so accepted that few take the time to study it and set up firm procedures for installing the packing. Unfortunately, procedures vary from application to application, and from product to product. Therefore, untrained maintenance personnel frequently use one technique for all packing applications—thus getting a variety of results. Properly trained maintenance personnel can avoid many failures due to installation procedures.

One guideline on the packing procedure for pumps uses four pages of data and recommends 44 steps. Another technical article on packing uses 19 different steps for packing installation and, at the end, implies that these instructions apply to approximately 90% of such installations—but that there are many exceptions to the 19 steps.

Let us try to reduce the number of steps to a workable amount, and simplify the explanations:

1. Measure shaft run-out and end-play. These should not exceed the pump manufacturer's tolerances. On some old equipment, the packing acted as a form of bearing and its life under these conditions was reduced. When checking equipment, you should keep in mind the requirements that the shaft run smooth, be free of

burrs, and operate without vibration or whip. Always check the bearings; if in doubt, replace them. For good results, make certain that the equipment is in good condition and does not cause packing failures.

2. Check the condition of the inside surface of the stuffing box. This is more important than most people realize. There can be substantial leakage in the static seal that is formed between the outside diameter of the packing and the inside (i.e., bore) of the stuffing box. The bore should be smooth and have a finish not to exceed 70 $\mu$in. If the O.D. of the stuffing box is rough, it may "hang up" the packing, and require excessive gland pressure to correct it. Such pressure frequently results in packing failure.

3. Locate a lantern ring properly if flushing is required. This is critical. Check the pump manufacturer's details for the number of rings of packing installed beyond the lantern ring. Consider die-formed rings to help position the lantern ring. See Fig. 7 for lantern-ring arrangements.

4. Determine proper packing and size for the application. Every mechanic knows that the packing can be made to fit. However, every manufacturer knows equally well that the best results are only obtained by proper selection and fit for the application.

5. Cut the rings on a mandrel, as shown in Fig. 8a. If a mandrel is not available, use the actual sleeve or shaft of the pump. Use a straight cut so that the rings are butt-jointed. Remove metallic rings from the mandrel, as shown in Fig. 8b. Cutting the rings on a packing cutter is recommended.

6. Remove the old rings from the pump by using the proper tools, and avoid metal-to-metal contact where possible. Make certain that all rings are removed. Leaving one ring in the bottom of the stuffing box is a frequent cause for misinstallation of the lantern ring. Be sure the stuffing box is clean. Use a degreaser or other cleaning agent. Be certain that no foreign matter or cleaning products get into the bearings.

7. Check the packing-manufacturer's instructions. Are there any special recommendations regarding lubricants? If the requirements for a packing lubricant are unknown and the interaction between the lubricant and pumped product is unknown, avoid using lubricant. Since 70% of the wear on packing takes place on the last two rings (installed next to the gland), lubricant on them may be critical—provided that a lubricant can be added.

8. Open the rings with a twisting motion when they are installed on pump shafts (Fig. 8b).

9. Tamp each ring as it is installed. Each ring is then individually positioned and compressed into place. This should be done with a tool designed for this purpose, or with a split-holder cylinder. The shaft should be turned occasionally to be certain that no binding occurs due to overtamping. Stagger the joints, normally at 120-deg positions. After wrapping the ring on the shaft, one should avoid gaps in the cut ends.

10. Tighten the gland finger-tight after the packing is totally installed. If possible, jog the pump. The initial leakage should be generous to the point where it is a small stream rather than a slow drip. With TFE packings, it is necessary to back-off the gland an additional amount. If the packing happens to be 100% TFE, this becomes critical. In this step, the installer should follow the packing-manufacturer's instructions. If the packing starts to smoke, stop the pump and loosen the packing set. Obtain a generous leakage before restarting the pump.

## Standards for packings and seals

In actual practice, mechanical end-face seals are finding wide acceptance. In many instances, these seals are legally required for many services on suspected or actual carcinogens such as vinyl chloride and benzene.

On many pumps, mechanical packings made from modern materials and installed by trained personnel can achieve results that approach those of mechanical seals. Packing will never enjoy the success of mechanical seals in controlling the fluid being sealed—particularly as packing cannot be legally used with certain fluids. But packing is a viable alternative for a high percentage of the remaining applications.

In hazardous refinery services (for example, gasoline and propane), API (American Petroleum Inst.) Standard 610 requires the use of mechanical seals under strict arrangements. The EPA requires double mechanical seals for carcinogens. The International Standards Organization (ISO) and the American Soc. of Lubrication Engineers (ASLE) are setting standards for mechanical seals, while the Fluid Sealing Assn. (FSA) and the National Fluid Power Assn. (NFPA) set standards for mechanical and hydraulic packings. For example, the FSA has created standard test procedures for packing. Using these, anyone could establish packing-application parameters. One goal of FSA is to reduce the "art" of packing to a defined science.

What still predominates in choosing between seals and packings is simplicity of installation. Each technology has its own area of usage and our intent here has been to help the user make the decision.

## The author

**Richard Hoyle** is responsible for corporate development and engineering at the A. W. Chesterton Co., Stoneham, MA 02180. He joined the company in 1957, and has worked on plant expansions, acquisitions, and special technical problems. He serves on committees dealing with mechanical seals and packings in the following organizations: American Soc. of Lubrication Engineers, Fluid Sealing Assn., American National Standards Institute, and Technical Assn. of the Pulp and Paper Industry. He has a B.S. from Lowell Technological Institute.

# Pump Piping Design

Layout and sizing of pump piping and fittings are critically important to centrifugal pump performance. Suction and discharge piping present individual problems and must be designed with equal care.

ROBERT KERN, Design Engineer

When designing suction piping, the engineer's main concern is reliable operation. In general, suction piping consists of a draw-off nozzle, one or two sections of horizontal and vertical pipe runs, a block valve, strainer and connecting elbow section before the final connection to the pump suction nozzle. Each of these parts should be designed and located to avoid vaporization or air pockets.

The lowest pressure point in a pump is just before the impeller vanes. If the pressure here falls below the vapor pressure of the liquid, vapor bubbles will form. Then the bubbles collapse because of the rapid pressure increase within the impeller. This phenomenon (the rapid formation and collapse of vapor bubbles) is called cavitation. It can reduce pump efficiency and cause noise, vibration, impeller wear and breakage.

Cavitation can be avoided by providing sufficient static head before the pump suction nozzle to satisfy net positive suction head requirements and to overcome friction losses in the line. A simple suction pipe configuration is essential.

## Suction Line Layouts

Fig. 1 shows a typical example of suction piping for a cooling tower. Water level determines the cooling water basin height and its cost. Consequently, a minimum water level is preferred, but it must be sufficient for the pump's NPSH and suction pipe resistance.

The draw off nozzle in Fig. 1 is well rounded and tapered for least resistance. Pipe diameter is larger than the pump suction nozzle, and piping is kept simple and short as possible. The eccentric reducer is placed in the line flat side up to avoid an air pocket just before the pump entry.

The vortex breaker in the cooling water basin prevents funnel formation and air entry into the suction pipe. The butterfly valve in the line offers little resistance and has no air pocket in its housing. Because of the rigid pipe system, expansion joints are necessary. The bottom of the basin is designed to prevent excessive sediments from entering the pump. Screens are built into the wall between the basin and suction trough and a trench is provided for cleaning.

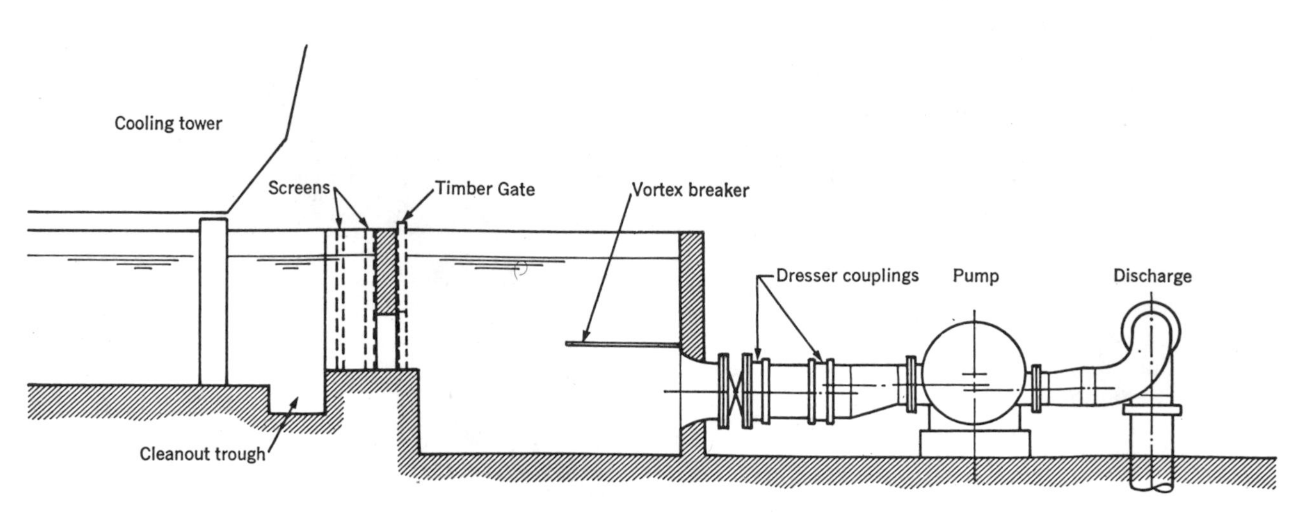


**SUCTION PIPING** for a cooling tower pump—Fig. 1

Originally published October 11, 1971

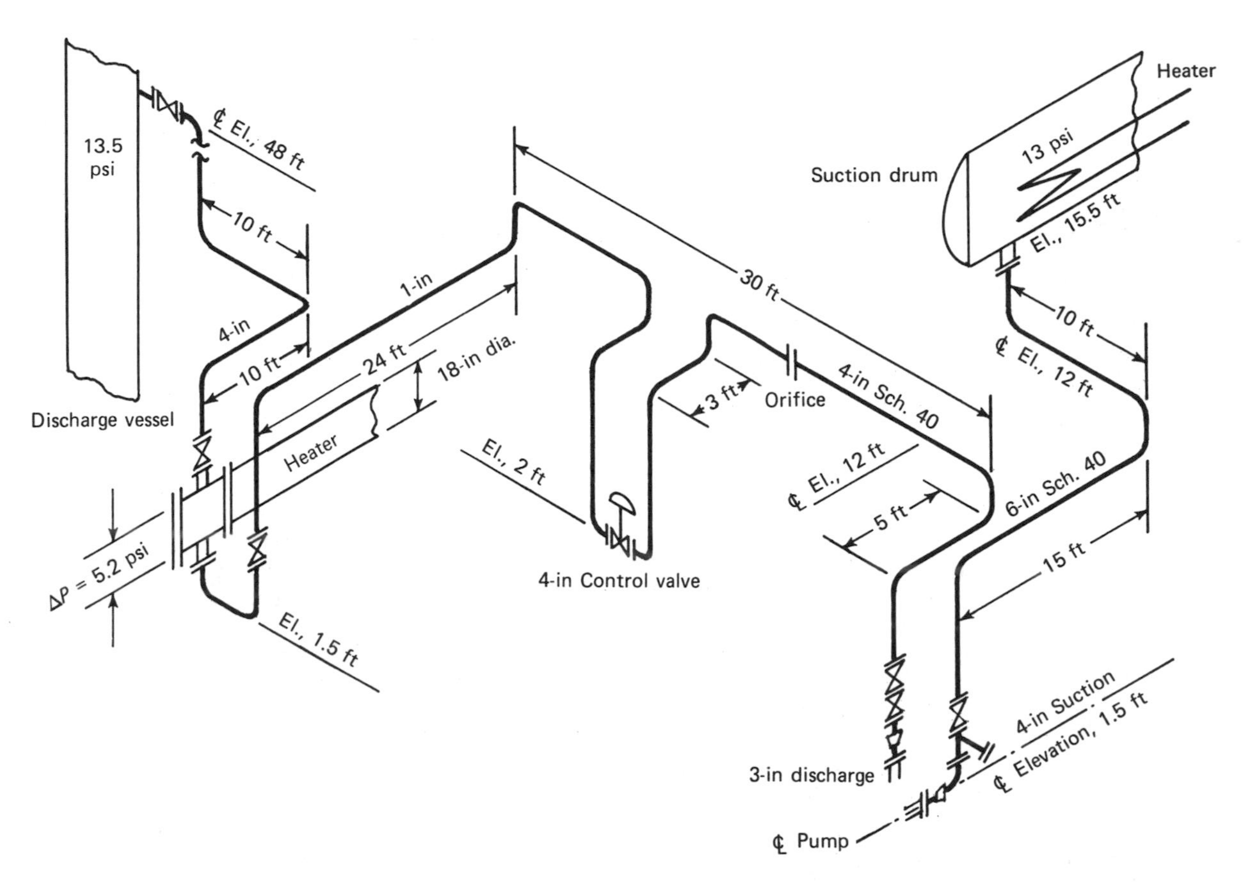


**PIPING** and equipment layout for the suction and discharge lines to the process pump as given for the problem—Fig. 5

25 psi, or expressed as head of liquid (the required total head):

$$H = (25)(144)/50 = 72 \text{ ft}$$

## Example Illustrates Detailed Design

A centrifugal pump having a 4-in suction nozzle and a 3-in discharge nozzle will handle a gas oil at a normal flowrate of 250 gpm through a piping and components system, as drawn in Fig. 5. Allowing for a safety factor of 1.1, we find that the maximum flowrate will be 1.1(250) = 275 gpm. Specific gravity and density are: $S_{60} = 1.18$, and $\rho_{60} = 73.6$ lb/ft$^3$, respectively. At the flowing temperature of 555° F, $S = 1.04$ and $\rho = 64.87$ lb/ft$^3$. Viscosity of the gas oil is 0.6 cp. There is a flow control valve in the discharge line.

Let us calculate the total head on the pump when it is expected to operate at a normal flowrate and at a maximum flowrate.

We will begin the analysis by first determining the loss in the suction line and then that in the discharge line. Pertinent data for the Schedule 40 pipe are:

| | Suction Line | Discharge Line |
|---|---|---|
| Nominal size, in | 6 | 4 |
| Inside dia., $d$, in | 6.065 | 4.026 |
| $d^5$ | 8,206 | 1,058 |

*Suction-Line*—We fine the loss in the suction line by initially calculating the Reynolds number from:

$$N_{Re} = 50.6(Q/d)(\rho/\mu)$$
$$N_{Re} = 50.6(250/6.065)(64.87/0.6) = 225{,}500$$

The friction factor, $f$, for this Reynolds number equals 0.0175 (from charts in Part 1 of this series, *Chem. Eng.*, Dec. 23, 1974, p. 65). By substituting in the following equation, we obtain the unit loss:

$$\Delta P_{100} = 1.35 fSQ^2/d^5$$
$$\Delta P_{100} = 1.35(0.0175)(1.04)(250)^2/8{,}206$$
$$\Delta P_{100} = 0.19 \text{ psi}/100 \text{ ft}$$

We then find the equivalent length for the suction line and its fittings from data in Part 2 of this series (*Chem. Eng.*, Jan. 6, 1975, Tables I to IV):

| | |
|---|---|
| Pipe length | 39 ft |
| 5 Short-radius elbows | 75 ft |
| 1 Reducer | 4 ft |
| 1 Strainer | 30 ft |
| 1 Gate valve | 6.5 ft |
| 1 Inlet to pipe | 18 ft |
| Total equivalent length, $L$, | 172.5 ft |

Hence, the overall loss for the line and its fittings at the normal flow of 250 gpm becomes;

$$\Delta P = \Delta P_{100}(L/100)$$
$$\Delta P = 0.19(172.5/100) = 0.33 \text{ psi at normal flow}$$
$$\Delta P = 0.33(250/275)^2 = 0.4 \text{ psi at maximum flow}$$

To find the pressure at the suction nozzle, we calculate the static-head pressure: (14)(64.87)/144 = 6.3 psi, and add this to the vessel pressure of 13 psi to get 19.3 psi. Since the pipe-friction loss at normal flow is 0.33 psi, the pressure at the suction nozzle becomes 19.3 − 0.33 = 18.97 psi. At maximum flow, the pressure at the suction nozzle is 19.3 − 0.4, or 18.9 psi.

*Discharge-Line*—We now perform similar computations to find the loss in the discharge line. Therefore, for the discharge line:

$$N_{Re} = 50.6(250/4.026)(64.87/0.6) = 340{,}000$$

The friction factor, $f$, for this Reynolds number is 0.0178 from charts in Part 1 of this series. We now use Fig. 4 to find the unit loss, $\Delta P_{100}$, as 1.32 psi for totally turbulent conditions at the normal flow of 250 gpm. Since the flow for this Reynolds number is in the transitional region (see Part 1 of this series, *Chem. Eng.*, Dec. 23, 1974, p. 65), we must correct the unit loss as follows:

$$\Delta P_{100} = 1.32(1.04)(0.0178/0.0165) = 1.48 \text{ psi}/100 \text{ ft}$$

We then find the equivalent length for the discharge and its fittings from data in Part 2 of this series:

| | |
|---|---|
| Pipe length | 156 ft |
| 20 Short-radius elbows | 210 ft |
| 4 Gate valves | 18 ft |
| 1 Reducer | 3 ft |
| 2 Exits | 40 ft |
| 1 Inlet | 10 ft |
| Total equivalent length, $L$ | 437 ft |

Hence, the overall loss for the line and its fittings at the normal flow of 250 gpm becomes:

$$\Delta P = \Delta P_{100}(L/100)$$
$$\Delta P = 1.48(437/100) = 6.47 \text{ psi}$$

And at maximum flow,

$$\Delta P = 6.47(275/250)^2 = 7.83 \text{ psi}$$

A detailed, specific and systematic procedure is shown in Table I for calculating the total head on the pump when it operates at a normal flowrate and at the maximum flowrate.

For the calculations presented in Table I, the following steps are recommended:

Lines 1 to 6 can be worked out simultaneously for normal and maximum flow. Lines 1 and 2: Discharge-vessel pressure and discharge static head do not change with alternative flow capacities. Lines 3, 4 and 5: Orifice $\Delta P$, pipe-friction loss, and equipment-friction loss increase with capacity. Hence:

$$\Delta P_{max} = (Pump\ Safety\ Factor)^2 \Delta P_{normal}$$

Line 3: Orifice pressure drop depends on the manometer deflection and on the orifice diameter/pipe diameter ratio.

The manometer deflection for an orifice at a 250-gpm flow will be:

$$h_w = [0.176Q\sqrt{S}/(d_1{}^2\beta^2C)]^2$$
$$h_w = [0.176(250)\sqrt{1.04}/(16.21 \times 0.339)]^2$$
$$h_w = 80.7 \text{ in, or } 6.73 \text{ ft}$$

For this deflection, we can use a 100- or 125-in-long manometer. The orifice $\Delta P_o$ will be:

$$\Delta P_o = 6.73(62.37)/144 = 2.92 \text{ psi}$$

With $\beta = 0.7$, the permanent loss will be 52% of the orifice pressure differential.*

$$\Delta P_o(\text{loss}) = 0.52(2.92) = 1.52 \text{ psi}$$

Line 4: Friction loss in the discharge (and suction) line has been previously computed as 6.47 psi. Line 5: Pressure drop through the exchanger (and other equipment) can be obtained from the manufacturers. Line 6 is the sum of Lines 3 to 5 at the normal flowrate. For a 10% greater flow and a pump safety factor of 1.1, resistance of the discharge line will increase by $(1.1)^2$. Line 7 is the subtotal of Lines 1, 2 and 6 at normal flow and at maximum flow.

At this point, we will continue the calculations at the maximum flowrate for reasons that will become evident.

Line 8: For the control valve to operate in an optimum range at normal flow, we usually consider the valve plug in a fully open position at maximum flow. This also gives a minimum pressure drop through the control valve†. A 4-in single-seat control valve has a valve coefficient† $C_v = 124$. And, with $C_{vc}/C_v = 1$:

$$\Delta P_{(min)} = \left[\frac{Q}{(C_{vc}/C_v)C_v}\right]^2 S$$

$$\Delta P_{(min)} = \left[\frac{275}{1 \times 124}\right]^2 1.04 = 5.12 \text{ psi}$$

Line 9 is the required discharge pressure, including control-valve $\Delta P$ at maximum flow. Suction-nozzle pres-

†For details, *Chem. Eng.*, Apr. 14, 1975, p. 89.

**Calculations for Total Head of Pump — Table I**

| Line | Variable | | Normal Flow, Psi | | Maximum Fl Psi |
|---|---|---|---|---|---|
| 1. | Discharge-vessel pressure | | 13.5 | | 13.5 |
| 2. | Discharge static head | | 20.94 | | 20.9 |
| 3. | Orifice $\Delta P_o$ | 1.52 | | | |
| 4. | Discharge pipe-friction loss | 6.47 | | | |
| 5. | Exchanger $\Delta P$ | 5.2 | | | |
| | [a] | + | | | |
| 6. | Discharge-system resistance | → | 13.19 | x (1.1)[2] = | 15.96 |
| 7. | Subtotals | | 47.63 | | 50.4 |
| 8. | Control valve $\Delta P$ | | 10.71 [d] | | + 5.12 |
| 9. | Discharge pressure | | + 58.34 | | 55.52 |
| 10. | Suction-nozzle pressure | | 18.97 | | − 18.9 |
| 11. | Pump differential pressure | | 39.37 | | 36.62 |
| 12. | Required total head, $(144 \times \Delta P)/\rho$ | | 87.4 [e] ft | ← | 81.3 |
| 13. | Total head from head-capacity curve | | 95 ft | | 92 |
| 14. | Total-head safety margin | | 7.6 ft | | 10.7 |

[a]Other equipment resistance.
[b]Pump safety factor, $s.f.$ = 1.1
[c]With $C_{vc}/C_v$ = 1.
[d]With $C_{vc}/C_v$ = 0.5 to 0.8
[e]$\Delta H$(normal flow) = $\Delta H$ (maximum flow) x 1.075
= 81.3 x 1.075 = 87.4

sure has been previously computed. By deducting the suction-inlet pressure from the required discharge pressure, we obtain the pump's differential pressure at maximum flow (Line 11). This is converted to the equivalent head (Line 12) at maximum flow by using the previously determined value of 36.62 psi from Table I:

$$\frac{144(\Delta P)}{\rho} = \frac{144(36.62)}{64.87} = 81.3 \text{ ft}$$

Let us now summarize some of these results. The pressure needed at the pump-discharge nozzle to overcome backpressure in the discharge line is the sum (Line 9 in Table I) of actual pressure in the discharge vessel, static-head lift up to the terminating nozzle (or liquid level in discharge drum), control-valve $\Delta P$, and total discharge-pipe and equipment resistances. Pump differential pressure (Line 11) equals discharge pressure (Line 9) minus suction-nozzle pressure (Line 10).

We can now estimate the total head at normal flowrate. Total head will increase by an amount ranging from 0.5 to 1 of the percentage decrease in capacity. In this example, there is a 10% decrease in capacity. Hence, for a single-impeller pump, we will assume an increase in the total head of about 7.5% (i.e., 0.75 of 10%). The computed value is on Line 12 for the normal flow. Calculated values for the total head for this example are shown in Fig. 1b.

Suction-nozzle pressure at normal flow is on Line 10. Line 9 = Line 10 + Line 11.

The available pressure differential at normal flow for control-valve sizing (Line 8) equals discharge pressure (Line 9) minus the line backpressure without the control valve (Line 7). This $\Delta P$ should give a control-valve coefficient falling within the recommended ranges of $C_{vc}/C_v = 0.5$ to 0.8 for equal-percentage contoured plugs:

$$C_{vc} = Q\sqrt{S}/\sqrt{\Delta P} = 250(\sqrt{1.04})/\sqrt{10.71}) = 77.9$$

For the selected 4-in control valve, $C_{vc}/C_v = 77.9/124$, or 0.63, which is acceptable.

For a normal flow of 250 gpm and 87.4 ft total head on the one hand, and a maximum flow of 275 gpm and 81.3 ft on the other, we can now select the pump, as shown in Fig. 1. Impeller diameter for the selected pump is 10 in, and a standard motor of 10 bhp is required. The motor will work with a better than 65% efficiency.

The calculated total-head points fall between the 9- and 10-in impellers. The pump will operate at 95-ft total head at 250 gpm, and at 92-ft at 275 gpm. The extra head (7.6 and 10.7 ft, respectively, here) provides a safety margin compensating for inaccuracies in the assessment of the flow-properties and line data. These additional pressure differentials can be absorbed by the control valve. Or, the block valve in the discharge line can be slightly closed to bring the operating point up to the head-capacity curve. Also, the motor will be able to drive the pump when the liquid is colder and specific gravities are greater than at operating conditions.

If a pump has not been selected and the head-capacity curve is not available, a safety margin of 5 to 15% can be estimated and added to the required total head of Line 12.

## Economy of Pump Piping

As the data in Fig. 1b show, several sizes of impellers can be placed in one pump case. The cost difference between impeller sizes is negligible. Motor sizes are usually well determined. A difference in the cost of the pump and motor occurs for pumps falling in adjacent envelopes of the composite rating chart. In some borderline cases, it may be more economical to redesign the discharge piping for lower pressure drop rather than to invest in a larger pump and motor.

For economy in utility cost, the pump should work at its highest efficiency. High pump efficiency results in minimum horsepower input, and minimum wear and maintenance. High-efficiency pumps last longer, are quieter and vibrate less than low-efficiency ones [*2*].

Small pumps should not be oversized. The total of oversized small motors in a plant can add to substantial waste in energy usage.

For pipe diameters above 12 in, more than one size may be selected initially because capacity increments in large-size pipe are very close. Piping costs, of course, increase with diameter, while utility costs decrease because of smaller pipe and components resistances. The best size can be determined by adding the total cost of utilities over the period of capital payout to the cost of the mechanical and electrical installation. The lowest total cost calculated for a 2-, 5- or 10-yr amortization will give the most economical design.

A detailed investigation for the most economical pipe size is justified if line sizes are large, pipe configurations are long or complicated, or if the piping material is expensive. Pipe friction must contribute a major portion of the discharge pressure—otherwise, there will be little difference in total heads between alternative designs. Actual vessel pressures and static liquid heights usually cannot be altered.

For reciprocating pumps, the available *NPSH*, pump differential pressure, suction-line and discharge-line resistances cannot be calculated in the same way as for centrifugals. Because of pulsating flow, the minimum-pressure levels should not fall below the vapor pressure when saturated liquid is pumped. For identical flowrates, pressure losses in suction and discharge lines of reciprocating pumps are greater than in those of centrifugal pumps. These principles have been adequately discussed by Hattiangadi [3]. For information on power ratings, installation and operation of reciprocating pumps, consult the "Hydraulic Institute Standards" [*4*]. The Standards [*4*] also contain information on electric-motor-driven and steam-driven reciprocating pumps.

The next article in this series will appear in the issue of June 23, 1975, and will review piping design for two-phase (i.e., vapor-liquid) flow.

## References

1. "Hydraulic Institute Standards," 12th ed., p. 81, Hydraulic Institute, New York, 1969.
2. Marischen, J. P., "Critical Centrifugal Pump Information," Ampco Metal Inc., Milwaukee, WI 53201.
3. Hattiangadi, U. S., Specifying Centrifugal and Reciprocating Pumps, *Chem. Eng.*, Feb. 23, 1970, pp. 101–108.
4. "Hydraulic Institute Standards," 12th ed. pp. 165–166, 181–182, 204–205, Hydraulic Institute, New York, 1969.

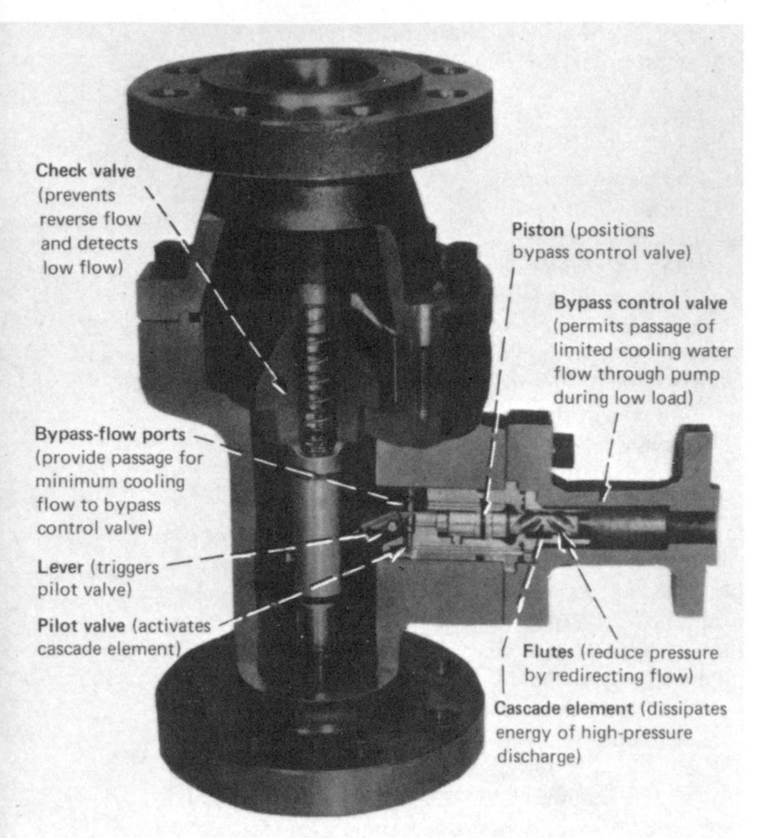

◀ **AUTOMATIC** recirculation control valve.

# Bypass Systems For Centrifugal Pumps

To prevent overheating and consequent seizure of centrifugal pumps, it is most important that a reliable bypass recirculating arrangement be designed into the pumping system.

PETER P. VAN BLARCOM, Yarway Corp.

For cooling, every centrifugal pump depends largely on the fluid being pumped. The heat generated by the rotating members of the pump and by the pump bearings is carried away as the fluid discharges.

This is most critical at low flow when the high horsepower of the electric motor or turbine driver is converted to heat. Overheating the pump can cause its failure from flashing of the fluid, which in turn produces bearing seizure or contact of the rotating elements with the stationary casing. To assure safe operation at all flows, but particularly those of less than 25% of pump capacity, a bypass or recirculation system is used.

All factors of a pump bypass system must be considered to assure positive pump protection at low flows. These factors include: capacity; positive, reliable operation; long life; quiet operation; power requirements; system simplicity; system design time; pressure breakdown ability; and installed cost.

Originally published February 4, 1974

To ensure proper sizing of a centrifugal-pump bypass system, regardless of the fluid being pumped, it is best to rely on the specific pump curve and recommendations of the manufacturer. As a minimum, this information should be provided to the prospective supplier of a bypass system: (1) maximum pump capacity; (2) bypass flow quantity: (3) pump shutoff head; (4) sump or reservoir pressure; (5) fluid to be pumped; and (6) fluid temperature and specific gravity.

Three basic types of recirculation systems are used in industry today: continuous recirculation, flow-controlled recirculation, and automatic recirculation control.

## Continuous Recirculation Systems

In a continuous recirculation system, the pump and its prime mover are oversized to provide enough fluid to keep the pump cool by continuous recirculation from the discharge of the pump, back to the sump (or reservoir) on the suction side.

A fixed orifice in the recirculation line breaks down the pressure differential between the pump discharge and the reservoir. This orifice is sized to continuously recirculate sufficient fluid to keep the pump cool. But because of this continuous recirculation, this fluid never adds to the value of the final product. The system is, therefore, inefficient and very costly because:

- Both the pump and its prime mover must be oversized to handle the recirculated quantity even at full load on the pump, where there is sufficient flow to prevent overheating of the pump.
- The additional power costs to run the pump driver are higher than normally realized. For example, a pump in a 0.9¢/kWh area discharging 300 gpm at a 2,500-ft head, requires 50 gpm recirculated flow to keep it cool. Assuming the water to be at 330°F, extra annual power costs may be calculated as follows:

$$\text{hp} = \frac{(\text{gpm})(\text{discharge pressure, ft})(\text{sp gr})}{3{,}960(\text{pump efficiency})}$$

$$= [(50)(6{,}535)(\text{power cost})]/[(3{,}690)(0.75)]$$

$$= 37.5 \text{ hp over that needed for process demands (see Fig. 1).}$$

$$\text{Annual cost} = [(\text{hp})(6{,}535)(0.9)]/(\text{driver efficiency})$$

$$= [(37.5)(6{,}535)(0.9)]/0.85 = \$2{,}600$$

Thus, $2,600 would be the cost to recirculate sufficient fluid to cool the pump (Fig. 2). This cost would gradually increase as the fixed orifice began to wear and enlarge due to the high fluid-velocity through it, which would then cause an increase in the quantity of fluid passing through it. This, plus the original purchase costs for oversized pumps and drivers, make a continuous-recirculation system prohibitively expensive except on the smallest of pumps.

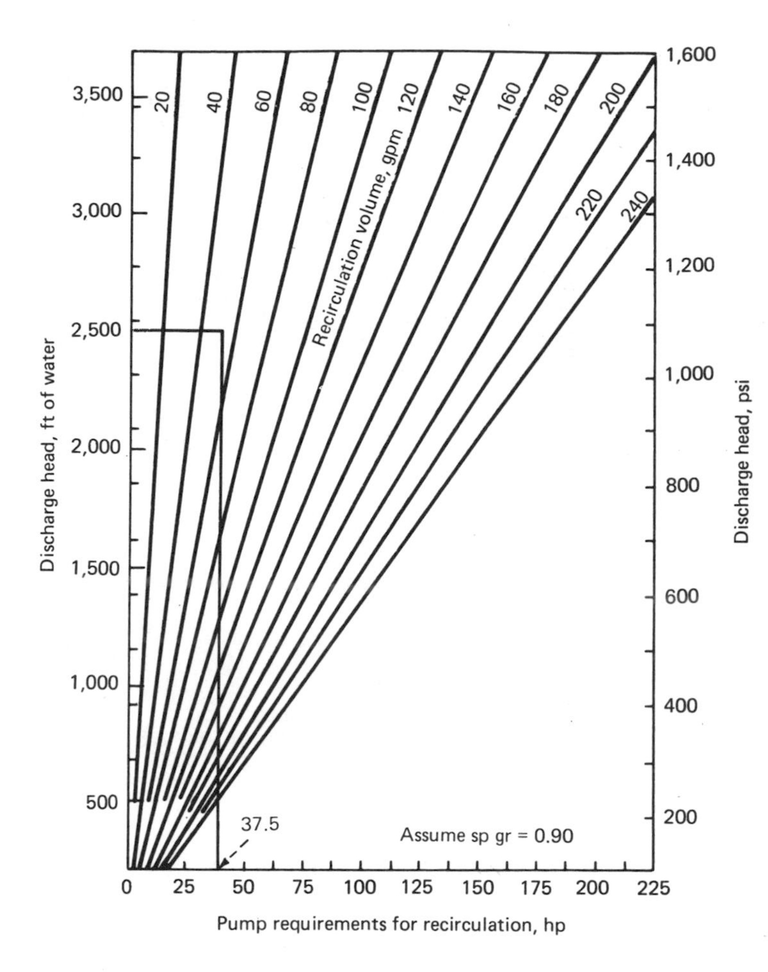


**HORSEPOWER REQUIRED** for pump recirculation—Fig. 1

## Flow-Controlled Recirculation

In a flow-controlled system, (Fig. 3a), a fluid is recirculated only when flow through the pump approaches the minimum specified by the pump manufacturer.

As the orifice on the suction (or discharge) side of the pump senses that flow through the unit nears the minimum requirement, the flow transmitter sends an electrical signal to open the solenoid valve, thus opening the recirculation-control valve. Fluid is now recirculated to the storage vessel on the suction side of the pump in a quantity governed by the size of the recirculation-control valve and the pressure drop across it. Fluid may also be going to the process.

A multiple pressure-reducing orifice, located after the recirculation-control valve, reduces the pressure drop from the discharge of the pump to the reservoir. This pressure drop can be very large at high pump-discharge pressures. At the instant of opening and the instant of closing, the recirculation-control valve receives the entire pressure drop from the pump discharge to the reservoir, and because of this is subject to severe wire-drawing and erosion.

The control valve must also shut tightly against full pump-discharge pressure. Any leakage through the valve should be subtracted from the total pump or system capacity. At times of severe leakage, the pump may not be able to satisfy the system at full-flow requirements.

A recent survey of electric utilities indicated that the single most troublesome control valve in their system was the recirculation-control valve. Repair is frequently due to the high pressure-drop, and high inlet or shutoff pressure. In any type of recirculation system, the recirculation-control valve should fail open to completely assure protection of the pump, in the event of air or electrical system failure.

Variations of the flow-controlled system include a temperature-differential system wherein the temperature rise across the pump is used to open and close the recirculation-control valve. Thermocouples are used to measure suction and discharge temperatures. There have been difficulties, however, in determining the proper placement of sensing elements.

When electric motors are used to drive pumps, it is possible to control the recirculation system by measuring motor amperage, which is a function of the work, and therefore of the flow through the pump. But neither of these latter systems are in frequent use today.

All the systems so far discussed include a check valve on the discharge side of the pump to prevent reverse flow

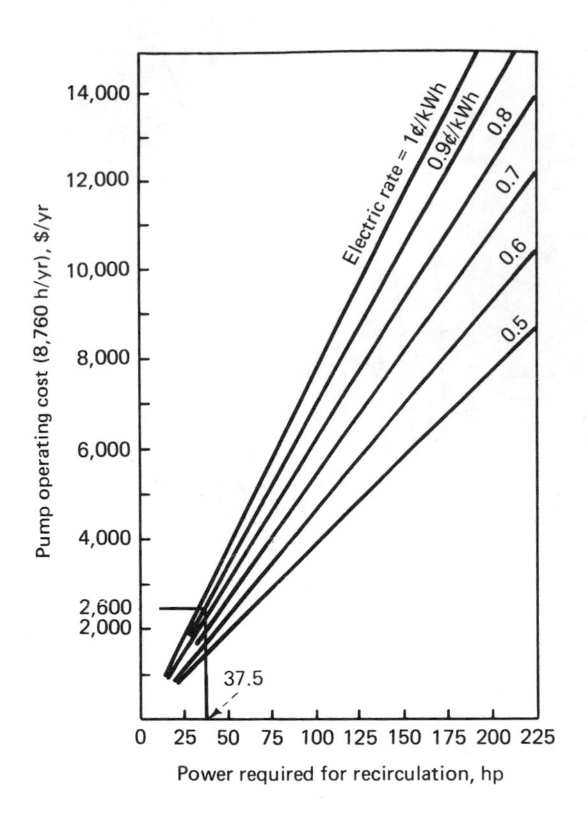


**ANNUAL COST** of recirculation (continuous flow)—Fig. 2

through the pump. Flow- and temperature-controlled systems also require both electric power and a pneumatic system to actuate the control valves.

## Automatic Recirculation Control

Automatic-recirculation control can be provided in a system that combines the functions of the pump discharge check valve—and the recirculation and pressure breakdown elements—in a single, selfpowered unit (as shown in Fig. 3b).

In this system, the rising disk-type of check or nonreturn valve acts as a flow-sensing element. This opens and closes a small pilot valve, which triggers the opening and closing of the recirculation-control valve. This is a tight shutoff element, whose unique cascade design dissipates the high-pressure energy prior to passing the recirculated fluid back to the low-pressure sump or reservoir. A cutaway photo of the recirculation-control valve is shown on p. xxx.

With normal main flow (Fig. 4a), the check valve rises off its seat and floats on the load or flow discharging from the pump. The lower extension of the check valve lifts the left end of the lever arm, thus allowing the pilot valve to seat to prevent flow through the lower passage to the low-pressure system.

Full pump-discharge pressure on the head of the pilot valve keeps it well seated. This pressure is also exerted on the head of the cascade piston—via the annular clearance around the pilot-valve stem—which closes the cascade valve and prevents bypass flow. Pump pressure is also applied on the opposite side of the piston, but to a smaller area. Thus, the cascade recirculation valve remains closed.

As process requirements decrease and pump capacity is reduced, the spring-loaded check valve begins to descend toward its seat (Fig. 4b). The lever arm, through its pivot point, now can open the pilot valve. When the pilot valve opens, the high pressure on the head of the cascade-valve piston is vented downstream to the low-pressure bypass portion of the system. As this happens, the piston moves to the right—due to pressure imbalance—and recirculation flow begins.

The point at which the pilot valve opens to place the system in the bypass mode is carefully calculated to match individual pump characteristics. It is controlled by the annular clearances between the tapered lower portion of the check-valve disk and the surrounding body. The check-valve disk becomes, in effect, a variable-area

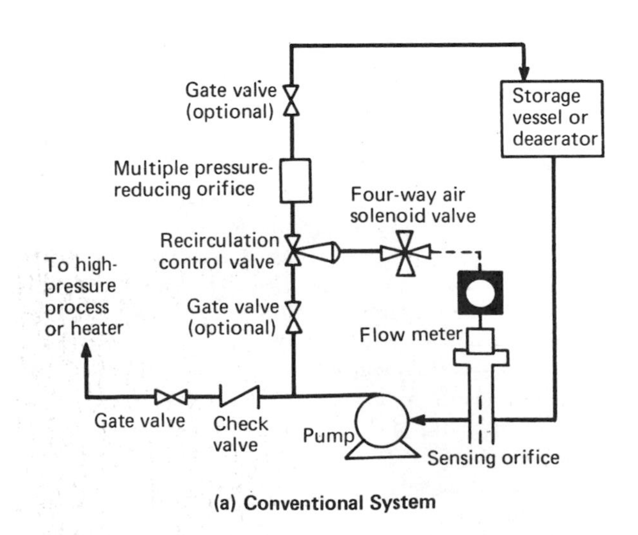


(a) Conventional System

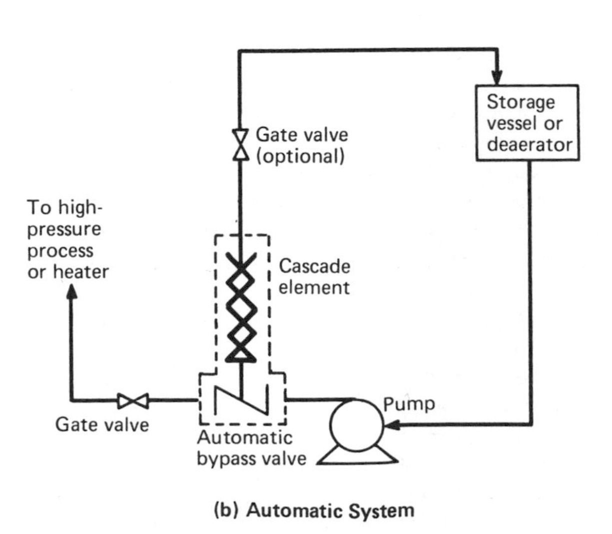


(b) Automatic System

**RECIRCULATION-CONTROL SYSTEMS.** Conventional arrangement is shown in (a), the automatic kind in (b)—Fig. 3

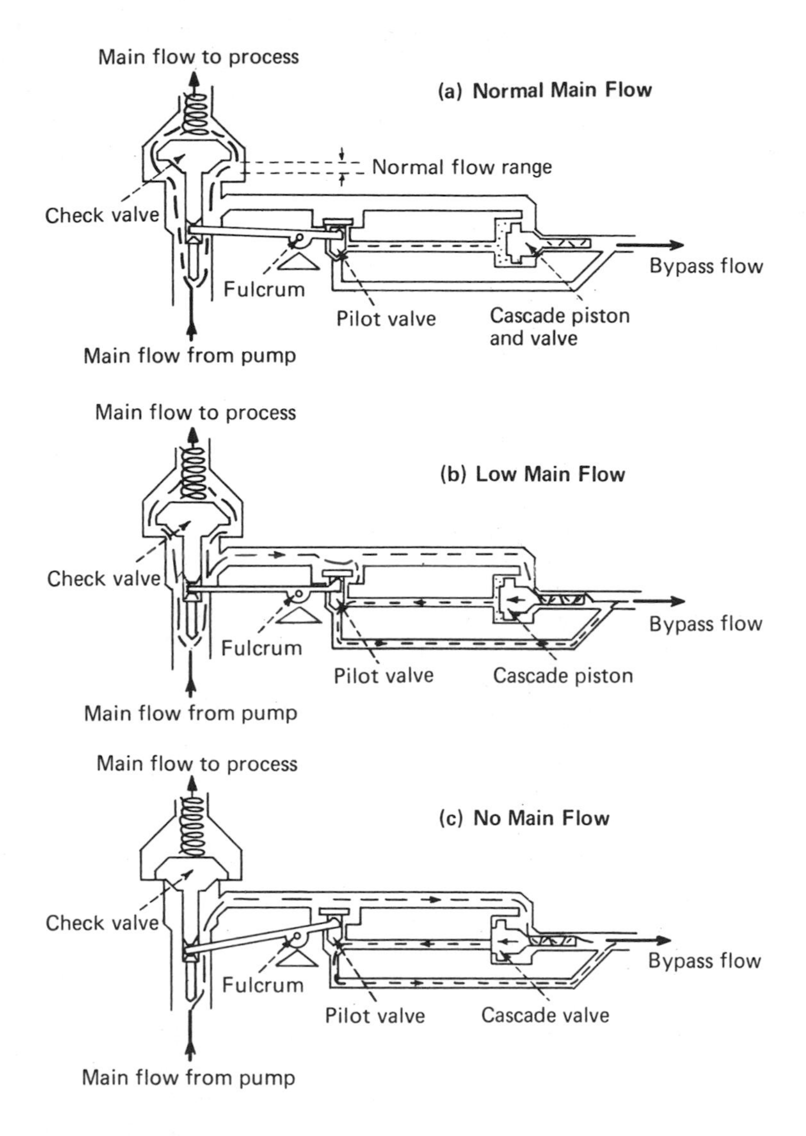


**BYPASS-CONTROL VALVE** mode of operation—Fig. 4

flowmeter when pump-discharge quantities fall below 40% of pump capacity.

When the check valve is fully seated due to lack of process-flow requirements (as shown in Fig. 4c), the cascade piston and valve are fully open. Bypass flow, which then is at its maximum, is returned to the low-pressure sump or reservoir.

Although the tightly seated check valve prevents reverse flow through the pump when the pump is out of service, reverse flow through the bypass system can be used to keep the pump casing and internals warm when the pump is not being used.

The cascade valve controls the bypass flow and dissipates the high-pressure energy differential from the pump discharge to the sump. This valve splits the main velocity flow into multiple streams.

Parallel flutes, milled into the cascade cylinder, direct fluid flow through a series of 90-deg turns as it cascades through the valve. The flutes constitute a series of variable orifices, and each set of flutes (or "stages") absorbs part of the fluid energy. The number of stages is governed by the magnitude of the pressure drop across the valve. The size or depth of the milled flutes governs flow capacity.

Because the seating surfaces are not exposed to high fluid velocity, they maintain tight shutoff of the bypass system for long periods of time. And since no external source of power—electrical or pneumatic—is needed, the system is designed to fail-safe, guaranteeing minimum flow through the pump. Virtually any magnitude of pressure drop can be handled by the system.

## Meet the Author

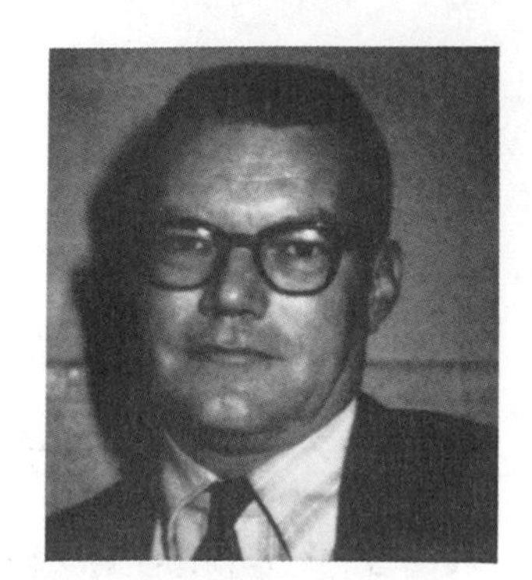

**Peter P. Van Blarcom** is assistant product sales manager for control valves and liquid-level measuring devices with Yarway Corp., Blue Bell, PA 19422. Before joining Yarway in 1962, he was with the Industrial Div. of Armstrong Cork Co. He holds a degree in mechanical engineering from Duke University, and is a member of the American Soc. of Mechanical Engineers.

# Pump Installation and Maintenance

Much of the effort devoted to careful selection and sizing of a process pump may be wasted if the unit is not installed and maintained properly. Here is a practical guide for the production engineer that details the right procedures and possible pitfalls.

JOHN A. REYNOLDS, Union Carbide Corp.

This article is directed toward the practicing engineer who may be involved in design or startup and operation of a chemical production unit. Most of the discussion will deal with the horizontal end-suction centrifugal pump, by far the workhorse of the CPI.

We will begin with a typical pump upon its receipt at a new plant site and assume that it has been specified, purchased, manufactured, inspected, packaged, and delivered properly. Give the pump a quick check for shipping damage and then a more detailed examination for corrosion, poor workmanship, dirt or metal chips in sensitive areas, and poor fits at seams and gaskets.

Be sure all loose items (coupling halves, uninstalled mechanical seals, piping, etc.) are either labeled and stored or are well sealed and wired to the pump. Also, verify that the pump is weather proofed since months may elapse before the pump is finally installed and started up. Lack of attention in the beginning can mean endless headaches during the critical startup period, when time is of the essence.

Drivers should be given the same careful attention as pumps. Most vertical turbine pumps and some large horizontal pumps are shipped with the drivers unmounted. Check these for shaft damage; they are quite vulnerable to being battered or bent. Also, plan to hook up space heaters on motors so equipped, even at the field storage stage. Large steam turbines should contain desiccant to prevent corrosion of the internals during storage.

Finally, prepare and carry out a plan for routine inspection of all mechanical equipment being held for installation. This includes manual rotation of the shafts to wet the bearings with oil, changing desiccant bags, and generally keeping things clean and weather-tight. Spare parts should also be at the jobsite during equipment installation because they may be needed to replace those damaged by corrosion or rough handling. A summary of installation checkpoints is presented in Table I.

### Pump Installation Check List—Table I

**1.** The foundation should be substantial to support the pump rigidly and absorb vibration.

**2.** The base plate must be leveled on the foundation so the oil in the pump bearing housing will be at the correct level and the oiler on the housing will feed properly.

**3.** There are two types of coupling misalignment—angular and parallel. Wedges, feeler gages or dial indicators are generally used to check alignment, as shown in Fig. 1.

**4.** Factory alignment of the pump and driver must be checked after the complete unit has been leveled on the foundation, and again after the grout has set and the foundation bolts tightened.

**5.** Suction piping is especially critical for proper pump operation and should not have pockets that may collect air or have sharp turns at the suction nozzle, which could cause an uneven flow to the impeller.

**6.** Very little pipe loading should be imposed on the pump. Expansion joints and pipe supports should be used judiciously.

**7.** The direction of rotation should be checked by jogging the motor before the coupling halves are joined.

**8.** A final alignment should be made after the pump has run long enough for temperatures to stabilize. Some manufacturers' manuals give cold alignment data that indicate the amount of misalignment the couplings should have when cold in order to be correctly aligned at the normal running temperature.

Originally published October 11, 1971

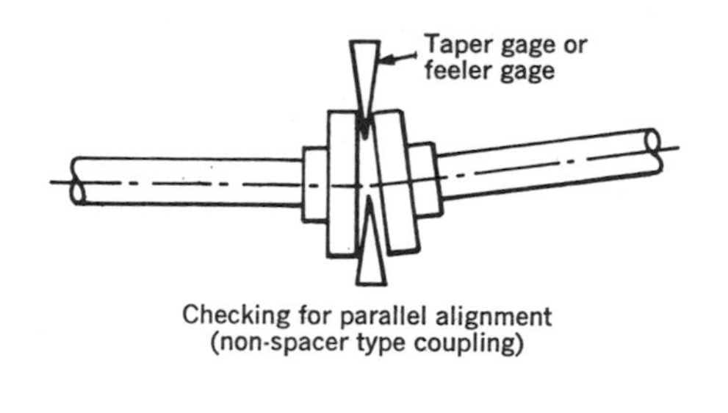


Checking for parallel alignment
(non-spacer type coupling)

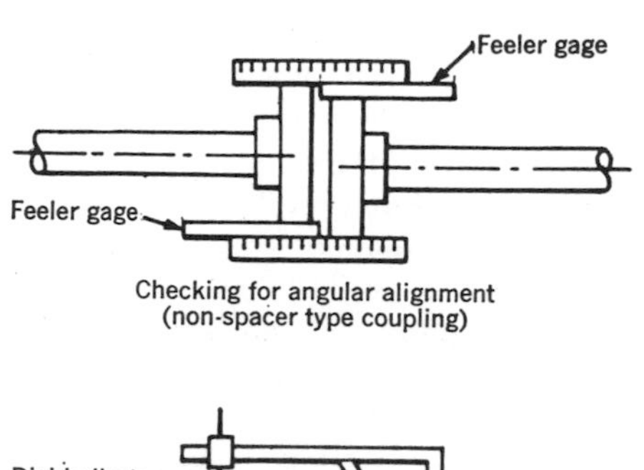


Checking for angular alignment
(non-spacer type coupling)

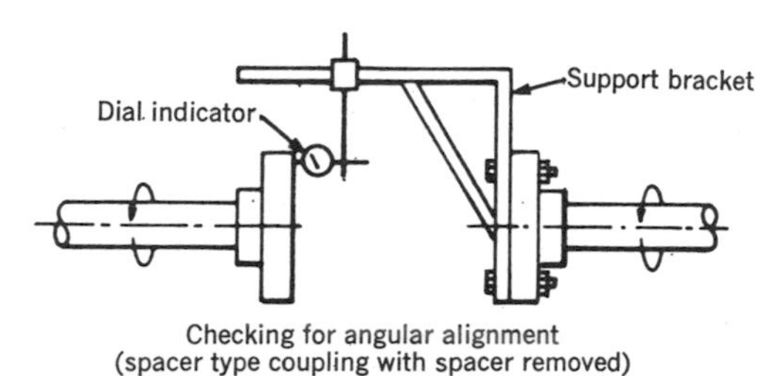


Checking for angular alignment
(spacer type coupling with spacer removed)

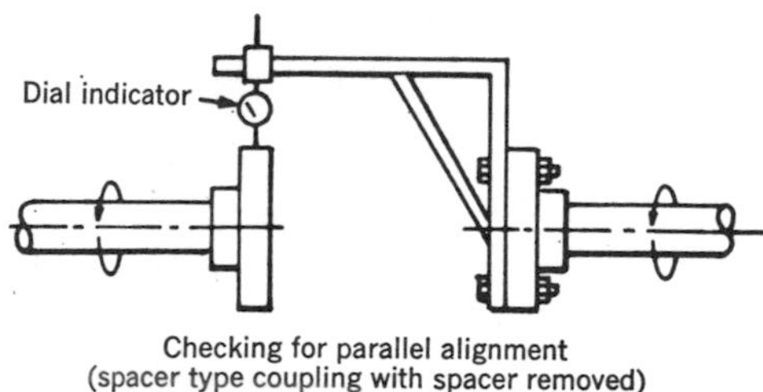


Checking for parallel alignment
(spacer type coupling with spacer removed)

**COUPLING ALIGNMENT** can be checked by several methods—Fig. 1

## Base Plates and Piping

Large pumps and drivers (above 100 hp.) generally have common base plates that are mounted on concrete foundations. These are rigid anchor points and every consideration must be given to forces imposed by the piping. Most manufacturers will furnish the allowable forces and moments permitted on their pump's nozzles. Beware of the supplier who does not give this information and states flatly that no pipe loading may be imposed on the nozzle. This is just not practical.

The pipe stress analyst must have some leeway to come up with an economical design. Otherwise, the result will be long and tortuous piping runs or extra expansion joints. The design engineer should always check the final suction piping layout. He may find that the available NPSH has shrunk to an impractical level and that the pump, as originally specified, will no longer work.

Large, low-head circulating pumps can be especially troublesome during piping layout because the available NPSH is usually low, the liquid is often near its boiling point, and the pump's NPSH requirements are usually high since it's well out toward the end of its performance curve. All this demands an elevated suction vessel with large diameter piping to reduce suction line pressure drop. In this situation, a spring-loaded and/or a sliding base plate may be the best way to cope with thermal expansion.

The piping layout may be improved by rotating the pump discharge nozzle to some position other than vertically upward. Many of the large, pedestal-supported pumps have casings that make this possible. One arrangement, in conjunction with a sliding base plate to better position another piece of equipment, is shown in Fig. 2.

Some very large pumps and drivers are installed without base plates, especially if a "train" of equipment is involved such as a turbine driver, a high speed pump, a gear box and a low speed pump. Here the equipment is bolted to metal sole plates that are positioned and anchored in the concrete foundation. The sole plates are used so that the individual machines can be easily removed for maintenance. This shortens downtime.

Positioning lugs with jack screws, welded onto base plates or sole plates, are most helpful to position heavy pumps and drivers axially and laterally. Preferably, these should be specified as a part of the original equipment package, but usually they can be added later in the field without warping or damaging the base plate. It's less expensive to have the equipment manufacturer furnish them.

**SLIDING BASEPLATE** is useful to position other equipment—Fig. 2

## In-line Pumps

In-line pumps can be used to advantage where space is at a premium or where elimination of a foundation is economically attractive. Many are installed with some type of concrete or steel pedestal under the pump, but this is still less expensive than a massive concrete foundation, with its base well below the frost line. One definite asset of this design is that it does not require detailed leveling, shimming and grouting for coupling alignment. Since all the bearings are in the motor, which is usually grease lubricated, there is no oil sump to fill.

Some of the off-setting factors for this type of pump are generally higher first cost (vertical motors more expensive) and lack of dimensional standardization, in contrast to the AVS horizontal end-suction pump. (Now American National Standards Institute Standard B123.1-1971). However, the Manufacturing Chemists Association has recently issued a proposed standard.

Standards are most useful because once the hydraulics of an application are established, the pump can be selected and the layout and piping drafting done before equipment procurement. Later in the field, dimensional standardization aids pump maintenance and replacement.

The pipe loading is also important with smaller pumps, but usually is not as critical. Some manufacturers offer foot-supported base plates that do not require separate foundations. These rest on concrete padding without anchor bolts, shims, or grouting (see Fig. 3). Only the piping holds them in place and thus they offer a greater degree of design flexibility. These pumps are generally limited to 75 hp. at 3,500 rpm. or 40 hp. at 1,750 rpm.

## Other Precautions

Piping spool pieces should be included in suction and discharge lines to make the pump accessible for maintenance unless it has the "back pull-out" feature in which the bearing housing, rear cover, and impeller can be removed without disturbing the piping or the motor. Refinery pumps, AVS pumps, and most vertical in-line pumps are designed this way. Be sure, though, that the maintenance shops have procedures to utilize this feature; otherwise, spool pieces are advisable.

Elbows should not be installed against the nozzles of end-suction centrifugal pumps that have suction nozzles over 4 in. dia. The elbow can cause liquid to enter the impeller eye with an uneven velocity distribution, which results in turbulence. One manufacturer recommends 10 diameters of straight piping immediately before the suction nozzle. This is most important where the NPSH available is close to that required by the pump.

A word about check valves. Don't depend on them to completely stop liquid flow or leakage. Always use regular valves to isolate the pump so it can be accessible for maintenance. If a check valve is also required in the system, place it between the pump and the regular valve so it too can be removed for repair.

Where parallel pumps are installed, check valves should be on the discharge lines of both pumps, so one cannot pump back through the other. The most common problem is the case of a spare pump that may be called upon at a moment's notice. If it has been primed and both suction and discharge valves opened, it should be ready to assume the load at the push of a button, without pumping back through the other. This can also happen where two identical pumps are opèrating in parallel and the wear rings of one become worn or broken, which causes its effective head capacity curve to drop.

## Auxiliary Piping

One of the thorniest pump problems is the hook up of the auxiliary piping, especially around the mechanical seal area. Many manufacturers' drawings give little detail here and most mechanical seal manufacturers' drawings show the connections rather schematically, especially with reference to those provided by the pump supplier. One practical solution is to sketch the auxiliary piping configurations in use and then tabulate the pumps that have each type.

The pump supplier can provide detail for cooling water connections to the bearings and stuffing box jacket, as well as bypass piping. Certain seal gland connections are supplied by the seal manufacturer. By combining the details supplied by both, a good set of sketches can be made up for the field. One plant went to the extreme of photographing each pump from both sides as it was received, and drew on the photo all the piping and labeled all connections that had to be made up. It was rather time consuming initially but simplified field installation. The photographs were of later benefit during maintenance.

Some seal glands are furnished with labels or tags that describe the function of the connection. Both have their shortcomings because they may be either painted over or pulled off. A practical scheme to help field personnel with auxiliary piping would be a

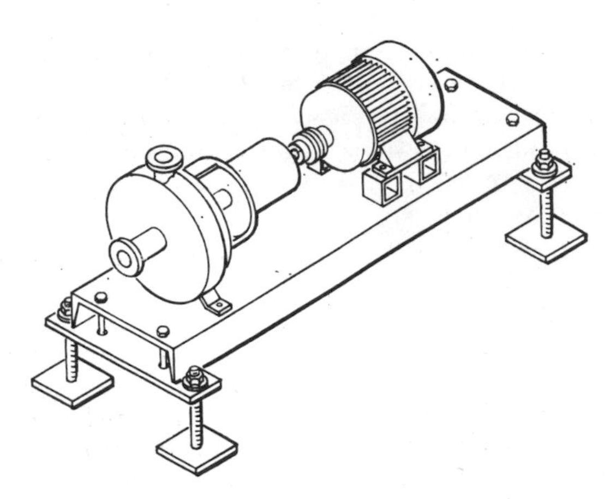

**FOOT-SUPPORTED** base plate does not require separate foundation—Fig. 3

system of standard letter abbreviations stamped next to each connection. Small plastic cards listing these symbols could be supplied to the construction and maintenance crews for ready reference. See Table II for a suggested list of abbreviations.

## Vents and Drains

Vent and drain connections are also an important part of pump installation. Drain connections should always be specified unless the liquid is hazardous or corrosive. Mechanics have a great aversion to opening a pump casing and having a deluge of smelly liquid drenching them and their tools. In situations where the process fluid is toxic or corrosive, a blowout with air or nitrogen, followed by a neutralization rinse and a final blow, is most feasible. A vacuum purge can be used instead of a neutralization rinse with low boiling liquids.

Vent connections are necessary for priming centrifugal pumps unless the unit has a top discharge and is self-venting. Even some self-venting pumps may require a vent if they can become filled with air or vapor and have to develop full head to open a check valve in the discharge line. A similar situation can occur in vacuum service. Even though the pump is at the lowest point in the system, it can become air or vapor filled and not prime properly without a bypass line from the discharge back to the suction vessel.

High points in the suction piping can become a trap for air or other entrained gases. These can vapor bind a pump, even after it has been vented and primed initially. Attention should be given to this possibility during piping layout for the project. A particularly bad point is where a liquid leaves a vessel, rises to a pipe rack, and then later drops down to a pump installed at grade level. If the material has a low boiling point, warming may generate vapor that will blind the pump. Concentric pipe reducers can also produce vapor pockets in horizontal piping.

Operating centrifugal pumps at greatly reduced flows will cause a heat rise that may degrade or vaporize the liquid. Heat rise may be calculated by:

$$\text{Temp. rise, F.} = \frac{\text{Head}}{778 \times \text{Sp. heat}}\left[\frac{1.0}{\text{Eff.}} - 1.0\right]$$

In the above equation, head is the total head of the pump in feet of liquid, and efficiency is hydraulic efficiency expressed as a fraction, both at the reduced flow condition.

The common solution is to install a bypass line to the suction vessel or through a heat exchanger. A fixed orifice in the bypass line can give sufficient flow to control the heat problem and still not reduce the pump's effective capacity. These are usually used on large, high-head pumps, such as boiler feed pumps, to permit some circulation at startup rather than have the pump dead headed against a check valve. Such devices are sometimes referred to as warm-up orifices and are used when running the pump at reduced speed or flow until its temperature has stabilized and its prime well established.

**Abbreviations for seal connections—Table II**

| | |
|---|---|
| CWI | —Cooling water in |
| CWO | —Cooling water out |
| D | —Drain |
| FI | —Flush in |
| FO | —Flush out |
| G | —Gage connection |
| HI | —Heat in (steam or hot water) |
| HO | —Heat out (steam or hot water) |
| LI | —Lubricant in |
| LO | —Lubricant out |
| Q | —Quench |
| SFI | —Sealing fluid in |
| SFO | —Sealing fluid out |
| V | —Venti |

For hazardous liquids, casing vent connections have to be more than just a threaded pipe plug or nipple and valve with open discharge to the atmosphere. Additional piping must be added to send the vapors and liquid to special drains or scrubbers. A flow indicator in the line is helpful to let the operator know when all liquid is vented.

Canned-rotor centrifugal pumps, which are generally used for hazardous liquids, require special venting if the liquid is near its boiling point. This design has journal bearings that support both the electrical rotor and the pump impeller. These bearings are lubricated by the pump liquid, which flows through a bypass pipe to the rear of the pump. The liquid passes through the outboard bearing, over the rotor, through the inboard bearing, and returns to the casing near the impeller hub.

Near-boiling liquid in the bypass line may be warmed by the pump motor and vaporize, causing the pump to lose prime. One solution is to add a heat exchanger to cool the liquid well below its boiling point before it's injected back into the motor cavity. Another technique is to specify that the pump have reverse circulation, so that the lubricating liquid flows through the pump and then discharges to the suction vessel. If some vapor is generated, it is generally of no consequence if injected back into the suction vessel.

## Pressure Relief

All pumping systems need to be checked carefully for placement of pressure relief devices. All positive displacement pumps (rotary and reciprocating) should have relief valves in the discharge piping since any blockage in the line will cause the pressure to rise until the motor stalls, the piping ruptures, or the pump breaks. Some of these pumps have built-in relief valves that are primarily for protecting the pump and driver since they only permit internal bypassing. As such, they do not protect the piping system as a whole.

For complete protection, the relief device must be able to discharge some liquid out of the system, or at least out of that part which may be blocked or isolated with valves. Also, an internal bypass does not dissipate heat that will develop within a pump running against a closed line. Liquid being recycled within the pump can reach temperatures that can ruin the pump, its seal, and the liquid itself.

There are a few rather rare cases where rotary pumps may be exposed to reverse rotation and have foot valves, check or hand valves to block the reverse flow. Pressure relief should be provided in the suction line. Damage to an unprotected pump may be of little consequence compared to the effect on a large compressor that loses its lube oil supply because a small pump fails.

Regular centrifugal pumps generally don't require pressure relief protection because they reach a shutoff head condition and cannot develop more pressure (see Fig. 4). Since the pump curve is rather flat, a relief valve usually cannot be used to prevent excessive temperatures when the pump is operating at shutoff.

Regenerative turbines, which are also kinetic machines, have such steep performance curves (Fig. 5) that a relief valve is usually in order. This type of pump should not be operated at shutoff under any circumstance because of high unbalanced radial loads, rapid temperature rise, and possible motor overload. In general, any piping system should have pressure relief if it may be blocked and experience a temperature rise. This phenomenon has caused much grief for operating personnel by damaging equipment and producing packing, mechanical seal, and flange leaks.

## Direction of Rotation

Direction of pump rotation should first be considered during the detailed layout stages in engineering. Many rotary and horizontal split case pumps can be manufactured to rotate in either direction, and generally one direction is more suitable for the layout than the other. By taking this into account, the most economical piping arrangement will result.

Later, during the installation phase, all pumps and drivers should be carefully checked for direction of rotation before joining the coupling halves. Many centrifugal pumps have impellers that are screwed on the shaft. Just jogging the motor to check rotation with the coupling connected can spin the impeller loose, usually jamming it into the casing. Canned rotor pumps, which have no exposed shaft or coupling, can be checked with a very inexpensive phase indicator.

Centrifugal pumps will develop some head and flow while running backward, though much less than their normal rating. This has given more than one operating person some concern. It seems to happen more after maintenance than during the initial startup, because of motor leads incorrectly connected.

Some reciprocating pumps will operate in either direction, but most have a preferred direction. With a shaft-driven lube oil pump, direction of rotation is critical.

Reverse rotation due to back flow can be damaging to centrifugal pumps, and especially their drivers, if it is sustained to cause excess equipment speeds. This can easily happen to vertical turbine pumps, where there is a long uphill line or an elevated reservoir that can drain back through the pump if it stops. Vertical motors can be furnished with antireverse ratchets.

Reverse rotation can also occur with process pumps where there is a large vessel of pressurized liquid that can feed back through the pump if it stops. Check valves are mostly used to prevent this, but they are not infallible. Electric motors will usually start destructing at a much lower rpm. than pumps because of their intricate rotors, especially in large form-wound machines.

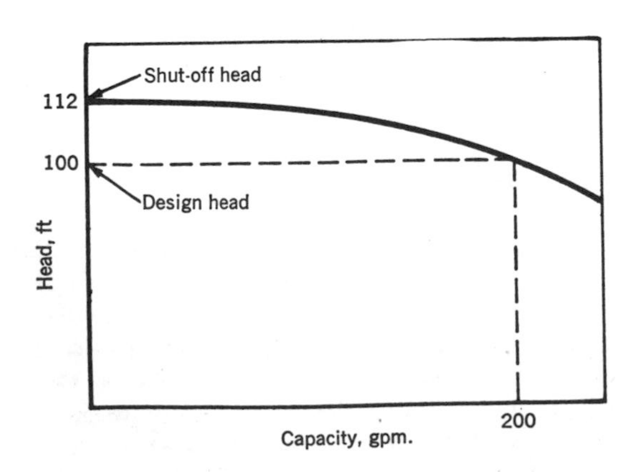


**PRESSURE RELIEF** is not required with flat pump curve —Fig. 4

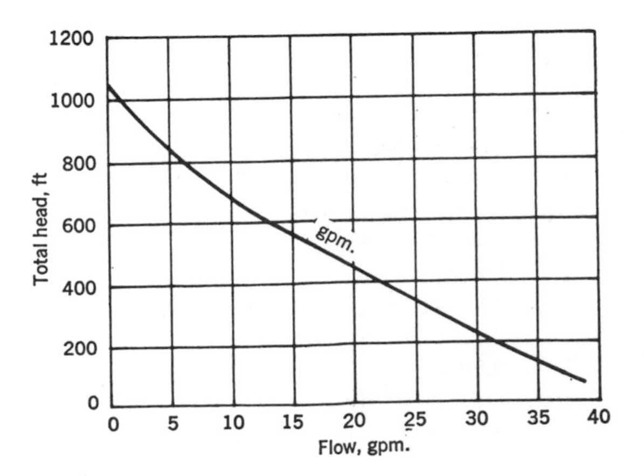


**REGENERATIVE TURBINES** have steep performance curve—Fig. 5

## Couplings

Spacer-type couplings should be used with all horizontal "back pull-out" design pumps. They permit removal of the bearing housing, along with the shaft, rear cover, mechanical seal and impeller without disturbing the pump casing, the piping or the motor. Thus, the motor-pump alignment is not disturbed and after the bearings or mechanical seal have been repaired the pump is easily reassembled without shifting large heavy pieces of equipment.

Spacer-type couplings are also quite useful with horizontal split-case pumps that are equipped with mechanical seals, providing the seals are mounted on a separate sleeve or cartridge that can be removed without opening the casing. The spacer coupling can be removed, providing space to remove the inboard bearing housing and the seal cartridge. This shortens and simplifies the maintenance cycle and should pay for itself many times over.

There are several elastomeric couplings, both spacer and non-spacer types, which have the advantage of requiring no lubrication (grit and dirt adhere to grease, causing wear), being very flexible and economical. These should be considered for services up to 50 hp. at 3,500 rpm. For larger sizes, these couplings tend to be bulky and expensive. Of course, there are some services where high temperature, high speed, solvent atmosphere or high horsepower make them impractical. In these situations, all metal couplings are the best choice.

Slip-type couplings are used successfully to protect rotary pumps from overload where the product is gummy and may foul the relief valve. In this case, pressure relief valves are also used, but are normally never actuated, when set for maximum line pressure.

## Pulsation Dampeners

Pulsation dampeners should be considered with reciprocating pumps and quick-closing valves. Both can cause severe hydraulic pulsations, which try to straighten out piping elbows and cause piping to vibrate. This, in turn, causes flanges to leak and piping to fail in fatigue or even rupture from overpressure.

There are two basic kinds of pulsation dampeners: the displacement type and the phase damping type. The first has several variations, as illustrated in Fig. 6. The weighted type, in which a piston is displaced in a reservoir, has a rather long response time, perhaps one cycle-per-second or longer. By substituting a spring for the weight, which reduces the mass of the piston, the response time is shortened. Another variation is the pneumatic or gas-loaded reservoir, usually having an elastomeric bladder to prevent loss of the gas charge. It is very responsive because of the bladder's low mass and can accommodate pulsations of 10 cycles-per-second and higher.

The second type, also shown in Fig. 6, utilizes phase dampening to attenuate pulsations, similar to the action of an automotive muffler. It works most efficiently above five cycles-per-second. This device has no moving parts but it does present some pressure drop. Dampeners should be located as close to the source of the pulsation as possible, generally at the pump discharge.

Fig. 7 illustrates the fluid velocity on the discharge of a simplex plunger-type metering pump. The same profile exists on the suction side if it is undamped, which means that the fluid is being stopped and started, accelerated and decelerated. With inertia and pipe friction at work, the liquid may flash at the suction side, resulting in poor pump operation.

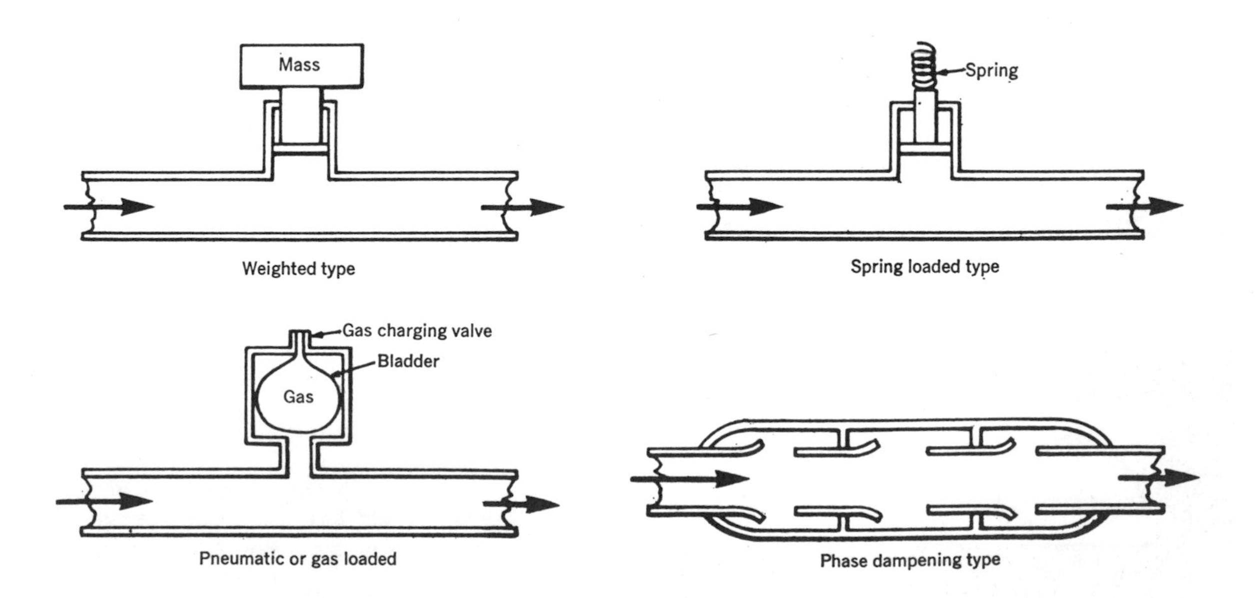


**PULSATION DAMPENERS** reduce hydraulic hammer—Fig. 6

## Catch Basins and Shields

The most common pump failure is leakage through the seal or packing. Leakage should be recognized from the beginning and steps taken to control, collect, and protect the equipment from it. There are many seal devices, such as the throttle bushing in the seal gland and the quench collar for packing, that can restrict leakage and direct it through piping to a collection point.

Most pump manufacturers offer drip pans that fit between the bearing housing and the casing, directly under the shaft at the point of possible leakage. These usually have a threaded connection that can be piped to drain away leakage. They have one shortcoming though; maintenance personnel have a tendency to lose them, especially if the pump is removed to the shop. Some production personnel prefer the larger catch basins that are mounted between the pump and the base plate and catch leakage from flanges and sample connections as well as the stuffing box. These are available in stainless steel and offer better protection for a steel or cast iron base plate. In fact, they may not cost any more than a drip-lip base plate, which is fine for catching noncorrosive leakage but can have a very limited life with aggressive liquids.

Stainless steel drip shields have also been used to protect base plates where the entire pump is surrounded by an acid brick dike. The shield is thin sheet metal, mounted directly between the pump and the base plate. It covers all or most of the basic plate and overhangs an inch or two around the edges. In a few extreme corrosive situations, the entire bearing housing has been fabricated of stainless steel. There are some paints and plastic coatings that can greatly increase the life of cast iron and steel parts in corrosive service, although nicks and scratches can detract from their utility.

## Startup Screens

All piping should be thoroughly cleaned before startup. This involves pumping a cleansing liquid through the lines, usually water, and sometimes blowing them clean with compressed air. Even after this has been done, it's sometimes amazing what appears in pump impellers. Thus, it is good practice to install ¼-in. mesh startup screens before the suction nozzles of rotating equipment.

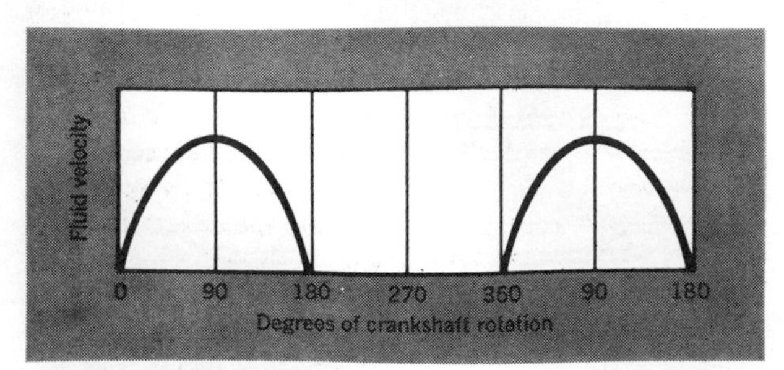


**FLUID VELOCITY** profile for a plunger pump—Fig. 7

For pumps with critical NPSH, startup screens should be the long conical type with low pressure drop. Flat screens blank off about 50% of the pipe area, which could cause cavitation. Sometimes it is wise to have screens near suction nozzles during normal operation, especially for two parallel pumps in a closed loop system or for pumps in series. There have been many cases where a pump has failed, dumping broken wear rings and impeller bits into the system and damaging the installed spare immediately upon its startup or causing the failure of other pumps. One wreck is bad enough; don't compound it by lack of protection.

## Startup

Even after much engineering effort has been expended in design, specification, and installation, it is good practice to call in the manufacturer's field or erection engineer before pushing the start button. For large, complicated pumps this can be particularly valuable and time saving. It gives the manufacturer direct information on how the installation was handled and he has assurance that his instructions were followed, both to the intent and the letter of the instruction manuals. This will save time in getting to the crux of the problem, should one arise during startup or initial operation of the pump.

While the erection engineer is on site he should also go over maintenance details and instruction books with plant personnel and even help identify and catalog spare parts. This can be done most efficiently if advance plans are agreed upon. In any case, there should always be some independent audit of the installation unless the construction crew has a performance warranty to meet.

Now, be sure the pump is filled with liquid! Most pumps do a reasonable job of handling liquids, but pumping air causes all kinds of misfortunes. Even self primers need an initial charge of liquid. This can be accomplished with an air, water or steam ejector; with a vacuum pump connected to the vent connection; or even with a bucket of liquid poured in the priming chamber.

Running a centrifugal pump dry can cause the wear rings to seize, mechanical seal faces to rapidly wear and score, packing to burn, shaft sleeves to score and any interstage bushings or internal bearings to wear rapidly or seize. Reciprocating pumps usually score their plungers or pistons and cylinders, as well as burn the packing. Rotaries will wear or score gear teeth and wear plates, if so equipped, and ruin the seals or packing.

A seasoned maintenance man once quipped, "If you operators could just remember to keep liquids in pumps and gases in compressors, my job would be so much easier."

When starting centrifugals, have the discharge valve almost closed to prevent the pump from "running off the end of its curve" and cavitating. Unless the discharge piping is completely filled, there isn't enough head or friction to keep the pump at its design flow. It will operate at end of its curve where the NPSH requirement exceeds that available and cavitation will result (this generally sounds like pumping gravel).

As the flow fills the system, the control valve can be regulated to move the pump output to its design condition. With a regenerative turbine, don't restrict the discharge valve too much because excessive pressures are generated at shutoff, as mentioned earlier. For positive displacement pumps, just completely open the discharge valve and leave it in that position.

## Operation and Maintenance

If all the foregoing precautions are followed, there is very little left to discuss about pump operation except for turning them on and off and doing a little trouble shooting. The first operating consideration is, who will do the trouble shooting? Generally, production personnel are so concerned with other routine problems that they are hard pressed to become knowledgeable about the intricacies of pumps. For this reason, it is well to designate an individual or team, in either the maintenance or the engineering department, to do the trouble shooting.

The specialist, by training, daily exposure and perhaps native ability, can help arrive at the nub of a problem without delay, trial and error methods and false starts. He can also serve as a contact point for the various manufacturers' representatives. This position can be used as a training ground for maintenance engineers. In this case, be sure to keep an individual in the slot long enough to really become an expert and become known as such.

Of course, production operators should be trained in pump fundamentals. This is a must for good operation and low maintenance costs. Some operators are accused of being careless with equipment when, in truth, they have never had proper training. The specialist can be used to provide training sessions for plant operators.

## Trouble Shooting

Loss of prime is one of the common problems that will be encountered during daily trouble shooting. This may be caused by incomplete venting, a blocked suction line or entrained gases in the liquid. Entrainment can result from a number of situations including: having incoming lines in the suction vessel above the liquid's surface; vortexing in the suction vessel; air leaks through joints, valve packing or pump packing; and flashing of low boiling liquids.

The reason a centrifugal stops pumping liquid when the impeller becomes filled with air is that its output pressure in psi. is drastically reduced. The impeller may be creating about the same head in feet of "fluid," but the specific gravity of the "fluid" has changed from that of the liquid (1.0 for water at 20 C) to that of air (0.00123). The conversion to psi. from feet of liquid is:

$$\text{psi} = (\text{ft. of fluid})\ (\text{sp. gr.})/2.31$$

From this we can see that a water pump generating 100 ft. of head or 43.3 psi. will only generate about 0.053 psi when it becomes "air bound." And since the liquid in the discharge line needs a given pressure to move it at a certain velocity, the suddenly reduced pressure allows flow to stop.

With reciprocating pumps the reason for air binding is somewhat different. The air is drawn into the pumping chamber where the plunger or piston tries to compress it. If there is pressure in the discharge line to keep the discharge valves seated, then the air is simply compressed without being expelled. As the plunger is withdrawn from the chamber, the air within expands to the suction pressure without drawing in more. This is repeated until the packing burns and the plunger becomes scored.

Rotary pumps become air bound because air is so much less viscous than liquid that it simply bypasses from the discharge back around the rotor side and peripheral clearances to the suction. It is recycled again and again. This, too, causes burned packing and scored shafts since they are being lubricated with air instead of liquid.

## Motor Overloading

Motor overloading can be a little trickier to solve than air binding. With a centrifugal pump, look for hot bearings or vibration. Internal parts may be loose and rubbing, or the bearings may be failing and causing an overload on the motor. A reduced discharge pressure may indicate that the pump is operating far out on its performance curve, which generally requires more horsepower.

If the pump has worn internals there may be internal bypassing. The output flow may be at the design point and a reduced discharge pressure may be the only indication that extra fluid is passing through the impeller, requiring more horsepower. In this case, shut off the pump and turn the shaft by hand. The motor may have bad bearings or a rubbing rotor. For smaller pumps, tight packing can use as much power as the pump itself.

If the pump duty has changed, the specific gravity of the new fluid may be greater, calling for more horsepower. Or a new, larger impeller may have been installed, also requiring more power. Pump speed is another important factor since horsepower varies as the cube of the speed ratios. A change in liquid viscosity can require more power, even with a decrease in both head and capacity. Refer to the correction chart in the Hydraulic Institute Standards to determine the effect of viscosity on pump performance. If a positive displacement pump is overloading, check for tight packing, high discharge pressure, worn bearings and gears or an increase in fluid viscosity. Also have a qualified electrician check the motor and the power source.

## Vibration

Vibration problems in a pump may be due to a bearing failing or coupling misalignment. First turn off the pump and check for coupling misalignment. This should be done as soon as possible after the pump stops in order to get a hot reading. Then let the pump cool down to ambient temperature and check the coupling again. Record both readings so that cold alignment can be made to compensate for temperature changes.

For large pumps the alignment as well as the vibration level is very important. Some sophisticated techniques have evolved, using inductive proximity probes to check the hot alignment and monitor vibration[1]. Many large or critical pumps are equipped with resistance temperature detectors or thermocouples to monitor bearing temperatures and give warning of incipient failure so the pump can be shut down before a wreck occurs.

Other points that should be checked when vibration is a problem include:

1. Check for a bent shaft. A dial indicator used along the shaft can easily identify this problem.
2. Inspect the coupling for abnormal wear or looseness. If the coupling is large, check it for balance.
3. Check the impeller for damage and for foreign material lodged between its vanes.
4. Check the piping for vibration. Loose pipe supports can permit piping vibration to be passed on to the pump. Perhaps more pipe supports are needed.
5. Check the hydraulic conditions of the suction piping. Is the NPSH adequate? Is there vortexing or entrained gasses in the suction line? (Rumbling and surging of the pump.)

## Bearing Failure

Bearing failure may be caused by water or product contamination in the bearing housing. Some of the older designs of centrifugal pumps do not have lip-type oil seals in their bearing housings and a seal failure usually sprays enough liquid to ruin the bearings. Slingers on the shaft can help exclude liquids (while rotating) but on non-running spare pumps they don't give much protection. Even in-breathing of moisture laden air during ambient temperature changes can cause water to condense and dilute the lubricant.

Oil mist lubrication has helped solve some of these problems. It keeps the bearing housing pressurized with clean dry air and a "mist" of oil. The mist is usually created by a centrally located generator feeding to overhead distribution headers, to branch lines, and finally to drop lines delivering the mist to individual pumps. For remotely located pumps, grease lubrication appears to be superior to the standard oil sump lubrication.

High suction pressures may create unusual thrust loads on bearings. Also, be sure the thrust bearings are installed in the correct position. Angular contact bearings can be accidently switched from a tandem arrangement to face-to-face or back-to-back, which reduces their load carrying capacity. Check with the pump manufacturer to see if the proper thrust bearings were furnished.

Inboard bearing failure may be caused by high unbalanced radial loads, which can result from operating at or near shutoff. A bypass line to move the pump out on its curve can help reduce this load. (For volute type centrifugal pumps this unbalanced radial load approaches zero at the maximum efficiency point.)[2]

Coupling alignment can also cause bearing failure. Although misalignment always creates vibration it may not be noticeable, especially at very low speeds or in a particularly rigid system, and yet tremendous forces may be imposed on the bearings. Most manufacturers offer a bearing failure analysis service that can help pinpoint the reason for failure.

## Mechanical Seal Failure

A survey of maintenance records would reveal that mechanical seal failures account for the bulk of pump repairs. Thus, improvement in this area can have significant results. Of course, even if the number of repairs are not reduced, it is a step in the right direction to cut the cost per failure. If current practice is to remove an entire pump to the maintenance shop, then the use of the "field seal repair" feature can save time and dollars. It is generally applied to AVS end-suction horizontal pumps, but can be applied to other "back pull-out" desings. The procedure is illustrated in Fig. 8.

The pump must be equipped with a hook-type shaft sleeve whose internal diameter is relieved in the area of the seal clamp collar so that the sleeve can be slipped on or off the shaft with the seal clamped to it. Without the internal relief, the force of the clamp collar set screw can distort the sleeve enough to prevent its installation or removal. There must be some provision to rigidly support the bearing housing on the base plate after it has been "back-pulled" from the casing. For larger pumps, a special rear-foot fixture is required so that the housing can be slid toward the driver, rotated for access and locked in position without being completely free to topple over.

Once in this position, the impeller and rear cover can be removed to expose the sleeve-seal unit, which can be removed and replaced. All this is accomplished without the pump leaving its base plate. The old seal is returned to the shop for repair.

The two mating faces of a mechanical seal must be treated as if they were a sliding bearing (which they are) that must be lubricated, cooled if necessary and kept in good alignment. Because of the heat generated by the rubbing faces of the seal (½ hp. or more) the liquid entering the stuffing box from the pump casing may flash and form a vapor pocket. Lubrication is lost and the faces run dry causing scoring and heat-checking. Bypass or external flushes are used to remove heat.

Dry running of seal faces is common in tank transfers where an operator starts the pump and returns at

**1.** Remove spacer coupling as shown above. Then remove casing nuts, and loosen cap screw under bearing housing.

**2.** Slide bearing housing unit back and to one side, turning 90°. Then retighten cap screw under the bearing housing.

**3.** Remove seal gland bolts and slide gland back. Remove impeller and rear cover plate.

**4.** Remove old seal and replace with new or rebuilt one. Illustration shows pump with replaceable sleeve.

**FIELD SEAL** repair technique saves time and money—Fig. 8.

some later time to check its progress. Often the tank is empty and pump is running dry. A flow detection switch, either in the discharge line or in a small bypass to the seal, can save many costly repairs.

Pressure balanced seals must be used for higher stuffing box pressures (roughly over 75 or 100 psi, depending upon liquid lubricity) to keep the face loading pressure from getting too high. Some very high head pumps have stuffing box pressures very near suction pressure. Several books have been written about the design and use of mechanical seals but if the above fundamentals are kept in mind, many seal problems can be solved.

## References


1. Jackson, Charles, Successful Shaft Hot Alignment, *Hydrocarbon Processing*, Jan. 1969, Vol. 48, No. 1.
2. Stepanoff, A. J., Centrifugal and Axial Flow Pumps, 2nd Edition, N. Y., Wiley, 1957.


## Meet the Author

**John A. Reynolds** is a senior engineer in the Chemicals and Plastics Engineering Dept., Union Carbide Corp., South Charleston, WV 25303. He is responsible for specification, procurement and consultation to plants on pumps, seals and related equipment. Mr. Reynolds holds a B.S. degree in mechanical engineering from West Virginia University and has been active in mechanical evaluation of machinery since 1951. He has specialized in pumping equipment since 1962.

# How to get the best process-plant layouts for pumps and compressors

Economy of piping and structures, along with ease of operation and maintenance, are the principal aims when making arrangements for pumps and compressors, their drivers and auxiliary components.

***Robert Kern,*** *Hoffmann - La Roche Inc.*

☐ Layout design and piping configurations affect the capital cost of, and energy used by, pumps and compressors. We will combine the discussion for this equipment because the requirements for layout and piping design often overlap. Significant differences will be taken up separately.

Piping design can be influenced only to a limited degree by the choice of alternatives. Since pumps and compressors have been extensively described we will review only the pertinent details of this equipment as they affect plant design, layout, and piping design.

## Centrifugal pumps for process plants

If a pressure difference does not exist between two points in a piping system, a pump (or compressor) has to be used to provide the necessary flow. Centrifugal machines are the types most frequently found in process plants.

The discharge head of a centrifugal pump (or compressor) depends on the type of impeller, its diameter and speed. Impellers (Fig. 1a) have three basic forms:

*1. Radial Flow*—In the radial-flow pump, the inlet is axial and the outlet radial. The discharge head is provided entirely by centrifugal force. This impeller is suitable for developing high heads. Most process pumps and compressors belong in this category. Many horizontal and vertical pumps have this type of impeller.

*2. Mixed Flow*—In the mixed-flow pump, the liquid enters axially and the impeller discharges at an angle to the pump shaft. In this design, the energy imparted to the liquid is a combination of centrifugal force and axial displacement. Some vertical pumps have this type of impeller.

*3. Axial Flow*—In the axial-flow pump, the liquid enters and leaves axially. All the energy imparted to the liquid is by the lifting action of the impeller. There is virtually no centrifugal force. Some vertical pumps have this type of impeller. Axial-flow compressors or blowers have low heads and high impeller speeds.

Before considering layout and piping design for these machines, we will examine the mechanical-design characteristics of centrifugal pumps.

The single-stage pump has one impeller. In response to the needs of the chemical process industries, a "standard" design (Fig. 1b) was developed. This standard pump was originally known as the American Voluntary Standard, AVS, pump, but is now made in accordance with ANSI B73.1-1974. The pump has a single-end suction. A wide capacity range is available from a few interchangeable impellers. Rotating parts can be removed without disturbing piping, casing or motors. This pump can be used when the suction vessel is at grade, the suction line is near grade, the liquid is subcooled, or available NPSH is low.

A variation of the horizontal end-suction pump has the inlet located at the top of the pump casing, as shown in Fig. 1c. Here, the piping and valving will occupy less space, but the suction vessel must be elevated. The horizontal-inlet pump can also be used, but piping and valving will occupy space in front of the pump suction. Estimates of the floor space required for these pumps are given in Table I.

Multistage pumps have two or more impellers in

Originally published December 5, 1977

Pipe sizes and floor space for single-impeller centrifugal pumps
(Total head 40 to 400 ft)

Table I

| Pump type | Capacity, gpm | Pump nozzles: Suction, in. | Pump nozzles: Discharge, in. | Pipe sizes: Suction, in. | Pipe sizes: Discharge, in. | Floor space ft |
|---|---|---|---|---|---|---|
| **Single-inlet impeller** (End suction, top discharge) or (Top suction, top discharge) | up to 100 | 2 | 1 | 2–3 | 1–2 | 1.5 × 4 |
| | 100 to 200 | 3 | 1½ | 4 | 2–3 | 1.5 × 5 |
| | 200 to 300 | 3 | 2 | 4–6 | 3–4 | 2 × 5.5 |
| | 300 to 700 | 4 | 3 | 6–8 | 3–6 | 2 × 6 |
| **Double-inlet impeller** (Side suction, side discharge) | 700 to 1,000 | 6 | 4 | 8 | 6 | 2 × 6 |
| | 1,000 to 1,500 | 8 | 6 | 10–12 | 6–8 | 2.5 × 6.5 |

Nozzle and pipe dimensions are nominal pipe sizes.

series. The discharge of one impeller is the suction of the next one. The heads developed in all stages are totaled. These pumps are for working against high discharge pressure—for example, heads provided by medium-pressure and high-pressure boiler-feed pumps.

Depending on the number of stages, the casing can be long (Fig. 1d) for multistage pumps, and more space will be needed. Suction and discharge nozzles are usually vertical on these pumps. Pumps having horizontally split casings should have access from both sides for convenience in maintenance. The barrel-type arrangements (vertically split casings) require removal space in front of the pump for pulling the shaft and impeller during inplace maintenance.

## Impeller and casing design

A variety of centrifugal-type impellers exist. Liquid accelerates through the cavities of the enclosed impeller. The wide impeller provides high capacities and low head—usually for pumping slurries, sewage and water. The semi-enclosed impeller has one side open; the pump casing encloses the open side of the impeller vanes. The open impeller is enclosed on both sides by the pump casing. The double-inlet impeller has two impeller eyes, and is usually found in relatively high-head and high-capacity pumps used for cooling-water and fire-water services.

Large-capacity water pumps usually have horizontally split casings with a double-inlet impeller. Inlet and outlet are horizontal (i.e., at right angles to the pump shaft), and suction piping is simple—often a short, straight line with one or two expansion joints. A horizontal elbow at the pump suction is undesirable because it supplies uneven flow to the double-inlet impeller. This can affect the life of pump bearings.

Maintenance is simple and can be done without disconnecting the piping. Considerable space is required around these pumps because of the large-size lines, fittings and valves, and to meet space requirements for mobile-maintenance facilities.

Vertically split casing (perpendicular to the pump shaft) lends itself to good maintenance access. For large pumps, clearance has to be provided in front of the pump for casing-head removal, and shaft and impeller pulling. (In some of these pumps, the impeller can be removed toward the driver.)

Most pumps are horizontal and must be mounted with the motor shaft and pump shaft carefully aligned.

Inline pumps are compact, and economical from the standpoint of capital cost, layout, piping design and maintenance. According to their manufacturers, inline pumps (5 to 100-gpm capacity) can be mounted horizontally, vertically up or down, or at any angular position. If necessary, inline pumps can be located overhead without looping the piping to grade. However, it is important that temporary or permanent access be provided to these pumps and their associated valves and instruments.

Large inline pumps have a pedestal and have to be supported on a vertical plinth. Such pumps are generally located at grade or floor level.

## Vertical-shaft pumps

Vertical pumps occupy small areas but require removal access and vertical space for lifting the motor, pump shaft and impellers (Fig. 1e and 1f). On many types, motor and pump shaft can be separately removed. Several types of vertical pumps are available. Their principal features are:

■ The submerged pump is a single radial-impeller pump with a long vertical shaft (Fig. 1e). Required submergence and inlet dimensions are specified by the manufacturers. Suction-pit size and depth must satisfy these conditions.

■ The wet-pit propeller pump has a mixed-flow impeller. The impeller's lifting action is usually applied in several stages. In some pumps, the propeller blades can be adjusted to suit operating conditions.

■ Deepwell pumps have impellers with radial or mixed-flow discharge. Each impeller is in a semi-enclosed housing with discharge vanes. A great number of impellers can be mounted in series. Consequently, these pumps can develop high heads, and they find use for pumping well water. A long shaft (and the corresponding removal height) can be avoided by using a submersible motor.

A foot valve at the inlet keeps vertical pumps primed. One of the maintenance operations is to clear the foot

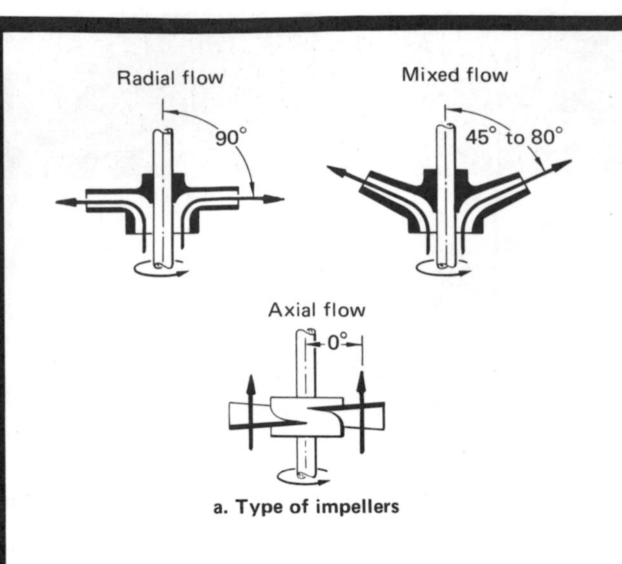


a. Type of impellers

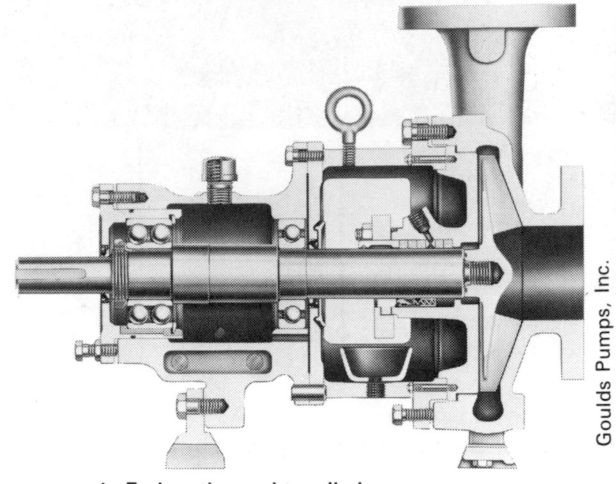


b. End-suction and top-discharge pump
(For elevated vessels and vessels at grade)

c. Top-suction and top-discharge pump
(For elevated suction vessels)

d. Multiimpeller pump
(Requires more space)

e. Sump pump
(Requires well-dimensioned pit)

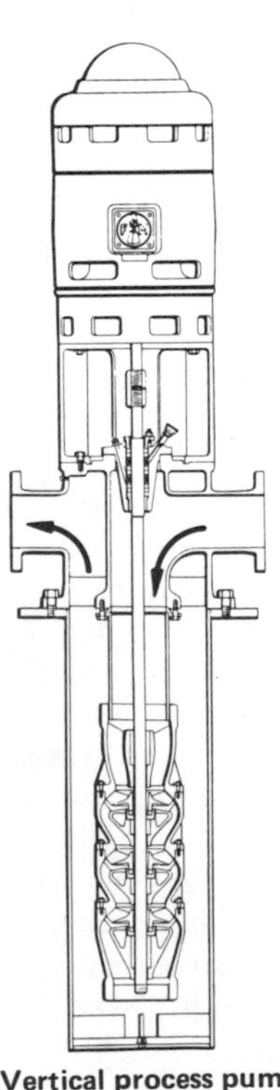


f. Vertical process pump
Flexibility in inlet and outlet orientation)

g. Close-coupled pump
(Mounts in any position)

Centrifugal pumps Fig. 1

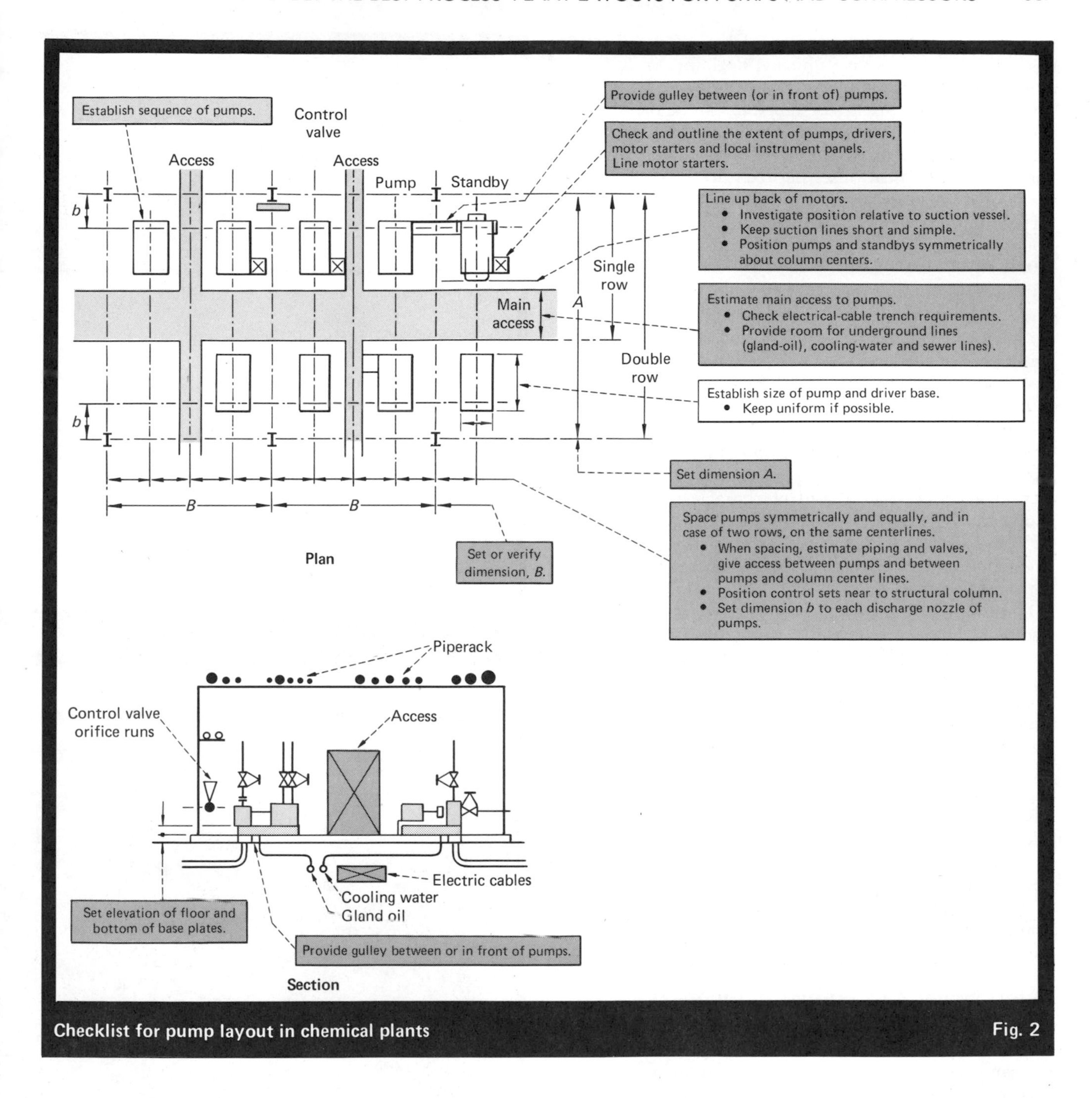


**Checklist for pump layout in chemical plants** Fig. 2

valve and inlet screen by removing the pump from its pit. For this purpose, valves, piping and electrical conduits should not interfere with the operation.

■ The dry-well pump is the most commonly used vertical pump in process plants (Fig. 1f). It is a shorter and enclosed version of the deepwell pump. Its suction and discharge nozzles are above grade, but the impeller inlet is below grade. Priming is not required. The pump is removed with its casing.

To exploit the capabilities of a vertical process pump, it should be located right under, or alongside, the vessel it takes suction from. Provision can be made for mobile-crane removal space or hitching points.

For dry-well pumps, the suction and discharge nozzles can have any practical horizontal angle to each other to suit required piping arrangements. Off-the-shelf pumps have suction and discharge nozzles at 180° to each other.

## Types of pump drives

Cradle-mounted pumps having coupled drivers are the most common units (Fig. 1b, 1c and 1d). Most centrifugal pumps are electric-motor driven. Turbine-driven (steam, gas or hydraulic) pumps need substantially more space for the gearbox, valves, piping and instrumentation than motor-driven pumps. Maintenance access and possible overhead trolley beams have to be planned for.

Close-coupled pumps are mounted directly on the driver-motor shaft (Fig. 1g). These pumps are compact and can work in any shaft position. Due to their design, capacities are limited. The discharge nozzle can be rotated in bolt-hole increments relative to the motor base. These features give unusual flexibility in layout and piping design.

Gear-driven pumps are designed to handle the different requirements between driver and pump speeds. These pumps are usually installed horizontally, with a horizontal or vertical offset between driver and pump shafts. This increases overall width and length and, for vertical offset, height.

## Cost, operation and maintenance

Centrifugal pumps pay for themselves in a short time. The low-investment cost is due to simple designs, direct-coupled motor arrangements, and a wide range of material, size, performance and operating characteristics. Such pumps occupy small space without shelter. Piping arrangements are simple.

Centrifugal pumps are dependable and long lasting. They can tolerate internal corrosion and erosion without substantially decreasing performance. Flexible operation gives good flow-control characteristics over a wide capacity range at constant speed. Capacities can be conveniently varied by throttling the discharge. These pumps are quiet, need little attention, and operate pulsation-free.

Due to the long life of the pumps, maintenance cost is also low. They can be easily disassembled, few parts have close tolerances, and worn parts can be quickly replaced.

Capacity, head and efficiency rapidly decrease with changing viscosity. The centrifugal pump (except for the regenerative type) cannot transfer liquid having a vapor content. At low flows (below 15 to 20% of design capacity), the centrifugal pump becomes unstable. Thus, a minimum flow is necessary. This means an additional bypass line to the suction vessel.

Pumps are selected by specialists, and the piping designer has little influence on the basic selection. However, the layout designer can request preferred orientation for suction and discharge, and NPSH limitations (to meet required equipment elevations).

## Pump layout and piping: outdoors

Pumps rarely influence plant layout except where a common standby for two services might require the rearrangement of process equipment. Pumps are placed close to process vessels. A number of pumps should be lined up and esthetically well arranged.

In a chemical or petrochemical plant, most pumps are located in two rows under the piperack. Lined-up motor ends define the access space in the center of the two rows. Pump suction and discharge face toward the process vessels. Fig. 2 illustrates one evaluation for a pump (or pump house) arrangement. Single pumps have access all around. With few exceptions, suction lines can be routed overhead. Discharge lines with flowmeters run just above headroom under the piperack. Control valves are usually looped to grade preferably at structural columns to facilitate good support. The preferred pump-nozzle arrangement (for vessels at high elevations) is top suction and top discharge.

Where space is restricted, or the pumps are small, or in the expensive space of a structure, two pumps can be

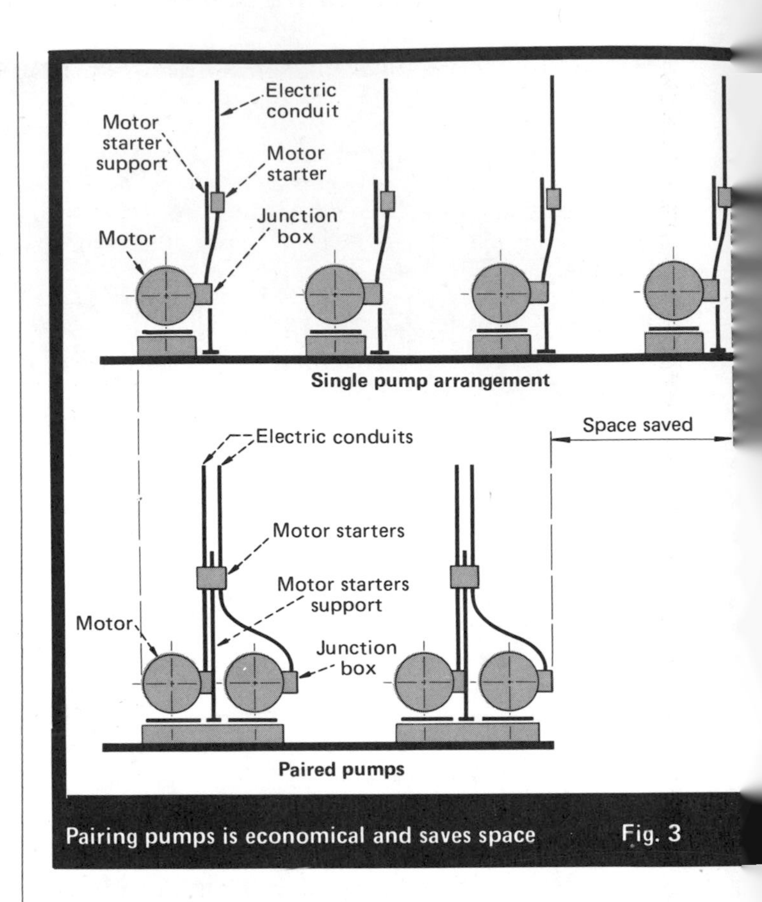


Pairing pumps is economical and saves space Fig. 3

Arrangements showing pump and vessel relationships inside a building Fig. 4

placed on a common concrete plinth, with one starter support for the two. Space usage and arrangements for such conditions are shown in Fig. 3.

For safety and operators' convenience, pumps to vessels containing flammable liquids should be located outside the dikes. (An exception occurs with API tanks having inline pumps mounted on the tanks' suction nozzles.)

## Pump layout and piping: indoors

Pumps that deliver subcooled liquid from process vessels usually take suction almost at floor level. These pumps must have an end-suction or side-suction inlet, and usually top discharge. Inline pumps are often chosen for such services. Looking from the access aisle, the motor-starter supports should be placed behind the pump and motor, away from pump flanges and maintenance clearances.

Saturated-liquid, steam-condensate and vacuum conditions require elevated suction vessels. These vessels are placed under the ceiling, elevated through the floor, or upon the floor above. Fig. 4 shows the lined-up pumps for various arrangements.

Hydraulically, a long vertical drop for the suction lines is preferred. Because of building restrictions, a horizontal run is not often provided in the suction line. For thermal-expansion purposes, an expansion joint must be installed in the vertical leg, close to the pump. Piping at the expansion joint must be well guided so that side-deflection of the vertical line does not tear the expansion joint apart. Loads (deadweight and pipe expansion) should not be imposed on the flanges of pumps, or on any rotating machinery.

A space-saving arrangement is possible if the pumps can be placed along structural walls. Motor starters, control valves and utility manifolds can be supported on the wall over the pump space. For example, the work tanks may be outside the building, while the pumps are inside. In general, existing structures, walls and building columns should be used for supporting instrumentation and electrical hardware.

Auxiliary pipe manifolds should be located overhead above the pump. These manifolds may supply cooling water to bearings, or gland oil, or fluids for pump-jacket heating. Necessary spool pieces, break flanges, or couplings should be provided for convenient access to the pump, and for easy removal of valves and auxiliary-piping manifolds.

## Positive-displacement pumps

Rotary pumps (and compressors) work with forced-volume displacement and can deliver constant pulsation-free flow against much higher pressures than can centrifugal pumps. The layout and piping design does not differ from that of centrifugal machines. Rotary pumps must have built-in or piped-up relief valves.

Reciprocating pumps pump liquids containing vapor. For the same flowrate, resistance through the machine because of nonreturn inlet and outlet valves is much greater than through a centrifugal pump. Consequently, when pumping a saturated liquid, a static head is required in front of the pump suction for vapor-free flow to the pump cylinder. A vapor-liquid mixture flowing to the cylinder considerably reduces the pump's volumetric efficiency. In practice, this creates a flowrate that is much lower than the pump's rated capacity. In layout, this means an elevated suction drum and simple, direct, restriction-free piping from the vessel to the pump's suction.

Reciprocating pumps are classified by their type of cylinder and plunger arrangements. The principal pumps and their features:

- The single-acting pump is a single-piston plunger pump. Pulsation is high in these pumps. It can be reduced by using an air chamber at the discharge, or by combining two or more pump cylinders in parallel and out of phase. Proportioning pumps belong to this group and are usually motor driven. Capacity control is obtained by adjusting the length of stroke. Normal capacity ranges from 0.15 to 10 gpm.
- The simplex pump is a direct-coupled machine having a steam cylinder at one end and a double-acting pump at the other. Because of its simplicity, this pump provides reliable service and is used in smaller boilers as a feedwater pump. It can handle volatile and viscous liquids. Power pumps of this type are driven through a crankshaft by an engine or electric motor.

Many variations of cylinder, piston, valving and driving mechanisms have been designed for these pumps. The most common are: single pump (one cylinder), duplex, triplex and quadruplex. These pumps can be built with the cylinders in horizontal or vertical arrays. Horizontals usually require more floor space.

- The diaphragm pump handles precise quantities of fluid. A reciprocating plunger forces a motive liquid to one side of a diaphragm to create pulsating motion. This causes the process liquid to be conveyed into the pumping chamber (which forms the other side of the diaphragm). Suction and discharge valves regulate the movement of the pumped liquid. This pump is not suitable for high-viscosity liquids. Diaphragm pumps are compact and occupy a square-shaped floor area.

Reciprocating pumps are used where high head is needed. Capacity stays constant for varying discharge pressures and viscosities. Discharge cannot be throttled to obtain capacity control as for centrifugal pumps. Instead, a variable-speed drive or stroke adjustment is used.

Due to the characteristics of their suction and discharge valves, these pumps are not suitable for pumping liquid containing solids or dirt and are not built for corrosive or abrasive services.

The alternating action of reciprocating pumps produces pulsating flow. The extent and frequency of pulsation depends on the number of cylinders in parallel and whether the cylinders are single- or double-acting. Increasing the number of cylinders reduces the amplitude but increases the frequency of pulsation. To dampen pulsation, suction and discharge chambers are either built as part of the pump or are included in the piping. Without pulsation chambers, flow metering is difficult.

Due to pulsating operation, these pumps are bulky, and require large foundations and substantial pipe-supporting structures. Cushion-chamber sizes and locations are usually obtained from the pump manufacturer. Suction and discharge chambers often require more space in layout than does the pump itself.

Streamlined piping is desirable at the suction and discharge lines of reciprocating pumps. Avoid dead ends, opposing flows and sudden changes of direction. Use long-radius elbows and angular branch connections. After the discharge nozzle, a check valve and a block valve are usually installed. For short discharge lines, a check valve is not necessary.

Reciprocating pumps must have pressure-relieving arrangements. The following relief circuits are possible: (1) built-in relief valve integral with the pump casing; (2) closed-circuit relief line connecting the suction and discharge pipes immediately after the pump nozzles; (3) relief-valve discharge connecting to the suction drum (in this case, the block valve in the suction line should be locked open); and (4) relief line discharging to sewer.

Auxiliary-piping systems are usually associated with reciprocating pumps. These may provide cooling water, steam or heat-transfer media to jackets; seal oil to bearing seals; lubricating oil to pump bearings; and vent and drain connections.

Piping should not interfere with pump maintenance. Access to valve covers, case covers and shaft packing must be provided, as well as space for cylinder and shaft removal. Spool pieces or flanged elbows facilitate pump removal. Because maintenance of these pumps is relatively frequent, a trolley beam or manual traveling crane and weather protection are desirable.

## Centrifugal compressors

Layout and piping considerations and plant-design principles for arranging small centrifugal compressors (or blowers) do not differ in concept from the arrangements for centrifugal pumps. However, pipe sizes and components for centrifugal compressors are much larger than for centrifugal pumps.

Large centrifugal compressors are extensively used in process plants, e.g. in catalytic-cracking, ethylene and ammonia units. The advantages of these machines: (a) ability to handle large volumes in relatively small-size equipment, (b) mechanical simplicity of one rotating assembly, and (c) adaptability to various drivers, i.e., electric motor, steam turbine and gas turbine.

Large-size multistage compressors usually have horizontally split casings. Compressors with top connections can be arranged close to grade. The machine cover can only be lifted after piping has been disconnected and removed. Suitable break-flanges must be provided in the piping design.

Large compressors having bottom connections are elevated (Fig. 5). Piping can stay in place, and the machine cover can be conveniently removed. In outdoor arrangements, a number of concrete supporting columns are arranged close to the machine. Access is provided by a cantilevered platform surrounding the machine.

Concrete columns and a table-top arrangement are equally suitable. The wide span of support between two columns will not become vibration-prone. Compressors inside a building usually have foundations that are independent of the building's foundation.

If a roof or housing is required over a compressor, a trolley beam (located over the compressor's centerline) is usually provided. The trolley beam extends over the removal space that may be inside or outside the compressor shelter. If a compressor house contains several machines, a hand-operated traveling crane is normally supplied. The height of this crane has to be carefully estimated so that machine covers and rotors can be lifted over adjacent compressors and motors. Laydown space has to be allowed for.

Knockout drums and interstage exchangers are usu-

Bottom-outlet centrifugal-compressor arrangement Fig. 5

Lube-oil console occupies large floor space Fig. 6

Complexity of compressor, piping, intercoolers and silencer show importance of access. Fig. 7

Worthington Corp.

ally placed at grade, adjacent to the compressor platform or housing for short and simple interconnecting piping. Long-radius elbows or long-radius reducing elbows should be used immediately before the compressor's suction inlet. For air compressors, the vertical open-air intake has a strainer at the inlet that is protected from rain water, and, if required, a venturi meter to measure flow. The venturi requires a straight length of pipe on the upstream side.

## Auxiliaries for centrifugal compressors

Lubricating-oil and seal-oil consoles occupy large areas near or under these compressors.

The lube-oil console (Fig. 6) supplies lubricating oil to the compressor bearings. It is a constant-pressure, forced-circulation system with gear pumps, oil coolers and filters. These units are neither small nor simple. From the compressor, oil flows by gravity to the storage tank. The inlet has to be located below the compressor (see Fig. 5). A great number of valves must be accessible; clearances for pulling the exchanger bundle must be provided; and access for operations and maintenance is essential.

In some cases, the lube-oil pump is driven off the compressor, with an auxiliary pump provided for startup and shutdown.

The seal-oil console (similar to Fig. 6) supplies clean, filtered oil to the hydraulic seals of the compressor at constant pressure and temperature. The shaft end-seals prevent leakage of compressed gas to the atmosphere. Seal oil is forced between the seal rings at both ends of the compressor shaft at slightly higher pressures than that of the gas in the compressor case. The oil escaping on the low-pressure side of the seal returns to the reservoir and is recirculated. Oil escaping through the high-pressure side passes through automatic traps. Entry of oil into the compressor gas is prevented by labyrinth seals located between the oil seal and compressor impeller case.

Both lube-oil and seal-oil components are integrally mounted on the respective oil reservoirs. One console can serve several compressors. Manufacturers' recommendations should be followed.

## Reciprocating compressors

Reciprocating compressors should be arranged as close to grade as possible. Smaller machines, with or without pulsation dampeners, have foundations about 1.5 to 2 ft above grade. Larger machines, or groups of them, have foundations about 4 to 5 ft above grade. Sizable reciprocating compressors should not be supported higher than grade, and not on a concrete table top.

Most of these compressors have a shelter or compressor house whose floor level is near the top of the compressor foundation. Building and compressor foundations should be separate to avoid transmitting vibrations to the building structure. The floor has openings for access to pulsation dampeners and valves that are placed below floor level. Roof structure or columns support a traveling crane or trolley beams (over the cylinders) for maintenance. Sides of the shelter can be open, with a curtain wall above headroom, or completely enclosed. If it is enclosed, doors and stairways must be provided as required. Laydown space for compressor parts should also be planned.

The compressor piping interconnects pulsation dampeners, knockout pots, intercoolers and aftercoolers, possibly reactors and other process equipment, valves, and measuring and controlling components. A compact arrangement with short and simple piping will be less vibration-prone than widely spaced equipment interconnected by long slender pipelines. Knockout pots and intercoolers should be as close as possible to the pulsation dampeners, which are placed on or below the compressor cylinders. Knockout pots either in the compressor house or lined up outside the edge of buildings and supported at grade are common arrangements.

Compressor manufacturers can supply intercoolers that connect two cylinders as an integral part of the machine. Double-pipe intercoolers can be placed below the floor level or right outside the compressor house. To avoid vibration, it is advisable to provide three supports for a group of slender double-pipe exchangers 16 to 20 ft long.

Fig. 7 emphasizes these principles, and points out the importance of access needed around and above reciprocating compressors. Overhead traveling crane and housing are essential.

## Piping for reciprocating compressors

Pipe runs are grouped and placed just below or just outside the compressor cylinders. If below, the piping goes under the compressor floor; if outside, the piping is adjacent to the compressor house. Preferred location of all such piping is at grade.

The central question arising at every detail of compressor piping is: Will it or will it not induce vibration in the piping system? Because of the potential for vibration, customary piping details are modified. Even so, during final design it is impossible to predict which portion of a pipeline will have the potential for sympathetic vibration that is induced by flow and pressure pulsations.

Discharge dampeners and piping can be well supported below the compressor cylinder or at grade. Pipe turns, elevation changes and pipe junctions should be streamlined, as shown in Fig. 8. The following designs will help in avoiding pulsating flow: bends instead of elbows (Fig. 8a), angular inlets instead of laterals (Fig. 8b), one-plane turns instead of double offsets (Fig. 8c), smooth junctions instead of head-on opposing flow (Fig. 8d), streamlined end-of-header arrangement instead of dead-end header (Fig. 8e). Obstructions should be kept to a minimum; components having large pressure losses should be avoided.

It is advisable to check the process-flow-system diagram against the final piping design. The layout is not always connected in the same junction sequence as a flow diagram shows. An unexpected surge of suddenly changing flowrate and increased velocities, coupled with obstructions in the pipeline, can cause pulsating flow and vibration.

Valves should be placed in the piping without altering the simple pipe configurations. Vibration-free pipelines are more important than lined-up valve handwheels.

Supports for compressor piping should be independent of building structures, and building and compressor foundations. These supports also control piping movement. Expansion joints, anchors and guides are usually placed to support pipes and restrict pipe movements.

Supports are also located at directional changes, at valves, and generally where external or internal forces act and might induce vibration. Because of its mass, a valve placed in the center of a pipeline between two pipe supports can have a larger amplitude of vibration than the bare pipe alone. A valve placed close to a support is less likely to vibrate. Long compressor piping is more likely to vibrate sympathetically if supported and anchored at equal intervals than piping with irregular spacing of supports and anchors.

## Pulsation dampeners

Pulsating fluid flow can be seen at reciprocating compressors and pumps; at very large high-pressure centrifugal compressors and pumps; at blowers, rotary compressors and pumps; and at high-pressure letdown valves. If pulsating flow is transmitted to piping, structures and process equipment, material fatigue can occur. This results in failures and breakage, requiring frequent maintenance and even causing shutdowns. In piping, metering inaccuracies will occur, and unpleasant vibrations and considerable noise will arise.

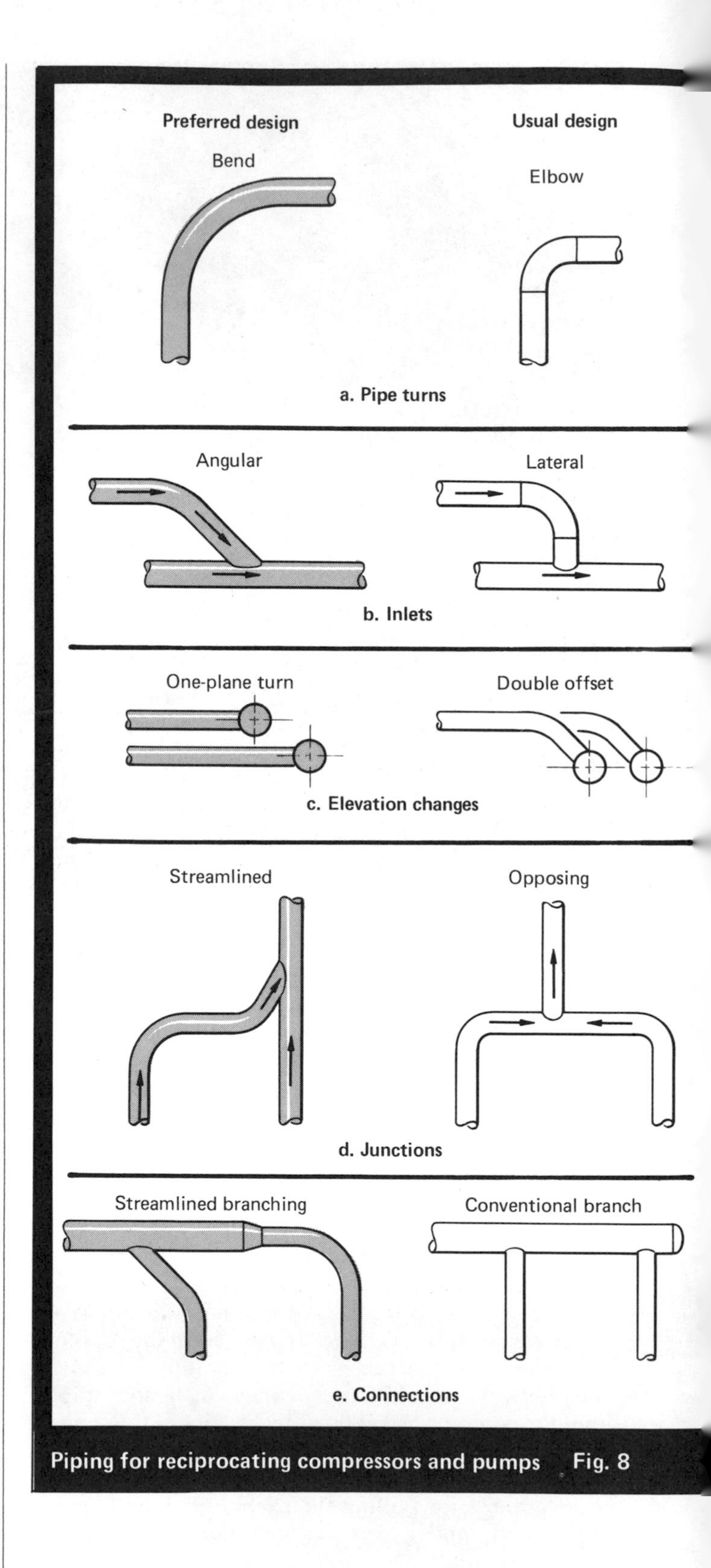

Piping for reciprocating compressors and pumps Fig. 8

Vibration in piping and heavy machinery becomes

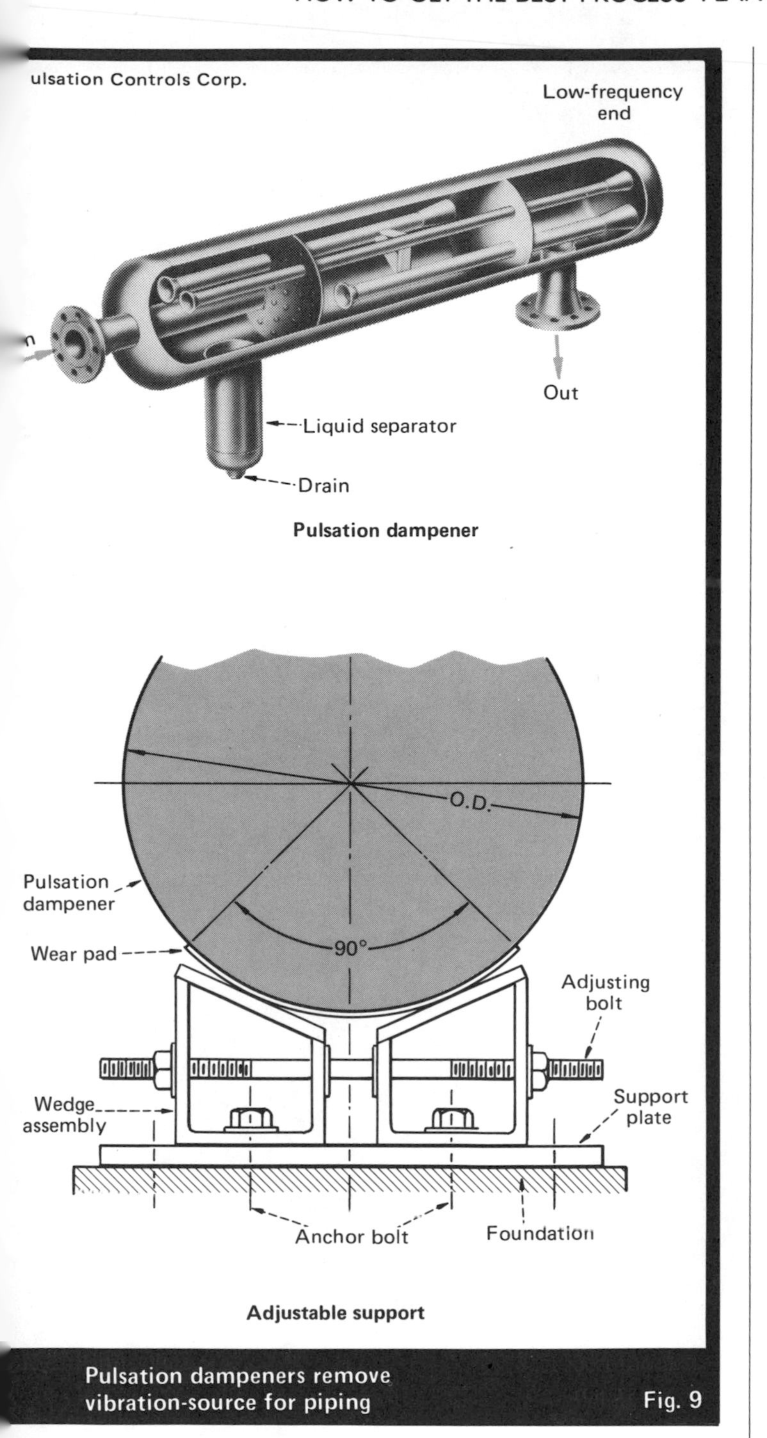


Pulsation dampeners remove vibration-source for piping Fig. 9

more hazardous as frequency and amplitude increase. High-speed pistons and/or high compressor-inlet velocities increase pulsation frequencies. Pulsations from a single compressor cylinder can be as hazardous as those from several compressor cylinders in parallel that discharge into the same pipe system. Fluids with higher densities will produce higher pressure pulsations than those with lower densities. Therefore, high-speed and high-pressure compressors must have pulsation-control devices.

Pulsation dampeners are used to eliminate pulsation in suction and discharge piping, to separate the source of vibration from the piping system, and to increase the volumetric efficiency of the compressor. The greatest hazard from vibration fatigue is at the dampener. Consequently, dampeners are rugged and heavy.

A dampener is an elongated vessel with expansion chambers, connected by a network of venturi tubes, and so designed as to break up volume and velocities (Fig. 9). Dampener size depends on flowrate, frequency of gas pulses, and pressure, temperature and composition of the gas. Inlet and outlet locations depend on the dampener's design, compressor-cylinder outlets, and piping arrangement. Horizontal and vertical dampener arrangements exist. At large centrifugal compressors, they greatly reduce noise.

Dampener design is highly specialized. In order to secure operating guarantees from manufacturers and design organizations, dampeners must be installed as specified. In general:

- Dampeners should be located as close to the compressor nozzles as possible (each dampener has a maximum distance limitation). Inlet and outlet nozzles are noted and flow direction indicated on the dampener's body.
- Dampeners should be securely anchored to a foundation and held down with straps or braces. Compressor cylinders should not be supported on a dampener unless designed for. Adjustable wedge support for precise load distribution is shown in Fig. 9. Heat expansion of supports after adjustment should be avoided.

For parallel compressor cylinders and a multiple-inlet single dampener, inlet flanges on the dampener should be field welded to the nozzle necks. In addition to design data, nozzle-orientation preferences must be given before final design and manufacture of pulsation dampeners.

During startup, screens and strainers should be installed in the suction line to prevent the drawing of foreign materials into the compressor and dampener.

Pulsation dampeners have drains and vents. Relief-valve nozzles can be located on the shell, preferably on the low-frequency end. Pressure loss across a pulsation dampener is small. Nozzle velocities at inlet and outlet are limited to a maximum of 50 ft/s. This value permits estimation of reasonable pipe sizes.

While construction access is necessary to install pulsation dampeners, servicing or maintenance access is usually not required.

## References


1. "Hydraulic Institute Standards," Hydraulic Institute, Cleveland, OH 44115, 1969.
2. Neerken, R. F., Pump Selection for the Chemical Process Industries; Birk, J. R. and Peacock, J. H., Pump Requirements for the Chemical Process Industries, *Chem. Eng.*, Feb. 18, 1974, pp. 104–124.*
3. Holland, F. A. and Chapman, F. S., Positive-Displacement Pumps, *Chem. Eng.*, Feb. 14, 1966, pp. 129–152.
4. Neerken, R. F., Compressor Selection for the Chemical Process Industries; Lapina, R. P., Can You Rerate Your Centrifugal Compressor?, *Chem. Eng.*, Jan. 20, 1975, pp. 78–98.*
5. Pollak, R., Selecting Fans and Blowers, *Chem. Eng.*, Jan. 22, 1973, pp. 86–100.*
6. Pump and Valve Selector, *Chem. Eng. Deskbook*, Oct. 11, 1971.*

*These articles are available as reprints. Ref. 2 is reprint number 200; Ref. 4 is number 220; Ref. 5 is number 173; and Ref. 6 is number 135. To order, check the appropriate number on the order form in the back of any issue. Prices are shown on order form.

# Index